Graduate Texts in Mathematics 218

Graduate Texts in Mathematics

Graduate Texts in Mathematics bridge the gap between passive study and creative understanding, offering graduate-level introductions to advanced topics in mathematics. The volumes are carefully written as teaching aids and highlight characteristic features of the theory. Although these books are frequently used as textbooks in graduate courses, they are also suitable for individual study.

For further volumes:
www.springer.com/series/136

John M. Lee

Introduction to
Smooth Manifolds

Second Edition

 Springer

John M. Lee
Department of Mathematics
University of Washington
Seattle, WA, USA

ISSN 0072-5285
ISBN 978-1-4899-9475-2   ISBN 978-1-4419-9982-5 (eBook)
DOI 10.1007/978-1-4419-9982-5
Springer New York Heidelberg Dordrecht London

Mathematics Subject Classification: 53-01, 58-01, 57-01

Printed on acid-free paper

Springer is part of Springer Science+Business Media (www.springer.com)

Preface

Manifolds crop up everywhere in mathematics. These generalizations of curves and surfaces to arbitrarily many dimensions provide the mathematical context for understanding "space" in all of its manifestations. Today, the tools of manifold theory are indispensable in most major subfields of pure mathematics, and are becoming increasingly important in such diverse fields as genetics, robotics, econometrics, statistics, computer graphics, biomedical imaging, and, of course, the undisputed leader among consumers (and inspirers) of mathematics—theoretical physics. No longer the province of differential geometers alone, smooth manifold technology is now a basic skill that all mathematics students should acquire as early as possible.

Over the past century or two, mathematicians have developed a wondrous collection of conceptual machines that enable us to peer ever more deeply into the invisible world of geometry in higher dimensions. Once their operation is mastered, these powerful machines enable us to think geometrically about the 6-dimensional solution set of a polynomial equation in four complex variables, or the 10-dimensional manifold of 5×5 orthogonal matrices, as easily as we think about the familiar 2-dimensional sphere in $\mathbb{R}^3$. The price we pay for this power, however, is that the machines are assembled from layer upon layer of abstract structure. Starting with the familiar raw materials of Euclidean spaces, linear algebra, multivariable calculus, and differential equations, one must progress through topological spaces, smooth atlases, tangent bundles, immersed and embedded submanifolds, vector fields, flows, cotangent bundles, tensors, Riemannian metrics, differential forms, foliations, Lie derivatives, Lie groups, Lie algebras, and more—just to get to the point where one can even think about studying specialized applications of manifold theory such as comparison theory, gauge theory, symplectic topology, or Ricci flow.

This book is designed as a first-year graduate text on manifold theory, for students who already have a solid acquaintance with undergraduate linear algebra, real analysis, and topology. I have tried to focus on the portions of manifold theory that will be needed by most people who go on to use manifolds in mathematical or scientific research. I introduce and use all of the standard tools of the subject, and prove most of its fundamental theorems, while avoiding unnecessary generalization or specialization. I try to keep the approach as concrete as possible, with pictures and intuitive discussions of how one should think geometrically about the abstract concepts, but without shying away from the powerful tools that modern mathematics has to offer. To fit in all of the basics and still maintain a reasonably sane pace, I have had to omit or barely touch on a number of important topics, such as complex manifolds, infinite-dimensional manifolds, connections, geodesics, curvature, fiber bundles, sheaves, characteristic classes, and Hodge theory. Think of them as dessert, to be savored after completing this book as the main course.

To convey the book's compass, it is easiest to describe where it starts and where it ends. The starting line is drawn just after topology: I assume that the reader has had a rigorous introduction to general topology, including the fundamental group and covering spaces. One convenient source for this material is my *Introduction to Topological Manifolds* [LeeTM], which I wrote partly with the aim of providing the topological background needed for this book. There are other books that cover similar material well; I am especially fond of the second edition of Munkres's *Topology* [Mun00]. The finish line is drawn just after a broad and solid background has been established, but before getting into the more specialized aspects of any particular subject. In particular, I introduce Riemannian metrics, but I do not go into connections, geodesics, or curvature. There are many Riemannian geometry books for the interested student to take up next, including one that I wrote [LeeRM] with the goal of moving expediently in a one-quarter course from basic smooth manifold theory to nontrivial geometric theorems about curvature and topology. Similar material is covered in the last two chapters of the recent book by Jeffrey Lee (no relation) [LeeJeff09], and do Carmo [dC92] covers a bit more. For more ambitious readers, I recommend the beautiful books by Petersen [Pet06], Sharpe [Sha97], and Chavel [Cha06].

This subject is often called "differential geometry." I have deliberately avoided using that term to describe what this book is about, however, because the term applies more properly to the study of smooth manifolds endowed with some extra structure—such as Lie groups, Riemannian manifolds, symplectic manifolds, vector bundles, foliations—and of their properties that are invariant under structure-preserving maps. Although I do give all of these geometric structures their due (after all, smooth manifold theory is pretty sterile without some geometric applications), I felt that it was more honest not to suggest that the book is primarily about one or all of these geometries. Instead, it is about developing the general tools for working with smooth manifolds, so that the reader can go on to work in whatever field of differential geometry or its cousins he or she feels drawn to.

There is no canonical linear path through this material. I have chosen an ordering of topics designed to establish a good technical foundation in the first half of the book, so that I can discuss interesting applications in the second half. Once the first twelve chapters have been completed, there is some flexibility in ordering the remaining chapters. For example, Chapter 13 (Riemannian Metrics) can be postponed if desired, although some sections of Chapters 15 and 16 would have to be postponed as well. On the other hand, Chapters 19–21 (Distributions and Foliations, The Exponential Map, and Quotient Manifolds, respectively) could in principle be inserted any time after Chapter 14, and much of the material can be covered even earlier if you are willing to skip over the references to differential forms. And the final chapter (Symplectic Manifolds) would make sense any time after Chapter 17, or even after Chapter 14 if you skip the references to de Rham cohomology.

As you might have guessed from the size of the book, and will quickly confirm when you start reading it, my style tends toward more detailed explanations and proofs than one typically finds in graduate textbooks. I realize this is not to every instructor's taste, but in my experience most students appreciate having the details spelled out when they are first learning the subject. The detailed proofs in the book provide students with useful models of rigor, and can free up class time for discussion of the meanings and motivations behind the definitions as well as the "big ideas" underlying some of the more difficult proofs. There are plenty of opportunities in the exercises and problems for students to provide arguments of their own.

I should say something about my choices of conventions and notations. The old joke that "differential geometry is the study of properties that are invariant under change of notation" is funny primarily because it is alarmingly close to the truth. Every geometer has his or her favorite system of notation, and while the systems are all in some sense formally isomorphic, the transformations required to get from one to another are often not at all obvious to students. Because one of my central goals is to prepare students to read advanced texts and research articles in differential geometry, I have tried to choose notations and conventions that are as close to the mainstream as I can make them without sacrificing too much internal consistency. (One difference between this edition and the previous one is that I have changed a number of my notational conventions to make them more consistent with mainstream mathematical usage.) When there are multiple conventions in common use (such as for the wedge product or the Laplace operator), I explain what the alternatives are and alert the student to be aware of which convention is in use by any given writer. Striving for too much consistency in this subject can be a mistake, however, and I have eschewed absolute consistency whenever I felt it would get in the way of ease of understanding. I have also introduced some common shortcuts at an early stage, such as the Einstein summation convention and the systematic confounding of maps with their coordinate representations, both of which tend to drive students crazy at first, but pay off enormously in efficiency later.

Prerequisites

This subject draws on most of the topics that are covered in a typical undergraduate mathematics education. The appendices (which most readers should read, or at least skim, first) contain a cursory summary of prerequisite material on topology, linear algebra, calculus, and differential equations. Although students who have not seen this material before will not learn it from reading the appendices, I hope readers will appreciate having all of the background material collected in one place. Besides giving me a convenient way to refer to results that I want to assume as known, it also gives the reader a splendid opportunity to brush up on topics that were once (hopefully) understood but may have faded.

Exercises and Problems

This book has a rather large number of exercises and problems for the student to work out. Embedded in the text of each chapter are questions labeled as "Exercises." These are (mostly) short opportunities to fill in gaps in the text. Some of them are routine verifications that would be tedious to write out in full, but are not quite trivial enough to warrant tossing off as obvious. I recommend that serious readers take the time at least to stop and convince themselves that they fully understand what is involved in doing each exercise, if not to write out a complete solution, because it will make their reading of the text far more fruitful.

At the end of each chapter is a collection of (mostly) longer and harder questions labeled as "Problems." These are the ones from which I select written homework assignments when I teach this material. Many of them will take hours for students to work through. Only by doing a substantial number of these problems can one hope to absorb this material deeply. I have tried insofar as possible to choose problems that are enlightening in some way and have interesting consequences in their own right. When the result of a problem is used in an essential way in the text, the page where it is used is noted at the end of the problem statement.

I have deliberately not provided written solutions to any of the problems, either in the back of the book or on the Internet. In my experience, if written solutions to problems are available, even the most conscientious students find it very hard to resist the temptation to look at the solutions as soon as they get stuck. But it is exactly at that stage of being stuck that students learn most effectively, by struggling to get unstuck and eventually finding a path through the thicket. Reading someone else's solution too early can give one a comforting, but ultimately misleading, sense of understanding. If you really feel you have run out of ideas, talk with an instructor, a fellow student, or one of the online mathematical discussion communities such as *math.stackexchange.com*. Even if someone else gives you a suggestion that turns out to be the key to getting unstuck, you will still learn much more from absorbing the suggestion and working out the details on your own than you would from reading someone else's polished proof.

About the Second Edition

Those who are familiar with the first edition of this book will notice first that the topics have been substantially rearranged. This is primarily because I decided it was worthwhile to introduce the two most important analytic tools (the rank theorem and the fundamental theorem on flows) much earlier, so that they can be used throughout the book rather than being relegated to later chapters.

A few new topics have been added, notably Sard's theorem, some transversality theorems, a proof that infinitesimal Lie group actions generate global group actions, a more thorough study of first-order partial differential equations, a brief treatment of degree theory for smooth maps between compact manifolds, and an introduction to contact structures. I have consolidated the introductory treatments of Lie groups,

Riemannian metrics, and symplectic manifolds in chapters of their own, to make it easier to concentrate on the special features of those subjects when they are first introduced (although Lie groups and Riemannian metrics still appear repeatedly in later chapters). In addition, manifolds with boundary are now treated much more systematically throughout the book.

Apart from additions and rearrangement, there are thousands of small changes and also some large ones. Parts of every chapter have been substantially rewritten to improve clarity. Some proofs that seemed too labored in the original have been streamlined, while others that seemed unclear have been expanded. I have modified some of my notations, usually moving toward more consistency with common notations in the literature. There is a new notation index just before the subject index.

There are also some typographical improvements in this edition. Most importantly, mathematical terms are now typeset in ***bold italics*** when they are officially defined, to reflect the fact that definitions are just as important as theorems and proofs but fit better into the flow of paragraphs rather than being called out with special headings. The exercises in the text are now indicated more clearly with a special symbol (▶), and numbered consecutively with the theorems to make them easier to find. The symbol □, in addition to marking the ends of proofs, now also marks the ends of statements of corollaries that follow so easily that they do not need proofs; and I have introduced the symbol // to mark the ends of numbered examples. The entire book is now set in Times Roman, supplemented by the excellent *MathTime Professional II* mathematics fonts from Personal TₑX, Inc.

Acknowledgments

Many people have contributed to the development of this book in indispensable ways. I would like to mention Tom Duchamp, Jim Isenberg, and Steve Mitchell, all of whom generously shared their own notes and ideas about teaching this subject; and Gary Sandine, who made lots of helpful suggestions and created more than a third of the illustrations in the book. In addition, I would like to thank the many others who have read the book and sent their corrections and suggestions to me. (In the Internet age, textbook writing becomes ever more a collaborative venture.) And most of all, I owe a debt of gratitude to Judith Arms, who has improved the book in countless ways with her thoughtful and penetrating suggestions.

For the sake of future readers, I hope each reader will take the time to keep notes of any mistakes or passages that are awkward or unclear, and let me know about them as soon as it is convenient for you. I will keep an up-to-date list of corrections on my website, whose address is listed below. (Sad experience suggests that there will be plenty of corrections despite my best efforts to root them out in advance.) If that site becomes unavailable for any reason, the publisher will know where to find me. Happy reading!

Seattle, Washington, USA   John M. Lee
                                                    www.math.washington.edu/~lee

Contents

Chapter 1
Smooth Manifolds

This book is about *smooth manifolds*. In the simplest terms, these are spaces that locally look like some Euclidean space $\mathbb{R}^n$, and on which one can do calculus. The most familiar examples, aside from Euclidean spaces themselves, are smooth plane curves such as circles and parabolas, and smooth surfaces such as spheres, tori, paraboloids, ellipsoids, and hyperboloids. Higher-dimensional examples include the set of points in $\mathbb{R}^{n+1}$ at a constant distance from the origin (an *n-sphere*) and graphs of smooth maps between Euclidean spaces.

The simplest manifolds are the topological manifolds, which are topological spaces with certain properties that encode what we mean when we say that they "locally look like" $\mathbb{R}^n$. Such spaces are studied intensively by topologists.

However, many (perhaps most) important applications of manifolds involve calculus. For example, applications of manifold theory to geometry involve such properties as volume and curvature. Typically, volumes are computed by integration, and curvatures are computed by differentiation, so to extend these ideas to manifolds would require some means of making sense of integration and differentiation on a manifold. Applications to classical mechanics involve solving systems of ordinary differential equations on manifolds, and the applications to general relativity (the theory of gravitation) involve solving a system of partial differential equations.

The first requirement for transferring the ideas of calculus to manifolds is some notion of "smoothness." For the simple examples of manifolds we described above, all of which are subsets of Euclidean spaces, it is fairly easy to describe the meaning of smoothness on an intuitive level. For example, we might want to call a curve "smooth" if it has a tangent line that varies continuously from point to point, and similarly a "smooth surface" should be one that has a tangent plane that varies continuously. But for more sophisticated applications it is an undue restriction to require smooth manifolds to be subsets of some ambient Euclidean space. The ambient coordinates and the vector space structure of $\mathbb{R}^n$ are superfluous data that often have nothing to do with the problem at hand. It is a tremendous advantage to be able to work with manifolds as abstract topological spaces, without the excess baggage of such an ambient space. For example, in general relativity, spacetime is modeled as a 4-dimensional smooth manifold that carries a certain geometric structure, called a

J.M. Lee, *Introduction to Smooth Manifolds*, Graduate Texts in Mathematics 218, DOI 10.1007/978-1-4419-9982-5_1, © Springer Science+Business Media New York 2013

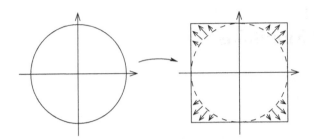

Fig. 1.1 A homeomorphism from a circle to a square

Lorentz metric, whose curvature results in gravitational phenomena. In such a model there is no physical meaning that can be assigned to any higher-dimensional ambient space in which the manifold lives, and including such a space in the model would complicate it needlessly. For such reasons, we need to think of smooth manifolds as abstract topological spaces, not necessarily as subsets of larger spaces.

It is not hard to see that there is no way to define a purely topological property that would serve as a criterion for "smoothness," because it cannot be invariant under homeomorphisms. For example, a circle and a square in the plane are homeomorphic topological spaces (Fig. 1.1), but we would probably all agree that the circle is "smooth," while the square is not. Thus, topological manifolds will not suffice for our purposes. Instead, we will think of a smooth manifold as a set with two layers of structure: first a topology, then a smooth structure.

In the first section of this chapter we describe the first of these structures. A *topological manifold* is a topological space with three special properties that express the notion of being locally like Euclidean space. These properties are shared by Euclidean spaces and by all of the familiar geometric objects that look locally like Euclidean spaces, such as curves and surfaces. We then prove some important topological properties of manifolds that we use throughout the book.

In the next section we introduce an additional structure, called a *smooth structure*, that can be added to a topological manifold to enable us to make sense of derivatives.

Following the basic definitions, we introduce a number of examples of manifolds, so you can have something concrete in mind as you read the general theory. At the end of the chapter we introduce the concept of a *smooth manifold with boundary*, an important generalization of smooth manifolds that will have numerous applications throughout the book, especially in our study of integration in Chapter 16.

Topological Manifolds

In this section we introduce topological manifolds, the most basic type of manifolds. We assume that the reader is familiar with the definition and basic properties of topological spaces, as summarized in Appendix A.

Suppose M is a topological space. We say that M is a ***topological manifold of dimension n*** or a ***topological n-manifold*** if it has the following properties:

- M is a **Hausdorff space**: for every pair of distinct points $p, q \in M$, there are disjoint open subsets $U, V \subseteq M$ such that $p \in U$ and $q \in V$.
- M is **second-countable**: there exists a countable basis for the topology of M.
- M is **locally Euclidean of dimension n**: each point of M has a neighborhood that is homeomorphic to an open subset of $\mathbb{R}^n$.

The third property means, more specifically, that for each $p \in M$ we can find

- an open subset $U \subseteq M$ containing p,
- an open subset $\hat{U} \subseteq \mathbb{R}^n$, and
- a homeomorphism $\varphi: U \to \hat{U}$.

▶ **Exercise 1.1.** Show that equivalent definitions of manifolds are obtained if instead of allowing U to be homeomorphic to *any* open subset of $\mathbb{R}^n$, we require it to be homeomorphic to an open ball in $\mathbb{R}^n$, or to $\mathbb{R}^n$ itself.

If M is a topological manifold, we often abbreviate the dimension of M as $\dim M$. Informally, one sometimes writes "Let M^n be a manifold" as shorthand for "Let M be a manifold of dimension n." The superscript n is not part of the name of the manifold, and is usually not included in the notation after the first occurrence.

It is important to note that every topological manifold has, by definition, a specific, well-defined dimension. Thus, we do not consider spaces of mixed dimension, such as the disjoint union of a plane and a line, to be manifolds at all. In Chapter 17, we will use the theory of de Rham cohomology to prove the following theorem, which shows that the dimension of a (nonempty) topological manifold is in fact a topological invariant.

Theorem 1.2 (Topological Invariance of Dimension). *A nonempty n-dimensional topological manifold cannot be homeomorphic to an m-dimensional manifold unless $m = n$.*

For the proof, see Theorem 17.26. In Chapter 2, we will also prove a related but weaker theorem (diffeomorphism invariance of dimension, Theorem 2.17). See also [LeeTM, Chap. 13] for a different proof of Theorem 1.2 using singular homology theory.

The empty set satisfies the definition of a topological n-manifold for every n. For the most part, we will ignore this special case (sometimes without remembering to say so). But because it is useful in certain contexts to allow the empty manifold, we choose not to exclude it from the definition.

The basic example of a topological n-manifold is $\mathbb{R}^n$ itself. It is Hausdorff because it is a metric space, and it is second-countable because the set of all open balls with rational centers and rational radii is a countable basis for its topology.

Requiring that manifolds share these properties helps to ensure that manifolds behave in the ways we expect from our experience with Euclidean spaces. For example, it is easy to verify that in a Hausdorff space, finite subsets are closed and limits of convergent sequences are unique (see Exercise A.11 in Appendix A). The motivation for second-countability is a bit less evident, but it will have important

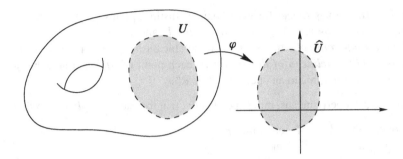

Fig. 1.2 A coordinate chart

consequences throughout the book, mostly based on the existence of partitions of unity (see Chapter 2).

In practice, both the Hausdorff and second-countability properties are usually easy to check, especially for spaces that are built out of other manifolds, because both properties are inherited by subspaces and finite products (Propositions A.17 and A.23). In particular, it follows that every open subset of a topological n-manifold is itself a topological n-manifold (with the subspace topology, of course).

We should note that some authors choose to omit the Hausdorff property or second-countability or both from the definition of manifolds. However, most of the interesting results about manifolds do in fact require these properties, and it is exceedingly rare to encounter a space "in nature" that would be a manifold except for the failure of one or the other of these hypotheses. For a couple of simple examples, see Problems 1-1 and 1-2; for a more involved example (a connected, locally Euclidean, Hausdorff space that is not second-countable), see [LeeTM, Problem 4-6].

Coordinate Charts

Let M be a topological n-manifold. A **coordinate chart** (or just a **chart**) on M is a pair (U, φ), where U is an open subset of M and $\varphi \colon U \to \hat{U}$ is a homeomorphism from U to an open subset $\hat{U} = \varphi(U) \subseteq \mathbb{R}^n$ (Fig. 1.2). By definition of a topological manifold, each point $p \in M$ is contained in the domain of some chart (U, φ). If $\varphi(p) = 0$, we say that the chart is **centered at p**. If (U, φ) is any chart whose domain contains p, it is easy to obtain a new chart centered at p by subtracting the constant vector $\varphi(p)$.

Given a chart (U, φ), we call the set U a **coordinate domain**, or a **coordinate neighborhood** of each of its points. If, in addition, $\varphi(U)$ is an open ball in $\mathbb{R}^n$, then U is called a **coordinate ball**; if $\varphi(U)$ is an open cube, U is a **coordinate cube**. The map φ is called a **(local) coordinate map**, and the component functions $(x^1, \ldots, x^n)$ of φ, defined by $\varphi(p) = (x^1(p), \ldots, x^n(p))$, are called **local coordinates** on U. We sometimes write things such as "(U, φ) is a chart containing p" as shorthand for "(U, φ) is a chart whose domain U contains p." If we wish to emphasize the coordinate functions $(x^1, \ldots, x^n)$ instead of the coordinate map φ, we sometimes denote the chart by $(U, (x^1, \ldots, x^n))$ or $(U, (x^i))$.

Examples of Topological Manifolds

Here are some simple examples.

Example 1.3 (Graphs of Continuous Functions). Let $U \subseteq \mathbb{R}^n$ be an open subset, and let $f: U \to \mathbb{R}^k$ be a continuous function. The **graph of f** is the subset of $\mathbb{R}^n \times \mathbb{R}^k$ defined by

$$\Gamma(f) = \{(x, y) \in \mathbb{R}^n \times \mathbb{R}^k : x \in U \text{ and } y = f(x)\},$$

with the subspace topology. Let $\pi_1: \mathbb{R}^n \times \mathbb{R}^k \to \mathbb{R}^n$ denote the projection onto the first factor, and let $\varphi: \Gamma(f) \to U$ be the restriction of π_1 to $\Gamma(f)$:

$$\varphi(x, y) = x, \quad (x, y) \in \Gamma(f).$$

Because φ is the restriction of a continuous map, it is continuous; and it is a homeomorphism because it has a continuous inverse given by $\varphi^{-1}(x) = (x, f(x))$. Thus $\Gamma(f)$ is a topological manifold of dimension n. In fact, $\Gamma(f)$ is homeomorphic to U itself, and $(\Gamma(f), \varphi)$ is a global coordinate chart, called **graph coordinates**. The same observation applies to any subset of $\mathbb{R}^{n+k}$ defined by setting any k of the coordinates (not necessarily the last k) equal to some continuous function of the other n, which are restricted to lie in an open subset of $\mathbb{R}^n$. //

Example 1.4 (Spheres). For each integer $n \geq 0$, the unit n-sphere $\mathbb{S}^n$ is Hausdorff and second-countable because it is a topological subspace of $\mathbb{R}^{n+1}$. To show that it is locally Euclidean, for each index $i = 1, \ldots, n+1$ let U_i^+ denote the subset of $\mathbb{R}^{n+1}$ where the ith coordinate is positive:

$$U_i^+ = \{(x^1, \ldots, x^{n+1}) \in \mathbb{R}^{n+1} : x^i > 0\}.$$

(See Fig. 1.3.) Similarly, U_i^- is the set where $x^i < 0$.

Let $f: \mathbb{B}^n \to \mathbb{R}$ be the continuous function

$$f(u) = \sqrt{1 - |u|^2}.$$

Then for each $i = 1, \ldots, n+1$, it is easy to check that $U_i^+ \cap \mathbb{S}^n$ is the graph of the function

$$x^i = f(x^1, \ldots, \widehat{x^i}, \ldots, x^{n+1}),$$

where the hat indicates that x^i is omitted. Similarly, $U_i^- \cap \mathbb{S}^n$ is the graph of

$$x^i = -f(x^1, \ldots, \widehat{x^i}, \ldots, x^{n+1}).$$

Thus, each subset $U_i^\pm \cap \mathbb{S}^n$ is locally Euclidean of dimension n, and the maps $\varphi_i^\pm: U_i^\pm \cap \mathbb{S}^n \to \mathbb{B}^n$ given by

$$\varphi_i^\pm(x^1, \ldots, x^{n+1}) = (x^1, \ldots, \widehat{x^i}, \ldots, x^{n+1})$$

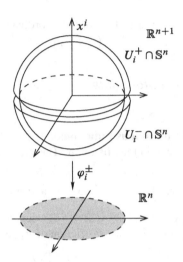

Fig. 1.3 Charts for $\mathbb{S}^n$

are graph coordinates for $\mathbb{S}^n$. Since each point of $\mathbb{S}^n$ is in the domain of at least one of these $2n + 2$ charts, $\mathbb{S}^n$ is a topological n-manifold.   //

Example 1.5 (Projective Spaces). The *n-dimensional real projective space*, denoted by $\mathbb{RP}^n$ (or sometimes just $\mathbb{P}^n$), is defined as the set of 1-dimensional linear subspaces of $\mathbb{R}^{n+1}$, with the quotient topology determined by the natural map $\pi \colon \mathbb{R}^{n+1} \smallsetminus \{0\} \to \mathbb{RP}^n$ sending each point $x \in \mathbb{R}^{n+1} \smallsetminus \{0\}$ to the subspace spanned by x. The 2-dimensional projective space $\mathbb{RP}^2$ is called the *projective plane*. For any point $x \in \mathbb{R}^{n+1} \smallsetminus \{0\}$, let $[x] = \pi(x) \in \mathbb{RP}^n$ denote the line spanned by x.

For each $i = 1, \ldots, n + 1$, let $\widetilde{U}_i \subseteq \mathbb{R}^{n+1} \smallsetminus \{0\}$ be the set where $x^i \neq 0$, and let $U_i = \pi(\widetilde{U}_i) \subseteq \mathbb{RP}^n$. Since $\widetilde{U}_i$ is a saturated open subset, U_i is open and $\pi|_{\widetilde{U}_i} \colon \widetilde{U}_i \to U_i$ is a quotient map (see Theorem A.27). Define a map $\varphi_i \colon U_i \to \mathbb{R}^n$ by

$$\varphi_i \left[x^1, \ldots, x^{n+1} \right] = \left(\frac{x^1}{x^i}, \ldots, \frac{x^{i-1}}{x^i}, \frac{x^{i+1}}{x^i}, \ldots, \frac{x^{n+1}}{x^i} \right).$$

This map is well defined because its value is unchanged by multiplying x by a nonzero constant. Because $\varphi_i \circ \pi$ is continuous, φ_i is continuous by the characteristic property of quotient maps (Theorem A.27). In fact, φ_i is a homeomorphism, because it has a continuous inverse given by

$$\varphi_i^{-1} \left(u^1, \ldots, u^n \right) = \left[u^1, \ldots, u^{i-1}, 1, u^i, \ldots, u^n \right],$$

as you can check. Geometrically, $\varphi([x]) = u$ means $(u, 1)$ is the point in $\mathbb{R}^{n+1}$ where the line $[x]$ intersects the affine hyperplane where $x^i = 1$ (Fig. 1.4). Because the sets $U_1, \ldots, U_{n+1}$ cover $\mathbb{RP}^n$, this shows that $\mathbb{RP}^n$ is locally Euclidean of dimension n. The Hausdorff and second-countability properties are left as exercises.   //

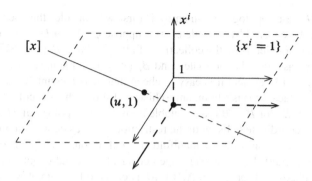

Fig. 1.4 A chart for $\mathbb{RP}^n$

▶ **Exercise 1.6.** Show that $\mathbb{RP}^n$ is Hausdorff and second-countable, and is therefore a topological n-manifold.

▶ **Exercise 1.7.** Show that $\mathbb{RP}^n$ is compact. [Hint: show that the restriction of π to $\mathbb{S}^n$ is surjective.]

Example 1.8 (Product Manifolds). Suppose $M_1, \ldots, M_k$ are topological manifolds of dimensions $n_1, \ldots, n_k$, respectively. The product space $M_1 \times \cdots \times M_k$ is shown to be a topological manifold of dimension $n_1 + \cdots + n_k$ as follows. It is Hausdorff and second-countable by Propositions A.17 and A.23, so only the locally Euclidean property needs to be checked. Given any point $(p_1, \ldots, p_k) \in M_1 \times \cdots \times M_k$, we can choose a coordinate chart (U_i, φ_i) for each M_i with $p_i \in U_i$. The product map

$$\varphi_1 \times \cdots \times \varphi_k : U_1 \times \cdots \times U_k \to \mathbb{R}^{n_1 + \cdots + n_k}$$

is a homeomorphism onto its image, which is a product open subset of $\mathbb{R}^{n_1 + \cdots + n_k}$. Thus, $M_1 \times \cdots \times M_k$ is a topological manifold of dimension $n_1 + \cdots + n_k$, with charts of the form $(U_1 \times \cdots \times U_k, \varphi_1 \times \cdots \times \varphi_k)$. //

Example 1.9 (Tori). For a positive integer n, the **n-torus** (plural: **tori**) is the product space $\mathbb{T}^n = \mathbb{S}^1 \times \cdots \times \mathbb{S}^1$. By the discussion above, it is a topological n-manifold. (The 2-torus is usually called simply **the torus**.) //

Topological Properties of Manifolds

As topological spaces go, manifolds are quite special, because they share so many important properties with Euclidean spaces. Here we discuss a few such properties that will be of use to us throughout the book.

Most of the properties we discuss in this section depend on the fact that every manifold possesses a particularly well-behaved basis for its topology.

Lemma 1.10. *Every topological manifold has a countable basis of precompact coordinate balls.*

Proof. Let M be a topological n-manifold. First we consider the special case in which M can be covered by a single chart. Suppose $\varphi \colon M \to \hat{U} \subseteq \mathbb{R}^n$ is a global coordinate map, and let $\mathscr{B}$ be the collection of all open balls $B_r(x) \subseteq \mathbb{R}^n$ such that r is rational, x has rational coordinates, and $B_{r'}(x) \subseteq \hat{U}$ for some $r' > r$. Each such ball is precompact in $\hat{U}$, and it is easy to check that $\mathscr{B}$ is a countable basis for the topology of $\hat{U}$. Because φ is a homeomorphism, it follows that the collection of sets of the form $\varphi^{-1}(B)$ for $B \in \mathscr{B}$ is a countable basis for the topology of M, consisting of precompact coordinate balls, with the restrictions of φ as coordinate maps.

Now let M be an arbitrary n-manifold. By definition, each point of M is in the domain of a chart. Because every open cover of a second-countable space has a countable subcover (Proposition A.16), M is covered by countably many charts $\{(U_i, \varphi_i)\}$. By the argument in the preceding paragraph, each coordinate domain U_i has a countable basis of coordinate balls that are precompact in U_i, and the union of all these countable bases is a countable basis for the topology of M. If $V \subseteq U_i$ is one of these balls, then the closure of V in U_i is compact, and because M is Hausdorff, it is closed in M. It follows that the closure of V in M is the same as its closure in U_i, so V is precompact in M as well. $\qquad \square$

Connectivity

The existence of a basis of coordinate balls has important consequences for the connectivity properties of manifolds. Recall that a topological space X is

- **connected** if there do not exist two disjoint, nonempty, open subsets of X whose union is X;
- **path-connected** if every pair of points in X can be joined by a path in X; and
- **locally path-connected** if X has a basis of path-connected open subsets.

(See Appendix A.) The following proposition shows that connectivity and path connectivity coincide for manifolds.

Proposition 1.11. *Let M be a topological manifold.*

(a) *M is locally path-connected.*
(b) *M is connected if and only if it is path-connected.*
(c) *The components of M are the same as its path components.*
(d) *M has countably many components, each of which is an open subset of M and a connected topological manifold.*

Proof. Since each coordinate ball is path-connected, (a) follows from the fact that M has a basis of coordinate balls. Parts (b) and (c) are immediate consequences of (a) and Proposition A.43. To prove (d), note that each component is open in M by Proposition A.43, so the collection of components is an open cover of M. Because M is second-countable, this cover must have a countable subcover. But since the components are all disjoint, the cover must have been countable to begin with, which is to say that M has only countably many components. Because the components are open, they are connected topological manifolds in the subspace topology. $\qquad \square$

Local Compactness and Paracompactness

The next topological property of manifolds that we need is local compactness (see Appendix A for the definition).

Proposition 1.12 (Manifolds Are Locally Compact). *Every topological manifold is locally compact.*

Proof. Lemma 1.10 showed that every manifold has a basis of precompact open subsets. ☐

Another key topological property possessed by manifolds is called *paracompactness*. It is a consequence of local compactness and second-countability, and in fact is one of the main reasons why second-countability is included in the definition of manifolds.

Let M be a topological space. A collection $\mathcal{X}$ of subsets of M is said to be **locally finite** if each point of M has a neighborhood that intersects at most finitely many of the sets in $\mathcal{X}$. Given a cover $\mathcal{U}$ of M, another cover $\mathcal{V}$ is called a **refinement of** $\mathcal{U}$ if for each $V \in \mathcal{V}$ there exists some $U \in \mathcal{U}$ such that $V \subseteq U$. We say that M is **paracompact** if every open cover of M admits an open, locally finite refinement.

Lemma 1.13. *Suppose $\mathcal{X}$ is a locally finite collection of subsets of a topological space M.*

(a) *The collection $\{\overline{X} : X \in \mathcal{X}\}$ is also locally finite.*
(b) $\overline{\bigcup_{X \in \mathcal{X}} X} = \bigcup_{X \in \mathcal{X}} \overline{X}.$

▶ **Exercise 1.14.** Prove the preceding lemma.

Theorem 1.15 (Manifolds Are Paracompact). *Every topological manifold is paracompact. In fact, given a topological manifold M, an open cover $\mathcal{X}$ of M, and any basis $\mathcal{B}$ for the topology of M, there exists a countable, locally finite open refinement of $\mathcal{X}$ consisting of elements of $\mathcal{B}$.*

Proof. Given M, $\mathcal{X}$, and $\mathcal{B}$ as in the hypothesis of the theorem, let $(K_j)_{j=1}^{\infty}$ be an exhaustion of M by compact sets (Proposition A.60). For each j, let $V_j = K_{j+1} \smallsetminus \text{Int } K_j$ and $W_j = \text{Int } K_{j+2} \smallsetminus K_{j-1}$ (where we interpret K_j as $\varnothing$ if $j < 1$). Then V_j is a compact set contained in the open subset W_j. For each $x \in V_j$, there is some $X_x \in \mathcal{X}$ containing x, and because $\mathcal{B}$ is a basis, there exists $B_x \in \mathcal{B}$ such that $x \in B_x \subseteq X_x \cap W_j$. The collection of all such sets B_x as x ranges over V_j is an open cover of V_j, and thus has a finite subcover. The union of all such finite subcovers as j ranges over the positive integers is a countable open cover of M that refines $\mathcal{X}$. Because the finite subcover of V_j consists of sets contained in W_j, and $W_j \cap W_{j'} = \varnothing$ except when $j - 2 \le j' \le j + 2$, the resulting cover is locally finite. ☐

Problem 1-5 shows that, at least for connected spaces, paracompactness can be used as a substitute for second-countability in the definition of manifolds.

Fundamental Groups of Manifolds

The following result about fundamental groups of manifolds will be important in our study of covering manifolds in Chapter 4. For a brief review of the fundamental group, see Appendix A.

Proposition 1.16. *The fundamental group of a topological manifold is countable.*

Proof. Let M be a topological manifold. By Lemma 1.10, there is a countable collection $\mathcal{B}$ of coordinate balls covering M. For any pair of coordinate balls $B, B' \in \mathcal{B}$, the intersection $B \cap B'$ has at most countably many components, each of which is path-connected. Let $\mathcal{X}$ be a countable set containing a point from each component of $B \cap B'$ for each $B, B' \in \mathcal{B}$ (including $B = B'$). For each $B \in \mathcal{B}$ and each $x, x' \in \mathcal{X}$ such that $x, x' \in B$, let $h^B_{x,x'}$ be some path from x to x' in B.

Since the fundamental groups based at any two points in the same component of M are isomorphic, and $\mathcal{X}$ contains at least one point in each component of M, we may as well choose a point $p \in \mathcal{X}$ as base point. Define a *special loop* to be a loop based at p that is equal to a finite product of paths of the form $h^B_{x,x'}$. Clearly, the set of special loops is countable, and each special loop determines an element of $\pi_1(M, p)$. To show that $\pi_1(M, p)$ is countable, therefore, it suffices to show that each element of $\pi_1(M, p)$ is represented by a special loop.

Suppose $f : [0, 1] \to M$ is a loop based at p. The collection of components of sets of the form $f^{-1}(B)$ as B ranges over $\mathcal{B}$ is an open cover of $[0, 1]$, so by compactness it has a finite subcover. Thus, there are finitely many numbers $0 = a_0 < a_1 < \cdots < a_k = 1$ such that $[a_{i-1}, a_i] \subseteq f^{-1}(B)$ for some $B \subseteq \mathcal{B}$. For each i, let f_i be the restriction of f to the interval $[a_{i-1}, a_i]$, reparametrized so that its domain is $[0, 1]$, and let $B_i \in \mathcal{B}$ be a coordinate ball containing the image of f_i. For each i, we have $f(a_i) \in B_i \cap B_{i+1}$, and there is some $x_i \in \mathcal{X}$ that lies in the same component of $B_i \cap B_{i+1}$ as $f(a_i)$. Let g_i be a path in $B_i \cap B_{i+1}$ from x_i to $f(a_i)$ (Fig. 1.5), with the understanding that $x_0 = x_k = p$, and g_0 and g_k are both equal to the constant path c_p based at p. Then, because $\bar{g}_i \cdot g_i$ is path-homotopic to a constant path (where $\bar{g}_i(t) = g_i(1 - t)$ is the reverse path of g_i),

$$f \sim f_1 \cdot \cdots \cdot f_k$$
$$\sim g_0 \cdot f_1 \cdot \bar{g}_1 \cdot g_1 \cdot f_2 \cdot \bar{g}_2 \cdot \cdots \cdot \bar{g}_{k-1} \cdot g_{k-1} \cdot f_k \cdot \bar{g}_k$$
$$\sim \tilde{f}_1 \cdot \tilde{f}_2 \cdot \cdots \cdot \tilde{f}_k,$$

where $\tilde{f}_i = g_{i-1} \cdot f_i \cdot \bar{g}_i$. For each i, $\tilde{f}_i$ is a path in B_i from x_{i-1} to x_i. Since B_i is simply connected, $\tilde{f}_i$ is path-homotopic to $h^{B_i}_{x_{i-1}, x_i}$. It follows that f is path-homotopic to a special loop, as claimed.   □

Smooth Structures

The definition of manifolds that we gave in the preceding section is sufficient for studying topological properties of manifolds, such as compactness, connectedness,

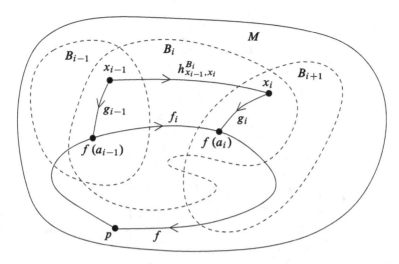

Fig. 1.5 The fundamental group of a manifold is countable

simple connectivity, and the problem of classifying manifolds up to homeomorphism. However, in the entire theory of topological manifolds there is no mention of calculus. There is a good reason for this: however we might try to make sense of derivatives of functions on a manifold, such derivatives cannot be invariant under homeomorphisms. For example, the map $\varphi\colon \mathbb{R}^2 \to \mathbb{R}^2$ given by $\varphi(u, v) = \left(u^{1/3}, v^{1/3}\right)$ is a homeomorphism, and it is easy to construct differentiable functions $f\colon \mathbb{R}^2 \to \mathbb{R}$ such that $f \circ \varphi$ is not differentiable at the origin. (The function $f(x, y) = x$ is one such.)

To make sense of derivatives of real-valued functions, curves, or maps between manifolds, we need to introduce a new kind of manifold called a *smooth manifold*. It will be a topological manifold with some extra structure in addition to its topology, which will allow us to decide which functions to or from the manifold are smooth.

The definition will be based on the calculus of maps between Euclidean spaces, so let us begin by reviewing some basic terminology about such maps. If U and V are open subsets of Euclidean spaces $\mathbb{R}^n$ and $\mathbb{R}^m$, respectively, a function $F\colon U \to V$ is said to be **smooth** (or C^∞, or **infinitely differentiable**) if each of its component functions has continuous partial derivatives of all orders. If in addition F is bijective and has a smooth inverse map, it is called a **diffeomorphism**. A diffeomorphism is, in particular, a homeomorphism.

A review of some important properties of smooth maps is given in Appendix C. You should be aware that some authors define the word *smooth* differently—for example, to mean continuously differentiable or merely differentiable. On the other hand, some use the word *differentiable* to mean what we call *smooth*. Throughout this book, *smooth* is synonymous with C^∞.

To see what additional structure on a topological manifold might be appropriate for discerning which maps are smooth, consider an arbitrary topological n-manifold M. Each point in M is in the domain of a coordinate map $\varphi\colon U \to \hat{U} \subseteq \mathbb{R}^n$.

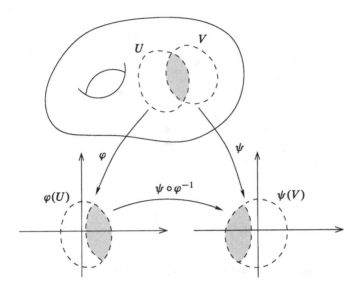

Fig. 1.6   A transition map

A plausible definition of a smooth function on M would be to say that $f : M \to \mathbb{R}$ is smooth if and only if the composite function $f \circ \varphi^{-1} : \widehat{U} \to \mathbb{R}$ is smooth in the sense of ordinary calculus. But this will make sense only if this property is independent of the choice of coordinate chart. To guarantee this independence, we will restrict our attention to "smooth charts." Since smoothness is not a homeomorphism-invariant property, the way to do this is to consider the collection of all smooth charts as a new kind of structure on M.

With this motivation in mind, we now describe the details of the construction.

Let M be a topological n-manifold. If (U, φ), (V, ψ) are two charts such that $U \cap V \neq \varnothing$, the composite map $\psi \circ \varphi^{-1} : \varphi(U \cap V) \to \psi(U \cap V)$ is called the ***transition map from φ to ψ*** (Fig. 1.6). It is a composition of homeomorphisms, and is therefore itself a homeomorphism. Two charts (U, φ) and (V, ψ) are said to be ***smoothly compatible*** if either $U \cap V = \varnothing$ or the transition map $\psi \circ \varphi^{-1}$ is a diffeomorphism. Since $\varphi(U \cap V)$ and $\psi(U \cap V)$ are open subsets of $\mathbb{R}^n$, smoothness of this map is to be interpreted in the ordinary sense of having continuous partial derivatives of all orders.

We define an ***atlas for M*** to be a collection of charts whose domains cover M. An atlas $\mathcal{A}$ is called a ***smooth atlas*** if any two charts in $\mathcal{A}$ are smoothly compatible with each other.

To show that an atlas is smooth, we need only verify that each transition map $\psi \circ \varphi^{-1}$ is smooth whenever (U, φ) and (V, ψ) are charts in $\mathcal{A}$; once we have proved this, it follows that $\psi \circ \varphi^{-1}$ is a diffeomorphism because its inverse $\left(\psi \circ \varphi^{-1}\right)^{-1} = \varphi \circ \psi^{-1}$ is one of the transition maps we have already shown to be smooth. Alternatively, given two particular charts (U, φ) and (V, ψ), it is often easiest to show that they are smoothly compatible by verifying that $\psi \circ \varphi^{-1}$ is smooth and injective with nonsingular Jacobian at each point, and appealing to Corollary C.36.

Our plan is to define a "smooth structure" on M by giving a smooth atlas, and to define a function $f : M \to \mathbb{R}$ to be smooth if and only if $f \circ \varphi^{-1}$ is smooth in the sense of ordinary calculus for each coordinate chart (U, φ) in the atlas. There is one minor technical problem with this approach: in general, there will be many possible atlases that give the "same" smooth structure, in that they all determine the same collection of smooth functions on M. For example, consider the following pair of atlases on $\mathbb{R}^n$:

$$\mathcal{A}_1 = \left\{ \left(\mathbb{R}^n, \mathrm{Id}_{\mathbb{R}^n} \right) \right\},$$

$$\mathcal{A}_2 = \left\{ \left(B_1(x), \mathrm{Id}_{B_1(x)} \right) : x \in \mathbb{R}^n \right\}.$$

Although these are different smooth atlases, clearly a function $f : \mathbb{R}^n \to \mathbb{R}$ is smooth with respect to either atlas if and only if it is smooth in the sense of ordinary calculus.

We could choose to define a smooth structure as an equivalence class of smooth atlases under an appropriate equivalence relation. However, it is more straightforward to make the following definition: a smooth atlas $\mathcal{A}$ on M is *maximal* if it is not properly contained in any larger smooth atlas. This just means that any chart that is smoothly compatible with every chart in $\mathcal{A}$ is already in $\mathcal{A}$. (Such a smooth atlas is also said to be *complete*.)

Now we can define the main concept of this chapter. If M is a topological manifold, a *smooth structure on M* is a maximal smooth atlas. A *smooth manifold* is a pair $(M, \mathcal{A})$, where M is a topological manifold and $\mathcal{A}$ is a smooth structure on M. When the smooth structure is understood, we usually omit mention of it and just say "M is a smooth manifold." Smooth structures are also called *differentiable structures* or C^∞ *structures* by some authors. We also use the term *smooth manifold structure* to mean a manifold topology together with a smooth structure.

We emphasize that a smooth structure is an additional piece of data that must be added to a topological manifold before we are entitled to talk about a "smooth manifold." In fact, a given topological manifold may have many different smooth structures (see Example 1.23 and Problem 1-6). On the other hand, it is not always possible to find a smooth structure on a given topological manifold: there exist topological manifolds that admit no smooth structures at all. (The first example was a compact 10-dimensional manifold found in 1960 by Michel Kervaire [Ker60].)

It is generally not very convenient to define a smooth structure by explicitly describing a maximal smooth atlas, because such an atlas contains very many charts. Fortunately, we need only specify *some* smooth atlas, as the next proposition shows.

Proposition 1.17. *Let M be a topological manifold.*

(a) *Every smooth atlas $\mathcal{A}$ for M is contained in a unique maximal smooth atlas, called the **smooth structure determined by** $\mathcal{A}$.*

(b) *Two smooth atlases for M determine the same smooth structure if and only if their union is a smooth atlas.*

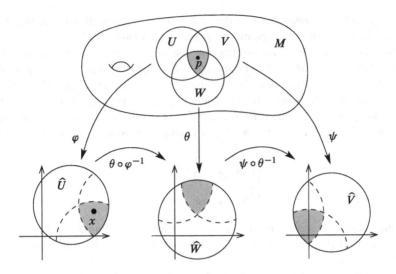

Fig. 1.7   Proof of Proposition 1.17(a)

Proof. Let $\mathcal{A}$ be a smooth atlas for M, and let $\overline{\mathcal{A}}$ denote the set of all charts that are smoothly compatible with every chart in $\mathcal{A}$. To show that $\overline{\mathcal{A}}$ is a smooth atlas, we need to show that any two charts of $\overline{\mathcal{A}}$ are smoothly compatible with each other, which is to say that for any (U, φ), $(V, \psi) \in \overline{\mathcal{A}}$, the map $\psi \circ \varphi^{-1} \colon \varphi(U \cap V) \to \psi(U \cap V)$ is smooth.

Let $x = \varphi(p) \in \varphi(U \cap V)$ be arbitrary. Because the domains of the charts in $\mathcal{A}$ cover M, there is some chart $(W, \theta) \in \mathcal{A}$ such that $p \in W$ (Fig. 1.7). Since every chart in $\overline{\mathcal{A}}$ is smoothly compatible with (W, θ), both of the maps $\theta \circ \varphi^{-1}$ and $\psi \circ \theta^{-1}$ are smooth where they are defined. Since $p \in U \cap V \cap W$, it follows that $\psi \circ \varphi^{-1} = \left(\psi \circ \theta^{-1}\right) \circ \left(\theta \circ \varphi^{-1}\right)$ is smooth on a neighborhood of x. Thus, $\psi \circ \varphi^{-1}$ is smooth in a neighborhood of each point in $\varphi(U \cap V)$. Therefore, $\overline{\mathcal{A}}$ is a smooth atlas. To check that it is maximal, just note that any chart that is smoothly compatible with every chart in $\overline{\mathcal{A}}$ must in particular be smoothly compatible with every chart in $\mathcal{A}$, so it is already in $\overline{\mathcal{A}}$. This proves the existence of a maximal smooth atlas containing $\mathcal{A}$. If $\mathcal{B}$ is any other maximal smooth atlas containing $\mathcal{A}$, each of its charts is smoothly compatible with each chart in $\mathcal{A}$, so $\mathcal{B} \subseteq \overline{\mathcal{A}}$. By maximality of $\mathcal{B}$, $\mathcal{B} = \overline{\mathcal{A}}$.

The proof of (b) is left as an exercise.   □

▶ **Exercise 1.18.** Prove Proposition 1.17(b).

For example, if a topological manifold M can be covered by a single chart, the smooth compatibility condition is trivially satisfied, so any such chart automatically determines a smooth structure on M.

It is worth mentioning that the notion of smooth structure can be generalized in several different ways by changing the compatibility requirement for charts. For example, if we replace the requirement that charts be smoothly compatible by the weaker requirement that each transition map $\psi \circ \varphi^{-1}$ (and its inverse) be of class C^k, we obtain the definition of a C^k **structure**. Similarly, if we require that each transition map be real-analytic (i.e., expressible as a convergent power series in a neighborhood of each point), we obtain the definition of a **real-analytic structure**, also called a C^ω **structure**. If M has even dimension $n = 2m$, we can identify $\mathbb{R}^{2m}$ with $\mathbb{C}^m$ and require that the transition maps be complex-analytic; this determines a **complex-analytic structure**. A manifold endowed with one of these structures is called a C^k **manifold**, **real-analytic manifold**, or **complex manifold**, respectively. (Note that a C^0 manifold is just a topological manifold.) We do not treat any of these other kinds of manifolds in this book, but they play important roles in analysis, so it is useful to know the definitions.

Local Coordinate Representations

If M is a smooth manifold, any chart (U, φ) contained in the given maximal smooth atlas is called a **smooth chart**, and the corresponding coordinate map φ is called a **smooth coordinate map**. It is useful also to introduce the terms **smooth coordinate domain** or **smooth coordinate neighborhood** for the domain of a smooth coordinate chart. A **smooth coordinate ball** means a smooth coordinate domain whose image under a smooth coordinate map is a ball in Euclidean space. A **smooth coordinate cube** is defined similarly.

It is often useful to restrict attention to coordinate balls whose closures sit nicely inside larger coordinate balls. We say a set $B \subseteq M$ is a **regular coordinate ball** if there is a smooth coordinate ball $B' \supseteq \bar{B}$ and a smooth coordinate map $\varphi \colon B' \to \mathbb{R}^n$ such that for some positive real numbers $r < r'$,

$$\varphi(B) = B_r(0), \qquad \varphi(\bar{B}) = \bar{B}_r(0), \quad \text{and} \quad \varphi(B') = B_{r'}(0).$$

Because $\bar{B}$ is homeomorphic to $\bar{B}_r(0)$, it is compact, and thus every regular coordinate ball is precompact in M. The next proposition gives a slight improvement on Lemma 1.10 for smooth manifolds. Its proof is a straightforward adaptation of the proof of that lemma.

Proposition 1.19. *Every smooth manifold has a countable basis of regular coordinate balls.*

▶ **Exercise 1.20.** Prove Proposition 1.19.

Here is how one usually thinks about coordinate charts on a smooth manifold. Once we choose a smooth chart (U, φ) on M, the coordinate map $\varphi \colon U \to \hat{U} \subseteq \mathbb{R}^n$ can be thought of as giving a temporary *identification* between U and $\hat{U}$. Using this identification, while we work in this chart, we can think of U simultaneously as an open subset of M and as an open subset of $\mathbb{R}^n$. You can visualize this identification by thinking of a "grid" drawn on U representing the preimages of the coordinate lines under φ (Fig. 1.8). Under this identification, we can represent a point $p \in U$ by its coordinates $(x^1, \ldots, x^n) = \varphi(p)$, and think of this n-tuple as *being* the

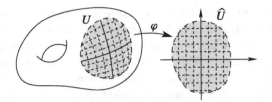

Fig. 1.8 A coordinate grid

point p. We typically express this by saying "$(x^1,\ldots,x^n)$ is the (local) coordinate representation for p" or "$p = (x^1,\ldots,x^n)$ in local coordinates."

Another way to look at it is that by means of our identification $U \leftrightarrow \hat{U}$, we can think of φ as the identity map and suppress it from the notation. This takes a bit of getting used to, but the payoff is a huge simplification of the notation in many situations. You just need to remember that the identification is in general only local, and depends heavily on the choice of coordinate chart.

You are probably already used to such identifications from your study of multivariable calculus. The most common example is **polar coordinates** (r, θ) in the plane, defined implicitly by the relation $(x, y) = (r \cos \theta, r \sin \theta)$ (see Example C.37). On an appropriate open subset such as $U = \{(x, y) : x > 0\} \subseteq \mathbb{R}^2$, (r, θ) can be expressed as smooth functions of (x, y), and the map that sends (x, y) to the corresponding (r, θ) is a smooth coordinate map with respect to the standard smooth structure on $\mathbb{R}^2$. Using this map, we can write a given point $p \in U$ either as $p = (x, y)$ in standard coordinates or as $p = (r, \theta)$ in polar coordinates, where the two coordinate representations are related by $(r, \theta) = \left(\sqrt{x^2 + y^2}, \tan^{-1} y/x\right)$ and $(x, y) = (r \cos \theta, r \sin \theta)$. Other polar coordinate charts can be obtained by restricting (r, θ) to other open subsets of $\mathbb{R}^2 \smallsetminus \{0\}$.

The fact that manifolds do not come with any predetermined choice of coordinates is both a blessing and a curse. The flexibility to choose coordinates more or less arbitrarily can be a big advantage in approaching problems in manifold theory, because the coordinates can often be chosen to simplify some aspect of the problem at hand. But we pay for this flexibility by being obliged to ensure that any objects we wish to define globally on a manifold are not dependent on a particular choice of coordinates. There are generally two ways of doing this: either by writing down a coordinate-dependent definition and then proving that the definition gives the same results in any coordinate chart, or by writing down a definition that is manifestly coordinate-independent (often called an *invariant definition*). We will use the coordinate-dependent approach in a few circumstances where it is notably simpler, but for the most part we will give coordinate-free definitions whenever possible. The need for such definitions accounts for much of the abstraction of modern manifold theory. One of the most important skills you will need to acquire in order to use manifold theory effectively is an ability to switch back and forth easily between invariant descriptions and their coordinate counterparts.

Examples of Smooth Manifolds

Before proceeding further with the general theory, let us survey some examples of smooth manifolds.

Example 1.21 (0-Dimensional Manifolds). A topological manifold M of dimension 0 is just a countable discrete space. For each point $p \in M$, the only neighborhood of p that is homeomorphic to an open subset of $\mathbb{R}^0$ is $\{p\}$ itself, and there is exactly one coordinate map $\varphi \colon \{p\} \to \mathbb{R}^0$. Thus, the set of all charts on M trivially satisfies the smooth compatibility condition, and each 0-dimensional manifold has a unique smooth structure. //

Example 1.22 (Euclidean Spaces). For each nonnegative integer n, the Euclidean space $\mathbb{R}^n$ is a smooth n-manifold with the smooth structure determined by the atlas consisting of the single chart $(\mathbb{R}^n, \mathrm{Id}_{\mathbb{R}^n})$. We call this the *standard smooth structure on* $\mathbb{R}^n$ and the resulting coordinate map *standard coordinates*. Unless we explicitly specify otherwise, we always use this smooth structure on $\mathbb{R}^n$. With respect to this smooth structure, the smooth coordinate charts for $\mathbb{R}^n$ are exactly those charts (U, φ) such that φ is a diffeomorphism (in the sense of ordinary calculus) from U to another open subset $\hat{U} \subseteq \mathbb{R}^n$. //

Example 1.23 (Another Smooth Structure on $\mathbb{R}$). Consider the homeomorphism $\psi \colon \mathbb{R} \to \mathbb{R}$ given by

$$\psi(x) = x^3. \tag{1.1}$$

The atlas consisting of the single chart $(\mathbb{R}, \psi)$ defines a smooth structure on $\mathbb{R}$. This chart is not smoothly compatible with the standard smooth structure, because the transition map $\mathrm{Id}_{\mathbb{R}} \circ \psi^{-1}(y) = y^{1/3}$ is not smooth at the origin. Therefore, the smooth structure defined on $\mathbb{R}$ by ψ is not the same as the standard one. Using similar ideas, it is not hard to construct many distinct smooth structures on any given positive-dimensional topological manifold, as long as it has one smooth structure to begin with (see Problem 1-6). //

Example 1.24 (Finite-Dimensional Vector Spaces). Let V be a finite-dimensional real vector space. Any norm on V determines a topology, which is independent of the choice of norm (Exercise B.49). With this topology, V is a topological n-manifold, and has a natural smooth structure defined as follows. Each (ordered) basis $(E_1, \ldots, E_n)$ for V defines a basis isomorphism $E \colon \mathbb{R}^n \to V$ by

$$E(x) = \sum_{i=1}^{n} x^i E_i.$$

This map is a homeomorphism, so (V, E^{-1}) is a chart. If $(\tilde{E}_1, \ldots, \tilde{E}_n)$ is any other basis and $\tilde{E}(x) = \sum_j x^j \tilde{E}_j$ is the corresponding isomorphism, then there is some invertible matrix (A_i^j) such that $E_i = \sum_j A_i^j \tilde{E}_j$ for each i. The transition map between the two charts is then given by $\tilde{E}^{-1} \circ E(x) = \tilde{x}$, where $\tilde{x} = (\tilde{x}^1, \ldots, \tilde{x}^n)$

is determined by

$$\sum_{j=1}^{n} \tilde{x}^j \tilde{E}_j = \sum_{i=1}^{n} x^i E_i = \sum_{i,j=1}^{n} x^i A_i^j \tilde{E}_j.$$

It follows that $\tilde{x}^j = \sum_i A_i^j x^i$. Thus, the map sending x to $\tilde{x}$ is an invertible linear map and hence a diffeomorphism, so any two such charts are smoothly compatible. The collection of all such charts thus defines a smooth structure, called the *standard smooth structure on V*.   //

The Einstein Summation Convention

This is a good place to pause and introduce an important notational convention that is commonly used in the study of smooth manifolds. Because of the proliferation of summations such as $\sum_i x^i E_i$ in this subject, we often abbreviate such a sum by omitting the summation sign, as in

$$E(x) = x^i E_i, \quad \text{an abbreviation for } E(x) = \sum_{i=1}^{n} x^i E_i.$$

We interpret any such expression according to the following rule, called the *Einstein summation convention*: if the same index name (such as i in the expression above) appears exactly twice in any monomial term, once as an upper index and once as a lower index, that term is understood to be summed over all possible values of that index, generally from 1 to the dimension of the space in question. This simple idea was introduced by Einstein to reduce the complexity of expressions arising in the study of smooth manifolds by eliminating the necessity of explicitly writing summation signs. We use the summation convention systematically throughout the book (except in the appendices, which many readers will look at before the rest of the book).

Another important aspect of the summation convention is the positions of the indices. We always write basis vectors (such as E_i) with lower indices, and components of a vector with respect to a basis (such as x^i) with upper indices. These index conventions help to ensure that, in summations that make mathematical sense, each index to be summed over typically appears twice in any given term, once as a lower index and once as an upper index. Any index that is implicitly summed over is a "dummy index," meaning that the value of such an expression is unchanged if a different name is substituted for each dummy index. For example, $x^i E_i$ and $x^j E_j$ mean exactly the same thing.

Since the coordinates of a point $(x^1, \ldots, x^n) \in \mathbb{R}^n$ are also its components with respect to the standard basis, in order to be consistent with our convention of writing components of vectors with upper indices, we need to use upper indices for these coordinates, and we do so throughout this book. Although this may seem awkward at first, in combination with the summation convention it offers enormous advantages when we work with complicated indexed sums, not the least of which is that expressions that are not mathematically meaningful often betray themselves quickly by violating the index convention. (The main exceptions are expressions involving the Euclidean dot product $x \cdot y = \sum_i x^i y^i$, in which the same index appears twice in the upper position, and the standard symplectic form on $\mathbb{R}^{2n}$, which we will define in Chapter 22. We always explicitly write summation signs in such expressions.)

More Examples

Now we continue with our examples of smooth manifolds.

Example 1.25 (Spaces of Matrices). Let $\mathrm{M}(m \times n, \mathbb{R})$ denote the set of $m \times n$ matrices with real entries. Because it is a real vector space of dimension mn under matrix addition and scalar multiplication, $\mathrm{M}(m \times n, \mathbb{R})$ is a smooth mn-dimensional manifold. (In fact, it is often useful to *identify* $\mathrm{M}(m \times n, \mathbb{R})$ with $\mathbb{R}^{mn}$, just by stringing all the matrix entries out in a single row.) Similarly, the space $\mathrm{M}(m \times n, \mathbb{C})$ of $m \times n$ complex matrices is a vector space of dimension $2mn$ over $\mathbb{R}$, and thus a smooth manifold of dimension $2mn$. In the special case in which $m = n$ (square matrices), we abbreviate $\mathrm{M}(n \times n, \mathbb{R})$ and $\mathrm{M}(n \times n, \mathbb{C})$ by $\mathrm{M}(n, \mathbb{R})$ and $\mathrm{M}(n, \mathbb{C})$, respectively.   //

Example 1.26 (Open Submanifolds). Let U be any open subset of $\mathbb{R}^n$. Then U is a topological n-manifold, and the single chart (U, Id_U) defines a smooth structure on U.

More generally, let M be a smooth n-manifold and let $U \subseteq M$ be any open subset. Define an atlas on U by

$$\mathcal{A}_U = \{\text{smooth charts } (V, \varphi) \text{ for } M \text{ such that } V \subseteq U\}.$$

Every point $p \in U$ is contained in the domain of some chart (W, φ) for M; if we set $V = W \cap U$, then $(V, \varphi|_V)$ is a chart in $\mathcal{A}_U$ whose domain contains p. Therefore, U is covered by the domains of charts in $\mathcal{A}_U$, and it is easy to verify that this is a smooth atlas for U. Thus any open subset of M is itself a smooth n-manifold in a natural way. Endowed with this smooth structure, we call any open subset an **open submanifold of M.** (We will define a more general class of submanifolds in Chapter 5.)   //

Example 1.27 (The General Linear Group). The *general linear group* $\mathrm{GL}(n, \mathbb{R})$ is the set of invertible $n \times n$ matrices with real entries. It is a smooth n^2-dimensional manifold because it is an open subset of the n^2-dimensional vector space $\mathrm{M}(n, \mathbb{R})$, namely the set where the (continuous) determinant function is nonzero.   //

Example 1.28 (Matrices of Full Rank). The previous example has a natural generalization to rectangular matrices of full rank. Suppose $m < n$, and let $\mathrm{M}_m(m \times n, \mathbb{R})$ denote the subset of $\mathrm{M}(m \times n, \mathbb{R})$ consisting of matrices of rank m. If A is an arbitrary such matrix, the fact that rank $A = m$ means that A has some nonsingular $m \times m$ submatrix. By continuity of the determinant function, this same submatrix has nonzero determinant on a neighborhood of A in $M(m \times n, \mathbb{R})$, which implies that A has a neighborhood contained in $M_m(m \times n, \mathbb{R})$. Thus, $M_m(m \times n, \mathbb{R})$ is an open subset of $M(m \times n, \mathbb{R})$, and therefore is itself a smooth mn-dimensional manifold. A similar argument shows that $M_n(m \times n, \mathbb{R})$ is a smooth mn-manifold when $n < m$. //

Example 1.29 (Spaces of Linear Maps). Suppose V and W are finite-dimensional real vector spaces, and let $L(V; W)$ denote the set of linear maps from V to W. Then because $L(V; W)$ is itself a finite-dimensional vector space (whose dimension is the product of the dimensions of V and W), it has a natural smooth manifold structure as in Example 1.24. One way to put global coordinates on it is to choose bases for V and W, and represent each $T \in L(V; W)$ by its matrix, which yields an isomorphism of $L(V; W)$ with $M(m \times n, \mathbb{R})$ for $m = \dim W$ and $n = \dim V$. //

Example 1.30 (Graphs of Smooth Functions). If $U \subseteq \mathbb{R}^n$ is an open subset and $f: U \to \mathbb{R}^k$ is a smooth function, we have already observed above (Example 1.3) that the graph of f is a topological n-manifold in the subspace topology. Since $\Gamma(f)$ is covered by the single graph coordinate chart $\varphi: \Gamma(f) \to U$ (the restriction of π_1), we can put a canonical smooth structure on $\Gamma(f)$ by declaring the graph coordinate chart $(\Gamma(f), \varphi)$ to be a smooth chart. //

Example 1.31 (Spheres). We showed in Example 1.4 that the n-sphere $\mathbb{S}^n \subseteq \mathbb{R}^{n+1}$ is a topological n-manifold. We put a smooth structure on $\mathbb{S}^n$ as follows. For each $i = 1, \ldots, n + 1$, let $(U_i^{\pm}, \varphi_i^{\pm})$ denote the graph coordinate charts we constructed in Example 1.4. For any distinct indices i and j, the transition map $\varphi_i^{\pm} \circ (\varphi_j^{\pm})^{-1}$ is easily computed. In the case $i < j$, we get

$$\varphi_i^{\pm} \circ (\varphi_j^{\pm})^{-1} (u^1, \ldots, u^n) = \left(u^1, \ldots, \widehat{u^i}, \ldots, \pm\sqrt{1 - |u|^2}, \ldots, u^n \right)$$

(with the square root in the jth position), and a similar formula holds when $i > j$. When $i = j$, an even simpler computation gives $\varphi_i^+ \circ (\varphi_i^-)^{-1} = \varphi_i^- \circ (\varphi_i^+)^{-1} = \mathrm{Id}_{\mathbb{B}^n}$. Thus, the collection of charts $\{(U_i^{\pm}, \varphi_i^{\pm})\}$ is a smooth atlas, and so defines a smooth structure on $\mathbb{S}^n$. We call this its **standard smooth structure**. //

Example 1.32 (Level Sets). The preceding example can be generalized as follows. Suppose $U \subseteq \mathbb{R}^n$ is an open subset and $\Phi: U \to \mathbb{R}$ is a smooth function. For any $c \in \mathbb{R}$, the set $\Phi^{-1}(c)$ is called a **level set of** Φ. Choose some $c \in \mathbb{R}$, let $M = \Phi^{-1}(c)$, and suppose it happens that the total derivative $D\Phi(a)$ is nonzero for each $a \in \Phi^{-1}(c)$. Because $D\Phi(a)$ is a row matrix whose entries are the partial derivatives $(\partial \Phi / \partial x^1(a), \ldots, \partial \Phi / \partial x^n(a))$, for each $a \in M$ there is some i such that $\partial \Phi / \partial x^i(a) \neq 0$. It follows from the implicit function theorem (Theorem C.40, with x^i playing the role of y) that there is a neighborhood U_0 of a such that $M \cap U_0$ can be expressed as a graph of an equation of the form

$$x^i = f\left(x^1, \ldots, \widehat{x^i}, \ldots, x^n\right),$$

for some smooth real-valued function f defined on an open subset of $\mathbb{R}^{n-1}$. Therefore, arguing just as in the case of the n-sphere, we see that M is a topological manifold of dimension $(n-1)$, and has a smooth structure such that each of the graph coordinate charts associated with a choice of f as above is a smooth chart. In Chapter 5, we will develop the theory of smooth submanifolds, which is a far-reaching generalization of this construction. //

Example 1.33 (Projective Spaces). The n-dimensional real projective space $\mathbb{R}\mathbb{P}^n$ is a topological n-manifold by Example 1.5. Let us check that the coordinate charts (U_i, φ_i) constructed in that example are all smoothly compatible. Assuming for convenience that $i > j$, it is straightforward to compute that

$$\varphi_j \circ \varphi_i^{-1}(u^1,\ldots,u^n) = \left(\frac{u^1}{u^j},\ldots,\frac{u^{j-1}}{u^j},\frac{u^{j+1}}{u^j},\ldots,\frac{u^{i-1}}{u^j},\frac{1}{u^j},\frac{u^i}{u^j},\ldots,\frac{u^n}{u^j}\right),$$

which is a diffeomorphism from $\varphi_i(U_i \cap U_j)$ to $\varphi_j(U_i \cap U_j)$. //

Example 1.34 (Smooth Product Manifolds). If $M_1,\ldots,M_k$ are smooth manifolds of dimensions $n_1,\ldots,n_k$, respectively, we showed in Example 1.8 that the product space $M_1 \times \cdots \times M_k$ is a topological manifold of dimension $n_1 + \cdots + n_k$, with charts of the form $(U_1 \times \cdots \times U_k, \varphi_1 \times \cdots \times \varphi_k)$. Any two such charts are smoothly compatible because, as is easily verified,

$$(\psi_1 \times \cdots \times \psi_k) \circ (\varphi_1 \times \cdots \times \varphi_k)^{-1} = \left(\psi_1 \circ \varphi_1^{-1}\right) \times \cdots \times \left(\psi_k \circ \varphi_k^{-1}\right),$$

which is a smooth map. This defines a natural smooth manifold structure on the product, called the ***product smooth manifold structure***. For example, this yields a smooth manifold structure on the n-torus $\mathbb{T}^n = \mathbb{S}^1 \times \cdots \times \mathbb{S}^1$. //

In each of the examples we have seen so far, we constructed a smooth manifold structure in two stages: we started with a topological space and checked that it was a topological manifold, and then we specified a smooth structure. It is often more convenient to combine these two steps into a single construction, especially if we start with a set that is not already equipped with a topology. The following lemma provides a shortcut—it shows how, given a set with suitable "charts" that overlap smoothly, we can use the charts to define both a topology and a smooth structure on the set.

Lemma 1.35 (Smooth Manifold Chart Lemma). *Let M be a set, and suppose we are given a collection $\{U_\alpha\}$ of subsets of M together with maps $\varphi_\alpha : U_\alpha \to \mathbb{R}^n$, such that the following properties are satisfied:*

(i) *For each α, φ_α is a bijection between U_α and an open subset $\varphi_\alpha(U_\alpha) \subseteq \mathbb{R}^n$.*

(ii) *For each α and β, the sets $\varphi_\alpha(U_\alpha \cap U_\beta)$ and $\varphi_\beta(U_\alpha \cap U_\beta)$ are open in $\mathbb{R}^n$.*

(iii) *Whenever $U_\alpha \cap U_\beta \neq \varnothing$, the map $\varphi_\beta \circ \varphi_\alpha^{-1} : \varphi_\alpha(U_\alpha \cap U_\beta) \to \varphi_\beta(U_\alpha \cap U_\beta)$ is smooth.*

(iv) *Countably many of the sets U_α cover M.*

(v) *Whenever p, q are distinct points in M, either there exists some U_α containing both p and q or there exist disjoint sets U_α, U_β with $p \in U_\alpha$ and $q \in U_\beta$.*

Then M has a unique smooth manifold structure such that each $(U_\alpha, \varphi_\alpha)$ is a smooth chart.

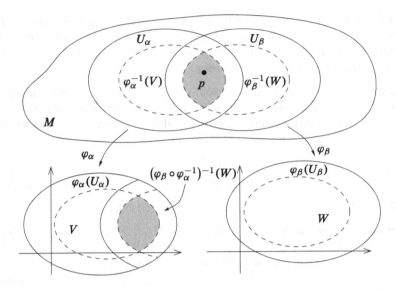

Fig. 1.9 The smooth manifold chart lemma

Proof. We define the topology by taking all sets of the form $\varphi_\alpha^{-1}(V)$, with V an open subset of $\mathbb{R}^n$, as a basis. To prove that this is a basis for a topology, we need to show that for any point p in the intersection of two basis sets $\varphi_\alpha^{-1}(V)$ and $\varphi_\beta^{-1}(W)$, there is a third basis set containing p and contained in the intersection. It suffices to show that $\varphi_\alpha^{-1}(V) \cap \varphi_\beta^{-1}(W)$ is itself a basis set (Fig. 1.9). To see this, observe that (iii) implies that $\left(\varphi_\beta \circ \varphi_\alpha^{-1}\right)^{-1}(W)$ is an open subset of $\varphi_\alpha(U_\alpha \cap U_\beta)$, and (ii) implies that this set is also open in $\mathbb{R}^n$. It follows that

$$\varphi_\alpha^{-1}(V) \cap \varphi_\beta^{-1}(W) = \varphi_\alpha^{-1}\left(V \cap \left(\varphi_\beta \circ \varphi_\alpha^{-1}\right)^{-1}(W)\right)$$

is also a basis set, as claimed.

Each map φ_α is then a homeomorphism onto its image (essentially by definition), so M is locally Euclidean of dimension n. The Hausdorff property follows easily from (v), and second-countability follows from (iv) and the result of Exercise A.22, because each U_α is second-countable. Finally, (iii) guarantees that the collection $\{(U_\alpha, \varphi_\alpha)\}$ is a smooth atlas. It is clear that this topology and smooth structure are the unique ones satisfying the conclusions of the lemma. $\qquad\square$

Example 1.36 (Grassmann Manifolds). Let V be an n-dimensional real vector space. For any integer $0 \leq k \leq n$, we let $G_k(V)$ denote the set of all k-dimensional linear subspaces of V. We will show that $G_k(V)$ can be naturally given the structure of a smooth manifold of dimension $k(n-k)$. With this structure, it is called a **Grassmann manifold**, or simply a **Grassmannian**. In the special case $V = \mathbb{R}^n$, the Grassmannian $G_k\left(\mathbb{R}^n\right)$ is often denoted by some simpler notation such as $G_{k,n}$ or $G(k,n)$. Note that $G_1\left(\mathbb{R}^{n+1}\right)$ is exactly the n-dimensional projective space $\mathbb{R}P^n$.

The construction of a smooth structure on $G_k(V)$ is somewhat more involved than the ones we have done so far, but the basic idea is just to use linear algebra to construct charts for $G_k(V)$, and then apply the smooth manifold chart lemma. We will give a shorter proof that $G_k(V)$ is a smooth manifold in Chapter 21 (see Example 21.21).

Let P and Q be any complementary subspaces of V of dimensions k and $n-k$, respectively, so that V decomposes as a direct sum: $V = P \oplus Q$. The graph of any linear map $X: P \to Q$ can be identified with a k-dimensional subspace $\Gamma(X) \subseteq V$, defined by

$$\Gamma(X) = \{v + Xv : v \in P\}.$$

Any such subspace has the property that its intersection with Q is the zero subspace. Conversely, any subspace $S \subseteq V$ that intersects Q trivially is the graph of a unique linear map $X: P \to Q$, which can be constructed as follows: let $\pi_P: V \to P$ and $\pi_Q: V \to Q$ be the projections determined by the direct sum decomposition; then the hypothesis implies that $\pi_P|_S$ is an isomorphism from S to P. Therefore, $X = (\pi_Q|_S) \circ (\pi_P|_S)^{-1}$ is a well-defined linear map from P to Q, and it is straightforward to check that S is its graph.

Let $L(P; Q)$ denote the vector space of linear maps from P to Q, and let U_Q denote the subset of $G_k(V)$ consisting of k-dimensional subspaces whose intersections with Q are trivial. The assignment $X \mapsto \Gamma(X)$ defines a map $\Gamma: L(P; Q) \to U_Q$, and the discussion above shows that Γ is a bijection. Let $\varphi = \Gamma^{-1}: U_Q \to L(P; Q)$. By choosing bases for P and Q, we can identify $L(P; Q)$ with $M((n-k) \times k, \mathbb{R})$ and hence with $\mathbb{R}^{k(n-k)}$, and thus we can think of (U_Q, φ) as a coordinate chart. Since the image of each such chart is all of $L(P; Q)$, condition (i) of Lemma 1.35 is clearly satisfied.

Now let (P', Q') be any other such pair of subspaces, and let $\pi_{P'}, \pi_{Q'}$ be the corresponding projections and $\varphi': U_{Q'} \to L(P'; Q')$ the corresponding map. The set $\varphi(U_Q \cap U_{Q'}) \subseteq L(P; Q)$ consists of all linear maps $X: P \to Q$ whose graphs intersect Q' trivially. To see that this set is open in $L(P; Q)$, for each $X \in L(P; Q)$, let $I_X: P \to V$ be the map $I_X(v) = v + Xv$, which is a bijection from P to the graph of X. Because $\Gamma(X) = \operatorname{Im} I_X$ and $Q' = \operatorname{Ker} \pi_{P'}$, it follows from Exercise B.22(d) that the graph of X intersects Q' trivially if and only if $\pi_{P'} \circ I_X$ has full rank. Because the matrix entries of $\pi_{P'} \circ I_X$ (with respect to any bases) depend continuously on X, the result of Example 1.28 shows that the set of all such X is open in $L(P; Q)$. Thus property (ii) in the smooth manifold chart lemma holds.

We need to show that the transition map $\varphi' \circ \varphi^{-1}$ is smooth on $\varphi(U_Q \cap U_{Q'})$. Suppose $X \in \varphi(U_Q \cap U_{Q'}) \subseteq L(P; Q)$ is arbitrary, and let S denote the subspace $\Gamma(X) \subseteq V$. If we put $X' = \varphi' \circ \varphi^{-1}(X)$, then as above, $X' = (\pi_{Q'}|_S) \circ (\pi_{P'}|_S)^{-1}$ (see Fig. 1.10). To relate this map to X, note that $I_X: P \to S$ is an isomorphism, so we can write

$$X' = (\pi_{Q'}|_S) \circ I_X \circ (I_X)^{-1} \circ (\pi_{P'}|_S)^{-1} = (\pi_{Q'} \circ I_X) \circ (\pi_{P'} \circ I_X)^{-1}.$$

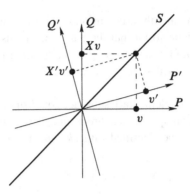

Fig. 1.10 Smooth compatibility of coordinates on $G_k(V)$

To show that this depends smoothly on X, define linear maps $A \colon P \to P'$, $B \colon P \to Q'$, $C \colon Q \to P'$, and $D \colon Q \to Q'$ as follows:

$$A = \pi_{P'}\big|_P, \qquad B = \pi_{Q'}\big|_P, \qquad C = \pi_{P'}\big|_Q, \qquad D = \pi_{Q'}\big|_Q.$$

Then for $v \in P$, we have

$$(\pi_{P'} \circ I_X)v = (A + CX)v, \qquad (\pi_{Q'} \circ I_X)v = (B + DX)v,$$

from which it follows that $X' = (B + DX)(A + CX)^{-1}$. Once we choose bases for P, Q, P', and Q', all of these linear maps are represented by matrices. Because the matrix entries of $(A + CX)^{-1}$ are rational functions of those of $A + CX$ by Cramer's rule, it follows that the matrix entries of X' depend smoothly on those of X. This proves that $\varphi' \circ \varphi^{-1}$ is a smooth map, so the charts we have constructed satisfy condition (iii) of Lemma 1.35.

To check condition (iv), we just note that $G_k(V)$ can in fact be covered by *finitely* many of the sets U_Q: for example, if $(E_1, \ldots, E_n)$ is any fixed basis for V, any partition of the basis elements into two subsets containing k and $n - k$ elements determines appropriate subspaces P and Q, and any subspace S must have trivial intersection with Q for at least one of these partitions (see Exercise B.9). Thus, $G_k(V)$ is covered by the finitely many charts determined by all possible partitions of a fixed basis.

Finally, the Hausdorff condition (v) is easily verified by noting that for any two k-dimensional subspaces $P, P' \subseteq V$, it is possible to find a subspace Q of dimension $n - k$ whose intersections with both P and P' are trivial, and then P and P' are both contained in the domain of the chart determined by, say, (P, Q).   //

Manifolds with Boundary

In many important applications of manifolds, most notably those involving integration, we will encounter spaces that would be smooth manifolds except that they

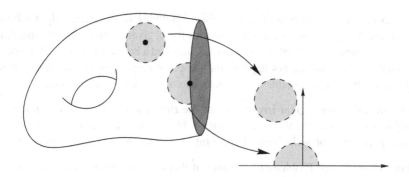

Fig. 1.11 A manifold with boundary

have a "boundary" of some sort. Simple examples of such spaces include closed intervals in $\mathbb{R}$, closed balls in $\mathbb{R}^n$, and closed hemispheres in $\mathbb{S}^n$. To accommodate such spaces, we need to extend our definition of manifolds.

Points in these spaces will have neighborhoods modeled either on open subsets of $\mathbb{R}^n$ or on open subsets of the *closed n-dimensional upper half-space* $\mathbb{H}^n \subseteq \mathbb{R}^n$, defined as

$$\mathbb{H}^n = \left\{ \left(x^1, \ldots, x^n \right) \in \mathbb{R}^n : x^n \geq 0 \right\}.$$

We will use the notations $\operatorname{Int} \mathbb{H}^n$ and $\partial \mathbb{H}^n$ to denote the interior and boundary of $\mathbb{H}^n$, respectively, as a subset of $\mathbb{R}^n$. When $n > 0$, this means

$$\operatorname{Int} \mathbb{H}^n = \left\{ \left(x^1, \ldots, x^n \right) \in \mathbb{R}^n : x^n > 0 \right\},$$

$$\partial \mathbb{H}^n = \left\{ \left(x^1, \ldots, x^n \right) \in \mathbb{R}^n : x^n = 0 \right\}.$$

In the $n = 0$ case, $\mathbb{H}^0 = \mathbb{R}^0 = \{0\}$, so $\operatorname{Int} \mathbb{H}^0 = \mathbb{R}^0$ and $\partial \mathbb{H}^0 = \varnothing$.

An *n-dimensional topological manifold with boundary* is a second-countable Hausdorff space M in which every point has a neighborhood homeomorphic either to an open subset of $\mathbb{R}^n$ or to a (relatively) open subset of $\mathbb{H}^n$ (Fig. 1.11). An open subset $U \subseteq M$ together with a map $\varphi \colon U \to \mathbb{R}^n$ that is a homeomorphism onto an open subset of $\mathbb{R}^n$ or $\mathbb{H}^n$ will be called a *chart for M*, just as in the case of manifolds. When it is necessary to make the distinction, we will call (U, φ) an *interior chart* if $\varphi(U)$ is an open subset of $\mathbb{R}^n$ (which includes the case of an open subset of $\mathbb{H}^n$ that does not intersect $\partial \mathbb{H}^n$), and a *boundary chart* if $\varphi(U)$ is an open subset of $\mathbb{H}^n$ such that $\varphi(U) \cap \partial \mathbb{H}^n \neq \varnothing$. A boundary chart whose image is a set of the form $B_r(x) \cap \mathbb{H}^n$ for some $x \in \partial \mathbb{H}^n$ and $r > 0$ is called a *coordinate half-ball*.

A point $p \in M$ is called an *interior point of M* if it is in the domain of some interior chart. It is a *boundary point of M* if it is in the domain of a boundary chart that sends p to $\partial \mathbb{H}^n$. The *boundary of M* (the set of all its boundary points) is denoted by ∂M; similarly, its *interior*, the set of all its interior points, is denoted by $\operatorname{Int} M$.

It follows from the definition that each point $p \in M$ is either an interior point or a boundary point: if p is not a boundary point, then either it is in the domain of an interior chart or it is in the domain of a boundary chart (U, φ) such that $\varphi(p) \notin \partial \mathbb{H}^n$, in which case the restriction of φ to $U \cap \varphi^{-1}(\operatorname{Int} \mathbb{H}^n)$ is an interior chart whose domain contains p. However, it is not obvious that a given point cannot be simultaneously an interior point with respect to one chart and a boundary point with respect to another. In fact, this cannot happen, but the proof requires more machinery than we have available at this point. For convenience, we state the theorem here.

Theorem 1.37 (Topological Invariance of the Boundary). *If M is a topological manifold with boundary, then each point of M is either a boundary point or an interior point, but not both. Thus ∂M and* Int *M are disjoint sets whose union is M.*

For the proof, see Problem 17-9. Later in this chapter, we will prove a weaker version of this result for smooth manifolds with boundary (Theorem 1.46), which will be sufficient for most of our purposes.

Be careful to observe the distinction between these new definitions of the terms *boundary* and *interior* and their usage to refer to the boundary and interior of a subset of a topological space. A manifold with boundary may have nonempty boundary in this new sense, irrespective of whether it has a boundary as a subset of some other topological space. If we need to emphasize the difference between the two notions of boundary, we will use the terms ***topological boundary*** and ***manifold boundary*** as appropriate. For example, the closed unit ball $\overline{\mathbb{B}}^n$ is a manifold with boundary (see Problem 1-11), whose manifold boundary is $\mathbb{S}^{n-1}$. Its topological boundary as a subset of $\mathbb{R}^n$ happens to be the sphere as well. However, if we think of $\overline{\mathbb{B}}^n$ as a topological space in its own right, then as a subset of itself, it has empty topological boundary. And if we think of it as a subset of $\mathbb{R}^{n+1}$ (considering $\mathbb{R}^n$ as a subset of $\mathbb{R}^{n+1}$ in the obvious way), its topological boundary is all of $\overline{\mathbb{B}}^n$. Note that $\mathbb{H}^n$ is itself a manifold with boundary, and its manifold boundary is the same as its topological boundary as a subset of $\mathbb{R}^n$. Every interval in $\mathbb{R}$ is a 1-manifold with boundary, whose manifold boundary consists of its endpoints (if any).

The nomenclature for manifolds with boundary is traditional and well established, but it must be used with care. Despite their name, manifolds with boundary are *not* in general manifolds, because boundary points do not have locally Euclidean neighborhoods. (This is a consequence of the theorem on invariance of the boundary.) Moreover, a manifold with boundary might have empty boundary—there is nothing in the definition that requires the boundary to be a nonempty set. On the other hand, a manifold is also a manifold with boundary, whose boundary is empty. Thus, every manifold is a manifold with boundary, but a manifold with boundary is a manifold if and only if its boundary is empty (see Proposition 1.38 below).

Even though the term *manifold with boundary* encompasses manifolds as well, we will often use redundant phrases such as ***manifold without boundary*** if we wish to emphasize that we are talking about a manifold in the original sense, and ***manifold with or without boundary*** to refer to a manifold with boundary if we wish emphasize that the boundary might be empty. (The latter phrase will often appear when our primary interest is in manifolds, but the results being discussed are just as easy to state and prove in the more general case of manifolds with boundary.) Note that the word "manifold" without further qualification always means a manifold without boundary. In the literature, you will also encounter the terms ***closed manifold*** to mean a compact manifold without boundary, and ***open manifold*** to mean a noncompact manifold without boundary.

Proposition 1.38. *Let M be a topological n-manifold with boundary.*

(a) *Int M is an open subset of M and a topological n-manifold without boundary.*
(b) *∂M is a closed subset of M and a topological $(n - 1)$-manifold without boundary.*
(c) *M is a topological manifold if and only if $\partial M = \varnothing$.*
(d) *If $n = 0$, then $\partial M = \varnothing$ and M is a 0-manifold.*

▶ **Exercise 1.39.** Prove the preceding proposition. For this proof, you may use the theorem on topological invariance of the boundary when necessary. Which parts require it?

The topological properties of manifolds that we proved earlier in the chapter have natural extensions to manifolds with boundary, with essentially the same proofs as in the manifold case. For the record, we state them here.

Proposition 1.40. *Let M be a topological manifold with boundary.*

(a) *M has a countable basis of precompact coordinate balls and half-balls.*
(b) *M is locally compact.*
(c) *M is paracompact.*
(d) *M is locally path-connected.*
(e) *M has countably many components, each of which is an open subset of M and a connected topological manifold with boundary.*
(f) *The fundamental group of M is countable.*

▶ **Exercise 1.41.** Prove the preceding proposition.

Smooth Structures on Manifolds with Boundary

To see how to define a smooth structure on a manifold with boundary, recall that a map from an arbitrary subset $A \subseteq \mathbb{R}^n$ to $\mathbb{R}^k$ is said to be smooth if in a neighborhood of each point of A it admits an extension to a smooth map defined on an open subset of $\mathbb{R}^n$ (see Appendix C, p. 645). Thus, if U is an open subset of $\mathbb{H}^n$, a map $F: U \to \mathbb{R}^k$ is smooth if for each $x \in U$, there exists an open subset $\widetilde{U} \subseteq \mathbb{R}^n$ containing x and a smooth map $\widetilde{F}: \widetilde{U} \to \mathbb{R}^k$ that agrees with F on $\widetilde{U} \cap \mathbb{H}^n$ (Fig. 1.12). If F is such a map, the restriction of F to $U \cap \operatorname{Int} \mathbb{H}^n$ is smooth in the usual sense. By continuity, all partial derivatives of F at points of $U \cap \partial \mathbb{H}^n$ are determined by their values in $\operatorname{Int} \mathbb{H}^n$, and therefore in particular are independent of the choice of extension. It is a fact (which we will neither prove nor use) that $F: U \to \mathbb{R}^k$ is smooth in this sense if and only if F is continuous, $F|_{U \cap \operatorname{Int} \mathbb{H}^n}$ is smooth, and the partial derivatives of $F|_{U \cap \operatorname{Int} \mathbb{H}^n}$ of all orders have continuous extensions to all of U. (One direction is obvious; the other direction depends on a lemma of Émile Borel, which shows that there is a smooth function defined in the lower half-space whose derivatives all match those of F on $U \cap \partial \mathbb{H}^n$. See, e.g., [Hör90, Thm. 1.2.6].)

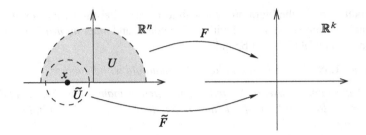

Fig. 1.12 Smoothness of maps on open subsets of $\mathbb{H}^n$

For example, let $\mathbb{B}^2 \subseteq \mathbb{R}^2$ be the open unit disk, let $U = \mathbb{B}^2 \cap \mathbb{H}^2$, and define $f \colon U \to \mathbb{R}$ by $f(x, y) = \sqrt{1 - x^2 - y^2}$. Because f extends smoothly to all of $\mathbb{B}^2$ (by the same formula), f is a smooth function on U. On the other hand, although $g(x, y) = \sqrt{y}$ is continuous on U and smooth in $U \cap \operatorname{Int} \mathbb{H}^2$, it has no smooth extension to any neighborhood of the origin in $\mathbb{R}^2$ because $\partial g / \partial y \to \infty$ as $y \to 0$. Thus g is not smooth on U.

Now let M be a topological manifold with boundary. As in the manifold case, a **smooth structure for M** is defined to be a maximal smooth atlas—a collection of charts whose domains cover M and whose transition maps (and their inverses) are smooth in the sense just described. With such a structure, M is called a **smooth manifold with boundary**. Every smooth manifold is automatically a smooth manifold with boundary (whose boundary is empty).

Just as for smooth manifolds, if M is a smooth manifold with boundary, any chart in the given smooth atlas is called a **smooth chart for M**. **Smooth coordinate balls**, **smooth coordinate half-balls**, and **regular coordinate balls in M** are defined in the obvious ways. In addition, a subset $B \subseteq M$ is called a **regular coordinate half-ball** if there is a smooth coordinate half-ball $B' \supseteq \bar{B}$ and a smooth coordinate map $\varphi \colon B' \to \mathbb{H}^n$ such that for some $r' > r > 0$ we have

$$\varphi(B) = B_r(0) \cap \mathbb{H}^n, \qquad \varphi(\bar{B}) = \bar{B}_r(0) \cap \mathbb{H}^n, \quad \text{and} \quad \varphi(B') = B_{r'}(0) \cap \mathbb{H}^n.$$

▶ **Exercise 1.42.** Show that every smooth manifold with boundary has a countable basis consisting of regular coordinate balls and half-balls.

▶ **Exercise 1.43.** Show that the smooth manifold chart lemma (Lemma 1.35) holds with "$\mathbb{R}^n$" replaced by "$\mathbb{R}^n$ or $\mathbb{H}^n$" and "smooth manifold" replaced by "smooth manifold with boundary."

▶ **Exercise 1.44.** Suppose M is a smooth n-manifold with boundary and U is an open subset of M. Prove the following statements:

(a) U is a topological n-manifold with boundary, and the atlas consisting of all smooth charts (V, φ) for M such that $V \subseteq U$ defines a smooth structure on U. With this topology and smooth structure, U is called an **open submanifold with boundary**.
(b) If $U \subseteq \operatorname{Int} M$, then U is actually a smooth manifold (without boundary); in this case we call it an **open submanifold of M**.
(c) $\operatorname{Int} M$ is an open submanifold of M (without boundary).

One important result about smooth manifolds that does *not* extend directly to smooth manifolds with boundary is the construction of smooth structures on finite products (see Example 1.8). Because a product of half-spaces $\mathbb{H}^n \times \mathbb{H}^m$ is not itself a half-space, a finite product of smooth manifolds with boundary cannot generally be considered as a smooth manifold with boundary. (Instead, it is an example of a *smooth manifold with corners*, which we will study in Chapter 16.) However, we do have the following result.

Proposition 1.45. *Suppose $M_1, \ldots, M_k$ are smooth manifolds and N is a smooth manifold with boundary. Then $M_1 \times \cdots \times M_k \times N$ is a smooth manifold with boundary, and $\partial(M_1 \times \cdots \times M_k \times N) = M_1 \times \cdots \times M_k \times \partial N$.*

Proof. Problem 1-12.   □

For smooth manifolds with boundary, the following result is often an adequate substitute for the theorem on invariance of the boundary.

Theorem 1.46 (Smooth Invariance of the Boundary). *Suppose M is a smooth manifold with boundary and $p \in M$. If there is some smooth chart (U, φ) for M such that $\varphi(U) \subseteq \mathbb{H}^n$ and $\varphi(p) \in \partial \mathbb{H}^n$, then the same is true for every smooth chart whose domain contains p.*

Proof. Suppose on the contrary that p is in the domain of a smooth interior chart (U, ψ) and also in the domain of a smooth boundary chart (V, φ) such that $\varphi(p) \in \partial \mathbb{H}^n$. Let $\tau = \varphi \circ \psi^{-1}$ denote the transition map; it is a homeomorphism from $\psi(U \cap V)$ to $\varphi(U \cap V)$. The smooth compatibility of the charts ensures that both τ and τ^{-1} are smooth, in the sense that locally they can be extended, if necessary, to smooth maps defined on open subsets of $\mathbb{R}^n$.

Write $x_0 = \psi(p)$ and $y_0 = \varphi(p) = \tau(x_0)$. There is some neighborhood W of y_0 in $\mathbb{R}^n$ and a smooth function $\eta \colon W \to \mathbb{R}^n$ that agrees with τ^{-1} on $W \cap \varphi(U \cap V)$. On the other hand, because we are assuming that ψ is an interior chart, there is an open Euclidean ball B that is centered at x_0 and contained in $\varphi(U \cap V)$, so τ itself is smooth on B in the ordinary sense. After shrinking B if necessary, we may assume that $B \subseteq \tau^{-1}(W)$. Then $\eta \circ \tau|_B = \tau^{-1} \circ \tau|_B = \mathrm{Id}_B$, so it follows from the chain rule that $D\eta(\tau(x)) \circ D\tau(x)$ is the identity map for each $x \in B$. Since $D\tau(x)$ is a square matrix, this implies that it is nonsingular. It follows from Corollary C.36 that τ (considered as a map from B to $\mathbb{R}^n$) is an open map, so $\tau(B)$ is an open subset of $\mathbb{R}^n$ that contains $y_0 = \varphi(p)$ and is contained in $\varphi(V)$. This contradicts the assumption that $\varphi(V) \subseteq \mathbb{H}^n$ and $\varphi(p) \in \partial \mathbb{H}^n$.   □

Problems

1-1. Let X be the set of all points $(x, y) \in \mathbb{R}^2$ such that $y = \pm 1$, and let M be the quotient of X by the equivalence relation generated by $(x, -1) \sim (x, 1)$ for all $x \neq 0$. Show that M is locally Euclidean and second-countable, but not Hausdorff. (This space is called the ***line with two origins***.)

1-2. Show that a disjoint union of uncountably many copies of $\mathbb{R}$ is locally Euclidean and Hausdorff, but not second-countable.

1-3. A topological space is said to be σ-*compact* if it can be expressed as a union of countably many compact subspaces. Show that a locally Euclidean Hausdorff space is a topological manifold if and only if it is σ-compact.

1-4. Let M be a topological manifold, and let $\mathcal{U}$ be an open cover of M.
   (a) Assuming that each set in $\mathcal{U}$ intersects only finitely many others, show that $\mathcal{U}$ is locally finite.
   (b) Give an example to show that the converse to (a) may be false.
   (c) Now assume that the sets in $\mathcal{U}$ are precompact in M, and prove the converse: if $\mathcal{U}$ is locally finite, then each set in $\mathcal{U}$ intersects only finitely many others.

1-5. Suppose M is a locally Euclidean Hausdorff space. Show that M is second-countable if and only if it is paracompact and has countably many connected components. [Hint: assuming M is paracompact, show that each component of M has a locally finite cover by precompact coordinate domains, and extract from this a countable subcover.]

1-6. Let M be a nonempty topological manifold of dimension $n \geq 1$. If M has a smooth structure, show that it has uncountably many distinct ones. [Hint: first show that for any $s > 0$, $F_s(x) = |x|^{s-1}x$ defines a homeomorphism from $\mathbb{B}^n$ to itself, which is a diffeomorphism if and only if $s = 1$.]

1-7. Let N denote the *north pole* $(0,\ldots,0,1) \in \mathbb{S}^n \subseteq \mathbb{R}^{n+1}$, and let S denote the *south pole* $(0,\ldots,0,-1)$. Define the *stereographic projection* $\sigma : \mathbb{S}^n \smallsetminus \{N\} \to \mathbb{R}^n$ by

$$\sigma\left(x^1,\ldots,x^{n+1}\right) = \frac{(x^1,\ldots,x^n)}{1-x^{n+1}}.$$

Let $\tilde{\sigma}(x) = -\sigma(-x)$ for $x \in \mathbb{S}^n \smallsetminus \{S\}$.
   (a) For any $x \in \mathbb{S}^n \smallsetminus \{N\}$, show that $\sigma(x) = u$, where $(u,0)$ is the point where the line through N and x intersects the linear subspace where $x^{n+1} = 0$ (Fig. 1.13). Similarly, show that $\tilde{\sigma}(x)$ is the point where the line through S and x intersects the same subspace. (For this reason, $\tilde{\sigma}$ is called *stereographic projection from the south pole*.)
   (b) Show that σ is bijective, and

$$\sigma^{-1}\left(u^1,\ldots,u^n\right) = \frac{(2u^1,\ldots,2u^n,|u|^2-1)}{|u|^2+1}.$$

   (c) Compute the transition map $\tilde{\sigma} \circ \sigma^{-1}$ and verify that the atlas consisting of the two charts $(\mathbb{S}^n \smallsetminus \{N\}, \sigma)$ and $(\mathbb{S}^n \smallsetminus \{S\}, \tilde{\sigma})$ defines a smooth structure on $\mathbb{S}^n$. (The coordinates defined by σ or $\tilde{\sigma}$ are called *stereographic coordinates*.)
   (d) Show that this smooth structure is the same as the one defined in Example 1.31.
   (*Used on pp. 201, 269, 301, 345, 347, 450.*)

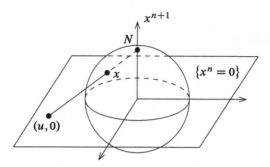

Fig. 1.13 Stereographic projection

1-8. By identifying $\mathbb{R}^2$ with $\mathbb{C}$, we can think of the unit circle $\mathbb{S}^1$ as a subset of the complex plane. An **angle function** on a subset $U \subseteq \mathbb{S}^1$ is a continuous function $\theta: U \to \mathbb{R}$ such that $e^{i\theta(z)} = z$ for all $z \in U$. Show that there exists an angle function θ on an open subset $U \subseteq \mathbb{S}^1$ if and only if $U \neq \mathbb{S}^1$. For any such angle function, show that (U, θ) is a smooth coordinate chart for $\mathbb{S}^1$ with its standard smooth structure. (*Used on pp. 37, 152, 176.*)

1-9. **Complex projective n-space**, denoted by $\mathbb{CP}^n$, is the set of all 1-dimensional complex-linear subspaces of $\mathbb{C}^{n+1}$, with the quotient topology inherited from the natural projection $\pi: \mathbb{C}^{n+1} \smallsetminus \{0\} \to \mathbb{CP}^n$. Show that $\mathbb{CP}^n$ is a compact $2n$-dimensional topological manifold, and show how to give it a smooth structure analogous to the one we constructed for $\mathbb{RP}^n$. (We use the correspondence

$$\left(x^1 + iy^1, \ldots, x^{n+1} + iy^{n+1}\right) \leftrightarrow \left(x^1, y^1, \ldots, x^{n+1}, y^{n+1}\right)$$

to identify $\mathbb{C}^{n+1}$ with $\mathbb{R}^{2n+2}$.) (*Used on pp. 48, 96, 172, 560, 561.*)

1-10. Let k and n be integers satisfying $0 < k < n$, and let $P, Q \subseteq \mathbb{R}^n$ be the linear subspaces spanned by $(e_1, \ldots, e_k)$ and $(e_{k+1}, \ldots, e_n)$, respectively, where e_i is the ith standard basis vector for $\mathbb{R}^n$. For any k-dimensional subspace $S \subseteq \mathbb{R}^n$ that has trivial intersection with Q, show that the coordinate representation $\varphi(S)$ constructed in Example 1.36 is the unique $(n-k) \times k$ matrix B such that S is spanned by the columns of the matrix $\binom{I_k}{B}$, where I_k denotes the $k \times k$ identity matrix.

1-11. Let $M = \overline{\mathbb{B}}^n$, the closed unit ball in $\mathbb{R}^n$. Show that M is a topological manifold with boundary in which each point in $\mathbb{S}^{n-1}$ is a boundary point and each point in $\mathbb{B}^n$ is an interior point. Show how to give it a smooth structure such that every smooth interior chart is a smooth chart for the standard smooth structure on $\mathbb{B}^n$. [Hint: consider the map $\pi \circ \sigma^{-1}: \mathbb{R}^n \to \mathbb{R}^n$, where $\sigma: \mathbb{S}^n \to \mathbb{R}^n$ is the stereographic projection (Problem 1-7) and π is a projection from $\mathbb{R}^{n+1}$ to $\mathbb{R}^n$ that omits some coordinate other than the last.]

1-12. Prove Proposition 1.45 (a product of smooth manifolds together with one smooth manifold with boundary is a smooth manifold with boundary).

Chapter 2
Smooth Maps

The main reason for introducing smooth structures was to enable us to define smooth functions on manifolds and smooth maps between manifolds. In this chapter we carry out that project.

We begin by defining smooth real-valued and vector-valued functions, and then generalize this to smooth maps between manifolds. We then focus our attention for a while on the special case of *diffeomorphisms*, which are bijective smooth maps with smooth inverses. If there is a diffeomorphism between two smooth manifolds, we say that they are *diffeomorphic*. The main objects of study in smooth manifold theory are properties that are invariant under diffeomorphisms.

At the end of the chapter, we introduce a powerful tool for blending together locally defined smooth objects, called *partitions of unity*. They are used throughout smooth manifold theory for building global smooth objects out of local ones.

Smooth Functions and Smooth Maps

Although the terms *function* and *map* are technically synonymous, in studying smooth manifolds it is often convenient to make a slight distinction between them. Throughout this book we generally reserve the term **function** for a map whose codomain is $\mathbb{R}$ (a **real-valued function**) or $\mathbb{R}^k$ for some $k > 1$ (a **vector-valued function**). Either of the words **map** or **mapping** can mean any type of map, such as a map between arbitrary manifolds.

Smooth Functions on Manifolds

Suppose M is a smooth n-manifold, k is a nonnegative integer, and $f : M \to \mathbb{R}^k$ is any function. We say that f is a **smooth function** if for every $p \in M$, there exists a smooth chart (U, φ) for M whose domain contains p and such that the composite function $f \circ \varphi^{-1}$ is smooth on the open subset $\widehat{U} = \varphi(U) \subseteq \mathbb{R}^n$ (Fig. 2.1). If M is a smooth manifold with boundary, the definition is exactly the same, except that

J.M. Lee, *Introduction to Smooth Manifolds*, Graduate Texts in Mathematics 218,
DOI 10.1007/978-1-4419-9982-5_2, © Springer Science+Business Media New York 2013

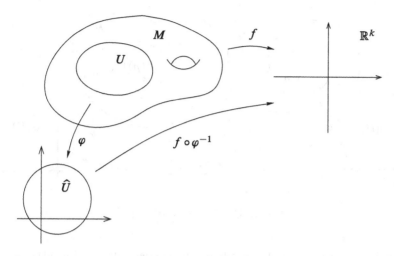

Fig. 2.1 Definition of smooth functions

$\varphi(U)$ is now an open subset of either $\mathbb{R}^n$ or $\mathbb{H}^n$, and in the latter case we interpret smoothness of $f \circ \varphi^{-1}$ to mean that each point of $\varphi(U)$ has a neighborhood (in $\mathbb{R}^n$) on which $f \circ \varphi^{-1}$ extends to a smooth function in the ordinary sense.

The most important special case is that of smooth real-valued functions $f : M \to \mathbb{R}$; the set of all such functions is denoted by $C^\infty(M)$. Because sums and constant multiples of smooth functions are smooth, $C^\infty(M)$ is a vector space over $\mathbb{R}$.

▶ **Exercise 2.1.** Let M be a smooth manifold with or without boundary. Show that pointwise multiplication turns $C^\infty(M)$ into a commutative ring and a commutative and associative algebra over $\mathbb{R}$. (See Appendix B, p. 624, for the definition of an algebra.)

▶ **Exercise 2.2.** Let U be an open submanifold of $\mathbb{R}^n$ with its standard smooth manifold structure. Show that a function $f : U \to \mathbb{R}^k$ is smooth in the sense just defined if and only if it is smooth in the sense of ordinary calculus. Do the same for an open submanifold with boundary in $\mathbb{H}^n$ (see Exercise 1.44).

▶ **Exercise 2.3.** Let M be a smooth manifold with or without boundary, and suppose $f : M \to \mathbb{R}^k$ is a smooth function. Show that $f \circ \varphi^{-1} : \varphi(U) \to \mathbb{R}^k$ is smooth for *every* smooth chart (U, φ) for M.

Given a function $f : M \to \mathbb{R}^k$ and a chart (U, φ) for M, the function $\hat{f} : \varphi(U) \to \mathbb{R}^k$ defined by $\hat{f}(x) = f \circ \varphi^{-1}(x)$ is called the ***coordinate representation of*** f. By definition, f is smooth if and only if its coordinate representation is smooth in some smooth chart around each point. By the preceding exercise, smooth functions have smooth coordinate representations in every smooth chart.

For example, consider the real-valued function $f(x, y) = x^2 + y^2$ defined on the plane. In polar coordinates on, say, the set $U = \{(x, y) : x > 0\}$, it has the coordinate representation $\hat{f}(r, \theta) = r^2$. In keeping with our practice of using local coordinates

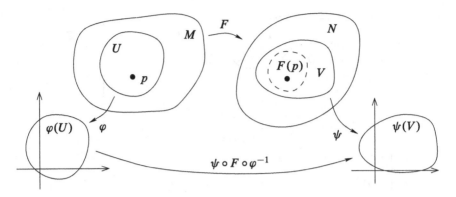

Fig. 2.2 Definition of smooth maps

to identify an open subset of a manifold with an open subset of Euclidean space, in cases where it causes no confusion we often do not even observe the distinction between $\hat{f}$ and f itself, and instead say something like "f is smooth on U because its coordinate representation $f(r, \theta) = r^2$ is smooth."

Smooth Maps Between Manifolds

The definition of smooth functions generalizes easily to maps between manifolds. Let M, N be smooth manifolds, and let $F \colon M \to N$ be any map. We say that F is a ***smooth map*** if for every $p \in M$, there exist smooth charts (U, φ) containing p and (V, ψ) containing $F(p)$ such that $F(U) \subseteq V$ and the composite map $\psi \circ F \circ \varphi^{-1}$ is smooth from $\varphi(U)$ to $\psi(V)$ (Fig. 2.2). If M and N are smooth manifolds with boundary, smoothness of F is defined in exactly the same way, with the usual understanding that a map whose domain is a subset of $\mathbb{H}^n$ is smooth if it admits an extension to a smooth map in a neighborhood of each point, and a map whose codomain is a subset of $\mathbb{H}^n$ is smooth if it is smooth as a map into $\mathbb{R}^n$. Note that our previous definition of smoothness of real-valued or vector-valued functions can be viewed as a special case of this one, by taking $N = V = \mathbb{R}^k$ and $\psi = \mathrm{Id} \colon \mathbb{R}^k \to \mathbb{R}^k$.

The first important observation about our definition of smooth maps is that, as one might expect, smoothness implies continuity.

Proposition 2.4. *Every smooth map is continuous.*

Proof. Suppose M and N are smooth manifolds with or without boundary, and $F \colon M \to N$ is smooth. Given $p \in M$, smoothness of F means there are smooth charts (U, φ) containing p and (V, ψ) containing $F(p)$, such that $F(U) \subseteq V$ and $\psi \circ F \circ \varphi^{-1} \colon \varphi(U) \to \psi(V)$ is smooth, hence continuous. Since $\varphi \colon U \to \varphi(U)$ and $\psi \colon V \to \psi(V)$ are homeomorphisms, this implies in turn that

$$F|_U = \psi^{-1} \circ \left(\psi \circ F \circ \varphi^{-1} \right) \circ \varphi \colon U \to V,$$

which is a composition of continuous maps. Since F is continuous in a neighborhood of each point, it is continuous on M.   □

To prove that a map $F: M \to N$ is smooth directly from the definition requires, in part, that for each $p \in M$ we prove the existence of coordinate domains U containing p and V containing $F(p)$ such that $F(U) \subseteq V$. This requirement is included in the definition precisely so that smoothness automatically implies continuity. (Problem 2-1 illustrates what can go wrong if this requirement is omitted.) There are other ways of characterizing smoothness of maps between manifolds that accomplish the same thing. Here are two of them.

Proposition 2.5 (Equivalent Characterizations of Smoothness). *Suppose M and N are smooth manifolds with or without boundary, and $F: M \to N$ is a map. Then F is smooth if and only if either of the following conditions is satisfied:*

(a) *For every $p \in M$, there exist smooth charts (U, φ) containing p and (V, ψ) containing $F(p)$ such that $U \cap F^{-1}(V)$ is open in M and the composite map $\psi \circ F \circ \varphi^{-1}$ is smooth from $\varphi(U \cap F^{-1}(V))$ to $\psi(V)$.*
(b) *F is continuous and there exist smooth atlases $\{(U_\alpha, \varphi_\alpha)\}$ and $\{(V_\beta, \psi_\beta)\}$ for M and N, respectively, such that for each α and β, $\psi_\beta \circ F \circ \varphi_\alpha^{-1}$ is a smooth map from $\varphi_\alpha(U_\alpha \cap F^{-1}(V_\beta))$ to $\psi_\beta(V_\beta)$.*

Proposition 2.6 (Smoothness Is Local). *Let M and N be smooth manifolds with or without boundary, and let $F: M \to N$ be a map.*

(a) *If every point $p \in M$ has a neighborhood U such that the restriction $F|_U$ is smooth, then F is smooth.*
(b) *Conversely, if F is smooth, then its restriction to every open subset is smooth.*

▶ **Exercise 2.7.** Prove the preceding two propositions.

The next corollary is essentially just a restatement of the previous proposition, but it gives a highly useful way of constructing smooth maps.

Corollary 2.8 (Gluing Lemma for Smooth Maps). *Let M and N be smooth manifolds with or without boundary, and let $\{U_\alpha\}_{\alpha \in A}$ be an open cover of M. Suppose that for each $\alpha \in A$, we are given a smooth map $F_\alpha: U_\alpha \to N$ such that the maps agree on overlaps: $F_\alpha|_{U_\alpha \cap U_\beta} = F_\beta|_{U_\alpha \cap U_\beta}$ for all α and β. Then there exists a unique smooth map $F: M \to N$ such that $F|_{U_\alpha} = F_\alpha$ for each $\alpha \in A$.*   □

If $F: M \to N$ is a smooth map, and (U, φ) and (V, ψ) are any smooth charts for M and N, respectively, we call $\widehat{F} = \psi \circ F \circ \varphi^{-1}$ the **coordinate representation of F** with respect to the given coordinates. It maps the set $\varphi(U \cap F^{-1}(V))$ to $\psi(V)$.

▶ **Exercise 2.9.** Suppose $F: M \to N$ is a smooth map between smooth manifolds with or without boundary. Show that the coordinate representation of F with respect to *every* pair of smooth charts for M and N is smooth.

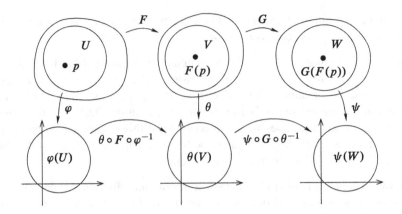

Fig. 2.3 A composition of smooth maps is smooth

As with real-valued or vector-valued functions, once we have chosen specific local coordinates in both the domain and codomain, we can often ignore the distinction between F and $\widehat{F}$.

Next we examine some simple classes of maps that are automatically smooth.

Proposition 2.10. *Let M, N, and P be smooth manifolds with or without boundary.*

(a) *Every constant map $c \colon M \to N$ is smooth.*
(b) *The identity map of M is smooth.*
(c) *If $U \subseteq M$ is an open submanifold with or without boundary, then the inclusion map $U \hookrightarrow M$ is smooth.*
(d) *If $F \colon M \to N$ and $G \colon N \to P$ are smooth, then so is $G \circ F \colon M \to P$.*

Proof. We prove (d) and leave the rest as exercises. Let $F \colon M \to N$ and $G \colon N \to P$ be smooth maps, and let $p \in M$. By definition of smoothness of G, there exist smooth charts (V, θ) containing $F(p)$ and (W, ψ) containing $G(F(p))$ such that $G(V) \subseteq W$ and $\psi \circ G \circ \theta^{-1} \colon \theta(V) \to \psi(W)$ is smooth. Since F is continuous, $F^{-1}(V)$ is a neighborhood of p in M, so there is a smooth chart (U, φ) for M such that $p \in U \subseteq F^{-1}(V)$ (Fig. 2.3). By Exercise 2.9, $\theta \circ F \circ \varphi^{-1}$ is smooth from $\varphi(U)$ to $\theta(V)$. Then we have $G \circ F(U) \subseteq G(V) \subseteq W$, and $\psi \circ (G \circ F) \circ \varphi^{-1} = \left(\psi \circ G \circ \theta^{-1} \right) \circ \left(\theta \circ F \circ \varphi^{-1} \right) \colon \varphi(U) \to \psi(W)$ is smooth because it is a composition of smooth maps between subsets of Euclidean spaces. □

▶ **Exercise 2.11.** Prove parts (a)–(c) of the preceding proposition.

Proposition 2.12. *Suppose $M_1, \dots, M_k$ and N are smooth manifolds with or without boundary, such that at most one of $M_1, \dots, M_k$ has nonempty boundary. For each i, let $\pi_i \colon M_1 \times \cdots \times M_k \to M_i$ denote the projection onto the M_i factor. A map $F \colon N \to M_1 \times \cdots \times M_k$ is smooth if and only if each of the component maps $F_i = \pi_i \circ F \colon N \to M_i$ is smooth.*

Proof. Problem 2-2.   □

Although most of our efforts in this book are devoted to the study of smooth manifolds and smooth maps, we also need to work with topological manifolds and continuous maps on occasion. For the sake of consistency, we adopt the following conventions: without further qualification, the words "function" and "map" are to be understood purely in the set-theoretic sense, and carry no assumptions of continuity or smoothness. Most other objects we study, however, will be understood to carry some minimal topological structure by default. Unless otherwise specified, a "manifold" or "manifold with boundary" is always to be understood as a topological one, and a "coordinate chart" is to be understood in the topological sense, as a homeomorphism from an open subset of the manifold to an open subset of $\mathbb{R}^n$ or $\mathbb{H}^n$. If we wish to restrict attention to smooth manifolds or smooth coordinate charts, we will say so. Similarly, our default assumptions for many other specific types of geometric objects and the maps between them will be continuity at most; smoothness will not be assumed unless explicitly specified. The only exceptions will be a few concepts that require smoothness for their very definitions.

This convention requires a certain discipline, in that we have to remember to state the smoothness hypothesis whenever it is needed; but its advantage is that it frees us (for the most part) from having to remember which types of maps are assumed to be smooth and which are not.

On the other hand, because the definition of a smooth map requires smooth structures in the domain and codomain, if we say "$F\colon M \to N$ is a smooth map" without specifying what M and N are, it should always be understood that they are smooth manifolds with or without boundaries.

We now have enough information to produce a number of interesting examples of smooth maps. In spite of the apparent complexity of the definition, it is usually not hard to prove that a particular map is smooth. There are basically only three common ways to do so:

- Write the map in smooth local coordinates and recognize its component functions as compositions of smooth elementary functions.
- Exhibit the map as a composition of maps that are known to be smooth.
- Use some special-purpose theorem that applies to the particular case under consideration.

Example 2.13 (Smooth Maps).

(a) Any map from a zero-dimensional manifold into a smooth manifold with or without boundary is automatically smooth, because each coordinate representation is constant.

(b) If the circle $\mathbb{S}^1$ is given its standard smooth structure, the map $\varepsilon\colon \mathbb{R} \to \mathbb{S}^1$ defined by $\varepsilon(t) = e^{2\pi i t}$ is smooth, because with respect to any angle coordinate θ for $\mathbb{S}^1$ (see Problem 1-8) it has a coordinate representation of the form $\widehat{\varepsilon}(t) = 2\pi t + c$ for some constant c, as you can check.

(c) The map $\varepsilon^n\colon \mathbb{R}^n \to \mathbb{T}^n$ defined by $\varepsilon^n\left(x^1,\dots,x^n\right) = \left(e^{2\pi i x^1},\dots,e^{2\pi i x^n}\right)$ is smooth by Proposition 2.12.

(d) Now consider the n-sphere $\mathbb{S}^n$ with its standard smooth structure. The inclusion map $\iota \colon \mathbb{S}^n \hookrightarrow \mathbb{R}^{n+1}$ is certainly continuous, because it is the inclusion map of a topological subspace. It is a smooth map because its coordinate representation with respect to any of the graph coordinates of Example 1.31 is

$$\hat{\iota}\left(u^1, \ldots, u^n\right) = \iota \circ \left(\varphi_i^{\pm}\right)^{-1}\left(u^1, \ldots, u^n\right)$$
$$= \left(u^1, \ldots, u^{i-1}, \pm\sqrt{1-|u|^2}, u^i, \ldots, u^n\right),$$

which is smooth on its domain (the set where $|u|^2 < 1$).

(e) The quotient map $\pi \colon \mathbb{R}^{n+1} \smallsetminus \{0\} \to \mathbb{RP}^n$ used to define $\mathbb{RP}^n$ is smooth, because its coordinate representation in terms of any of the coordinates for $\mathbb{RP}^n$ constructed in Example 1.33 and standard coordinates on $\mathbb{R}^{n+1} \smallsetminus \{0\}$ is

$$\hat{\pi}\left(x^1, \ldots, x^{n+1}\right) = \varphi_i \circ \pi\left(x^1, \ldots, x^{n+1}\right) = \varphi_i\left[x^1, \ldots, x^{n+1}\right]$$
$$= \left(\frac{x^1}{x^i}, \ldots, \frac{x^{i-1}}{x^i}, \frac{x^{i+1}}{x^i}, \ldots, \frac{x^{n+1}}{x^i}\right).$$

(f) Define $q \colon \mathbb{S}^n \to \mathbb{RP}^n$ as the restriction of $\pi \colon \mathbb{R}^{n+1} \smallsetminus \{0\} \to \mathbb{RP}^n$ to $\mathbb{S}^n \subseteq \mathbb{R}^{n+1} \smallsetminus \{0\}$. It is a smooth map, because it is the composition $q = \pi \circ \iota$ of the maps in the preceding two examples.

(g) If $M_1, \ldots, M_k$ are smooth manifolds, then each projection map $\pi_i \colon M_1 \times \cdots \times M_k \to M_i$ is smooth, because its coordinate representation with respect to any of the product charts of Example 1.8 is just a coordinate projection.   //

Diffeomorphisms

If M and N are smooth manifolds with or without boundary, a **diffeomorphism from M to N** is a smooth bijective map $F \colon M \to N$ that has a smooth inverse. We say that **M and N are diffeomorphic** if there exists a diffeomorphism between them. Sometimes this is symbolized by $M \approx N$.

Example 2.14 (Diffeomorphisms).

(a) Consider the maps $F \colon \mathbb{B}^n \to \mathbb{R}^n$ and $G \colon \mathbb{R}^n \to \mathbb{B}^n$ given by

$$F(x) = \frac{x}{\sqrt{1-|x|^2}}, \qquad G(y) = \frac{y}{\sqrt{1+|y|^2}}. \qquad (2.1)$$

These maps are smooth, and it is straightforward to compute that they are inverses of each other. Thus they are both diffeomorphisms, and therefore $\mathbb{B}^n$ is diffeomorphic to $\mathbb{R}^n$.

(b) If M is any smooth manifold and (U, φ) is a smooth coordinate chart on M, then $\varphi \colon U \to \varphi(U) \subseteq \mathbb{R}^n$ is a diffeomorphism. (In fact, it has an identity map as a coordinate representation.)   //

Proposition 2.15 (Properties of Diffeomorphisms).

(a) *Every composition of diffeomorphisms is a diffeomorphism.*
(b) *Every finite product of diffeomorphisms between smooth manifolds is a diffeomorphism.*
(c) *Every diffeomorphism is a homeomorphism and an open map.*
(d) *The restriction of a diffeomorphism to an open submanifold with or without boundary is a diffeomorphism onto its image.*
(e) *"Diffeomorphic" is an equivalence relation on the class of all smooth manifolds with or without boundary.*

▶ **Exercise 2.16.** Prove the preceding proposition.

The following theorem is a weak version of invariance of dimension, which suffices for many purposes.

Theorem 2.17 (Diffeomorphism Invariance of Dimension). *A nonempty smooth manifold of dimension m cannot be diffeomorphic to an n-dimensional smooth manifold unless $m = n$.*

Proof. Suppose M is a nonempty smooth m-manifold, N is a nonempty smooth n-manifold, and $F: M \to N$ is a diffeomorphism. Choose any point $p \in M$, and let (U, φ) and (V, ψ) be smooth coordinate charts containing p and $F(p)$, respectively. Then (the restriction of) $\widehat{F} = \psi \circ F \circ \varphi^{-1}$ is a diffeomorphism from an open subset of $\mathbb{R}^m$ to an open subset of $\mathbb{R}^n$, so it follows from Proposition C.4 that $m = n$. $\square$

There is a similar invariance statement for boundaries.

Theorem 2.18 (Diffeomorphism Invariance of the Boundary). *Suppose M and N are smooth manifolds with boundary and $F: M \to N$ is a diffeomorphism. Then $F(\partial M) = \partial N$, and F restricts to a diffeomorphism from Int M to Int N.*

▶ **Exercise 2.19.** Use Theorem 1.46 to prove the preceding theorem.

Just as two topological spaces are considered to be "the same" if they are homeomorphic, two smooth manifolds with or without boundary are essentially indistinguishable if they are diffeomorphic. The central concern of smooth manifold theory is the study of properties of smooth manifolds that are preserved by diffeomorphisms. Theorem 2.17 shows that dimension is one such property.

It is natural to wonder whether the smooth structure on a given topological manifold is unique. This straightforward version of the question is easy to answer: we observed in Example 1.21 that every zero-dimensional manifold has a unique smooth structure, but as Problem 1-6 showed, each positive-dimensional manifold admits many distinct smooth structures as soon as it admits one.

A more subtle and interesting question is whether a given topological manifold admits smooth structures that are not diffeomorphic to each other. For example, let $\widetilde{\mathbb{R}}$ denote the topological manifold $\mathbb{R}$, but endowed with the smooth structure described in Example 1.23 (defined by the global chart $\psi(x) = x^3$). It turns out that $\widetilde{\mathbb{R}}$ is diffeomorphic to $\mathbb{R}$ with its standard smooth structure. Define a map $F: \mathbb{R} \to \tilde{\mathbb{R}}$ by $F(x) = x^{1/3}$. The coordinate representation of this map is $\hat{F}(t) = \psi \circ F \circ \mathrm{Id}_{\mathbb{R}}^{-1}(t) = t$, which is clearly smooth. Moreover, the coordinate representation of its inverse is

$$\widehat{F^{-1}}(y) = \mathrm{Id}_{\mathbb{R}} \circ F^{-1} \circ \psi^{-1}(y) = y,$$

which is also smooth, so F is a diffeomorphism. (This is a case in which it *is* important to maintain the distinction between a map and its coordinate representation!)

In fact, as you will see later, there is only one smooth structure on $\mathbb{R}$ up to diffeomorphism (see Problem 15-13). More precisely, if $\mathcal{A}_1$ and $\mathcal{A}_2$ are any two smooth structures on $\mathbb{R}$, there exists a diffeomorphism $F: (\mathbb{R}, \mathcal{A}_1) \to (\mathbb{R}, \mathcal{A}_2)$. In fact, it follows from work of James Munkres [Mun60] and Edwin Moise [Moi77] that every topological manifold of dimension less than or equal to 3 has a smooth structure that is unique up to diffeomorphism. The analogous question in higher dimensions turns out to be quite deep, and is still largely unanswered. Even for Euclidean spaces, the question of uniqueness of smooth structures was not completely settled until late in the twentieth century. The answer is surprising: as long as $n \neq 4$, $\mathbb{R}^n$ has a unique smooth structure (up to diffeomorphism); but $\mathbb{R}^4$ has uncountably many distinct smooth structures, no two of which are diffeomorphic to each other! The existence of nonstandard smooth structures on $\mathbb{R}^4$ (called *fake* $\mathbb{R}^4$'s) was first proved by Simon Donaldson and Michael Freedman in 1984 as a consequence of their work on the geometry and topology of compact 4-manifolds; the results are described in [DK90] and [FQ90].

For compact manifolds, the situation is even more fascinating. In 1956, John Milnor [Mil56] showed that there are smooth structures on $\mathbb{S}^7$ that are not diffeomorphic to the standard one. Later, he and Michel Kervaire [KM63] showed (using a deep theorem of Steve Smale [Sma62]) that there are exactly 15 diffeomorphism classes of such structures (or 28 classes if you restrict to diffeomorphisms that preserve a property called *orientation*, which will be discussed in Chapter 15).

On the other hand, in all dimensions greater than 3 there are compact topological manifolds that have no smooth structures at all. The problem of identifying the number of smooth structures (if any) on topological 4-manifolds is an active subject of current research.

Partitions of Unity

A frequently used tool in topology is the gluing lemma (Lemma A.20), which shows how to construct continuous maps by "gluing together" maps defined on open or closed subsets. We have a version of the gluing lemma for smooth maps defined on *open* subsets (Corollary 2.8), but we cannot expect to glue together smooth maps defined on *closed* subsets and obtain a smooth result. For example, the two functions $f_+: [0, \infty) \to \mathbb{R}$ and $f_-: (-\infty, 0] \to \mathbb{R}$ defined by

$$f_+(x) = +x, \quad x \in [0, \infty),$$
$$f_-(x) = -x, \quad x \in (-\infty, 0],$$

are both smooth and agree at the point 0 where they overlap, but the continuous function $f : \mathbb{R} \to \mathbb{R}$ that they define, namely $f(x) = |x|$, is not smooth at the origin.

A disadvantage of Corollary 2.8 is that in order to use it, we must construct maps that agree exactly on relatively large subsets of the manifold, which is too restrictive for some purposes. In this section we introduce *partitions of unity*, which are tools for "blending together" local smooth objects into global ones without necessarily assuming that they agree on overlaps. They are indispensable in smooth manifold theory and will reappear throughout the book.

All of our constructions in this section are based on the existence of smooth functions that are positive in a specified part of a manifold and identically zero in some other part. We begin by defining a smooth function on the real line that is zero for $t \leq 0$ and positive for $t > 0$.

Lemma 2.20. *The function $f : \mathbb{R} \to \mathbb{R}$ defined by*

$$f(t) = \begin{cases} e^{-1/t}, & t > 0, \\ 0, & t \leq 0, \end{cases}$$

is smooth.

Proof. The function in question is pictured in Fig. 2.4. It is smooth on $\mathbb{R} \smallsetminus \{0\}$ by composition, so we need only show f has continuous derivatives of all orders at the origin. Because existence of the $(k + 1)$st derivative implies continuity of the kth, it suffices to show that each such derivative exists. We begin by noting that f is continuous at 0 because $\lim_{t \searrow 0} e^{-1/t} = 0$. In fact, a standard application of l'Hôpital's rule and induction shows that for any integer $k \geq 0$,

$$\lim_{t \searrow 0} \frac{e^{-1/t}}{t^k} = \lim_{t \searrow 0} \frac{t^{-k}}{e^{1/t}} = 0. \tag{2.2}$$

We show by induction that for $t > 0$, the kth derivative of f is of the form

$$f^{(k)}(t) = p_k(t) \frac{e^{-1/t}}{t^{2k}} \tag{2.3}$$

for some polynomial p_k of degree at most k. This is clearly true (with $p_0(t) = 1$) for $k = 0$, so suppose it is true for some $k \geq 0$. By the product rule,

$$f^{(k+1)}(t) = p_k'(t) \frac{e^{-1/t}}{t^{2k}} + p_k(t) \frac{t^{-2} e^{-1/t}}{t^{2k}} - 2k p_k(t) \frac{e^{-1/t}}{t^{2k+1}}$$

$$= \left(t^2 p_k'(t) + p_k(t) - 2kt p_k(t) \right) \frac{e^{-1/t}}{t^{2(k+1)}},$$

which is of the required form.

Finally, we prove by induction that $f^{(k)}(0) = 0$ for each integer $k \geq 0$. For $k = 0$ this is true by definition, so assume that it is true for some $k \geq 0$. To prove that $f^{(k+1)}(0)$ exists, it suffices to show that $f^{(k)}$ has one-sided derivatives from both sides at $t = 0$ and that they are equal. Clearly, the derivative from the left is zero.

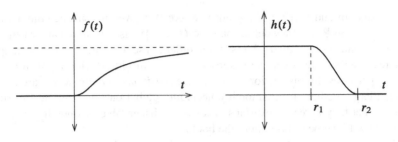

Fig. 2.4 $f(t) = e^{-1/t}$ **Fig. 2.5** A cutoff function

Using (2.3) and (2.2) again, we find that the derivative of $f^{(k)}$ from the right at $t = 0$ is equal to

$$\lim_{t \searrow 0} \frac{p_k(t)\dfrac{e^{-1/t}}{t^{2k}} - 0}{t} = \lim_{t \searrow 0} p_k(t)\frac{e^{-1/t}}{t^{2k+1}} = p_k(0)\lim_{t \searrow 0}\frac{e^{-1/t}}{t^{2k+1}} = 0.$$

Thus $f^{(k+1)}(0) = 0$. $\square$

Lemma 2.21. *Given any real numbers r_1 and r_2 such that $r_1 < r_2$, there exists a smooth function $h \colon \mathbb{R} \to \mathbb{R}$ such that $h(t) \equiv 1$ for $t \le r_1$, $0 < h(t) < 1$ for $r_1 < t < r_2$, and $h(t) \equiv 0$ for $t \ge r_2$.*

Proof. Let f be the function of the previous lemma, and set

$$h(t) = \frac{f(r_2 - t)}{f(r_2 - t) + f(t - r_1)}.$$

(See Fig. 2.5.) Note that the denominator is positive for all t, because at least one of the expressions $r_2 - t$ and $t - r_1$ is always positive. The desired properties of h follow easily from those of f. $\square$

A function with the properties of h in the preceding lemma is usually called a *cutoff function*.

Lemma 2.22. *Given any positive real numbers $r_1 < r_2$, there is a smooth function $H \colon \mathbb{R}^n \to \mathbb{R}$ such that $H \equiv 1$ on $\bar{B}_{r_1}(0)$, $0 < H(x) < 1$ for all $x \in B_{r_2}(0) \smallsetminus \bar{B}_{r_1}(0)$, and $H \equiv 0$ on $\mathbb{R}^n \smallsetminus B_{r_2}(0)$.*

Proof. Just set $H(x) = h(|x|)$, where h is the function of the preceding lemma. Clearly, H is smooth on $\mathbb{R}^n \smallsetminus \{0\}$, because it is a composition of smooth functions there. Since it is identically equal to 1 on $B_{r_1}(0)$, it is smooth there too. $\square$

The function H constructed in this lemma is an example of a *smooth bump function*, a smooth real-valued function that is equal to 1 on a specified set and is zero outside a specified neighborhood of that set. Later in this chapter, we will generalize this notion to manifolds.

If f is any real-valued or vector-valued function on a topological space M, the **support of** f, denoted by supp f, is the closure of the set of points where f is nonzero:

$$\text{supp } f = \overline{\{p \in M : f(p) \neq 0\}}.$$

(For example, if H is the function constructed in the preceding lemma, then supp $H = \bar{B}_{r_2}(0)$.) If supp f is contained in some set $U \subseteq M$, we say that f is **supported in** U. A function f is said to be **compactly supported** if supp f is a compact set. Clearly, every function on a compact space is compactly supported.

The next construction is the most important application of paracompactness. Suppose M is a topological space, and let $\mathcal{X} = (X_\alpha)_{\alpha \in A}$ be an arbitrary open cover of M, indexed by a set A. A **partition of unity subordinate to** $\mathcal{X}$ is an indexed family $(\psi_\alpha)_{\alpha \in A}$ of continuous functions $\psi_\alpha : M \to \mathbb{R}$ with the following properties:

(i) $0 \leq \psi_\alpha(x) \leq 1$ for all $\alpha \in A$ and all $x \in M$.
(ii) supp $\psi_\alpha \subseteq X_\alpha$ for each $\alpha \in A$.
(iii) The family of supports $(\text{supp } \psi_\alpha)_{\alpha \in A}$ is locally finite, meaning that every point has a neighborhood that intersects supp ψ_α for only finitely many values of α.
(iv) $\sum_{\alpha \in A} \psi_\alpha(x) = 1$ for all $x \in M$.

Because of the local finiteness condition (iii), the sum in (iv) actually has only finitely many nonzero terms in a neighborhood of each point, so there is no issue of convergence. If M is a smooth manifold with or without boundary, a **smooth partition of unity** is one for which each of the functions ψ_α is smooth.

Theorem 2.23 (Existence of Partitions of Unity). *Suppose M is a smooth manifold with or without boundary, and $\mathcal{X} = (X_\alpha)_{\alpha \in A}$ is any indexed open cover of M. Then there exists a smooth partition of unity subordinate to $\mathcal{X}$.*

Proof. For simplicity, suppose for this proof that M is a smooth manifold without boundary; the general case is left as an exercise. Each set X_α is a smooth manifold in its own right, and thus has a basis $\mathcal{B}_\alpha$ of regular coordinate balls by Proposition 1.19, and it is easy to check that $\mathcal{B} = \bigcup_\alpha \mathcal{B}_\alpha$ is a basis for the topology of M. It follows from Theorem 1.15 that $\mathcal{X}$ has a countable, locally finite refinement $\{B_i\}$ consisting of elements of $\mathcal{B}$. By Lemma 1.13(a), the cover $\{\bar{B}_i\}$ is also locally finite.

For each i, the fact that B_i is a regular coordinate ball in some X_α guarantees that there is a coordinate ball $B_i' \subseteq X_\alpha$ such that $B_i' \supseteq \bar{B}_i$, and a smooth coordinate map $\varphi_i : B_i' \to \mathbb{R}^n$ such that $\varphi_i(\bar{B}_i) = \bar{B}_{r_i}(0)$ and $\varphi_i(B_i') = B_{r_i'}(0)$ for some $r_i < r_i'$. For each i, define a function $f_i : M \to \mathbb{R}$ by

$$f_i = \begin{cases} H_i \circ \varphi_i & \text{on } B_i', \\ 0 & \text{on } M \smallsetminus \bar{B}_i, \end{cases}$$

where $H_i : \mathbb{R}^n \to \mathbb{R}$ is a smooth function that is positive in $B_{r_i}(0)$ and zero elsewhere, as in Lemma 2.22. On the set $B_i' \smallsetminus \bar{B}_i$ where the two definitions overlap, both definitions yield the zero function, so f_i is well defined and smooth, and supp $f_i = \bar{B}_i$.

Define $f: M \to \mathbb{R}$ by $f(x) = \sum_i f_i(x)$. Because of the local finiteness of the cover $\{\overline{B}_i\}$, this sum has only finitely many nonzero terms in a neighborhood of each point and thus defines a smooth function. Because each f_i is nonnegative everywhere and positive on B_i, and every point of M is in some B_i, it follows that $f(x) > 0$ everywhere on M. Thus, the functions $g_i: M \to \mathbb{R}$ defined by $g_i(x) = f_i(x)/f(x)$ are also smooth. It is immediate from the definition that $0 \le g_i \le 1$ and $\sum_i g_i \equiv 1$.

Finally, we need to reindex our functions so that they are indexed by the same set A as our open cover. Because the cover $\{B_i'\}$ is a refinement of $\mathcal{X}$, for each i we can choose some index $a(i) \in A$ such that $B_i' \subseteq X_{a(i)}$. For each $\alpha \in A$, define $\psi_\alpha: M \to \mathbb{R}$ by

$$\psi_\alpha = \sum_{i:a(i)=\alpha} g_i.$$

If there are no indices i for which $a(i) = \alpha$, then this sum should be interpreted as the zero function. It follows from Lemma 1.13(b) that

$$\operatorname{supp}\psi_\alpha = \overline{\bigcup_{i:a(i)=\alpha} B_i} = \bigcup_{i:a(i)=\alpha} \overline{B}_i \subseteq X_\alpha.$$

Each ψ_α is a smooth function that satisfies $0 \le \psi_\alpha \le 1$. Moreover, the family of supports $(\operatorname{supp}\psi_\alpha)_{\alpha \in A}$ is still locally finite, and $\sum_\alpha \psi_\alpha \equiv \sum_i g_i \equiv 1$, so this is the desired partition of unity. $\square$

▶ **Exercise 2.24.** Show how the preceding proof needs to be modified for the case in which M has nonempty boundary.

There are basically two different strategies for patching together locally defined smooth maps to obtain a global one. If you can define a map in a neighborhood of each point in such a way that the locally defined maps all agree where they overlap, then the local definitions piece together to yield a global smooth map by Corollary 2.8. (This usually requires some sort of uniqueness result.) But if the local definitions are not guaranteed to agree, then you usually have to resort to a partition of unity. The trick then is showing that the patched-together objects still have the required properties. We use both strategies repeatedly throughout the book.

Applications of Partitions of Unity

As our first application of partitions of unity, we extend the notion of bump functions to arbitrary closed subsets of manifolds. If M is a topological space, $A \subseteq M$ is a closed subset, and $U \subseteq M$ is an open subset containing A, a continuous function $\psi: M \to \mathbb{R}$ is called a ***bump function for A supported in U*** if $0 \le \psi \le 1$ on M, $\psi \equiv 1$ on A, and $\operatorname{supp}\psi \subseteq U$.

Proposition 2.25 (Existence of Smooth Bump Functions). *Let M be a smooth manifold with or without boundary. For any closed subset $A \subseteq M$ and any open subset U containing A, there exists a smooth bump function for A supported in U.*

Proof. Let $U_0 = U$ and $U_1 = M \smallsetminus A$, and let $\{\psi_0, \psi_1\}$ be a smooth partition of unity subordinate to the open cover $\{U_0, U_1\}$. Because $\psi_1 \equiv 0$ on A and thus $\psi_0 = \sum_i \psi_i = 1$ there, the function ψ_0 has the required properties. $\qquad\square$

Our second application is an important result concerning the possibility of extending smooth functions from closed subsets. Suppose M and N are smooth manifolds with or without boundary, and $A \subseteq M$ is an arbitrary subset. We say that a map $F: A \to N$ is **smooth on** A if it has a smooth extension in a neighborhood of each point: that is, if for every $p \in A$ there is an open subset $W \subseteq M$ containing p and a smooth map $\widetilde{F}: W \to N$ whose restriction to $W \cap A$ agrees with F.

Lemma 2.26 (Extension Lemma for Smooth Functions). *Suppose M is a smooth manifold with or without boundary, $A \subseteq M$ is a closed subset, and $f: A \to \mathbb{R}^k$ is a smooth function. For any open subset U containing A, there exists a smooth function $\widetilde{f}: M \to \mathbb{R}^k$ such that $\widetilde{f}|_A = f$ and $\operatorname{supp} \widetilde{f} \subseteq U$.*

Proof. For each $p \in A$, choose a neighborhood W_p of p and a smooth function $\widetilde{f_p}: W_p \to \mathbb{R}^k$ that agrees with f on $W_p \cap A$. Replacing W_p by $W_p \cap U$, we may assume that $W_p \subseteq U$. The family of sets $\{W_p : p \in A\} \cup \{M \smallsetminus A\}$ is an open cover of M. Let $\{\psi_p : p \in A\} \cup \{\psi_0\}$ be a smooth partition of unity subordinate to this cover, with $\operatorname{supp} \psi_p \subseteq W_p$ and $\operatorname{supp} \psi_0 \subseteq M \smallsetminus A$.

For each $p \in A$, the product $\psi_p \widetilde{f_p}$ is smooth on W_p, and has a smooth extension to all of M if we interpret it to be zero on $M \smallsetminus \operatorname{supp} \psi_p$. (The extended function is smooth because the two definitions agree on the open subset $W_p \smallsetminus \operatorname{supp} \psi_p$ where they overlap.) Thus we can define $\widetilde{f}: M \to \mathbb{R}^k$ by

$$\widetilde{f}(x) = \sum_{p \in A} \psi_p(x) \widetilde{f_p}(x).$$

Because the collection of supports $\{\operatorname{supp} \psi_p\}$ is locally finite, this sum actually has only a finite number of nonzero terms in a neighborhood of any point of M, and therefore defines a smooth function. If $x \in A$, then $\psi_0(x) = 0$ and $\widetilde{f_p}(x) = f(x)$ for each p such that $\psi_p(x) \neq 0$, so

$$\widetilde{f}(x) = \sum_{p \in A} \psi_p(x) f(x) = \left(\psi_0(x) + \sum_{p \in A} \psi_p(x) \right) f(x) = f(x),$$

so $\widetilde{f}$ is indeed an extension of f. It follows from Lemma 1.13(b) that

$$\operatorname{supp} \widetilde{f} = \overline{\bigcup_{p \in A} \operatorname{supp} \psi_p} = \bigcup_{p \in A} \operatorname{supp} \psi_p \subseteq U. \qquad\square$$

▶ **Exercise 2.27.** Give a counterexample to show that the conclusion of the extension lemma can be false if A is not closed.

The assumption in the extension lemma that the codomain of f is $\mathbb{R}^k$, and not some other smooth manifold, is needed: for other codomains, extensions can fail to exist for topological reasons. (For example, the identity map $\mathbb{S}^1 \to \mathbb{S}^1$ is smooth, but does not have even a *continuous* extension to a map from $\mathbb{R}^2$ to $\mathbb{S}^1$.) Later we will show that a smooth map from a closed subset of a smooth manifold into a smooth manifold has a smooth extension if and only if it has a continuous one (see Corollary 6.27).

This extension lemma, by the way, illustrates an essential difference between smooth manifolds and real-analytic manifolds. The analogue of the extension lemma for real-analytic functions on real-analytic manifolds is decidedly false, because a real-analytic function that is defined on a connected domain and vanishes on an open subset must be identically zero.

Next, we use partitions of unity to construct a special kind of smooth function. If M is a topological space, an **exhaustion function for** M is a continuous function $f : M \to \mathbb{R}$ with the property that the set $f^{-1}((-\infty, c])$ (called a **sublevel set of** f) is compact for each $c \in \mathbb{R}$. The name comes from the fact that as n ranges over the positive integers, the sublevel sets $f^{-1}((-\infty, n])$ form an exhaustion of M by compact sets; thus an exhaustion function provides a sort of continuous version of an exhaustion by compact sets. For example, the functions $f : \mathbb{R}^n \to \mathbb{R}$ and $g : \mathbb{B}^n \to \mathbb{R}$ given by

$$f(x) = |x|^2, \qquad g(x) = \frac{1}{1 - |x|^2}$$

are smooth exhaustion functions. Of course, if M is compact, any continuous real-valued function on M is an exhaustion function, so such functions are interesting only for noncompact manifolds.

Proposition 2.28 (Existence of Smooth Exhaustion Functions). *Every smooth manifold with or without boundary admits a smooth positive exhaustion function.*

Proof. Let M be a smooth manifold with or without boundary, let $\{V_j\}_{j=1}^{\infty}$ be any countable open cover of M by precompact open subsets, and let $\{\psi_j\}$ be a smooth partition of unity subordinate to this cover. Define $f \in C^{\infty}(M)$ by

$$f(p) = \sum_{j=1}^{\infty} j \psi_j(p).$$

Then f is smooth because only finitely many terms are nonzero in a neighborhood of any point, and positive because $f(p) \geq \sum_j \psi_j(p) = 1$.

To see that f is an exhaustion function, let $c \in \mathbb{R}$ be arbitrary, and choose a positive integer $N > c$. If $p \notin \bigcup_{j=1}^{N} \bar{V}_j$, then $\psi_j(p) = 0$ for $1 \leq j \leq N$, so

$$f(p) = \sum_{j=N+1}^{\infty} j \psi_j(p) \geq \sum_{j=N+1}^{\infty} N \psi_j(p) = N \sum_{j=1}^{\infty} \psi_j(p) = N > c.$$

Equivalently, if $f(p) \leq c$, then $p \in \bigcup_{j=1}^{N} \bar{V}_j$. Thus $f^{-1}((-\infty, c])$ is a closed subset of the compact set $\bigcup_{j=1}^{N} \bar{V}_j$ and is therefore compact. $\square$

As our final application of partitions of unity, we will prove the remarkable fact that every closed subset of a manifold can be expressed as a level set of some smooth real-valued function. We will not use this result in this book (except in a few of the problems), but it provides an interesting contrast with the result of Example 1.32.

Theorem 2.29 (Level Sets of Smooth Functions). *Let M be a smooth manifold. If K is any closed subset of M, there is a smooth nonnegative function $f : M \to \mathbb{R}$ such that $f^{-1}(0) = K$.*

Proof. We begin with the special case in which $M = \mathbb{R}^n$ and $K \subseteq \mathbb{R}^n$ is a closed subset. For each $x \in M \smallsetminus K$, there is a positive number $r \leq 1$ such that $B_r(x) \subseteq M \smallsetminus K$. By Proposition A.16, $M \smallsetminus K$ is the union of countably many such balls $\{B_{r_i}(x_i)\}_{i=1}^{\infty}$.

Let $h : \mathbb{R}^n \to \mathbb{R}$ be a smooth bump function that is equal to 1 on $\bar{B}_{1/2}(0)$ and supported in $B_1(0)$. For each positive integer i, let $C_i \geq 1$ be a constant that bounds the absolute values of h and all of its partial derivatives up through order i. Define $f : \mathbb{R}^n \to \mathbb{R}$ by

$$f(x) = \sum_{i=1}^{\infty} \frac{(r_i)^i}{2^i C_i} h\left(\frac{x - x_i}{r_i}\right).$$

The terms of the series are bounded in absolute value by those of the convergent series $\sum_i 1/2^i$, so the entire series converges uniformly to a continuous function by the Weierstrass M-test. Because the ith term is positive exactly when $x \in B_{r_i}(x_i)$, it follows that f is zero in K and positive elsewhere.

It remains only to show that f is smooth. We have already shown that it is continuous, so suppose $k \geq 1$ and assume by induction that all partial derivatives of f of order less than k exist and are continuous. By the chain rule and induction, every kth partial derivative of the ith term in the series can be written in the form

$$\frac{(r_i)^{i-k}}{2^i C_i} D_k h\left(\frac{x - x_i}{r_i}\right),$$

where $D_k h$ is some kth partial derivative of h. By our choices of r_i and C_i, as soon as $i \geq k$, each of these terms is bounded in absolute value by $1/2^i$, so the differentiated series also converges uniformly to a continuous function. It then follows from Theorem C.31 that the kth partial derivatives of f exist and are continuous. This completes the induction, and shows that f is smooth.

Now let M be an arbitrary smooth manifold, and $K \subseteq M$ be any closed subset. Let $\{B_\alpha\}$ be an open cover of M by smooth coordinate balls, and let $\{\psi_\alpha\}$ be a subordinate partition of unity. Since each B_α is diffeomorphic to $\mathbb{R}^n$, the preceding argument shows that for each α there is a smooth nonnegative function $f_\alpha : B_\alpha \to \mathbb{R}$ such that $f_\alpha^{-1}(0) = B_\alpha \cap K$. The function $f = \sum_\alpha \psi_\alpha f_\alpha$ does the trick. $\square$

Problems

2-1. Define $f: \mathbb{R} \to \mathbb{R}$ by

$$f(x) = \begin{cases} 1, & x \geq 0, \\ 0, & x < 0. \end{cases}$$

Show that for every $x \in \mathbb{R}$, there are smooth coordinate charts (U, φ) containing x and (V, ψ) containing $f(x)$ such that $\psi \circ f \circ \varphi^{-1}$ is smooth as a map from $\varphi(U \cap f^{-1}(V))$ to $\psi(V)$, but f is not smooth in the sense we have defined in this chapter.

2-2. Prove Proposition 2.12 (smoothness of maps into product manifolds).

2-3. For each of the following maps between spheres, compute sufficiently many coordinate representations to prove that it is smooth.
  (a) $p_n: \mathbb{S}^1 \to \mathbb{S}^1$ is the **nth power map** for $n \in \mathbb{Z}$, given in complex notation by $p_n(z) = z^n$.
  (b) $\alpha: \mathbb{S}^n \to \mathbb{S}^n$ is the **antipodal map** $\alpha(x) = -x$.
  (c) $F: \mathbb{S}^3 \to \mathbb{S}^2$ is given by $F(w, z) = (z\bar{w} + w\bar{z}, iw\bar{z} - iz\bar{w}, z\bar{z} - w\bar{w})$, where we think of $\mathbb{S}^3$ as the subset $\{(w, z) : |w|^2 + |z|^2 = 1\}$ of $\mathbb{C}^2$.

2-4. Show that the inclusion map $\overline{\mathbb{B}}^n \hookrightarrow \mathbb{R}^n$ is smooth when $\overline{\mathbb{B}}^n$ is regarded as a smooth manifold with boundary.

2-5. Let $\mathbb{R}$ be the real line with its standard smooth structure, and let $\widetilde{\mathbb{R}}$ denote the same topological manifold with the smooth structure defined in Example 1.23. Let $f: \mathbb{R} \to \mathbb{R}$ be a function that is smooth in the usual sense.
  (a) Show that f is also smooth as a map from $\mathbb{R}$ to $\widetilde{\mathbb{R}}$.
  (b) Show that f is smooth as a map from $\widetilde{\mathbb{R}}$ to $\mathbb{R}$ if and only if $f^{(n)}(0) = 0$ whenever n is not an integral multiple of 3.

2-6. Let $P: \mathbb{R}^{n+1} \smallsetminus \{0\} \to \mathbb{R}^{k+1} \smallsetminus \{0\}$ be a smooth function, and suppose that for some $d \in \mathbb{Z}$, $P(\lambda x) = \lambda^d P(x)$ for all $\lambda \in \mathbb{R} \smallsetminus \{0\}$ and $x \in \mathbb{R}^{n+1} \smallsetminus \{0\}$. (Such a function is said to be **homogeneous of degree** d.) Show that the map $\widetilde{P}: \mathbb{RP}^n \to \mathbb{RP}^k$ defined by $\widetilde{P}([x]) = [P(x)]$ is well defined and smooth.

2-7. Let M be a nonempty smooth n-manifold with or without boundary, and suppose $n \geq 1$. Show that the vector space $C^\infty(M)$ is infinite-dimensional. [Hint: show that if $f_1, \ldots, f_k$ are elements of $C^\infty(M)$ with nonempty disjoint supports, then they are linearly independent.]

2-8. Define $F: \mathbb{R}^n \to \mathbb{RP}^n$ by $F(x^1, \ldots, x^n) = [x^1, \ldots, x^n, 1]$. Show that F is a diffeomorphism onto a dense open subset of $\mathbb{RP}^n$. Do the same for $G: \mathbb{C}^n \to \mathbb{CP}^n$ defined by $G(z^1, \ldots, z^n) = [z^1, \ldots, z^n, 1]$ (see Problem 1-9).

2-9. Given a polynomial p in one variable with complex coefficients, not identically zero, show that there is a unique smooth map $\widetilde{p}: \mathbb{CP}^1 \to \mathbb{CP}^1$ that makes the following diagram commute, where $\mathbb{CP}^1$ is 1-dimensional complex projective space and $G: \mathbb{C} \to \mathbb{CP}^1$ is the map of Problem 2-8:

$$
\begin{array}{ccc}
\mathbb{C} & \xrightarrow{\ G\ } & \mathbb{CP}^1 \\
p \downarrow & & \downarrow \tilde{p} \\
\mathbb{C} & \xrightarrow[\ G\ ]{} & \mathbb{CP}^1.
\end{array}
$$

(Used on p. 465.)

2-10. For any topological space M, let $C(M)$ denote the algebra of continuous functions $f: M \to \mathbb{R}$. Given a continuous map $F: M \to N$, define $F^*: C(N) \to C(M)$ by $F^*(f) = f \circ F$.
   (a) Show that F^* is a linear map.
   (b) Suppose M and N are smooth manifolds. Show that $F: M \to N$ is smooth if and only if $F^*(C^\infty(N)) \subseteq C^\infty(M)$.
   (c) Suppose $F: M \to N$ is a homeomorphism between smooth manifolds. Show that it is a diffeomorphism if and only if F^* restricts to an isomorphism from $C^\infty(N)$ to $C^\infty(M)$.
   [Remark: this result shows that in a certain sense, the entire smooth structure of M is encoded in the subset $C^\infty(M) \subseteq C(M)$. In fact, some authors *define* a smooth structure on a topological manifold M to be a subalgebra of $C(M)$ with certain properties; see, e.g., [Nes03].] *(Used on p. 75.)*

2-11. Suppose V is a real vector space of dimension $n \geq 1$. Define the *projectivization of V*, denoted by $\mathbb{P}(V)$, to be the set of 1-dimensional linear subspaces of V, with the quotient topology induced by the map $\pi: V \smallsetminus \{0\} \to \mathbb{P}(V)$ that sends x to its span. (Thus $\mathbb{P}(\mathbb{R}^n) = \mathbb{RP}^{n-1}$.) Show that $\mathbb{P}(V)$ is a topological $(n-1)$-manifold, and has a unique smooth structure with the property that for each basis $(E_1, \ldots, E_n)$ for V, the map $E: \mathbb{RP}^{n-1} \to \mathbb{P}(V)$ defined by $E[v^1, \ldots, v^n] = [v^i E_i]$ (where brackets denote equivalence classes) is a diffeomorphism. *(Used on p. 561.)*

2-12. State and prove an analogue of Problem 2-11 for complex vector spaces.

2-13. Suppose M is a topological space with the property that for every indexed open cover $\mathcal{X}$ of M, there exists a partition of unity subordinate to $\mathcal{X}$. Show that M is paracompact.

2-14. Suppose A and B are disjoint closed subsets of a smooth manifold M. Show that there exists $f \in C^\infty(M)$ such that $0 \leq f(x) \leq 1$ for all $x \in M$, $f^{-1}(0) = A$, and $f^{-1}(1) = B$.

Chapter 3
Tangent Vectors

The central idea of calculus is *linear approximation*. This arises repeatedly in the study of calculus in Euclidean spaces, where, for example, a function of one variable can be approximated by its tangent line, a parametrized curve in $\mathbb{R}^n$ by its velocity vector, a surface in $\mathbb{R}^3$ by its tangent plane, or a map from $\mathbb{R}^n$ to $\mathbb{R}^m$ by its total derivative (see Appendix C).

In order to make sense of calculus on manifolds, we need to introduce the *tangent space to a manifold at a point*, which we can think of as a sort of "linear model" for the manifold near the point. Because of the abstractness of the definition of a smooth manifold, this takes some work, which we carry out in this chapter.

We begin by studying much more concrete objects: *geometric tangent vectors* in $\mathbb{R}^n$, which can be visualized as "arrows" attached to points. Because the definition of smooth manifolds is built around the idea of identifying which functions are smooth, the property of a geometric tangent vector that is amenable to generalization is its action on smooth functions as a "directional derivative." The key observation, which we prove in the first section of this chapter, is that the process of taking directional derivatives gives a natural one-to-one correspondence between geometric tangent vectors and linear maps from $C^\infty(\mathbb{R}^n)$ to $\mathbb{R}$ satisfying the product rule. (Such maps are called *derivations*.) With this as motivation, we then *define* a tangent vector on a smooth manifold as a derivation of $C^\infty(M)$ at a point.

In the second section of the chapter, we show how a smooth map between manifolds yields a linear map between tangent spaces, called the *differential* of the map, which generalizes the total derivative of a map between Euclidean spaces. This allows us to connect the abstract definition of tangent vectors to our concrete geometric picture by showing that any smooth coordinate chart (U, φ) gives a natural isomorphism from the space of tangent vectors to M at p to the space of tangent vectors to $\mathbb{R}^n$ at $\varphi(p)$, which in turn is isomorphic to the space of geometric tangent vectors at $\varphi(p)$. Thus, any smooth coordinate chart yields a basis for each tangent space. Using this isomorphism, we describe how to do concrete computations in such a basis. Based on these coordinate computations, we show how the union of all the tangent spaces at all points of a smooth manifold can be "glued together" to form a new manifold, called the *tangent bundle* of the original manifold.

J.M. Lee, *Introduction to Smooth Manifolds*, Graduate Texts in Mathematics 218,
DOI 10.1007/978-1-4419-9982-5_3, © Springer Science+Business Media New York 2013

Next we show how a smooth curve determines a tangent vector at each point, called its *velocity*, which can be regarded as the derivation of $C^\infty(M)$ that takes the derivative of each function along the curve.

In the final two sections we discuss and compare several other approaches to defining tangent spaces, and give a brief overview of the terminology of *category theory*, which puts the tangent space and differentials in a larger context.

Tangent Vectors

Imagine a manifold in Euclidean space—for example, the unit sphere $\mathbb{S}^{n-1} \subseteq \mathbb{R}^n$. What do we mean by a "tangent vector" at a point of $\mathbb{S}^{n-1}$? Before we can answer this question, we have to come to terms with a dichotomy in the way we think about elements of $\mathbb{R}^n$. On the one hand, we usually think of them as *points* in space, whose only property is location, expressed by the coordinates $(x^1, \ldots, x^n)$. On the other hand, when doing calculus we sometimes think of them instead as *vectors*, which are objects that have magnitude and direction, but whose location is irrelevant. A vector $v = v^i e_i$ (where e_i denotes the ith standard basis vector) can be visualized as an arrow with its initial point anywhere in $\mathbb{R}^n$; what is relevant from the vector point of view is only which direction it points and how long it is.

What we really have in mind here is a separate copy of $\mathbb{R}^n$ at each point. When we talk about vectors tangent to the sphere at a point a, for example, we imagine them as living in a copy of $\mathbb{R}^n$ with its origin translated to a.

Geometric Tangent Vectors

Here is a preliminary definition of tangent vectors in Euclidean space. Given a point $a \in \mathbb{R}^n$, let us define the **geometric tangent space to $\mathbb{R}^n$ at a**, denoted by $\mathbb{R}^n_a$, to be the set $\{a\} \times \mathbb{R}^n = \{(a, v) : v \in \mathbb{R}^n\}$. A **geometric tangent vector** in $\mathbb{R}^n$ is an element of $\mathbb{R}^n_a$ for some $a \in \mathbb{R}^n$. As a matter of notation, we abbreviate (a, v) as v_a (or sometimes $v|_a$ if it is clearer, for example if v itself has a subscript). We think of v_a as the vector v with its initial point at a (Fig. 3.1). The set $\mathbb{R}^n_a$ is a real vector space under the natural operations

$$v_a + w_a = (v + w)_a, \qquad c(v_a) = (cv)_a.$$

The vectors $e_i|_a$, $i = 1, \ldots, n$, are a basis for $\mathbb{R}^n_a$. In fact, as a vector space, $\mathbb{R}^n_a$ is essentially the same as $\mathbb{R}^n$ itself; the only reason we add the index a is so that the geometric tangent spaces $\mathbb{R}^n_a$ and $\mathbb{R}^n_b$ at distinct points a and b will be disjoint sets.

With this definition we could think of the tangent space to $\mathbb{S}^{n-1}$ at a point $a \in \mathbb{S}^{n-1}$ as a certain subspace of $\mathbb{R}^n_a$ (Fig. 3.2), namely the space of vectors that are orthogonal to the radial unit vector through a, using the inner product that $\mathbb{R}^n_a$ inherits from $\mathbb{R}^n$ via the natural isomorphism $\mathbb{R}^n \cong \mathbb{R}^n_a$. The problem with this definition, however, is that it gives us no clue as to how we might define tangent vectors on an arbitrary smooth manifold, where there is no ambient Euclidean space. So we

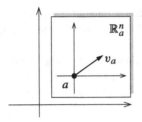
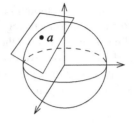

Fig. 3.1 Geometric tangent space   **Fig. 3.2** Tangent space to $\mathbb{S}^{n-1}$

need to look for another characterization of tangent vectors that might make sense on a manifold.

The only things we have to work with on smooth manifolds so far are smooth functions, smooth maps, and smooth coordinate charts. One thing that a geometric tangent vector provides is a means of taking directional derivatives of functions. For example, any geometric tangent vector $v_a \in \mathbb{R}_a^n$ yields a map $D_v|_a \colon C^\infty(\mathbb{R}^n) \to \mathbb{R}$, which takes the directional derivative in the direction v at a:

$$D_v|_a f = D_v f(a) = \left.\frac{d}{dt}\right|_{t=0} f(a+tv). \tag{3.1}$$

This operation is linear over $\mathbb{R}$ and satisfies the product rule:

$$D_v|_a (fg) = f(a)D_v|_a g + g(a)D_v|_a f. \tag{3.2}$$

If $v_a = v^i e_i|_a$ in terms of the standard basis, then by the chain rule $D_v|_a f$ can be written more concretely as

$$D_v|_a f = v^i \frac{\partial f}{\partial x^i}(a).$$

(Here we are using the summation convention as usual, so the expression on the right-hand side is understood to be summed over $i = 1, \ldots, n$. This sum is consistent with our index convention if we stipulate that an upper index "in the denominator" is to be regarded as a lower index.) For example, if $v_a = e_j|_a$, then

$$D_v|_a f = \frac{\partial f}{\partial x^j}(a).$$

With this construction in mind, we make the following definition. If a is a point of $\mathbb{R}^n$, a map $w \colon C^\infty(\mathbb{R}^n) \to \mathbb{R}$ is called a **derivation at a** if it is linear over $\mathbb{R}$ and satisfies the following product rule:

$$w(fg) = f(a)wg + g(a)wf. \tag{3.3}$$

Let $T_a \mathbb{R}^n$ denote the set of all derivations of $C^\infty(\mathbb{R}^n)$ at a. Clearly, $T_a \mathbb{R}^n$ is a vector space under the operations

$$(w_1 + w_2)f = w_1 f + w_2 f, \qquad (cw)f = c(wf).$$

The most important (and perhaps somewhat surprising) fact about $T_a\mathbb{R}^n$ is that it is finite-dimensional, and in fact is naturally isomorphic to the geometric tangent space $\mathbb{R}_a^n$ that we defined above. The proof will be based on the following lemma.

Lemma 3.1 (Properties of Derivations). *Suppose $a \in \mathbb{R}^n$, $w \in T_a\mathbb{R}^n$, and $f, g \in C^\infty(\mathbb{R}^n)$.*

(a) *If f is a constant function, then $wf = 0$.*
(b) *If $f(a) = g(a) = 0$, then $w(fg) = 0$.*

Proof. It suffices to prove (a) for the constant function $f_1(x) \equiv 1$, for then $f(x) \equiv c$ implies $wf = w(cf_1) = cwf_1 = 0$ by linearity. For f_1, the product rule gives

$$wf_1 = w(f_1 f_1) = f_1(a)wf_1 + f_1(a)wf_1 = 2wf_1,$$

which implies that $wf_1 = 0$. Similarly, (b) also follows from the product rule:

$$w(fg) = f(a)wg + g(a)wf = 0 + 0 = 0. \qquad \square$$

The next proposition shows that derivations at a are in one-to-one correspondence with geometric tangent vectors.

Proposition 3.2. *Let $a \in \mathbb{R}^n$.*

(a) *For each geometric tangent vector $v_a \in \mathbb{R}_a^n$, the map $D_v|_a \colon C^\infty(\mathbb{R}^n) \to \mathbb{R}$ defined by (3.1) is a derivation at a.*
(b) *The map $v_a \mapsto D_v|_a$ is an isomorphism from $\mathbb{R}_a^n$ onto $T_a\mathbb{R}^n$.*

Proof. The fact that $D_v|_a$ is a derivation at a is an immediate consequence of the product rule (3.2).

To prove that the map $v_a \mapsto D_v|_a$ is an isomorphism, we note first that it is linear, as is easily checked. To see that it is injective, suppose $v_a \in \mathbb{R}_a^n$ has the property that $D_v|_a$ is the zero derivation. Writing $v_a = v^i e_i|_a$ in terms of the standard basis, and taking f to be the jth coordinate function $x^j \colon \mathbb{R}^n \to \mathbb{R}$, thought of as a smooth function on $\mathbb{R}^n$, we obtain

$$0 = D_v|_a\left(x^j\right) = v^i \frac{\partial}{\partial x^i}\left(x^j\right)\bigg|_{x=a} = v^j,$$

where the last equality follows because $\partial x^j / \partial x^i = 0$ except when $i = j$, in which case it is equal to 1. Since this is true for each j, it follows that v_a is the zero vector.

To prove surjectivity, let $w \in T_a\mathbb{R}^n$ be arbitrary. Motivated by the computation in the preceding paragraph, we define $v = v^i e_i$, where the real numbers $v^1, \dots, v^n$ are given by $v^i = w(x^i)$. We will show that $w = D_v|_a$.

To see this, let f be any smooth real-valued function on $\mathbb{R}^n$. By Taylor's theorem (Theorem C.15), we can write

$$f(x) = f(a) + \sum_{i=1}^n \frac{\partial f}{\partial x^i}(a)\left(x^i - a^i\right)$$

$$+ \sum_{i,j=1}^n \left(x^i - a^i\right)\left(x^j - a^j\right) \int_0^1 (1-t)\frac{\partial^2 f}{\partial x^i \partial x^j}\left(a + t(x-a)\right) dt.$$

Note that each term in the last sum above is a product of two smooth functions of x that vanish at $x = a$: one is $(x^i - a^i)$, and the other is $(x^j - a^j)$ times the integral. The derivation w annihilates this entire sum by Lemma 3.1(b). Thus

$$wf = w(f(a)) + \sum_{i=1}^{n} w\left(\frac{\partial f}{\partial x^i}(a)(x^i - a^i)\right)$$

$$= 0 + \sum_{i=1}^{n} \frac{\partial f}{\partial x^i}(a)\left(w(x^i) - w(a^i)\right)$$

$$= \sum_{i=1}^{n} \frac{\partial f}{\partial x^i}(a)v^i = D_v|_a f.$$

$\square$

Corollary 3.3. *For any $a \in \mathbb{R}^n$, the n derivations*

$$\frac{\partial}{\partial x^1}\bigg|_a , \ldots , \frac{\partial}{\partial x^n}\bigg|_a \quad \text{defined by} \quad \frac{\partial}{\partial x^i}\bigg|_a f = \frac{\partial f}{\partial x^i}(a)$$

form a basis for $T_a\mathbb{R}^n$, which therefore has dimension n.

Proof. Apply the previous proposition and note that $\partial/\partial x^i|_a = D_{e_i}|_a$. $\square$

Tangent Vectors on Manifolds

Now we are in a position to define tangent vectors on manifolds and manifolds with boundary. The definition is the same in both cases. Let M be a smooth manifold with or without boundary, and let p be a point of M. A linear map $v: C^\infty(M) \to \mathbb{R}$ is called a **derivation at p** if it satisfies

$$v(fg) = f(p)vg + g(p)vf \quad \text{for all } f, g \in C^\infty(M). \tag{3.4}$$

The set of all derivations of $C^\infty(M)$ at p, denoted by T_pM, is a vector space called the **tangent space to M at p**. An element of T_pM is called a **tangent vector at p**.

The following lemma is the analogue of Lemma 3.1 for manifolds.

Lemma 3.4 (Properties of Tangent Vectors on Manifolds). *Suppose M is a smooth manifold with or without boundary, $p \in M$, $v \in T_pM$, and $f, g \in C^\infty(M)$.*

(a) *If f is a constant function, then $vf = 0$.*
(b) *If $f(p) = g(p) = 0$, then $v(fg) = 0$.*

▶ **Exercise 3.5.** Prove Lemma 3.4.

With the motivation of geometric tangent vectors in $\mathbb{R}^n$ in mind, you should visualize tangent vectors to M as "arrows" that are tangent to M and whose base points are attached to M at the given point. Proofs of theorems about tangent vectors must, of course, be based on the abstract definition in terms of derivations, but your intuition should be guided as much as possible by the geometric picture.

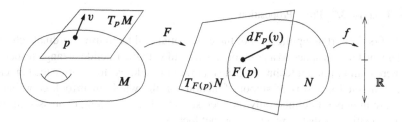

Fig. 3.3 The differential

The Differential of a Smooth Map

To relate the abstract tangent spaces we have defined on manifolds to geometric tangent spaces in $\mathbb{R}^n$, we have to explore the way smooth maps affect tangent vectors. In the case of a smooth map between Euclidean spaces, the total derivative of the map at a point (represented by its Jacobian matrix) is a linear map that represents the "best linear approximation" to the map near the given point. In the manifold case there is a similar linear map, but it makes no sense to talk about a linear map between manifolds. Instead, it will be a linear map between tangent spaces.

If M and N are smooth manifolds with or without boundary and $F: M \to N$ is a smooth map, for each $p \in M$ we define a map

$$dF_p: T_pM \to T_{F(p)}N,$$

called the **differential of F at p** (Fig. 3.3), as follows. Given $v \in T_pM$, we let $dF_p(v)$ be the derivation at $F(p)$ that acts on $f \in C^\infty(N)$ by the rule

$$dF_p(v)(f) = v(f \circ F).$$

Note that if $f \in C^\infty(N)$, then $f \circ F \in C^\infty(M)$, so $v(f \circ F)$ makes sense. The operator $dF_p(v): C^\infty(N) \to \mathbb{R}$ is linear because v is, and is a derivation at $F(p)$ because for any $f, g \in C^\infty(N)$ we have

$$\begin{aligned} dF_p(v)(fg) &= v\big((fg) \circ F\big) = v\big((f \circ F)(g \circ F)\big) \\ &= f \circ F(p)v(g \circ F) + g \circ F(p)v(f \circ F) \\ &= f\big(F(p)\big)dF_p(v)(g) + g\big(F(p)\big)dF_p(v)(f). \end{aligned}$$

Proposition 3.6 (Properties of Differentials). *Let M, N, and P be smooth manifolds with or without boundary, let $F: M \to N$ and $G: N \to P$ be smooth maps, and let $p \in M$.*

(a) $dF_p: T_pM \to T_{F(p)}N$ *is linear.*
(b) $d(G \circ F)_p = dG_{F(p)} \circ dF_p: T_pM \to T_{G \circ F(p)}P.$
(c) $d(\mathrm{Id}_M)_p = \mathrm{Id}_{T_pM}: T_pM \to T_pM.$
(d) *If F is a diffeomorphism, then $dF_p: T_pM \to T_{F(p)}N$ is an isomorphism, and* $(dF_p)^{-1} = d\big(F^{-1}\big)_{F(p)}.$

▶ **Exercise 3.7.** Prove Proposition 3.6.

Our first important application of the differential will be to use coordinate charts to relate the tangent space to a point on a manifold with the Euclidean tangent space. But there is an important technical issue that we must address first: while the tangent space is defined in terms of smooth functions on the whole manifold, coordinate charts are in general defined only on open subsets. The key point, expressed in the next proposition, is that tangent vectors act locally.

Proposition 3.8. *Let M be a smooth manifold with or without boundary, $p \in M$, and $v \in T_pM$. If $f, g \in C^\infty(M)$ agree on some neighborhood of p, then $vf = vg$.*

Proof. Let $h = f - g$, so that h is a smooth function that vanishes in a neighborhood of p. Let $\psi \in C^\infty(M)$ be a smooth bump function that is identically equal to 1 on the support of h and is supported in $M \smallsetminus \{p\}$. Because $\psi \equiv 1$ where h is nonzero, the product ψh is identically equal to h. Since $h(p) = \psi(p) = 0$, Lemma 3.4 implies that $vh = v(\psi h) = 0$. By linearity, this implies $vf = vg$.   □

Using this proposition, we can identify the tangent space to an open submanifold with the tangent space to the whole manifold.

Proposition 3.9 (The Tangent Space to an Open Submanifold). *Let M be a smooth manifold with or without boundary, let $U \subseteq M$ be an open subset, and let $\iota: U \hookrightarrow M$ be the inclusion map. For every $p \in U$, the differential $d\iota_p: T_pU \to T_pM$ is an isomorphism.*

Proof. To prove injectivity, suppose $v \in T_pU$ and $d\iota_p(v) = 0 \in T_pM$. Let B be a neighborhood of p such that $\bar{B} \subseteq U$. If $f \in C^\infty(U)$ is arbitrary, the extension lemma for smooth functions guarantees that there exists $\tilde{f} \in C^\infty(M)$ such that $\tilde{f} \equiv f$ on $\bar{B}$. Then since f and $\tilde{f}|_U$ are smooth functions on U that agree in a neighborhood of p, Proposition 3.8 implies

$$vf = v(\tilde{f}|_U) = v(\tilde{f} \circ \iota) = d\iota(v)_p \tilde{f} = 0.$$

Since this holds for every $f \in C^\infty(U)$, it follows that $v = 0$, so $d\iota_p$ is injective.

On the other hand, to prove surjectivity, suppose $w \in T_pM$ is arbitrary. Define an operator $v: C^\infty(U) \to \mathbb{R}$ by setting $vf = w\tilde{f}$, where $\tilde{f}$ is any smooth function on all of M that agrees with f on $\bar{B}$. By Proposition 3.8, vf is independent of the choice of $\tilde{f}$, so v is well defined, and it is easy to check that it is a derivation of $C^\infty(U)$ at p. For any $g \in C^\infty(M)$,

$$d\iota_p(v)g = v(g \circ \iota) = w(\widetilde{g \circ \iota}) = wg,$$

where the last two equalities follow from the facts that $g \circ \iota$, $\widetilde{g \circ \iota}$, and g all agree on B. Therefore, $d\iota_p$ is also surjective.   □

Given an open subset $U \subseteq M$, the isomorphism $d\iota_p$ between T_pU and T_pM is canonically defined, independently of any choices. From now on we *identify* T_pU with T_pM for any point $p \in U$. This identification just amounts to the observation

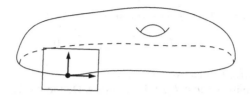

Fig. 3.4 The tangent space to a manifold with boundary

that $d\iota_p(v)$ is the *same derivation as* v, thought of as acting on functions on the bigger manifold M instead of functions on U. Since the action of a derivation on a function depends only on the values of the function in an arbitrarily small neighborhood, this is a harmless identification. In particular, this means that any tangent vector $v \in T_pM$ can be unambiguously applied to functions defined only in a neighborhood of p, not necessarily on all of M.

Proposition 3.10 (Dimension of the Tangent Space). *If M is an n-dimensional smooth manifold, then for each $p \in M$, the tangent space T_pM is an n-dimensional vector space.*

Proof. Given $p \in M$, let (U, φ) be a smooth coordinate chart containing p. Because φ is a diffeomorphism from U onto an open subset $\hat{U} \subseteq \mathbb{R}^n$, it follows from Proposition 3.6(d) that $d\varphi_p$ is an isomorphism from T_pU to $T_{\varphi(p)}\hat{U}$. Since Proposition 3.9 guarantees that $T_pM \cong T_pU$ and $T_{\varphi(p)}\hat{U} \cong T_{\varphi(p)}\mathbb{R}^n$, it follows that $\dim T_pM = \dim T_{\varphi(p)}\mathbb{R}^n = n$. $\qquad\square$

Next we need to prove an analogous result for manifolds with boundary. In fact, if M is an n-manifold with boundary, it might not be immediately clear what one should expect the tangent space at a boundary point of M to look like. Should it be an n-dimensional vector space, like the tangent space at an interior point? Or should it be $(n - 1)$-dimensional, like the boundary? Or should it be an n-dimensional half-space, like the space $\mathbb{H}^n$ on which M is modeled locally?

As we will show below, our definition implies that the tangent space at a boundary point is an n-dimensional vector space (Fig. 3.4), just like the tangent spaces at interior points. This may or may not seem like the most geometrically intuitive choice, but it has the advantage of making most of the definitions of geometric objects on a manifold with boundary look exactly the same as those on a manifold.

First, we need to relate the tangent spaces $T_a\mathbb{H}^n$ and $T_a\mathbb{R}^n$ for points $a \in \partial\mathbb{H}^n$. Since $\mathbb{H}^n$ is not an open subset of $\mathbb{R}^n$, Proposition 3.9 does not apply. As a substitute, we have the following lemma.

Lemma 3.11. *Let $\iota: \mathbb{H}^n \hookrightarrow \mathbb{R}^n$ denote the inclusion map. For any $a \in \partial\mathbb{H}^n$, the differential $d\iota_a: T_a\mathbb{H}^n \to T_a\mathbb{R}^n$ is an isomorphism.*

Proof. Suppose $a \in \partial\mathbb{H}^n$. To show that $d\iota_a$ is injective, assume $d\iota_a(v) = 0$. Suppose $f: \mathbb{H}^n \to \mathbb{R}$ is smooth, and let $\tilde{f}$ be any extension of f to a smooth function defined on all of $\mathbb{R}^n$. (Such an extension exists by the extension lemma for smooth functions, Lemma 2.26.) Then $\tilde{f} \circ \iota = f$, so

$$vf = v(\tilde{f} \circ \iota) = d\iota_a(v)\tilde{f} = 0,$$

which implies that $d\iota_a$ is injective.

To show surjectivity, let $w \in T_a\mathbb{R}^n$ be arbitrary. Define $v \in T_a\mathbb{H}^n$ by

$$vf = w\tilde{f},$$

where $\tilde{f}$ is any smooth extension of f. Writing $w = w^i \partial/\partial x^i|_a$ in terms of the standard basis for $T_a\mathbb{R}^n$, this means that

$$vf = w^i \frac{\partial \tilde{f}}{\partial x^i}(a).$$

This is independent of the choice of $\tilde{f}$, because by continuity the derivatives of $\tilde{f}$ at a are determined by those of f in $\mathbb{H}^n$. It is easy to check that v is a derivation at a and that $w = d\iota_a(v)$, so $d\iota_a$ is surjective. $\square$

Just as we use Proposition 3.9 to identify T_pU with T_pM when U is an open subset of M, we use this lemma to identify $T_a\mathbb{H}^n$ with $T_a\mathbb{R}^n$ when $a \in \partial\mathbb{H}^n$, and we do not distinguish notationally between an element of $T_a\mathbb{H}^n$ and its image in $T_a\mathbb{R}^n$.

Proposition 3.12 (Dimension of Tangent Spaces on a Manifold with Boundary).
Suppose M is an n-dimensional smooth manifold with boundary. For each $p \in M$, T_pM is an n-dimensional vector space.

Proof. Let $p \in M$ be arbitrary. If p is an interior point, then because $\operatorname{Int} M$ is an open submanifold of M, Proposition 3.9 implies that $T_p(\operatorname{Int} M) \cong T_pM$. Since $\operatorname{Int} M$ is a smooth n-manifold without boundary, its tangent spaces all have dimension n.

On the other hand, if $p \in \partial M$, let (U, φ) be a smooth boundary chart containing p, and let $\hat{U} = \varphi(U) \subseteq \mathbb{H}^n$. There are isomorphisms $T_pM \cong T_pU$ (by Proposition 3.9); $T_pU \cong T_{\varphi(p)}\hat{U}$ (by Proposition 3.6(d), because φ is a diffeomorphism); $T_{\varphi(p)}\hat{U} \cong T_{\varphi(p)}\mathbb{H}^n$ (by Proposition 3.9 again); and $T_{\varphi(p)}\mathbb{H}^n \cong T_{\varphi(p)}\mathbb{R}^n$ (by Lemma 3.11). The result follows. $\square$

Recall from Example 1.24 that every finite-dimensional vector space has a natural smooth manifold structure that is independent of any choice of basis or norm. The following proposition shows that the tangent space to a vector space can be naturally identified with the vector space itself.

Suppose V is a finite-dimensional vector space and $a \in V$. Just as we did earlier in the case of $\mathbb{R}^n$, for any vector $v \in V$, we define a map $D_v|_a : C^\infty(V) \to \mathbb{R}$ by

$$D_v|_a f = \left.\frac{d}{dt}\right|_{t=0} f(a + tv). \tag{3.5}$$

Proposition 3.13 (The Tangent Space to a Vector Space). *Suppose V is a finite-dimensional vector space with its standard smooth manifold structure. For each point $a \in V$, the map $v \mapsto D_v|_a$ defined by (3.5) is a canonical isomorphism from V to $T_a V$, such that for any linear map $L: V \to W$, the following diagram commutes:*

$$
\begin{array}{ccc}
V & \overset{\cong}{\longrightarrow} & T_a V \\
{\scriptstyle L}\big\downarrow & & \big\downarrow{\scriptstyle dL_a} \\
W & \underset{\cong}{\longrightarrow} & T_{La} W.
\end{array}
\qquad (3.6)
$$

Proof. Once we choose a basis for V, we can use the same argument as in the proof of Proposition 3.2 to show that $D_v|_a$ is indeed a derivation at a, and that the map $v \mapsto D_v|_a$ is an isomorphism.

Now suppose $L: V \to W$ is a linear map. Because its components with respect to any choices of bases for V and W are linear functions of the coordinates, L is smooth. Unwinding the definitions and using the linearity of L, we compute

$$
\begin{aligned}
dL_a\big(D_v|_a\big) f &= D_v|_a (f \circ L) \\
&= \frac{d}{dt}\bigg|_{t=0} f\big(L(a + tv)\big) = \frac{d}{dt}\bigg|_{t=0} f(La + tLv) \\
&= D_{Lv}|_{La} f. \qquad \qquad \square
\end{aligned}
$$

It is important to understand that each isomorphism $V \cong T_a V$ is canonically defined, independently of any choice of basis (notwithstanding the fact that we used a choice of basis to prove that it is an isomorphism). Because of this result, we can routinely *identify* tangent vectors to a finite-dimensional vector space with elements of the space itself. More generally, if M is an open submanifold of a vector space V, we can combine our identifications $T_p M \leftrightarrow T_p V \leftrightarrow V$ to obtain a canonical identification of each tangent space to M with V. For example, since $\mathrm{GL}(n, \mathbb{R})$ is an open submanifold of the vector space $M(n, \mathbb{R})$, we can identify its tangent space at each point $X \in \mathrm{GL}(n, \mathbb{R})$ with the full space of matrices $M(n, \mathbb{R})$.

There is another natural identification for tangent spaces to a product manifold.

Proposition 3.14 (The Tangent Space to a Product Manifold). *Let $M_1, \ldots, M_k$ be smooth manifolds, and for each j, let $\pi_j : M_1 \times \cdots \times M_k \to M_j$ be the projection onto the M_j factor. For any point $p = (p_1, \ldots, p_k) \in M_1 \times \cdots \times M_k$, the map*

$$
\alpha: T_p(M_1 \times \cdots \times M_k) \to T_{p_1} M_1 \oplus \cdots \oplus T_{p_k} M_k
$$

defined by

$$
\alpha(v) = \big(d(\pi_1)_p(v), \ldots, d(\pi_k)_p(v)\big) \qquad (3.7)
$$

is an isomorphism. The same is true if one of the spaces M_i is a smooth manifold with boundary.

Proof. See Problem 3-2. $\qquad \square$

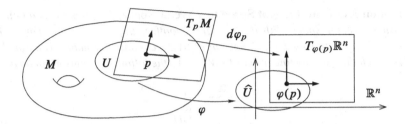

Fig. 3.5 Tangent vectors in coordinates

Once again, because the isomorphism (3.7) is canonically defined, independently of any choice of coordinates, we can consider it as a canonical identification, and we will always do so. Thus, for example, we identify $T_{(p,q)}(M \times N)$ with $T_p M \oplus T_q N$, and treat $T_p M$ and $T_q N$ as subspaces of $T_{(p,q)}(M \times N)$.

Computations in Coordinates

Our treatment of the tangent space to a manifold so far might seem hopelessly abstract. To bring it down to earth, we will show how to do computations with tangent vectors and differentials in local coordinates.

First, suppose M is a smooth manifold (without boundary), and let (U, φ) be a smooth coordinate chart on M. Then φ is, in particular, a diffeomorphism from U to an open subset $\hat{U} \subseteq \mathbb{R}^n$. Combining Propositions 3.9 and 3.6(d), we see that $d\varphi_p \colon T_p M \to T_{\varphi(p)} \mathbb{R}^n$ is an isomorphism.

By Corollary 3.3, the derivations $\partial/\partial x^1|_{\varphi(p)}, \dots, \partial/\partial x^n|_{\varphi(p)}$ form a basis for $T_{\varphi(p)} \mathbb{R}^n$. Therefore, the preimages of these vectors under the isomorphism $d\varphi_p$ form a basis for $T_p M$ (Fig. 3.5). In keeping with our standard practice of treating coordinate maps as identifications whenever possible, we use the notation $\partial/\partial x^i|_p$ for these vectors, characterized by either of the following expressions:

$$\frac{\partial}{\partial x^i}\bigg|_p = (d\varphi_p)^{-1} \left(\frac{\partial}{\partial x^i}\bigg|_{\varphi(p)} \right) = d(\varphi^{-1})_{\varphi(p)} \left(\frac{\partial}{\partial x^i}\bigg|_{\varphi(p)} \right). \tag{3.8}$$

Unwinding the definitions, we see that $\partial/\partial x^i|_p$ acts on a function $f \in C^\infty(U)$ by

$$\frac{\partial}{\partial x^i}\bigg|_p f = \frac{\partial}{\partial x^i}\bigg|_{\varphi(p)} (f \circ \varphi^{-1}) = \frac{\partial \hat{f}}{\partial x^i}(\hat{p}),$$

where $\hat{f} = f \circ \varphi^{-1}$ is the coordinate representation of f, and $\hat{p} = (p^1, \dots, p^n) = \varphi(p)$ is the coordinate representation of p. In other words, $\partial/\partial x^i|_p$ is just the derivation that takes the ith partial derivative of (the coordinate representation of) f at (the coordinate representation of) p. The vectors $\partial/\partial x^i|_p$ are called the **coordinate vectors at** p associated with the given coordinate system. In the special case of standard coordinates on $\mathbb{R}^n$, the vectors $\partial/\partial x^i|_p$ are literally the partial derivative operators.

When M is a smooth manifold with boundary and p is an interior point, the discussion above applies verbatim. For $p \in \partial M$, the only change that needs to be made is to substitute $\mathbb{H}^n$ for $\mathbb{R}^n$, with the understanding that the notation $\partial/\partial x^i|_{\varphi(p)}$ can be used interchangeably to denote either an element of $T_{\varphi(p)}\mathbb{R}^n$ or an element of $T_{\varphi(p)}\mathbb{H}^n$, in keeping with our convention of considering the isomorphism $d\iota_{\varphi(p)} : T_{\varphi(p)}\mathbb{H}^n \to T_{\varphi(p)}\mathbb{R}^n$ as an identification. The nth coordinate vector $\partial/\partial x^n|_p$ should be interpreted as a one-sided derivative in this case.

The following proposition summarizes the discussion so far.

Proposition 3.15. *Let M be a smooth n-manifold with or without boundary, and let $p \in M$. Then T_pM is an n-dimensional vector space, and for any smooth chart $(U, (x^i))$ containing p, the coordinate vectors $\partial/\partial x^1|_p, \dots, \partial/\partial x^n|_p$ form a basis for T_pM.* $\square$

Thus, a tangent vector $v \in T_pM$ can be written uniquely as a linear combination

$$v = v^i \frac{\partial}{\partial x^i}\bigg|_p,$$

where we use the summation convention as usual, with an upper index in the denominator being considered as a lower index, as explained on p. 52. The ordered basis $(\partial/\partial x^i|_p)$ is called a *coordinate basis for T_pM*, and the numbers $(v^1, \dots, v^n)$ are called the *components of v* with respect to the coordinate basis. If v is known, its components can be computed easily from its action on the coordinate functions. For each j, the components of v are given by $v^j = v(x^j)$ (where we think of x^j as a smooth real-valued function on U), because

$$v(x^j) = \left(v^i \frac{\partial}{\partial x^i}\bigg|_p \right)(x^j) = v^i \frac{\partial x^j}{\partial x^i}(p) = v^j.$$

The Differential in Coordinates

Next we explore how differentials look in coordinates. We begin by considering the special case of a smooth map $F : U \to V$, where $U \subseteq \mathbb{R}^n$ and $V \subseteq \mathbb{R}^m$ are open subsets of Euclidean spaces. For any $p \in U$, we will determine the matrix of $dF_p : T_p\mathbb{R}^n \to T_{F(p)}\mathbb{R}^m$ in terms of the standard coordinate bases. Using $(x^1, \dots, x^n)$ to denote the coordinates in the domain and $(y^1, \dots, y^m)$ to denote those in the codomain, we use the chain rule to compute the action of dF_p on a typical basis vector as follows:

$$dF_p\left(\frac{\partial}{\partial x^i}\bigg|_p \right)f = \frac{\partial}{\partial x^i}\bigg|_p (f \circ F) = \frac{\partial f}{\partial y^j}(F(p)) \frac{\partial F^j}{\partial x^i}(p)$$

$$= \left(\frac{\partial F^j}{\partial x^i}(p) \frac{\partial}{\partial y^j}\bigg|_{F(p)} \right)f.$$

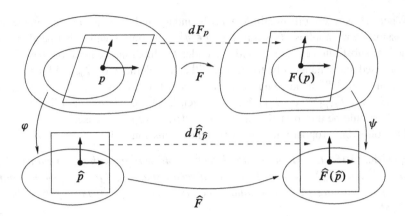

Fig. 3.6 The differential in coordinates

Thus

$$dF_p\left(\frac{\partial}{\partial x^i}\bigg|_p\right) = \frac{\partial F^j}{\partial x^i}(p)\,\frac{\partial}{\partial y^j}\bigg|_{F(p)}. \tag{3.9}$$

In other words, the matrix of dF_p in terms of the coordinate bases is

$$\begin{pmatrix} \dfrac{\partial F^1}{\partial x^1}(p) & \cdots & \dfrac{\partial F^1}{\partial x^n}(p) \\ \vdots & \ddots & \vdots \\ \dfrac{\partial F^m}{\partial x^1}(p) & \cdots & \dfrac{\partial F^m}{\partial x^n}(p) \end{pmatrix}.$$

(Recall that the columns of the matrix are the components of the images of the basis vectors.) This matrix is none other than the Jacobian matrix of F at p, which is the matrix representation of the total derivative $DF(p)\colon \mathbb{R}^n \to \mathbb{R}^m$. Therefore, in this case, $dF_p\colon T_p\mathbb{R}^n \to T_{F(p)}\mathbb{R}^m$ corresponds to the total derivative $DF(p)\colon \mathbb{R}^n \to \mathbb{R}^m$, under our usual identification of Euclidean spaces with their tangent spaces. The same calculation applies if U is an open subset of $\mathbb{H}^n$ and V is an open subset of $\mathbb{H}^m$.

Now consider the more general case of a smooth map $F\colon M \to N$ between smooth manifolds with or without boundary. Choosing smooth coordinate charts (U,φ) for M containing p and (V,ψ) for N containing $F(p)$, we obtain the coordinate representation $\widehat{F} = \psi \circ F \circ \varphi^{-1}\colon \varphi(U \cap F^{-1}(V)) \to \psi(V)$ (Fig. 3.6). Let $\widehat{p} = \varphi(p)$ denote the coordinate representation of p. By the computation above, $d\widehat{F}_{\widehat{p}}$ is represented with respect to the standard coordinate bases by the Jacobian matrix of $\widehat{F}$ at $\widehat{p}$. Using the fact that $F \circ \varphi^{-1} = \psi^{-1} \circ \widehat{F}$, we compute

$$dF_p\left(\frac{\partial}{\partial x^i}\bigg|_p\right) = dF_p\left(d(\varphi^{-1})_{\widehat{p}}\left(\frac{\partial}{\partial x^i}\bigg|_{\widehat{p}}\right)\right) = d(\psi^{-1})_{\widehat{F}(\widehat{p})}\left(d\widehat{F}_{\widehat{p}}\left(\frac{\partial}{\partial x^i}\bigg|_{\widehat{p}}\right)\right)$$

$$= d\left(\psi^{-1}\right)_{\widehat{F}(\widehat{p})}\left(\frac{\partial \widehat{F}^{j}}{\partial x^{i}}(\widehat{p}) \left.\frac{\partial}{\partial y^{j}}\right|_{\widehat{F}(\widehat{p})}\right)$$

$$= \frac{\partial \widehat{F}^{j}}{\partial x^{i}}(\widehat{p}) \left.\frac{\partial}{\partial y^{j}}\right|_{F(p)}. \tag{3.10}$$

Thus, dF_p is represented in coordinate bases by the Jacobian matrix of (the coordinate representative of) F. In fact, the definition of the differential was cooked up precisely to give a coordinate-independent meaning to the Jacobian matrix.

In the differential geometry literature, the differential is sometimes called the *tangent map*, the *total derivative*, or simply the *derivative of F*. Because it "pushes" tangent vectors forward from the domain manifold to the codomain, it is also called the *(pointwise) pushforward*. Different authors denote it by symbols such as

$$F'(p), \quad DF, \quad DF(p), \quad F_*, \quad TF, \quad T_pF.$$

We will stick with the notation dF_p for the differential of a smooth map between manifolds, and reserve $DF(p)$ for the total derivative of a map between finite-dimensional vector spaces, which in the case of Euclidean spaces we identify with the Jacobian matrix of F.

Change of Coordinates

Suppose (U,φ) and (V,ψ) are two smooth charts on M, and $p \in U \cap V$. Let us denote the coordinate functions of φ by $\left(x^i\right)$ and those of ψ by $\left(\widetilde{x}^i\right)$. Any tangent vector at p can be represented with respect to either basis $\left(\partial/\partial x^i|_p\right)$ or $\left(\partial/\partial \widetilde{x}^i|_p\right)$. How are the two representations related?

In this situation, it is customary to write the transition map $\psi \circ \varphi^{-1} : \varphi(U \cap V) \rightarrow \psi(U \cap V)$ in the following shorthand notation:

$$\psi \circ \varphi^{-1}(x) = \left(\widetilde{x}^1(x), \ldots, \widetilde{x}^n(x)\right).$$

Here we are indulging in a typical abuse of notation: in the expression $\widetilde{x}^i(x)$, we think of $\widetilde{x}^i$ as a coordinate *function* (whose domain is an open subset of M, identified with an open subset of $\mathbb{R}^n$ or $\mathbb{H}^n$); but we think of x as representing a *point* (in this case, in $\varphi(U \cap V)$). By (3.9), the differential $d\left(\psi \circ \varphi^{-1}\right)_{\varphi(p)}$ can be written

$$d\left(\psi \circ \varphi^{-1}\right)_{\varphi(p)}\left(\left.\frac{\partial}{\partial x^{i}}\right|_{\varphi(p)}\right) = \frac{\partial \widetilde{x}^{j}}{\partial x^{i}}(\varphi(p)) \left.\frac{\partial}{\partial \widetilde{x}^{j}}\right|_{\psi(p)}.$$

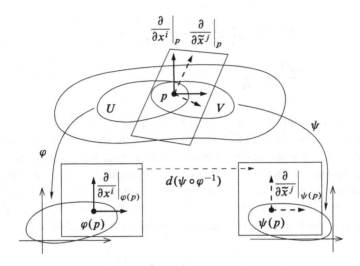

Fig. 3.7 Change of coordinates

(See Fig. 3.7.) Using the definition of coordinate vectors, we obtain

$$\left.\frac{\partial}{\partial x^i}\right|_p = d\left(\varphi^{-1}\right)_{\varphi(p)}\left(\left.\frac{\partial}{\partial x^i}\right|_{\varphi(p)}\right)$$

$$= d\left(\psi^{-1}\right)_{\psi(p)} \circ d\left(\psi \circ \varphi^{-1}\right)_{\varphi(p)}\left(\left.\frac{\partial}{\partial x^i}\right|_{\varphi(p)}\right)$$

$$= d\left(\psi^{-1}\right)_{\psi(p)}\left(\frac{\partial \widetilde{x}^j}{\partial x^i}(\varphi(p)) \left.\frac{\partial}{\partial \widetilde{x}^j}\right|_{\psi(p)}\right) = \frac{\partial \widetilde{x}^j}{\partial x^i}(\widehat{p}) \left.\frac{\partial}{\partial \widetilde{x}^j}\right|_p, \qquad (3.11)$$

where again we have written $\widehat{p} = \varphi(p)$. (This formula is easy to remember, because it looks exactly the same as the chain rule for partial derivatives in $\mathbb{R}^n$.) Applying this to the components of a vector $v = v^i \partial/\partial x^i|_p = \widetilde{v}^j \partial/\partial \widetilde{x}^j|_p$, we find that the components of v transform by the rule

$$\widetilde{v}^j = \frac{\partial \widetilde{x}^j}{\partial x^i}(\widehat{p}) v^i. \qquad (3.12)$$

Example 3.16. The transition map between polar coordinates and standard coordinates in suitable open subsets of the plane is given by $(x, y) = (r \cos \theta, r \sin \theta)$. Let p be the point in $\mathbb{R}^2$ whose polar coordinate representation is $(r, \theta) = (2, \pi/2)$, and let $v \in T_p\mathbb{R}^2$ be the tangent vector whose polar coordinate representation is

$$v = 3\left.\frac{\partial}{\partial r}\right|_p - \left.\frac{\partial}{\partial \theta}\right|_p.$$

Applying (3.11) to the coordinate vectors, we find

$$\frac{\partial}{\partial r}\bigg|_p = \cos\left(\frac{\pi}{2}\right)\frac{\partial}{\partial x}\bigg|_p + \sin\left(\frac{\pi}{2}\right)\frac{\partial}{\partial y}\bigg|_p = \frac{\partial}{\partial y}\bigg|_p,$$

$$\frac{\partial}{\partial \theta}\bigg|_p = -2\sin\left(\frac{\pi}{2}\right)\frac{\partial}{\partial x}\bigg|_p + 2\cos\left(\frac{\pi}{2}\right)\frac{\partial}{\partial y}\bigg|_p = -2\frac{\partial}{\partial x}\bigg|_p,$$

and thus v has the following coordinate representation in standard coordinates:

$$v = 3\frac{\partial}{\partial y}\bigg|_p + 2\frac{\partial}{\partial x}\bigg|_p. \qquad\qquad //$$

One important fact to bear in mind is that each coordinate vector $\partial/\partial x^i|_p$ depends on the entire *coordinate system*, not just on the single coordinate function x^i. Geometrically, this reflects the fact that $\partial/\partial x^i|_p$ is the derivation obtained by differentiating with respect to x^i while *all the other coordinates are held constant*. If the coordinate functions other than x^i are changed, then the direction of this coordinate derivative can change. The next exercise illustrates how this can happen.

▶ **Exercise 3.17.** Let (x, y) denote the standard coordinates on $\mathbb{R}^2$. Verify that $(\tilde{x}, \tilde{y})$ are global smooth coordinates on $\mathbb{R}^2$, where

$$\tilde{x} = x, \qquad \tilde{y} = y + x^3.$$

Let p be the point $(1, 0) \in \mathbb{R}^2$ (in standard coordinates), and show that

$$\frac{\partial}{\partial x}\bigg|_p \neq \frac{\partial}{\partial \tilde{x}}\bigg|_p,$$

even though the coordinate functions x and $\tilde{x}$ are identically equal.

The Tangent Bundle

Often it is useful to consider the set of all tangent vectors at all points of a manifold. Given a smooth manifold M with or without boundary, we define the *tangent bundle of M*, denoted by TM, to be the disjoint union of the tangent spaces at all points of M:

$$TM = \coprod_{p \in M} T_p M.$$

We usually write an element of this disjoint union as an ordered pair (p, v), with $p \in M$ and $v \in T_p M$ (instead of putting the point p in the second position, as elements of a disjoint union are more commonly written). The tangent bundle comes equipped with a natural *projection map* $\pi \colon TM \to M$, which sends each vector in $T_p M$ to the point p at which it is tangent: $\pi(p, v) = p$. We will often commit the usual mild sin of identifying $T_p M$ with its image under the canonical injection $v \mapsto (p, v)$, and will use any of the notations (p, v), v_p, and v for a tangent vector in $T_p M$, depending on how much emphasis we wish to give to the point p.

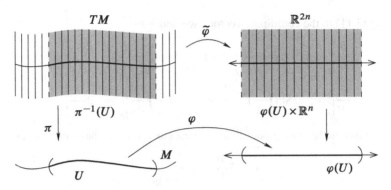

Fig. 3.8 Coordinates for the tangent bundle

For example, in the special case $M = \mathbb{R}^n$, using Proposition 3.2, we see that the tangent bundle of $\mathbb{R}^n$ can be canonically identified with the union of its geometric tangent spaces, which in turn is just the Cartesian product of $\mathbb{R}^n$ with itself:

$$T\mathbb{R}^n = \coprod_{a \in \mathbb{R}^n} T_a\mathbb{R}^n \cong \coprod_{a \in \mathbb{R}^n} \mathbb{R}^n_a = \coprod_{a \in \mathbb{R}^n} \{a\} \times \mathbb{R}^n = \mathbb{R}^n \times \mathbb{R}^n.$$

An element (a, v) of this Cartesian product can be thought of as representing either the geometric tangent vector v_a or the derivation $D_v|_a$ defined by (3.1). Be warned, however, that in general the tangent bundle of a smooth manifold cannot be identified in any natural way with a Cartesian product, because there is no canonical way to identify tangent spaces at different points with each other. We will have more to say about this below.

If M is a smooth manifold, the tangent bundle TM can be thought of simply as a disjoint union of vector spaces; but it is much more than that. The next proposition shows that TM can be considered as a smooth manifold in its own right.

Proposition 3.18. *For any smooth n-manifold M, the tangent bundle TM has a natural topology and smooth structure that make it into a 2n-dimensional smooth manifold. With respect to this structure, the projection $\pi\colon TM \to M$ is smooth.*

Proof. We begin by defining the maps that will become our smooth charts. Given any smooth chart (U, φ) for M, note that $\pi^{-1}(U) \subseteq TM$ is the set of all tangent vectors to M at all points of U. Let $(x^1, \dots, x^n)$ denote the coordinate functions of φ, and define a map $\widetilde{\varphi}\colon \pi^{-1}(U) \to \mathbb{R}^{2n}$ by

$$\widetilde{\varphi}\left(v^i \frac{\partial}{\partial x^i}\bigg|_p\right) = \left(x^1(p), \dots, x^n(p), v^1, \dots, v^n\right). \tag{3.13}$$

(See Fig. 3.8.) Its image set is $\varphi(U) \times \mathbb{R}^n$, which is an open subset of $\mathbb{R}^{2n}$. It is a bijection onto its image, because its inverse can be written explicitly as

$$\widetilde{\varphi}^{-1}\left(x^1, \dots, x^n, v^1, \dots, v^n\right) = v^i \frac{\partial}{\partial x^i}\bigg|_{\varphi^{-1}(x)}.$$

Now suppose we are given two smooth charts (U, φ) and (V, ψ) for M, and let $\left(\pi^{-1}(U), \widetilde{\varphi}\right), \left(\pi^{-1}(V), \widetilde{\psi}\right)$ be the corresponding charts on TM. The sets

$$\widetilde{\varphi}\left(\pi^{-1}(U) \cap \pi^{-1}(V)\right) = \varphi(U \cap V) \times \mathbb{R}^n \quad \text{and}$$

$$\widetilde{\psi}\left(\pi^{-1}(U) \cap \pi^{-1}(V)\right) = \psi(U \cap V) \times \mathbb{R}^n$$

are open in $\mathbb{R}^{2n}$, and the transition map $\widetilde{\psi} \circ \widetilde{\varphi}^{-1} \colon \varphi(U \cap V) \times \mathbb{R}^n \to \psi(U \cap V) \times \mathbb{R}^n$ can be written explicitly using (3.12) as

$$\widetilde{\psi} \circ \widetilde{\varphi}^{-1}\left(x^1, \ldots, x^n, v^1, \ldots, v^n\right)$$

$$= \left(\widetilde{x}^1(x), \ldots, \widetilde{x}^n(x), \frac{\partial \widetilde{x}^1}{\partial x^j}(x)v^j, \ldots, \frac{\partial \widetilde{x}^n}{\partial x^j}(x)v^j\right).$$

This is clearly smooth.

Choosing a countable cover $\{U_i\}$ of M by smooth coordinate domains, we obtain a countable cover of TM by coordinate domains $\{\pi^{-1}(U_i)\}$ satisfying conditions (i)–(iv) of the smooth manifold chart lemma (Lemma 1.35). To check the Hausdorff condition (v), just note that any two points in the same fiber of π lie in one chart, while if (p, v) and (q, w) lie in different fibers, there exist disjoint smooth coordinate domains U, V for M such that $p \in U$ and $q \in V$, and then $\pi^{-1}(U)$ and $\pi^{-1}(V)$ are disjoint coordinate neighborhoods containing (p, v) and (q, w), respectively.

To see that π is smooth, note that with respect to charts (U, φ) for M and $\left(\pi^{-1}(U), \widetilde{\varphi}\right)$ for TM, its coordinate representation is $\pi(x, v) = x$. $\qquad \square$

The coordinates $\left(x^i, v^i\right)$ given by (3.13) are called **natural coordinates on TM**.

▶ **Exercise 3.19.** Suppose M is a smooth manifold with boundary. Show that TM has a natural topology and smooth structure making it into a smooth manifold with boundary, such that if $\left(U, \left(x^i\right)\right)$ is any smooth boundary chart for M, then rearranging the coordinates in the natural chart $\left(\pi^{-1}(U), \left(x^i, v^i\right)\right)$ for TM yields a boundary chart $\left(\pi^{-1}(U), \left(v^i, x^i\right)\right)$.

Proposition 3.20. *If M is a smooth n-manifold with or without boundary, and M can be covered by a single smooth chart, then TM is diffeomorphic to $M \times \mathbb{R}^n$.*

Proof. If (U, φ) is a global smooth chart for M, then φ is, in particular, a diffeomorphism from $U = M$ to an open subset $\widehat{U} \subseteq \mathbb{R}^n$ or $\mathbb{H}^n$. The proof of the previous proposition showed that the natural coordinate chart $\widetilde{\varphi}$ is a bijection from TM to $\widehat{U} \times \mathbb{R}^n$, and the smooth structure on TM is defined essentially by declaring $\widetilde{\varphi}$ to be a diffeomorphism. $\qquad \square$

Although the picture of a product $U \times \mathbb{R}^n$ is a useful way to visualize the smooth structure on a tangent bundle locally as in Fig. 3.8, do not be misled into imagining that every tangent bundle is *globally* diffeomorphic (or even homeomorphic) to a product of the manifold with $\mathbb{R}^n$. This is not the case for most smooth manifolds. We will revisit this question in Chapters 8, 10, and 16.

By putting together the differentials of F at all points of M, we obtain a globally defined map between tangent bundles, called the **global differential** or **global tangent map** and denoted by $dF: TM \to TN$. This is just the map whose restriction to each tangent space $T_p M \subseteq TM$ is dF_p. When we apply the differential of F to a specific vector $v \in T_p M$, we can write either $dF_p(v)$ or $dF(v)$, depending on how much emphasis we wish to give to the point p. The former notation is more informative, while the second is more concise.

One important feature of the smooth structure we have defined on TM is that it makes the differential of a smooth map into a smooth map between tangent bundles.

Proposition 3.21. *If* $F: M \to N$ *is a smooth map, then its global differential* $dF: TM \to TN$ *is a smooth map.*

Proof. From the local expression (3.9) for dF_p in coordinates, it follows that dF has the following coordinate representation in terms of natural coordinates for TM and TN:

$$dF\left(x^1, \ldots, x^n, v^1, \ldots, v^n\right) = \left(F^1(x), \ldots, F^n(x), \frac{\partial F^1}{\partial x^i}(x)v^i, \ldots, \frac{\partial F^n}{\partial x^i}(x)v^i\right).$$

This is smooth because F is. $\square$

The following properties of the global differential follow immediately from Proposition 3.6.

Corollary 3.22 (Properties of the Global Differential). *Suppose* $F: M \to N$ *and* $G: N \to P$ *are smooth maps.*

(a) $d(G \circ F) = dG \circ dF$.
(b) $d(\mathrm{Id}_M) = \mathrm{Id}_{TM}$.
(c) *If* F *is a diffeomorphism, then* $dF: TM \to TN$ *is also a diffeomorphism, and* $(dF)^{-1} = d\left(F^{-1}\right)$. $\square$

Because of part (c) of this corollary, when F is a diffeomorphism we can use the notation dF^{-1} unambiguously to mean either $(dF)^{-1}$ or $d\left(F^{-1}\right)$.

Velocity Vectors of Curves

The *velocity* of a smooth parametrized curve in $\mathbb{R}^n$ is familiar from elementary calculus. It is just the vector whose components are the derivatives of the component functions of the curve. In this section we extend this notion to curves in manifolds.

If M is a manifold with or without boundary, we define a **curve in** M to be a continuous map $\gamma: J \to M$, where $J \subseteq \mathbb{R}$ is an interval. (Most of the time, we will be interested in curves whose domains are open intervals, but for some purposes it is useful to allow J to have one or two endpoints; the definitions all make sense with minor modifications in that case, either by considering J as a manifold with boundary or by interpreting derivatives as one-sided derivatives.) Note that in this

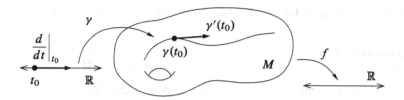

Fig. 3.9 The velocity of a curve

book the term *curve* always refers to a map from an interval into M (a parametrized curve), not just a set of points in M.

Now let M be a smooth manifold, still with or without boundary. Our definition of tangent spaces leads to a natural interpretation of velocity vectors: given a smooth curve $\gamma: J \to M$ and $t_0 \in J$, we define the **velocity of γ at t_0** (Fig. 3.9), denoted by $\gamma'(t_0)$, to be the vector

$$\gamma'(t_0) = d\gamma\left(\left.\frac{d}{dt}\right|_{t_0}\right) \in T_{\gamma(t_0)}M,$$

where $d/dt|_{t_0}$ is the standard coordinate basis vector in $T_{t_0}\mathbb{R}$. (As in ordinary calculus, it is customary to use d/dt instead of $\partial/\partial t$ when the manifold is 1-dimensional.) Other common notations for the velocity are

$$\dot{\gamma}(t_0), \quad \frac{d\gamma}{dt}(t_0), \quad \text{and} \quad \left.\frac{d\gamma}{dt}\right|_{t=t_0}.$$

This tangent vector acts on functions by

$$\gamma'(t_0)f = d\gamma\left(\left.\frac{d}{dt}\right|_{t_0}\right)f = \left.\frac{d}{dt}\right|_{t_0}(f \circ \gamma) = (f \circ \gamma)'(t_0).$$

In other words, $\gamma'(t_0)$ is the derivation at $\gamma(t_0)$ obtained by taking the derivative of a function along γ. (If t_0 is an endpoint of J, this still holds, provided that we interpret the derivative with respect to t as a one-sided derivative, or equivalently as the derivative of any smooth extension of $f \circ \gamma$ to an open subset of $\mathbb{R}$.)

Now let (U, φ) be a smooth chart with coordinate functions (x^i). If $\gamma(t_0) \in U$, we can write the coordinate representation of γ as $\gamma(t) = (\gamma^1(t), \dots, \gamma^n(t))$, at least for t sufficiently close to t_0, and then the coordinate formula for the differential yields

$$\gamma'(t_0) = \frac{d\gamma^i}{dt}(t_0) \left.\frac{\partial}{\partial x^i}\right|_{\gamma(t_0)}.$$

This means that $\gamma'(t_0)$ is given by essentially the same formula as it would be in Euclidean space: it is the tangent vector whose components in a coordinate basis are the derivatives of the component functions of γ.

The next proposition shows that every tangent vector on a manifold is the velocity vector of some curve. This gives a different and somewhat more geometric way to think about the tangent bundle: it is just the set of all velocity vectors of smooth curves in M.

Proposition 3.23. *Suppose M is a smooth manifold with or without boundary and $p \in M$. Every $v \in T_p M$ is the velocity of some smooth curve in M.*

Proof. First suppose that $p \in \operatorname{Int} M$ (which includes the case $\partial M = \varnothing$). Let (U, φ) be a smooth coordinate chart centered at p, and write $v = v^i \partial/\partial x^i|_p$ in terms of the coordinate basis. For sufficiently small $\varepsilon > 0$, let $\gamma \colon (-\varepsilon, \varepsilon) \to U$ be the curve whose coordinate representation is

$$\gamma(t) = \left(t v^1, \dots, t v^n\right). \tag{3.14}$$

(Remember, this really means $\gamma(t) = \varphi^{-1}\left(t v^1, \dots, t v^n\right)$.) This is a smooth curve with $\gamma(0) = p$, and the computation above shows that $\gamma'(0) = v^i \partial/\partial x^i|_{\gamma(0)} = v$.

Now suppose $p \in \partial M$. Let (U, φ) be a smooth boundary chart centered at p, and write $v = v^i \partial/\partial x^i|_p$ as before. We wish to let γ be the curve whose coordinate representation is (3.14), but this formula represents a point of M only when $t v^n \geq 0$. We can accommodate this requirement by suitably restricting the domain of γ: if $v^n = 0$, we define $\gamma \colon (-\varepsilon, \varepsilon) \to U$ as before; if $v^n > 0$, we let the domain be $[0, \varepsilon)$; and if $v^n < 0$, we let it be $(-\varepsilon, 0]$. In each case, γ is a smooth curve in M with $\gamma(0) = p$ and $\gamma'(0) = v$.   □

The next proposition shows that velocity vectors behave well under composition with smooth maps.

Proposition 3.24 (The Velocity of a Composite Curve). *Let $F \colon M \to N$ be a smooth map, and let $\gamma \colon J \to M$ be a smooth curve. For any $t_0 \in J$, the velocity at $t = t_0$ of the composite curve $F \circ \gamma \colon J \to N$ is given by*

$$(F \circ \gamma)'(t_0) = dF\left(\gamma'(t_0)\right).$$

Proof. Just go back to the definition of the velocity of a curve:

$$(F \circ \gamma)'(t_0) = d(F \circ \gamma)\left(\frac{d}{dt}\bigg|_{t_0}\right) = dF \circ d\gamma \left(\frac{d}{dt}\bigg|_{t_0}\right) = dF\left(\gamma'(t_0)\right). \qquad □$$

On the face of it, the preceding proposition tells us how to compute the velocity of a composite curve in terms of the differential. However, it is often much more useful to turn it around the other way, and use it as a streamlined way to compute differentials. Suppose $F \colon M \to N$ is a smooth map, and we need to compute the differential dF_p at some point $p \in M$. We can compute $dF_p(v)$ for any $v \in T_p M$ by choosing a smooth curve γ whose initial tangent vector is v, and then applying Proposition 3.23 to the composite curve $F \circ \gamma$. The next corollary summarizes the result.

Corollary 3.25 (Computing the Differential Using a Velocity Vector). *Suppose $F \colon M \to N$ is a smooth map, $p \in M$, and $v \in T_p M$. Then*

$$dF_p(v) = (F \circ \gamma)'(0)$$

for any smooth curve $\gamma \colon J \to M$ such that $0 \in J$, $\gamma(0) = p$, and $\gamma'(0) = v$.   □

This corollary frequently yields a much more succinct computation of dF, especially if F is presented in some form other than an explicit coordinate representation. We will see many examples of this technique in later chapters.

Alternative Definitions of the Tangent Space

In the literature you will find tangent vectors to a smooth manifold defined in several different ways. Here we describe the most common ones. (Yet another definition is suggested in the remark following Problem 11-4.) It is good to be conversant with all of them. Throughout this section, M represents an arbitrary smooth manifold with or without boundary.

Tangent Vectors as Derivations of the Space of Germs

The most common alternative definition is based on the notion of "germs" of smooth functions, which we now define.

A *smooth function element* on M is an ordered pair (f, U), where U is an open subset of M and $f : U \to \mathbb{R}$ is a smooth function. Given a point $p \in M$, let us define an equivalence relation on the set of all smooth function elements whose domains contain p by setting $(f, U) \sim (g, V)$ if $f \equiv g$ on some neighborhood of p. The equivalence class of a function element (f, U) is called the **germ of f at p**. The set of all germs of smooth functions at p is denoted by $C_p^\infty(M)$. It is a real vector space and an associative algebra under the operations

$$c[(f, U)] = [(cf, U)],$$
$$[(f, U)] + [(g, V)] = [(f + g, U \cap V)],$$
$$[(f, U)][(g, V)] = [(fg, U \cap V)].$$

(The zero element of this algebra is the equivalence class of the zero function on M.) Let us denote the germ at p of the function element (f, U) simply by $[f]_p$; there is no need to include the domain U in the notation, because the same germ is represented by the restriction of f to any neighborhood of p. To say that two germs $[f]_p$ and $[g]_p$ are equal is simply to say that $f \equiv g$ on some neighborhood of p, however small.

A **derivation of $C_p^\infty(M)$** is a linear map $v : C_p^\infty(M) \to \mathbb{R}$ satisfying the following product rule analogous to (3.4):

$$v[fg]_p = f(p)v[g]_p + g(p)v[f]_p.$$

It is common to define the tangent space to M at p as the vector space $\mathcal{D}_p M$ of derivations of $C_p^\infty(M)$. Thanks to Proposition 3.8, it is a simple matter to prove that $\mathcal{D}_p M$ is naturally isomorphic to the tangent space as we have defined it (see Problem 3-7).

The germ definition has a number of advantages. One of the most significant is that it makes the local nature of the tangent space clearer, without requiring the use of bump functions. Because there do not exist analytic bump functions, the germ definition of tangent vectors is the only one available on real-analytic or complex-analytic manifolds. The chief disadvantage of the germ approach is simply that it adds an additional level of complication to an already highly abstract definition.

Tangent Vectors as Equivalence Classes of Curves

Another common approach to tangent vectors is to define an intrinsic equivalence relation on the set of smooth curves with the same starting point, which captures the idea of "having the same velocity," and to define a tangent vector as an equivalence class of curves. Here we describe one such equivalence relation.

Suppose p is a point of M. We wish to define an equivalence relation on the set of all smooth curves of the form $\gamma: J \to M$, where J is an interval containing 0 and $\gamma(0) = p$. Given two such curves $\gamma_1: J_1 \to M$ and $\gamma_2: J_2 \to M$, let us say that $\gamma_1 \sim \gamma_2$ if $(f \circ \gamma_1)'(0) = (f \circ \gamma_2)'(0)$ for every smooth real-valued function f defined in a neighborhood of p. Let $\mathcal{V}_p M$ denote the set of equivalence classes. The tangent space to M at p is often defined to be the set $\mathcal{V}_p M$.

Using this definition, it is very easy to define the differential of a smooth map $F: M \to N$ as the map that sends $[\gamma] \in \mathcal{V}_p M$ to $[F \circ \gamma] \in \mathcal{V}_{F(p)} N$. Velocity vectors of smooth curves are almost as easy to define. Suppose $\gamma: J \to M$ is any smooth curve. If $0 \in J$, then the velocity of γ at 0 is just the equivalence class of γ in $\mathcal{V}_{\gamma(0)} M$. The velocity at any other point $t_0 \in J$ can be defined as the equivalence class in $\mathcal{V}_{\gamma(t_0)} M$ of the curve γ_{t_0} defined by $\gamma_{t_0}(t) = \gamma(t_0 + t)$.

Problem 3-8 shows that there is a natural one-to-one correspondence between $\mathcal{V}_p M$ and $T_p M$. This definition has the advantage of being geometrically more intuitive, but it has the serious drawback that the existence of a vector space structure on $\mathcal{V}_p M$ is not at all obvious.

Tangent Vectors as Equivalence Classes of n-Tuples

Yet another approach to defining the tangent space is based on the transformation rule (3.12) for the components of tangent vectors in coordinates. One defines a tangent vector at a point $p \in M$ to be a rule that assigns an ordered n-tuple $(v^1, \ldots, v^n) \in \mathbb{R}^n$ to each smooth coordinate chart containing p, with the property that the n-tuples assigned to overlapping charts transform according to (3.12). (This is, in fact, the oldest definition of all, and many physicists are still apt to think of tangent vectors this way.)

In this approach, the velocity of a curve is defined by the usual Euclidean formula in coordinates, and the differential of $F: M \to N$ is defined as the linear map determined by the Jacobian matrix of F in coordinates. One then has to show, by means of tedious computations involving the chain rule, that these operations are well defined, independently of the choices of coordinates.

It is a matter of individual taste which of the various characterizations of $T_p M$ one chooses to take as the definition. The definition we have chosen, however abstract it may seem at first, has several advantages: it is relatively concrete (tangent vectors are actual derivations of $C^\infty(M)$, with no equivalence classes involved); it makes the vector space structure on $T_p M$ obvious; and it leads to straightforward coordinate-independent definitions of differentials, velocities, and many of the other geometric objects we will be studying.

Categories and Functors

Another useful perspective on tangent spaces and differentials is provided by the theory of categories. In this section we summarize the basic definitions of category theory. We do not do much with the theory in this book, but we mention it because it provides a convenient and powerful language for talking about many of the mathematical structures we will meet.

A *category* C consists of the following things:

- a class Ob(C), whose elements are called *objects of* C,
- a class Hom(C), whose elements are called *morphisms of* C,
- for each morphism $f \in$ Hom(C), two objects $X, Y \in$ Ob(C) called the *source* and *target of* f, respectively,
- for each triple $X, Y, Z \in$ Ob(C), a mapping called *composition*:

$$\mathrm{Hom}_C(X, Y) \times \mathrm{Hom}_C(Y, Z) \to \mathrm{Hom}_C(X, Z),$$

written $(f, g) \mapsto g \circ f$, where $\mathrm{Hom}_C(X, Y)$ denotes the class of all morphisms with source X and target Y.

The morphisms are required to satisfy the following axioms:

(i) ASSOCIATIVITY: $(f \circ g) \circ h = f \circ (g \circ h)$.
(ii) EXISTENCE OF IDENTITIES: For each object $X \in$ Ob(C), there exists an *identity morphism* $\mathrm{Id}_X \in \mathrm{Hom}_C(X, X)$, such that $\mathrm{Id}_Y \circ f = f = f \circ \mathrm{Id}_X$ for all $f \in \mathrm{Hom}_C(X, Y)$.

A morphism $f \in \mathrm{Hom}_C(X, Y)$ is called an *isomorphism in* C if there exists a morphism $g \in \mathrm{Hom}_C(Y, X)$ such that $f \circ g = \mathrm{Id}_Y$ and $g \circ f = \mathrm{Id}_X$.

Example 3.26 (Categories). In most of the categories that one meets "in nature," the objects are sets with some extra structure, the morphisms are maps that preserve that structure, and the composition laws and identity morphisms are the obvious ones. Some of the categories of this type that appear in this book (implicitly or explicitly) are listed below. In each case, we describe the category by giving its objects and its morphisms.

- Set: sets and maps
- Top: topological spaces and continuous maps
- Man: topological manifolds and continuous maps

- $\mathsf{Man_b}$: topological manifolds with boundary and continuous maps
- Diff: smooth manifolds and smooth maps
- $\mathsf{Diff_b}$: smooth manifolds with boundary and smooth maps
- $\mathsf{Vec_{\mathbb{R}}}$: real vector spaces and real-linear maps
- $\mathsf{Vec_{\mathbb{C}}}$: complex vector spaces and complex-linear maps
- Grp: groups and group homomorphisms
- Ab: abelian groups and group homomorphisms
- Rng: rings and ring homomorphisms
- CRng: commutative rings and ring homomorphisms

There are also important categories whose objects are sets with distinguished base points, in addition to (possibly) other structures. A **pointed set** is an ordered pair (X, p), where X is a set and p is an element of X. Other pointed objects such as **pointed topological spaces** or **pointed smooth manifolds** are defined similarly. If (X, p) and (X', p') are pointed sets (or topological spaces, etc.), a map $F: X \to X'$ is said to be a **pointed map** if $F(p) = p'$; in this case, we write $F: (X, p) \to (X', p')$. Here are some important examples of categories of pointed objects.

- $\mathsf{Set_*}$: pointed sets and pointed maps
- $\mathsf{Top_*}$: pointed topological spaces and pointed continuous maps
- $\mathsf{Man_*}$: pointed topological manifolds and pointed continuous maps
- $\mathsf{Diff_*}$: pointed smooth manifolds and pointed smooth maps   //

We use the word *class* instead of *set* for the collections of objects and morphisms in a category because in some categories they are "too large" to be considered sets. For example, in the category Set, $\mathrm{Ob}(\mathsf{Set})$ is the class of all sets; any attempt to treat it as a set in its own right leads to the well-known Russell paradox of set theory. (See [LeeTM, Appendix A] or almost any book on set theory for more.) Even though the classes of objects and morphisms might not constitute sets, we still use notations such as $X \in \mathrm{Ob}(\mathsf{C})$ and $f \in \mathrm{Hom}(\mathsf{C})$ to indicate that X is an object and f is a morphism in C. A category in which both $\mathrm{Ob}(\mathsf{C})$ and $\mathrm{Hom}(\mathsf{C})$ are sets is called a **small category**, and one in which each class of morphisms $\mathrm{Hom}_\mathsf{C}(X, Y)$ is a set is called **locally small**. All the categories listed above are locally small but not small.

If C and D are categories, a **covariant functor from C to D** is a rule $\mathscr{F}$ that assigns to each object $X \in \mathrm{Ob}(\mathsf{C})$ an object $\mathscr{F}(X) \in \mathrm{Ob}(\mathsf{D})$, and to each morphism $f \in \mathrm{Hom}_\mathsf{C}(X, Y)$ a morphism $\mathscr{F}(f) \in \mathrm{Hom}_\mathsf{D}(\mathscr{F}(X), \mathscr{F}(Y))$, so that identities and composition are preserved:

$$\mathscr{F}(\mathrm{Id}_X) = \mathrm{Id}_{\mathscr{F}(X)}; \qquad \mathscr{F}(g \circ h) = \mathscr{F}(g) \circ \mathscr{F}(h).$$

We also need to consider functors that reverse morphisms: a **contravariant functor from C to D** is a rule $\mathscr{F}$ that assigns to each object $X \in \mathrm{Ob}(\mathsf{C})$ an object $\mathscr{F}(X) \in \mathrm{Ob}(\mathsf{D})$, and to each morphism $f \in \mathrm{Hom}_\mathsf{C}(X, Y)$ a morphism $\mathscr{F}(f) \in \mathrm{Hom}_\mathsf{D}(\mathscr{F}(Y), \mathscr{F}(X))$, such that

$$\mathscr{F}(\mathrm{Id}_X) = \mathrm{Id}_{\mathscr{F}(X)}; \qquad \mathscr{F}(g \circ h) = \mathscr{F}(h) \circ \mathscr{F}(g).$$

▶ **Exercise 3.27.** Show that any (covariant or contravariant) functor from C to D takes isomorphisms in C to isomorphisms in D.

One trivial example of a covariant functor is the *identity functor* from any category to itself: it takes each object and each morphism to itself. Another example is the *forgetful functor*: if C is a category whose objects are sets with some additional structure and whose morphisms are maps preserving that structure (as are all the categories listed in the first part of Example 3.26 except Set itself), the forgetful functor $\mathcal{F}$: C $\to$ Set assigns to each object its underlying set, and to each morphism the same map thought of as a map between sets.

More interesting functors arise when we associate "invariants" to classes of mathematical objects. For example, the fundamental group is a covariant functor from Top$_*$ to Grp. The results of Problem 2-10 show that there is a contravariant functor from Diff to Vec$_\mathbb{R}$ defined by assigning to each smooth manifold M the vector space $C^\infty(M)$, and to each smooth map F: $M \to N$ the linear map F^*: $C^\infty(N) \to C^\infty(M)$ defined by $F^*(f) = f \circ F$.

The discussion in this chapter has given us some other important examples of functors. First, the *tangent space functor* is a covariant functor from the category Diff$_*$ of pointed smooth manifolds to the category Vec$_\mathbb{R}$ of real vector spaces. To each pointed smooth manifold (M, p) it assigns the tangent space $T_p M$, and to each pointed smooth map F: $(M, p) \to (N, F(p))$ it assigns the differential dF_p. The fact that this is a functor is the content of parts (b) and (c) of Proposition 3.6.

Similarly, we can think of the assignments $M \mapsto TM$ and $F \mapsto dF$ (sending each smooth manifold to its tangent bundle and each smooth map to its global differential) as a covariant functor from Diff to itself, called the *tangent functor*.

Problems

3-1. Suppose M and N are smooth manifolds with or without boundary, and F: $M \to N$ is a smooth map. Show that dF_p: $T_p M \to T_{F(p)} N$ is the zero map for each $p \in M$ if and only if F is constant on each component of M.

3-2. Prove Proposition 3.14 (the tangent space to a product manifold).

3-3. Prove that if M and N are smooth manifolds, then $T(M \times N)$ is diffeomorphic to $TM \times TN$.

3-4. Show that $T\mathbb{S}^1$ is diffeomorphic to $\mathbb{S}^1 \times \mathbb{R}$.

3-5. Let $\mathbb{S}^1 \subseteq \mathbb{R}^2$ be the unit circle, and let $K \subseteq \mathbb{R}^2$ be the boundary of the square of side 2 centered at the origin: $K = \{(x, y) : \max(|x|, |y|) = 1\}$. Show that there is a homeomorphism F: $\mathbb{R}^2 \to \mathbb{R}^2$ such that $F(\mathbb{S}^1) = K$, but there is no *diffeomorphism* with the same property. [Hint: let γ be a smooth curve whose image lies in $\mathbb{S}^1$, and consider the action of $dF(\gamma'(t))$ on the coordinate functions x and y.] *(Used on p. 123.)*

3-6. Consider $\mathbb{S}^3$ as the unit sphere in $\mathbb{C}^2$ under the usual identification $\mathbb{C}^2 \leftrightarrow \mathbb{R}^4$. For each $z = (z^1, z^2) \in \mathbb{S}^3$, define a curve γ_z: $\mathbb{R} \to \mathbb{S}^3$ by $\gamma_z(t) = (e^{it}z^1, e^{it}z^2)$. Show that γ_z is a smooth curve whose velocity is never zero.

3-7. Let M be a smooth manifold with or without boundary and p be a point of
M. Let $C_p^\infty(M)$ denote the algebra of germs of smooth real-valued functions at p, and let $\mathcal{D}_p M$ denote the vector space of derivations of $C_p^\infty(M)$.
Define a map $\Phi\colon \mathcal{D}_p M \to T_p M$ by $(\Phi v) f = v([f]_p)$. Show that Φ is an
isomorphism. *(Used on p. 71.)*

3-8. Let M be a smooth manifold with or without boundary and $p \in M$. Let $\mathcal{V}_p M$
denote the set of equivalence classes of smooth curves starting at p under the
relation $\gamma_1 \sim \gamma_2$ if $(f \circ \gamma_1)'(0) = (f \circ \gamma_2)'(0)$ for every smooth real-valued
function f defined in a neighborhood of p. Show that the map $\Psi\colon \mathcal{V}_p M \to$
$T_p M$ defined by $\Psi[\gamma] = \gamma'(0)$ is well defined and bijective. *(Used on p. 72.)*

Chapter 4
Submersions, Immersions, and Embeddings

Because the differential of a smooth map is supposed to represent the "best linear approximation" to the map near a given point, we can learn a great deal about a map by studying linear-algebraic properties of its differential. The most essential property of the differential—in fact, just about the only property that can be defined independently of choices of bases—is its *rank* (the dimension of its image).

In this chapter we undertake a detailed study of the ways in which geometric properties of smooth maps can be detected from their differentials. The maps for which differentials give good local models turn out to be the ones whose differentials have constant rank. Three categories of such maps play special roles: *smooth submersions* (whose differentials are surjective everywhere), *smooth immersions* (whose differentials are injective everywhere), and *smooth embeddings* (injective smooth immersions that are also homeomorphisms onto their images). Smooth immersions and embeddings, as we will see in the next chapter, are essential ingredients in the theory of submanifolds, while smooth submersions play a role in smooth manifold theory closely analogous to the role played by quotient maps in topology.

The engine that powers this discussion is the *rank theorem*, a corollary of the inverse function theorem. In the first section of the chapter, we prove the rank theorem and some of its important consequences. Then we delve more deeply into smooth embeddings and smooth submersions, and apply the theory to a particularly useful class of smooth submersions, the *smooth covering maps*.

Maps of Constant Rank

The key linear-algebraic property of a linear map is its rank. In fact, as Theorem B.20 shows, the rank is the *only* property that distinguishes different linear maps if we are free to choose bases independently for the domain and codomain.

Suppose M and N are smooth manifolds with or without boundary. Given a smooth map $F : M \to N$ and a point $p \in M$, we define the **rank of F at p** to be the rank of the linear map $dF_p : T_p M \to T_{F(p)} N$; it is the rank of the Jacobian matrix of F in any smooth chart, or the dimension of $\operatorname{Im} dF_p \subseteq T_{F(p)}N$. If F has the same rank r at every point, we say that it has **constant rank**, and write rank $F = r$.

Because the rank of a linear map is never higher than the dimension of either its domain or its codomain (Exercise B.22), the rank of F at each point is bounded above by the minimum of $\{\dim M, \dim N\}$. If the rank of dF_p is equal to this upper bound, we say that F **has full rank at** p, and if F has full rank everywhere, we say F **has full rank**.

The most important constant-rank maps are those of full rank. A smooth map $F: M \to N$ is called a **smooth submersion** if its differential is surjective at each point (or equivalently, if rank $F = \dim N$). It is called a **smooth immersion** if its differential is injective at each point (equivalently, rank $F = \dim M$).

Proposition 4.1. *Suppose* $F: M \to N$ *is a smooth map and* $p \in M$. *If* dF_p *is surjective, then* p *has a neighborhood* U *such that* $F|_U$ *is a submersion. If* dF_p *is injective, then* p *has a neighborhood* U *such that* $F|_U$ *is an immersion.*

Proof. If we choose any smooth coordinates for M near p and for N near $F(p)$, either hypothesis means that Jacobian matrix of F in coordinates has full rank at p. Example 1.28 shows that the set of $m \times n$ matrices of full rank is an open subset of $\mathrm{M}(m \times n, \mathbb{R})$ (where $m = \dim M$ and $n = \dim N$), so by continuity, the Jacobian of F has full rank in some neighborhood of p. $\square$

As we will see in this chapter, smooth submersions and immersions behave locally like surjective and injective linear maps, respectively. (There are also analogous notions of *topological submersions* and *topological immersions*, which apply to maps that are merely continuous. We do not have any need to use these, but for the sake of completeness, we describe them later in the chapter.)

Example 4.2 (Submersions and Immersions).

(a) Suppose $M_1, \ldots, M_k$ are smooth manifolds. Then each of the projection maps $\pi_i: M_1 \times \cdots \times M_k \to M_i$ is a smooth submersion. In particular, the projection $\pi: \mathbb{R}^{n+k} \to \mathbb{R}^n$ onto the first n coordinates is a smooth submersion.

(b) If $\gamma: J \to M$ is a smooth curve in a smooth manifold M with or without boundary, then γ is a smooth immersion if and only if $\gamma'(t) \neq 0$ for all $t \in J$.

(c) If M is a smooth manifold and its tangent bundle TM is given the smooth manifold structure described in Proposition 3.18, the projection $\pi: TM \to M$ is a smooth submersion. To verify this, just note that with respect to any smooth local coordinates (x^i) on an open subset $U \subseteq M$ and the corresponding natural coordinates (x^i, v^i) on $\pi^{-1}(U) \subseteq TM$ (see Proposition 3.18), the coordinate representation of π is $\widehat{\pi}(x, v) = x$.

(d) The smooth map $X: \mathbb{R}^2 \to \mathbb{R}^3$ given by

$$X(u, v) = \big((2 + \cos 2\pi u) \cos 2\pi v, (2 + \cos 2\pi u) \sin 2\pi v, \sin 2\pi u\big)$$

is a smooth immersion of $\mathbb{R}^2$ into $\mathbb{R}^3$ whose image is the doughnut-shaped surface obtained by revolving the circle $(y - 2)^2 + z^2 = 1$ in the (y, z)-plane about the z-axis (Fig. 4.1).   //

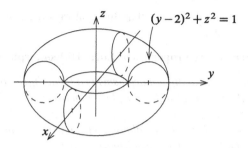

Fig. 4.1 A torus of revolution in $\mathbb{R}^3$

▶ **Exercise 4.3.** Verify the claims made in the preceding example.

▶ **Exercise 4.4.** Show that a composition of smooth submersions is a smooth submersion, and a composition of smooth immersions is a smooth immersion. Give a counterexample to show that a composition of maps of constant rank need not have constant rank.

Local Diffeomorphisms

If M and N are smooth manifolds with or without boundary, a map $F : M \to N$ is called a *local diffeomorphism* if every point $p \in M$ has a neighborhood U such that $F(U)$ is open in N and $F|_U : U \to F(U)$ is a diffeomorphism. The next theorem is the key to the most important properties of local diffeomorphisms.

Theorem 4.5 (Inverse Function Theorem for Manifolds). *Suppose M and N are smooth manifolds, and $F : M \to N$ is a smooth map. If $p \in M$ is a point such that dF_p is invertible, then there are connected neighborhoods U_0 of p and V_0 of $F(p)$ such that $F|_{U_0} : U_0 \to V_0$ is a diffeomorphism.*

Proof. The fact that dF_p is bijective implies that M and N have the same dimension, say n. Choose smooth charts (U, φ) centered at p and (V, ψ) centered at $F(p)$, with $F(U) \subseteq V$. Then $\widehat{F} = \psi \circ F \circ \varphi^{-1}$ is a smooth map from the open subset $\widehat{U} = \varphi(U) \subseteq \mathbb{R}^n$ into $\widehat{V} = \psi(V) \subseteq \mathbb{R}^n$, with $\widehat{F}(p) = 0$. Because φ and ψ are diffeomorphisms, the differential $d\widehat{F}_0 = d\psi_{F(p)} \circ dF_p \circ d(\varphi^{-1})_0$ is nonsingular. The ordinary inverse function theorem (Theorem C.34) shows that there are connected open subsets $\widehat{U}_0 \subseteq \widehat{U}$ and $\widehat{V}_0 \subseteq \widehat{V}$ containing 0 such that $\widehat{F}$ restricts to a diffeomorphism from $\widehat{U}_0$ to $\widehat{V}_0$. Then $U_0 = \varphi^{-1}(\widehat{U}_0)$ and $V_0 = \psi^{-1}(\widehat{V}_0)$ are connected neighborhoods of p and $F(p)$, respectively, and it follows by composition that $F|_{U_0}$ is a diffeomorphism from U_0 to V_0. $\square$

It is important to notice that we have stated Theorem 4.5 only for manifolds without boundary. In fact, it can fail for a map whose domain has nonempty boundary (see Problem 4-1). However, when the *codomain* has nonempty boundary, there is something useful that can be said: provided the map takes its values in the interior of the codomain, the same conclusion holds because the interior is a smooth manifold without boundary. Problem 4-2 shows that this is always the case at points where the differential is invertible.

Proposition 4.6 (Elementary Properties of Local Diffeomorphisms).

(a) *Every composition of local diffeomorphisms is a local diffeomorphism.*
(b) *Every finite product of local diffeomorphisms between smooth manifolds is a local diffeomorphism.*
(c) *Every local diffeomorphism is a local homeomorphism and an open map.*
(d) *The restriction of a local diffeomorphism to an open submanifold with or without boundary is a local diffeomorphism.*
(e) *Every diffeomorphism is a local diffeomorphism.*
(f) *Every bijective local diffeomorphism is a diffeomorphism.*
(g) *A map between smooth manifolds with or without boundary is a local diffeomorphism if and only if in a neighborhood of each point of its domain, it has a coordinate representation that is a local diffeomorphism.*

▶ **Exercise 4.7.** Prove the preceding proposition.

Proposition 4.8. *Suppose M and N are smooth manifolds (without boundary), and $F: M \to N$ is a map.*

(a) *F is a local diffeomorphism if and only if it is both a smooth immersion and a smooth submersion.*
(b) *If $\dim M = \dim N$ and F is either a smooth immersion or a smooth submersion, then it is a local diffeomorphism.*

Proof. Suppose first that F is a local diffeomorphism. Given $p \in M$, there is a neighborhood U of p such that F is a diffeomorphism from U to $F(U)$. It then follows from Proposition 3.6(d) that $dF_p: T_pM \to T_{F(p)}N$ is an isomorphism. Thus rank $F = \dim M = \dim N$, so F is both a smooth immersion and a smooth submersion. Conversely, if F is both a smooth immersion and a smooth submersion, then dF_p is an isomorphism at each $p \in M$, and the inverse function theorem for manifolds (Theorem 4.5) shows that p has a neighborhood on which F restricts to a diffeomorphism onto its image. This proves (a).

To prove (b), note that if M and N have the same dimension, then either injectivity or surjectivity of dF_p implies bijectivity, so F is a smooth submersion if and only if it is a smooth immersion, and thus (b) follows from (a).   □

▶ **Exercise 4.9.** Show that the conclusions of Proposition 4.8 still hold if N is allowed to be a smooth manifold with boundary, but not if M is. (See Problems 4-1 and 4-2.)

▶ **Exercise 4.10.** Suppose M, N, P are smooth manifolds with or without boundary, and $F: M \to N$ is a local diffeomorphism. Prove the following:

(a) If $G: P \to M$ is continuous, then G is smooth if and only if $F \circ G$ is smooth.
(b) If in addition F is surjective and $G: N \to P$ is any map, then G is smooth if and only if $G \circ F$ is smooth.

Example 4.11 (Local Diffeomorphisms). The map $\varepsilon\colon \mathbb{R} \to \mathbb{S}^1$ defined in Example 2.13(b) is a local diffeomorphism because in a neighborhood of each point it has a coordinate representation of the form $t \mapsto 2\pi t + c$, which is a local diffeomorphism. Similarly, the map $\varepsilon^n\colon \mathbb{R}^n \to \mathbb{T}^n$ defined in Example 2.13(c) is a local diffeomorphism because it is a product of local diffeomorphisms.   //

At the end of this chapter, we will explore an important special class of local diffeomorphisms, the *smooth covering maps*.

The Rank Theorem

The most important fact about constant-rank maps is the following consequence of the inverse function theorem, which says that a constant-rank smooth map can be placed locally into a particularly simple canonical form by a change of coordinates. It is a nonlinear version of the canonical form theorem for linear maps given in Theorem B.20.

Theorem 4.12 (Rank Theorem). *Suppose M and N are smooth manifolds of dimensions m and n, respectively, and $F\colon M \to N$ is a smooth map with constant rank r. For each $p \in M$ there exist smooth charts (U, φ) for M centered at p and (V, ψ) for N centered at $F(p)$ such that $F(U) \subseteq V$, in which F has a coordinate representation of the form*

$$\widehat{F}\left(x^1, \ldots, x^r, x^{r+1}, \ldots, x^m\right) = \left(x^1, \ldots, x^r, 0, \ldots, 0\right). \tag{4.1}$$

In particular, if F is a smooth submersion, this becomes

$$\widehat{F}\left(x^1, \ldots, x^n, x^{n+1}, \ldots, x^m\right) = \left(x^1, \ldots, x^n\right), \tag{4.2}$$

and if F is a smooth immersion, it is

$$\widehat{F}\left(x^1, \ldots, x^m\right) = \left(x^1, \ldots, x^m, 0, \ldots, 0\right). \tag{4.3}$$

Proof. Because the theorem is local, after choosing smooth coordinates we can replace M and N by open subsets $U \subseteq \mathbb{R}^m$ and $V \subseteq \mathbb{R}^n$. The fact that $DF(p)$ has rank r implies that its matrix has some $r \times r$ submatrix with nonzero determinant. By reordering the coordinates, we may assume that it is the upper left submatrix, $\left(\partial F^i/\partial x^j\right)$ for $i, j = 1, \ldots, r$. Let us relabel the standard coordinates as $(x, y) = \left(x^1, \ldots, x^r, y^1, \ldots, y^{m-r}\right)$ in $\mathbb{R}^m$ and $(v, w) = \left(v^1, \ldots, v^r, w^1, \ldots, w^{n-r}\right)$ in $\mathbb{R}^n$. By initial translations of the coordinates, we may assume without loss of generality that $p = (0,0)$ and $F(p) = (0,0)$. If we write $F(x, y) = \left(Q(x, y), R(x, y)\right)$ for some smooth maps $Q\colon U \to \mathbb{R}^r$ and $R\colon U \to \mathbb{R}^{n-r}$, then our hypothesis is that $\left(\partial Q^i/\partial x^j\right)$ is nonsingular at $(0,0)$.

Define $\varphi\colon U \to \mathbb{R}^m$ by $\varphi(x, y) = \left(Q(x, y), y\right)$. Its total derivative at $(0,0)$ is

$$D\varphi(0,0) = \begin{pmatrix} \dfrac{\partial Q^i}{\partial x^j}(0,0) & \dfrac{\partial Q^i}{\partial y^j}(0,0) \\ 0 & \delta^i_j \end{pmatrix},$$

where we have used the following standard notation: for positive integers i and j, the symbol δ^i_j, called the **Kronecker delta**, is defined by

$$\delta^i_j = \begin{cases} 1 & \text{if } i = j, \\ 0 & \text{if } i \neq j. \end{cases} \tag{4.4}$$

The matrix $D\varphi(0,0)$ is nonsingular by virtue of the hypothesis. Therefore, by the inverse function theorem, there are connected neighborhoods U_0 of $(0,0)$ and $\tilde{U}_0$ of $\varphi(0,0) = (0,0)$ such that $\varphi \colon U_0 \to \tilde{U}_0$ is a diffeomorphism. By shrinking U_0 and $\tilde{U}_0$ if necessary, we may assume that $\tilde{U}_0$ is an open cube. Writing the inverse map as $\varphi^{-1}(x,y) = \big(A(x,y), B(x,y)\big)$ for some smooth functions $A \colon \tilde{U}_0 \to \mathbb{R}^r$ and $B \colon \tilde{U}_0 \to \mathbb{R}^{m-r}$, we compute

$$(x,y) = \varphi\big(A(x,y), B(x,y)\big) = \big(Q\big(A(x,y), B(x,y)\big), B(x,y)\big). \tag{4.5}$$

Comparing y components shows that $B(x,y) = y$, and therefore φ^{-1} has the form

$$\varphi^{-1}(x,y) = \big(A(x,y), y\big).$$

On the other hand, $\varphi \circ \varphi^{-1} = \text{Id}$ implies $Q\big(A(x,y), y\big) = x$, and therefore $F \circ \varphi^{-1}$ has the form

$$F \circ \varphi^{-1}(x,y) = \big(x, \tilde{R}(x,y)\big),$$

where $\tilde{R} \colon \tilde{U}_0 \to \mathbb{R}^{n-r}$ is defined by $\tilde{R}(x,y) = R\big(A(x,y), y\big)$. The Jacobian matrix of this composite map at an arbitrary point $(x,y) \in \tilde{U}_0$ is

$$D\big(F \circ \varphi^{-1}\big)(x,y) = \begin{pmatrix} \delta^i_j & 0 \\ \dfrac{\partial \tilde{R}^i}{\partial x^j}(x,y) & \dfrac{\partial \tilde{R}^i}{\partial y^j}(x,y) \end{pmatrix}.$$

Since composing with a diffeomorphism does not change the rank of a map, this matrix has rank r everywhere in $\tilde{U}_0$. The first r columns are obviously linearly independent, so the rank can be r only if the derivatives $\partial \tilde{R}^i / \partial y^j$ vanish identically on $\tilde{U}_0$, which implies that $\tilde{R}$ is actually independent of $(y^1, \dots, y^{m-r})$. (This is one reason we arranged for $\tilde{U}_0$ to be a cube.) Thus, if we let $S(x) = \tilde{R}(x,0)$, then we have

$$F \circ \varphi^{-1}(x,y) = \big(x, S(x)\big). \tag{4.6}$$

To complete the proof, we need to define an appropriate smooth chart in some neighborhood of $(0,0) \in V$. Let $V_0 \subseteq V$ be the open subset defined by $V_0 = \{(v,w) \in V : (v,0) \in \tilde{U}_0\}$. Then V_0 is a neighborhood of $(0,0)$. Because $\tilde{U}_0$ is a cube and $F \circ \varphi^{-1}$ has the form (4.6), it follows that $F \circ \varphi^{-1}\big(\tilde{U}_0\big) \subseteq V_0$, and therefore $F(U_0) \subseteq V_0$. Define $\psi \colon V_0 \to \mathbb{R}^n$ by $\psi(v,w) = \big(v, w - S(v)\big)$. This is a diffeomorphism onto its image, because its inverse is given explicitly by $\psi^{-1}(s,t) = \big(s, t + S(s)\big)$; thus (V_0, ψ) is a smooth chart. It follows from (4.6) that

$$\psi \circ F \circ \varphi^{-1}(x,y) = \psi\big(x, S(x)\big) = \big(x, S(x) - S(x)\big) = (x,0),$$

which was to be proved. $\square$

The next corollary can be viewed as a more invariant statement of the rank theorem. It says that constant-rank maps are precisely the ones whose local behavior is the same as that of their differentials.

Corollary 4.13. *Let M and N be smooth manifolds, let $F: M \to N$ be a smooth map, and suppose M is connected. Then the following are equivalent:*

(a) *For each $p \in M$ there exist smooth charts containing p and $F(p)$ in which the coordinate representation of F is linear.*

(b) *F has constant rank.*

Proof. First suppose F has a linear coordinate representation in a neighborhood of each point. Since every linear map has constant rank, it follows that the rank of F is constant in a neighborhood of each point, and thus by connectedness it is constant on all of M. Conversely, if F has constant rank, the rank theorem shows that it has the linear coordinate representation (4.1) in a neighborhood of each point. $\square$

The rank theorem is a purely local statement. However, it has the following powerful global consequence.

Theorem 4.14 (Global Rank Theorem). *Let M and N be smooth manifolds, and suppose $F: M \to N$ is a smooth map of constant rank.*

(a) *If F is surjective, then it is a smooth submersion.*

(b) *If F is injective, then it is a smooth immersion.*

(c) *If F is bijective, then it is a diffeomorphism.*

Proof. Let $m = \dim M$, $n = \dim N$, and suppose F has constant rank r. To prove (a), assume that F is not a smooth submersion, which means that $r < n$. By the rank theorem, for each $p \in M$ there are smooth charts (U, φ) for M centered at p and (V, ψ) for N centered at $F(p)$ such that $F(U) \subseteq V$ and the coordinate representation of F is given by (4.1). (See Fig. 4.2.) Shrinking U if necessary, we may assume that it is a regular coordinate ball and $F(\bar{U}) \subseteq V$. This implies that $F(\bar{U})$ is a compact subset of the set $\{y \in V : y^{r+1} = \cdots = y^n = 0\}$, so it is closed in N and contains no open subset of N; hence it is nowhere dense in N. Since every open cover of a manifold has a countable subcover, we can choose countably many such charts $\{(U_i, \varphi_i)\}$ covering M, with corresponding charts $\{(V_i, \psi_i)\}$ covering $F(M)$. Because $F(M)$ is equal to the countable union of the nowhere dense sets $F(\bar{U}_i)$, it follows from the Baire category theorem (Theorem A.58) that $F(M)$ has empty interior in N, which means F cannot be surjective.

To prove (b), assume that F is not a smooth immersion, so that $r < m$. By the rank theorem, for each $p \in M$ we can choose charts on neighborhoods of p and $F(p)$ in which F has the coordinate representation (4.1). It follows that $F(0, \ldots, 0, \varepsilon) = F(0, \ldots, 0, 0)$ for any sufficiently small ε, so F is not injective.

Finally, (c) follows from (a) and (b), because a bijective smooth map of constant rank is a smooth submersion by part (a) and a smooth immersion by part (b); so Proposition 4.8 implies that F is a local diffeomorphism, and because it is bijective, it is a diffeomorphism. $\square$

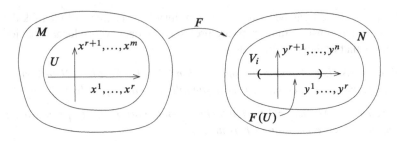

Fig. 4.2 Proof of Theorem 4.14(a)

The Rank Theorem for Manifolds with Boundary

In the context of manifolds with boundary, we need the rank theorem only in one special case: that of a smooth immersion whose domain is a smooth manifold with boundary. Of course, since the interior of a smooth manifold with boundary is a smooth manifold, near any interior point of the domain the ordinary rank theorem applies. For boundary points, we have the following substitute for the rank theorem.

Theorem 4.15 (Local Immersion Theorem for Manifolds with Boundary). *Suppose M is a smooth m-manifold with boundary, N is a smooth n-manifold, and $F: M \to N$ is a smooth immersion. For any $p \in \partial M$, there exist a smooth boundary chart (U, φ) for M centered at p and a smooth coordinate chart (V, ψ) for N centered at $F(p)$ with $F(U) \subseteq V$, in which F has the coordinate representation*

$$\widehat{F}\left(x^1, \ldots, x^m\right) = \left(x^1, \ldots, x^m, 0, \ldots, 0\right). \tag{4.7}$$

Proof. By choosing initial smooth charts for M and N, we may assume that M and N are open subsets of $\mathbb{H}^m$ and $\mathbb{R}^n$, respectively, and also that $p = 0 \in \mathbb{H}^m$, and $F(p) = 0 \in \mathbb{R}^n$. By definition of smoothness for functions on $\mathbb{H}^m$, F extends to a smooth map $\widetilde{F}: W \to \mathbb{R}^n$, where W is some open subset of $\mathbb{R}^m$ containing 0. Because $d\widetilde{F}_0 = dF_0$ is injective, by shrinking W if necessary, we may assume that $\widetilde{F}$ is a smooth immersion. Let us write the coordinates on $\mathbb{R}^m$ as $x = \left(x^1, \ldots, x^m\right)$, and those on $\mathbb{R}^n$ as $(v, w) = \left(v^1, \ldots, v^m, w^1, \ldots, w^{n-m}\right)$.

By the rank theorem, there exist smooth charts (U_0, φ_0) for $\mathbb{R}^m$ centered at 0 and (V_0, ψ_0) for $\mathbb{R}^n$ centered at 0 such that $\widehat{F} = \psi_0 \circ \widetilde{F} \circ \varphi_0^{-1}$ is given by (4.7). The only problem with these coordinates is that φ_0 might not restrict to a boundary chart for M. But we can correct this easily as follows. Because φ_0 is a diffeomorphism from U_0 to an open subset $\widehat{U}_0 = \varphi_0(U_0) \subseteq \mathbb{R}^m$, the map $\varphi_0^{-1} \times \mathrm{Id}_{\mathbb{R}^{n-m}}$ is a diffeomorphism from $\widehat{U}_0 \times \mathbb{R}^{n-m}$ to $U_0 \times \mathbb{R}^{n-m}$. Let $\psi = \left(\varphi_0^{-1} \times \mathrm{Id}_{\mathbb{R}^{n-m}}\right) \circ \psi_0$, which is a diffeomorphism from some open subset $V \subseteq V_0$ containing 0 to a neighborhood of 0 in $\mathbb{R}^n$. Using (4.7), we compute

$$\psi \circ F(x) = \left(\varphi_0^{-1} \times \mathrm{Id}_{\mathbb{R}^{n-m}}\right) \circ \psi_0 \circ F \circ \varphi_0^{-1} \circ \varphi_0(x)$$

$$= \left(\varphi_0^{-1} \times \mathrm{Id}_{\mathbb{R}^{n-m}}\right) \circ \widehat{F}\left(\varphi_0(x)\right)$$

$$= \left(\varphi_0^{-1} \times \mathrm{Id}_{\mathbb{R}^{n-m}} \right) \left(\varphi_0(x), 0 \right) = (x, 0).$$

Thus, the original coordinates for M (restricted to a sufficiently small neighborhood of 0) and the chart (V, ψ) for N satisfy the desired conditions. □

It is possible to prove a similar theorem for more general maps with constant rank out of manifolds with boundary, but the proof is more elaborate because an extension of F to an open subset does not automatically have constant rank. Since we have no need for this more general result, we leave it to the interested reader to pursue (Problem 4-3). On the other hand, the situation is considerably more complicated for a map whose *codomain* is a manifold with boundary: since the image of the map could intersect the boundary in unpredictable ways, there is no way to put such a map into a simple canonical form without strong restrictions on the map.

Embeddings

One special kind of immersion is particularly important. If M and N are smooth manifolds with or without boundary, a *smooth embedding of M into N* is a smooth immersion $F: M \to N$ that is also a topological embedding, i.e., a homeomorphism onto its image $F(M) \subseteq N$ in the subspace topology. A smooth embedding is a map that is both a topological embedding and a smooth immersion, not just a topological embedding that happens to be smooth.

▶ **Exercise 4.16.** Show that every composition of smooth embeddings is a smooth embedding.

Example 4.17 (Smooth Embeddings).

(a) If M is a smooth manifold with or without boundary and $U \subseteq M$ is an open submanifold, the inclusion map $U \hookrightarrow M$ is a smooth embedding.

(b) If $M_1, \dots, M_k$ are smooth manifolds and $p_i \in M_i$ are arbitrarily chosen points, each of the maps $\iota_j: M_j \to M_1 \times \cdots \times M_k$ given by

$$\iota_j(q) = (p_1, \dots, p_{j-1}, q, p_{j+1}, \dots, p_k)$$

is a smooth embedding. In particular, the inclusion map $\mathbb{R}^n \hookrightarrow \mathbb{R}^{n+k}$ given by sending $(x^1, \dots, x^n)$ to $(x^1, \dots, x^n, 0, \dots, 0)$ is a smooth embedding.

(c) Problem 4-12 shows that the map $X: \mathbb{R}^2 \to \mathbb{R}^3$ of Example 4.2(d) descends to a smooth embedding of the torus $\mathbb{S}^1 \times \mathbb{S}^1$ into $\mathbb{R}^3$. //

To understand more fully what it means for a map to be a smooth embedding, it is useful to bear in mind some examples of injective smooth maps that are *not* smooth embeddings. The next three examples illustrate three rather different ways in which this can happen.

Example 4.18 (A Smooth Topological Embedding). The map $\gamma: \mathbb{R} \to \mathbb{R}^2$ given by $\gamma(t) = (t^3, 0)$ is a smooth map and a topological embedding, but it is not a smooth embedding because $\gamma'(0) = 0$. //

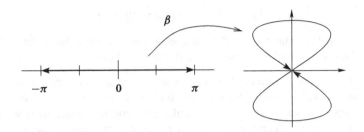

Fig. 4.3 The figure-eight curve of Example 4.19

Example 4.19 (The Figure-Eight Curve). Consider the curve $\beta\colon (-\pi,\pi) \to \mathbb{R}^2$ defined by

$$\beta(t) = (\sin 2t, \sin t).$$

Its image is a set that looks like a figure-eight in the plane (Fig. 4.3), sometimes called a *lemniscate*. (It is the locus of points (x,y) where $x^2 = 4y^2\left(1-y^2\right)$, as you can check.) It is easy to see that β is an injective smooth immersion because $\beta'(t)$ never vanishes; but it is not a topological embedding, because its image is compact in the subspace topology, while its domain is not.   //

Example 4.20 (A Dense Curve on the Torus). Let $\mathbb{T}^2 = \mathbb{S}^1 \times \mathbb{S}^1 \subseteq \mathbb{C}^2$ denote the torus, and let α be any irrational number. The map $\gamma\colon \mathbb{R} \to \mathbb{T}^2$ given by

$$\gamma(t) = \left(e^{2\pi i t}, e^{2\pi i \alpha t}\right)$$

is a smooth immersion because $\gamma'(t)$ never vanishes. It is also injective, because $\gamma(t_1) = \gamma(t_2)$ implies that both $t_1 - t_2$ and $\alpha t_1 - \alpha t_2$ are integers, which is impossible unless $t_1 = t_2$.

Consider the set $\gamma(\mathbb{Z}) = \{\gamma(n) : n \in \mathbb{Z}\}$. It follows from Dirichlet's approximation theorem (see below) that for every $\varepsilon > 0$, there are integers n, m such that $|\alpha n - m| < \varepsilon$. Using the fact that $\left|e^{it_1} - e^{it_2}\right| \le |t_1 - t_2|$ for $t_1, t_2 \in \mathbb{R}$ (because the line segment from e^{it_1} to e^{it_2} is shorter than the circular arc of length $|t_1 - t_2|$), we have $\left|e^{2\pi i \alpha n} - 1\right| = \left|e^{2\pi i \alpha n} - e^{2\pi i m}\right| \le \left|2\pi(\alpha n - m)\right| < 2\pi\varepsilon$. Therefore,

$$\left|\gamma(n) - \gamma(0)\right| = \left|\left(e^{2\pi i n}, e^{2\pi i \alpha n}\right) - (1,1)\right| = \left|\left(1, e^{2\pi i \alpha n}\right) - (1,1)\right| < 2\pi\varepsilon.$$

Thus, $\gamma(0)$ is a limit point of $\gamma(\mathbb{Z})$. But this means that γ is not a homeomorphism onto its image, because $\mathbb{Z}$ has no limit point in $\mathbb{R}$. In fact, it is not hard to show that the image set $\gamma(\mathbb{R})$ is actually dense in $\mathbb{T}^2$ (see Problem 4-4).   //

The preceding example and Problem 4-4 depend on the following elementary result from number theory.

Lemma 4.21 (Dirichlet's Approximation Theorem). *Given $\alpha \in \mathbb{R}$ and any positive integer N, there exist integers n, m with $1 \le n \le N$ such that $|n\alpha - m| < 1/N$.*

Proof. For any real number x, let $f(x) = x - \lfloor x \rfloor$, where $\lfloor x \rfloor$ is the greatest integer less than or equal to x. Since the $N + 1$ numbers $\{f(i\alpha) : i = 0,\ldots,N\}$ all lie in the interval $[0, 1)$, by the pigeonhole principle there must exist integers i and j with $0 \le i < j \le N$ such that both $f(i\alpha)$ and $f(j\alpha)$ lie in one of the N subintervals $[0, 1/N), [1/N, 2/N), \ldots, [(N-1)/N, 1)$. This means that $|f(j\alpha) - f(i\alpha)| < 1/N$, so we can take $n = j - i$ and $m = \lfloor j\alpha \rfloor - \lfloor i\alpha \rfloor$.   □

The following proposition gives a few simple sufficient criteria for an injective immersion to be an embedding.

Proposition 4.22. *Suppose M and N are smooth manifolds with or without boundary, and $F: M \to N$ is an injective smooth immersion. If any of the following holds, then F is a smooth embedding.*

(a) *F is an open or closed map.*
(b) *F is a proper map.*
(c) *M is compact.*
(d) *M has empty boundary and $\dim M = \dim N$.*

Proof. If F is open or closed, then it is a topological embedding by Theorem A.38, so it is a smooth embedding. Either (b) or (c) implies that F is closed: if F is proper, then it is closed by Theorem A.57, and if M is compact, then F is closed by the closed map lemma. Finally, assume M has empty boundary and $\dim M = \dim N$. Then dF_p is nonsingular everywhere, and Problem 4-2 shows that $F(M) \subseteq \operatorname{Int} N$. Proposition 4.8(b) shows that $F: M \to \operatorname{Int} N$ is a local diffeomorphism, so it is an open map. It follows that $F: M \to N$ is a composition of open maps $M \to \operatorname{Int} N \hookrightarrow N$, so it is an embedding.   □

Example 4.23. Let $\iota: \mathbb{S}^n \hookrightarrow \mathbb{R}^{n+1}$ be the inclusion map. We showed in Example 2.13(d) that ι is smooth by computing its coordinate representation with respect to graph coordinates. It is easy to verify in the same coordinates that its differential is injective at each point, so it is an injective smooth immersion. Because $\mathbb{S}^n$ is compact, ι is a smooth embedding by Proposition 4.22.   //

▶ **Exercise 4.24.** Give an example of a smooth embedding that is neither an open map nor a closed map.

Theorem 4.25 (Local Embedding Theorem). *Suppose M and N are smooth manifolds with or without boundary, and $F: M \to N$ is a smooth map. Then F is a smooth immersion if and only if every point in M has a neighborhood $U \subseteq M$ such that $F|_U: U \to N$ is a smooth embedding.*

Proof. One direction is immediate: if every point has a neighborhood on which F is a smooth embedding, then F has full rank everywhere, so it is a smooth immersion.

Conversely, suppose F is a smooth immersion, and let $p \in M$. We show first that p has a neighborhood on which F is injective. If $F(p) \notin \partial N$, then either the rank theorem (if $p \notin \partial M$) or Theorem 4.15 (if $p \in \partial M$) implies that there is a neighborhood U_1 of p on which F has a coordinate representation of the form (4.3). It follows from this formula that $F|_{U_1}$ is injective. On the other hand, suppose $F(p) \in \partial N$, and let (W, ψ) be any smooth boundary chart for N centered at $F(p)$. If we let $U_0 = F^{-1}(W)$, which is a neighborhood of p, and let $\iota: \mathbb{H}^n \hookrightarrow \mathbb{R}^n$ be the inclusion map, then the preceding argument can be applied to the composite map $\iota \circ \psi \circ F|_{U_0} \colon U_0 \to \mathbb{R}^n$, to show that p has a neighborhood $U_1 \subseteq U_0$ such that $\iota \circ \psi \circ F|_{U_1}$ is injective, from which it follows that $F|_{U_1}$ is injective.

Now let $p \in M$ be arbitrary, and let U_1 be a neighborhood of p on which F is injective. There exists a precompact neighborhood U of p such that $\bar{U} \subseteq U_1$. The restriction of F to $\bar{U}$ is an injective continuous map with compact domain, so it is a topological embedding by the closed map lemma. Because any restriction of a topological embedding is again a topological embedding, $F|_U$ is both a topological embedding and a smooth immersion, hence a smooth embedding. $\square$

Theorem 4.25 points the way to a notion of immersions that makes sense for arbitrary topological spaces: if X and Y are topological spaces, a continuous map $F \colon X \to Y$ is called a **topological immersion** if every point of X has a neighborhood U such that $F|_U$ is a topological embedding. Thus, every smooth immersion is a topological immersion; but, just as with embeddings, a topological immersion that happens to be smooth need not be a smooth immersion (cf. Example 4.18).

Submersions

One of the most important applications of the rank theorem is to vastly expand our understanding of the properties of submersions. If $\pi \colon M \to N$ is any continuous map, a **section of** π is a continuous right inverse for π, i.e., a continuous map $\sigma \colon N \to M$ such that $\pi \circ \sigma = \mathrm{Id}_N$:

$$
\begin{array}{c}
M \\
\pi \Big\downarrow \;\; \Big\uparrow \sigma \\
N.
\end{array}
$$

A **local section of** π is a continuous map $\sigma \colon U \to M$ defined on some open subset $U \subseteq N$ and satisfying the analogous relation $\pi \circ \sigma = \mathrm{Id}_U$. Many of the important properties of smooth submersions follow from the fact that they admit an abundance of smooth local sections.

Theorem 4.26 (Local Section Theorem). *Suppose M and N are smooth manifolds and $\pi \colon M \to N$ is a smooth map. Then π is a smooth submersion if and only if every point of M is in the image of a smooth local section of π.*

Proof. First suppose that π is a smooth submersion. Given $p \in M$, let $q = \pi(p) \in N$. By the rank theorem, we can choose smooth coordinates $(x^1, \dots, x^m)$ centered at p and $(y^1, \dots, y^n)$ centered at q in which π has the coordinate representation $\pi(x^1, \dots, x^n, x^{n+1}, \dots, x^m) = (x^1, \dots, x^n)$. If ε is a sufficiently small positive number, the coordinate cube

$$
C_\varepsilon = \{x : |x^i| < \varepsilon \text{ for } i = 1, \dots, m\}
$$

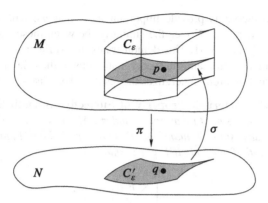

Fig. 4.4 Local section of a submersion

is a neighborhood of p whose image under π is the cube

$$C'_\varepsilon = \left\{ y : \left| y^i \right| < \varepsilon \text{ for } i = 1, \dots, n \right\}.$$

The map $\sigma : C'_\varepsilon \to C_\varepsilon$ whose coordinate representation is

$$\sigma \left(x^1, \dots, x^n \right) = \left(x^1, \dots, x^n, 0, \dots, 0 \right)$$

is a smooth local section of π satisfying $\sigma(q) = p$ (Fig. 4.4).

Conversely, assume each point of M is in the image of a smooth local section. Given $p \in M$, let $\sigma : U \to M$ be a smooth local section such that $\sigma(q) = p$, where $q = \pi\big(\sigma(q)\big) = \pi(p) \in N$. The equation $\pi \circ \sigma = \mathrm{Id}_U$ implies that $d\pi_p \circ d\sigma_q = \mathrm{Id}_{T_q N}$, which in turn implies that $d\pi_p$ is surjective. $\qquad \square$

This theorem motivates the following definition: if $\pi : X \to Y$ is a continuous map, we say π is a ***topological submersion*** if every point of X is in the image of a (continuous) local section of π. The preceding theorem shows that every smooth submersion is a topological submersion.

▶ **Exercise 4.27.** Give an example of a smooth map that is a topological submersion but not a smooth submersion.

Proposition 4.28 (Properties of Smooth Submersions). *Let M and N be smooth manifolds, and suppose $\pi : M \to N$ is a smooth submersion. Then π is an open map, and if it is surjective it is a quotient map.*

Proof. Suppose W is an open subset of M and q is a point of $\pi(W)$. For any $p \in W$ such that $\pi(p) = q$, there is a neighborhood U of q on which there exists a smooth local section $\sigma : U \to M$ of π satisfying $\sigma(q) = p$. For each $y \in \sigma^{-1}(W)$, the fact that $\sigma(y) \in W$ implies $y = \pi\big(\sigma(y)\big) \in \pi(W)$. Thus $\sigma^{-1}(W)$ is a neighborhood of q contained in $\pi(W)$, which implies that $\pi(W)$ is open. The second assertion follows from the first because every surjective open continuous map is a quotient map. $\qquad \square$

The next three theorems provide important tools that we will use frequently when studying submersions. Notice the similarity between these results and Theorems A.27(a), A.30, and A.31. This demonstrates that surjective smooth submersions play a role in smooth manifold theory analogous to the role of quotient maps in topology. The first theorem generalizes the result of Exercise 4.10(b).

Theorem 4.29 (Characteristic Property of Surjective Smooth Submersions).
Suppose M and N are smooth manifolds, and $\pi \colon M \to N$ is a surjective smooth submersion. For any smooth manifold P with or without boundary, a map $F \colon N \to P$ is smooth if and only if $F \circ \pi$ is smooth:

$$
\begin{array}{ccc}
M & & \\
\pi \downarrow & \searrow^{F \circ \pi} & \\
N & \xrightarrow{F} & P.
\end{array}
$$

Proof. If F is smooth, then $F \circ \pi$ is smooth by composition. Conversely, suppose that $F \circ \pi$ is smooth, and let $q \in N$ be arbitrary. Since π is surjective, there is a point $p \in \pi^{-1}(q)$, and then the local section theorem guarantees the existence of a neighborhood U of q and a smooth local section $\sigma \colon U \to M$ of π such that $\sigma(q) = p$. Then $\pi \circ \sigma = \mathrm{Id}_U$ implies

$$
F|_U = F|_U \circ \mathrm{Id}_U = F|_U \circ (\pi \circ \sigma) = (F \circ \pi) \circ \sigma,
$$

which is a composition of smooth maps. This shows that F is smooth in a neighborhood of each point, so it is smooth.   □

Problem 4-7 explains the sense in which this property is "characteristic."

The next theorem gives a very general sufficient condition under which a smooth map can be "pushed down" by a submersion.

Theorem 4.30 (Passing Smoothly to the Quotient). *Suppose M and N are smooth manifolds and $\pi \colon M \to N$ is a surjective smooth submersion. If P is a smooth manifold with or without boundary and $F \colon M \to P$ is a smooth map that is constant on the fibers of π, then there exists a unique smooth map $\widetilde{F} \colon N \to P$ such that $\widetilde{F} \circ \pi = F$:*

Proof. Because a surjective smooth submersion is a quotient map, Theorem A.30 shows that there exists a unique continuous map $\widetilde{F} \colon N \to P$ satisfying $\widetilde{F} \circ \pi = F$. It is smooth by Theorem 4.29.   □

Finally, we have the following uniqueness result.

Theorem 4.31 (Uniqueness of Smooth Quotients). *Suppose that M, N_1, and N_2 are smooth manifolds, and $\pi_1 \colon M \to N_1$ and $\pi_2 \colon M \to N_2$ are surjective smooth* *submersions that are constant on each other's fibers. Then there exists a unique diffeomorphism* $F \colon N_1 \to N_2$ *such that* $F \circ \pi_1 = \pi_2$:

$$
\begin{array}{ccc}
 & M & \\
\pi_1 \swarrow & & \searrow \pi_2 \\
N_1 & \dashrightarrow & N_2. \\
 & F &
\end{array}
$$

▶ **Exercise 4.32.** Prove Theorem 4.31.

Smooth Covering Maps

In this section, we introduce a class of local diffeomorphisms that play a significant role in smooth manifold theory. You are probably already familiar with the notion of a ***covering map*** between topological spaces: this is a surjective continuous map $\pi \colon E \to M$ between connected, locally path-connected spaces with the property that each point of M has a neighborhood U that is ***evenly covered***, meaning that each component of $\pi^{-1}(U)$ is mapped homeomorphically onto U by π. The basic properties of covering maps are summarized in Appendix A (pp. 615–616).

In the context of smooth manifolds, it is useful to introduce a slightly more restrictive type of covering map. If E and M are connected smooth manifolds with or without boundary, a map $\pi \colon E \to M$ is called a ***smooth covering map*** if π is smooth and surjective, and each point in M has a neighborhood U such that each component of $\pi^{-1}(U)$ is mapped *diffeomorphically* onto U by π. In this context we also say that U is evenly covered. The space M is called the ***base of the covering***, and E is called a ***covering manifold of*** M. If E is simply connected, it is called the ***universal covering manifold of*** M.

To distinguish this new definition from the previous one, we often call an ordinary (not necessarily smooth) covering map a ***topological covering map***. A smooth covering map is, in particular, a topological covering map. But as with other types of maps we have studied in this chapter, a smooth covering map is more than just a topological covering map that happens to be smooth: the definition requires in addition that the restriction of π to each component of the preimage of an evenly covered set be a diffeomorphism, not just a smooth homeomorphism.

Proposition 4.33 (Properties of Smooth Coverings).

(a) *Every smooth covering map is a local diffeomorphism, a smooth submersion, an open map, and a quotient map.*
(b) *An injective smooth covering map is a diffeomorphism.*
(c) *A topological covering map is a smooth covering map if and only if it is a local diffeomorphism.*

▶ **Exercise 4.34.** Prove Proposition 4.33.

Example 4.35 (Smooth Covering Maps). The map $\varepsilon \colon \mathbb{R} \to \mathbb{S}^1$ defined in Example 2.13(b) is a topological covering map and a local diffeomorphism (see also

Example 4.11), so it is a smooth covering map. Similarly, the map $\varepsilon^n \colon \mathbb{R}^n \to \mathbb{T}^n$ of Example 2.13(c) is a smooth covering map. For each $n \geq 1$, the map $q \colon \mathbb{S}^n \to \mathbb{R}\mathbb{P}^n$ defined in Example 2.13(f) is a two-sheeted smooth covering map (see Problem 4-10). //

Because smooth covering maps are surjective smooth submersions, all of the results in the preceding section about smooth submersions can be applied to them. For example, Theorem 4.30 is a particularly useful tool for defining a smooth map out of the base of a covering space. See Problems 4-12 and 4-13 for examples of this technique.

For smooth covering maps, the local section theorem can be strengthened.

Proposition 4.36 (Local Section Theorem for Smooth Covering Maps). *Suppose E and M are smooth manifolds with or without boundary, and $\pi \colon E \to M$ is a smooth covering map. Given any evenly covered open subset $U \subseteq M$, any $q \in U$, and any p in the fiber of π over q, there exists a unique smooth local section $\sigma \colon U \to E$ such that $\sigma(q) = p$.*

Proof. Suppose $U \subseteq M$ is evenly covered, $q \in U$, and $p \in \pi^{-1}(q)$. Let $\tilde{U}_0$ be the component of $\pi^{-1}(U)$ containing p. Since the restriction of π to $\tilde{U}_0$ is a diffeomorphism onto U, the map $\sigma = \left(\pi|_{\tilde{U}_0}\right)^{-1}$ is the required smooth local section.

To prove uniqueness, suppose $\sigma' \colon U \to E$ is any other smooth local section satisfying $\sigma'(q) = p$. Since U is connected, $\sigma'(U)$ is contained in the component $\tilde{U}_0$ containing p. Because σ' is a right inverse for the bijective map $\pi|_{\tilde{U}_0}$, it must be equal to its inverse, and therefore equal to σ. $\square$

▶ **Exercise 4.37.** Suppose $\pi \colon E \to M$ is a smooth covering map. Show that every local section of π is smooth.

▶ **Exercise 4.38.** Suppose $E_1, \ldots, E_k$ and $M_1, \ldots, M_k$ are smooth manifolds (without boundary), and $\pi_i \colon E_i \to M_i$ is a smooth covering map for each $i = 1, \ldots, k$. Show that $\pi_1 \times \cdots \times \pi_k \colon E_1 \times \cdots \times E_k \to M_1 \times \cdots \times M_k$ is a smooth covering map.

▶ **Exercise 4.39.** Suppose $\pi \colon E \to M$ is a smooth covering map. Since π is also a topological covering map, there is a potential ambiguity about what it means for a subset $U \subseteq M$ to be evenly covered: does π map the components of $\pi^{-1}(U)$ diffeomorphically onto U, or merely homeomorphically? Show that the two concepts are equivalent: if $U \subseteq M$ is evenly covered in the topological sense, then π maps each component of $\pi^{-1}(U)$ diffeomorphically onto U.

Proposition 4.40 (Covering Spaces of Smooth Manifolds). *Suppose M is a connected smooth n-manifold, and $\pi \colon E \to M$ is a topological covering map. Then E is a topological n-manifold, and has a unique smooth structure such that π is a smooth covering map.*

Proof. Because π is a local homeomorphism, E is locally Euclidean. To show that it is Hausdorff, let p_1 and p_2 be distinct points in E. If $\pi(p_1) = \pi(p_2)$ and $U \subseteq M$ is an evenly covered open subset containing $\pi(p_1)$, then the components of $\pi^{-1}(U)$ containing p_1 and p_2 are disjoint open subsets of E that separate p_1 and p_2. On

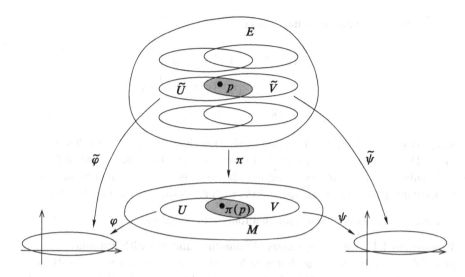

Fig. 4.5 Smooth compatibility of charts on a covering manifold

the other hand, if $\pi(p_1) \neq \pi(p_2)$, there are disjoint open subsets $U_1, U_2 \subseteq M$ containing $\pi(p_1)$ and $\pi(p_2)$, respectively, and then $\pi^{-1}(U_1)$ and $\pi^{-1}(U_2)$ are disjoint open subsets of E containing p_1 and p_2. Thus E is Hausdorff.

To show that E is second-countable, we will show first that each fiber of π is countable. Given $q \in M$ and an arbitrary point $p_0 \in \pi^{-1}(q)$, we will construct a surjective map $\beta : \pi_1(M, q) \to \pi^{-1}(q)$; since $\pi_1(M, q)$ is countable by Proposition 1.16, this suffices. Let $[f] \in \pi_1(M, q)$ be the path class of an arbitrary loop $f : [0, 1] \to M$ based at q. The path-lifting property of covering maps (Proposition A.77(b)) guarantees that there is a lift $\tilde{f} : [0, 1] \to E$ of f starting at p_0, and the monodromy theorem (Proposition A.77(c)) shows that the endpoint $\tilde{f}(1) \in \pi^{-1}(q)$ depends only on the path class of f, so it makes sense to define $\beta[f] = \tilde{f}(1)$. To see that β is surjective, just note that for any point $p \in \pi^{-1}(q)$, there is a path $\tilde{f}$ in E from p_0 to p, and then $f = \pi \circ \tilde{f}$ is a loop in M such that $p = \beta[f]$.

The collection of all evenly covered open subsets is an open cover of M, and therefore has a countable subcover $\{U_i\}$. For any given i, each component of $\pi^{-1}(U_i)$ contains exactly one point in each fiber over U_i, so $\pi^{-1}(U_i)$ has countably many components. The collection of all components of all sets of the form $\pi^{-1}(U_i)$ is thus a countable open cover of E; since each such component is second-countable, it follows from Exercise A.22 that E is second-countable. This completes the proof that E is a topological manifold.

To construct a smooth structure on E, suppose p is any point in E, and let U be an evenly covered neighborhood of $\pi(p)$. After shrinking U if necessary, we may assume also that it is the domain of a smooth coordinate map $\varphi : U \to \mathbb{R}^n$ (see Fig. 4.5).If $\tilde{U}$ is the component of $\pi^{-1}(U)$ containing p, and $\tilde{\varphi} = \varphi \circ \pi|_{\tilde{U}} : \tilde{U} \to \mathbb{R}^n$, then $(\tilde{U}, \tilde{\varphi})$ is a chart on E. If two such charts $(\tilde{U}, \tilde{\varphi})$ and

$\left(\widetilde{V}, \widetilde{\psi}\right)$ overlap, the transition map can be written

$$
\begin{aligned}
\widetilde{\psi} \circ \widetilde{\varphi}^{-1} &= \left(\psi \circ \pi|_{\widetilde{U} \cap \widetilde{V}}\right) \circ \left(\varphi \circ \pi|_{\widetilde{U} \cap \widetilde{V}}\right)^{-1} \\
&= \psi \circ \left(\pi|_{\widetilde{U} \cap \widetilde{V}}\right) \circ \left(\pi|_{\widetilde{U} \cap \widetilde{V}}\right)^{-1} \circ \varphi^{-1} \\
&= \psi \circ \varphi^{-1},
\end{aligned}
$$

which is smooth. Thus the collection of all such charts defines a smooth structure on E. The uniqueness of this smooth structure is left to the reader (Problem 4-9).

Finally, π is a smooth covering map because its coordinate representation in terms of any pair of charts $\left(\widetilde{U}, \widetilde{\varphi}\right)$ and (U, φ) constructed above is the identity. $\square$

Here is the analogous result for manifolds with boundary.

Proposition 4.41 (Covering Spaces of Smooth Manifolds with Boundary). *Suppose M is a connected smooth n-manifold with boundary, and $\pi \colon E \to M$ is a topological covering map. Then E is a topological n-manifold with boundary such that $\partial E = \pi^{-1}(\partial M)$, and it has a unique smooth structure such that π is a smooth covering map.*

▶ **Exercise 4.42.** Prove the preceding proposition.

Corollary 4.43 (Existence of a Universal Covering Manifold). *If M is a connected smooth manifold, there exists a simply connected smooth manifold $\widetilde{M}$, called the **universal covering manifold of** M, and a smooth covering map $\pi \colon \widetilde{M} \to M$. The universal covering manifold is unique in the following sense: if $\widetilde{M}'$ is any other simply connected smooth manifold that admits a smooth covering map $\pi' \colon \widetilde{M}' \to M$, then there exists a diffeomorphism $\Phi \colon \widetilde{M} \to \widetilde{M}'$ such that $\pi' \circ \Phi = \pi$.*

▶ **Exercise 4.44.** Prove the preceding corollary.

▶ **Exercise 4.45.** Generalize the preceding corollary to smooth manifolds with boundary.

There are not many simple criteria for determining whether a given map is a smooth covering map, even if it is known to be a surjective local diffeomorphism. The following proposition gives one useful sufficient criterion. (It is not necessary, however; see Problem 4-11.)

Proposition 4.46. *Suppose E and M are nonempty connected smooth manifolds with or without boundary. If $\pi \colon E \to M$ is a proper local diffeomorphism, then π is a smooth covering map.*

Proof. Because π is a local diffeomorphism, it is an open map, and because it is proper, it is a closed map (Theorem A.57). Thus $\pi(E)$ is both open and closed in M. Since it is obviously nonempty, it is all of M, so π is surjective.

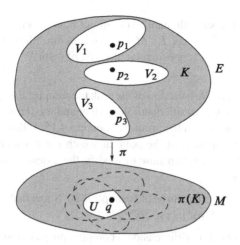

Fig. 4.6 A proper local diffeomorphism is a covering map

Let $q \in M$ be arbitrary. Since π is a local diffeomorphism, each point of $\pi^{-1}(q)$ has a neighborhood on which π is injective, so $\pi^{-1}(q)$ is a discrete subset of E. Since π is proper, $\pi^{-1}(q)$ is also compact, so it is finite. Write $\pi^{-1}(q) = \{p_1, \dots, p_k\}$. For each i, there exists a neighborhood V_i of p_i on which π is a diffeomorphism onto an open subset $U_i \subseteq M$. Shrinking each V_i if necessary, we may assume also that $V_i \cap V_j = \varnothing$ for $i \neq j$.

Set $U = U_1 \cap \dots \cap U_k$ (Fig. 4.6), which is a neighborhood of q. Then U satisfies

$$U \subseteq U_i \quad \text{for each } i. \tag{4.8}$$

Because $K = E \smallsetminus (V_1 \cup \dots \cup V_k)$ is closed in E and π is a closed map, $\pi(K)$ is closed in M. Replacing U by $U \smallsetminus \pi(K)$, we can assume that U also satisfies

$$\pi^{-1}(U) \subseteq V_1 \cup \dots \cup V_k. \tag{4.9}$$

Finally, after replacing U by the connected component of U containing q, we can assume that U is connected and still satisfies (4.8) and (4.9). We will show that U is evenly covered.

Let $\widetilde{V}_i = \pi^{-1}(U) \cap V_i$. By virtue of (4.9), $\pi^{-1}(U) = \widetilde{V}_1 \cup \dots \cup \widetilde{V}_k$. Because $\pi \colon V_i \to U_i$ is a diffeomorphism, (4.8) implies that $\pi \colon \widetilde{V}_i \to U$ is still a diffeomorphism, and in particular $\widetilde{V}_i$ is connected. Because $\widetilde{V}_1, \dots, \widetilde{V}_k$ are disjoint connected open subsets of $\pi^{-1}(U)$, they are exactly the components of $\pi^{-1}(U)$. $\square$

Problems

4-1. Use the inclusion map $\mathbb{H}^n \hookrightarrow \mathbb{R}^n$ to show that Theorem 4.5 does not extend to the case in which M is a manifold with boundary. (*Used on p. 80.*)

4-2. Suppose M is a smooth manifold (without boundary), N is a smooth manifold with boundary, and $F: M \to N$ is smooth. Show that if $p \in M$ is a point such that dF_p is nonsingular, then $F(p) \in \operatorname{Int} N$. (*Used on pp. 80, 87.*)

4-3. Formulate and prove a version of the rank theorem for a map of constant rank whose domain is a smooth manifold with boundary. [Hint: after extending F arbitrarily as we did in the proof of Theorem 4.15, follow through the proof of the rank theorem until the point at which the constant-rank hypothesis is used, and then explain how to modify the extended map so that it has constant rank.]

4-4. Let $\gamma: \mathbb{R} \to \mathbb{T}^2$ be the curve of Example 4.20. Show that the image set $\gamma(\mathbb{R})$ is dense in $\mathbb{T}^2$. (*Used on pp. 502, 542.*)

4-5. Let $\mathbb{CP}^n$ denote the n-dimensional complex projective space, as defined in Problem 1-9.
    (a) Show that the quotient map $\pi: \mathbb{C}^{n+1} \smallsetminus \{0\} \to \mathbb{CP}^n$ is a surjective smooth submersion.
    (b) Show that $\mathbb{CP}^1$ is diffeomorphic to $\mathbb{S}^2$.
    (*Used on pp. 172, 560.*)

4-6. Let M be a nonempty smooth compact manifold. Show that there is no smooth submersion $F: M \to \mathbb{R}^k$ for any $k > 0$.

4-7. Suppose M and N are smooth manifolds, and $\pi: M \to N$ is a surjective smooth submersion. Show that there is no other smooth manifold structure on N that satisfies the conclusion of Theorem 4.29; in other words, assuming that $\widetilde{N}$ represents the same set as N with a possibly different topology and smooth structure, and that for every smooth manifold P with or without boundary, a map $F: \widetilde{N} \to P$ is smooth if and only if $F \circ \pi$ is smooth, show that Id_N is a diffeomorphism between N and $\widetilde{N}$. [Remark: this shows that the property described in Theorem 4.29 is "characteristic" in the same sense as that in which Theorem A.27(a) is characteristic of the quotient topology.]

4-8. This problem shows that the converse of Theorem 4.29 is false. Let $\pi: \mathbb{R}^2 \to \mathbb{R}$ be defined by $\pi(x, y) = xy$. Show that π is surjective and smooth, and for each smooth manifold P, a map $F: \mathbb{R} \to P$ is smooth if and only if $F \circ \pi$ is smooth; but π is not a smooth submersion.

4-9. Let M be a connected smooth manifold, and let $\pi: E \to M$ be a topological covering map. Complete the proof of Proposition 4.40 by showing that there is only one smooth structure on E such that π is a smooth covering map. [Hint: use the existence of smooth local sections.]

4-10. Show that the map $q: \mathbb{S}^n \to \mathbb{RP}^n$ defined in Example 2.13(f) is a smooth covering map. (*Used on p. 550.*)

4-11. Show that a topological covering map is proper if and only if its fibers are finite, and therefore the converse of Proposition 4.46 is false.

4-12. Using the covering map $\varepsilon^2 \colon \mathbb{R}^2 \to \mathbb{T}^2$ (see Example 4.35), show that the immersion $X \colon \mathbb{R}^2 \to \mathbb{R}^3$ defined in Example 4.2(d) descends to a smooth embedding of $\mathbb{T}^2$ into $\mathbb{R}^3$. Specifically, show that X passes to the quotient to define a smooth map $\widetilde{X} \colon \mathbb{T}^2 \to \mathbb{R}^3$, and then show that $\widetilde{X}$ is a smooth embedding whose image is the given surface of revolution.

4-13. Define a map $F \colon \mathbb{S}^2 \to \mathbb{R}^4$ by $F(x, y, z) = (x^2 - y^2, xy, xz, yz)$. Using the smooth covering map of Example 2.13(f) and Problem 4-10, show that F descends to a smooth embedding of $\mathbb{RP}^2$ into $\mathbb{R}^4$.

Chapter 5
Submanifolds

Many familiar manifolds appear naturally as subsets of other manifolds. We have already seen that open subsets of smooth manifolds can be viewed as smooth manifolds in their own right; but there are many interesting examples beyond the open ones. In this chapter we explore *smooth submanifolds*, which are smooth manifolds that are subsets of other smooth manifolds. As you will soon discover, the situation is quite a bit more subtle than the analogous theory of topological subspaces.

We begin by defining the most important type of smooth submanifolds, called *embedded submanifolds*. These have the subspace topology inherited from their containing manifold, and turn out to be exactly the images of smooth embeddings. As we will see in this chapter, they are modeled locally on linear subspaces of Euclidean spaces. Because embedded submanifolds are most often presented as level sets of smooth maps, we devote some time to analyzing the conditions under which level sets are smooth submanifolds. We will see, for example, that level sets of constant-rank maps (in particular, smooth submersions) are always embedded submanifolds.

Next, we introduce a more general kind of submanifolds, called *immersed submanifolds*, which turn out to be the images of injective immersions. An immersed submanifold looks locally like an embedded one, but globally it may have a topology that is different from the subspace topology.

After introducing these basic concepts, we address two crucial technical questions about submanifolds: When is it possible to restrict the domain or codomain of a smooth map to a smooth submanifold and still retain smoothness? How can we identify the tangent space to a smooth submanifold as a subspace of the tangent space of its ambient manifold? Then we show how the theory of submanifolds can be generalized to the case of submanifolds with boundary.

Embedded Submanifolds

Suppose M is a smooth manifold with or without boundary. An ***embedded submanifold of*** M is a subset $S \subseteq M$ that is a manifold (without boundary) in the subspace topology, endowed with a smooth structure with respect to which the inclusion map

J.M. Lee, *Introduction to Smooth Manifolds*, Graduate Texts in Mathematics 218, DOI 10.1007/978-1-4419-9982-5_5, © Springer Science+Business Media New York 2013

$S \hookrightarrow M$ is a smooth embedding. Embedded submanifolds are also called *regular submanifolds* by some authors.

If S is an embedded submanifold of M, the difference $\dim M - \dim S$ is called the *codimension of S in M*, and the containing manifold M is called the *ambient manifold* for S. An *embedded hypersurface* is an embedded submanifold of codimension 1. The empty set is an embedded submanifold of any dimension.

The easiest embedded submanifolds to understand are those of codimension 0. Recall that in Example 1.26, for any smooth manifold M we defined an *open submanifold of M* to be any open subset with the subspace topology and with the smooth charts obtained by restricting those of M.

Proposition 5.1 (Open Submanifolds). *Suppose M is a smooth manifold. The embedded submanifolds of codimension 0 in M are exactly the open submanifolds.*

Proof. Suppose $U \subseteq M$ is an open submanifold, and let $\iota \colon U \hookrightarrow M$ be the inclusion map. Example 1.26 showed that U is a smooth manifold of the same dimension as M, so it has codimension 0. In terms of the smooth charts for U constructed in Example 1.26, ι is represented in coordinates by an identity map, so it is a smooth immersion; and because U has the subspace topology, ι is a smooth embedding. Thus U is an embedded submanifold. Conversely, suppose U is any codimension-0 embedded submanifold of M. Then inclusion $\iota \colon U \hookrightarrow M$ is a smooth embedding by definition, and therefore it is a local diffeomorphism by Proposition 4.8, and an open map by Proposition 4.6. Thus U is an open subset of M. $\square$

The next few propositions demonstrate several other ways to produce embedded submanifolds.

Proposition 5.2 (Images of Embeddings as Submanifolds). *Suppose M is a smooth manifold with or without boundary, N is a smooth manifold, and $F \colon N \to M$ is a smooth embedding. Let $S = F(N)$. With the subspace topology, S is a topological manifold, and it has a unique smooth structure making it into an embedded submanifold of M with the property that F is a diffeomorphism onto its image.*

Proof. If we give S the subspace topology that it inherits from M, then the assumption that F is an embedding means that F can be considered as a homeomorphism from N onto S, and thus S is a topological manifold. We give S a smooth structure by taking the smooth charts to be those of the form $\left(F(U), \varphi \circ F^{-1}\right)$, where (U, φ) is any smooth chart for N; smooth compatibility of these charts follows immediately from the smooth compatibility of the corresponding charts for N. With this smooth structure on S, the map F is a diffeomorphism onto its image (essentially by definition), and this is obviously the only smooth structure with this property. The inclusion map $S \hookrightarrow M$ is equal to the composition of a diffeomorphism followed by a smooth embedding:

$$S \xrightarrow{\ F^{-1}\ } N \xrightarrow{\ F\ } M, \tag{5.1}$$

and therefore it is a smooth embedding. $\square$

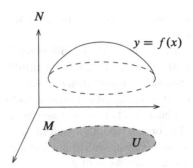

Fig. 5.1 A graph is an embedded submanifold

Since every embedded submanifold is the image of a smooth embedding (namely its own inclusion map), the previous proposition shows that embedded submanifolds are exactly the images of smooth embeddings.

Proposition 5.3 (Slices of Product Manifolds). *Suppose M and N are smooth manifolds. For each $p \in N$, the subset $M \times \{p\}$ (called a **slice** of the product manifold) is an embedded submanifold of $M \times N$ diffeomorphic to M.*

Proof. The set $M \times \{p\}$ is the image of the smooth embedding $x \mapsto (x, p)$. $\qquad\square$

Proposition 5.4 (Graphs as Submanifolds). *Suppose M is a smooth m-manifold (without boundary), N is a smooth n-manifold with or without boundary, $U \subseteq M$ is open, and $f : U \to N$ is a smooth map. Let $\Gamma(f) \subseteq M \times N$ denote the graph of f:*

$$\Gamma(f) = \{(x, y) \in M \times N : x \in U, \ y = f(x)\}.$$

Then $\Gamma(f)$ is an embedded m-dimensional submanifold of $M \times N$ (see Fig. 5.1).

Proof. Define a map $\gamma_f : U \to M \times N$ by

$$\gamma_f(x) = (x, f(x)). \tag{5.2}$$

It is a smooth map whose image is $\Gamma(f)$. Because the projection $\pi_M : M \times N \to M$ satisfies $\pi_M \circ \gamma_f(x) = x$ for $x \in U$, the composition $d(\pi_M)_{(x, f(x))} \circ d(\gamma_f)_x$ is the identity on $T_x M$ for each $x \in U$. Thus, $d(\gamma_f)_x$ is injective, so γ_f is a smooth immersion. It a homeomorphism onto its image because $\pi_M|_{\Gamma(f)}$ is a continuous inverse for it. Thus, $\Gamma(f)$ is an embedded submanifold diffeomorphic to U. $\qquad\square$

For some purposes, merely being an embedded submanifold is not quite a strong enough condition. (See, e.g., Lemma 5.34 below.) An embedded submanifold $S \subseteq M$ is said to be ***properly embedded*** if the inclusion $S \hookrightarrow M$ is a proper map.

Proposition 5.5. *Suppose M is a smooth manifold with or without boundary and $S \subseteq M$ is an embedded submanifold. Then S is properly embedded if and only if it is a closed subset of M.*

Proof. If S is properly embedded, then it is closed by Theorem A.57. Conversely, if S is closed in M, then Proposition A.53(c) shows that the inclusion map $S \hookrightarrow M$ is proper. $\qquad \square$

Corollary 5.6. *Every compact embedded submanifold is properly embedded.*

Proof. Compact subsets of Hausdorff spaces are closed. $\qquad \square$

Graphs of globally defined functions are common examples of properly embedded submanifolds.

Proposition 5.7 (Global Graphs Are Properly Embedded). *Suppose M is a smooth manifold, N is a smooth manifold with or without boundary, and $f : M \to N$ is a smooth map. With the smooth manifold structure of Proposition 5.4, $\Gamma(f)$ is properly embedded in $M \times N$.*

Proof. In this case, the projection $\pi_M : M \times N \to M$ is a smooth left inverse for the embedding $\gamma_f : M \to M \times N$ defined by (5.2). Thus γ_f is proper by Proposition A.53. $\qquad \square$

Slice Charts for Embedded Submanifolds

As our next theorem will show, embedded submanifolds are modeled locally on the standard embedding of $\mathbb{R}^k$ into $\mathbb{R}^n$, identifying $\mathbb{R}^k$ with the subspace

$$\{(x^1, \ldots, x^k, x^{k+1}, \ldots, x^n) : x^{k+1} = \cdots = x^n = 0\} \subseteq \mathbb{R}^n.$$

Somewhat more generally, if U is an open subset of $\mathbb{R}^n$ and $k \in \{0, \ldots, n\}$, a *k-dimensional slice of U* (or simply a *k-slice*) is any subset of the form

$$S = \{(x^1, \ldots, x^k, x^{k+1}, \ldots, x^n) \in U : x^{k+1} = c^{k+1}, \ldots, x^n = c^n\}$$

for some constants $c^{k+1}, \ldots, c^n$. (When $k = n$, this just means $S = U$.) Clearly, every k-slice is homeomorphic to an open subset of $\mathbb{R}^k$. (Sometimes it is convenient to consider slices defined by setting some subset of the coordinates other than the last ones equal to constants. The meaning should be clear from the context.)

Let M be a smooth n-manifold, and let (U, φ) be a smooth chart on M. If S is a subset of U such that $\varphi(S)$ is a k-slice of $\varphi(U)$, then we say that *S is a k-slice of U*. (Although in general we allow our slices to be defined by arbitrary constants $c^{k+1}, \ldots, c^n$, it is sometimes useful to have slice coordinates for which the constants are all zero, which can easily be achieved by subtracting a constant from each coordinate function.) Given a subset $S \subseteq M$ and a nonnegative integer k, we say that S satisfies the *local k-slice condition* if each point of S is contained in the domain of a smooth chart (U, φ) for M such that $S \cap U$ is a single k-slice in U. Any such chart is called a *slice chart for S in M*, and the corresponding coordinates $(x^1, \ldots, x^n)$ are called *slice coordinates*.

Theorem 5.8 (Local Slice Criterion for Embedded Submanifolds). *Let M be a smooth n-manifold. If $S \subseteq M$ is an embedded k-dimensional submanifold, then S*

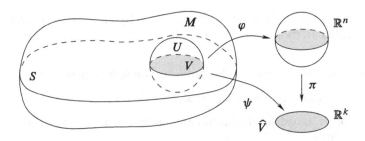

Fig. 5.2 A chart for a subset satisfying the k-slice condition

satisfies the local k-slice condition. Conversely, if $S \subseteq M$ is a subset that satisfies the local k-slice condition, then with the subspace topology, S is a topological manifold of dimension k, and it has a smooth structure making it into a k-dimensional embedded submanifold of M.

Proof. First suppose that $S \subseteq M$ is an embedded k-dimensional submanifold. Since the inclusion map $S \hookrightarrow M$ is an immersion, the rank theorem shows that for any $p \in S$ there are smooth charts (U, φ) for S (in its given smooth manifold structure) and (V, ψ) for M, both centered at p, in which the inclusion map $\iota|_U \colon U \to V$ has the coordinate representation

$$\left(x^1, \dots, x^k\right) \mapsto \left(x^1, \dots, x^k, 0, \dots, 0\right).$$

Choose $\varepsilon > 0$ small enough that both U and V contain coordinate balls of radius ε centered at p, and denote these coordinate balls by $U_0 \subseteq U$ and $V_0 \subseteq V$. It follows that $U_0 = \iota(U_0)$ is exactly a single slice in V_0. Because S has the subspace topology, the fact that U_0 is open in S means that there is an open subset $W \subseteq M$ such that $U_0 = W \cap S$. Setting $V_1 = V_0 \cap W$, we obtain a smooth chart $\left(V_1, \psi|_{V_1}\right)$ for M containing p such that $V_1 \cap S = U_0$, which is a single slice of V_1.

Conversely, suppose S satisfies the local k-slice condition. With the subspace topology, S is Hausdorff and second-countable, because both properties are inherited by subspaces. To see that S is locally Euclidean, we construct an atlas. The basic idea of the construction is that if $\left(x^1, \dots, x^n\right)$ are slice coordinates for S in M, we can use $\left(x^1, \dots, x^k\right)$ as local coordinates for S.

For this proof, let $\pi \colon \mathbb{R}^n \to \mathbb{R}^k$ denote the projection onto the first k coordinates. Let (U, φ) be any slice chart for S in M (Fig. 5.2), and define

$$V = U \cap S, \qquad \widehat{V} = \pi \circ \varphi(V), \qquad \psi = \pi \circ \varphi|_V \colon V \to \widehat{V}.$$

By definition of slice charts, $\varphi(V)$ is the intersection of $\varphi(U)$ with a certain k-slice $A \subseteq \mathbb{R}^n$ defined by setting $x^{k+1} = c^{k+1}, \dots, x^n = c^n$, and therefore $\varphi(V)$ is open in A. Since $\pi|_A$ is a diffeomorphism from A to $\mathbb{R}^k$, it follows that $\widehat{V}$ is open in $\mathbb{R}^k$. Moreover, ψ is a homeomorphism because it has a continuous inverse given by $\varphi^{-1} \circ j|_{\widehat{V}}$, where $j \colon \mathbb{R}^k \to \mathbb{R}^n$ is the map

$$j\left(x^1, \dots, x^k\right) = \left(x^1, \dots, x^k, c^{k+1}, \dots, c^n\right).$$

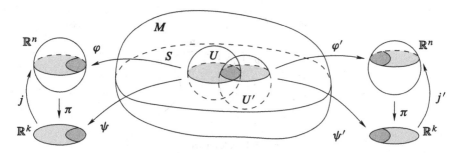

Fig. 5.3 Smooth compatibility of slice charts

Thus S is a topological k-manifold, and the inclusion map $\iota\colon S \hookrightarrow M$ is a topological embedding.

To put a smooth structure on S, we need to verify that the charts constructed above are smoothly compatible. Suppose (U,φ) and (U',φ') are two slice charts for S in M, and let (V,ψ), (V',ψ') be the corresponding charts for S. The transition map is given by $\psi' \circ \psi^{-1} = \pi \circ \varphi' \circ \varphi^{-1} \circ j$, which is a composition of four smooth maps (Fig. 5.3). Thus the atlas we have constructed is in fact a smooth atlas, and it defines a smooth structure on S. In terms of a slice chart (U,φ) for M and the corresponding chart (V,ψ) for S, the inclusion map $S \hookrightarrow M$ has a coordinate representation of the form

$$\left(x^1,\ldots,x^k\right) \mapsto \left(x^1,\ldots,x^k,c^{k+1},\ldots,c^n\right),$$

which is a smooth immersion. Since the inclusion is a smooth immersion and a topological embedding, S is an embedded submanifold.   □

Notice that the local slice condition for $S \subseteq M$ is a condition on the *subset* S only; it does not presuppose any particular topology or smooth structure on S. As we will see later (Theorem 5.31), the smooth manifold structure constructed in the preceding theorem is the *unique* one in which S can be considered as a submanifold, so a subset satisfying the local slice condition is an embedded submanifold in only one way.

Example 5.9 (Spheres as Submanifolds). For any $n \geq 0$, $\mathbb{S}^n$ is an embedded submanifold of $\mathbb{R}^{n+1}$, because it is locally the graph of a smooth function: as we showed in Example 1.4, the intersection of $\mathbb{S}^n$ with the open subset $\{x : x^i > 0\}$ is the graph of the smooth function

$$x^i = f\left(x^1,\ldots,x^{i-1},x^{i+1},\ldots,x^{n+1}\right),$$

where $f\colon \mathbb{B}^n \to \mathbb{R}$ is given by $f(u) = \sqrt{1 - |u|^2}$. Similarly, the intersection of $\mathbb{S}^n$ with $\{x : x^i < 0\}$ is the graph of $-f$. Since every point in $\mathbb{S}^n$ is in one of these sets, $\mathbb{S}^n$ satisfies the local n-slice condition and is thus an embedded submanifold of $\mathbb{R}^{n+1}$. The smooth structure thus induced on $\mathbb{S}^n$ is the same as the one we defined in Chapter 1: in fact, the coordinates for $\mathbb{S}^n$ determined by these slice charts are exactly the graph coordinates defined in Example 1.31.   //

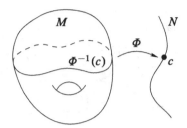

Fig. 5.4 A level set

▶ **Exercise 5.10.** Show that spherical coordinates (Example C.38) form a slice chart for S^2 in $\mathbb{R}^3$ on any open subset where they are defined.

If M is a smooth manifold with nonempty boundary and $S \subseteq M$ is an embedded submanifold, then S might intersect ∂M in very complicated ways, so we will not attempt to prove any general results about the existence of slice charts for S in M in that case.However, in the special case in which the submanifold is the boundary of M itself, the boundary charts for M play the role of slice charts for ∂M in M, and we do have the following result.

Theorem 5.11. *If M is a smooth n-manifold with boundary, then with the subspace topology, ∂M is a topological $(n-1)$-dimensional manifold (without boundary), and has a smooth structure such that it is a properly embedded submanifold of M.*

Proof. See Problem 5-2.   □

We will see later that the smooth structure on ∂M is unique. (See Theorem 5.31.)

In order to analyze more general submanifolds of M when M has a boundary, the most effective technique is often to find an embedding of M into a larger smooth manifold $\widetilde{M}$ without boundary, and apply the preceding results in that context. Example 9.32 will show that every smooth manifold with boundary can be embedded in such a way.

Level Sets

In practice, embedded submanifolds are most often presented as solution sets of equations or systems of equations. Extending the terminology we introduced in Example 1.32, if $\Phi: M \to N$ is any map and c is any point of N, we call the set $\Phi^{-1}(c)$ a *level set of* Φ (Fig. 5.4). (In the special case $N = \mathbb{R}^k$ and $c = 0$, the level set $\Phi^{-1}(0)$ is usually called the *zero set of* Φ.)

It is easy to find level sets of smooth functions that are not smooth submanifolds. For instance, consider the three smooth functions $\Theta, \Phi, \Psi: \mathbb{R}^2 \to \mathbb{R}$ defined by

$$\Theta(x, y) = x^2 - y, \qquad \Phi(x, y) = x^2 - y^2, \qquad \Psi(x, y) = x^2 - y^3.$$

(See Fig. 5.5.) Although the zero set of Θ (a parabola) is an embedded submanifold of $\mathbb{R}^2$ because it is the graph of the smooth function $f(x) = x^2$, Problem 5-11 asks you to show that neither the zero set of Φ nor that of Ψ is an embedded submanifold. In fact, without further assumptions on the smooth function, the situation is about as bad as could be imagined: as Theorem 2.29 showed, *every* closed subset of M can be expressed as the zero set of some smooth real-valued function.

The argument we used in Example 1.32 (based on the implicit function theorem) to show that certain level sets in $\mathbb{R}^n$ are smooth manifolds can be adapted to show that those level sets are in fact embedded submanifolds of $\mathbb{R}^n$. But using the rank theorem, we can prove something much stronger.

Theorem 5.12 (Constant-Rank Level Set Theorem). *Let M and N be smooth manifolds, and let $\Phi \colon M \to N$ be a smooth map with constant rank r. Each level set of Φ is a properly embedded submanifold of codimension r in M.*

Proof. Write $m = \dim M$, $n = \dim N$, and $k = m - r$. Let $c \in N$ be arbitrary, and let S denote the level set $\Phi^{-1}(c) \subseteq M$. From the rank theorem, for each $p \in S$ there are smooth charts (U, φ) centered at p and (V, ψ) centered at $c = \Phi(p)$ in which Φ has a coordinate representation of the form (4.1), and therefore $S \cap U$ is the slice

$$\left\{ \left(x^1, \dots, x^r, x^{r+1}, \dots, x^m \right) \in U : x^1 = \cdots = x^r = 0 \right\}.$$

Thus S satisfies the local k-slice condition, so it is an embedded submanifold of dimension k. It is closed in M by continuity, so it is properly embedded by Proposition 5.5. $\qquad\square$

Corollary 5.13 (Submersion Level Set Theorem). *If M and N are smooth manifolds and $\Phi \colon M \to N$ is a smooth submersion, then each level set of Φ is a properly embedded submanifold whose codimension is equal to the dimension of N.*

Proof. Every smooth submersion has constant rank equal to the dimension of its codomain. $\qquad\square$

This result should be compared to the corresponding result in linear algebra: if $L \colon \mathbb{R}^m \to \mathbb{R}^r$ is a surjective linear map, then the kernel of L is a linear subspace of codimension r by the rank-nullity law. The vector equation $Lx = 0$ is equivalent to r linearly independent scalar equations, each of which can be thought of as cutting down one of the degrees of freedom in $\mathbb{R}^m$, leaving a subspace of codimension r. In the context of smooth manifolds, the analogue of a surjective linear map is a smooth submersion, each of whose (local) component functions cuts down the dimension by one.

Corollary 5.13 can be strengthened considerably, because we need only check the submersion condition on the level set we are interested in. If $\Phi \colon M \to N$ is a smooth map, a point $p \in M$ is said to be a **regular point of Φ** if $d\Phi_p \colon T_p M \to T_{\Phi(p)} N$ is surjective; it is a **critical point of Φ** otherwise. This means, in particular, that every point of M is critical if $\dim M < \dim N$, and every point is regular if and only if F is a submersion. Note that the set of regular points of Φ is always an open subset of M by Proposition 4.1. A point $c \in N$ is said to be a **regular value of Φ** if every point of the level set $\Phi^{-1}(c)$ is a regular point, and a **critical value** otherwise. In particular, if $\Phi^{-1}(c) = \varnothing$, then c is a regular value. Finally, a level set $\Phi^{-1}(c)$ is

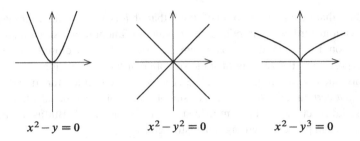

$$x^2 - y = 0 \qquad\qquad x^2 - y^2 = 0 \qquad\qquad x^2 - y^3 = 0$$

Fig. 5.5 Level sets may or may not be embedded submanifolds

called a *regular level set* if c is a regular value of Φ; in other words, a regular level set is a level set consisting entirely of regular points of Φ (points p such that $d\Phi_p$ is surjective).

Corollary 5.14 (Regular Level Set Theorem). *Every regular level set of a smooth map between smooth manifolds is a properly embedded submanifold whose codimension is equal to the dimension of the codomain.*

Proof. Let $\Phi\colon M \to N$ be a smooth map and let $c \in N$ be a regular value. The set U of points $p \in M$ where $\operatorname{rank} d\Phi_p = \dim N$ is open in M by Proposition 4.1, and contains $\Phi^{-1}(c)$ because of the assumption that c is a regular value. It follows that $\Phi|_U\colon U \to N$ is a smooth submersion, and the preceding corollary shows that $\Phi^{-1}(c)$ is an embedded submanifold of U. Since the composition of smooth embeddings $\Phi^{-1}(c) \hookrightarrow U \hookrightarrow M$ is again a smooth embedding, it follows that $\Phi^{-1}(c)$ is an embedded submanifold of M, and it is closed by continuity. $\qquad\square$

It is worth noting that the previous corollary also applies to empty level sets, which are both regular level sets and properly embedded submanifolds.

Example 5.15 (Spheres). Now we can give a much easier proof that $\mathbb{S}^n$ is an embedded submanifold of $\mathbb{R}^{n+1}$. The sphere is a regular level set of the smooth function $f\colon \mathbb{R}^{n+1} \to \mathbb{R}$ given by $f(x) = |x|^2$, since $df_x(v) = 2\sum_i x^i v^i$, which is surjective except at the origin.   //

Not all embedded submanifolds can be expressed as level sets of smooth submersions. However, the next proposition shows that every embedded submanifold is at least locally of this form.

Proposition 5.16. *Let S be a subset of a smooth m-manifold M. Then S is an embedded k-submanifold of M if and only if every point of S has a neighborhood U in M such that $U \cap S$ is a level set of a smooth submersion $\Phi\colon U \to \mathbb{R}^{m-k}$.*

Proof. First suppose S is an embedded k-submanifold. If $(x^1,\dots,x^m)$ are slice coordinates for S on an open subset $U \subseteq M$, the map $\Phi\colon U \to \mathbb{R}^{m-k}$ given in coordinates by $\Phi(x) = (x^{k+1},\dots,x^m)$ is easily seen to be a smooth submersion, one of whose level sets is $S \cap U$ (Fig. 5.6). Conversely, suppose that around every point $p \in S$ there is a neighborhood U and a smooth submersion $\Phi\colon U \to \mathbb{R}^{m-k}$

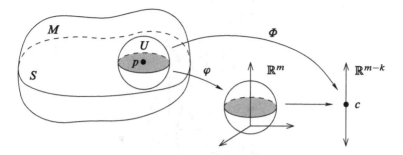

Fig. 5.6 An embedded submanifold is locally a level set

such that $S \cap U$ is a level set of Φ. By the submersion level set theorem, $S \cap U$ is an embedded submanifold of U, so it satisfies the local slice condition; it follows that S is itself an embedded submanifold of M. □

If $S \subseteq M$ is an embedded submanifold, a smooth map $\Phi \colon M \to N$ such that S is a regular level set of Φ is called a **defining map for S**. In the special case $N = \mathbb{R}^{m-k}$ (so that Φ is a real-valued or vector-valued function), it is usually called a **defining function**. Example 5.15 shows that $f(x) = |x|^2$ is a defining function for the sphere. More generally, if U is an open subset of M and $\Phi \colon U \to N$ is a smooth map such that $S \cap U$ is a regular level set of Φ, then Φ is called a **local defining map** (or **local defining function**) **for S**. Proposition 5.16 says that every embedded submanifold admits a local defining function in a neighborhood of each of its points.

In specific examples, finding a (local or global) defining function for a submanifold is usually just a matter of using geometric information about how the submanifold is defined together with some computational ingenuity. Here is an example.

Example 5.17 (Surfaces of Revolution). Let H be the half-plane $\{(r,z) : r > 0\}$, and suppose $C \subseteq H$ is an embedded 1-dimensional submanifold. The **surface of revolution** determined by C is the subset $S_C \subseteq \mathbb{R}^3$ given by

$$S_C = \left\{ (x,y,z) : \left(\sqrt{x^2 + y^2}, z \right) \in C \right\}.$$

The set C is called its **generating curve** (see Fig. 5.7). If $\varphi \colon U \to \mathbb{R}$ is any local defining function for C in H, we get a local defining function Φ for S_C by

$$\Phi(x,y,z) = \varphi\left(\sqrt{x^2 + y^2}, z \right),$$

defined on the open subset

$$\tilde{U} = \left\{ (x,y,z) : \left(\sqrt{x^2 + y^2}, z \right) \in U \right\} \subseteq \mathbb{R}^3.$$

A computation shows that the Jacobian matrix of Φ is

$$D\Phi(x,y,z) = \left(\frac{x}{r} \frac{\partial \varphi}{\partial r}(r,z) \quad \frac{y}{r} \frac{\partial \varphi}{\partial r}(r,z) \quad \frac{\partial \varphi}{\partial z}(r,z) \right),$$

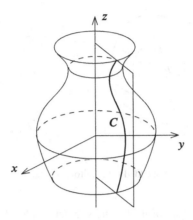

Fig. 5.7 A surface of revolution

where we have written $r = \sqrt{x^2 + y^2}$. At any point $(x, y, z) \in S_C$, at least one of the components of $D\Phi(x, y, z)$ is nonzero, so S_C is a regular level set of Φ and is thus an embedded 2-dimensional submanifold of $\mathbb{R}^3$.

For a specific example, the doughnut-shaped torus of revolution D described in Example 4.2(d) is the surface of revolution obtained from the circle $(r - 2)^2 + z^2 = 1$. It is a regular level set of the function $\Phi(x, y, z) = \left(\sqrt{x^2 + y^2} - 2\right)^2 + z^2$, which is smooth on $\mathbb{R}^3$ minus the z-axis. //

Immersed Submanifolds

Although embedded submanifolds are the most natural and common submanifolds and suffice for most purposes, it is sometimes important to consider a more general notion of submanifold. In particular, when we study Lie subgroups in Chapter 7 and foliations in Chapter 19, we will encounter subsets of smooth manifolds that are images of injective immersions, but not necessarily of embeddings. To see some of the kinds of phenomena that occur, look back at the two examples we introduced in Chapter 4 of sets that are images of injective immersions that are not embeddings: the figure-eight curve of Example 4.19 and the dense curve on the torus of Example 4.20. Neither of these sets is an embedded submanifold (see Problems 5-4 and 5-5).

So as to have a convenient language for talking about examples like these, we introduce the following definition. Let M be a smooth manifold with or without boundary. An *immersed submanifold of M* is a subset $S \subseteq M$ endowed with a topology (not necessarily the subspace topology) with respect to which it is a topological manifold (without boundary), and a smooth structure with respect to which the inclusion map $S \hookrightarrow M$ is a smooth immersion. As for embedded submanifolds, we define the *codimension of S in M* to be $\dim M - \dim S$.

Every embedded submanifold is also an immersed submanifold. Because immersed submanifolds are the more general of the two types of submanifolds, we

adopt the convention that the term *smooth submanifold* without further qualification means an immersed one, which includes an embedded submanifold as a special
case. Similarly, the term *smooth hypersurface* without qualification means an immersed submanifold of codimension 1.

You should be aware that there are variations in how smooth submanifolds are
defined in the literature. Some authors reserve the unqualified term "submanifold"
to mean what we call an embedded submanifold. If there is room for confusion, it is
safest to specify explicitly which type of submanifold—embedded or immersed—is
meant. Even though both terms "smooth submanifold" and "immersed submanifold" encompass embedded ones as well, when we are considering general submanifolds we sometimes use the phrase *immersed or embedded submanifold* as a
reminder that the discussion applies equally to the embedded case.

(Some authors define immersed submanifolds even more generally than we have,
as images of smooth immersions with no injectivity requirement. Such a submanifold can have "self-crossings" at points where the immersion fails to be injective.
We do not consider such sets as submanifolds, but it is good to be aware that some
authors do.)

There are also various notions of submanifolds in the topological category. For
example, if M is a topological manifold, one could define an *immersed topological
submanifold of M* to be a subset $S \subseteq M$ endowed with a topology such that it
is a topological manifold and such that the inclusion map is a topological immersion. It is an *embedded topological submanifold* if the inclusion is a topological
embedding. To be entirely consistent with our convention of assuming by default
only continuity rather than smoothness, we would have to distinguish the types of
submanifolds we have defined in this chapter by calling them *smooth embedded
submanifolds* and *smooth immersed submanifolds*, respectively; but since we have
no reason to treat topological submanifolds in this book, for the sake of simplicity
let us agree that the terms *embedded submanifold* and *immersed submanifold* always
refer to the smooth kind.

Immersed submanifolds often arise in the following way.

Proposition 5.18 (Images of Immersions as Submanifolds). *Suppose M is a
smooth manifold with or without boundary, N is a smooth manifold, and $F: N \to
M$ is an injective smooth immersion. Let $S = F(N)$. Then S has a unique topol-
ogy and smooth structure such that it is a smooth submanifold of M and such that
$F: N \to S$ is a diffeomorphism onto its image.*

Proof. The proof is very similar to that of Proposition 5.2, except that now we also
have to define the topology on S. We give S a topology by declaring a set $U \subseteq S$ to
be open if and only if $F^{-1}(U) \subseteq N$ is open, and then give it a smooth structure by
taking the smooth charts to be those of the form $\left(F(U), \varphi \circ F^{-1}\right)$, where (U, φ) is
any smooth chart for N. As in the proof of Proposition 5.2, the smooth compatibility
condition follows from that for N. With this topology and smooth structure on S,
the map F is a diffeomorphism onto its image, and these are the only topology and
smooth structure on S with this property. As in the embedding case, the inclusion

$S \hookrightarrow M$ can be written as the composition

$$ S \xrightarrow{F^{-1}} N \xrightarrow{F} M; $$

in this case, the first map is a diffeomorphism and the second is a smooth immersion, so the composition is a smooth immersion. □

Example 5.19 (The Figure-Eight and the Dense Curve on the Torus). Look back at the two examples we introduced in Chapter 4 of injective smooth immersions that are not embeddings: because the figure-eight of Example 4.19 and the dense curve of Example 4.20 are images of injective smooth immersions, they are immersed submanifolds when given appropriate topologies and smooth structures. As smooth manifolds, they are diffeomorphic to $\mathbb{R}$. They are *not* embedded submanifolds, because neither one has the subspace topology. In fact, their image sets cannot be made into embedded submanifolds even if we are allowed to change their topologies and smooth structures (see Problems 5-4 and 5-5). //

The following observation is sometimes useful when thinking about the topology of an immersed submanifold.

▶ **Exercise 5.20.** Suppose M is a smooth manifold and $S \subseteq M$ is an immersed submanifold. Show that every subset of S that is open in the subspace topology is also open in its given submanifold topology; and the converse is true if and only if S is embedded.

Given a smooth submanifold that is known only to be immersed, it is often useful to have simple criteria that guarantee that it is embedded. The next proposition gives several such criteria.

Proposition 5.21. *Suppose M is a smooth manifold with or without boundary, and $S \subseteq M$ is an immersed submanifold. If any of the following holds, then S is embedded.*

(a) *S has codimension 0 in M.*
(b) *The inclusion map $S \subseteq M$ is proper.*
(c) *S is compact.*

Proof. Problem 5-3. □

Although many immersed submanifolds are not embedded, the next proposition shows that the *local* structure of an immersed submanifold is the same as that of an embedded one.

Proposition 5.22 (Immersed Submanifolds Are Locally Embedded). *If M is a smooth manifold with or without boundary, and $S \subseteq M$ is an immersed submanifold, then for each $p \in S$ there exists a neighborhood U of p in S that is an embedded submanifold of M.*

Proof. Theorem 4.25 shows that each $p \in S$ has a neighborhood U in S such that the inclusion $\iota|_U : U \hookrightarrow M$ is an embedding. □

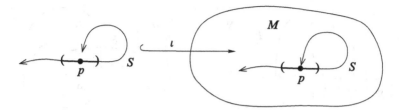

Fig. 5.8 An immersed submanifold is locally embedded

It is important to be clear about what this proposition does and does not say: given an immersed submanifold $S \subseteq M$ and a point $p \in S$, it is possible to find a neighborhood U of p (in S) such that U is embedded; but it may not be possible to find a neighborhood V of p in M such that $V \cap S$ is embedded (see Fig. 5.8).

Suppose $S \subseteq M$ is an immersed k-dimensional submanifold. A *local parametrization of S* is a continuous map $X \colon U \to M$ whose domain is an open subset $U \subseteq \mathbb{R}^k$, whose image is an open subset of S, and which, considered as a map into S, is a homeomorphism onto its image. It is called a *smooth local parametrization* if it is a diffeomorphism onto its image (with respect to S's smooth manifold structure). If the image of X is all of S, it is called a *global parametrization*.

Proposition 5.23. *Suppose M is a smooth manifold with or without boundary, $S \subseteq M$ is an immersed k-submanifold, $\iota \colon S \hookrightarrow M$ is the inclusion map, and U is an open subset of $\mathbb{R}^k$. A map $X \colon U \to M$ is a smooth local parametrization of S if and only if there is a smooth coordinate chart (V, φ) for S such that $X = \iota \circ \varphi^{-1}$. Therefore, every point of S is in the image of some local parametrization.*

▶ **Exercise 5.24.** Prove the preceding proposition.

Example 5.25 (Graph Parametrizations). Suppose $U \subseteq \mathbb{R}^n$ is an open subset and $f \colon U \to \mathbb{R}^k$ is a smooth function. The map $\gamma_f \colon U \to \mathbb{R}^n \times \mathbb{R}^k$ given by $\gamma_f(u) = (u, f(u))$ is a smooth global parametrization of $\Gamma(f)$, called a *graph parametrization*. Its inverse is the graph coordinate map constructed in Example 1.3. For example, the map $F \colon \mathbb{B}^2 \to \mathbb{R}^3$ given by

$$F(u, v) = \left(u, v, \sqrt{1 - u^2 - v^2} \right)$$

is a smooth local parametrization of $\mathbb{S}^2$ whose image is the open upper hemisphere, and whose inverse is one of the graph coordinate maps described in Example 1.4. //

Example 5.26 (Parametrization of the Figure-Eight Curve). Let $S \subseteq \mathbb{R}^2$ be the figure-eight curve of Example 5.19, considered as an immersed submanifold of $\mathbb{R}^2$. The map $\beta \colon (-\pi, \pi) \to \mathbb{R}^2$ of Example 4.19 is a smooth global parametrization of S. //

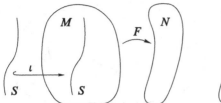 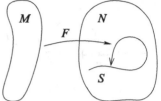

Fig. 5.9 Restricting the domain   **Fig. 5.10** Restricting the codomain

Restricting Maps to Submanifolds

Given a smooth map $F: M \to N$, it is important to know whether F is still smooth when its domain or codomain is restricted to a submanifold. In the case of restricting the domain, the answer is easy.

Theorem 5.27 (Restricting the Domain of a Smooth Map). *If M and N are smooth manifolds with or without boundary, $F: M \to N$ is a smooth map, and $S \subseteq M$ is an immersed or embedded submanifold (Fig. 5.9), then $F|_S: S \to N$ is smooth.*

Proof. The inclusion map $\iota: S \hookrightarrow M$ is smooth by definition of an immersed submanifold. Since $F|_S = F \circ \iota$, the result follows.   □

When the codomain is restricted, however, the resulting map may not be smooth, as the following example shows.

Example 5.28. Let $S \subseteq \mathbb{R}^2$ be the figure-eight submanifold, with the topology and smooth structure induced by the immersion β of Example 4.19. Define a smooth map $G: \mathbb{R} \to \mathbb{R}^2$ by

$$G(t) = (\sin 2t, \sin t).$$

(This is the same formula that we used to define β, but now the domain is extended to the whole real line instead of being just a subinterval.) It is easy to check that the image of G lies in S. However, as a map from $\mathbb{R}$ to S, G is not even continuous, because $\beta^{-1} \circ G$ is not continuous at $t = \pi$.   //

The next theorem gives sufficient conditions for a map to be smooth when its codomain is restricted to an immersed submanifold. It shows that the failure of continuity is the only thing that can go wrong.

Theorem 5.29 (Restricting the Codomain of a Smooth Map). *Suppose M is a smooth manifold (without boundary), $S \subseteq M$ is an immersed submanifold, and $F: N \to M$ is a smooth map whose image is contained in S (Fig. 5.10). If F is continuous as a map from N to S, then $F: N \to S$ is smooth.*

Remark. This theorem is stated only for the case in which the ambient manifold M is a manifold without boundary, because it is only in that case that we have constructed slice charts for embedded submanifolds of M. But the conclusion of the theorem is still true when M has nonempty boundary; see Problem 9-13.

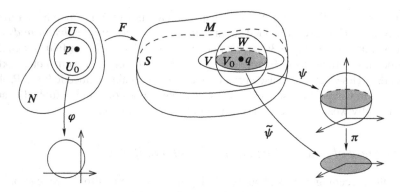

Fig. 5.11 Proof of Theorem 5.29

Proof. Let $p \in N$ be arbitrary and let $q = F(p) \in S$. Proposition 5.22 guarantees that there is a neighborhood V of q in S such that $\iota|_V : V \hookrightarrow M$ is a smooth embedding. Thus there exists a smooth chart (W, ψ) for M that is a slice chart for V in M centered at q (Fig. 5.11). (It might not be a slice chart for S in M.) The fact that (W, ψ) is a slice chart means that $(V_0, \widetilde{\psi})$ is a smooth chart for V, where $V_0 = W \cap V$ and $\widetilde{\psi} = \pi \circ \psi$, with $\pi \colon \mathbb{R}^n \to \mathbb{R}^k$ the projection onto the first $k = \dim S$ coordinates. Since $V_0 = (\iota|_V)^{-1}(W)$ is open in V, it is open in S in its given topology, and so $(V_0, \widetilde{\psi})$ is also a smooth chart for S.

Let $U = F^{-1}(V_0)$, which is an open subset of N containing p. (Here is where we use the hypothesis that F is continuous into S.) Choose a smooth chart (U_0, φ) for N such that $p \in U_0 \subseteq U$. Then the coordinate representation of $F \colon N \to S$ with respect to the charts (U_0, φ) and $(V_0, \widetilde{\psi})$ is

$$\widetilde{\psi} \circ F \circ \varphi^{-1} = \pi \circ \left(\psi \circ F \circ \varphi^{-1} \right),$$

which is smooth because $F \colon N \to M$ is smooth. $\qquad \square$

In the special case in which the submanifold S is embedded, the continuity hypothesis is always satisfied.

Corollary 5.30 (Embedded Case). *Let M be a smooth manifold and $S \subseteq M$ be an embedded submanifold. Then every smooth map $F \colon N \to M$ whose image is contained in S is also smooth as a map from N to S.*

Proof. Since $S \subseteq M$ has the subspace topology, a continuous map $F \colon N \to M$ whose image is contained in S is automatically continuous into S, by the characteristic property of the subspace topology (Proposition A.17(a)). $\qquad \square$

Although the conclusion of the preceding corollary fails for some immersed submanifolds such as the figure-eight curve (see Example 5.28), it turns out that there are certain immersed but nonembedded submanifolds for which it holds. To distinguish them, we introduce the following definition. If M is a smooth manifold and $S \subseteq M$ is an immersed submanifold, then S is said to be *weakly embedded in*

M if every smooth map $F : N \to M$ whose image lies in S is smooth as a map from N to S. (Weakly embedded submanifolds are called *initial submanifolds* by some authors.) Corollary 5.30 shows that every embedded submanifold is weakly embedded. It follows from Example 5.28 that the figure-eight curve is not weakly embedded. However, the dense curve on the torus *is* weakly embedded; see Problem 5-13. In Chapter 19, we will encounter some classes of submanifolds that are automatically weakly embedded (see Theorems 19.17 and 19.25).

Uniqueness of Smooth Structures on Submanifolds

Using the preceding results about restricting maps to submanifolds, we can prove the promised uniqueness theorem for the smooth manifold structure on an embedded submanifold.

Theorem 5.31. *Suppose M is a smooth manifold and $S \subseteq M$ is an embedded submanifold. The subspace topology on S and the smooth structure described in Theorem 5.8 are the only topology and smooth structure with respect to which S is an embedded or immersed submanifold.*

Proof. Suppose $S \subseteq M$ is an embedded k-dimensional submanifold. Theorem 5.8 shows that it satisfies the local k-slice condition, so it is an embedded submanifold with the subspace topology and the smooth structure of Theorem 5.8. Suppose there were some other topology and smooth structure on S making it into an immersed submanifold of some dimension. Let $\widetilde{S}$ denote the same set S, considered as a smooth manifold with the non-standard topology and smooth structure, and let $\widetilde{\iota} : \widetilde{S} \hookrightarrow M$ denote the inclusion map, which by assumption is an injective immersion (but not necessarily an embedding). Because $\widetilde{\iota}(\widetilde{S}) = S$, Corollary 5.30 implies that $\widetilde{\iota}$ is also smooth when considered as a map from $\widetilde{S}$ to S. For each $p \in \widetilde{S}$, the differential $d\widetilde{\iota}_p : T_p\widetilde{S} \to T_pM$ is equal to the composition

$$T_p\widetilde{S} \xrightarrow{\ d\widetilde{\iota}_p\ } T_pS \xrightarrow{\ d\iota_p\ } T_pM,$$

where $\iota : S \hookrightarrow M$ is also inclusion. Because this composition is injective (since $\widetilde{S}$ is assumed to be a smooth submanifold of M), $d\widetilde{\iota}_p$ must be injective. In particular, this means that $\widetilde{\iota} : \widetilde{S} \to S$ is an immersion. Because it is bijective, it follows from the global rank theorem that it is a diffeomorphism. In other words, the topology and smooth manifold structure of $\widetilde{S}$ are the same as those of S. $\qquad\square$

Thanks to this uniqueness result, we now know that a subset $S \subseteq M$ is an embedded submanifold if and only if it satisfies the local slice condition, and if so, its topology and smooth structure are uniquely determined. Because the local slice condition is a local condition, if every point $p \in S$ has a neighborhood $U \subseteq M$ such that $U \cap S$ is an embedded k-submanifold of U, then S is an embedded k-submanifold of M.

The preceding theorem is false in general if S is merely immersed; but we do have the following uniqueness theorem for the smooth structure of an immersed submanifold once the topology is known.

Theorem 5.32. *Suppose M is a smooth manifold and $S \subseteq M$ is an immersed sub-manifold. For the given topology on S, there is only one smooth structure making S into an immersed submanifold.*

Proof. See Problem 5-14.   □

It is certainly possible for a given subset of M to have more than one topology making it into an immersed submanifold (see Problem 5-15). However, for weakly embedded submanifolds we have a stronger uniqueness result.

Theorem 5.33. *If M is a smooth manifold and $S \subseteq M$ is a weakly embedded sub-manifold, then S has only one topology and smooth structure with respect to which it is an immersed submanifold.*

Proof. See Problem 5-16.   □

Extending Functions from Submanifolds

Complementary to the restriction problem is the problem of extending smooth functions from a submanifold to the ambient manifold. Let M be a smooth manifold with or without boundary, and let $S \subseteq M$ be a smooth submanifold. If $f : S \to \mathbb{R}$ is a function, there are two ways we might interpret the statement "f is smooth": it might mean that f is smooth as a function on the smooth manifold S (i.e., each coordinate representation is smooth), or it might mean that it is smooth as a function on the subset $S \subseteq M$ (i.e., it admits a smooth extension to a neighborhood of each point). We adopt the convention that the notation $f \in C^\infty(S)$ always means that f is smooth in the former sense (as a function on the manifold S).

Lemma 5.34 (Extension Lemma for Functions on Submanifolds). *Suppose M is a smooth manifold, $S \subseteq M$ is a smooth submanifold, and $f \in C^\infty(S)$.*

(a) *If S is embedded, then there exist a neighborhood U of S in M and a smooth function $\tilde{f} \in C^\infty(U)$ such that $\tilde{f}|_S = f$.*
(b) *If S is properly embedded, then the neighborhood U in part (a) can be taken to be all of M.*

Proof. Problem 5-17.   □

Problem 5-18 shows that the hypotheses in both (a) and (b) are necessary.

The Tangent Space to a Submanifold

If S is a smooth submanifold of $\mathbb{R}^n$, we intuitively think of the tangent space $T_p S$ at a point of S as a subspace of the tangent space $T_p \mathbb{R}^n$. Similarly, the tangent space to a smooth submanifold of an abstract smooth manifold can be viewed as a subspace of the tangent space to the ambient manifold, once we make appropriate identifications.

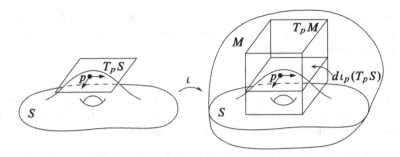

Fig. 5.12 The tangent space to an embedded submanifold

Let M be a smooth manifold with or without boundary, and let $S \subseteq M$ be an immersed or embedded submanifold. Since the inclusion map $\iota: S \hookrightarrow M$ is a smooth immersion, at each point $p \in S$ we have an injective linear map $d\iota_p: T_pS \to T_pM$. In terms of derivations, this injection works in the following way: for any vector $v \in T_pS$, the image vector $\tilde{v} = d\iota_p(v) \in T_pM$ acts on smooth functions on M by

$$\tilde{v}f = d\iota_p(v)f = v(f \circ \iota) = v(f|_S).$$

We adopt the convention of *identifying* T_pS with its image under this map, thereby thinking of T_pS as a certain linear subspace of T_pM (Fig. 5.12). This identification makes sense regardless of whether S is embedded or immersed.

There are several alternative ways of characterizing T_pS as a subspace of T_pM. The first one is the most general; it is just a straightforward generalization of Proposition 3.23.

Proposition 5.35. *Suppose M is a smooth manifold with or without boundary, $S \subseteq M$ is an immersed or embedded submanifold, and $p \in S$. A vector $v \in T_pM$ is in T_pS if and only if there is a smooth curve $\gamma: J \to M$ whose image is contained in S, and which is also smooth as a map into S, such that $0 \in J$, $\gamma(0) = p$, and $\gamma'(0) = v$.*

▶ **Exercise 5.36.** Prove the preceding proposition.

The next proposition gives a useful way to characterize T_pS in the embedded case. (Problem 5-20 shows that this does not work in the nonembedded case.)

Proposition 5.37. *Suppose M is a smooth manifold, $S \subseteq M$ is an embedded submanifold, and $p \in S$. As a subspace of T_pM, the tangent space T_pS is characterized by*

$$T_pS = \{v \in T_pM : vf = 0 \text{ whenever } f \in C^\infty(M) \text{ and } f|_S = 0\}.$$

Proof. First suppose $v \in T_pS \subseteq T_pM$. This means, more precisely, that $v = d\iota_p(w)$ for some $w \in T_pS$, where $\iota: S \to M$ is inclusion. If f is any smooth real-valued function on M that vanishes on S, then $f \circ \iota \equiv 0$, so

$$vf = d\iota_p(w)f = w(f \circ \iota) = 0.$$

Conversely, if $v \in T_p M$ satisfies $vf = 0$ whenever f vanishes on S, we need to show that there is a vector $w \in T_p S$ such that $v = d\iota_p(w)$. Let $(x^1, \ldots, x^n)$ be slice coordinates for S in some neighborhood U of p, so that $U \cap S$ is the subset of U where $x^{k+1} = \cdots = x^n = 0$, and $(x^1, \ldots, x^k)$ are coordinates for $U \cap S$. Because the inclusion map $\iota \colon S \cap U \hookrightarrow M$ has the coordinate representation

$$\iota(x^1, \ldots, x^k) = (x^1, \ldots, x^k, 0, \ldots, 0)$$

in these coordinates, it follows that $T_p S$ (that is, $d\iota_p(T_p S)$) is exactly the subspace of $T_p M$ spanned by $\partial/\partial x^1|_p, \ldots, \partial/\partial x^k|_p$. If we write the coordinate representation of v as

$$v = \sum_{i=1}^n v^i \left. \frac{\partial}{\partial x^i} \right|_p,$$

we see that $v \in T_p S$ if and only if $v^i = 0$ for $i > k$.

Let φ be a smooth bump function supported in U that is equal to 1 in a neighborhood of p. Choose an index $j > k$, and consider the function $f(x) = \varphi(x) x^j$, extended to be zero on $M \smallsetminus \operatorname{supp} \varphi$. Then f vanishes identically on S, so

$$0 = vf = \sum_{i=1}^n v^i \frac{\partial \left(\varphi(x) x^j \right)}{\partial x^i}(p) = v^j.$$

Thus $v \in T_p S$ as desired. $\qquad\qquad\square$

If an embedded submanifold is characterized by a defining map, the defining map gives a concise characterization of its tangent space at each point, as the next proposition shows.

Proposition 5.38. *Suppose M is a smooth manifold and $S \subseteq M$ is an embedded submanifold. If $\Phi \colon U \to N$ is any local defining map for S, then $T_p S = \operatorname{Ker} d\Phi_p \colon T_p M \to T_{\Phi(p)} N$ for each $p \in S \cap U$.*

Proof. Recall that we identify $T_p S$ with the subspace $d\iota_p(T_p S) \subseteq T_p M$, where $\iota \colon S \hookrightarrow M$ is the inclusion map. Because $\Phi \circ \iota$ is constant on $S \cap U$, it follows that $d\Phi_p \circ d\iota_p$ is the zero map from $T_p S$ to $T_{\Phi(p)} N$, and therefore $\operatorname{Im} d\iota_p \subseteq \operatorname{Ker} d\Phi_p$. On the other hand, $d\Phi_p$ is surjective by the definition of a defining map, so the rank–nullity law implies that

$$\dim \operatorname{Ker} d\Phi_p = \dim T_p M - \dim T_{\Phi(p)} N = \dim T_p S = \dim \operatorname{Im} d\iota_p,$$

which implies that $\operatorname{Im} d\iota_p = \operatorname{Ker} d\Phi_p$. $\qquad\qquad\square$

When the defining function Φ takes its values in $\mathbb{R}^k$, it is useful to restate the proposition in terms of component functions of Φ. The proof of the next corollary is immediate.

Corollary 5.39. *Suppose $S \subseteq M$ is a level set of a smooth submersion $\Phi = (\Phi^1, \ldots, \Phi^k) \colon M \to \mathbb{R}^k$. A vector $v \in T_p M$ is tangent to S if and only if $v\Phi^1 = \cdots = v\Phi^k = 0$.* $\qquad\qquad\square$

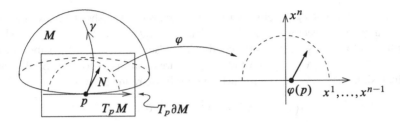

Fig. 5.13 An inward-pointing vector

▶ **Exercise 5.40.** Suppose $S \subseteq M$ is a level set of a smooth map $\Phi \colon M \to N$ with constant rank. Show that $T_p S = \operatorname{Ker} d\Phi_p$ for each $p \in S$.

If M is a smooth manifold with boundary and $p \in \partial M$, it is intuitively evident that the vectors in $T_p M$ can be separated into three classes: those tangent to the boundary, those pointing inward, and those pointing outward. Formally, we make the following definition. If $p \in \partial M$, a vector $v \in T_p M \smallsetminus T_p \partial M$ is said to be **inward-pointing** if for some $\varepsilon > 0$ there exists a smooth curve $\gamma \colon [0, \varepsilon) \to M$ such that $\gamma(0) = p$ and $\gamma'(0) = v$, and it is **outward-pointing** if there exists such a curve whose domain is $(-\varepsilon, 0]$. The following proposition gives another characterization of inward-pointing and outward-pointing vectors, which is usually much easier to check. (See Fig. 5.13.)

Proposition 5.41. *Suppose M is a smooth n-dimensional manifold with boundary, $p \in \partial M$, and (x^i) are any smooth boundary coordinates defined on a neighborhood of p. The inward-pointing vectors in $T_p M$ are precisely those with positive x^n-component, the outward-pointing ones are those with negative x^n-component, and the ones tangent to ∂M are those with zero x^n-component. Thus, $T_p M$ is the disjoint union of $T_p \partial M$, the set of inward-pointing vectors, and the set of outward-pointing vectors, and $v \in T_p M$ is inward-pointing if and only if $-v$ is outward-pointing.*

▶ **Exercise 5.42.** Prove Proposition 5.41.

If M is a smooth manifold with boundary, a **boundary defining function** for M is a smooth function $f \colon M \to [0, \infty)$ such that $f^{-1}(0) = \partial M$ and $df_p \neq 0$ for all $p \in \partial M$. For example, $f(x) = 1 - |x|^2$ is a boundary defining function for the closed unit ball $\overline{\mathbb{B}}^n$.

Proposition 5.43. *Every smooth manifold with boundary admits a boundary defining function.*

Proof. Let $\{(U_\alpha, \varphi_\alpha)\}$ be a collection of smooth charts whose domains cover M. For each α, define a smooth function $f_\alpha \colon U_\alpha \to [0, \infty)$ as follows: if U_α is an interior chart, let $f_\alpha \equiv 1$; while if U_α is a boundary chart, let $f_\alpha(x^1, \ldots, x^n) = x^n$ (the nth coordinate function in that chart). Thus, $f_\alpha(p)$ is positive if $p \in \operatorname{Int} M$ and zero if $p \in \partial M$. Let $\{\psi_\alpha\}$ be a partition of unity subordinate to this cover, and let

$f = \sum_\alpha \psi_\alpha f_\alpha$. Then f is smooth, identically zero on ∂M, and strictly positive in $\operatorname{Int} M$. To see that df does not vanish on ∂M, suppose $p \in \partial M$ and v is an inward-pointing vector at p. For each α such that $p \in U_\alpha$, we have $f_\alpha(p) = 0$ and $df_\alpha|_p(v) = dx^n|_p(v) > 0$ by Proposition 5.41. Thus

$$df_p(v) = \sum_\alpha \left(f_\alpha(p) d\psi_\alpha|_p(v) + \psi_\alpha(p) df_\alpha|_p(v) \right).$$

For each α, the first term in parentheses is zero and the second is nonnegative, and there is at least one α for which the second term is positive. Thus $df_p(v) > 0$, which implies that $df_p \neq 0$. $\qquad\square$

▶ **Exercise 5.44.** Suppose M is a smooth manifold with boundary, f is a boundary defining function, and $p \in \partial M$. Show that a vector $v \in T_p M$ is inward-pointing if and only if $vf > 0$, outward-pointing if and only if $vf < 0$, and tangent to ∂M if and only if $vf = 0$.

The results of this section have important applications to the problem of deciding whether a given subset of a smooth manifold is a submanifold. Given a smooth manifold M and a subset $S \subseteq M$, it is important to bear in mind that there are two very different questions one can ask. The simplest question is whether S is an embedded submanifold. Because embedded submanifolds are exactly those subsets satisfying the local slice condition, this is simply a question about the subset S itself: either it is an embedded submanifold or it is not, and if so, the topology and smooth structure making it into an embedded submanifold are uniquely determined (Theorem 5.31).

A more subtle question is whether S can be an immersed submanifold. In this case, neither the topology nor the smooth structure is known in advance, so one needs to ask whether there exist *any* topology and smooth structure on S making it into an immersed submanifold. This question is not always straightforward to answer, and it can be especially tricky to prove that S is *not* a smooth submanifold. A typical approach is to assume that it is, and then use one or more of the following phenomena to derive a contradiction:

- At each $p \in S$, the tangent space $T_p S$ is a linear subspace of $T_p M$, with the same dimension at each point.
- Each point of S is in the image of a local parametrization of S.
- Each vector tangent to S is the velocity vector of some smooth curve in S.
- Each vector tangent to S annihilates every smooth function that is constant on S.

Here is one example of how this can be done; others can be found in Problems 5-4 through 5-11.

Example 5.45. Consider the subset $S = \{(x, y) : y = |x|\} \subseteq \mathbb{R}^2$. It is easy to check that $S \smallsetminus \{(0,0)\}$ is an embedded 1-dimensional submanifold of $\mathbb{R}^2$, so if S itself is a smooth submanifold at all, it must be 1-dimensional. Suppose there were some smooth manifold structure on S making it into an immersed submanifold. Then $T_{(0,0)} S$ would be a 1-dimensional subspace of $T_{(0,0)} \mathbb{R}^2$, so by Proposition 5.35,

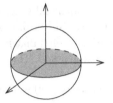

Fig. 5.14 A submanifold with boundary in a manifold with boundary

there would be a smooth curve $\gamma\colon (-\varepsilon, \varepsilon) \to \mathbb{R}^2$ whose image lies in S, and that satisfies $\gamma(0) = (0,0)$ and $\gamma'(0) \neq 0$. Writing $\gamma(t) = \big(x(t), y(t)\big)$, we see that $y(t)$ takes a global minimum at $t = 0$, so $y'(0) = 0$. On the other hand, because every point $(x, y) \in S$ satisfies $x^2 = y^2$, we have $x(t)^2 = y(t)^2$ for all t. Differentiating twice and setting $t = 0$, we conclude that $2x'(0)^2 = 2y'(0)^2 = 0$, which is a contradiction. Thus, there is no such smooth manifold structure.   //

Submanifolds with Boundary

So far in this chapter, all of our submanifolds have been manifolds without boundary. For some purposes (notably in the theory of integration), it is important also to consider submanifolds that have boundaries. The definitions are straightforward generalizations of the ones for ordinary submanifolds. If M is a smooth manifold with or without boundary, a *smooth submanifold with boundary in M* is a subset $S \subseteq M$ endowed with a topology and smooth structure making it into a smooth manifold with boundary such that the inclusion map is a smooth immersion. If the inclusion map is an embedding, then it is called an *embedded submanifold with boundary*; in the general case, it is an *immersed submanifold with boundary*. The terms *codimension* and *properly embedded* are defined just as in the submanifold case.

For example, for any positive integers $k \leq n$, the closed unit k-dimensional ball $\overline{\mathbb{B}}^k$ is a properly embedded submanifold with boundary in $\overline{\mathbb{B}}^n$, because the inclusion map $\overline{\mathbb{B}}^k \hookrightarrow \overline{\mathbb{B}}^n$ is easily seen to be a proper smooth embedding (Fig. 5.14).

One particular type of submanifold with boundary is especially important. If M is a smooth manifold with or without boundary, a *regular domain in M* is a properly embedded codimension-0 submanifold with boundary. Familiar examples are the closed upper half space $\mathbb{H}^n \subseteq \mathbb{R}^n$, the closed unit ball $\overline{\mathbb{B}}^n \subseteq \mathbb{R}^n$, and the closed upper hemisphere in $\mathbb{S}^n$.

Proposition 5.46. *Suppose M is a smooth manifold without boundary and $D \subseteq M$ is a regular domain. The topological interior and boundary of D are equal to its manifold interior and boundary, respectively.*

Proof. Suppose $p \in D$ is arbitrary. If p is in the manifold boundary of D, Theorem 4.15 shows that there exist a smooth boundary chart (U, φ) for D centered at p

and a smooth chart (V, ψ) for M centered at p in which F has the coordinate representation $F\left(x^1, \ldots, x^n\right) = \left(x^1, \ldots, x^n\right)$, where $n = \dim M = \dim D$. Since D has the subspace topology, $U = D \cap W$ for some open subset $W \subseteq M$, so $V_0 = V \cap W$ is a neighborhood of p in M such that $V_0 \cap D$ consists of all the points in V_0 whose x^m coordinate is nonnegative. Thus every neighborhood of p intersects both D and $M \setminus D$, so p is in the topological boundary of D.

On the other hand, suppose p is in the manifold interior of D. The manifold interior is a smooth embedded codimension-0 submanifold without boundary in M, so it is an open subset by Proposition 5.1. Thus p is in the topological interior of D.

Conversely, if p is in the topological interior of D, then it is not in the topological boundary, so the preceding argument shows that it is not in the manifold boundary and hence must be in the manifold interior. Similarly, if p is in the topological boundary, it is also in the manifold boundary. $\qquad\square$

Here are some ways in which regular domains often arise.

Proposition 5.47. *Suppose M is a smooth manifold and $f \in C^\infty(M)$.*

(a) *For each regular value b of f, the sublevel set $f^{-1}\left((-\infty, b]\right)$ is a regular domain in M.*

(b) *If a and b are two regular values of f with $a < b$, then $f^{-1}\left([a, b]\right)$ is a regular domain in M.*

Proof. Problem 5-21. $\qquad\square$

A set of the form $f^{-1}\left((-\infty, b]\right)$ for b a regular value of f is called a ***regular sublevel set of f***. Part (a) of the preceding theorem shows that every regular sublevel set of a smooth real-valued function is a regular domain. If $D \subseteq M$ is a regular domain and $f \in C^\infty(M)$ is a smooth function such that D is a regular sublevel set of f, then f is called a ***defining function for D***.

Theorem 5.48. *If M is a smooth manifold and $D \subseteq M$ is a regular domain, then there exists a defining function for D. If D is compact, then f can be taken to be a smooth exhaustion function for M.*

Proof. Problem 5-22. $\qquad\square$

Many (though not all) of the earlier results in this chapter have analogues for submanifolds with boundary. Since we will have little reason to consider nonembedded submanifolds with boundary, we focus primarily on the embedded case. The statements in the following proposition can be proved in the same way as their submanifold counterparts.

Proposition 5.49 (Properties of Submanifolds with Boundary). *Suppose M is a smooth manifold with or without boundary.*

(a) *Every open subset of M is an embedded codimension-0 submanifold with (possibly empty) boundary.*

(b) *If N is a smooth manifold with boundary and $F \colon N \to M$ is a smooth embedding, then with the subspace topology $F(N)$ is a topological manifold with* *boundary, and it has a smooth structure making it into an embedded submanifold with boundary in M.*

(c) *An embedded submanifold with boundary in M is properly embedded if and only if it is closed.*

(d) *If $S \subseteq M$ is an immersed submanifold with boundary, then for each $p \in S$ there exists a neighborhood U of p in S that is embedded in M.*

▶ **Exercise 5.50.** Prove the preceding proposition.

In order to adapt the results that depended on the existence of local slice charts, we have to generalize the local k-slice condition as follows. Suppose M is a smooth manifold (without boundary). If $(U, (x^i))$ is a chart for M, a **k-dimensional half-slice of** U is a subset of the following form for some constants $c^{k+1}, \ldots, c^n$:

$$\left\{ (x^1, \ldots, x^n) \in U : x^{k+1} = c^{k+1}, \ldots, x^n = c^n, \text{ and } x^k \geq 0 \right\}.$$

We say that a subset $S \subseteq M$ satisfies the **local k-slice condition for submanifolds with boundary** if each point of S is contained in the domain of a smooth chart $(U, (x^i))$ such that $S \cap U$ is either an ordinary k-dimensional slice or a k-dimensional half-slice. In the former case, the chart is called an **interior slice chart for S in M**, and in the latter, it is a **boundary slice chart for S in M**.

Theorem 5.51. *Let M be a smooth n-manifold without boundary. If $S \subseteq M$ is an embedded k-dimensional submanifold with boundary, then S satisfies the local k-slice condition for submanifolds with boundary. Conversely, if $S \subseteq M$ is a subset that satisfies the local k-slice condition for submanifolds with boundary, then with the subspace topology, S is a topological k-manifold with boundary, and it has a smooth structure making it into an embedded submanifold with boundary in M.*

▶ **Exercise 5.52.** Prove the preceding theorem.

Using the preceding theorem in place of Theorem 5.8, one can readily prove the following theorem.

Theorem 5.53 (Restricting Maps to Submanifolds with Boundary). *Suppose M and N are smooth manifolds with boundary and $S \subseteq M$ is an embedded submanifold with boundary.*

(a) RESTRICTING THE DOMAIN: *If $F: M \to N$ is a smooth map, then $F|_S: S \to N$ is smooth.*

(b) RESTRICTING THE CODOMAIN: *If $\partial M = \varnothing$ and $F: N \to M$ is a smooth map whose image is contained in S, then F is smooth as a map from N to S.*

Remark. The requirement that $\partial M = \varnothing$ can be removed in part (b) just as for Theorem 5.29; see Problem 9-13.

▶ **Exercise 5.54.** Prove Theorem 5.53.

Problems

5-1. Consider the map $\Phi\colon \mathbb{R}^4 \to \mathbb{R}^2$ defined by
$$\Phi(x, y, s, t) = \left(x^2 + y, x^2 + y^2 + s^2 + t^2 + y\right).$$
Show that $(0, 1)$ is a regular value of Φ, and that the level set $\Phi^{-1}(0, 1)$ is diffeomorphic to $\mathbb{S}^2$.

5-2. Prove Theorem 5.11 (the boundary of a manifold with boundary is an embedded submanifold).

5-3. Prove Proposition 5.21 (sufficient conditions for immersed submanifolds to be embedded).

5-4. Show that the image of the curve $\beta\colon (-\pi, \pi) \to \mathbb{R}^2$ of Example 4.19 is not an embedded submanifold of $\mathbb{R}^2$. [Be careful: this is not the same as showing that β is not an embedding.]

5-5. Let $\gamma\colon \mathbb{R} \to \mathbb{T}^2$ be the curve of Example 4.20. Show that $\gamma(\mathbb{R})$ is not an embedded submanifold of the torus. [Remark: the warning in Problem 5-4 applies in this case as well.]

5-6. Suppose $M \subseteq \mathbb{R}^n$ is an embedded m-dimensional submanifold, and let $UM \subseteq T\mathbb{R}^n$ be the set of all unit tangent vectors to M:
$$UM = \left\{(x, v) \in T\mathbb{R}^n : x \in M, \ v \in T_x M, \ |v| = 1\right\}.$$
It is called the **unit tangent bundle of M**. Prove that UM is an embedded $(2m - 1)$-dimensional submanifold of $T\mathbb{R}^n \approx \mathbb{R}^n \times \mathbb{R}^n$. (*Used on p. 147.*)

5-7. Let $F\colon \mathbb{R}^2 \to \mathbb{R}$ be defined by $F(x, y) = x^3 + xy + y^3$. Which level sets of F are embedded submanifolds of $\mathbb{R}^2$? For each level set, prove either that it is or that it is not an embedded submanifold.

5-8. Suppose M is a smooth n-manifold and $B \subseteq M$ is a regular coordinate ball. Show that $M \smallsetminus B$ is a smooth manifold with boundary, whose boundary is diffeomorphic to $\mathbb{S}^{n-1}$. (*Used on p. 225.*)

5-9. Let $S \subseteq \mathbb{R}^2$ be the boundary of the square of side 2 centered at the origin (see Problem 3-5). Show that S does not have a topology and smooth structure in which it is an immersed submanifold of $\mathbb{R}^2$.

5-10. For each $a \in \mathbb{R}$, let M_a be the subset of $\mathbb{R}^2$ defined by
$$M_a = \left\{(x, y) : y^2 = x(x - 1)(x - a)\right\}.$$
For which values of a is M_a an embedded submanifold of $\mathbb{R}^2$? For which values can M_a be given a topology and smooth structure making it into an immersed submanifold?

5-11. Let $\Phi\colon \mathbb{R}^2 \to \mathbb{R}$ be defined by $\Phi(x, y) = x^2 - y^2$.
(a) Show that $\Phi^{-1}(0)$ is not an embedded submanifold of $\mathbb{R}^2$.
(b) Can $\Phi^{-1}(0)$ be given a topology and smooth structure making it into an immersed submanifold of $\mathbb{R}^2$?

(c) Answer the same two questions for $\Psi \colon \mathbb{R}^2 \to \mathbb{R}$ defined by $\Psi(x, y) = x^2 - y^3$.

5-12. Suppose E and M are smooth manifolds with boundary, and $\pi \colon E \to M$ is a smooth covering map. Show that the restriction of π to each connected component of ∂E is a smooth covering map onto a component of ∂M. (*Used on p. 433.*)

5-13. Prove that the image of the dense curve on the torus described in Example 4.20 is a weakly embedded submanifold of $\mathbb{T}^2$.

5-14. Prove Theorem 5.32 (uniqueness of the smooth structure on an immersed submanifold once the topology is given).

5-15. Show by example that an immersed submanifold $S \subseteq M$ might have more than one topology and smooth structure with respect to which it is an immersed submanifold.

5-16. Prove Theorem 5.33 (uniqueness of the topology and smooth structure of a weakly embedded submanifold).

5-17. Prove Lemma 5.34 (the extension lemma for functions on submanifolds).

5-18. Suppose M is a smooth manifold and $S \subseteq M$ is a smooth submanifold.
   (a) Show that S is embedded if and only if every $f \in C^\infty(S)$ has a smooth extension to a neighborhood of S in M. [Hint: if S is not embedded, let $p \in S$ be a point that is not in the domain of any slice chart. Let U be a neighborhood of p in S that is embedded, and consider a function $f \in C^\infty(S)$ that is supported in U and equal to 1 at p.]
   (b) Show that S is properly embedded if and only if every $f \in C^\infty(S)$ has a smooth extension to all of M.

5-19. Suppose $S \subseteq M$ is an embedded submanifold and $\gamma \colon J \to M$ is a smooth curve whose image happens to lie in S. Show that $\gamma'(t)$ is in the subspace $T_{\gamma(t)}S$ of $T_{\gamma(t)}M$ for all $t \in J$. Give a counterexample if S is not embedded.

5-20. Show by giving a counterexample that the conclusion of Proposition 5.37 may be false if S is merely immersed.

5-21. Prove Proposition 5.47 (regular domains defined by smooth functions).

5-22. Prove Theorem 5.48 (existence of defining functions for regular domains).

5-23. Suppose M is a smooth manifold with boundary, N is a smooth manifold, and $F \colon M \to N$ is a smooth map. Let $S = F^{-1}(c)$, where $c \in N$ is a regular value for both F and $F|_{\partial M}$. Prove that S is a smooth submanifold with boundary in M, with $\partial S = S \cap \partial M$.

Chapter 6
Sard's Theorem

This chapter introduces a powerful tool in smooth manifold theory, *Sard's theorem*, which says that the set of critical values of a smooth function has measure zero. This theorem is fundamental in differential topology (the study of properties of smooth manifolds that are preserved by diffeomorphisms or by smooth deformations).

Before we begin, we need to extend the notion of sets of measure zero to manifolds. These are sets that are "small" in a sense that is closely related to having zero volume (even though we do not yet have a way to measure volume quantitatively on manifolds); they include such things as countable unions of submanifolds of positive codimension. With this tool in hand, we then prove Sard's theorem itself.

After proving Sard's theorem, we use it to prove three important results about smooth manifolds. The first result is the *Whitney embedding theorem*, which says that every smooth manifold can be smoothly embedded in some Euclidean space. (This justifies our habit of visualizing manifolds as subsets of $\mathbb{R}^n$.) The second result is the *Whitney approximation theorem*, which comes in two versions: every continuous real-valued or vector-valued function can be uniformly approximated by smooth ones, and every continuous map between smooth manifolds is homotopic to a smooth map. The third result is the *transversality homotopy theorem*, which says, among other things, that embedded submanifolds can always be deformed slightly so that they intersect "nicely" in a certain sense that we will make precise.

We will use some basic properties of sets of measure zero in the theory of integration in Chapter 16, and we will use the Whitney approximation theorems in our treatment of line integrals and de Rham cohomology in Chapters 16–18.

Sets of Measure Zero

An important notion in integration theory is that certain subsets of $\mathbb{R}^n$, called *sets of measure zero*, are so "thin" that they are negligible in integrals. In this section, we show how to define sets of measure zero in manifolds, and show that smooth maps between manifolds of the same dimension take sets of measure zero to sets of measure zero.

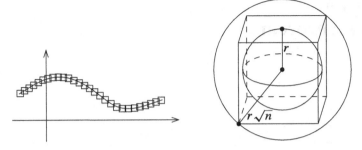

Fig. 6.1 A set of measure zero   **Fig. 6.2** *Balls* and *cubes*

Recall what it means for a set $A \subseteq \mathbb{R}^n$ to have measure zero (see Appendix C): for any $\delta > 0$, A can be covered by a countable collection of open rectangles, the sum of whose volumes is less than δ (Fig. 6.1).

▶ **Exercise 6.1.** Show that open rectangles can be replaced by open balls or open cubes in the definition of subsets of measure zero. [Hint: for cubes, first show that every open rectangle $R \subseteq \mathbb{R}^n$ can be covered by finitely many open cubes, the sum of whose volumes is no more than 2^n times the volume of R. For balls, Fig. 6.2 suggests the main idea.]

We need the following technical lemma about sets of measure zero. If you are familiar with the theory of Lebesgue measure, you will notice that this result follows easily from Fubini's theorem for integrals of measurable functions over product sets; but this is an elementary proof that does not depend on measure theory.

Lemma 6.2. *Suppose $A \subseteq \mathbb{R}^n$ is a compact subset whose intersection with $\{c\} \times \mathbb{R}^{n-1}$ has $(n-1)$-dimensional measure zero for every $c \in \mathbb{R}$. Then A has n-dimensional measure zero.*

Proof. Choose an interval $[a,b] \subseteq \mathbb{R}$ such that $A \subseteq [a,b] \times \mathbb{R}^{n-1}$. For each $c \in [a,b]$, let $A_c \subseteq \mathbb{R}^{n-1}$ denote the compact subset $\{x \in \mathbb{R}^{n-1} : (c,x) \in A\}$.

Let $\delta > 0$ be given. Our hypothesis implies that for each $c \in [a,b]$, the set A_c is covered by finitely many $(n-1)$-dimensional open cubes $C_1, \dots, C_k$ with total volume less than δ. Let U_c be the open subset $C_1 \cup \cdots \cup C_k \subseteq \mathbb{R}^{n-1}$. Because A is compact, there must be an open interval J_c containing c such that the intersection of A with $J_c \times \mathbb{R}^{n-1}$ is contained in $J_c \times U_c$, for otherwise there would be a sequence of points $(c_i, x_i) \in A$ such that $c_i \to c$ and $x_i \notin U_c$; but then passing to a convergent subsequence we obtain $x_i \to x \in A_c \smallsetminus U_c$, which contradicts the fact that $A_c \subseteq U_c$.

The intervals $\{J_c : c \in [a,b]\}$ form an open cover of $[a,b]$, so there are finitely many numbers $c_1 < \cdots < c_m$ such that the intervals $J_{c_1}, \dots, J_{c_m}$ cover $[a,b]$. By shrinking the intervals J_{c_i} where they overlap if necessary, we can arrange that the combined lengths of $J_{c_1}, \dots, J_{c_m}$ add up to no more than $2|b-a|$. It follows that A is contained in $\left(J_{c_1} \times U_{c_1}\right) \cup \cdots \cup \left(J_{c_m} \times U_{c_m}\right)$, which is a union of finitely many open rectangles with combined volume less than $2\delta|b-a|$. Since this can be made as small as desired, it follows that A has n-dimensional measure zero.   □

The most important sets of measure zero are graphs of continuous functions.

Proposition 6.3. *Suppose A is an open or closed subset of $\mathbb{R}^{n-1}$ or $\mathbb{H}^{n-1}$, and $f: A \to \mathbb{R}$ is a continuous function. Then the graph of f has measure zero in $\mathbb{R}^n$.*

Proof. First assume A is compact. We prove the theorem in this case by induction on n. When $n = 1$, it is trivial because the graph of f is at most a single point. To prove the inductive step, we appeal to Lemma 6.2. For each $c \in \mathbb{R}$, the intersection of the graph of f with $\{c\} \times \mathbb{R}^{n-1}$ is just the graph of the restriction of f to $\{x \in A : x^1 = c\}$, which is in turn the graph of a continuous function of $n - 2$ variables. It follows by induction that each such graph has $(n-1)$-dimensional measure zero, and thus by Lemma 6.2, the graph of f itself has n-dimensional measure zero.

If A is noncompact, it is a countable union of compact subsets by Proposition A.60, so the graph of f is a countable union of sets of measure zero. □

Corollary 6.4. *Every proper affine subspace of $\mathbb{R}^n$ has measure zero in $\mathbb{R}^n$.*

Proof. Let $S \subseteq \mathbb{R}^n$ be a proper affine subspace. Suppose first that $\dim S = n - 1$. Then there is at least one coordinate axis, say the x^i-axis, that is not parallel to S, and in that case S is the graph of an affine function of the form $x^i = F\left(x^1, \dots, x^{i-1}, x^{i+1}, \dots, x^n\right)$, so it has measure zero by Proposition 6.3. If $\dim S < n - 1$, then S is contained in some affine subspace of dimension $n - 1$, so it follows from Proposition C.18(b) that S has measure zero. □

Our goal is to extend the notion of measure zero in a diffeomorphism-invariant fashion to subsets of manifolds. Because a manifold does not come with a metric, volumes of cubes or balls do not make sense, so we cannot simply use the same definition. However, the key is provided by the next proposition, which implies that the condition of having measure zero is diffeomorphism-invariant for subsets of $\mathbb{R}^n$.

Proposition 6.5. *Suppose $A \subseteq \mathbb{R}^n$ has measure zero and $F: A \to \mathbb{R}^n$ is a smooth map. Then $F(A)$ has measure zero.*

Proof. By definition, for each $p \in A$, F has an extension to a smooth map, which we still denote by F, on a neighborhood of p in $\mathbb{R}^n$. Shrinking this neighborhood if necessary, we may assume that there is an open ball U containing p such that F extends smoothly to $\bar{U}$. By Proposition A.16, A is covered by countably many such precompact open subsets, so $F(A)$ is the union of countably many sets of the form $F\left(A \cap \bar{U}\right)$. Thus, it suffices to show that each such set has measure zero.

Since $\bar{U}$ is compact, there is a constant C such that $|DF(x)| \leq C$ for all $x \in \bar{U}$. Using the Lipschitz estimate for smooth functions (Proposition C.29), we have

$$\left|F(x) - F(x')\right| \leq C|x - x'| \tag{6.1}$$

for all $x, x' \in \bar{U}$.

Given $\delta > 0$, choose a countable cover $\{B_j\}$ of $A \cap \bar{U}$ by open balls satisfying $\sum_j \mathrm{Vol}(B_j) < \delta$. Then by (6.1), $F\left(\bar{U} \cap B_j\right)$ is contained in a ball $\tilde{B}_j$ whose radius is no more than C times that of B_j (Fig. 6.3). We conclude that $F\left(A \cap \bar{U}\right)$ is contained

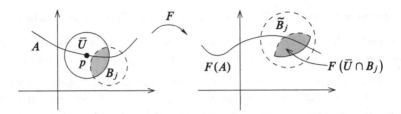

Fig. 6.3 The image of a set of measure zero

in the collection of balls $\{\widetilde{B}_j\}$, the sum of whose volumes is at most $C^n\delta$. Since this can be made as small as desired, it follows that $F(A \cap \bar{U})$ has measure zero. $\square$

If M is a smooth n-manifold with or without boundary, we say that a subset $A \subseteq M$ has **measure zero in** M if for every smooth chart (U, φ) for M, the subset $\varphi(A \cap U) \subseteq \mathbb{R}^n$ has n-dimensional measure zero. The following lemma shows that we need only check this condition for a single collection of smooth charts whose domains cover A.

Lemma 6.6. *Let M be a smooth n-manifold with or without boundary and $A \subseteq M$. Suppose that for some collection $\{(U_\alpha, \varphi_\alpha)\}$ of smooth charts whose domains cover A, $\varphi_\alpha(A \cap U_\alpha)$ has measure zero in $\mathbb{R}^n$ for each α. Then A has measure zero in M.*

Proof. Let (V, ψ) be an arbitrary smooth chart. We need to show that $\psi(A \cap V)$ has measure zero. Some countable collection of the U_α's covers $A \cap V$. For each such U_α, we have

$$\psi(A \cap V \cap U_\alpha) = (\psi \circ \varphi_\alpha^{-1}) \circ \varphi_\alpha(A \cap V \cap U_\alpha).$$

(See Fig. 6.4.)Now, $\varphi_\alpha(A \cap V \cap U_\alpha)$ is a subset of $\varphi_\alpha(A \cap U_\alpha)$, which has measure zero in $\mathbb{R}^n$ by hypothesis. By Proposition 6.5 applied to $\psi \circ \varphi_\alpha^{-1}$, therefore, $\psi(A \cap V \cap U_\alpha)$ has measure zero. Since $\psi(A \cap V)$ is the union of countably many such sets, it too has measure zero. $\square$

▶ **Exercise 6.7.** Let M be a smooth manifold with or without boundary. Show that a countable union of sets of measure zero in M has measure zero.

Proposition 6.8. *Suppose M is a smooth manifold with or without boundary and $A \subseteq M$ has measure zero in M. Then $M \smallsetminus A$ is dense in M.*

Proof. If $M \smallsetminus A$ is not dense, then A contains a nonempty open subset of M, which implies that there is a smooth chart (V, ψ) such that $\psi(A \cap V)$ contains a nonempty open subset of $\mathbb{R}^n$ (where $n = \dim M$). Because $\psi(A \cap V)$ has measure zero in $\mathbb{R}^n$, this contradicts Corollary C.25. $\square$

Theorem 6.9. *Suppose M and N are smooth n-manifolds with or without boundary, $F: M \to N$ is a smooth map, and $A \subseteq M$ is a subset of measure zero. Then $F(A)$ has measure zero in N.*

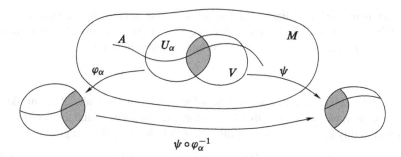

Fig. 6.4 Proof of Lemma 6.6

Proof. Let $\{(U_i, \varphi_i)\}$ be a countable cover of M by smooth charts. We need to show that for each smooth chart (V, ψ) for N, the set $\psi(F(A) \cap V)$ has measure zero in $\mathbb{R}^n$. Note that this set is the union of countably many subsets of the form $\psi \circ F \circ \varphi_i^{-1}(\varphi_i(A \cap U_i \cap F^{-1}(V)))$, each of which has measure zero by the result of Proposition 6.5. $\qquad\square$

Sard's Theorem

Here is the theorem that underlies all of our results about embedding, approximation, and transversality.

Theorem 6.10 (Sard's Theorem). *Suppose M and N are smooth manifolds with or without boundary and $F: M \to N$ is a smooth map. Then the set of critical values of F has measure zero in N.*

Proof. Let $m = \dim M$ and $n = \dim N$. We prove the theorem by induction on m. For $m = 0$, the result is immediate, because if $n = 0$, F has no critical points, while if $n > 0$, the entire image of F has measure zero because it is countable.

Now suppose $m \geq 1$, and assume the theorem holds for maps whose domains have dimensions less than m. By covering M and N with countably many smooth charts, we can reduce to the case in which F is a smooth map from an open subset $U \subseteq \mathbb{R}^m$ or $\mathbb{H}^m$ to $\mathbb{R}^n$. Write the coordinates in the domain U as $(x^1, \ldots, x^m)$, and those in the codomain as $(y^1, \ldots, y^n)$.

Let $C \subseteq U$ denote the set of critical points of F. We define a decreasing sequence of subsets $C \supseteq C_1 \supseteq C_2 \supseteq \cdots$ as follows:

$$C_k = \{x \in C : \text{for } 1 \leq i \leq k, \text{ all } i\text{th partial derivatives of } F \text{ vanish at } x\}.$$

By continuity, C and all of the C_k's are closed in U. We will prove in three steps that $F(C)$ has measure zero.

STEP 1: $F(C \smallsetminus C_1)$ *has measure zero.* Because C_1 is closed in U, we might as well replace U by $U \smallsetminus C_1$, and assume that $C_1 = \varnothing$. Let a be a point of C. Our assumption means that some first partial derivative of F is not zero at a.

By rearranging the coordinates in the domain and codomain, we may assume that $\partial F^1 / \partial x^1 (a) \neq 0$. This means that we can define new smooth coordinates $(u, v) = (u, v^2, \ldots, v^m)$ in some neighborhood V_a of a in U by

$$u = F^1, \quad v^2 = x^2, \quad \ldots, \quad v^m = x^m,$$

because the Jacobian of the coordinate transformation is nonsingular at a. Shrinking V_a if necessary, we can assume that $\overline{V}_a$ is a compact subset of U and the coordinates extend smoothly to $\overline{V}_a$. In these coordinates, F has the coordinate representation

$$F(u, v^2, \ldots, v^m) = (u, F^2(u, v), \ldots, F^n(u, v)), \tag{6.2}$$

and its Jacobian is

$$DF(u, v) = \begin{pmatrix} 1 & 0 \\ * & \dfrac{\partial F^i}{\partial v^j} \end{pmatrix}.$$

Therefore, $C \cap \overline{V}_a$ consists of exactly those points where the $(n-1) \times (m-1)$ matrix $(\partial F^i / \partial v^j)$ has rank less than $n - 1$.

We wish to show that the set $F(C \cap \overline{V}_a)$ has measure zero in $\mathbb{R}^n$. Because this set is compact, by Lemma 6.2 it suffices to show that its intersection with each hyperplane $y^1 = c$ has $(n-1)$-dimensional measure zero.

For $c \in \mathbb{R}$, let $B_c = \{v : (c, v) \in \overline{V}_a\} \subseteq \mathbb{R}^{m-1}$, and define $F_c : B_c \to \mathbb{R}^{n-1}$ by

$$F_c(v) = (F^2(c, v), \ldots, F^n(c, v)).$$

Because $F(c, v) = (c, F_c(v))$, the critical values of $F|_{\overline{V}_a}$ that lie in the hyperplane $y^1 = c$ are exactly the points of the form (c, w) with w a critical value of F_c. By the induction hypothesis, the set of critical values of each F_c has $(n-1)$-dimensional measure zero. Thus by Lemma 6.2, $F(C \cap \overline{V}_a)$ has measure zero.

Because U is covered by countably many sets of the form $\overline{V}_a$, it follows that $F(C \cap U)$ is a countable union of sets of measure zero and thus has measure zero. This completes the proof of Step 1.

STEP 2: *For each k, $F(C_k \smallsetminus C_{k+1})$ has measure zero.* Again, since C_{k+1} is closed in U, we can discard it and assume that at every point of C_k there is some $(k+1)$st partial derivative of F that does not vanish.

Let $a \in C_k$ be arbitrary, and let $y : U \to \mathbb{R}$ denote some kth partial derivative of F that has at least one nonvanishing first partial derivative at a. Then a is a regular point of the smooth map y, so there is a neighborhood V_a of a consisting entirely of regular points of y. Let Y be the zero set of y in V_a, which is a smooth hypersurface by the regular level set theorem. By definition of C_k, all kth derivatives of F (including y) vanish on C_k, so $C_k \cap V_a$ is contained in Y. At any $p \in C_k \cap V_a$, dF_p is not surjective, so certainly $d(F|_Y)_p = (dF_p)|_{T_p Y}$ is not surjective. Thus, $F(C_k \cap V_a)$ is contained in the set of critical values of $F|_Y : Y \to \mathbb{R}^n$, which has measure zero by the induction hypothesis. Since U can be covered by countably many neighborhoods like V_a, it follows that $F(C_k \smallsetminus C_{k+1})$ is contained in a countable union of sets of the form $F(C_k \cap V_a)$, and thus has measure zero.

We are not yet finished, because there may be points of C at which all partial derivatives of F vanish, which means that they are neither in $C \smallsetminus C_1$ nor in $C_k \smallsetminus C_{k+1}$ for any k. This possibility is taken care of by the final step.

STEP 3: *For $k > m/n - 1$, $F(C_k)$ has measure zero.* For each $a \in U$, there is a closed cube E such that $a \in E \subseteq U$. Since U can be covered by countably many such cubes, it suffices to show that $F(C_k \cap E)$ has measure zero whenever E is a closed cube contained in U. Let E be such a cube, and let A be a constant that bounds the absolute values of all of the $(k + 1)$st derivatives of F in E. Let R denote the side length of E, and let K be a large integer to be chosen later. We can subdivide E into K^m cubes of side length R/K, denoted by $(E_1, \dots, E_{K^m})$. If E_i is one of these cubes and there is a point $a_i \in C_k \cap E_i$, then Corollary C.16 to Taylor's theorem implies that for all $x \in E_i$ we have

$$\left| F(x) - F(a_i) \right| \le A' |x - a_i|^{k+1},$$

for some constant A' that depends only on A, k, and m. Thus, $F(E_i)$ is contained in a ball of radius $A'(R/K)^{k+1}$. This implies that $F(C_k \cap E)$ is contained in a union of K^m balls, the sum of whose n-dimensional volumes is no more than

$$K^m (A')^n (R/K)^{n(k+1)} = A'' K^{m-nk-n},$$

where $A'' = (A')^n R^{n(k+1)}$. Since $k > m/n - 1$, this can be made as small as desired by choosing K large, so $F(C_k \cap E)$ has measure zero. $\square$

Corollary 6.11. *Suppose M and N are smooth manifolds with or without boundary, and $F: M \to N$ is a smooth map. If $\dim M < \dim N$, then $F(M)$ has measure zero in N.*

Proof. In this case, each point of M is a critical point for F. $\square$

Problem 6-1 outlines a simple proof of the preceding corollary that does not depend on the full strength of Sard's theorem.

It is important to be aware that Corollary 6.11 is false if F is merely assumed to be continuous. For example, there is a continuous map $F: [0,1] \to \mathbb{R}^2$ whose image is the entire unit square $[0,1] \times [0,1]$. (Such a map is called a ***space-filling curve***. See [Rud76, p. 168] for an example.)

Corollary 6.12. *Suppose M is a smooth manifold with or without boundary, and $S \subseteq M$ is an immersed submanifold with or without boundary. If $\dim S < \dim M$, then S has measure zero in M.*

Proof. Apply Corollary 6.11 to the inclusion map $S \hookrightarrow M$. $\square$

The Whitney Embedding Theorem

Our first application of Sard's theorem is to show that every smooth manifold can be embedded into a Euclidean space. In fact, we will show that every smooth n-manifold with or without boundary is diffeomorphic to a properly embedded submanifold (with or without boundary) of $\mathbb{R}^{2n+1}$.

The first step is to show that an injective immersion of an n-manifold into $\mathbb{R}^N$ can be turned into an injective immersion into a lower-dimensional Euclidean space if $N > 2n + 1$.

Lemma 6.13. *Suppose $M \subseteq \mathbb{R}^N$ is a smooth n-dimensional submanifold with or without boundary. For any $v \in \mathbb{R}^N \smallsetminus \mathbb{R}^{N-1}$, let $\pi_v : \mathbb{R}^N \to \mathbb{R}^{N-1}$ be the projection with kernel $\mathbb{R}v$ (where we identify $\mathbb{R}^{N-1}$ with the subspace of $\mathbb{R}^N$ consisting of points with last coordinate zero). If $N > 2n + 1$, then there is a dense set of vectors $v \in \mathbb{R}^N \smallsetminus \mathbb{R}^{N-1}$ for which $\pi_v|_M$ is an injective immersion of M into $\mathbb{R}^{N-1}$.*

Proof. In order for $\pi_v|_M$ to be injective, it is necessary and sufficient that $p - q$ never be parallel to v when p and q are distinct points in M. Similarly, in order for $\pi_v|_M$ to be a smooth immersion, it is necessary and sufficient that $T_p M$ not contain any nonzero vectors in $\operatorname{Ker} d(\pi_v)_p$ for any $p \in M$. Because π_v is linear, its differential is the same linear map (under the usual identification $T_p \mathbb{R}^N \cong \mathbb{R}^N$), so this condition is equivalent to the requirement that $T_p M$ not contain any nonzero vectors parallel to v.

Let $\Delta_M \subseteq M \times M$ denote the closed set $\Delta_M = \{(p,p) : p \in M\}$ (called the **diagonal of $M \times M$**), and let $M_0 \subseteq TM$ denote the closed set $M_0 = \{(p,0) \in TM : p \in M\}$ (the set of zero vectors at all points of M). Consider the following two maps into the real projective space $\mathbb{RP}^{N-1}$:

$$\kappa : (M \times M) \smallsetminus \Delta_M \to \mathbb{RP}^{N-1}, \quad \kappa(p,q) = [p - q],$$
$$\tau : TM \smallsetminus M_0 \to \mathbb{RP}^{N-1}, \qquad \tau(p,w) = [w],$$

where the brackets mean the equivalence class of a vector in $\mathbb{R}^N \smallsetminus \{0\}$ considered as a point in $\mathbb{RP}^{N-1}$. These are both smooth because they are compositions of smooth maps with the projection $\mathbb{R}^N \smallsetminus \{0\} \to \mathbb{RP}^{N-1}$, and the condition that $\pi_v|_M$ be an injective smooth immersion is precisely the condition that $[v]$ not be in the image of either κ or τ. Because the domains of both κ and τ have dimension $2n < N - 1 = \dim \mathbb{RP}^{N-1}$, Corollary 6.11 to Sard's theorem implies that the image of each map has measure zero, and so their union has measure zero as well. Thus, the set of vectors whose equivalence classes are not in either image is dense.   □

By applying the preceding lemma repeatedly, we can conclude that if an n-manifold admits an injective immersion into *some* Euclidean space, then it admits one into $\mathbb{R}^{2n+1}$. When M is compact, this map is actually an embedding by Proposition 4.22(c); but if M is not compact, we need to work a little harder to ensure that our injective immersions are also embeddings.

Lemma 6.14. *Let M be a smooth n-manifold with or without boundary. If M admits a smooth embedding into $\mathbb{R}^N$ for some N, then it admits a proper smooth embedding into $\mathbb{R}^{2n+1}$.*

Proof. For this proof, given a one-dimensional linear subspace $S \subseteq \mathbb{R}^N$ and a positive number R, let us define the **tube with axis S and radius R** to be the open subset $T_R(S) \subseteq \mathbb{R}^N$ consisting of points whose distance from S is less than R:

$$T_R(S) = \{x \in \mathbb{R}^N : |x - y| < R \text{ for some } y \in S\}.$$

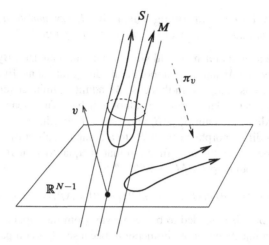

Fig. 6.5 Reducing the codimension of a proper embedding

Suppose $F: M \to \mathbb{R}^N$ is an arbitrary smooth embedding. Let $G: \mathbb{R}^N \to \mathbb{B}^N$ be a diffeomorphism, and let $f: M \to \mathbb{R}$ be a smooth exhaustion function (see Proposition 2.28). Define $\Psi: M \to \mathbb{R}^N \times \mathbb{R}$ by $\Psi(p) = \big(G \circ F(p), f(p)\big)$. Because $G \circ F$ is an embedding, it follows that Ψ is injective and $d\Psi_p$ is injective for each p, so Ψ is an injective immersion. It is proper because the preimage of any compact set is a closed subset of the compact set $f^{-1}\big((-\infty, c]\big)$ for some c, so Ψ is a smooth embedding by Proposition 4.22(b). By construction, the image of Ψ is contained in the tube $\mathbb{B}^N \times \mathbb{R}$.

Henceforth (after replacing $N + 1$ by N), we assume that M admits a proper smooth embedding into $\mathbb{R}^N$ that takes its values in some tube $T_R(S)$ (Fig. 6.5). Identifying M with its image, we may consider M as a properly embedded submanifold of $\mathbb{R}^N$ contained in the tube.

If $N > 2n + 1$, Lemma 6.13 shows that there exists $v \in \mathbb{R}^N \smallsetminus \mathbb{R}^{N-1}$ so that $\pi_v|_M$ is an injective immersion of M into $\mathbb{R}^{N-1}$. Moreover, we may choose v so that it does not lie in the subspace S; it follows that $\pi_v(S)$ is a one-dimensional subspace of $\mathbb{R}^{N-1}$, and $\pi_v(M)$ lies in a tube around $\pi_v(S)$ because π_v is a bounded linear map. We will show that $\pi_v|_M$ is proper, so it is an embedding by Proposition 4.22(b).

Suppose $K \subseteq \mathbb{R}^{N-1}$ is a compact set. Then K is contained in the open ball around 0 of some radius R_1. For any $x \in \pi_v^{-1}(K)$, there is some $c \in \mathbb{R}$ such that $\pi_v(x) = x - cv$; since $|\pi_v(x)| < R_1$, this means that x is in the tube of radius R_1 around the line $\mathbb{R}v$ spanned by v. It follows that $M \cap \pi_v^{-1}(K)$ is contained in two tubes, one around S and the other around $\mathbb{R}v$. A simple geometric argument shows that the intersection of two tubes is bounded when their axes are not parallel, so $M \cap \pi_v^{-1}(K)$ is compact. Thus $\pi_v|_M$ is proper, which implies that $\pi_v(M)$ is a properly embedded submanifold of $\mathbb{R}^{N-1}$ contained in a tube. We can now iterate this argument until we achieve a proper smooth embedding of M into $\mathbb{R}^{2n+1}$. $\square$

Theorem 6.15 (Whitney Embedding Theorem). *Every smooth n-manifold with or without boundary admits a proper smooth embedding into* $\mathbb{R}^{2n+1}$.

Proof. Let M be a smooth n-manifold with or without boundary. By Lemma 6.14, it suffices to show that M admits a smooth embedding into some Euclidean space.

First suppose M is compact. In this case M admits a finite cover $\{B_1, \ldots, B_m\}$ in which each B_i is a regular coordinate ball or half-ball. This means that for each i there exist a coordinate domain $B_i' \supseteq \bar{B}_i$ and a smooth coordinate map $\varphi_i \colon B_i' \to \mathbb{R}^n$ that restricts to a diffeomorphism from $\bar{B}_i$ to a compact subset of $\mathbb{R}^n$. For each i, let $\rho_i \colon M \to \mathbb{R}$ be a smooth bump function that is equal to 1 on $\bar{B}_i$ and supported in B_i'. Define a smooth map $F \colon M \to \mathbb{R}^{nm+m}$ by

$$F(p) = \big(\rho_1(p)\varphi_1(p), \ldots, \rho_m(p)\varphi_m(p), \rho_1(p), \ldots, \rho_m(p)\big),$$

where, as usual, $\rho_i \varphi_i$ is extended to be zero away from the support of ρ_i. We will show that F is an injective smooth immersion; because M is compact, this implies that F is a smooth embedding.

To see that F is injective, suppose $F(p) = F(q)$. Because the sets B_i cover M, there is some i such that $p \in B_i$. Then $\rho_i(p) = 1$, and the fact that $\rho_i(q) = \rho_i(p) = 1$ implies that $q \in \operatorname{supp} \rho_i \subseteq B_i'$, and

$$\varphi_i(q) = \rho_i(q)\varphi_i(q) = \rho_i(p)\varphi_i(p) = \varphi_i(p).$$

Since φ_i is injective on B_i', it follows that $p = q$.

Next, to see that F is a smooth immersion, let $p \in M$ be arbitrary and choose i such that $p \in B_i$. Because $\rho_i \equiv 1$ on a neighborhood of p, we have $d(\rho_i \varphi_i)_p = d(\varphi_i)_p$, which is injective. It follows easily that dF_p is injective. Thus, F is an injective smooth immersion and hence an embedding.

Now suppose M is noncompact. Let $f \colon M \to \mathbb{R}$ be a smooth exhaustion function. Sard's theorem shows that for each nonnegative integer i, there are regular values a_i, b_i of f such that $i < a_i < b_i < i + 1$. Define subsets $D_i, E_i \subseteq M$ by

$$D_0 = f^{-1}\big((-\infty, 1]\big), \quad E_0 = f^{-1}\big((-\infty, a_1]\big);$$
$$D_i = f^{-1}\big([i, i+1]\big), \quad E_i = f^{-1}\big([b_{i-i}, a_{i+1}]\big), \quad i \ge 1.$$

By Proposition 5.47, each E_i is a compact regular domain. We have $D_i \subseteq \operatorname{Int} E_i$, $M = \bigcup_i D_i$, and $E_i \cap E_j = \varnothing$ unless $j = i - 1, i$, or $i + 1$. The first part of the proof shows that for each i there is a smooth embedding of E_i into some Euclidean space, and therefore by Lemma 6.14 there is an embedding $\varphi_i \colon E_i \to \mathbb{R}^{2n+1}$. For each i, let $\rho_i \colon M \to \mathbb{R}$ be a smooth bump function that is equal to 1 on a neighborhood of D_i and supported in $\operatorname{Int} E_i$, and define $F \colon M \to \mathbb{R}^{2n+1} \times \mathbb{R}^{2n+1} \times \mathbb{R}$ by

$$F(p) = \left(\sum_{i \text{ even}} \rho_i(p)\varphi_i(p), \sum_{i \text{ odd}} \rho_i(p)\varphi_i(p), f(p) \right).$$

Then F is smooth because only one term in each sum is nonzero in a neighborhood of each point, and F is proper because f is. We will show that F is also an injective smooth immersion, hence a smooth embedding.

Suppose $F(p) = F(q)$. Then $p \in D_j$ for some j, and $f(q) = f(p)$ implies that $q \in D_j$ as well. Arguing just as in the compact case above, we conclude that $p = q$.

Now let $p \in M$ be arbitrary, and choose j such that $p \in D_j$. Then $\rho_j \equiv 1$ on a neighborhood of p. Assuming j is odd, for all q in that neighborhood we have

$$F(q) = (\varphi_j(q), \ldots, \ldots),$$

which implies that dF_p is injective. A similar argument applies when j is even. $\square$

Corollary 6.16. *Every smooth n-dimensional manifold with or without boundary is diffeomorphic to a properly embedded submanifold (with or without boundary) of $\mathbb{R}^{2n+1}$.* $\square$

Corollary 6.17. *Suppose M is a compact smooth n-manifold with or without boundary. If $N \geq 2n + 1$, then every smooth map from M to $\mathbb{R}^N$ can be uniformly approximated by embeddings.*

Proof. Assume $N \geq 2n + 1$, and let $f : M \to \mathbb{R}^N$ be a smooth map. By the Whitney embedding theorem, there is a smooth embedding $F : M \to \mathbb{R}^{2n+1}$. The map $G = f \times F : M \to \mathbb{R}^N \times \mathbb{R}^{2n+1}$ is also a smooth embedding, and f is equal to the composition $\pi \circ G$, where $\pi : \mathbb{R}^N \times \mathbb{R}^{2n+1} \to \mathbb{R}^N$ is the projection. Let $\widetilde{M} = G(M) \subseteq \mathbb{R}^N \times \mathbb{R}^{2n+1}$. Lemma 6.13 shows that there is a vector $v_{N+2n+1} \in \mathbb{R}^N \times \mathbb{R}^{2n+1}$ arbitrarily close to $e_{N+2n+1} = (0, \ldots, 0, 1)$ such that $\pi_{v_{N+2n+1}}|_{\widetilde{M}}$ is an embedding. This implies that $\pi_{v_{N+2n+1}}$ is arbitrarily close to $\pi_{e_{N+2n+1}}$. Iterating this, we obtain vectors $v_{N+2n+1}, v_{N+2n}, \ldots, v_{N+1}$ such that $\pi_{v_{N+1}} \circ \cdots \circ \pi_{v_{N+2n+1}}$ restricts to an embedding of $\widetilde{M}$ that is arbitrarily close to $\pi_{e_{N+1}} \circ \cdots \circ \pi_{e_{N+2n+1}} = \pi$, and therefore $\pi_{v_{N+1}} \circ \cdots \circ \pi_{v_{N+2n+1}} \circ G$ is an embedding of M into $\mathbb{R}^N$ that is arbitrarily close to f. $\square$

If we require only immersions rather than embeddings, we can lower the dimension by one.

Theorem 6.18 (Whitney Immersion Theorem). *Every smooth n-manifold with or without boundary admits a smooth immersion into $\mathbb{R}^{2n}$.*

Proof. See Problem 6-2 for the case $\partial M = \varnothing$, and Problem 9-14 for the general case. $\square$

Theorem 6.15, first proved by Hassler Whitney (in the case of empty boundary) in 1936 [Whi36], answered a question that had been nagging mathematicians since the notion of an abstract manifold was first introduced: Are there abstract smooth manifolds that are not diffeomorphic to embedded submanifolds of Euclidean space? Now we know that there are not.

Although the version of the embedding theorem that we have proved is quite sufficient for our purposes, it is interesting to note that eight years later, using much more sophisticated techniques of algebraic topology, Whitney was able to obtain the following improvements [Whi44a, Whi44b].

Theorem 6.19 (Strong Whitney Embedding Theorem). *If $n > 0$, every smooth n-manifold admits a smooth embedding into $\mathbb{R}^{2n}$.*

Theorem 6.20 (Strong Whitney Immersion Theorem). *If $n > 1$, every smooth n-manifold admits a smooth immersion into $\mathbb{R}^{2n-1}$.*

Because of these results, Theorems 6.15 and 6.18 are sometimes called the *easy* or *weak Whitney embedding and immersion theorems.*

In fact, not even the strong Whitney theorems are sharp in all dimensions. In 1985, Ralph Cohen proved that every compact smooth n-manifold can be immersed in $\mathbb{R}^{2n-a(n)}$, where $a(n)$ is the number of 1's in the binary expression for n. Thus, for example, every 3-manifold can be immersed in $\mathbb{R}^4$, while 4-manifolds require $\mathbb{R}^7$. This result is the best possible in every dimension. On the other hand, the best possible embedding dimension is known only for certain dimensions. For example, Whitney's dimension $2n$ is optimal for manifolds of dimensions $n = 1$ and $n = 2$, but C.T.C. Wall showed in 1965 [Wal65] that every 3-manifold can be embedded in $\mathbb{R}^5$. A good summary of the state of the art with references can be found in [Osb82].

The Whitney Approximation Theorems

In this section we prove the two theorems mentioned at the beginning of the chapter on approximation of continuous maps by smooth ones. Both of these theorems, like the embedding theorem we just proved, are due to Hassler Whitney [Whi36].

We begin with a theorem about smoothly approximating functions into Euclidean spaces. Our first theorem shows, in particular, that any continuous function from a smooth manifold M into $\mathbb{R}^k$ can be uniformly approximated by a smooth function. In fact, we will prove something stronger. If $\delta: M \to \mathbb{R}$ is a positive continuous function, we say that two functions $F, \widetilde{F}: M \to \mathbb{R}^k$ are **δ-close** if $\left| F(x) - \widetilde{F}(x) \right| < \delta(x)$ for all $x \in M$.

Theorem 6.21 (Whitney Approximation Theorem for Functions). *Suppose M is a smooth manifold with or without boundary, and $F: M \to \mathbb{R}^k$ is a continuous function. Given any positive continuous function $\delta: M \to \mathbb{R}$, there exists a smooth function $\widetilde{F}: M \to \mathbb{R}^k$ that is δ-close to F. If F is smooth on a closed subset $A \subseteq M$, then $\widetilde{F}$ can be chosen to be equal to F on A.*

Proof. If F is smooth on the closed subset A, then by the extension lemma for smooth functions (Lemma 2.26), there is a smooth function $F_0: M \to \mathbb{R}^k$ that agrees with F on A. Let

$$U_0 = \left\{ y \in M : \left| F_0(y) - F(y) \right| < \delta(y) \right\}.$$

Then U_0 is an open subset containing A. (If there is no such set A, we just take $U_0 = A = \varnothing$ and $F_0 \equiv 0$.)

We will show that there are countably many points $\{x_i\}_{i=1}^{\infty}$ in $M \smallsetminus A$ and neighborhoods U_i of x_i in $M \smallsetminus A$ such that $\{U_i\}_{i=1}^{\infty}$ is an open cover of $M \smallsetminus A$ and

$$\left| F(y) - F(x_i) \right| < \delta(y) \quad \text{for all } y \in U_i. \tag{6.3}$$

To see this, for any $x \in M \smallsetminus A$, let U_x be a neighborhood of x contained in $M \smallsetminus A$ and small enough that

$$\delta(y) > \tfrac{1}{2}\delta(x) \quad \text{and} \quad |F(y) - F(x)| < \tfrac{1}{2}\delta(x)$$

for all $y \in U_x$. (Such a neighborhood exists by continuity of δ and F.) Then if $y \in U_x$, we have

$$|F(y) - F(x)| < \tfrac{1}{2}\delta(x) < \delta(y).$$

The collection $\{U_x : x \in M \smallsetminus A\}$ is an open cover of $M \smallsetminus A$. Choosing a countable subcover $\{U_{x_i}\}_{i=1}^{\infty}$ and setting $U_i = U_{x_i}$, we have (6.3).

Let $\{\varphi_0, \varphi_i\}$ be a smooth partition of unity subordinate to the cover $\{U_0, U_i\}$ of M, and define $\widetilde{F} \colon M \to \mathbb{R}^k$ by

$$\widetilde{F}(y) = \varphi_0(y)F_0(y) + \sum_{i \geq 1} \varphi_i(y)F(x_i).$$

Then clearly $\widetilde{F}$ is smooth, and is equal to F on A. For any $y \in M$, the fact that $\sum_{i \geq 0} \varphi_i \equiv 1$ implies that

$$\left|\widetilde{F}(y) - F(y)\right| = \left|\varphi_0(y)F_0(y) + \sum_{i \geq 1} \varphi_i(y)F(x_i) - \left(\varphi_0(y) + \sum_{i \geq 1} \varphi_i(y)\right)F(y)\right|$$

$$\leq \varphi_0(y)\left|F_0(y) - F(y)\right| + \sum_{i \geq 1} \varphi_i(y)\left|F(x_i) - F(y)\right|$$

$$< \varphi_0(y)\delta(y) + \sum_{i \geq 1} \varphi_i(y)\delta(y) = \delta(y). \qquad \square$$

Corollary 6.22. *If M is a smooth manifold with or without boundary and $\delta \colon M \to \mathbb{R}$ is a positive continuous function, there is a smooth function $e \colon M \to \mathbb{R}$ such that $0 < e(x) < \delta(x)$ for all $x \in M$.*

Proof. Use the Whitney approximation theorem to construct a smooth function $e \colon M \to \mathbb{R}$ that satisfies $\left|e(x) - \tfrac{1}{2}\delta(x)\right| < \tfrac{1}{2}\delta(x)$ for all $x \in M$. $\qquad \square$

Tubular Neighborhoods

We would like to find a way to apply the Whitney approximation theorem to produce smooth approximations to continuous maps between smooth manifolds. If $F \colon N \to M$ is such a map, then by the Whitney embedding theorem we can consider M as an embedded submanifold of some Euclidean space $\mathbb{R}^n$, and approximate F by a smooth map into R^n. However, in general, the image of this smooth map will not lie in M. To correct for this, we need to know that there is a smooth retraction from some neighborhood of M onto M. For this purpose, we introduce tubular neighborhoods.

For each $x \in \mathbb{R}^n$, the tangent space $T_x\mathbb{R}^n$ is canonically identified with $\mathbb{R}^n$ itself, and the tangent bundle $T\mathbb{R}^n$ is canonically diffeomorphic to $\mathbb{R}^n \times \mathbb{R}^n$. By virtue of this identification, each tangent space $T_x \mathbb{R}^n$ inherits a Euclidean dot product. Suppose $M \subseteq \mathbb{R}^n$ is an embedded m-dimensional submanifold. For each $x \in M$, we define the **normal space to M at x** to be the $(n-m)$-dimensional subspace $N_x M \subseteq T_x \mathbb{R}^n$ consisting of all vectors that are orthogonal to $T_x M$ with respect to the Euclidean dot product. The **normal bundle of M**, denoted by NM, is the subset of $T\mathbb{R}^n \approx \mathbb{R}^n \times \mathbb{R}^n$ consisting of vectors that are normal to M:

$$NM = \{(x, v) \in \mathbb{R}^n \times \mathbb{R}^n : x \in M, \ v \in N_x M\}.$$

There is a natural projection $\pi_{NM} : NM \to M$ defined as the restriction to NM of $\pi : T\mathbb{R}^n \to \mathbb{R}^n$.

Theorem 6.23. *If $M \subseteq \mathbb{R}^n$ is an embedded m-dimensional submanifold, then NM is an embedded n-dimensional submanifold of $T\mathbb{R}^n \approx \mathbb{R}^n \times \mathbb{R}^n$.*

Proof. Let x_0 be any point of M, and let (U, φ) be a slice chart for M in $\mathbb{R}^n$ centered at x_0. Write $\widehat{U} = \varphi(U) \subseteq \mathbb{R}^n$, and write the coordinate functions of φ as $(u^1, \ldots, u^n)$, so that $M \cap U$ is the set where $u^{m+1} = \cdots = u^n = 0$. At each point $x \in U$, the vectors $E_j|_x = (d\varphi_x)^{-1}(\partial/\partial u^j|_{\varphi(x)})$ form a basis for $T_x \mathbb{R}^n$. We can expand each $E_j|_x$ in terms of the standard coordinate frame as

$$E_j\big|_x = E_j^i(x) \left.\frac{\partial}{\partial x^i}\right|_x,$$

where each $E_j^i(x)$ is a partial derivative of φ^{-1} evaluated at $\varphi(x)$, and thus is a smooth function of x.

Define a smooth function $\Phi : U \times \mathbb{R}^n \to \widehat{U} \times \mathbb{R}^n$ by

$$\Phi(x, v) = \left(u^1(x), \ldots, u^n(x), v \cdot E_1\big|_x, \ldots, v \cdot E_n\big|_x\right).$$

The total derivative of Φ at a point (x, v) is

$$D\Phi_{(x,v)} = \begin{pmatrix} \dfrac{\partial u^i}{\partial x^j}(x) & 0 \\[2mm] * & E_j^i(x) \end{pmatrix},$$

which is invertible, so Φ is a local diffeomorphism. If $\Phi(x, v) = \Phi(x', v')$, then $x = x'$ because φ is injective, and then the fact that $v \cdot E_i|_x = v' \cdot E_i|_x$ for each i implies that $v - v'$ is orthogonal to the span of $(E_1|_x, \ldots, E_n|_x)$ and is therefore zero. Thus Φ is injective, so it defines a smooth coordinate chart on $U \times \mathbb{R}^n$. The definitions imply that $(x, v) \in NM$ if and only if $\Phi(x, v)$ is in the slice

$$\{(y, z) \in \mathbb{R}^n \times \mathbb{R}^n : y^{m+1} = \cdots = y^n = 0, \ z^1 = \cdots = z^m = 0\}.$$

Thus Φ is a slice chart for NM in $\mathbb{R}^n \times \mathbb{R}^n$. $\square$

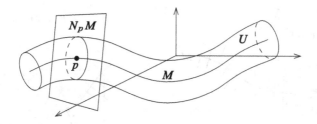

Fig. 6.6 A tubular neighborhood

(We will be able to give a shorter proof of this theorem in Chapter 10; see Corollary 10.36.)

Thinking of NM as a submanifold of $\mathbb{R}^n \times \mathbb{R}^n$, we define $E \colon NM \to \mathbb{R}^n$ by

$$E(x, v) = x + v.$$

This maps each normal space $N_x M$ affinely onto the affine subspace through x and orthogonal to $T_x M$. Clearly, E is smooth, because it is the restriction to NM of the addition map $\mathbb{R}^n \times \mathbb{R}^n \to \mathbb{R}^n$. A *tubular neighborhood of* M is a neighborhood U of M in $\mathbb{R}^n$ that is the diffeomorphic image under E of an open subset $V \subseteq NM$ of the form

$$V = \{(x, v) \in NM : |v| < \delta(x)\}, \tag{6.4}$$

for some positive continuous function $\delta \colon M \to \mathbb{R}$ (Fig. 6.6).

Theorem 6.24 (Tubular Neighborhood Theorem). *Every embedded submanifold of $\mathbb{R}^n$ has a tubular neighborhood.*

Proof. Let $M_0 \subseteq NM$ be the subset $\{(x, 0) : x \in M\}$. We begin by showing that E is a local diffeomorphism on a neighborhood of M_0. By the inverse function theorem, it suffices to show that the differential $dE_{(x,0)}$ is bijective at each point $(x, 0) \in M_0$. This follows easily from the following two facts: First, the restriction of E to M_0 is the obvious diffeomorphism $M_0 \to M$, so $dE_{(x,0)}$ maps the subspace $T_{(x,0)}M_0 \subseteq T_{(x,0)}NM$ isomorphically onto $T_x M$. Second, the restriction of E to the fiber $N_x M$ is the affine map $w \mapsto x + w$, so $dE_{(x,0)}$ maps $T_{(x,0)}(N_x M) \subseteq T_{(x,0)}NM$ isomorphically onto $N_x M$. Since $T_x \mathbb{R}^n = T_x M \oplus N_x M$, this shows that $dE_{(x,0)}$ is surjective, and hence is bijective for dimensional reasons. Thus, E is a diffeomorphism on a neighborhood of $(x, 0)$ in NM, which we can take to be of the form $V_\delta(x) = \{(x', v') \in NM : |x - x'| < \delta, |v'| < \delta\}$ for some $\delta > 0$. (This uses the fact that NM is embedded in $\mathbb{R}^n \times \mathbb{R}^n$, and therefore its topology is induced by the Euclidean metric.)

To complete the proof, we need to show that there is an open subset V of the form (6.4) on which E is a global diffeomorphism. For each point $x \in M$, let $\rho(x)$ be the supremum of all $\delta \leq 1$ such that E is a diffeomorphism from $V_\delta(x)$ to its image. The argument in the preceding paragraph implies that $\rho \colon M \to \mathbb{R}$ is positive. To show it is continuous, let $x, x' \in M$ be arbitrary, and suppose first that $|x - x'| <$

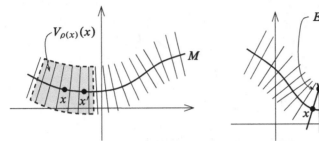

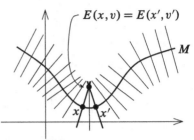

Fig. 6.7 Continuity of ρ **Fig. 6.8** Injectivity of E

$\rho(x)$. By the triangle inequality, $V_\delta(x')$ is contained in $V_{\rho(x)}(x)$ for $\delta = \rho(x) - |x - x'|$ (Fig. 6.7), which implies that $\rho(x') \geq \rho(x) - |x - x'|$, or

$$\rho(x) - \rho(x') \leq |x - x'|. \tag{6.5}$$

On the other hand, if $|x - x'| \geq \rho(x)$, then (6.5) holds for trivial reasons. Reversing the roles of x and x' yields an analogous inequality, which shows that $|\rho(x) - \rho(x')| \leq |x - x'|$, so ρ is continuous. Note that E is injective on the entire set $V_{\rho(x)}(x)$, because any two points $(x_1, v_1), (x_2, v_2)$ in this set are in $V_\delta(x)$ for some $\delta < \rho(x)$.

Now let $V = \{(x, v) \in NM : |v| < \tfrac{1}{2}\rho(x)\}$. We will show that E is injective on V. Suppose that (x, v) and (x', v') are points in V such that $E(x, v) = E(x', v')$ (Fig. 6.8). Assume without loss of generality that $\rho(x') \leq \rho(x)$. It follows from $x + v = x' + v'$ that

$$|x - x'| = |v - v'| \leq |v| + |v'| < \tfrac{1}{2}\rho(x) + \tfrac{1}{2}\rho(x') \leq \rho(x).$$

Therefore, both (x, v) and (x', v') are in $V_{\rho(x)}(x)$. Since E is injective on this set, this implies $(x, v) = (x', v')$.

The set $U = E(V)$ is open in $\mathbb{R}^n$ because $E|_V$ is a local diffeomorphism and thus an open map. It follows that $E : V \to U$ is a smooth bijection and a local diffeomorphism, hence a diffeomorphism by Proposition 4.6. Therefore, U is a tubular neighborhood of M.   □

One of the most useful features of tubular neighborhoods is expressed in the next proposition. A ***retraction*** of a topological space X onto a subspace $M \subseteq X$ is a continuous map $r : X \to M$ such that $r|_M$ is the identity map of M.

Proposition 6.25. *Let $M \subseteq \mathbb{R}^n$ be an embedded submanifold. If U is any tubular neighborhood of M, there exists a smooth map $r : U \to M$ that is both a retraction and a smooth submersion.*

Proof. Let $NM \subseteq T\mathbb{R}^n$ be the normal bundle of M, and let $M_0 \subseteq NM$ be the set $M_0 = \{(x, 0) : x \in M\}$. By definition of a tubular neighborhood, there is an open subset $V \subseteq NM$ containing M_0 such that $E : V \to U$ is a diffeomorphism.

Define $r: U \to M$ by $r = \pi_{NM} \circ E^{-1}$, where $\pi_{NM}: NM \to M$ is the natural projection. Then r is smooth by composition. For $x \in M$, note that $E(x,0) = x$, so $r(x) = \pi \circ E^{-1}(x) = \pi(x,0) = x$, which shows that r is a retraction. Since π is a smooth submersion and E^{-1} is a diffeomorphism, it follows that r is a smooth submersion. $\qquad\qquad\square$

Smooth Approximation of Maps Between Manifolds

Now we can extend the Whitney approximation theorem to maps between manifolds. This extension will have important applications to line integrals in Chapter 16 and to de Rham cohomology in Chapters 17–18.

Theorem 6.26 (Whitney Approximation Theorem). *Suppose N is a smooth manifold with or without boundary, M is a smooth manifold (without boundary), and $F: N \to M$ is a continuous map. Then F is homotopic to a smooth map. If F is already smooth on a closed subset $A \subseteq N$, then the homotopy can be taken to be relative to A.*

Proof. By the Whitney embedding theorem, we may as well assume that M is a properly embedded submanifold of $\mathbb{R}^n$. Let U be a tubular neighborhood of M in $\mathbb{R}^n$, and let $r: U \to M$ be the smooth retraction given by Proposition 6.25. For any $x \in M$, let

$$\delta(x) = \sup\{\varepsilon \leq 1 : B_\varepsilon(x) \subseteq U\}. \qquad (6.6)$$

By a triangle-inequality argument just like the one in the proof of the tubular neighborhood theorem, $\delta: M \to \mathbb{R}^+$ is continuous. Let $\tilde{\delta} = \delta \circ F: N \to \mathbb{R}^+$. By Theorem 6.21, there exists a smooth function $\tilde{F}: N \to \mathbb{R}^n$ that is $\tilde{\delta}$-close to F, and is equal to F on A (which might be the empty set). Let $H: N \times I \to M$ be the composition of r with the straight-line homotopy between F and $\tilde{F}$:

$$H(p,t) = r\big((1-t)F(p) + t\tilde{F}(p)\big).$$

This is well defined, because our condition on $\tilde{F}$ guarantees that for each $p \in N$, $\big|\tilde{F}(p) - F(p)\big| < \tilde{\delta}(p) = \delta(F(p))$, which means that $\tilde{F}(p)$ is contained in the ball of radius $\delta\big(F(p)\big)$ around $F(p)$; since this ball is contained in U, so is the entire line segment from $F(p)$ to $\tilde{F}(p)$.

Thus H is a homotopy between $H(p,0) = F(p)$ and $H(p,1) = r\big(\tilde{F}(p)\big)$, which is a smooth map by composition. It satisfies $H(p,t) = F(p)$ for all $p \in A$, since $F = \tilde{F}$ there. $\qquad\qquad\square$

Corollary 6.27 (Extension Lemma for Smooth Maps). *Suppose N is a smooth manifold with or without boundary, M is a smooth manifold, $A \subseteq N$ is a closed subset, and $f: A \to M$ is a smooth map. Then f has a smooth extension to N if and only if it has a continuous extension to N.*

Proof. If $F: N \to M$ is a continuous extension of f to all of N, the Whitney approximation theorem guarantees the existence of a smooth map $\tilde{F}$ (homotopic to F,

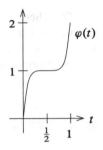

Fig. 6.9 The function φ in the proof of Lemma 6.28

in fact, though we do not need that here) that agrees with f on A; in other words, $\widetilde{F}$ is a smooth extension of f. The converse is obvious. $\square$

If N and M are two smooth manifolds with or without boundary, a homotopy $H: N \times I \to M$ is called a *smooth homotopy* if it is also a smooth map, in the sense that it extends to a smooth map on some neighborhood of $N \times I$ in $N \times \mathbb{R}$. Two maps are said to be *smoothly homotopic* if there is a smooth homotopy between them.

Lemma 6.28. *If N and M are smooth manifolds with or without boundary, smooth homotopy is an equivalence relation on the set of all smooth maps from N to M.*

Proof. Reflexivity and symmetry are proved just as for ordinary homotopy. To prove transitivity, suppose $F, G, K: N \to M$ are smooth maps, and $H_1, H_2: N \times I \to M$ are smooth homotopies from F to G and G to K, respectively. Let $\varphi: [0, 1] \to [0, 2]$ be a smooth map such that $0 \le \varphi(t) \le 1$ for $t \in \left[0, \frac{1}{2}\right]$, $1 \le \varphi(t) \le 2$ for $t \in \left[\frac{1}{2}, 1\right]$, $\varphi(0) = 0$, $\varphi(1) = 2$, and $\varphi(t) \equiv 1$ for t in a neighborhood of $\frac{1}{2}$ (see Fig. 6.9). Define $H: N \times I \to M$ by

$$H(x, t) = \begin{cases} H_1\big(x, \varphi(t)\big), & t \in \left[0, \frac{1}{2}\right], \\ H_2\big(x, \varphi(t) - 1\big), & t \in \left[\frac{1}{2}, 1\right]. \end{cases}$$

Then it is easy to check that H is a smooth homotopy from F to K. $\square$

Theorem 6.29. *Suppose N is a smooth manifold with or without boundary, M is a smooth manifold, and $F, G: N \to M$ are smooth maps. If F and G are homotopic, then they are smoothly homotopic. If F and G are homotopic relative to some closed subset $A \subseteq N$, then they are smoothly homotopic relative to A.*

Proof. Suppose $F, G: N \to M$ are smooth, and let $H: N \times I \to M$ be a homotopy from F to G (relative to A, which may be empty). We wish to show that H can be replaced by a smooth homotopy.

Define $\bar{H}\colon N \times \mathbb{R} \to M$ by

$$\bar{H}(x,t) = \begin{cases} H(x,t), & t \in [0,1], \\ H(x,0), & t \le 0, \\ H(x,1), & t \ge 1. \end{cases}$$

This is continuous by the gluing lemma. The restriction of $\bar{H}$ to $N \times \{0\} \cup N \times \{1\}$ is smooth, because it is equal to $F \circ \pi_1$ on $N \times \{0\}$ and $G \circ \pi_1$ on $N \times \{1\}$ (where $\pi_1\colon N \times \mathbb{R} \to N$ is the projection). If H is a homotopy relative to A, then $\bar{H}$ is also smooth on $A \times I$. Because $N \times \mathbb{R}$ is a smooth manifold with (possibly empty) boundary, the Whitney approximation theorem guarantees that there is a smooth map $\widetilde{H}\colon N \times \mathbb{R} \to M$ (homotopic to $\bar{H}$, but we do not need that here) whose restriction to $N \times \{0\} \cup N \times \{1\} \cup A \times I$ agrees with $\bar{H}$ (and therefore H). Restricting back to $N \times I$ again, we see that $\widetilde{H}|_{N \times I}$ is a smooth homotopy (relative to A) between F and G. $\qquad \square$

When the target manifold has nonempty boundary, the analogues of Theorems 6.26 and 6.29 do not hold, because it might not be possible to find a smooth map that agrees with F on A (see Problem 6-7). However, if we do not insist on homotopy relative to a subset, the rest of the results can be extended to maps into manifolds with boundary. The proofs will have to wait until Chapter 9 (see Theorems 9.27 and 9.28).

Transversality

As our final application of Sard's theorem, we show how submanifolds can be perturbed so that they intersect "nicely." To explain what this means, we introduce the concept of *transversality*.

The intersection of two linear subspaces of a vector space is always another linear subspace. The analogous statement for submanifolds is certainly not true: it is easy to come up with examples of smooth submanifolds whose intersection is not a submanifold. (See Problem 6-14.) But with an additional assumption about the submanifolds, it is possible to show that their intersection is again a submanifold.

Suppose M is a smooth manifold. Two embedded submanifolds $S, S' \subseteq M$ are said to **intersect transversely** if for each $p \in S \cap S'$, the tangent spaces $T_p S$ and $T_p S'$ together span $T_p M$ (where we consider $T_p S$ and $T_p S'$ as subspaces of $T_p M$).

For many purposes, it is more convenient to work with the following more general definition. If $F\colon N \to M$ is a smooth map and $S \subseteq M$ is an embedded submanifold, we say that F is **transverse to** S if for every $x \in F^{-1}(S)$, the spaces $T_{F(x)}S$ and $dF_x(T_x N)$ together span $T_{F(x)}M$. One special case is worth noting: if F is a smooth submersion, then it is automatically transverse to every embedded submanifold of M. Two embedded submanifolds intersect transversely if and only if the inclusion of either one is transverse to the other.

The next result, a generalization of the regular level set theorem, shows why transversality is desirable.

Theorem 6.30. *Suppose N and M are smooth manifolds and $S \subseteq M$ is an embedded submanifold.*

(a) *If $F: N \to M$ is a smooth map that is transverse to S, then $F^{-1}(S)$ is an embedded submanifold of N whose codimension is equal to the codimension of S in M.*

(b) *If $S' \subseteq M$ is an embedded submanifold that intersects S transversely, then $S \cap S'$ is an embedded submanifold of M whose codimension is equal to the sum of the codimensions of S and S'.*

Proof. The second statement follows easily from the first, simply by taking F to be the inclusion map $S' \hookrightarrow M$, and noting that a composition of smooth embeddings $S \cap S' \hookrightarrow S \hookrightarrow M$ is again a smooth embedding.

To prove (a), let m denote the dimension of M and k the codimension of S in M. Given $x \in F^{-1}(S)$, we can find a neighborhood U of $F(x)$ in M and a local defining function $\varphi: U \to \mathbb{R}^k$ for S, with $S \cap U = \varphi^{-1}(0)$. The theorem will be proved if we can show that 0 is a regular value of $\varphi \circ F$, because $F^{-1}(S) \cap F^{-1}(U)$ is the zero set of $\varphi \circ F|_{F^{-1}(U)}$.

Given $z \in T_0 \mathbb{R}^k$ and $p \in (\varphi \circ F)^{-1}(0)$, the fact that 0 is a regular value of φ means there is a vector $y \in T_{F(p)}M$ such that $d\varphi_{F(p)}(y) = z$. The fact that F is transverse to S means we can write $y = y_0 + dF_p(v)$ for some $y_0 \in T_{F(p)}S$ and some $v \in T_p N$. Because φ is constant on $S \cap U$, it follows that $d\varphi_{F(p)}(y_0) = 0$, so

$$d(\varphi \circ F)_p(v) = d\varphi_{F(p)}\big(dF_p(v)\big) = d\varphi_{F(p)}\big(y_0 + dF_p(v)\big) = d\varphi_{F(p)}(y) = z.$$

Thus $F^{-1}(S)$ is an embedded submanifold of codimension k. $\square$

For example, in $\mathbb{R}^3$, this theorem shows that a smooth curve and a smooth surface intersecting transversely have only isolated points in their intersection, while two smooth surfaces intersect transversely in a smooth curve. Two smooth curves in $\mathbb{R}^3$ intersect transversely if and only if their intersection is empty, because at any intersection point, the two one-dimensional tangent spaces to the curves would have to span the tangent space to $\mathbb{R}^3$.

Because a submersion is transverse to every embedded submanifold, the next corollary is immediate.

Corollary 6.31. *Suppose N and M are smooth manifolds, $S \subseteq M$ is an embedded submanifold of codimension k, and $F: N \to M$ is a submersion. Then $F^{-1}(S)$ is an embedded codimension-k submanifold of N.* $\square$

Transversality also provides a convenient criterion for recognizing a submanifold as a graph. The next theorem is a global version of the implicit function theorem.

Theorem 6.32 (Global Characterization of Graphs). *Suppose M and N are smooth manifolds and $S \subseteq M \times N$ is an immersed submanifold. Let π_M and π_N denote the projections from $M \times N$ onto M and N, respectively. The following are equivalent.*

(a) *S is the graph of a smooth map $f: M \to N$.*

(b) $\pi_M\big|_S$ is a diffeomorphism from S onto M.
(c) For each $p \in M$, the submanifolds S and $\{p\} \times N$ intersect transversely in exactly one point.

If these conditions hold, then S is the graph of the map $f : M \to N$ defined by $f = \pi_N \circ (\pi_M\big|_S)^{-1}$.

Proof. Problem 6-15.   □

Corollary 6.33 (Local Characterization of Graphs). *Suppose M and N are smooth manifolds, $S \subseteq M \times N$ is an immersed submanifold, and $(p, q) \in S$. If S intersects the submanifold $\{p\} \times N$ transversely at (p, q), then there exist a neighborhood U of p in M and a neighborhood V of (p, q) in S such that V is the graph of a smooth map $f : U \to N$.*

Proof. The hypothesis guarantees that $d(\pi_M)_{(p,q)} \colon T_{(p,q)}S \to T_p M$ is an isomorphism, so $\pi_M\big|_S$ restricts to a diffeomorphism from a neighborhood V of (p, q) in S to a neighborhood U of p. The result then follows from Theorem 6.32(b).   □

The surprising thing about transversely intersecting submanifolds and transverse maps is that they are "generic," as we will soon see. To set the stage, we need to consider families of maps that are somewhat more general than smooth homotopies.

Suppose N, M, and S are smooth manifolds, and for each $s \in S$ we are given a map $F_s \colon N \to M$. The collection $\{F_s : s \in S\}$ is called a ***smooth family of maps*** if the map $F \colon M \times S \to N$ defined by $F(x, s) = F_s(x)$ is smooth. You should think of such a family as a higher-dimensional analogue of a homotopy. The next proposition shows how such families are related to ordinary homotopies.

Proposition 6.34. *If $\{F_s : s \in S\}$ is a smooth family of maps from N to M and S is connected, then for any $s_1, s_2 \in S$, the maps $F_{s_1}, F_{s_2} \colon N \to M$ are homotopic.*

Proof. Because S is connected, it is path-connected. If $\gamma \colon [0, 1] \to S$ is any path from s_1 to s_2, then $H(x, s) = F(x, \gamma(s))$ is a homotopy from F_{s_1} to F_{s_2}.   □

The key to finding transverse maps is the following application of Sard's theorem, which gives a simple sufficient condition for a family of smooth maps to contain at least one map that is transverse to a given submanifold. If S is a smooth manifold and $B \subseteq S$ is a subset whose complement has measure zero in S, we say that B contains ***almost every element of S***.

Theorem 6.35 (Parametric Transversality Theorem). *Suppose N and M are smooth manifolds, $X \subseteq M$ is an embedded submanifold, and $\{F_s : s \in S\}$ is a smooth family of maps from N to M. If the map $F \colon N \times S \to M$ is transverse to X, then for almost every $s \in S$, the map $F_s \colon N \to M$ is transverse to X.*

Proof. The hypothesis implies that $W = F^{-1}(X)$ is an embedded submanifold of $N \times S$ by Theorem 6.30. Let $\pi \colon N \times S \to S$ be the projection onto the second factor. What we will actually show is that if $s \in S$ is a regular value of the restriction $\pi\big|_W$, then F_s is transverse to X. Since almost every s is a regular value by Sard's theorem, this proves the theorem.

Suppose $s \in S$ is a regular value of $\pi|_W$. Let $p \in F_s^{-1}(X)$ be arbitrary, and set $q = F_s(p) \in X$. We need to show that $T_q M = T_q X + d(F_s)(T_p N)$. Here is what we know. First, because of our hypothesis on F,

$$T_q M = T_q X + dF\big(T_{(p,s)}(N \times S)\big). \tag{6.7}$$

Second, because s is a regular value and $(p,s) \in W$,

$$T_s S = d\pi\big(T_{(p,s)} W\big). \tag{6.8}$$

Third, by the result of Problem 6-10, we have $T_{(p,s)} W = \big(dF_{(p,s)}\big)^{-1}(T_q X)$, which implies

$$dF\big(T_{(p,s)} W\big) = T_q X. \tag{6.9}$$

Now let $w \in T_q M$ be arbitrary. We need to find $v \in T_q X$ and $y \in T_p N$ such that

$$w = v + d(F_s)(y). \tag{6.10}$$

Because of (6.7), there exist $v_1 \in T_q X$ and $(y_1, z_1) \in T_p N \times T_s S \cong T_{(p,s)}(N \times S)$ such that

$$w = v_1 + dF(y_1, z_1). \tag{6.11}$$

By (6.8), there exists $(y_2, z_2) \in T_{(p,s)} W$ such that $d\pi(y_2, z_2) = z_1$. Since π is a projection, this means $z_2 = z_1$. By linearity, we can write

$$dF(y_1, z_1) = dF(y_2, z_1) + dF(y_1 - y_2, 0).$$

On the one hand, (6.9) implies $dF(y_2, z_1) = dF(y_2, z_2) \in dF\big(T_{(p,s)} W\big) = T_q X$. On the other hand, if $\iota_s : N \to N \times S$ is the map $\iota_s(p') = (p', s)$, then we have $F_s = F \circ \iota_s$ and $d(\iota_s)(y_1 - y_2) = (y_1 - y_2, 0)$, and therefore $dF(y_1 - y_2, 0) = dF \circ d(\iota_s)(y_1 - y_2) = d(F_s)(y_1 - y_2)$. By virtue of (6.11), therefore, (6.10) is satisfied with $v = v_1 + dF(y_2, z_1)$ and $y = y_1 - y_2$, and the proof is complete. $\square$

In order to make use of the parametric transversality theorem, we need to construct a smooth family of maps satisfying the hypothesis. The proof of the next theorem shows that it is always possible to do so.

Theorem 6.36 (Transversality Homotopy Theorem). *Suppose M and N are smooth manifolds and $X \subseteq M$ is an embedded submanifold. Every smooth map $f: N \to M$ is homotopic to a smooth map $g: N \to M$ that is transverse to X.*

Proof. The crux of the proof is constructing a smooth map $F: N \times S \to M$ that is transverse to X, where $S = \mathbb{B}^k$ for some k and $F_0 = f$. It then follows from the parametric transversality theorem that there is some $s \in S$ such that $F_s: N \to M$ is transverse to X, and from Proposition 6.34 that F_s is homotopic to f.

By the Whitney embedding theorem, we can assume that M is a properly embedded submanifold of $\mathbb{R}^k$ for some k. Let U be a tubular neighborhood of M in $\mathbb{R}^k$, and let $r: U \to M$ be a smooth retraction that is also a smooth submersion. If we define $\delta\colon M \to \mathbb{R}^+$ by (6.6), Corollary 6.22 shows that there exists a smooth function $e\colon N \to \mathbb{R}^+$ that satisfies $0 < e(p) < \delta(f(p))$ everywhere. Let S be the unit ball in $\mathbb{R}^k$, and define $F\colon N \times S \to M$ by

$$F(p,s) = r(f(p) + e(p)s).$$

Note that $|e(p)s| < e(p) < \delta(f(p))$, which implies that $f(p) + e(p)s \in U$, so F is well defined. Clearly, F is smooth, and $F_0 = f$ because r is a retraction.

For each $p \in N$, the restriction of F to $\{p\} \times S$ is the composition of the local diffeomorphism $s \mapsto f(p) + e(p)s$ followed by the smooth submersion r, so F is a smooth submersion and hence transverse to X. $\qquad\square$

Problems

6-1. Use Proposition 6.5 to give a simpler proof of Corollary 6.11 that does not use Sard's theorem. [Hint: given a smooth map $F\colon M \to N$, define a suitable map from $M \times \mathbb{R}^k$ to N, where $k = \dim N - \dim M$.]

6-2. Prove Theorem 6.18 (the Whitney immersion theorem) in the special case $\partial M = \varnothing$. [Hint: without loss of generality, assume that M is an embedded n-dimensional submanifold of $\mathbb{R}^{2n+1}$. Let $UM \subseteq T\mathbb{R}^{2n+1}$ be the unit tangent bundle of M (Problem 5-6), and let $G\colon UM \to \mathbb{RP}^{2n}$ be the map $G(x, v) = [v]$. Use Sard's theorem to conclude that there is some $v \in \mathbb{R}^{2n+1} \smallsetminus \mathbb{R}^{2n}$ such that $[v]$ is not in the image of G, and show that the projection from $\mathbb{R}^{2n+1}$ to $\mathbb{R}^{2n}$ with kernel $\mathbb{R}v$ restricts to an immersion of M into $\mathbb{R}^{2n}$.]

6-3. Let M be a smooth manifold, let $B \subseteq M$ be a closed subset, and let $\delta\colon M \to \mathbb{R}$ be a positive continuous function. Show that there is a smooth function $\tilde{\delta}\colon M \to \mathbb{R}$ that is zero on B, positive on $M \smallsetminus B$, and satisfies $\tilde{\delta}(x) < \delta(x)$ everywhere. [Hint: consider $f/(f + 1)$, where f is a smooth nonnegative function that vanishes exactly on B, and use Corollary 6.22.]

6-4. Let M be a smooth manifold, let B be a closed subset of M, and let $\delta\colon M \to \mathbb{R}$ be a positive continuous function.
   (a) Given any continuous function $f\colon M \to \mathbb{R}^k$, show that there is a continuous function $\tilde{f}\colon M \to \mathbb{R}^k$ that is smooth on $M \smallsetminus B$, agrees with f on B, and is δ-close to f. [Hint: use Problem 6-3.]
   (b) Given a smooth manifold N and a continuous map $F\colon M \to N$, show that F is homotopic relative to B to a map that is smooth on $M \smallsetminus B$.

6-5. Let $M \subseteq \mathbb{R}^n$ be an embedded submanifold. Show that M has a tubular neighborhood U with the following property: for each $y \in U$, $r(y)$ is the unique point in M closest to y, where $r\colon U \to M$ is the retraction defined in Proposition 6.25. [Hint: first show that if $y \in \mathbb{R}^n$ has a closest point $x \in M$, then $(y - x) \perp T_x M$. Then, using the notation of the proof of Theorem 6.24, show that for each $x \in M$, it is possible to choose $\delta > 0$ such that every $y \in E(V_\delta(x))$ has a closest point in M, and that point is equal to $r(y)$.]

6-6. Suppose $M \subseteq \mathbb{R}^n$ is a compact embedded submanifold. For any $\varepsilon > 0$, let M_ε be the set of points in $\mathbb{R}^n$ whose distance from M is less than ε. Show that for sufficiently small ε, ∂M_ε is a compact embedded hypersurface in $\mathbb{R}^n$, and $\overline{M}_\varepsilon$ is a compact regular domain in $\mathbb{R}^n$ whose interior contains M.

6-7. By considering the map $F\colon \mathbb{R} \to \mathbb{H}^2$ given by $F(t) = (t, |t|)$ and the subset $A = [0, \infty) \subseteq \mathbb{R}$, show that the conclusions of Theorem 6.26 and Corollary 6.27 can be false when M has nonempty boundary.

6-8. Prove that every proper continuous map between smooth manifolds is homotopic to a proper smooth map. [Hint: show that the map $\widetilde{F}$ constructed in the proof of Theorem 6.26 is proper if F is.]

6-9. Let $F\colon \mathbb{R}^2 \to \mathbb{R}^3$ be the map $F(x, y) = (e^y \cos x, e^y \sin x, e^{-y})$. For which positive numbers r is F transverse to the sphere $S_r(0) \subseteq \mathbb{R}^3$? For which positive numbers r is $F^{-1}(S_r(0))$ an embedded submanifold of $\mathbb{R}^2$?

6-10. Suppose $F\colon N \to M$ is a smooth map that is transverse to an embedded submanifold $X \subseteq M$, and let $W = F^{-1}(X)$. For each $p \in W$, show that $T_p W = (dF_p)^{-1}(T_{F(p)} X)$. Conclude that if two embedded submanifolds $X, X' \subseteq M$ intersect transversely, then $T_p(X \cap X') = T_p X \cap T_p X'$ for every $p \in X \cap X'$. (*Used on p. 146.*)

6-11. Suppose $F\colon M \to N$ and $G\colon N \to P$ are smooth maps, and G is transverse to an embedded submanifold $X \subseteq P$. Show that F is transverse to the submanifold $G^{-1}(X)$ if and only if $G \circ F$ is transverse to X.

6-12. Let M be a compact smooth n-manifold. Prove that if $N \geq 2n$, every smooth map from M to $\mathbb{R}^N$ can be uniformly approximated by smooth immersions.

6-13. Let M be a smooth manifold. In this chapter, we defined what it means for two embedded submanifolds of M to intersect transversely, and for a smooth map into M to be transverse to an embedded submanifold. More generally, if $F\colon N \to M$ and $F'\colon N' \to M$ are smooth maps into M, we say that F and F' are **transverse to each other** if for every $x \in N$ and $x' \in N'$ such that $F(x) = F'(x')$, the spaces $dF_x(T_x N)$ and $dF'_{x'}(T_{x'} N')$ together span $T_{F(x)} M$. Prove the following statements.

(a) With N, N', F, F' as above, F and F' are transverse to each other if and only if the map $F \times F'\colon N \times N' \to M \times M$ is transverse to the diagonal $\Delta_M = \{(x, x) : x \in M\}$.

(b) If S is an embedded submanifold of M, a smooth map $F\colon N \to M$ is transverse to S if and only if it is transverse to the inclusion $\iota\colon S \hookrightarrow M$.

(c) If $F\colon N \to M$ and $F'\colon N' \to M$ are smooth maps that are transverse to each other, then $F^{-1}(F'(N'))$ is an embedded submanifold of N of dimension equal to $\dim N + \dim N' - \dim M$.

6-14. This problem illustrates how badly Theorem 6.30 can fail if the transversality hypothesis is removed. Let $S = \mathbb{R}^n \times \{0\} \subseteq \mathbb{R}^{n+1}$, and suppose A is an arbitrary closed subset of S. Prove that there is a properly embedded hypersurface $S' \subseteq \mathbb{R}^{n+1}$ such that $S \cap S' = A$. [Hint: use Theorem 2.29.]

6-15. Prove Theorem 6.32 (global characterization of graphs).

6-16. Suppose M and N are smooth manifolds. A class $\mathcal{F}$ of smooth maps from N to M is said to be **stable** if it has the following property: whenever $\{F_s : s \in S\}$ is a smooth family of maps from N to M, and $F_{s_0} \in \mathcal{F}$ for some $s_0 \in S$, then there is a neighborhood U of s_0 in S such that $F_s \in \mathcal{F}$ for all $s \in U$. (Roughly speaking, a property of smooth maps is stable if it persists under small deformations.) Prove that if N is compact, then the following classes of smooth maps from N to M are stable:
   (a) immersions
   (b) submersions
   (c) embeddings
   (d) diffeomorphisms
   (e) local diffeomorphisms
   (f) maps that are transverse to a given properly embedded submanifold $X \subseteq M$

6-17. Let $\varphi \colon \mathbb{R} \to \mathbb{R}$ be a compactly supported smooth function such that $\varphi(0) = 1$. Use the family $\{F_s : s \in \mathbb{R}\}$ of maps from $\mathbb{R}$ to $\mathbb{R}$ given by $F_s(x) = x\varphi(sx)$ to show that the classes of maps described in Problem 6-16 need not be stable when N is not compact.

Chapter 7
Lie Groups

In this chapter we introduce *Lie groups*, which are smooth manifolds that are also groups in which multiplication and inversion are smooth maps. Besides providing many examples of interesting manifolds themselves, they are essential tools in the study of more general manifolds, primarily because of the role they play as groups of symmetries of other manifolds.

Our aim in this chapter is to introduce Lie groups and some of the tools for working with them, and to describe an abundant supply of examples. In subsequent chapters (especially Chapters 8, 20, and 21), we will develop many more properties and applications of Lie groups.

We begin with the definition of Lie groups and some of the basic structures associated with them, and then present a number of examples. Next we study *Lie group homomorphisms*, which are group homomorphisms that are also smooth maps. Then we introduce *Lie subgroups* (subgroups that are also smooth submanifolds), which lead to a number of new examples of Lie groups.

After explaining these basic ideas, we introduce actions of Lie groups on manifolds, which are the primary *raison d'être* of Lie groups. At the end of the chapter, we briefly touch on group representations.

The study of Lie groups was initiated in the late nineteenth century by the Norwegian mathematician Sophus Lie. Inspired by the way the French algebraist Évariste Galois had invented group theory and used it to analyze polynomial equations, Lie was interested in using symmetries, expressed in the form of group actions, to simplify problems in partial differential equations and geometry. However, Lie could not have conceived of the global objects that we now call Lie groups, for the simple reason that global topological notions such as manifolds (or even topological spaces!) had not yet been formulated. What Lie studied was essentially a local-coordinate version of Lie groups, now called *local Lie groups*. Despite the limitations imposed by the era in which he lived, he was able to lay much of the groundwork for our current understanding of Lie groups. We will describe his principal results in Chapter 20 (see Theorem 20.16).

J.M. Lee, *Introduction to Smooth Manifolds*, Graduate Texts in Mathematics 218, 150
DOI 10.1007/978-1-4419-9982-5_7, © Springer Science+Business Media New York 2013

Basic Definitions

A **Lie group** is a smooth manifold G (without boundary) that is also a group in the algebraic sense, with the property that the multiplication map $m\colon G \times G \to G$ and inversion map $i\colon G \to G$, given by

$$m(g,h) = gh, \qquad i(g) = g^{-1},$$

are both smooth. A Lie group is, in particular, a **topological group** (a topological space with a group structure such that the multiplication and inversion maps are continuous).

The group operation in an arbitrary Lie group is denoted by juxtaposition, except in certain abelian groups such as $\mathbb{R}^n$ in which the operation is usually written additively. It is traditional to denote the identity element of an arbitrary Lie group by the symbol e (for German *Einselement*, "unit element"), and we follow this convention, except in specific examples in which there are more common notations (such as I_n for the identity matrix in a matrix group, or 0 for the identity element in $\mathbb{R}^n$).

The following alternative characterization of the smoothness condition is sometimes useful. (See also Problem 7-3 for a stronger result.)

Proposition 7.1. *If G is a smooth manifold with a group structure such that the map $G \times G \to G$ given by $(g, h) \mapsto gh^{-1}$ is smooth, then G is a Lie group.*

▶ **Exercise 7.2.** Prove Proposition 7.1.

If G is a Lie group, any element $g \in G$ defines maps $L_g, R_g\colon G \to G$, called *left translation* and *right translation*, respectively, by

$$L_g(h) = gh, \qquad R_g(h) = hg.$$

Because L_g can be expressed as the composition of smooth maps

$$G \xrightarrow{\iota_g} G \times G \xrightarrow{m} G,$$

where $\iota_g(h) = (g, h)$ and m is multiplication, it follows that L_g is smooth. It is actually a diffeomorphism of G, because $L_{g^{-1}}$ is a smooth inverse for it. Similarly, $R_g\colon G \to G$ is a diffeomorphism. As we will see repeatedly below, many of the important properties of Lie groups follow from the fact that we can systematically map any point to any other by such a global diffeomorphism.

Example 7.3 (Lie Groups). Each of the following manifolds is a Lie group with the indicated group operation.

(a) The **general linear group** $\mathrm{GL}(n, \mathbb{R})$ is the set of invertible $n \times n$ matrices with real entries. It is a group under matrix multiplication, and it is an open submanifold of the vector space $\mathrm{M}(n, \mathbb{R})$, as we observed in Example 1.27. Multiplication is smooth because the matrix entries of a product matrix AB are polynomials in the entries of A and B. Inversion is smooth by Cramer's rule.

(b) Let $GL^+(n, \mathbb{R})$ denote the subset of $GL(n, \mathbb{R})$ consisting of matrices with positive determinant. Because $\det(AB) = (\det A)(\det B)$ and $\det(A^{-1}) = 1/\det A$, it is a subgroup of $GL(n, \mathbb{R})$; and because it is the preimage of $(0, \infty)$ under the continuous determinant function, it is an open subset of $GL(n, \mathbb{R})$ and therefore an n^2-dimensional manifold. The group operations are the restrictions of those of $GL(n, \mathbb{R})$, so they are smooth. Thus $GL^+(n, \mathbb{R})$ is a Lie group.

(c) Suppose G is an arbitrary Lie group and $H \subseteq G$ is an **open subgroup** (a subgroup that is also an open subset). By the same argument as in part (b), H is a Lie group with the inherited group structure and smooth manifold structure.

(d) The **complex general linear group** $GL(n, \mathbb{C})$ is the group of invertible complex $n \times n$ matrices under matrix multiplication. It is an open submanifold of $M(n, \mathbb{C})$ and thus a $2n^2$-dimensional smooth manifold, and it is a Lie group because matrix products and inverses are smooth functions of the real and imaginary parts of the matrix entries.

(e) If V is any real or complex vector space, $GL(V)$ denotes the set of invertible linear maps from V to itself. It is a group under composition. If V has finite dimension n, any basis for V determines an isomorphism of $GL(V)$ with $GL(n, \mathbb{R})$ or $GL(n, \mathbb{C})$, so $GL(V)$ is a Lie group. The transition map between any two such isomorphisms is given by a map of the form $A \mapsto BAB^{-1}$ (where B is the transition matrix between the two bases), which is smooth. Thus, the smooth manifold structure on $GL(V)$ is independent of the choice of basis.

(f) The real number field $\mathbb{R}$ and Euclidean space $\mathbb{R}^n$ are Lie groups under addition, because the coordinates of $x - y$ are smooth (linear!) functions of (x, y).

(g) Similarly, $\mathbb{C}$ and $\mathbb{C}^n$ are Lie groups under addition.

(h) The set $\mathbb{R}^*$ of nonzero real numbers is a 1-dimensional Lie group under multiplication. (In fact, it is exactly $GL(1, \mathbb{R})$ if we identify a 1×1 matrix with the corresponding real number.) The subset $\mathbb{R}^+$ of positive real numbers is an open subgroup, and is thus itself a 1-dimensional Lie group.

(i) The set $\mathbb{C}^*$ of nonzero complex numbers is a 2-dimensional Lie group under complex multiplication, which can be identified with $GL(1, \mathbb{C})$.

(j) The circle $\mathbb{S}^1 \subseteq \mathbb{C}^*$ is a smooth manifold and a group under complex multiplication. With appropriate angle functions as local coordinates on open subsets of $\mathbb{S}^1$ (see Problem 1-8), multiplication and inversion have the smooth coordinate expressions $(\theta_1, \theta_2) \mapsto \theta_1 + \theta_2$ and $\theta \mapsto -\theta$, and therefore $\mathbb{S}^1$ is a Lie group, called the **circle group**.

(k) Given Lie groups $G_1, \ldots, G_k$, their **direct product** is the product manifold $G_1 \times \cdots \times G_k$ with the group structure given by componentwise multiplication:

$$(g_1, \ldots, g_k)(g_1', \ldots, g_k') = (g_1 g_1', \ldots, g_k g_k').$$

It is a Lie group, as you can easily check.

(l) The n-torus $\mathbb{T}^n = \mathbb{S}^1 \times \cdots \times \mathbb{S}^1$ is an n-dimensional abelian Lie group.

(m) Any group with the discrete topology is a topological group, called a **discrete group**. If in addition the group is finite or countably infinite, then it is a zero-dimensional Lie group, called a **discrete Lie group**.   //

Lie Group Homomorphisms

If G and H are Lie groups, a **Lie group homomorphism from G to H** is a smooth map $F: G \to H$ that is also a group homomorphism. It is called a **Lie group isomorphism** if it is also a diffeomorphism, which implies that it has an inverse that is also a Lie group homomorphism. In this case we say that G and H are *isomorphic Lie groups*.

Example 7.4 (Lie Group Homomorphisms).

(a) The inclusion map $\mathbb{S}^1 \hookrightarrow \mathbb{C}^*$ is a Lie group homomorphism.

(b) Considering $\mathbb{R}$ as a Lie group under addition, and $\mathbb{R}^*$ as a Lie group under multiplication, the map $\exp: \mathbb{R} \to \mathbb{R}^*$ given by $\exp(t) = e^t$ is smooth, and is a Lie group homomorphism because $e^{(s+t)} = e^s e^t$. The image of $\exp$ is the open subgroup $\mathbb{R}^+$ consisting of positive real numbers, and $\exp: \mathbb{R} \to \mathbb{R}^+$ is a Lie group isomorphism with inverse $\log: \mathbb{R}^+ \to \mathbb{R}$.

(c) Similarly, $\exp: \mathbb{C} \to \mathbb{C}^*$ given by $\exp(z) = e^z$ is a Lie group homomorphism. It is surjective but not injective, because its kernel consists of the complex numbers of the form $2\pi i k$, where k is an integer.

(d) The map $\varepsilon: \mathbb{R} \to \mathbb{S}^1$ defined by $\varepsilon(t) = e^{2\pi i t}$ is a Lie group homomorphism whose kernel is the set $\mathbb{Z}$ of integers. Similarly, the map $\varepsilon^n: \mathbb{R}^n \to \mathbb{T}^n$ defined by $\varepsilon^n\left(x^1, \ldots, x^n\right) = \left(e^{2\pi i x^1}, \ldots, e^{2\pi i x^n}\right)$ is a Lie group homomorphism whose kernel is $\mathbb{Z}^n$.

(e) The determinant function $\det: \mathrm{GL}(n, \mathbb{R}) \to \mathbb{R}^*$ is smooth because $\det A$ is a polynomial in the matrix entries of A. It is a Lie group homomorphism because $\det(AB) = (\det A)(\det B)$. Similarly, $\det: \mathrm{GL}(n, \mathbb{C}) \to \mathbb{C}^*$ is a Lie group homomorphism.

(f) If G is a Lie group and $g \in G$, *conjugation by g* is the map $C_g: G \to G$ given by $C_g(h) = ghg^{-1}$. Because group multiplication and inversion are smooth, C_g is smooth, and a simple computation shows that it is a group homomorphism. In fact, it is an isomorphism, because it has $C_{g^{-1}}$ as an inverse. A subgroup $H \subseteq G$ is said to be *normal* if $C_g(H) = H$ for every $g \in G$. //

The next theorem is important for understanding many of the properties of Lie group homomorphisms.

Theorem 7.5. *Every Lie group homomorphism has constant rank.*

Proof. Let $F: G \to H$ be a Lie group homomorphism, and let e and $\tilde{e}$ denote the identity elements of G and H, respectively. Suppose g_0 is an arbitrary element of G. We will show that dF_{g_0} has the same rank as dF_e. The fact that F is a homomorphism means that for all $g \in G$,

$$F\left(L_{g_0}(g)\right) = F(g_0 g) = F(g_0)F(g) = L_{F(g_0)}\left(F(g)\right),$$

or in other words, $F \circ L_{g_0} = L_{F(g_0)} \circ F$. Taking differentials of both sides at the identity and using Proposition 3.6(b), we find that

$$dF_{g_0} \circ d\left(L_{g_0}\right)_e = d\left(L_{F(g_0)}\right)_{\tilde{e}} \circ dF_e.$$

Left multiplication by any element of a Lie group is a diffeomorphism, so both $d(L_{g_0})_e$ and $d(L_{F(g_0)})_{\tilde{e}}$ are isomorphisms. Because composing with an isomorphism does not change the rank of a linear map, it follows that dF_{g_0} and dF_e have the same rank. □

Corollary 7.6. *A Lie group homomorphism is a Lie group isomorphism if and only if it is bijective.*

Proof. The global rank theorem shows that a bijective Lie group homomorphism is a diffeomorphism. □

The Universal Covering Group

Covering space theory yields the following important result about Lie groups.

Theorem 7.7 (Existence of a Universal Covering Group). *Let G be a connected Lie group. There exists a simply connected Lie group $\widetilde{G}$, called the **universal covering group of G**, that admits a smooth covering map $\pi\colon \widetilde{G} \to G$ that is also a Lie group homomorphism.*

Proof. Let $\widetilde{G}$ be the universal covering manifold of G and $\pi\colon \widetilde{G} \to G$ be the corresponding smooth covering map. By Exercise 4.38, $\pi \times \pi\colon \widetilde{G} \times \widetilde{G} \to G \times G$ is also a smooth covering map.

Let $m\colon G \times G \to G$ and $i\colon G \to G$ denote the multiplication and inversion maps of G, respectively, and let $\tilde{e}$ be an arbitrary element of the fiber $\pi^{-1}(e) \subseteq \widetilde{G}$. Since $\widetilde{G}$ is simply connected, the lifting criterion for covering maps (Proposition A.78) guarantees that the map $m \circ (\pi \times \pi)\colon \widetilde{G} \times \widetilde{G} \to G$ has a unique continuous lift $\tilde{m}\colon \widetilde{G} \times \widetilde{G} \to \widetilde{G}$ satisfying $\tilde{m}(\tilde{e}, \tilde{e}) = \tilde{e}$ and $\pi \circ \tilde{m} = m \circ (\pi \times \pi)$:

$$
\begin{array}{ccc}
\widetilde{G} \times \widetilde{G} & \xrightarrow{\ \tilde{m}\ } & \widetilde{G} \\
{\scriptstyle \pi \times \pi}\big\downarrow & & \big\downarrow{\scriptstyle \pi} \\
G \times G & \xrightarrow[\ m\ ]{} & G.
\end{array}
\tag{7.1}
$$

Because π is a local diffeomorphism and $\pi \circ \tilde{m} = m \circ (\pi \times \pi)$ is smooth, it follows from Exercise 4.10(a) that $\tilde{m}$ is smooth. By the same reasoning, $i \circ \pi\colon \widetilde{G} \to G$ has a smooth lift $\tilde{\imath}\colon \widetilde{G} \to \widetilde{G}$ satisfying $\tilde{\imath}(\tilde{e}) = \tilde{e}$ and $\pi \circ \tilde{\imath} = i \circ \pi$:

$$
\begin{array}{ccc}
\widetilde{G} & \xrightarrow{\ \tilde{\imath}\ } & \widetilde{G} \\
{\scriptstyle \pi}\big\downarrow & & \big\downarrow{\scriptstyle \pi} \\
G & \xrightarrow[\ i\ ]{} & G.
\end{array}
\tag{7.2}
$$

We define multiplication and inversion in $\widetilde{G}$ by $xy = \tilde{m}(x, y)$ and $x^{-1} = \tilde{\imath}(x)$ for all $x, y \in \widetilde{G}$. Then (7.1) and (7.2) can be rewritten as

$$
\pi(xy) = \pi(x)\pi(y),
\tag{7.3}
$$

$$\pi\left(x^{-1}\right) = \pi(x)^{-1}. \tag{7.4}$$

It remains only to show that $\tilde{G}$ is a group with these operations, for then it is a Lie group because $\tilde{m}$ and $\tilde{\imath}$ are smooth, and (7.3) shows that π is a homomorphism.

First we show that $\tilde{e}$ is an identity for multiplication in $\tilde{G}$. Consider the map $f: \tilde{G} \to \tilde{G}$ defined by $f(x) = \tilde{e}x$. Then (7.3) implies that $\pi \circ f(x) = \pi(\tilde{e})\pi(x) = e\pi(x) = \pi(x)$, so f is a lift of $\pi: \tilde{G} \to G$. The identity map $\mathrm{Id}_{\tilde{G}}$ is another lift of π, and it agrees with f at a point because $f(\tilde{e}) = \tilde{m}(\tilde{e}, \tilde{e}) = \tilde{e}$, so the unique lifting property of covering maps (Proposition A.77(a)) implies that $f = \mathrm{Id}_{\tilde{G}}$, or equivalently, $\tilde{e}x = x$ for all $x \in \tilde{G}$. The same argument shows that $x\tilde{e} = x$.

Next, to show that multiplication in $\tilde{G}$ is associative, consider the two maps $\alpha_L, \alpha_R: \tilde{G} \times \tilde{G} \times \tilde{G} \to \tilde{G}$ defined by

$$\alpha_L(x, y, z) = (xy)z, \qquad \alpha_R(x, y, z) = x(yz).$$

Then (7.3) applied repeatedly implies that

$$\pi \circ \alpha_L(x, y, z) = \left(\pi(x)\pi(y)\right)\pi(z) = \pi(x)\left(\pi(y)\pi(z)\right) = \pi \circ \alpha_R(x, y, z),$$

so α_L and α_R are both lifts of the same map $\alpha(x, y, z) = \pi(x)\pi(y)\pi(z)$. Because α_L and α_R agree at $(\tilde{e}, \tilde{e}, \tilde{e})$, they are equal. A similar argument shows that $x^{-1}x = xx^{-1} = \tilde{e}$, so $\tilde{G}$ is a group. $\qquad\square$

▶ **Exercise 7.8.** Complete the proof of the preceding theorem by showing that $x^{-1}x = xx^{-1} = \tilde{e}$.

We also have the following uniqueness result.

Theorem 7.9 (Uniqueness of the Universal Covering Group). *For any connected Lie group G, the universal covering group is unique in the following sense: if $\tilde{G}$ and $\tilde{G}'$ are simply connected Lie groups that admit smooth covering maps $\pi: \tilde{G} \to G$ and $\pi': \tilde{G}' \to G$ that are also Lie group homomorphisms, then there exists a Lie group isomorphism $\Phi: \tilde{G} \to \tilde{G}'$ such that $\pi' \circ \Phi = \pi$.*

Proof. See Problem 7-5. $\qquad\square$

Example 7.10 (Universal Covering Groups).

(a) For each n, the map $\varepsilon^n: \mathbb{R}^n \to \mathbb{T}^n$ given by

$$\varepsilon^n\left(x^1, \ldots, x^n\right) = \left(e^{2\pi i x^1}, \ldots, e^{2\pi i x^n}\right)$$

is a Lie group homomorphism and a smooth covering map (see Example 7.4(d)). Since $\mathbb{R}^n$ is simply connected, this shows that the universal covering group of $\mathbb{T}^n$ is the additive Lie group $\mathbb{R}^n$.

(b) The Lie group homomorphism $\exp: \mathbb{C} \to \mathbb{C}^*$ described in Example 7.4(c) is also a smooth covering map, so $\mathbb{C}$ is the universal covering group of $\mathbb{C}^*$. $\qquad$ //

Lie Subgroups

Suppose G is a Lie group. A *Lie subgroup of G* is a subgroup of G endowed with a topology and smooth structure making it into a Lie group and an immersed submanifold of G. The following proposition shows that *embedded* subgroups are automatically Lie subgroups.

Proposition 7.11. *Let G be a Lie group, and suppose $H \subseteq G$ is a subgroup that is also an embedded submanifold. Then H is a Lie subgroup.*

Proof. We need only check that multiplication $H \times H \to H$ and inversion $H \to H$ are smooth maps. Because multiplication is a smooth map from $G \times G$ into G, its restriction is clearly smooth from $H \times H$ into G (this is true even if H is merely immersed). Because H is a subgroup, multiplication takes $H \times H$ into H, and since H is embedded, this is a smooth map into H by Corollary 5.30. A similar argument applies to inversion. This proves that H is a Lie subgroup. $\square$

The simplest examples of embedded Lie subgroups are the open subgroups. The following lemma shows that the possibilities for open subgroups are limited.

Lemma 7.12. *Suppose G is a Lie group and $H \subseteq G$ is an open subgroup. Then H is an embedded Lie subgroup. In addition, H is closed, so it is a union of connected components of G.*

Proof. If H is open in G, it is embedded by Proposition 5.1. In addition, every left coset $gH = \{gh : h \in H\}$ is open in G because it is the image of the open subset H under the diffeomorphism L_g. Because $G \smallsetminus H$ is the union of the cosets of H other than H itself, it is open, and therefore H is closed in G. Because H is both open and closed, it is a union of components. $\square$

If G is a group and $S \subseteq G$, the *subgroup generated by S* is the smallest subgroup containing S (i.e., the intersection of all subgroups containing S).

▶ **Exercise 7.13.** Given a group G and a subset $S \subseteq G$, show that the subgroup generated by S is equal to the set of all elements of G that can be expressed as finite products of elements of S and their inverses.

Proposition 7.14. *Suppose G is a Lie group, and $W \subseteq G$ is any neighborhood of the identity.*

(a) *W generates an open subgroup of G.*
(b) *If W is connected, it generates a connected open subgroup of G.*
(c) *If G is connected, then W generates G.*

Proof. Let $W \subseteq G$ be any neighborhood of the identity, and let H be the subgroup generated by W. As a matter of notation, if A and B are subsets of G, let us write

$$AB = \{ab : a \in A, b \in B\}, \qquad A^{-1} = \{a^{-1} : a \in A\}. \tag{7.5}$$

For each positive integer k, let W_k denote the set of all elements of G that can be expressed as products of k or fewer elements of $W \cup W^{-1}$. By Exercise 7.13, H is the union of all the sets W_k as k ranges over the positive integers. Now, W^{-1} is open because it is the image of W under the inversion map, which is a diffeomorphism. Thus, $W_1 = W \cup W^{-1}$ is open, and for each $k > 1$ we have

$$W_k = W_1 W_{k-1} = \bigcup_{g \in W_1} L_g(W_{k-1}).$$

Because each L_g is a diffeomorphism, it follows by induction that each W_k is open, and thus H is open.

Next suppose W is connected. Then W^{-1} is also connected because it is a diffeomorphic image of W, and $W_1 = W \cup W^{-1}$ is connected because it is a union of connected sets with the identity in common. Therefore, $W_2 = m(W_1 \times W_1)$ is connected because it is the image of a connected space under the continuous multiplication map m, and it follows by induction that $W_k = m(W_1 \times W_{k-1})$ is connected for each k. Thus, $H = \bigcup_k W_k$ is connected because it is a union of connected subsets with the identity in common.

Finally, assume G is connected. Since H is an open subgroup, it is also closed by Lemma 7.12, and it is not empty because it contains the identity. Thus $H = G$. $\square$

If G is a Lie group, the connected component of G containing the identity is called the ***identity component of G***.

Proposition 7.15. *Let G be a Lie group and let G_0 be its identity component. Then G_0 is a normal subgroup of G, and is the only connected open subgroup. Every connected component of G is diffeomorphic to G_0.*

Proof. Problem 7-7. $\square$

Now we move beyond the open subgroups to more general Lie subgroups. The following proposition shows how to produce many more examples of embedded Lie subgroups.

Proposition 7.16. *Let $F: G \to H$ be a Lie group homomorphism. The kernel of F is a properly embedded Lie subgroup of G, whose codimension is equal to the rank of F.*

Proof. Because F has constant rank, its kernel $F^{-1}(e)$ is a properly embedded submanifold of codimension equal to rank F. It is thus a Lie subgroup by Proposition 7.11. $\square$

Complementary to the preceding result about kernels is the following result about images. (In Chapter 21, we will prove the analogous result for images of arbitrary Lie group homomorphisms, not just injective ones; see Theorem 21.27.)

Proposition 7.17. *If $F: G \to H$ is an injective Lie group homomorphism, the image of F has a unique smooth manifold structure such that $F(G)$ is a Lie subgroup of H and $F: G \to F(G)$ is a Lie group isomorphism.*

Proof. Since a Lie group homomorphism has constant rank, it follows from the global rank theorem that F is a smooth immersion. Proposition 5.18 shows that

$F(G)$ has a unique smooth manifold structure such that it is an immersed submanifold of H and F is a diffeomorphism onto its image. It is a Lie group (because G is), and it is a subgroup for algebraic reasons, so it is a Lie subgroup. Because $F: G \to F(G)$ is a group isomorphism and a diffeomorphism, it is a Lie group isomorphism.   □

Example 7.18 (Embedded Lie Subgroups).

(a) The subgroup $GL^+(n, \mathbb{R}) \subseteq GL(n, \mathbb{R})$ described in Example 7.3(b) is an open subgroup and thus an embedded Lie subgroup.
(b) The circle $\mathbb{S}^1$ is an embedded Lie subgroup of $\mathbb{C}^*$ because it is a subgroup and an embedded submanifold.
(c) The set $SL(n, \mathbb{R})$ of $n \times n$ real matrices with determinant equal to 1 is called the **special linear group of degree n**. Because $SL(n, \mathbb{R})$ is the kernel of the Lie group homomorphism $\det: GL(n, \mathbb{R}) \to \mathbb{R}^*$, it is a properly embedded Lie subgroup. Because the determinant function is surjective, it is a smooth submersion by the global rank theorem, so $SL(n, \mathbb{R})$ has dimension $n^2 - 1$.
(d) Let n be a positive integer, and define a map $\beta: GL(n, \mathbb{C}) \to GL(2n, \mathbb{R})$ by replacing each complex matrix entry $a + ib$ with the 2×2 block $\begin{pmatrix} a & -b \\ b & a \end{pmatrix}$:

$$\beta \begin{pmatrix} a_1^1 + ib_1^1 \dots a_1^n + ib_1^n \\ \vdots \qquad\qquad \vdots \\ a_n^1 + ib_n^1 \dots a_n^n + ib_n^n \end{pmatrix} = \begin{pmatrix} a_1^1 & -b_1^1 & & a_1^n & -b_1^n \\ b_1^1 & a_1^1 & \cdots & b_1^n & a_1^n \\ \vdots & & & \vdots & \\ a_n^1 & -b_n^1 & \cdots & a_n^n & -b_n^n \\ b_n^1 & a_n^1 & & b_n^n & a_n^n \end{pmatrix}.$$

It is straightforward to verify that β is an injective Lie group homomorphism whose image is a properly embedded Lie subgroup of $GL(2n, \mathbb{R})$. Thus, $GL(n, \mathbb{C})$ is isomorphic to this Lie subgroup of $GL(2n, \mathbb{R})$. (You can check that β arises naturally from the identification of $\left(x^1 + iy^1, \dots, x^n + iy^n\right) \in \mathbb{C}^n$ with $\left(x^1, y^1, \dots, x^n, y^n\right) \in \mathbb{R}^{2n}$.)
(e) The subgroup $SL(n, \mathbb{C}) \subseteq GL(n, \mathbb{C})$ consisting of complex matrices of determinant 1 is called the **complex special linear group of degree n**. It is the kernel of the Lie group homomorphism $\det: GL(n, \mathbb{C}) \to \mathbb{C}^*$. This homomorphism is surjective, so it is a smooth submersion by the global rank theorem. Therefore, $SL(n, \mathbb{C}) = \mathrm{Ker}(\det)$ is a properly embedded Lie subgroup whose codimension is equal to $\dim \mathbb{C}^* = 2$ and whose dimension is therefore $2n^2 - 2$.   //

Finally, here is an example of a Lie subgroup that is not embedded.

Example 7.19 (A Dense Lie Subgroup of the Torus).
Let $H \subseteq \mathbb{T}^2$ be the dense submanifold of the torus that is the image of the immersion $\gamma: \mathbb{R} \to \mathbb{T}^2$ defined in Example 4.20. It is easy to check that γ is an injective Lie group homomorphism, and thus H is an immersed Lie subgroup of $\mathbb{T}^2$ by Proposition 7.17.   //

▶ **Exercise 7.20.** Let $S \subseteq \mathbb{T}^3$ be the image of the subgroup H of the preceding example under the obvious embedding $\mathbb{T}^2 \hookrightarrow \mathbb{T}^3$. Show that S is a Lie subgroup of

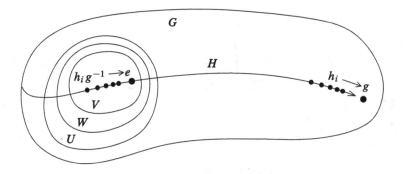

Fig. 7.1 An embedded Lie subgroup is closed

$\mathbb{T}^3$ that is not closed, embedded, or dense; but its closure is a properly embedded Lie subgroup of $\mathbb{T}^3$.

In Chapter 20, we will see that the subgroup $S \subseteq \mathbb{T}^3$ described in the preceding exercise is typical of nonembedded Lie subgroups: they are all dense subgroups of properly embedded Lie subgroups (see Problem 20-10).

In general, smooth submanifolds can be closed without being embedded (as is, for example, the figure-eight curve of Example 5.19) or embedded without being closed (as is the open unit ball in $\mathbb{R}^n$). However, as the next theorem shows, Lie subgroups have the remarkable property that closedness and embeddedness are not independent. This means that every embedded Lie subgroup is properly embedded.

Theorem 7.21. *Suppose G is a Lie group and $H \subseteq G$ is a Lie subgroup. Then H is closed in G if and only if it is embedded.*

Proof. Assume first that H is embedded in G. To prove that H is closed, let g be an arbitrary point of $\overline{H}$. Then there is a sequence (h_i) of points in H converging to g (Fig. 7.1). Let U be the domain of a slice chart for H containing the identity, and let W be a smaller neighborhood of e such that $\overline{W} \subseteq U$. By Problem 7-6, there is a neighborhood V of e with the property that $g_1 g_2^{-1} \in W$ whenever $g_1, g_2 \in V$.

Because $h_i g^{-1} \to e$, by discarding finitely many terms of the sequence we may assume that $h_i g^{-1} \in V$ for all i. This implies that

$$h_i h_j^{-1} = \left(h_i g^{-1}\right) \left(h_j g^{-1}\right)^{-1} \in W$$

for all i and j. Fixing j and letting $i \to \infty$, we find that $h_i h_j^{-1} \to g h_j^{-1} \in \overline{W} \subseteq U$. Since $H \cap U$ is a slice, it is closed in U, and therefore $g h_j^{-1} \in H$, which implies $g \in H$. Thus H is closed.

Conversely, assume H is a closed Lie subgroup, and let $m = \dim H$ and $n = \dim G$. We need to show that H is an embedded submanifold of G. If $m = n$, then H is embedded by Proposition 5.21(a), so we may assume henceforth that $m < n$.

It suffices to show that for some $h_1 \in H$, there is a neighborhood U_1 of h_1 in G such that $H \cap U_1$ is an embedded submanifold of U_1; for then if h is any other point

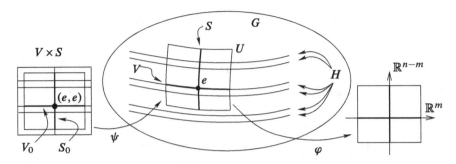

Fig. 7.2 Finding a slice chart

of H, right translation $R_{hh_1^{-1}}: G \to G$ is a diffeomorphism of G that takes H to H, and takes U_1 to a neighborhood U_1' of h such that $U_1' \cap H$ is embedded in U_1', so it follows from the local slice criterion that H is an embedded submanifold of G.

Because every immersed submanifold is locally embedded (Proposition 5.22), there exist a neighborhood V of e in H and a slice chart (U, φ) for V in G centered at e. By shrinking U if necessary, we may assume that it is a coordinate cube, and $U \cap V$ is the set of points whose coordinates are of the form $(x^1, \ldots, x^m, 0, \ldots, 0)$. Let $S \subseteq U$ be the set of points with coordinates of the form $(0, \ldots, 0, x^{m+1}, \ldots, x^n)$; it is the slice "perpendicular" to $U \cap V$ in these coordinates. Then S is an embedded submanifold of U and hence of G. Note that in these coordinates, $T_e V$ is spanned by the first m coordinate vectors and $T_e S$ by the last $n - m$, so $T_e G = T_e V \oplus T_e S$ (Fig. 7.2).

Now consider the map $\psi: V \times S \to G$ obtained by restricting group multiplication: $\psi(v, s) = vs$. Since $\psi(v, e) = v$ for $v \in V$ and $\psi(e, s) = s$ for $s \in S$, it follows easily that the differential of ψ at (e, e) satisfies $d\psi(X, 0) = X$ and $d\psi(0, Y) = Y$ for $X \in T_e V$, $Y \in T_e S$, and therefore $d\psi_{(e,e)}$ is bijective. By the inverse function theorem, there are connected neighborhoods W_0 of (e, e) in $V \times S$ and U_0 of e in G such that $\psi: W_0 \to U_0$ is a diffeomorphism. Shrinking the neighborhoods if necessary, we may assume that $W_0 = V_0 \times S_0$, where V_0 and S_0 are neighborhoods of e in V and S, respectively.

Let $K = S_0 \cap H$. There are two things we need to show about the set K:

(a) $\psi(V_0 \times K) = H \cap U_0$.
(b) K is a discrete set in the topology of H.

To prove (a), let $(v, s) \in V_0 \times S_0$ be arbitrary. Since H is a subgroup and $V_0 \subseteq H$, it follows that $vs \in H$ if and only if $s \in H$, which is to say that $\psi(v, s) \in H \cap U_0$ if and only if $(v, s) \in V_0 \times K$. To prove (b), suppose $h \in K$. Right translation $R_h: H \to H$ is a diffeomorphism of H taking e to h and taking V_0 to a neighborhood V_h of h in H. Note that $V_h = R_h(V_0) = \psi(V_0 \times \{h\})$, while $K = \psi(\{e\} \times K)$. Since ψ is injective on $V_0 \times S_0$, it follows that

$$V_h \cap K = \psi(\{e\} \times \{h\}) = \{h\}.$$

Thus each point $h \in K$ is isolated in H, which implies that K is discrete.

Since K is a discrete subset of the manifold H, it is countable, and since H is closed in G, it follows that $K = S_0 \cap H$ is closed in S_0. Thus, by Corollary A.59, there is a point $h_1 \in K$ that is isolated in S_0. (This step fails if H is not closed—for example, if H were a dense subgroup of the torus, then K would be dense in S_0.) This means there is a neighborhood S_1 of h_1 in S_0 such that $S_1 \cap H = \{h_1\}$. Then $U_1 = \psi(V_0 \times S_1)$ is a neighborhood of h_1 in G with the property that $U_1 \cap H$ is the slice $\psi(V_0 \times \{h_1\})$ in U_1. As explained at the beginning of the proof, the existence of such a neighborhood for one point of H implies that H is embedded. □

In Chapter 20, we will be able to prove a significantly strengthened form of this theorem, called the *closed subgroup theorem*, which asserts that every subgroup of a Lie group that is topologically a closed subset (but not assumed to be a submanifold) is automatically a properly embedded Lie subgroup.

Group Actions and Equivariant Maps

The most important applications of Lie groups to smooth manifold theory involve actions by Lie groups on other manifolds. If G is a group and M is a set, a *left action of G on M* is a map $G \times M \to M$, often written as $(g, p) \mapsto g \cdot p$, that satisfies

$$g_1 \cdot (g_2 \cdot p) = (g_1 g_2) \cdot p \quad \text{for all } g_1, g_2 \in G \text{ and } p \in M;$$
$$e \cdot p = p \qquad \text{for all } p \in M. \tag{7.6}$$

A *right action* is defined analogously as a map $M \times G \to M$ with the appropriate composition law:

$$(p \cdot g_1) \cdot g_2 = p \cdot (g_1 g_2) \quad \text{for all } g_1, g_2 \in G \text{ and } p \in M;$$
$$p \cdot e = p \qquad \text{for all } p \in M.$$

If M is a topological space and G is a topological group, an action of G on M is said to be a *continuous action* if the defining map $G \times M \to M$ or $M \times G \to M$ is continuous. In this case, M is said to be a (*left or right*) *G-space*. If in addition M is a smooth manifold with or without boundary, G is a Lie group, and the defining map is smooth, then the action is said to be a *smooth action*. We are primarily interested in smooth actions of Lie groups on smooth manifolds.

Sometimes it is useful to give a name to an action, such as $\theta \colon G \times M \to M$, with the action of a group element g on a point p usually written as $\theta_g(p)$. In terms of this notation, the conditions (7.6) for a left action read

$$\theta_{g_1} \circ \theta_{g_2} = \theta_{g_1 g_2},$$
$$\theta_e = \mathrm{Id}_M, \tag{7.7}$$

while for a right action the first equation is replaced by

$$\theta_{g_2} \circ \theta_{g_1} = \theta_{g_1 g_2}.$$

For a smooth action, each map $\theta_g \colon M \to M$ is a diffeomorphism, because $\theta_{g^{-1}}$ is a smooth inverse for it.

For left actions, we generally use the notations $g \cdot p$ and $\theta_g(p)$ interchangeably. The latter notation contains a bit more information, and is useful when it is important to specify the particular action under consideration, while the former is often more convenient when the action is understood. For right actions, the notation $p \cdot g$ is generally preferred because of the way composition works.

A right action can always be converted to a left action by the trick of defining $g \cdot p$ to be $p \cdot g^{-1}$, and a left action can similarly be converted to a right action. Thus, any results about left actions can be translated into results about right actions, and vice versa. We usually focus our attention on left actions, because their group law (7.7) has the property that multiplication of group elements corresponds to composition of maps. However, there are some circumstances in which right actions arise naturally; we will see several such actions later in the book.

Lie group actions typically arise in situations involving some kind of symmetry. For example, if M is a vector space or smooth manifold endowed with a metric or other geometric structure, the set of diffeomorphisms of M that preserve the structure (called the **symmetry group** of the structure) frequently turns out to be a Lie group acting smoothly on M.

Throughout the book, we use the following standard terminology regarding group actions. Suppose $\theta \colon G \times M \to M$ is a left action of a group G on a set M. (The definitions for right actions are similar. For these definitions, no continuity or smoothness assumption is necessary.)

- For each $p \in M$, the **orbit of** p, denoted by $G \cdot p$, is the set of all images of p under the action by elements of G:

$$G \cdot p = \{g \cdot p : g \in G\}.$$

- For each $p \in M$, the **isotropy group** or **stabilizer of** p, denoted by G_p, is the set of elements of G that fix p:

$$G_p = \{g \in G : g \cdot p = p\}.$$

The definition of a group action guarantees that G_p is a subgroup of G.
- The action is said to be **transitive** if for every pair of points $p, q \in M$, there exists $g \in G$ such that $g \cdot p = q$, or equivalently if the only orbit is all of M.
- The action is said to be **free** if the only element of G that fixes any element of M is the identity: $g \cdot p = p$ for some $p \in M$ implies $g = e$, or equivalently if every isotropy group is trivial.

Here are some examples of Lie group actions on manifolds. We will see more in Chapter 21.

Example 7.22 (Lie Group Actions).

(a) If G is any Lie group and M is any smooth manifold, the **trivial action of G on M** is defined by $g \cdot p = p$ for all $g \in G$ and $p \in M$. It is a smooth action, for which each orbit is a single point and each isotropy group is all of G.

(b) The **natural action of GL($n, \mathbb{R}$) on $\mathbb{R}^n$** is the left action given by matrix multiplication: $(A, x) \mapsto Ax$, considering $x \in \mathbb{R}^n$ as a column matrix. This is an action because $I_n x = x$ and matrix multiplication is associative: $(AB)x = A(Bx)$. It is smooth because the components of Ax depend polynomially on the matrix entries of A and the components of x. Because any nonzero vector can be taken to any other by some invertible linear transformation, there are exactly two orbits: $\{0\}$ and $\mathbb{R}^n \smallsetminus \{0\}$.

(c) Every Lie group G acts smoothly on itself by left translation. Given any two points $g_1, g_2 \in G$, there is a unique left translation of G taking g_1 to g_2, namely left translation by $g_2 g_1^{-1}$; thus the action is both free and transitive. More generally, if H is a Lie subgroup of G, then the restriction of the multiplication map to $H \times G \to G$ defines a smooth and free (but generally not transitive) left action of H on G. Similar observations apply to right translations.

(d) Every Lie group acts smoothly on itself by conjugation: $g \cdot h = ghg^{-1}$.

(e) An action of a discrete group Γ on a manifold M is smooth if and only if for each $g \in \Gamma$, the map $p \mapsto g \cdot p$ is a smooth map from M to itself. Thus, for example, $\mathbb{Z}^n$ acts smoothly and freely on $\mathbb{R}^n$ by left translation:

$$\left(m^1, \dots, m^n \right) \cdot \left(x^1, \dots, x^n \right) = \left(m^1 + x^1, \dots, m^n + x^n \right). \qquad \text{//}$$

Another important class of Lie group actions arises from covering maps. Suppose E and M are topological spaces, and $\pi \colon E \to M$ is a (topological) covering map. An **automorphism of π** (also called a **deck transformation** or **covering transformation**) is a homeomorphism $\varphi \colon E \to E$ such that $\pi \circ \varphi = \pi$:

$$\begin{array}{ccc} E & \overset{\varphi}{\longrightarrow} & E \\ & \searrow{\scriptstyle \pi} \quad {\scriptstyle \pi}\swarrow & \\ & M. & \end{array} \qquad (7.8)$$

The set $\mathrm{Aut}_\pi(E)$ of all automorphisms of π, called the **automorphism group of π**, is a group under composition, acting on E on the left. It can be shown that $\mathrm{Aut}_\pi(E)$ acts transitively on each fiber of π if and only if π is a **normal covering map**, which means that $\pi_*\left(\pi_1(E, q) \right)$ is a normal subgroup of $\pi_1(M, \pi(q))$ for every $q \in E$ (see, for example, [LeeTM, Cor. 12.5]).

Proposition 7.23. *Suppose E and M are smooth manifolds with or without boundary, and $\pi \colon E \to M$ is a smooth covering map. With the discrete topology, the automorphism group $\mathrm{Aut}_\pi(E)$ is a zero-dimensional Lie group acting smoothly and freely on E.*

Proof. Suppose $\varphi \in \mathrm{Aut}_\pi(E)$ is an automorphism that fixes a point $p \in E$. Simply by rotating diagram (7.8), we can consider φ as a lift of π:

$$
\begin{array}{ccc}
 & & E \\
 & \nearmid \varphi & \downarrow \pi \\
E & \xrightarrow{\ \ \pi\ \ } & M.
\end{array}
$$

Since the identity map of E is another such lift that agrees with φ at p, the unique lifting property of covering maps (Proposition A.77(a)) guarantees that $\varphi = \mathrm{Id}_E$. Thus, the action of $\mathrm{Aut}_\pi(E)$ is free.

To show that $\mathrm{Aut}_\pi(E)$ is a Lie group, we need only verify that it is countable. Let $q \in E$ be arbitrary, let $p = \pi(q) \in M$, and let $U \subseteq M$ be an evenly covered neighborhood of p. Because E is second-countable, $\pi^{-1}(U)$ has countably many components, and because each component contains exactly one point of $\pi^{-1}(p)$, it follows that $\pi^{-1}(p)$ is countable. Let $\theta^{(q)} \colon \mathrm{Aut}_\pi(E) \to E$ be the map $\theta^{(q)}(\varphi) = \varphi(q)$. Then $\theta^{(q)}$ maps $\mathrm{Aut}_\pi(E)$ into $\pi^{-1}(p)$, and the fact that the action is free implies that it is injective; thus $\mathrm{Aut}_\pi(E)$ is countable.

Smoothness of the action follows from Theorem 4.29. $\square$

Equivariant Maps

For some manifolds with group actions, there is an easily verified sufficient condition for a smooth map to have constant rank. Suppose G is a Lie group, and M and N are both smooth manifolds endowed with smooth left or right G-actions. A map $F \colon M \to N$ is said to be *equivariant* with respect to the given G-actions if for each $g \in G$,

$$
F(g \cdot p) = g \cdot F(p) \quad \text{(for left actions)},
$$
$$
F(p \cdot g) = F(p) \cdot g \quad \text{(for right actions)}.
$$

Equivalently, if θ and φ are the given actions on M and N, respectively, F is equivariant if the following diagram commutes for each $g \in G$:

$$
\begin{array}{ccc}
M & \xrightarrow{\ F\ } & N \\
\theta_g \downarrow & & \downarrow \varphi_g \\
M & \xrightarrow[\ F\ ]{} & N.
\end{array}
$$

This condition is also expressed by saying that *F intertwines θ and φ.*

Example 7.24. Let $v = (v^1, \dots, v^n) \in \mathbb{R}^n$ be any fixed nonzero vector. Define smooth left actions of $\mathbb{R}$ on $\mathbb{R}^n$ and $\mathbb{T}^n$ by

$$
t \cdot (x^1, \dots, x^n) = (x^1 + t v^1, \dots, x^n + t v^n),
$$
$$
t \cdot (z^1, \dots, z^n) = \left(e^{2\pi i t v^1} z^1, \dots, e^{2\pi i t v^n} z^n \right),
$$

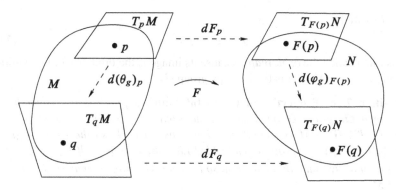

Fig. 7.3 The equivariant rank theorem

for $t \in \mathbb{R}$, $\left(x^1, \ldots, x^n\right) \in \mathbb{R}^n$, and $\left(z^1, \ldots, z^n\right) \in \mathbb{T}^n$. The smooth map $\varepsilon^n \colon \mathbb{R}^n \to \mathbb{T}^n$ given by $\varepsilon^n\left(x^1, \ldots, x^n\right) = \left(e^{2\pi i x^1}, \ldots, e^{2\pi i x^n}\right)$ is equivariant with respect to these actions. //

The following generalization of Theorem 7.5 is an extremely useful tool for proving that certain maps have constant rank.

Theorem 7.25 (Equivariant Rank Theorem). *Let M and N be smooth manifolds and let G be a Lie group. Suppose $F \colon M \to N$ is a smooth map that is equivariant with respect to a transitive smooth G-action on M and any smooth G-action on N. Then F has constant rank. Thus, if F is surjective, it is a smooth submersion; if it is injective, it is a smooth immersion; and if it is bijective, it is a diffeomorphism.*

Proof. Let θ and φ denote the G-actions on M and N, respectively, and let p and q be arbitrary points in M. Choose $g \in G$ such that $\theta_g(p) = q$. (Such a g exists because we are assuming that G acts transitively on M.) Because $\varphi_g \circ F = F \circ \theta_g$, the following diagram commutes (see Fig. 7.3):

$$
\begin{array}{ccc}
T_p M & \xrightarrow{\;dF_p\;} & T_{F(p)} N \\
{\scriptstyle d(\theta_g)_p} \downarrow & & \downarrow {\scriptstyle d(\varphi_g)_{F(p)}} \\
T_q M & \xrightarrow{\;dF_q\;} & T_{F(q)} N.
\end{array}
$$

Because the vertical linear maps in this diagram are isomorphisms, the horizontal ones have the same rank. In other words, the rank of F is the same at any two arbitrary points $p, q \in M$, so F has constant rank. The final statement follows from the global rank theorem. □

Here is an important application of the equivariant rank theorem. Suppose G is a Lie group, M is a smooth manifold, and $\theta \colon G \times M \to M$ is a smooth left action. (The definitions for right actions are analogous.) For each $p \in M$, define a map

$\theta^{(p)} \colon G \to M$ by

$$\theta^{(p)}(g) = g \cdot p. \tag{7.9}$$

This is often called the *orbit map*, because its image is the orbit $G \cdot p$. In addition, the preimage $(\theta^{(p)})^{-1}(p)$ is the isotropy group G_p.

Proposition 7.26 (Properties of the Orbit Map). *Suppose θ is a smooth left action of a Lie group G on a smooth manifold M. For each $p \in M$, the orbit map $\theta^{(p)} \colon G \to M$ is smooth and has constant rank, so the isotropy group $G_p = (\theta^{(p)})^{-1}(p)$ is a properly embedded Lie subgroup of G. If $G_p = \{e\}$, then $\theta^{(p)}$ is an injective smooth immersion, so the orbit $G \cdot p$ is an immersed submanifold of M.*

Remark. It is a fact that *every* orbit is an immersed submanifold of M, not just the ones corresponding to trivial isotropy groups; but the proof of that fact will have to wait until Chapter 21 (see Problem 21-17).

Proof. The orbit map is smooth because it is equal to the composition

$$G \approx G \times \{p\} \hookrightarrow G \times M \overset{\theta}{\to} M.$$

It follows from the definition of a group action that $\theta^{(p)}$ is equivariant with respect to the action of G on itself by left translation and the given action on M:

$$\theta^{(p)}(g'g) = (g'g) \cdot p = g' \cdot (g \cdot p) = g' \cdot \theta^{(p)}(g).$$

Since G acts transitively on itself, the equivariant rank theorem shows that $\theta^{(p)}$ has constant rank. Thus, G_p is a properly embedded submanifold by Theorem 5.12, and a Lie subgroup by Proposition 7.11.

Now suppose $G_p = \{e\}$. If $\theta^{(p)}(g') = \theta^{(p)}(g)$, then

$$g' \cdot p = g \cdot p \quad \Rightarrow \quad (g^{-1}g') \cdot p = p \quad \Rightarrow \quad g^{-1}g' = e \quad \Rightarrow \quad g = g',$$

showing that $\theta^{(p)}$ is injective. By the equivariant rank theorem, it is a smooth immersion, and thus the orbit (endowed with a suitable topology and smooth structure) is an immersed submanifold by Proposition 5.18. $\qquad\square$

Next we use the equivariant rank theorem to identify some important Lie subgroups of the general linear groups.

Example 7.27 (The Orthogonal Group). A real $n \times n$ matrix A is said to be *orthogonal* if as a linear map $A \colon \mathbb{R}^n \to \mathbb{R}^n$ it preserves the Euclidean dot product:

$$(Ax) \cdot (Ay) = x \cdot y \quad \text{for all } x, y \in \mathbb{R}^n.$$

The set $O(n)$ of all orthogonal $n \times n$ matrices is a subgroup of $GL(n, \mathbb{R})$, called the *orthogonal group of degree n*. It is easy to check that a matrix A is orthogonal if and only if it takes the standard basis of $\mathbb{R}^n$ to an orthonormal basis, which is equivalent to the columns of A being orthonormal. Since the (i, j)-entry of the matrix $A^T A$

(where A^T represents the transpose of A) is the dot product of the ith and jth columns of A, this condition is also equivalent to the requirement that $A^T A = I_n$.

Define a smooth map $\Phi \colon \mathrm{GL}(n, \mathbb{R}) \to \mathrm{M}(n, \mathbb{R})$ by $\Phi(A) = A^T A$. Then $\mathrm{O}(n)$ is equal to the level set $\Phi^{-1}(I_n)$. To show that Φ has constant rank and therefore that $\mathrm{O}(n)$ is an embedded Lie subgroup, we show that Φ is equivariant with respect to suitable right actions of $\mathrm{GL}(n, \mathbb{R})$. Let $\mathrm{GL}(n, \mathbb{R})$ act on itself by right multiplication, and define a right action of $\mathrm{GL}(n, \mathbb{R})$ on $\mathrm{M}(n, \mathbb{R})$ by

$$X \cdot B = B^T X B \quad \text{for } X \in \mathrm{M}(n, \mathbb{R}), B \in \mathrm{GL}(n, \mathbb{R}).$$

It is easy to check that this is a smooth action, and Φ is equivariant because

$$\Phi(AB) = (AB)^T (AB) = B^T A^T A B = B^T \Phi(A) B = \Phi(A) \cdot B.$$

Thus, $\mathrm{O}(n)$ is a properly embedded Lie subgroup of $\mathrm{GL}(n, \mathbb{R})$. It is compact because it is closed and bounded in $\mathrm{M}(n, \mathbb{R}) \cong \mathbb{R}^{n^2}$: closed because it is a level set of Φ, and bounded because every $A \in \mathrm{O}(n)$ has columns of norm 1, and therefore satisfies $|A| = \sqrt{n}$.

To determine the dimension of $\mathrm{O}(n)$, we need to compute the rank of Φ. Because the rank is constant, it suffices to compute it at the identity $I_n \in \mathrm{GL}(n, \mathbb{R})$. Thus for any $B \in T_{I_n} \mathrm{GL}(n, \mathbb{R}) = \mathrm{M}(n, \mathbb{R})$, let $\gamma \colon (-\varepsilon, \varepsilon) \to \mathrm{GL}(n, \mathbb{R})$ be the curve $\gamma(t) = I_n + tB$, and compute

$$d\Phi_{I_n}(B) = \frac{d}{dt}\bigg|_{t=0} \Phi \circ \gamma(t) = \frac{d}{dt}\bigg|_{t=0} (I_n + tB)^T (I_n + tB) = B^T + B.$$

From this formula, it is evident that the image of $d\Phi_{I_n}$ is contained in the vector space of *symmetric* matrices. Conversely, if $B \in \mathrm{M}(n, \mathbb{R})$ is an arbitrary symmetric $n \times n$ matrix, then $d\Phi_{I_n}\left(\frac{1}{2}B\right) = B$. It follows that the image of $d\Phi_{I_n}$ is exactly the space of symmetric matrices. This is a linear subspace of $\mathrm{M}(n, \mathbb{R})$ of dimension $n(n+1)/2$, because each symmetric matrix is uniquely determined by its values on and above the main diagonal. It follows that $\mathrm{O}(n)$ is an embedded Lie subgroup of dimension $n^2 - n(n+1)/2 = n(n-1)/2$. //

Example 7.28 (The Special Orthogonal Group). The *special orthogonal group of degree n* is defined as $\mathrm{SO}(n) = \mathrm{O}(n) \cap \mathrm{SL}(n, \mathbb{R}) \subseteq \mathrm{GL}(n, \mathbb{R})$. Because every matrix $A \in \mathrm{O}(n)$ satisfies

$$1 = \det I_n = \det(A^T A) = (\det A)(\det A^T) = (\det A)^2,$$

it follows that $\det A = \pm 1$ for all $A \in \mathrm{O}(n)$. Therefore, $\mathrm{SO}(n)$ is the open subgroup of $\mathrm{O}(n)$ consisting of matrices of positive determinant, and is therefore also an embedded Lie subgroup of dimension $n(n-1)/2$ in $\mathrm{GL}(n, \mathbb{R})$. It is a compact group because it is a closed subset of $\mathrm{O}(n)$. //

Example 7.29 (The Unitary Group). For any complex matrix A, the *adjoint of A* is the matrix A^* formed by conjugating the entries of A and taking the transpose: $A^* = \overline{A}^T$. Observe that $(AB)^* = (\overline{AB})^T = \overline{B}^T \overline{A}^T = B^* A^*$. For any positive integer n, the *unitary group of degree n* is the subgroup $\mathrm{U}(n) \subseteq \mathrm{GL}(n, \mathbb{C})$ consisting of complex $n \times n$ matrices whose columns form an orthonormal basis for $\mathbb{C}^n$ with respect to the Hermitian dot product $z \cdot w = \sum_i z^i \overline{w^i}$. It is straightforward to check that $U(n)$ consists of those matrices A such that $A^* A = I_n$. Problem 7-13 shows that it is a properly embedded Lie subgroup of $GL(n, \mathbb{C})$ of dimension n^2.   //

Example 7.30 (The Special Unitary Group). The group $SU(n) = U(n) \cap SL(n, \mathbb{C})$ is called the *complex special unitary group of degree n*. Problem 7-14 shows that it is a properly embedded $(n^2 - 1)$-dimensional Lie subgroup of $U(n)$. Since the composition of smooth embeddings $SU(n) \hookrightarrow U(n) \hookrightarrow GL(n, \mathbb{C})$ is again a smooth embedding, this implies that $SU(n)$ is also embedded in $GL(n, \mathbb{C})$.   //

Semidirect Products

Group actions give us a powerful new way to construct Lie groups. Suppose H and N are Lie groups, and $\theta \colon H \times N \to N$ is a smooth left action of H on N. It is said to be an *action by automorphisms* if for each $h \in H$, the map $\theta_h \colon N \to N$ is a group automorphism of N (i.e., an isomorphism from N to itself). Given such an action, we define a new Lie group $N \rtimes_\theta H$, called a *semidirect product of H and N*, as follows. As a smooth manifold, $N \rtimes_\theta H$ is just the Cartesian product $N \times H$; but the group multiplication is defined by

$$(n, h)(n', h') = \big(n\theta_h(n'), hh'\big). \tag{7.10}$$

Sometimes, if the action of H on N is understood or irrelevant, the semidirect product is denoted simply by $N \rtimes H$.

▶ **Exercise 7.31.** Verify that (7.10) does indeed define a Lie group structure on the manifold $N \times H$, with (e, e) as identity and $(n, h)^{-1} = \big(\theta_{h^{-1}}(n^{-1}), h^{-1}\big)$.

Example 7.32 (The Euclidean Group). If we consider $\mathbb{R}^n$ as a Lie group under addition, then the natural action of $O(n)$ on $\mathbb{R}^n$ is an action by automorphisms. The resulting semidirect product $E(n) = \mathbb{R}^n \rtimes O(n)$ is called the *Euclidean group*; its multiplication is given by $(b, A)(b', A') = (b + Ab', AA')$. It acts on $\mathbb{R}^n$ via

$$(b, A) \cdot x = b + Ax.$$

This action preserves lines, distances, and angle measures, and thus all of the relationships of Euclidean geometry.   //

The next proposition details some basic properties of the semidirect product. Recall the notation AB defined in (7.5) for subsets A, B of a group G.

Proposition 7.33 (Properties of Semidirect Products). *Suppose N and H are Lie groups, and θ is a smooth action of H on N by automorphisms. Let $G = N \rtimes_\theta H$.*

(a) *The subsets $\widetilde{N} = N \times \{e\}$ and $\widetilde{H} = \{e\} \times H$ are closed Lie subgroups of G isomorphic to N and H, respectively.*

(b) $\widetilde{N}$ is a normal subgroup of $\widetilde{G}$.
(c) $\widetilde{N} \cap \widetilde{H} = \{(e,e)\}$ and $\widetilde{N}\widetilde{H} = G$.

▶ **Exercise 7.34.** Prove the preceding proposition.

Thanks to the next theorem, many Lie groups can be realized as semidirect products of suitable subgroups.

Theorem 7.35 (Characterization of Semidirect Products). *Suppose G is a Lie group, and $N, H \subseteq G$ are closed Lie subgroups such that N is normal, $N \cap H = \{e\}$, and $NH = G$. Then the map $(n, h) \mapsto nh$ is a Lie group isomorphism between $N \rtimes_\theta H$ and G, where $\theta \colon H \times N \to N$ is the action by conjugation: $\theta_h(n) = hnh^{-1}$.*

Proof. Problem 7-18.   □

Under the hypotheses of Theorem (7.35), we say that G is the **internal semidirect product of N and H**. Some examples are described in Problem 7-20.

Representations

Most of the Lie groups we have seen so far can be realized as Lie subgroups of $GL(n, \mathbb{R})$ or $GL(n, \mathbb{C})$. It is natural to ask whether all Lie groups are of this form. The key to studying this question is the theory of *group representations*.

Recall that if V is a finite-dimensional real or complex vector space, $GL(V)$ denotes the group of invertible linear transformations of V, which is a Lie group isomorphic to $GL(n, \mathbb{R})$ or $GL(n, \mathbb{C})$ for $n = \dim V$. If G is a Lie group, a (*finite-dimensional*) *representation of G* is a Lie group homomorphism from G to $GL(V)$ for some V. (Although it is useful for many applications to consider also the case in which V is infinite-dimensional, in this book we consider only finite-dimensional representations.)

If a representation $\rho \colon G \to GL(V)$ is injective, it is said to be *faithful*. In that case, it follows from Proposition 7.17 that the image of ρ is a Lie subgroup of $GL(V)$, and ρ gives a Lie group isomorphism between G and $\rho(G) \subseteq GL(V) \cong GL(n, \mathbb{R})$ or $GL(n, \mathbb{C})$. Thus, a Lie group admits a faithful representation if and only if it is isomorphic to a Lie subgroup of $GL(n, \mathbb{R})$ or $GL(n, \mathbb{C})$ for some n. Not every Lie group admits such a representation. We do not yet have the technology to construct a counterexample, but Problem 21-26 asks you to prove that the universal covering group of $SL(2, \mathbb{R})$ has no faithful representation and therefore is not isomorphic to any matrix group.

Representation theory is a vast subject, with applications to fields as diverse as differential geometry, differential equations, harmonic analysis, number theory, quantum physics, and engineering; we can do no more than touch on it here.

Example 7.36 (Lie Group Representations).

(a) If G is any Lie subgroup of $GL(n, \mathbb{R})$, the inclusion map $G \hookrightarrow GL(n, \mathbb{R}) = GL(\mathbb{R}^n)$ is a faithful representation, called the *defining representation of G*. The defining representation of a Lie subgroup of $GL(n, \mathbb{C})$ is defined similarly.

(b) The inclusion map $S^1 \hookrightarrow \mathbb{C}^* \cong GL(1,\mathbb{C})$ is a faithful representation of the circle group. More generally, the map $\rho\colon \mathbb{T}^n \to GL(n,\mathbb{C})$ given by

$$\rho\left(z^1,\dots,z^n\right) = \begin{pmatrix} z^1 & 0 & \dots & 0 \\ 0 & z^2 & \dots & 0 \\ \vdots & \vdots & \ddots & \vdots \\ 0 & 0 & \dots & z^n \end{pmatrix}$$

is a faithful representation of $\mathbb{T}^n$.

(c) Let $\sigma\colon \mathbb{R}^n \to GL(n+1,\mathbb{R})$ be the map that sends $x \in \mathbb{R}^n$ to the matrix $\sigma(x)$ defined in block form by

$$\sigma(x) = \begin{pmatrix} I_n & x \\ 0 & 1 \end{pmatrix},$$

where I_n is the $n \times n$ identity matrix and x is regarded as an $n \times 1$ column matrix. A straightforward computation shows that σ is a faithful representation of the additive Lie group $\mathbb{R}^n$.

(d) Another faithful representation of $\mathbb{R}^n$ is the map $\mathbb{R}^n \to GL(n,\mathbb{R})$ that sends $\left(x^1,\dots,x^n\right)$ to the diagonal matrix whose diagonal entries are $\left(e^{x^1},\dots,e^{x^n}\right)$.

(e) Yet another representation of $\mathbb{R}^n$ is the map $\mathbb{R}^n \to GL(n,\mathbb{C})$ sending x to the diagonal matrix with diagonal entries $\left(e^{2\pi i x^1},\dots,e^{2\pi i x^n}\right)$. This one is not faithful, because its kernel is the subgroup $\mathbb{Z}^n \subseteq \mathbb{R}^n$.

(f) Let $E(n)$ be the Euclidean group (Example 7.32). A faithful representation of $E(n)$ is given by the map $\rho\colon E(n) \to GL(n+1,\mathbb{R})$ defined in block form by

$$\rho(b,A) = \begin{pmatrix} A & b \\ 0 & 1 \end{pmatrix},$$

where b is considered as a column matrix.

(g) For positive integers n and d, let $\mathcal{P}_d^n$ denote the vector space of real-valued polynomial functions $p\colon \mathbb{R}^n \to \mathbb{R}$ of degree at most d. For any matrix $A \in GL(n,\mathbb{R})$, define a linear map $\tau_d^n(A)\colon \mathcal{P}_d^n \to \mathcal{P}_d^n$ by

$$\tau_d^n(A)p = p \circ A^{-1}.$$

Problem 7-24 shows that the map $\tau_d^n\colon GL(n,\mathbb{R}) \to GL\left(\mathcal{P}_d^n\right)$ is a faithful representation.   //

There is a close connection between representations and group actions. Let G be a Lie group and V be a finite-dimensional vector space. An action of G on V is said to be a **linear action** if for each $g \in G$, the map from V to itself given by $x \mapsto g \cdot x$ is linear. For example, if $\rho\colon G \to GL(V)$ is a representation of G, there is an associated smooth linear action of G on V given by $g \cdot x = \rho(g)x$. The next proposition shows that every linear action is of this type.

Proposition 7.37. *Let G be a Lie group and V be a finite-dimensional vector space. A smooth left action of G on V is linear if and only if it is of the form $g \cdot x = \rho(g)x$ for some representation ρ of G.*

Proof. Every action induced by a representation is evidently linear. To prove the converse, assume that we are given a linear action of G on V. The hypothesis implies that for each $g \in G$ there is a linear map $\rho(g) \in GL(V)$ such that $g \cdot x = \rho(g)x$ for all $x \in V$. The fact that the action satisfies (7.6) translates to $\rho(g_1 g_2) = \rho(g_1)\rho(g_2)$, so $\rho \colon G \to GL(V)$ is a group homomorphism. Thus, to show that it is a Lie group representation, we need only show that it is smooth. Choose a basis (E_i) for V, and for each i let $\pi^i \colon V \to \mathbb{R}$ be the projection onto the ith coordinate with respect to this basis: $\pi^i (x^j E_j) = x^i$. If we let $\rho^i_j(g)$ denote the matrix entries of $\rho(g)$ with respect to this basis, it follows that $\rho^i_j(g) = \pi^i (g \cdot E_i)$, so each function ρ^i_j is a composition of smooth functions. Because the matrix entries form global smooth coordinates for $GL(V)$, this implies that ρ is smooth. □

Problems

7-1. Show that for any Lie group G, the multiplication map $m \colon G \times G \to G$ is a smooth submersion. [Hint: use local sections.]

7-2. Let G be a Lie group.
   (a) Let $m \colon G \times G \to G$ denote the multiplication map. Using Proposition 3.14 to identify $T_{(e,e)}(G \times G)$ with $T_e G \oplus T_e G$, show that the differential $dm_{(e,e)} \colon T_e G \oplus T_e G \to T_e G$ is given by

   $$dm_{(e,e)}(X, Y) = X + Y.$$

   [Hint: compute $dm_{(e,e)}(X, 0)$ and $dm_{(e,e)}(0, Y)$ separately.]
   (b) Let $i \colon G \to G$ denote the inversion map. Show that $di_e \colon T_e G \to T_e G$ is given by $di_e(X) = -X$.
   (Used on pp. 203, 522.)

7-3. Our definition of Lie groups includes the requirement that both the multiplication map and the inversion map are smooth. Show that smoothness of the inversion map is redundant: if G is a smooth manifold with a group structure such that the multiplication map $m \colon G \times G \to G$ is smooth, then G is a Lie group. [Hint: show that the map $F \colon G \times G \to G \times G$ defined by $F(g, h) = (g, gh)$ is a bijective local diffeomorphism.]

7-4. Let $\det \colon GL(n, \mathbb{R}) \to \mathbb{R}$ denote the determinant function. Use Corollary 3.25 to compute the differential of $\det$, as follows.
   (a) For any $A \in M(n, \mathbb{R})$, show that

   $$\frac{d}{dt}\bigg|_{t=0} \det(I_n + tA) = \operatorname{tr} A,$$

   where $\operatorname{tr}(A^i_j) = \sum_i A^i_i$ is the trace of A. [Hint: the defining equation (B.3) expresses $\det(I_n + tA)$ as a polynomial in t. What is the linear term?]

(b) For $X \in GL(n, \mathbb{R})$ and $B \in T_X GL(n, \mathbb{R}) \cong M(n, \mathbb{R})$, show that

$$d(\det)_X(B) = (\det X) \operatorname{tr}\left(X^{-1}B\right). \tag{7.11}$$

[Hint: $\det(X + tB) = \det(X)\det\left(I_n + tX^{-1}B\right)$.]
(*Used on p. 203.*)

7-5. Prove Theorem 7.9 (uniqueness of the universal covering group).

7-6. Suppose G is a Lie group and U is any neighborhood of the identity. Show that there exists a neighborhood V of the identity such that $V \subseteq U$ and $gh^{-1} \in U$ whenever $g, h \in V$. (*Used on pp. 159, 556.*)

7-7. Prove Proposition 7.15 (properties of the identity component of a Lie group).

7-8. Suppose a connected topological group G acts continuously on a discrete space K. Show that the action is trivial. (*Used on p. 562.*)

7-9. Show that the formula

$$A \cdot [x] = [Ax]$$

defines a smooth, transitive left action of $GL(n + 1, \mathbb{R})$ on $\mathbb{R}\mathbb{P}^n$.

7-10. Repeat Problem 7-9 for $GL(n + 1, \mathbb{C})$ and $\mathbb{C}\mathbb{P}^n$ (see Problems 1-9 and 4-5).

7-11. Considering $\mathbb{S}^{2n+1}$ as the unit sphere in $\mathbb{C}^{n+1}$, define an action of $\mathbb{S}^1$ on $\mathbb{S}^{2n+1}$, called the **Hopf action**, by

$$z \cdot \left(w^1, \ldots, w^{n+1}\right) = \left(zw^1, \ldots, zw^{n+1}\right).$$

Show that this action is smooth and its orbits are disjoint unit circles in $\mathbb{C}^{n+1}$ whose union is $\mathbb{S}^{2n+1}$. (*Used on p. 560.*)

7-12. Use the equivariant rank theorem to give another proof of Theorem 7.5 by showing that every Lie group homomorphism $F : G \to H$ is equivariant with respect to suitable smooth G-actions on G and H.

7-13. For each $n \geq 1$, prove that $U(n)$ is a properly embedded n^2-dimensional Lie subgroup of $GL(n, \mathbb{C})$. (See Example 7.29.)

7-14. For each $n \geq 1$, prove that $SU(n)$ is a properly embedded $(n^2 - 1)$-dimensional Lie subgroup of $U(n)$. (See Example 7.30.)

7-15. Show that $SO(2)$, $U(1)$, and $\mathbb{S}^1$ are all isomorphic as Lie groups.

7-16. Prove that $SU(2)$ is diffeomorphic to $\mathbb{S}^3$. (*Used on pp. 179, 563.*)

7-17. Determine which of the following Lie groups are compact:

$$GL(n, \mathbb{R}), \ SL(n, \mathbb{R}), \ GL(n, \mathbb{C}), \ SL(n, \mathbb{C}), \ U(n), \ SU(n).$$

7-18. Prove Theorem 7.35 (characterization of semidirect products).

7-19. Suppose G, N, and H are Lie groups. Prove that G is isomorphic to a semidirect product $N \rtimes H$ if and only if there are Lie group homomorphisms $\varphi : G \to H$ and $\psi : H \to G$ such that $\varphi \circ \psi = \operatorname{Id}_H$ and $\operatorname{Ker}\varphi \cong N$.

7-20. Prove that the following Lie groups are isomorphic to semidirect products as shown. [Hint: Use Problem 7-19.]

(a) $O(n) \cong SO(n) \rtimes O(1)$.

(b) $U(n) \cong SU(n) \rtimes U(1)$.

(c) $GL(n, \mathbb{R}) \cong SL(n, \mathbb{R}) \rtimes \mathbb{R}^*$.

(d) $GL(n, \mathbb{C}) \cong SL(n, \mathbb{C}) \rtimes \mathbb{C}^*$.

7-21. Prove that when $n > 1$, none of the groups in Problem 7-20 are isomorphic to direct products of the indicated groups. [Hint: the *center* of a group G is the set of all elements that commute with every element of G. Show that isomorphic groups have isomorphic centers.]

7-22. Let $\mathbb{H} = \mathbb{C} \times \mathbb{C}$ (considered as a real vector space), and define a bilinear product $\mathbb{H} \times \mathbb{H} \to \mathbb{H}$ by

$$(a, b)(c, d) = \left(ac - d\bar{b}, \ \bar{a}d + cb\right), \quad \text{for } a, b, c, d \in \mathbb{C}.$$

With this product, $\mathbb{H}$ is a 4-dimensional algebra over $\mathbb{R}$, called the algebra of *quaternions*. For each $p = (a, b) \in \mathbb{H}$, define $p^* = (\bar{a}, -b)$. It is useful to work with the basis $(1, i, j, k)$ for $\mathbb{H}$ defined by

$$1 = (1, 0), \quad i = (i, 0), \quad j = (0, 1), \quad k = (0, -i).$$

It is straightforward to verify that this basis satisfies

$$i^2 = j^2 = k^2 = -1, \qquad 1q = q1 = q \quad \text{for all } q \in \mathbb{H},$$

$$ij = -ji = k, \qquad jk = -kj = i, \qquad ki = -ik = j,$$

$$1^* = 1, \qquad i^* = -i, \qquad j^* = -j, \qquad k^* = -k.$$

A quaternion p is said to be *real* if $p^* = p$, and *imaginary* if $p^* = -p$. Real quaternions can be identified with real numbers via the correspondence $x \leftrightarrow x1$.

(a) Show that quaternionic multiplication is associative but not commutative.

(b) Show that $(pq)^* = q^* p^*$ for all $p, q \in \mathbb{H}$.

(c) Show that $\langle p, q \rangle = \frac{1}{2}(p^* q + q^* p)$ is an inner product on $\mathbb{H}$, whose associated norm satisfies $|pq| = |p| \, |q|$.

(d) Show that every nonzero quaternion has a two-sided multiplicative inverse given by $p^{-1} = |p|^{-2} p^*$.

(e) Show that the set $\mathbb{H}^*$ of nonzero quaternions is a Lie group under quaternionic multiplication.

(*Used on pp. 200, 200, 562.*)

7-23. Let $\mathbb{H}^*$ be the Lie group of nonzero quaternions (Problem 7-22), and let $S \subseteq \mathbb{H}^*$ be the set of unit quaternions. Show that S is a properly embedded Lie subgroup of $\mathbb{H}^*$, isomorphic to $SU(2)$. (*Used on pp. 200, 562.*)

7-24. Prove that each of the maps $\tau_d^n : GL(n, \mathbb{R}) \to GL\left(\mathscr{P}_d^n\right)$ described in Example 7.36(g) is a faithful representation of $GL(n, \mathbb{R})$.

Chapter 8
Vector Fields

Vector fields are familiar objects of study in multivariable calculus. In that setting, a vector field on an open subset $U \subseteq \mathbb{R}^n$ is simply a continuous map from U to $\mathbb{R}^n$, which can be visualized as attaching an "arrow" to each point of U. In this chapter we show how to extend this idea to smooth manifolds.

We think of a vector field on an abstract smooth manifold M as a particular kind of continuous map X from M to its tangent bundle—one that assigns to each point $p \in M$ a tangent vector $X_p \in T_p M$. After introducing the definitions, we explore the ways that vector fields behave under differentials of smooth maps.

In the next section we introduce the *Lie bracket* operation, which is a way of combining two smooth vector fields to obtain another. Then we describe the most important application of Lie brackets: the set of all smooth vector fields on a Lie group that are invariant under left multiplication is closed under Lie brackets, and thus forms an algebraic object naturally associated with the group, called the *Lie algebra of the Lie group*. We describe a few basic properties of Lie algebras, and compute the Lie algebras of some familiar groups. Then we show how Lie group homomorphisms induce homomorphisms of their Lie algebras, from which it follows that isomorphic Lie groups have isomorphic Lie algebras. Finally, at the end of the chapter we show how to identify Lie algebras of Lie subgroups.

Vector Fields on Manifolds

If M is a smooth manifold with or without boundary, a *vector field on M* is a section of the map $\pi \colon TM \to M$. More concretely, a vector field is a continuous map $X \colon M \to TM$, usually written $p \mapsto X_p$, with the property that

$$\pi \circ X = \mathrm{Id}_M, \tag{8.1}$$

or equivalently, $X_p \in T_p M$ for each $p \in M$. (We write the value of X at p as X_p instead of $X(p)$ to be consistent with our notation for elements of the tangent bundle, as well as to avoid conflict with the notation $v(f)$ for the action of a vector on a function.) You should visualize a vector field on M in the same way as you

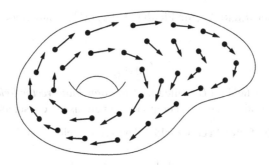

Fig. 8.1 A vector field

visualize vector fields in Euclidean space: as an arrow attached to each point of M, chosen to be tangent to M and to vary continuously from point to point (Fig. 8.1).

We are primarily interested in ***smooth vector fields***, the ones that are smooth as maps from M to TM, when TM is given the smooth manifold structure described in Proposition 3.18. In addition, for some purposes it is useful to consider maps from M to TM that would be vector fields except that they might not be continuous. A ***rough vector field on*** M is a (not necessarily continuous) map $X : M \to TM$ satisfying (8.1). Just as for functions, if X is a vector field on M, the ***support of*** X is defined to be the closure of the set $\{p \in M : X_p \neq 0\}$. A vector field is said to be ***compactly supported*** if its support is a compact set.

Suppose M is a smooth n-manifold (with or without boundary). If $X : M \to TM$ is a rough vector field and $(U, (x^i))$ is any smooth coordinate chart for M, we can write the value of X at any point $p \in U$ in terms of the coordinate basis vectors:

$$X_p = X^i(p) \left. \frac{\partial}{\partial x^i} \right|_p . \tag{8.2}$$

This defines n functions $X^i : U \to \mathbb{R}$, called the ***component functions of*** X in the given chart.

Proposition 8.1 (Smoothness Criterion for Vector Fields). *Let M be a smooth manifold with or without boundary, and let $X : M \to TM$ be a rough vector field. If $(U, (x^i))$ is any smooth coordinate chart on M, then the restriction of X to U is smooth if and only if its component functions with respect to this chart are smooth.*

Proof. Let (x^i, v^i) be the natural coordinates on $\pi^{-1}(U) \subseteq TM$ associated with the chart $(U, (x^i))$. By definition of natural coordinates, the coordinate representation of $X : M \to TM$ on U is

$$\widehat{X}(x) = \left(x^1, \dots, x^n, X^1(x), \dots, X^n(x) \right),$$

where X^i is the ith component function of X in x^i-coordinates. It follows immediately that smoothness of X in U is equivalent to smoothness of its component functions. $\qquad \square$

Example 8.2 (Coordinate Vector Fields). If $(U, (x^i))$ is any smooth chart on M, the assignment

$$p \mapsto \left. \frac{\partial}{\partial x^i} \right|_p$$

determines a vector field on U, called the *ith coordinate vector field* and denoted by $\partial / \partial x^i$. It is smooth because its component functions are constants.   //

Example 8.3 (The Euler Vector Field). The vector field V on $\mathbb{R}^n$ whose value at $x \in \mathbb{R}^n$ is

$$V_x = x^1 \left. \frac{\partial}{\partial x^1} \right|_x + \cdots + x^n \left. \frac{\partial}{\partial x^n} \right|_x$$

is smooth because its coordinate functions are linear. It vanishes at the origin, and points radially outward everywhere else. It is called the *Euler vector field* because of its appearance in Euler's homogeneous function theorem (see Problem 8-2).   //

Example 8.4 (The Angle Coordinate Vector Field on the Circle). Let θ be any angle coordinate on a proper open subset $U \subseteq \mathbb{S}^1$ (see Problem 1-8), and let $d/d\theta$ denote the corresponding coordinate vector field. Because any other angle coordinate $\widetilde{\theta}$ differs from θ by an additive constant in a neighborhood of each point, the transformation law for coordinate vector fields (3.11) shows that $d/d\theta = d/d\widetilde{\theta}$ on their common domain. For this reason, there is a globally defined vector field on $\mathbb{S}^1$ whose coordinate representation is $d/d\theta$ with respect to any angle coordinate. It is a smooth vector field because its component function is constant in any such chart. We denote this global vector field by $d/d\theta$, even though, strictly speaking, it cannot be considered as a coordinate vector field on the entire circle at once.   //

Example 8.5 (Angle Coordinate Vector Fields on Tori). On the n-dimensional torus $\mathbb{T}^n$, choosing an angle function θ^i for the ith circle factor, $i = 1, \ldots, n$, yields local coordinates $(\theta^1, \ldots, \theta^n)$ for $\mathbb{T}^n$. An analysis similar to that of the previous example shows that the coordinate vector fields $\partial / \partial \theta^1, \ldots, \partial / \partial \theta^n$ are smooth and globally defined on $\mathbb{T}^n$.   //

If $U \subseteq M$ is open, the fact that $T_p U$ is naturally identified with $T_p M$ for each $p \in U$ (Proposition 3.9) allows us to identify TU with the open subset $\pi^{-1}(U) \subseteq TM$. Therefore, a vector field on U can be thought of either as a map from U to TU or as a map from U to TM, whichever is more convenient. If X is a vector field on M, its restriction $X|_U$ is a vector field on U, which is smooth if X is.

The next lemma is a generalization of Lemma 2.26 to vector fields, and is proved in much the same way. If M is a smooth manifold with or without boundary and $A \subseteq M$ is an arbitrary subset, a *vector field along A* is a continuous map $X \colon A \to TM$ satisfying $\pi \circ X = \mathrm{Id}_A$ (or in other words $X_p \in T_p M$ for each $p \in A$). We call it a *smooth vector field along A* if for each $p \in A$, there is a neighborhood V of p in M and a smooth vector field $\widetilde{X}$ on V that agrees with X on $V \cap A$.

Lemma 8.6 (Extension Lemma for Vector Fields). *Let M be a smooth manifold with or without boundary, and let $A \subseteq M$ be a closed subset. Suppose X is a smooth* *vector field along A. Given any open subset U containing A, there exists a smooth global vector field $\widetilde{X}$ on M such that $\widetilde{X}|_A = X$ and supp $\widetilde{X} \subseteq U$.*

Proof. See Problem 8-1. □

As an important special case, any vector at a point can be extended to a smooth vector field on the entire manifold.

Proposition 8.7. *Let M be a smooth manifold with or without boundary. Given $p \in M$ and $v \in T_pM$, there is a smooth global vector field X on M such that $X_p = v$.*

Proof. The assignment $p \mapsto v$ is an example of a vector field along the set $\{p\}$ as defined above. It is smooth because it can be extended, say, to a constant-coefficient vector field in a coordinate neighborhood of p. Thus, the proposition follows from the extension lemma with $A = \{p\}$ and $U = M$. □

If M is a smooth manifold with or without boundary, it is standard to use the notation $\mathfrak{X}(M)$ to denote the set of all smooth vector fields on M. It is a vector space under pointwise addition and scalar multiplication:

$$(aX + bY)_p = aX_p + bY_p.$$

The zero element of this vector space is the zero vector field, whose value at each $p \in M$ is $0 \in T_pM$. In addition, smooth vector fields can be multiplied by smooth real-valued functions: if $f \in C^\infty(M)$ and $X \in \mathfrak{X}(M)$, we define $fX : M \to TM$ by

$$(fX)_p = f(p)X_p.$$

The next proposition shows that these operations yield smooth vector fields.

Proposition 8.8. *Let M be a smooth manifold with or without boundary.*

(a) *If X and Y are smooth vector fields on M and $f, g \in C^\infty(M)$, then $fX + gY$ is a smooth vector field.*

(b) *$\mathfrak{X}(M)$ is a module over the ring $C^\infty(M)$.*

▶ **Exercise 8.9.** Prove Proposition 8.8.

For example, the basis expression (8.2) for a vector field X can also be written as an equation between vector *fields* instead of an equation between vectors at a point:

$$X = X^i \frac{\partial}{\partial x^i},$$

where X^i is the ith component function of X in the given coordinates.

Local and Global Frames

Coordinate vector fields in a smooth chart provide a convenient way of representing vector fields, because their values form a basis for the tangent space at each point. However, they are not the only choices.

Suppose M is a smooth n-manifold with or without boundary. An ordered k-tuple $(X_1, \ldots, X_k)$ of vector fields defined on some subset $A \subseteq M$ is said to be **linearly independent** if $(X_1|_p, \ldots, X_k|_p)$ is a linearly independent k-tuple in $T_p M$ for each $p \in A$, and is said to **span the tangent bundle** if the k-tuple $(X_1|_p, \ldots, X_k|_p)$ spans $T_p M$ at each $p \in A$. A **local frame for** M is an ordered n-tuple of vector fields $(E_1, \ldots, E_n)$ defined on an open subset $U \subseteq M$ that is linearly independent and spans the tangent bundle; thus the vectors $(E_1|_p, \ldots, E_n|_p)$ form a basis for $T_p M$ at each $p \in U$. It is called a **global frame** if $U = M$, and a **smooth frame** if each of the vector fields E_i is smooth. We often use the shorthand notation (E_i) to denote a frame $(E_1, \ldots, E_n)$. If M has dimension n, then to check that an ordered n-tuple of vector fields $(E_1, \ldots, E_n)$ is a local frame, it suffices to check either that it is linearly independent or that it spans the tangent bundle.

Example 8.10 (Local and Global Frames).

(a) The standard coordinate vector fields form a smooth global frame for $\mathbb{R}^n$.

(b) If $(U, (x^i))$ is any smooth coordinate chart for a smooth manifold M (possibly with boundary), then the coordinate vector fields form a smooth local frame $(\partial/\partial x^i)$ on U, called a **coordinate frame**. Every point of M is in the domain of such a local frame.

(c) The vector field $d/d\theta$ defined in Example 8.4 constitutes a smooth global frame for the circle.

(d) The n-tuple of vector fields $(\partial/\partial\theta^1, \ldots, \partial/\partial\theta^n)$ on the n-torus, defined in Example 8.4, is a smooth global frame for $\mathbb{T}^n$.   //

The next proposition shows that local frames are easy to come by.

Proposition 8.11 (Completion of Local Frames). *Let M be a smooth n-manifold with or without boundary.*

(a) *If $(X_1, \ldots, X_k)$ is a linearly independent k-tuple of smooth vector fields on an open subset $U \subseteq M$, with $1 \le k < n$, then for each $p \in U$ there exist smooth vector fields $X_{k+1}, \ldots, X_n$ in a neighborhood V of p such that $(X_1, \ldots, X_n)$ is a smooth local frame for M on $U \cap V$.*

(b) *If $(v_1, \ldots, v_k)$ is a linearly independent k-tuple of vectors in $T_p M$ for some $p \in M$, with $1 \le k \le n$, then there exists a smooth local frame (X_i) on a neighborhood of p such that $X_i|_p = v_i$ for $i = 1, \ldots, k$.*

(c) *If $(X_1, \ldots, X_n)$ is a linearly independent n-tuple of smooth vector fields along a closed subset $A \subseteq M$, then there exists a smooth local frame $(\widetilde{X}_1, \ldots, \widetilde{X}_n)$ on some neighborhood of A such that $\widetilde{X}_i|_A = X_i$ for $i = 1, \ldots, n$.*

Proof. See Problem 8-5. $\square$

For subsets of $\mathbb{R}^n$, there is a special type of frame that is often more useful for geometric problems than arbitrary frames. A k-tuple of vector fields $(E_1, \ldots, E_k)$ defined on some subset $A \subseteq \mathbb{R}^n$ is said to be **orthonormal** if for each $p \in A$, the vectors $(E_1|_p, \ldots, E_k|_p)$ are orthonormal with respect to the Euclidean dot product (where we identify $T_p\mathbb{R}^n$ with $\mathbb{R}^n$ in the usual way). A (local or global) frame consisting of orthonormal vector fields is called an **orthonormal frame**.

Example 8.12. The standard coordinate frame is a global orthonormal frame on $\mathbb{R}^n$. For a less obvious example, consider the smooth vector fields defined on $\mathbb{R}^2 \smallsetminus \{0\}$ by

$$E_1 = \frac{x}{r}\frac{\partial}{\partial x} + \frac{y}{r}\frac{\partial}{\partial y}, \qquad E_2 = -\frac{y}{r}\frac{\partial}{\partial x} + \frac{x}{r}\frac{\partial}{\partial y}, \tag{8.3}$$

where $r = \sqrt{x^2 + y^2}$. A straightforward computation shows that (E_1, E_2) is an orthonormal frame for $\mathbb{R}^2$ over the open subset $\mathbb{R}^2 \smallsetminus \{0\}$. Geometrically, E_1 and E_2 are unit vector fields tangent to radial lines and circles centered at the origin, respectively. //

The next lemma describes a useful method for creating orthonormal frames.

Lemma 8.13 (Gram–Schmidt Algorithm for Frames). *Suppose (X_j) is a smooth local frame for $T\mathbb{R}^n$ over an open subset $U \subseteq \mathbb{R}^n$. Then there is a smooth orthonormal frame (E_j) over U such that* $\operatorname{span}(E_1|_p, \ldots, E_j|_p) = \operatorname{span}(X_1|_p, \ldots, X_j|_p)$ *for each $j = 1, \ldots, n$ and each $p \in U$.*

Proof. Applying the Gram–Schmidt algorithm to the vectors $(X_j|_p)$ at each $p \in U$, we obtain an n-tuple of rough vector fields $(E_1, \ldots, E_n)$ given inductively by

$$E_j = \frac{X_j - \sum_{i=1}^{j-1}(X_j \cdot E_i)E_i}{\left| X_j - \sum_{i=1}^{j-1}(X_j \cdot E_i)E_i \right|}.$$

For each $j = 1, \ldots, n$ and each $p \in U$, $X_j|_p \notin \operatorname{span}(E_1|_p, \ldots, E_{j-1}|_p)$ (which is equal to $\operatorname{span}(X_1|_p, \ldots, X_{j-1}|_p)$), so the denominator above is a nowhere-vanishing smooth function on U. Therefore, this formula defines (E_j) as a smooth orthonormal frame on U that satisfies the conclusion of the lemma. □

Although smooth local frames are plentiful, global ones are not. A smooth manifold with or without boundary is said to be **parallelizable** if it admits a smooth global frame. Example 8.10 shows that $\mathbb{R}^n$, $\mathbb{S}^1$, and $\mathbb{T}^n$ are all parallelizable. Problems 8-6 and 8-7 show that $\mathbb{S}^3$ and $\mathbb{S}^7$ are parallelizable. Later in this chapter, we will see that all Lie groups are parallelizable (see Corollary 8.39 below). However, despite the evidence of these examples, most smooth manifolds are not parallelizable. (As we will see in Chapter 10, parallelizability of M is intimately connected to the question of whether its tangent bundle is diffeomorphic to the product $M \times \mathbb{R}^n$.)

The simplest example of a nonparallelizable manifold is $\mathbb{S}^2$, but the proof of this fact will have to wait until we have developed more machinery (see Problem 16-6). In fact, using more advanced methods from algebraic topology, it was shown in 1958 by Raoul Bott and John Milnor [MB58] and independently by Michel Kervaire [Ker58] that $\mathbb{S}^1$, $\mathbb{S}^3$, and $\mathbb{S}^7$ are the *only* spheres that are parallelizable. Thus these are the only positive-dimensional spheres that can possibly admit Lie group structures. The first two do (see Example 7.3(j) and Problem 7-16), but it turns out that $\mathbb{S}^7$ has no Lie group structure (see [Bre93, p. 301]).

Vector Fields as Derivations of $C^\infty(M)$

An essential property of vector fields is that they define operators on the space of smooth real-valued functions. If $X \in \mathfrak{X}(M)$ and f is a smooth real-valued function defined on an open subset $U \subseteq M$, we obtain a new function $Xf : U \to \mathbb{R}$, defined by

$$(Xf)(p) = X_p f.$$

(Be careful not to confuse the notations fX and Xf: the former is the smooth *vector field* on U obtained by multiplying X by f, while the latter is the real-valued *function* on U obtained by applying the vector field X to the smooth function f.) Because the action of a tangent vector on a function is determined by the values of the function in an arbitrarily small neighborhood, it follows that Xf is locally determined. In particular, for any open subset $V \subseteq U$,

$$(Xf)|_V = X(f|_V). \tag{8.4}$$

This construction yields another useful smoothness criterion for vector fields.

Proposition 8.14. *Let M be a smooth manifold with or without boundary, and let $X : M \to TM$ be a rough vector field. The following are equivalent:*

(a) *X is smooth.*
(b) *For every $f \in C^\infty(M)$, the function Xf is smooth on M.*
(c) *For every open subset $U \subseteq M$ and every $f \in C^\infty(U)$, the function Xf is smooth on U.*

Proof. We will prove that (a) $\Rightarrow$ (b) $\Rightarrow$ (c) $\Rightarrow$ (a).

To prove (a) $\Rightarrow$ (b), assume X is smooth, and let $f \in C^\infty(M)$. For any $p \in M$, we can choose smooth coordinates (x^i) on a neighborhood U of p. Then for $x \in U$, we can write

$$Xf(x) = \left(X^i(x) \left. \frac{\partial}{\partial x^i} \right|_x \right) f = X^i(x) \frac{\partial f}{\partial x^i}(x).$$

Since the component functions X^i are smooth on U by Proposition 8.1, it follows that Xf is smooth in U. Since the same is true in a neighborhood of each point, Xf is smooth on M.

To prove (b) $\Rightarrow$ (c), suppose $U \subseteq M$ is open and $f \in C^\infty(U)$. For any $p \in U$, let ψ be a smooth bump function that is equal to 1 in a neighborhood of p and supported in U, and define $\tilde{f} = \psi f$, extended to be zero on $M \smallsetminus \operatorname{supp} \psi$. Then $X \tilde{f}$ is smooth by assumption, and is equal to Xf in a neighborhood of p by (8.4). This shows that Xf is smooth in a neighborhood of each point of U.

Finally, to prove (c) $\Rightarrow$ (a), suppose Xf is smooth whenever f is smooth on an open subset of M. If (x^i) are any smooth local coordinates on $U \subseteq M$, we can think of each coordinate x^i as a smooth function on U. Applying X to one of these functions, we obtain

$$Xx^i = X^j \frac{\partial}{\partial x^j}(x^i) = X^i.$$

Because Xx^i is smooth by assumption, it follows that the component functions of X are smooth, so X is smooth. $\square$

One consequence of the preceding proposition is that a smooth vector field $X \in \mathfrak{X}(M)$ defines a map from $C^\infty(M)$ to itself by $f \mapsto Xf$. This map is clearly linear over $\mathbb{R}$. Moreover, the product rule (3.4) for tangent vectors translates into the following product rule for vector fields:

$$X(fg) = f\,Xg + g\,Xf, \qquad (8.5)$$

as you can easily check by evaluating both sides at an arbitrary point $p \in M$. In general, a map $X\colon C^\infty(M) \to C^\infty(M)$ is called a **derivation** (as distinct from a *derivation at p*, defined in Chapter 3) if it is linear over $\mathbb{R}$ and satisfies (8.5) for all $f, g \in C^\infty(M)$.

The next proposition shows that derivations of $C^\infty(M)$ can be identified with smooth vector fields.

Proposition 8.15. *Let M be a smooth manifold with or without boundary. A map $D\colon C^\infty(M) \to C^\infty(M)$ is a derivation if and only if it is of the form $Df = Xf$ for some smooth vector field $X \in \mathfrak{X}(M)$.*

Proof. We just showed that every smooth vector field induces a derivation. Conversely, suppose $D\colon C^\infty(M) \to C^\infty(M)$ is a derivation. We need to concoct a vector field X such that $Df = Xf$ for all f. From the discussion above, it is clear that if there is such a vector field, its value at $p \in M$ must be the derivation at p whose action on any smooth real-valued function f is given by

$$X_p f = (Df)(p).$$

The linearity of D guarantees that this expression depends linearly on f, and the fact that D is a derivation yields the product rule (3.4) for tangent vectors. Thus, the map $X_p\colon C^\infty(M) \to \mathbb{R}$ so defined is indeed a tangent vector, that is, a derivation of $C^\infty(M)$ at p. This defines X as a rough vector field. Because $Xf = Df$ is smooth whenever $f \in C^\infty(M)$, this vector field is smooth by Proposition 8.14. $\square$

Because of this result, we sometimes *identify* smooth vector fields on M with derivations of $C^\infty(M)$, using the same letter for both the vector field (thought of as a smooth map from M to TM) and the derivation (thought of as a linear map from $C^\infty(M)$ to itself).

Vector Fields and Smooth Maps

If $F\colon M \to N$ is a smooth map and X is a vector field on M, then for each point $p \in M$, we obtain a vector $dF_p(X_p) \in T_{F(p)}N$ by applying the differential of F to X_p. However, this does not in general define a *vector field* on N. For example, if F is not surjective, there is no way to decide what vector to assign to a point $q \in N \smallsetminus F(M)$ (Fig. 8.2). If F is not injective, then for some points of N there

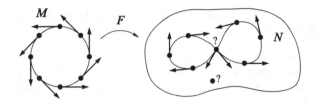

Fig. 8.2 The differential might not take vector fields to vector fields

may be several different vectors obtained by applying dF to X at different points of M.

Suppose $F: M \to N$ is smooth and X is a vector field on M, and suppose there happens to be a vector field Y on N with the property that for each $p \in M$, $dF_p(X_p) = Y_{F(p)}$. In this case, we say the vector fields X and Y are *F-related* (see Fig. 8.3). The next proposition shows how F-related vector fields act on smooth functions.

Proposition 8.16. *Suppose $F: M \to N$ is a smooth map between manifolds with or without boundary, $X \in \mathfrak{X}(M)$, and $Y \in \mathfrak{X}(N)$. Then X and Y are F-related if and only if for every smooth real-valued function f defined on an open subset of N,*

$$X(f \circ F) = (Yf) \circ F. \tag{8.6}$$

Proof. For any $p \in M$ and any smooth real-valued f defined in a neighborhood of $F(p)$,

$$X(f \circ F)(p) = X_p(f \circ F) = dF_p(X_p)f,$$

while

$$(Yf) \circ F(p) = (Yf)\big(F(p)\big) = Y_{F(p)}f.$$

Thus, (8.6) is true for all f if and only if $dF_p(X_p) = Y_{F(p)}$ for all p, i.e., if and only if X and Y are F-related. $\square$

Example 8.17. Let $F: \mathbb{R} \to \mathbb{R}^2$ be the smooth map $F(t) = (\cos t, \sin t)$. Then $d/dt \in \mathfrak{X}(\mathbb{R})$ is F-related to the vector field $Y \in \mathfrak{X}(\mathbb{R}^2)$ defined by

$$Y = x\frac{\partial}{\partial y} - y\frac{\partial}{\partial x}.$$ //

▶ **Exercise 8.18.** Prove the claim in the preceding example in two ways: directly from the definition, and by using Proposition 8.16.

It is important to remember that for a given smooth map $F: M \to N$ and vector field $X \in \mathfrak{X}(M)$, there may not be *any* vector field on N that is F-related to X. There is one special case, however, in which there is always such a vector field, as the next proposition shows.

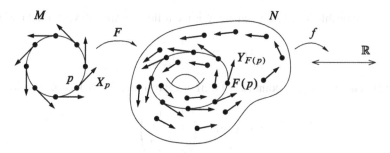

Fig. 8.3 F-related vector fields

Proposition 8.19. *Suppose M and N are smooth manifolds with or without boundary, and $F: M \to N$ is a diffeomorphism. For every $X \in \mathfrak{X}(M)$, there is a unique smooth vector field on N that is F-related to X.*

Proof. For $Y \in \mathfrak{X}(N)$ to be F-related to X means that $dF_p(X_p) = Y_{F(p)}$ for every $p \in M$. If F is a diffeomorphism, therefore, we *define* Y by

$$Y_q = dF_{F^{-1}(q)}\big(X_{F^{-1}(q)}\big).$$

It is clear that Y, so defined, is the unique (rough) vector field that is F-related to X. Note that $Y: N \to TN$ is the composition of the following smooth maps:

$$N \xrightarrow{\ F^{-1}\ } M \xrightarrow{\ X\ } TM \xrightarrow{\ dF\ } TN.$$

It follows that Y is smooth. $\square$

In the situation of the preceding proposition we denote the unique vector field that is F-related to X by F_*X, and call it the *pushforward of X by F*. Remember, it is only when F is a diffeomorphism that F_*X is defined. The proof of Proposition 8.19 shows that F_*X is defined explicitly by the formula

$$(F_*X)_q = dF_{F^{-1}(q)}\big(X_{F^{-1}(q)}\big). \tag{8.7}$$

As long as the inverse map F^{-1} can be computed explicitly, the pushforward of a vector field can be computed directly from this formula.

Example 8.20 (Computing the Pushforward of a Vector Field). Let M and N be the following open submanifolds of $\mathbb{R}^2$:

$$M = \{(x, y) : y > 0 \text{ and } x + y > 0\},$$

$$N = \{(u, v) : u > 0 \text{ and } v > 0\},$$

and define $F: M \to N$ by $F(x, y) = (x + y, \, x/y + 1)$. Then F is a diffeomorphism because its inverse is easily computed: just solve $(u, v) = (x + y, \, x/y + 1)$ for x and y to obtain the formula $(x, y) = F^{-1}(u, v) = (u - u/v, \, u/v)$. Let us compute the pushforward $F_* X$, where X is the following smooth vector field on M:

$$X_{(x,y)} = y^2 \frac{\partial}{\partial x}\bigg|_{(x,y)}.$$

The differential of F at a point $(x, y) \in M$ is represented by its Jacobian matrix,

$$DF(x, y) = \begin{pmatrix} 1 & 1 \\ \frac{1}{y} & -\frac{x}{y^2} \end{pmatrix},$$

and thus $dF_{F^{-1}(u,v)}$ is represented by the matrix

$$DF\left(u - \frac{u}{v}, \frac{u}{v}\right) = \begin{pmatrix} 1 & 1 \\ v & v - v^2 \\ \overline{u} & \overline{u} \end{pmatrix}.$$

For any $(u, v) \in N$,

$$X_{F^{-1}(u,v)} = \frac{u^2}{v^2} \frac{\partial}{\partial x}\bigg|_{F^{-1}(u,v)}.$$

Therefore, applying (8.7) with $p = (u, v)$ yields the formula for $F_* X$:

$$(F_* X)_{(u,v)} = \frac{u^2}{v^2} \frac{\partial}{\partial u}\bigg|_{(u,v)} + \frac{u}{v} \frac{\partial}{\partial v}\bigg|_{(u,v)}. \qquad\qquad /\!/$$

The next corollary follows directly from Proposition 8.16.

Corollary 8.21. *Suppose $F: M \to N$ is a diffeomorphism and $X \in \mathfrak{X}(M)$. For any $f \in C^\infty(N)$,*

$$\big((F_* X)f\big) \circ F = X(f \circ F). \qquad\qquad \square$$

Vector Fields and Submanifolds

If $S \subseteq M$ is an immersed or embedded submanifold (with or without boundary), a vector field X on M does not necessarily restrict to a vector field on S, because X_p may not lie in the subspace $T_p S \subseteq T_p M$ at a point $p \in S$. Given a point $p \in S$, a vector field X on M is said to be **tangent to S at p** if $X_p \in T_p S \subseteq T_p M$. It is **tangent to S** if it is tangent to S at every point of S (Fig. 8.4).

Proposition 8.22. *Let M be a smooth manifold, $S \subseteq M$ be an embedded submanifold with or without boundary, and X be a smooth vector field on M. Then X is tangent to S if and only if $(Xf)|_S = 0$ for every $f \in C^\infty(M)$ such that $f|_S \equiv 0$.*

Proof. This is an immediate consequence of Proposition 5.37. $\qquad\square$

Suppose $S \subseteq M$ is an immersed submanifold with or without boundary, and Y is a smooth vector field on M. If there is a vector field $X \in \mathfrak{X}(S)$ that is ι-related to Y, where $\iota: S \hookrightarrow M$ is the inclusion map, then clearly Y is tangent to S, because

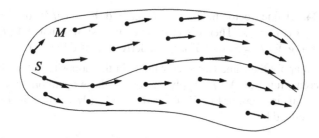

Fig. 8.4 A vector field tangent to a submanifold

$Y_p = d\iota_p(X_p)$ is in the image of $d\iota_p$ for each $p \in S$. The next proposition shows that the converse is true.

Proposition 8.23 (Restricting Vector Fields to Submanifolds). *Let M be a smooth manifold, let $S \subseteq M$ be an immersed submanifold with or without boundary, and let $\iota\colon S \hookrightarrow M$ denote the inclusion map. If $Y \in \mathfrak{X}(M)$ is tangent to S, then there is a unique smooth vector field on S, denoted by $Y|_S$, that is ι-related to Y.*

Proof. The fact that Y is tangent to S means by definition that Y_p is in the image of $d\iota_p$ for each p. Thus, for each p there is a vector $X_p \in T_pS$ such that $Y_p = d\iota_p(X_p)$. Since $d\iota_p$ is injective, X_p is unique, so this defines X as a rough vector field on S. If we can show that X is smooth, it is the unique vector field that is ι-related to Y. It suffices to show that it is smooth in a neighborhood of each point.

Let p be any point in S. Since an immersed submanifold (with or without boundary) is locally embedded, there is a neighborhood V of p in S that is embedded in M. Let $(U, (x^i))$ be a slice chart (or boundary slice chart) for V in M centered at p, so that $V \cap U$ is the subset where $x^{k+1} = \cdots = x^n = 0$ (and $x^k \geq 0$ if $p \in \partial S$), and $(x^1, \ldots, x^k)$ form local coordinates for S in $V \cap U$. If $Y = Y^1 \partial/\partial x^1 + \cdots + Y^n \partial/\partial x^n$ in these coordinates, it follows from our construction that X has the coordinate representation $Y^1 \partial/\partial x^1 + \cdots + Y^k \partial/\partial x^k$, which is clearly smooth on $V \cap U$. $\qquad\square$

Lie Brackets

In this section we introduce an important way of combining two smooth vector fields to obtain another vector field.

Let X and Y be smooth vector fields on a smooth manifold M. Given a smooth function $f\colon M \to \mathbb{R}$, we can apply X to f and obtain another smooth function Xf (see Proposition 8.14). In turn, we can apply Y to this function, and obtain yet another smooth function $YXf = Y(Xf)$. The operation $f \mapsto YXf$, however, does not in general satisfy the product rule and thus cannot be a vector field, as the following example shows.

Example 8.24. Define vector fields $X = \partial/\partial x$ and $Y = x\partial/\partial y$ on $\mathbb{R}^2$, and let $f(x, y) = x$, $g(x, y) = y$. Then direct computation shows that $XY(fg) = 2x$, while $f\, XYg + g\, XYf = x$, so XY is not a derivation of $C^\infty(\mathbb{R}^2)$.   //

We can also apply the same two vector fields in the opposite order, obtaining a (usually different) function XYf. Applying both of these operators to f and subtracting, we obtain an operator $[X, Y] \colon C^\infty(M) \to C^\infty(M)$, called the **Lie bracket of X and Y**, defined by

$$[X, Y]f = XYf - YXf.$$

The key fact is that this operator *is* a vector field.

Lemma 8.25. *The Lie bracket of any pair of smooth vector fields is a smooth vector field.*

Proof. By Proposition 8.15, it suffices to show that $[X, Y]$ is a derivation of $C^\infty(M)$. For arbitrary $f, g \in C^\infty(M)$, we compute

$$\begin{aligned}
[X, Y](fg) &= X\big(Y(fg)\big) - Y\big(X(fg)\big) \\
&= X(f\, Yg + g\, Yf) - Y(f\, Xg + g\, Xf) \\
&= Xf\, Yg + f\, XYg + Xg\, Yf + g\, XYf \\
&\quad - Yf\, Xg - f\, YXg - Yg\, Xf - g\, YXf \\
&= f\, XYg + g\, XYf - f\, YXg - g\, YXf \\
&= f[X, Y]g + g[X, Y]f. \qquad\qquad \square
\end{aligned}$$

We will describe one significant application of Lie brackets later in this chapter, and we will see many others in later chapters. Unfortunately, we are not yet in a position to give Lie brackets a geometric interpretation, but we will do so in Chapter 9. For now, we develop some of their basic properties.

The value of the vector field $[X, Y]$ at a point $p \in M$ is the derivation at p given by the formula

$$[X, Y]_p f = X_p(Yf) - Y_p(Xf).$$

However, this formula is of limited usefulness for computations, because it requires one to compute terms involving second derivatives of f that will always cancel each other out. The next proposition gives an extremely useful coordinate formula for the Lie bracket, in which the cancellations have already been accounted for.

Proposition 8.26 (Coordinate Formula for the Lie Bracket). *Let X, Y be smooth vector fields on a smooth manifold M with or without boundary, and let $X = X^i\, \partial/\partial x^i$ and $Y = Y^j\, \partial/\partial x^j$ be the coordinate expressions for X and Y in terms of some smooth local coordinates (x^i) for M. Then $[X, Y]$ has the following coordinate expression:*

$$[X, Y] = \left(X^i \frac{\partial Y^j}{\partial x^i} - Y^i \frac{\partial X^j}{\partial x^i} \right) \frac{\partial}{\partial x^j}, \tag{8.8}$$

or more concisely,

$$[X, Y] = \left(X Y^j - Y X^j\right) \frac{\partial}{\partial x^j}. \tag{8.9}$$

Proof. Because we know already that $[X, Y]$ is a smooth vector field, its action on a function is determined locally: $([X, Y]f)|_U = [X, Y](f|_U)$. Thus it suffices to compute in a single smooth chart, where we have

$$[X, Y]f = X^i \frac{\partial}{\partial x^i} \left(Y^j \frac{\partial f}{\partial x^j}\right) - Y^j \frac{\partial}{\partial x^j} \left(X^i \frac{\partial f}{\partial x^i}\right)$$

$$= X^i \frac{\partial Y^j}{\partial x^i} \frac{\partial f}{\partial x^j} + X^i Y^j \frac{\partial^2 f}{\partial x^i \partial x^j} - Y^j \frac{\partial X^i}{\partial x^j} \frac{\partial f}{\partial x^i} - Y^j X^i \frac{\partial^2 f}{\partial x^j \partial x^i}$$

$$= X^i \frac{\partial Y^j}{\partial x^i} \frac{\partial f}{\partial x^j} - Y^j \frac{\partial X^i}{\partial x^j} \frac{\partial f}{\partial x^i},$$

where in the last step we have used the fact that mixed partial derivatives of a smooth function can be taken in any order. Interchanging the roles of the dummy indices i and j in the second term, we obtain (8.8). $\qquad\square$

One trivial application of (8.8) is to compute the Lie brackets of the coordinate vector fields $(\partial/\partial x^i)$ in any smooth chart: because the component functions of the coordinate vector fields are all constants, it follows that

$$\left[\frac{\partial}{\partial x^i}, \frac{\partial}{\partial x^j}\right] \equiv 0 \quad \text{for all } i \text{ and } j. \tag{8.10}$$

(This also follows from the definition of the Lie bracket, and is essentially a restatement of the fact that mixed partial derivatives of smooth functions commute.) Here is a slightly less trivial computation.

Example 8.27. Define smooth vector fields $X, Y \in \mathfrak{X}\left(\mathbb{R}^3\right)$ by

$$X = x \frac{\partial}{\partial x} + \frac{\partial}{\partial y} + x(y + 1) \frac{\partial}{\partial z},$$

$$Y = \frac{\partial}{\partial x} + y \frac{\partial}{\partial z}.$$

Then (8.9) yields

$$[X, Y] = X(1) \frac{\partial}{\partial x} + X(y) \frac{\partial}{\partial z} - Y(x) \frac{\partial}{\partial x} - Y(1) \frac{\partial}{\partial y} - Y\left(x(y + 1)\right) \frac{\partial}{\partial z}$$

$$= 0 \frac{\partial}{\partial x} + 1 \frac{\partial}{\partial z} - 1 \frac{\partial}{\partial x} - 0 \frac{\partial}{\partial y} - (y + 1) \frac{\partial}{\partial z}$$

$$= -\frac{\partial}{\partial x} - y \frac{\partial}{\partial z}. \qquad /\!/$$

Proposition 8.28 (Properties of the Lie Bracket). *The Lie bracket satisfies the following identities for all $X, Y, Z \in \mathfrak{X}(M)$:*

(a) BILINEARITY: *For $a, b \in \mathbb{R}$,*

$$[aX + bY, Z] = a[X, Z] + b[Y, Z],$$
$$[Z, aX + bY] = a[Z, X] + b[Z, Y].$$

(b) ANTISYMMETRY:

$$[X, Y] = -[Y, X].$$

(c) JACOBI IDENTITY:

$$[X, [Y, Z]] + [Y, [Z, X]] + [Z, [X, Y]] = 0.$$

(d) *For $f, g \in C^\infty(M)$,*

$$[fX, gY] = fg[X, Y] + (fXg)Y - (gYf)X. \tag{8.11}$$

Proof. Bilinearity and antisymmetry are obvious consequences of the definition. The proof of the Jacobi identity is just a computation:

$$[X, [Y, Z]]f + [Y, [Z, X]]f + [Z, [X, Y]]f$$
$$= X[Y, Z]f - [Y, Z]Xf + Y[Z, X]f$$
$$\quad - [Z, X]Yf + Z[X, Y]f - [X, Y]Zf$$
$$= XYZf - XZYf - YZXf + ZYXf + YZXf - YXZf$$
$$\quad - ZXYf + XZYf + ZXYf - ZYXf - XYZf + YXZf.$$

In this last expression all the terms cancel in pairs. Part (d) is a direct computation from the definition of the Lie bracket, and is left as an exercise.   □

▶ **Exercise 8.29.** Prove part (d) of the preceding proposition.

The significance of part (d) of this proposition might not be evident at this point, but it will become clearer in the next chapter, where we will see that it expresses the fact that the Lie bracket satisfies product rules with respect to both of its arguments (see Corollary 9.39).

Proposition 8.30 (Naturality of the Lie Bracket). *Let $F: M \to N$ be a smooth map between manifolds with or without boundary, and let $X_1, X_2 \in \mathfrak{X}(M)$ and $Y_1, Y_2 \in \mathfrak{X}(N)$ be vector fields such that X_i is F-related to Y_i for $i = 1, 2$. Then $[X_1, X_2]$ is F-related to $[Y_1, Y_2]$.*

Proof. Using Proposition 8.16 and the fact that X_i and Y_i are F-related,

$$X_1 X_2 (f \circ F) = X_1 \big((Y_2 f) \circ F \big) = (Y_1 Y_2 f) \circ F.$$

Similarly,

$$X_2 X_1 (f \circ F) = (Y_2 Y_1 f) \circ F.$$

Therefore,

$$[X_1, X_2](f \circ F) = X_1 X_2 (f \circ F) - X_2 X_1 (f \circ F)$$

$$= (Y_1 Y_2 f) \circ F - (Y_2 Y_1 f) \circ F$$
$$= ([Y_1, Y_2] f) \circ F. \qquad\qquad \square$$

See Problem 11-18 for an indication of why this property is called "naturality." When applied in special cases, this result has the following important corollaries. First we consider the case in which the map is a diffeomorphism.

Corollary 8.31 (Pushforwards of Lie Brackets). *Suppose* $F: M \to N$ *is a diffeomorphism and* $X_1, X_2 \in \mathfrak{X}(M)$. *Then* $F_*[X_1, X_2] = [F_* X_1, F_* X_2]$.

Proof. This is just the special case of Proposition 8.30 in which F is a diffeomorphism and $Y_i = F_* X_i$. $\qquad\qquad \square$

The second special case is that of the inclusion of a submanifold.

Corollary 8.32 (Brackets of Vector Fields Tangent to Submanifolds). *Let* M *be a smooth manifold and let* S *be an immersed submanifold with or without boundary in* M. *If* Y_1 *and* Y_2 *are smooth vector fields on* M *that are tangent to* S, *then* $[Y_1, Y_2]$ *is also tangent to* S.

Proof. By Proposition 8.23, there exist smooth vector fields X_1 and X_2 on S such that X_i is ι-related to Y_i for $i = 1, 2$ (where $\iota: S \to M$ is the inclusion). By Proposition 8.30, $[X_1, X_2]$ is ι-related to $[Y_1, Y_2]$, which is therefore tangent to S. $\qquad\square$

The Lie Algebra of a Lie Group

One of the most important applications of Lie brackets occurs in the context of Lie groups. Suppose G is a Lie group. Recall that G acts smoothly and transitively on itself by left translation: $L_g(h) = gh$. (See Example 7.22(c).) A vector field X on G is said to be **left-invariant** if it is invariant under all left translations, in the sense that it is L_g-related to itself for every $g \in G$. More explicitly, this means

$$d(L_g)_{g'}(X_{g'}) = X_{gg'}, \quad \text{for all } g, g' \in G. \qquad (8.12)$$

Since L_g is a diffeomorphism, this can be abbreviated by writing $(L_g)_* X = X$ for every $g \in G$.

Because $(L_g)_*(aX + bY) = a(L_g)_* X + b(L_g)_* Y$, the set of all smooth left-invariant vector fields on G is a linear subspace of $\mathfrak{X}(G)$. But it is much more than that. The central fact is that it is closed under Lie brackets.

Proposition 8.33. *Let* G *be a Lie group, and suppose* X *and* Y *are smooth left-invariant vector fields on* G. *Then* $[X, Y]$ *is also left-invariant.*

Proof. Let $g \in G$ be arbitrary. Since $(L_g)_* X = X$ and $(L_g)_* Y = Y$ by definition of left-invariance, it follows from Corollary 8.31 that

$$(L_g)_*[X, Y] = [(L_g)_* X, (L_g)_* Y] = [X, Y].$$

Thus, $[X, Y]$ is L_g-related to itself for each g, which is to say it is left-invariant. $\square$

A *Lie algebra* (over $\mathbb{R}$) is a real vector space $\mathfrak{g}$ endowed with a map called the *bracket* from $\mathfrak{g} \times \mathfrak{g}$ to $\mathfrak{g}$, usually denoted by $(X, Y) \mapsto [X, Y]$, that satisfies the following properties for all $X, Y, Z \in \mathfrak{g}$:

(i) BILINEARITY: For $a, b \in \mathbb{R}$,

$$[aX + bY, Z] = a[X, Z] + b[Y, Z],$$
$$[Z, aX + bY] = a[Z, X] + b[Z, Y].$$

(ii) ANTISYMMETRY:

$$[X, Y] = -[Y, X].$$

(iii) JACOBI IDENTITY:

$$[X, [Y, Z]] + [Y, [Z, X]] + [Z, [X, Y]] = 0.$$

Notice that the Jacobi identity is a substitute for associativity, which does not hold in general for brackets in a Lie algebra. It is useful in some circumstances to define Lie algebras over $\mathbb{C}$ or other fields, but we do not have any reason to consider such Lie algebras; thus all of our Lie algebras are assumed without further comment to be real.

If $\mathfrak{g}$ is a Lie algebra, a linear subspace $\mathfrak{h} \subseteq \mathfrak{g}$ is called a *Lie subalgebra of* $\mathfrak{g}$ if it is closed under brackets. In this case $\mathfrak{h}$ is itself a Lie algebra with the restriction of the same bracket.

If $\mathfrak{g}$ and $\mathfrak{h}$ are Lie algebras, a linear map $A \colon \mathfrak{g} \to \mathfrak{h}$ is called a *Lie algebra homomorphism* if it preserves brackets: $A[X, Y] = [AX, AY]$. An invertible Lie algebra homomorphism is called a *Lie algebra isomorphism*. If there exists a Lie algebra isomorphism from $\mathfrak{g}$ to $\mathfrak{h}$, we say that they are *isomorphic* as Lie algebras.

▶ **Exercise 8.34.** Verify that the kernel and image of a Lie algebra homomorphism are Lie subalgebras.

▶ **Exercise 8.35.** Suppose $\mathfrak{g}$ and $\mathfrak{h}$ are finite-dimensional Lie algebras and $A \colon \mathfrak{g} \to \mathfrak{h}$ is a linear map. Show that A is a Lie algebra homomorphism if and only if $A[E_i, E_j] = [AE_i, AE_j]$ for some basis $(E_1, \ldots, E_n)$ of $\mathfrak{g}$.

Example 8.36 (Lie Algebras).

(a) The space $\mathfrak{X}(M)$ of all smooth vector fields on a smooth manifold M is a Lie algebra under the Lie bracket by Proposition 8.28.

(b) If G is a Lie group, the set of all smooth left-invariant vector fields on G is a Lie subalgebra of $\mathfrak{X}(G)$ and is therefore a Lie algebra.

(c) The vector space $\mathrm{M}(n, \mathbb{R})$ of $n \times n$ real matrices becomes an n^2-dimensional Lie algebra under the *commutator bracket*:

$$[A, B] = AB - BA.$$

Bilinearity and antisymmetry are obvious from the definition, and the Jacobi identity follows from a straightforward calculation. When we are regarding $\mathrm{M}(n, \mathbb{R})$ as a Lie algebra with this bracket, we denote it by $\mathfrak{gl}(n, \mathbb{R})$.

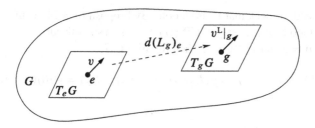

Fig. 8.5 Defining a left-invariant vector field

(d) Similarly, $\mathfrak{gl}(n,\mathbb{C})$ is the $2n^2$-dimensional (real) Lie algebra obtained by endowing $M(n,\mathbb{C})$ with the commutator bracket.

(e) If V is a vector space, the vector space of all linear maps from V to itself becomes a Lie algebra, which we denote by $\mathfrak{gl}(V)$, with the commutator bracket:

$$[A, B] = A \circ B - B \circ A.$$

Under our usual identification of $n \times n$ matrices with linear maps from $\mathbb{R}^n$ to itself, $\mathfrak{gl}(\mathbb{R}^n)$ is the same as $\mathfrak{gl}(n,\mathbb{R})$.

(f) Any vector space V becomes a Lie algebra if we define all brackets to be zero. Such a Lie algebra is said to be ***abelian***. (The name reflects the fact that brackets in most Lie algebras, as in the preceding examples, are defined as commutators in terms of underlying associative products, so all brackets are zero precisely when the underlying product is commutative; it also reflects the connection between abelian Lie algebras and abelian Lie groups, which you will explore in Problems 8-25 and 20-7.)   //

Example (b) is the most important one. The Lie algebra of all smooth left-invariant vector fields on a Lie group G is called the **Lie algebra of G**, and is denoted by $\mathrm{Lie}(G)$. (We will see below that the assumption of smoothness is redundant; see Corollary 8.38.) The fundamental fact is that $\mathrm{Lie}(G)$ is finite-dimensional, and in fact has the same dimension as G itself, as the following theorem shows.

Theorem 8.37. *Let G be a Lie group. The evaluation map $\varepsilon\colon \mathrm{Lie}(G) \to T_eG$, given by $\varepsilon(X) = X_e$, is a vector space isomorphism. Thus, $\mathrm{Lie}(G)$ is finite-dimensional, with dimension equal to $\dim G$.*

Proof. It is clear from the definition that ε is linear over $\mathbb{R}$. It is easy to prove that it is injective: if $\varepsilon(X) = X_e = 0$ for some $X \in \mathrm{Lie}(G)$, then left-invariance of X implies that $X_g = d(L_g)_e(X_e) = 0$ for every $g \in G$, so $X = 0$.

To show that ε is surjective, let $v \in T_eG$ be arbitrary, and define a (rough) vector field v^L on G by

$$v^L\big|_g = d(L_g)_e(v). \tag{8.13}$$

(See Fig. 8.5.) If there is a left-invariant vector field on G whose value at the identity is v, clearly it has to be given by this formula.

First we need to check that v^L is smooth. By Proposition 8.14, it suffices to show that $v^L f$ is smooth whenever $f \in C^\infty(G)$. Choose a smooth curve $\gamma \colon (-\delta, \delta) \to G$ such that $\gamma(0) = e$ and $\gamma'(0) = v$. Then for all $g \in G$,

$$\left(v^L f\right)(g) = v^L\big|_g f = d(L_g)_e(v) f = v(f \circ L_g) = \gamma'(0)(f \circ L_g)$$

$$= \frac{d}{dt}\bigg|_{t=0} (f \circ L_g \circ \gamma)(t).$$

If we define $\varphi \colon (-\delta, \delta) \times G \to \mathbb{R}$ by $\varphi(t, g) = f \circ L_g \circ \gamma(t) = f\big(g\gamma(t)\big)$, the computation above shows that $\left(v^L f\right)(g) = \partial\varphi/\partial t(0, g)$. Because φ is a composition of group multiplication, f, and γ, it is smooth. It follows that $\partial\varphi/\partial t(0, g)$ depends smoothly on g, so $v^L f$ is smooth.

Next we show that v^L is left-invariant, which is to say that $d(L_h)_g\left(v^L\big|_g\right) = v^L\big|_{hg}$ for all $g, h \in G$. This follows from the definition of v^L and the fact that $L_h \circ L_g = L_{hg}$:

$$d(L_h)_g\left(v^L\big|_g\right) = d(L_h)_g \circ d(L_g)_e(v) = d(L_h \circ L_g)_e(v) = d(L_{hg})_e(v) = v^L\big|_{hg}.$$

Thus $v^L \in \mathrm{Lie}(G)$. Since L_e (left translation by the identity) is the identity map of G, it follows that $\varepsilon\left(v^L\right) = v^L\big|_e = v$, so ε is surjective. $\square$

Given any vector $v \in T_e G$, we continue to use the notation v^L to denote the smooth left-invariant vector field defined by (8.13).

It is worth observing that the preceding proof also shows that the assumption of smoothness in the definition of $\mathrm{Lie}(G)$ is unnecessary.

Corollary 8.38. *Every left-invariant rough vector field on a Lie group is smooth.*

Proof. Let X be a left-invariant rough vector field on a Lie group G, and let $v = X_e$. The fact that X is left-invariant implies that $X = v^L$, which is smooth. $\square$

The existence of global left-invariant vector fields also yields the following important property of Lie groups. Recall that a smooth manifold is said to be *parallelizable* if it admits a smooth global frame. If G is a Lie group, a local or global frame consisting of left-invariant vector fields is called a **left-invariant frame**.

Corollary 8.39. *Every Lie group admits a left-invariant smooth global frame, and therefore every Lie group is parallelizable.*

Proof. If G is a Lie group, every basis for $\mathrm{Lie}(G)$ is a left-invariant smooth global frame for G. $\square$

Example 8.40. Let us determine the Lie algebras of some familiar Lie groups.

(a) EUCLIDEAN SPACE $\mathbb{R}^n$: If we consider $\mathbb{R}^n$ as a Lie group under addition, left translation by an element $b \in \mathbb{R}^n$ is given by the affine map $L_b(x) = b + x$, whose differential $d(L_b)$ is represented by the identity matrix in standard co-ordinates. Thus a vector field $X^i \partial/\partial x^i$ is left-invariant if and only if its coefficients X^i are constants. Because the Lie bracket of two constant-coefficient vector fields is zero by (8.8), the Lie algebra of $\mathbb{R}^n$ is abelian, and is isomorphic to $\mathbb{R}^n$ itself with the trivial bracket. In brief, $\text{Lie}(\mathbb{R}^n) \cong \mathbb{R}^n$.

(b) THE CIRCLE GROUP $\mathbb{S}^1$: In terms of appropriate angle coordinates, each left translation has a local coordinate representation of the form $\theta \mapsto \theta + c$. Since the differential of this map is the 1×1 identity matrix, it follows that the vector field $d/d\theta$ defined in Example 8.4 is left-invariant, and is therefore a basis for the Lie algebra of $\mathbb{S}^1$. This Lie algebra is 1-dimensional and abelian, and therefore $\text{Lie}(\mathbb{S}^1) \cong \mathbb{R}$.

(c) THE n-TORUS $\mathbb{T}^n = \mathbb{S}^1 \times \cdots \times \mathbb{S}^1$: An analysis similar to the one above shows that $(\partial/\partial\theta^1, \ldots, \partial/\partial\theta^n)$ is a basis for $\text{Lie}(\mathbb{T}^n)$, where $\partial/\partial\theta^i$ is the angle coordinate vector field on the ith $\mathbb{S}^1$ factor. Since the Lie brackets of these coordinate vector fields are all zero, $\text{Lie}(\mathbb{T}^n) \cong \mathbb{R}^n$. //

The Lie groups $\mathbb{R}^n$, $\mathbb{S}^1$, and $\mathbb{T}^n$ are abelian, and as the discussion above shows, their Lie algebras turn out also to be abelian. This is no accident: every abelian Lie group has an abelian Lie algebra (see Problem 8-25). Later, we will see that the converse is true provided that the group is connected (Problem 20-7).

Just as we can view the tangent space as a "linear model" of a smooth manifold near a point, the Lie algebra of a Lie group provides a "linear model" of the group, which reflects many of the properties of the group. Because Lie groups have more structure than ordinary smooth manifolds, it should come as no surprise that their linear models have more structure than ordinary vector spaces. Since a finite-dimensional Lie algebra is a purely linear-algebraic object, it is in many ways simpler to understand than the group itself. Much of the progress in the theory of Lie groups has come from a careful analysis of Lie algebras.

We conclude this section by analyzing the Lie algebra of the most important non-abelian Lie group of all, the general linear group. Theorem 8.37 gives a vector space isomorphism between $\text{Lie}(\text{GL}(n, \mathbb{R}))$ and the tangent space to $\text{GL}(n, \mathbb{R})$ at the identity matrix I_n. Because $\text{GL}(n, \mathbb{R})$ is an open subset of the vector space $\mathfrak{gl}(n, \mathbb{R})$, its tangent space is naturally isomorphic to $\mathfrak{gl}(n, \mathbb{R})$ itself. The composition of these two isomorphisms gives a vector space isomorphism $\text{Lie}(\text{GL}(n, \mathbb{R})) \cong \mathfrak{gl}(n, \mathbb{R})$.

The vector spaces $\text{Lie}(\text{GL}(n, \mathbb{R}))$ and $\mathfrak{gl}(n, \mathbb{R})$ have independently defined Lie algebra structures—the first coming from Lie brackets of vector fields, and the second from commutator brackets of matrices. The next proposition shows that the natural vector space isomorphism between these spaces is in fact a Lie algebra isomorphism.

Proposition 8.41 (Lie Algebra of the General Linear Group). *The composition of the natural maps*

$$\text{Lie}\left(\text{GL}(n, \mathbb{R})\right) \to T_{I_n} \text{GL}(n, \mathbb{R}) \to \mathfrak{gl}(n, \mathbb{R}) \tag{8.14}$$

gives a Lie algebra isomorphism between $\text{Lie}(\text{GL}(n, \mathbb{R}))$ *and the matrix algebra* $\mathfrak{gl}(n, \mathbb{R})$.

Proof. Using the matrix entries X^i_j as global coordinates on $GL(n, \mathbb{R}) \subseteq \mathfrak{gl}(n, \mathbb{R})$, the natural isomorphism $T_{I_n} GL(n, \mathbb{R}) \longleftrightarrow \mathfrak{gl}(n, \mathbb{R})$ takes the form

$$A^i_j \frac{\partial}{\partial X^i_j}\Big|_{I_n} \longleftrightarrow (A^i_j).$$

(Because of the dual role of the indices i, j as coordinate indices and matrix row and column indices, in this case it is impossible to maintain our convention that all coordinates have upper indices. However, we continue to observe the summation convention and the other index conventions associated with it. In particular, in the expression above, an upper index "in the denominator" is to be regarded as a lower index, and vice versa.)

Let $\mathfrak{g}$ denote the Lie algebra of $GL(n, \mathbb{R})$. Any matrix $A = (A^i_j) \in \mathfrak{gl}(n, \mathbb{R})$ determines a left-invariant vector field $A^L \in \mathfrak{g}$ defined by (8.13), which in this case becomes

$$A^L\big|_X = d(L_X)_{I_n}(A) = d(L_X)_{I_n}\left(A^i_j \frac{\partial}{\partial X^i_j}\Big|_{I_n}\right).$$

Since L_X is the restriction to $GL(n, \mathbb{R})$ of the linear map $A \mapsto XA$ on $\mathfrak{gl}(n, \mathbb{R})$, its differential is represented in coordinates by exactly the same linear map. In other words, the left-invariant vector field A^L determined by A is the one whose value at $X \in GL(n, \mathbb{R})$ is

$$A^L\big|_X = X^i_j A^j_k \frac{\partial}{\partial X^i_k}\Big|_X. \tag{8.15}$$

Given two matrices $A, B \in \mathfrak{gl}(n, \mathbb{R})$, the Lie bracket of the corresponding left-invariant vector fields is given by

$$[A^L, B^L] = \left[X^i_j A^j_k \frac{\partial}{\partial X^i_k}, X^p_q B^q_r \frac{\partial}{\partial X^p_r}\right]$$

$$= X^i_j A^j_k \frac{\partial}{\partial X^i_k}\left(X^p_q B^q_r\right)\frac{\partial}{\partial X^p_r} - X^p_q B^q_r \frac{\partial}{\partial X^p_r}\left(X^i_j A^j_k\right)\frac{\partial}{\partial X^i_k}$$

$$= X^i_j A^j_k B^k_r \frac{\partial}{\partial X^i_r} - X^p_q B^q_r A^r_k \frac{\partial}{\partial X^p_k}$$

$$= \left(X^i_j A^j_k B^k_r - X^i_j B^j_k A^k_r\right)\frac{\partial}{\partial X^i_r},$$

where we have used the fact that $\partial X^p_q / \partial X^i_k$ is equal to 1 if $p = i$ and $q = k$, and 0 otherwise, and A^i_j and B^i_j are constants. Evaluating this last expression when X is equal to the identity matrix, we get

$$[A^L, B^L]_{I_n} = \left(A^i_k B^k_r - B^i_k A^k_r\right)\frac{\partial}{\partial X^i_r}\Big|_{I_n}.$$

This is the vector corresponding to the matrix commutator bracket $[A, B]$. Since the left-invariant vector field $[A^L, B^L]$ is determined by its value at the identity, this implies that

$$[A^L, B^L] = [A, B]^L,$$

which is precisely the statement that the composite map (8.14) is a Lie algebra isomorphism.   □

There is an analogue of this result for abstract vector spaces. If V is any finite-dimensional real vector space, recall that we have defined $\mathrm{GL}(V)$ as the Lie group of invertible linear transformations of V, and $\mathfrak{gl}(V)$ as the Lie algebra of all linear transformations. Just as in the case of $\mathrm{GL}(n, \mathbb{R})$, we can regard $\mathrm{GL}(V)$ as an open submanifold of $\mathfrak{gl}(V)$, and thus there are canonical vector space isomorphisms

$$\mathrm{Lie}\left(\mathrm{GL}(V)\right) \to T_{\mathrm{Id}}\,\mathrm{GL}(V) \to \mathfrak{gl}(V). \tag{8.16}$$

Corollary 8.42. *If V is any finite-dimensional real vector space, the composition of the canonical isomorphisms in (8.16) yields a Lie algebra isomorphism between* $\mathrm{Lie}\left(\mathrm{GL}(V)\right)$ *and* $\mathfrak{gl}(V)$.

▶ **Exercise 8.43.** Prove the preceding corollary by choosing a basis for V and applying Proposition 8.41.

Induced Lie Algebra Homomorphisms

The importance of the Lie algebra of a Lie group stems, in large part, from the fact that each Lie group homomorphism induces a Lie algebra homomorphism, as the next theorem shows.

Theorem 8.44 (Induced Lie Algebra Homomorphisms). *Let G and H be Lie groups, and let $\mathfrak{g}$ and $\mathfrak{h}$ be their Lie algebras. Suppose $F : G \to H$ is a Lie group homomorphism. For every $X \in \mathfrak{g}$, there is a unique vector field in $\mathfrak{h}$ that is F-related to X. With this vector field denoted by $F_* X$, the map $F_* : \mathfrak{g} \to \mathfrak{h}$ so defined is a Lie algebra homomorphism.*

Proof. If there is any vector field $Y \in \mathfrak{h}$ that is F-related to X, it must satisfy $Y_e = dF_e(X_e)$, and thus it must be uniquely determined by

$$Y = \left(dF_e(X_e)\right)^L.$$

To show that this Y is F-related to X, we note that the fact that F is a homomorphism implies

$$F(gg') = F(g)F(g') \Rightarrow F(L_g g') = L_{F(g)} F(g')$$

$$\Rightarrow F \circ L_g = L_{F(g)} \circ F$$

$$\Rightarrow dF \circ d(L_g) = d(L_{F(g)}) \circ dF.$$

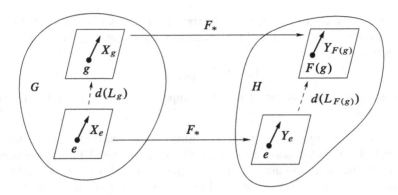

Fig. 8.6 The induced Lie algebra homomorphism

Thus,

$$dF(X_g) = dF\big(d(L_g)(X_e)\big) = d\left(L_{F(g)}\right)\big(dF(X_e)\big) = d\left(L_{F(g)}\right)(Y_e) = Y_{F(g)}.$$

(See Fig. 8.6.) This says precisely that X and Y are F-related.

For each $X \in \mathfrak{g}$, let F_*X denote the unique vector field in $\mathfrak{h}$ that is F-related to X. It then follows immediately from the naturality of Lie brackets that $F_*[X, Y] = [F_*X, F_*Y]$, so F_* is a Lie algebra homomorphism. □

The map $F_* \colon \mathfrak{g} \to \mathfrak{h}$ whose existence is asserted in this theorem is called the **induced Lie algebra homomorphism**. Note that the theorem implies that for any left-invariant vector field $X \in \mathfrak{g}$, F_*X is a well-defined smooth vector field on H, even though F may not be a diffeomorphism.

Proposition 8.45 (Properties of Induced Homomorphisms).

(a) *The homomorphism* $(\mathrm{Id}_G)_* \colon \mathrm{Lie}(G) \to \mathrm{Lie}(G)$ *induced by the identity map of* G *is the identity of* $\mathrm{Lie}(G)$.

(b) *If* $F_1 \colon G \to H$ *and* $F_2 \colon H \to K$ *are Lie group homomorphisms, then*

$$(F_2 \circ F_1)_* = (F_2)_* \circ (F_1)_* \colon \mathrm{Lie}(G) \to \mathrm{Lie}(K).$$

(c) *Isomorphic Lie groups have isomorphic Lie algebras.*

Proof. Both of the relations $d(\mathrm{Id}_G)_e = \mathrm{Id}_{T_eG}$ and $d(F_2 \circ F_1)_e = d(F_2)_e \circ d(F_1)_e$ hold for differentials. Since the induced homomorphism is determined by the differential at the identity, this proves (a) and (b). If $F \colon G \to H$ is an isomorphism, (a) and (b) together imply that $F_* \circ (F^{-1})_* = (F \circ F^{-1})_* = \mathrm{Id} = (F^{-1})_* \circ F_*$, so $F_* \colon \mathrm{Lie}(G) \to \mathrm{Lie}(H)$ is an isomorphism. □

The preceding proposition has a categorical interpretation, as you might have guessed. Let Lie denote the category whose objects are Lie groups and whose morphisms are Lie group homomorphisms, and lie the one whose objects are finite-dimensional Lie algebras and whose morphisms are Lie algebra homomorphisms. Proposition 8.45 can be interpreted as showing that the assignments $G \mapsto \mathrm{Lie}(G)$, $F \mapsto F_*$ define a covariant functor from Lie to lie.

The Lie Algebra of a Lie Subgroup

If G is a Lie group and $H \subseteq G$ is a Lie subgroup, we might hope that the Lie algebra of H would be a Lie subalgebra of that of G. However, elements of Lie(H) are vector fields on H, not G, and so, strictly speaking, are not elements of Lie(G). Nonetheless, the next proposition gives us a way to view Lie(H) as a subalgebra of Lie(G).

Theorem 8.46 (The Lie Algebra of a Lie Subgroup). *Suppose $H \subseteq G$ is a Lie subgroup, and $\iota\colon H \hookrightarrow G$ is the inclusion map. There is a Lie subalgebra $\mathfrak{h} \subseteq$ Lie(G) that is canonically isomorphic to Lie(H), characterized by either of the following descriptions:*

$$\mathfrak{h} = \iota_*\big(\mathrm{Lie}(H)\big)$$
$$= \{X \in \mathrm{Lie}(G) : X_e \in T_e H\}. \tag{8.17}$$

Proof. Because the inclusion map $\iota\colon H \hookrightarrow G$ is a Lie group homomorphism, $\iota_*\big(\mathrm{Lie}(H)\big)$ is a Lie subalgebra of Lie(G). By the way we defined the induced Lie algebra homomorphism, this subalgebra is precisely the set of left-invariant vector fields on G whose values at the identity are of the form $d\iota_e(v)$ for some $v \in T_e H$. Since the differential $d\iota_e\colon T_e H \to T_e G$ is the inclusion of $T_e H$ as a subspace in $T_e G$, the two characterizations of $\mathfrak{h}$ given in (8.17) are equal. Since $d\iota_e$ is injective on $T_e H$, it follows that ι_* is injective on Lie(H); since it is surjective by definition of $\mathfrak{h}$, it is an isomorphism between Lie(H) and $\mathfrak{h}$. $\square$

Using this proposition, whenever H is a Lie subgroup of G, we often *identify* Lie(H) as a subalgebra of Lie(G). As we mentioned above, elements of Lie(H) are not themselves left-invariant vector fields on G. But the preceding proposition shows that every element of Lie(H) corresponds to a unique element of Lie(G), determined by its value at the identity, and the injection of Lie(H) into Lie(G) thus determined respects Lie brackets; so by thinking of Lie(H) as a subalgebra of Lie(G) we are not committing a grave error.

This identification is especially illuminating in the case of Lie subgroups of $GL(n, \mathbb{R})$.

Example 8.47 (The Lie Algebra of O(n)). The orthogonal group O(n) is a Lie subgroup of $GL(n, \mathbb{R})$. By Example 7.27, it is equal to the level set $\Phi^{-1}(I_n)$, where $\Phi\colon GL(n, \mathbb{R}) \to M(n, \mathbb{R})$ is the map $\Phi(A) = A^T A$. By the result of Exercise 5.40, $T_{I_n} O(n)$ is equal to the kernel of $d\Phi_{I_n}\colon T_{I_n} GL(n, \mathbb{R}) \to T_{I_n} M(n, \mathbb{R})$. By the computation in Example 7.27, this differential is $d\Phi_{I_n}(B) = B^T + B$, so

$$T_{I_n} O(n) = \{B \in \mathfrak{gl}(n, \mathbb{R}) : B^T + B = 0\}$$
$$= \{\text{skew-symmetric } n \times n \text{ matrices}\}.$$

We denote this subspace of $\mathfrak{gl}(n, \mathbb{R})$ by $\mathfrak{o}(n)$. Theorem 8.46 then implies that $\mathfrak{o}(n)$ is a Lie subalgebra of $\mathfrak{gl}(n, \mathbb{R})$ that is canonically isomorphic to Lie$\big(O(n)\big)$. Notice that we did not even have to verify directly that $\mathfrak{o}(n)$ is a subalgebra.   //

We showed above that the Lie algebra of $GL(n, \mathbb{R})$ is naturally isomorphic to the matrix algebra $\mathfrak{gl}(n, \mathbb{R})$. We can now prove a similar result for $GL(n, \mathbb{C})$. Just as in the real case, our usual identification of $GL(n, \mathbb{C})$ as an open subset of $\mathfrak{gl}(n, \mathbb{C})$ yields a sequence of vector space isomorphisms

$$\operatorname{Lie}\left(GL(n, \mathbb{C})\right) \xrightarrow{\varepsilon} T_{I_n} GL(n, \mathbb{C}) \xrightarrow{\varphi} \mathfrak{gl}(n, \mathbb{C}), \tag{8.18}$$

where ε is the evaluation map and φ is the usual identification between the tangent space to an open subset of a vector space and the vector space itself. (Note that we are considering these as real vector spaces, not complex ones.)

Proposition 8.48 (The Lie Algebra of $GL(n, \mathbb{C})$). *The composition of the maps in (8.18) yields a Lie algebra isomorphism between* $\operatorname{Lie}\left(GL(n, \mathbb{C})\right)$ *and the matrix algebra* $\mathfrak{gl}(n, \mathbb{C})$.

Proof. The Lie group homomorphism $\beta \colon GL(n, \mathbb{C}) \to GL(2n, \mathbb{R})$ that we constructed in Example 7.18(d) induces a Lie algebra homomorphism

$$\beta_* \colon \operatorname{Lie}\left(GL(n, \mathbb{C})\right) \to \operatorname{Lie}\left(GL(2n, \mathbb{R})\right).$$

Composing β_* with our canonical isomorphisms yields a commutative diagram

$$
\begin{array}{ccccc}
\operatorname{Lie}\left(GL(n, \mathbb{C})\right) & \xrightarrow{\varepsilon} & T_{I_n} GL(n, \mathbb{C}) & \xrightarrow{\varphi} & \mathfrak{gl}(n, \mathbb{C}) \\
\downarrow{\beta_*} & & \downarrow{d\beta_{I_n}} & & \downarrow{\alpha} \\
\operatorname{Lie}\left(GL(2n, \mathbb{R})\right) & \xrightarrow{\varepsilon} & T_{I_{2n}} GL(2n, \mathbb{R}) & \xrightarrow{\varphi} & \mathfrak{gl}(2n, \mathbb{R}),
\end{array}
\tag{8.19}
$$

in which $\alpha = \varphi \circ d\beta_{I_n} \circ \varphi^{-1}$. Proposition 8.41 showed that the composition of the isomorphisms in the bottom row is a Lie algebra isomorphism; we need to show the same thing for the top row.

It is easy to see from the formula in Example 7.18(d) that β is (the restriction of) a linear map. It follows that $d\beta_{I_n} \colon T_{I_n} GL(n, \mathbb{C}) \to T_{I_{2n}} GL(2n, \mathbb{R})$ is given by exactly the same formula as β, as is $\alpha \colon \mathfrak{gl}(n, \mathbb{C}) \to \mathfrak{gl}(2n, \mathbb{R})$. Because $\beta(AB) = \beta(A)\beta(B)$, it follows that α preserves matrix commutators:

$$\alpha[A, B] = \alpha(AB - BA) = \alpha(A)\alpha(B) - \alpha(B)\alpha(A) = [\alpha(A), \alpha(B)].$$

Thus α is an injective Lie algebra homomorphism from $\mathfrak{gl}(n, \mathbb{C})$ to $\mathfrak{gl}(2n, \mathbb{R})$ (considering both as matrix algebras). Replacing the bottom row in (8.19) by the images of the vertical maps, we obtain a commutative diagram of vector space isomorphisms

$$
\begin{array}{ccc}
\operatorname{Lie}\left(GL(n, \mathbb{C})\right) & \xrightarrow{\cong} & \mathfrak{gl}(n, \mathbb{C}) \\
\downarrow{\beta_*} & & \downarrow{\alpha} \\
\beta_*\left(\operatorname{Lie}\left(GL(n, \mathbb{C})\right)\right) & \xrightarrow{\cong} & \alpha\left(\mathfrak{gl}(n, \mathbb{C})\right),
\end{array}
$$

in which the bottom map and the two vertical maps are Lie algebra isomorphisms; it follows that the top map is also a Lie algebra isomorphism. $\square$

Parallel to the notion of representations of Lie groups, there is also a notion of representations of Lie algebras. If $\mathfrak{g}$ is a finite-dimensional Lie algebra, a (*finite-dimensional*) *representation of* $\mathfrak{g}$ is a Lie algebra homomorphism $\varphi \colon \mathfrak{g} \to \mathfrak{gl}(V)$ for some finite-dimensional vector space V, where $\mathfrak{gl}(V)$ denotes the Lie algebra of linear maps from V to itself. If φ is injective, it is said to be *faithful*, in which case $\mathfrak{g}$ is isomorphic to the Lie subalgebra $\varphi(\mathfrak{g}) \subseteq \mathfrak{gl}(V) \cong \mathfrak{gl}(n, \mathbb{R})$.

There is a close connection between representations of Lie groups and representations of their Lie algebras. Suppose G is a Lie group and $\mathfrak{g}$ is its Lie algebra. If $\rho \colon G \to \mathrm{GL}(V)$ is any representation of G, then $\rho_* \colon \mathfrak{g} \to \mathfrak{gl}(V)$ is easily seen to be a representation of $\mathfrak{g}$.

We close this section by stating a deep algebraic result about Lie algebras, which we will use in Chapter 20. The proof of the following theorem requires far more algebra than we have at our disposal, so we refer the interested reader to the proof in [Var84].

Theorem 8.49 (Ado's Theorem). *Every finite-dimensional real Lie algebra admits a faithful finite-dimensional representation.*

Corollary 8.50. *Every finite-dimensional real Lie algebra is isomorphic to a Lie subalgebra of some matrix algebra* $\mathfrak{gl}(n, \mathbb{R})$ *with the commutator bracket.*

Proof. Let $\mathfrak{g}$ be a finite-dimensional real Lie algebra. By Ado's theorem, $\mathfrak{g}$ has a faithful representation $\rho \colon \mathfrak{g} \to \mathfrak{gl}(V)$ for some finite-dimensional real vector space V. Choosing a basis for V yields an isomorphism of $\mathfrak{gl}(V)$ with $\mathfrak{gl}(n, \mathbb{R})$ for some n, and composing ρ with this isomorphism yields an injective Lie algebra homomorphism from $\mathfrak{g}$ into $\mathfrak{gl}(n, \mathbb{R})$. Its image is a Lie subalgebra isomorphic to $\mathfrak{g}$. $\square$

As we mentioned in the previous chapter, it is important to remember that the analogous result for Lie groups is false: there are Lie groups that are not isomorphic to Lie subgroups of $\mathrm{GL}(n, \mathbb{R})$ (see Problem 21-26 for an example).

Problems

8-1. Prove Lemma 8.6 (the extension lemma for vector fields).

8-2. EULER'S HOMOGENEOUS FUNCTION THEOREM: Let c be a real number, and let $f \colon \mathbb{R}^n \smallsetminus \{0\} \to \mathbb{R}$ be a smooth function that is *positively homogeneous of degree* c, meaning that $f(\lambda x) = \lambda^c f(x)$ for all $\lambda > 0$ and $x \in \mathbb{R}^n \smallsetminus \{0\}$. Prove that $Vf = cf$, where V is the Euler vector field defined in Example 8.3. (*Used on p. 248.*)

8-3. Let M be a nonempty positive-dimensional smooth manifold with or without boundary. Show that $\mathfrak{X}(M)$ is infinite-dimensional.

8-4. Let M be a smooth manifold with boundary. Show that there exists a global smooth vector field on M whose restriction to ∂M is everywhere inward-pointing, and one whose restriction to ∂M is everywhere outward-pointing. (*Used on pp. 223, 386.*)

8-5. Prove Proposition 8.11 (completion of local frames).

8-6. Let $\mathbb{H}$ be the algebra of quaternions and let $\mathcal{S} \subseteq \mathbb{H}$ be the group of unit quaternions (see Problems 7-22 and 7-23).

    (a) Show that if $p \in \mathbb{H}$ is imaginary, then qp is tangent to $\mathcal{S}$ at each $q \in \mathcal{S}$. (Here we are identifying each tangent space to $\mathbb{H}$ with $\mathbb{H}$ itself in the usual way.)

    (b) Define vector fields X_1, X_2, X_3 on $\mathbb{H}$ by

$$X_1|_q = q\mathrm{i}, \qquad X_2|_q = q\mathrm{j}, \qquad X_3|_q = q\mathrm{k}.$$

Show that these vector fields restrict to a smooth left-invariant global frame on $\mathcal{S}$.

    (c) Under the isomorphism $(x^1, x^2, x^3, x^4) \leftrightarrow x^1\mathbb{1} + x^2\mathrm{i} + x^3\mathrm{j} + x^4\mathrm{k}$ between $\mathbb{R}^4$ and $\mathbb{H}$, show that these vector fields have the following coordinate representations:

$$X_1 = -x^2\frac{\partial}{\partial x^1} + x^1\frac{\partial}{\partial x^2} + x^4\frac{\partial}{\partial x^3} - x^3\frac{\partial}{\partial x^4},$$

$$X_2 = -x^3\frac{\partial}{\partial x^1} - x^4\frac{\partial}{\partial x^2} + x^1\frac{\partial}{\partial x^3} + x^2\frac{\partial}{\partial x^4},$$

$$X_3 = -x^4\frac{\partial}{\partial x^1} + x^3\frac{\partial}{\partial x^2} - x^2\frac{\partial}{\partial x^3} + x^1\frac{\partial}{\partial x^4}.$$

(*Used on pp. 179, 562.*)

8-7. The algebra of **octonions** (also called **Cayley numbers**) is the 8-dimensional real vector space $\mathbb{O} = \mathbb{H} \times \mathbb{H}$ (where $\mathbb{H}$ is the space of quaternions defined in Problem 7-22) with the following bilinear product:

$$(p, q)(r, s) = (pr - sq^*, p^*s + rq), \qquad \text{for } p, q, r, s \in \mathbb{H}. \tag{8.20}$$

Show that $\mathbb{O}$ is a noncommutative, nonassociative algebra over $\mathbb{R}$, and prove that there exists a smooth global frame on S^7 by imitating as much of Problem 8-6 as you can. [Hint: it might be helpful to prove that $(PQ^*)Q = P(Q^*Q)$ for all $P, Q \in \mathbb{O}$, where $(p, q)^* = (p^*, -q)$. For more about the octonions, see [Bae02].] (*Used on p. 179.*)

8-8. The algebra of **sedenions** is the 16-dimensional real vector space $\mathbb{S} = \mathbb{O} \times \mathbb{O}$ with the product defined by (8.20), but with p, q, r, and s interpreted as elements of $\mathbb{O}$. Why does sedenionic multiplication not yield a global frame for S^{15}? [Remark: the name "sedenion" comes from the Latin *sedecim*, meaning sixteen. A **division algebra** is an algebra with a multiplicative identity element and no zero divisors (i.e., $ab = 0$ if and only if $a = 0$ or $b = 0$).

It follows from the work of Bott, Milnor, and Kervaire on parallelizability of spheres [MB58, Ker58] that a finite-dimensional division algebra over $\mathbb{R}$ must have dimension 1, 2, 4, or 8.]

8-9. Show by finding a counterexample that Proposition 8.19 is false if we replace the assumption that F is a diffeomorphism by the weaker assumption that it is smooth and bijective.

8-10. Let M be the open submanifold of $\mathbb{R}^2$ where both x and y are positive, and let $F: M \to M$ be the map $F(x, y) = (xy, y/x)$. Show that F is a diffeomorphism, and compute F_*X and F_*Y, where

$$X = x\frac{\partial}{\partial x} + y\frac{\partial}{\partial y}, \qquad Y = y\frac{\partial}{\partial x}.$$

8-11. For each of the following vector fields on the plane, compute its coordinate representation in polar coordinates on the right half-plane $\{(x, y) : x > 0\}$.

(a) $X = x\dfrac{\partial}{\partial x} + y\dfrac{\partial}{\partial y}$.

(b) $Y = x\dfrac{\partial}{\partial y} - y\dfrac{\partial}{\partial x}$.

(c) $Z = (x^2 + y^2)\dfrac{\partial}{\partial x}$.

8-12. Let $F: \mathbb{R}^2 \to \mathbb{RP}^2$ be the smooth map $F(x, y) = [x, y, 1]$, and let $X \in \mathscr{X}(\mathbb{R}^2)$ be defined by $X = x\partial/\partial y - y\partial/\partial x$. Prove that there is a vector field $Y \in \mathscr{X}(\mathbb{RP}^2)$ that is F-related to X, and compute its coordinate representation in terms of each of the charts defined in Example 1.5.

8-13. Show that there is a smooth vector field on $\mathbb{S}^2$ that vanishes at exactly one point. [Hint: try using stereographic projection; see Problem 1-7.]

8-14. Let M be a smooth manifold with or without boundary, let N be a smooth manifold, and let $f: M \to N$ be a smooth map. Define $F: M \to M \times N$ by $F(x) = (x, f(x))$. Show that for every $X \in \mathscr{X}(M)$, there is a smooth vector field on $M \times N$ that is F-related to X.

8-15. EXTENSION LEMMA FOR VECTOR FIELDS ON SUBMANIFOLDS: Suppose M is a smooth manifold and $S \subseteq M$ is an embedded submanifold with or without boundary. Given $X \in \mathscr{X}(S)$, show that there is a smooth vector field Y on a neighborhood of S in M such that $X = Y|_S$. Show that every such vector field extends to all of M if and only if S is properly embedded.

8-16. For each of the following pairs of vector fields X, Y defined on $\mathbb{R}^3$, compute the Lie bracket $[X, Y]$.

(a) $X = y\dfrac{\partial}{\partial z} - 2xy^2\dfrac{\partial}{\partial y}; \quad Y = \dfrac{\partial}{\partial y}$.

(b) $X = x\dfrac{\partial}{\partial y} - y\dfrac{\partial}{\partial x}; \quad Y = y\dfrac{\partial}{\partial z} - z\dfrac{\partial}{\partial y}$.

(c) $X = x\dfrac{\partial}{\partial y} - y\dfrac{\partial}{\partial x};\quad Y = x\dfrac{\partial}{\partial y} + y\dfrac{\partial}{\partial x}.$

8-17. Let M and N be smooth manifolds. Given vector fields $X \in \mathfrak{X}(M)$ and $Y \in \mathfrak{X}(N)$, we can define a vector field $X \oplus Y$ on $M \times N$ by

$$(X \oplus Y)_{(p,q)} = (X_p, Y_q),$$

where we think of the right-hand side as an element of $T_p M \oplus T_q N$, which is naturally identified with $T_{p,q}(M \times N)$ as in Proposition 3.14. Prove that $X \oplus Y$ is smooth if X and Y are smooth, and $[X_1 \oplus Y_1, X_2 \oplus Y_2] = [X_1, X_2] \oplus [Y_1, Y_2]$. (*Used on pp. 346, 527.*)

8-18. Suppose $F \colon M \to N$ is a smooth submersion, where M and N are positive-dimensional smooth manifolds. Given $X \in \mathfrak{X}(M)$ and $Y \in \mathfrak{X}(N)$, we say that X is a **lift of** Y if X and Y are F-related. A vector field $V \in \mathfrak{X}(M)$ is said to be **vertical** if V is everywhere tangent to the fibers of F (or, equivalently, if V is F-related to the zero vector field on N).
   (a) Show that if $\dim M = \dim N$, then every smooth vector field on N has a unique lift.
   (b) Show that if $\dim M \neq \dim N$, then every smooth vector field on N has a lift, but that it is not unique.
   (c) Assume in addition that F is surjective. Given $X \in \mathfrak{X}(M)$, show that X is a lift of a smooth vector field on N if and only if $dF_p(X_p) = dF_q(X_q)$ whenever $F(p) = F(q)$. Show that if this is the case, then X is a lift of a *unique* smooth vector field.
   (d) Assume in addition that F is surjective with connected fibers. Show that a vector field $X \in \mathfrak{X}(M)$ is a lift of a smooth vector field on N if and only if $[V, X]$ is vertical whenever $V \in \mathfrak{X}(M)$ is vertical.
   (*Used on p. 434.*)

8-19. Show that $\mathbb{R}^3$ with the cross product is a Lie algebra.

8-20. Let $A \subseteq \mathfrak{X}\left(\mathbb{R}^3\right)$ be the subspace spanned by $\{X, Y, Z\}$, where

$$X = y\frac{\partial}{\partial z} - z\frac{\partial}{\partial y}, \qquad Y = z\frac{\partial}{\partial x} - x\frac{\partial}{\partial z}, \qquad Z = x\frac{\partial}{\partial y} - y\frac{\partial}{\partial x}.$$

Show that A is a Lie subalgebra of $\mathfrak{X}\left(\mathbb{R}^3\right)$, which is isomorphic to $\mathbb{R}^3$ with the cross product. (*Used on p. 538.*)

8-21. Prove that up to isomorphism, there are exactly one 1-dimensional Lie algebra and two 2-dimensional Lie algebras. Show that all three algebras are isomorphic to Lie subalgebras of $\mathfrak{gl}(2, \mathbb{R})$.

8-22. Let A be any algebra over $\mathbb{R}$. A **derivation of** A is a linear map $D \colon A \to A$ satisfying $D(xy) = (Dx)y + x(Dy)$ for all $x, y \in A$. Show that if D_1 and D_2 are derivations of A, then $[D_1, D_2] = D_1 \circ D_2 - D_2 \circ D_1$ is also a derivation. Show that the set of derivations of A is a Lie algebra with this bracket operation.

8-23. (a) Given Lie algebras $\mathfrak{g}$ and $\mathfrak{h}$, show that the direct sum $\mathfrak{g} \oplus \mathfrak{h}$ is a Lie algebra with the bracket defined by

$$[(X,Y),(X',Y')] = ([X,X'],[Y,Y']).$$

(b) Suppose G and H are Lie groups. Prove that $\mathrm{Lie}(G \times H)$ is isomorphic to $\mathrm{Lie}(G) \oplus \mathrm{Lie}(H)$.

8-24. Suppose G is a Lie group and $\mathfrak{g}$ is its Lie algebra. A vector field $X \in \mathfrak{X}(G)$ is said to be **right-invariant** if it is invariant under all right translations.

(a) Show that the set $\overline{\mathfrak{g}}$ of right-invariant vector fields on G is a Lie subalgebra of $\mathfrak{X}(G)$.

(b) Let $i : G \to G$ denote the inversion map $i(g) = g^{-1}$. Show that the pushforward $i_* : \mathfrak{X}(G) \to \mathfrak{X}(G)$ restricts to a Lie algebra isomorphism from $\mathfrak{g}$ to $\overline{\mathfrak{g}}$.

8-25. Prove that if G is an abelian Lie group, then $\mathrm{Lie}(G)$ is abelian. [Hint: show that the inversion map $i : G \to G$ is a group homomorphism, and use Problem 7-2.]

8-26. Suppose $F : G \to H$ is a Lie group homomorphism. Show that the kernel of $F_* : \mathrm{Lie}(G) \to \mathrm{Lie}(H)$ is the Lie algebra of $\mathrm{Ker}\, F$ (under the identification of the Lie algebra of a subgroup with a Lie subalgebra as in Theorem 8.46).

8-27. Let G and H be Lie groups, and suppose $F : G \to H$ is a Lie group homomorphism that is also a local diffeomorphism. Show that the induced homomorphism $F_* : \mathrm{Lie}(G) \to \mathrm{Lie}(H)$ is an isomorphism of Lie algebras. (*Used on pp. 531, 557.*)

8-28. Considering det: $\mathrm{GL}(n,\mathbb{R}) \to \mathbb{R}^*$ as a Lie group homomorphism, show that its induced Lie algebra homomorphism is tr: $\mathfrak{gl}(n,\mathbb{R}) \to \mathbb{R}$. [Hint: see Problem 7-4.]

8-29. Theorem 8.46 implies that the Lie algebra of any Lie subgroup of $\mathrm{GL}(n,\mathbb{R})$ is canonically isomorphic to a subalgebra of $\mathfrak{gl}(n,\mathbb{R})$, with a similar statement for Lie subgroups of $\mathrm{GL}(n,\mathbb{C})$. Under this isomorphism, show that

$$\mathrm{Lie}\left(\mathrm{SL}(n,\mathbb{R})\right) \cong \mathfrak{sl}(n,\mathbb{R}),$$

$$\mathrm{Lie}\left(\mathrm{SO}(n)\right) \cong \mathfrak{o}(n),$$

$$\mathrm{Lie}\left(\mathrm{SL}(n,\mathbb{C})\right) \cong \mathfrak{sl}(n,\mathbb{C}),$$

$$\mathrm{Lie}\left(\mathrm{U}(n)\right) \cong \mathfrak{u}(n),$$

$$\mathrm{Lie}\left(\mathrm{SU}(n)\right) \cong \mathfrak{su}(n),$$

where

$$\mathfrak{sl}(n,\mathbb{R}) = \left\{A \in \mathfrak{gl}(n,\mathbb{R}) : \mathrm{tr}\, A = 0\right\},$$

$$\mathfrak{o}(n) = \left\{A \in \mathfrak{gl}(n,\mathbb{R}) : A^T + A = 0\right\},$$

$$\mathfrak{sl}(n,\mathbb{C}) = \left\{A \in \mathfrak{gl}(n,\mathbb{C}) : \mathrm{tr}\, A = 0\right\},$$

$$\mathfrak{u}(n) = \{A \in \mathfrak{gl}(n,\mathbb{C}) : A^* + A = 0\},$$

$$\mathfrak{su}(n) = \mathfrak{u}(n) \cap \mathfrak{sl}(n,\mathbb{C}).$$

8-30. Show by giving an explicit isomorphism that $\mathfrak{su}(2)$ and $\mathfrak{o}(3)$ are isomorphic Lie algebras, and that both are isomorphic to $\mathbb{R}^3$ with the cross product.

8-31. Let $\mathfrak{g}$ be a Lie algebra. A linear subspace $\mathfrak{h} \subseteq \mathfrak{g}$ is called an **ideal in $\mathfrak{g}$** if $[X,Y] \in \mathfrak{h}$ whenever $X \in \mathfrak{h}$ and $Y \in \mathfrak{g}$.

   (a) Show that if $\mathfrak{h}$ is an ideal in $\mathfrak{g}$, then the quotient space $\mathfrak{g}/\mathfrak{h}$ has a unique Lie algebra structure such that the projection $\pi \colon \mathfrak{g} \to \mathfrak{g}/\mathfrak{h}$ is a Lie algebra homomorphism.

   (b) Show that a subspace $\mathfrak{h} \subseteq \mathfrak{g}$ is an ideal if and only if it is the kernel of a Lie algebra homomorphism.

   (*Used on p. 533.*)

Chapter 9
Integral Curves and Flows

In this chapter we continue our study of vector fields. The primary geometric objects associated with smooth vector fields are their *integral curves*, which are smooth curves whose velocity at each point is equal to the value of the vector field there. The collection of all integral curves of a given vector field on a manifold determines a family of diffeomorphisms of (open subsets of) the manifold, called a *flow*. Any smooth $\mathbb{R}$-action is a flow, for example; but there are flows that are not $\mathbb{R}$-actions because the diffeomorphisms may not be defined on the whole manifold for every $t \in \mathbb{R}$.

The main theorem of the chapter, the *fundamental theorem on flows*, asserts that every smooth vector field determines a unique maximal integral curve starting at each point, and the collection of all such integral curves determines a unique maximal flow. The proof is an application of the existence, uniqueness, and smoothness theorem for solutions of ordinary differential equations (see Appendix D).

After proving the fundamental theorem, we explore some of the properties of vector fields and flows. First, we investigate conditions under which a vector field generates a global flow. Then we show how "flowing out" from initial submanifolds along vector fields can be used to create useful parametrizations of larger submanifolds. Next we examine the local behavior of flows, and find that the behavior at points where the vector field vanishes, which correspond to *equilibrium points* of the flow, is very different from the behavior at points where it does not vanish, where the flow looks locally like translation along parallel coordinate lines.

We then introduce the *Lie derivative*, which is a coordinate-independent way of computing the rate of change of one vector field along the flow of another. It leads to some deep connections among vector fields, their Lie brackets, and their flows. In particular, we will prove that two vector fields have commuting flows if and only if their Lie bracket is zero. Based on this fact, we can prove a necessary and sufficient condition for a smooth local frame to be expressible as a coordinate frame. We then discuss how some of the results of this chapter can be generalized to *time-dependent vector fields* on manifolds.

In the last section of the chapter, we describe an important application of flows to the study of first-order partial differential equations.

J.M. Lee, *Introduction to Smooth Manifolds*, Graduate Texts in Mathematics 218,
DOI 10.1007/978-1-4419-9982-5_9, © Springer Science+Business Media New York 2013

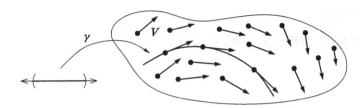

Fig. 9.1 An integral curve of a vector field

Integral Curves

Suppose M is a smooth manifold with or without boundary. If $\gamma\colon J \to M$ is a smooth curve, then for each $t \in J$, the velocity vector $\gamma'(t)$ is a vector in $T_{\gamma(t)}M$. In this section we describe a way to work backwards: given a tangent vector at each point, we seek a curve whose velocity at each point is equal to the given vector there.

If V is a vector field on M, an **integral curve of V** is a differentiable curve $\gamma\colon J \to M$ whose velocity at each point is equal to the value of V at that point:

$$\gamma'(t) = V_{\gamma(t)} \quad \text{for all } t \in J.$$

(See Fig. 9.1.) If $0 \in J$, the point $\gamma(0)$ is called the **starting point of γ**. (The reason for the term "integral curve" will be explained shortly. Note that this is one definition that requires some differentiability hypothesis, because the definition of an integral curve would make no sense for a curve that is merely continuous.)

Example 9.1 (Integral Curves).

(a) Let (x, y) be standard coordinates on $\mathbb{R}^2$, and let $V = \partial/\partial x$ be the first coordinate vector field. It is easy to check that the integral curves of V are precisely the straight lines parallel to the x-axis, with parametrizations of the form $\gamma(t) = (a + t, b)$ for constants a and b (Fig. 9.2(a)). Thus, there is a unique integral curve starting at each point of the plane, and the images of different integral curves are either identical or disjoint.

(b) Let $W = x\,\partial/\partial y - y\,\partial/\partial x$ on $\mathbb{R}^2$ (Fig. 9.2(b)). If $\gamma\colon \mathbb{R} \to \mathbb{R}^2$ is a smooth curve, written in standard coordinates as $\gamma(t) = (x(t), y(t))$, then the condition $\gamma'(t) = W_{\gamma(t)}$ for γ to be an integral curve translates to

$$x'(t)\,\frac{\partial}{\partial x}\bigg|_{\gamma(t)} + y'(t)\,\frac{\partial}{\partial y}\bigg|_{\gamma(t)} = x(t)\,\frac{\partial}{\partial y}\bigg|_{\gamma(t)} - y(t)\,\frac{\partial}{\partial x}\bigg|_{\gamma(t)}.$$

Comparing the components of these vectors, we see that this is equivalent to the system of ordinary differential equations

$$x'(t) = -y(t),$$
$$y'(t) = x(t).$$

These equations have the solutions

$$x(t) = a\cos t - b\sin t, \qquad y(t) = a\sin t + b\cos t,$$

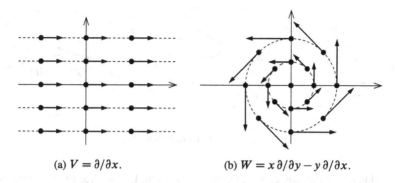

(a) $V = \partial/\partial x$.   (b) $W = x\,\partial/\partial y - y\,\partial/\partial x$.

Fig. 9.2 Vector fields and their integral curves in the plane

for arbitrary constants a and b, and thus each curve of the form $\gamma(t) = (a\cos t - b\sin t, a\sin t + b\cos t)$ is an integral curve of W. When $(a,b) = (0,0)$, this is the constant curve $\gamma(t) \equiv (0,0)$; otherwise, it is a circle traversed counterclockwise. Since $\gamma(0) = (a,b)$, we see once again that there is a unique integral curve starting at each point $(a,b) \in \mathbb{R}^2$, and the images of the various integral curves are either identical or disjoint.   //

As the second example above illustrates, finding integral curves boils down to solving a system of ordinary differential equations in a smooth chart. Suppose V is a smooth vector field on M and $\gamma\colon J \to M$ is a smooth curve. On a smooth coordinate domain $U \subseteq M$, we can write γ in local coordinates as $\gamma(t) = \big(\gamma^1(t),\dots,\gamma^n(t)\big)$. Then the condition $\gamma'(t) = V_{\gamma(t)}$ for γ to be an integral curve of V can be written

$$\dot{\gamma}^i(t)\,\frac{\partial}{\partial x^i}\bigg|_{\gamma(t)} = V^i\big(\gamma(t)\big)\,\frac{\partial}{\partial x^i}\bigg|_{\gamma(t)},$$

which reduces to the following autonomous system of ordinary differential equations (ODEs):

$$\dot{\gamma}^1(t) = V^1\big(\gamma^1(t),\dots,\gamma^n(t)\big),$$
$$\vdots \tag{9.1}$$
$$\dot{\gamma}^n(t) = V^n\big(\gamma^1(t),\dots,\gamma^n(t)\big).$$

(We use a dot to denote an ordinary derivative with respect to t when there are superscripts that would make primes hard to read.) The fundamental fact about such systems is the existence, uniqueness, and smoothness theorem, Theorem D.1. (This is the reason for the terminology "integral curves," because solving a system of ODEs is often referred to as "integrating" the system.) We will derive detailed consequences of that theorem later; for now, we just note the following simple result.

Proposition 9.2. *Let V be a smooth vector field on a smooth manifold M. For each point $p \in M$, there exist $\varepsilon > 0$ and a smooth curve $\gamma\colon (-\varepsilon,\varepsilon) \to M$ that is an integral curve of V starting at p.*

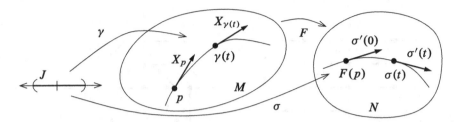

Fig. 9.3 Flows of F-related vector fields

Proof. This is just the existence statement of Theorem D.1 applied to the coordinate representation of V. □

The next two lemmas show how affine reparametrizations affect integral curves.

Lemma 9.3 (Rescaling Lemma). *Let V be a smooth vector field on a smooth manifold M, let $J \subseteq \mathbb{R}$ be an interval, and let $\gamma \colon J \to M$ be an integral curve of V. For any $a \in \mathbb{R}$, the curve $\tilde{\gamma} \colon \tilde{J} \to M$ defined by $\tilde{\gamma}(t) = \gamma(at)$ is an integral curve of the vector field aV, where $\tilde{J} = \{t : at \in J\}$.*

Proof. One way to see this is as a straightforward application of the chain rule in local coordinates. Somewhat more invariantly, we can examine the action of $\tilde{\gamma}'(t)$ on a smooth real-valued function f defined in a neighborhood of a point $\tilde{\gamma}(t_0)$. By the chain rule and the fact that γ is an integral curve of V,

$$\tilde{\gamma}'(t_0)f = \frac{d}{dt}\Big|_{t=t_0} (f \circ \tilde{\gamma})(t) = \frac{d}{dt}\Big|_{t=t_0} (f \circ \gamma)(at)$$

$$= a(f \circ \gamma)'(at_0) = a\gamma'(at_0)f = aV_{\tilde{\gamma}(t_0)}f. \qquad \square$$

Lemma 9.4 (Translation Lemma). *Let V, M, J, and γ be as in the preceding lemma. For any $b \in \mathbb{R}$, the curve $\hat{\gamma} \colon \hat{J} \to M$ defined by $\hat{\gamma}(t) = \gamma(t + b)$ is also an integral curve of V, where $\hat{J} = \{t : t + b \in J\}$.*

▶ **Exercise 9.5.** Prove the translation lemma.

Proposition 9.6 (Naturality of Integral Curves). *Suppose M and N are smooth manifolds and $F \colon M \to N$ is a smooth map. Then $X \in \mathfrak{X}(M)$ and $Y \in \mathfrak{X}(N)$ are F-related if and only if F takes integral curves of X to integral curves of Y, meaning that for each integral curve γ of X, $F \circ \gamma$ is an integral curve of Y.*

Proof. Suppose first that X and Y are F-related, and $\gamma \colon J \to M$ is an integral curve of X. If we define $\sigma \colon J \to N$ by $\sigma = F \circ \gamma$ (see Fig. 9.3), then

$$\sigma'(t) = (F \circ \gamma)'(t) = dF_{\gamma(t)}(\gamma'(t)) = dF_{\gamma(t)}(X_{\gamma(t)}) = Y_{F(\gamma(t))} = Y_{\sigma(t)},$$

so σ is an integral curve of Y.

Conversely, suppose F takes integral curves of X to integral curves of Y. Given $p \in M$, let $\gamma \colon (-\varepsilon, \varepsilon) \to M$ be an integral curve of X starting at p. Since $F \circ \gamma$ is an integral curve of Y starting at $F(p)$, we have

$$Y_{F(p)} = (F \circ \gamma)'(0) = dF_p\big(\gamma'(0)\big) = dF_p(X_p),$$

which shows that X and Y are F-related. $\square$

Flows

Here is another way to visualize the family of integral curves associated with a vector field. Let M be a smooth manifold and $V \in \mathfrak{X}(M)$, and suppose that for each point $p \in M$, V has a unique integral curve starting at p and defined for all $t \in \mathbb{R}$, which we denote by $\theta^{(p)} \colon \mathbb{R} \to M$. (It may not always be the case that every integral curve is defined for all t, but for purposes of illustration let us assume so for the time being.) For each $t \in \mathbb{R}$, we can define a map $\theta_t \colon M \to M$ by sending each $p \in M$ to the point obtained by following for time t the integral curve starting at p:

$$\theta_t(p) = \theta^{(p)}(t).$$

Each map θ_t "slides" the manifold along the integral curves for time t. The translation lemma implies that $t \mapsto \theta^{(p)}(t + s)$ is an integral curve of V starting at $q = \theta^{(p)}(s)$; since we are assuming uniqueness of integral curves, $\theta^{(q)}(t) = \theta^{(p)}(t + s)$. When we translate this into a statement about the maps θ_t, it becomes

$$\theta_t \circ \theta_s(p) = \theta_{t+s}(p).$$

Together with the equation $\theta_0(p) = \theta^{(p)}(0) = p$, which holds by definition, this implies that the map $\theta \colon \mathbb{R} \times M \to M$ is an action of the additive group $\mathbb{R}$ on M.

Motivated by these observations, we define a **global flow** on M (also called a **one-parameter group action**) to be a continuous left $\mathbb{R}$-action on M; that is, a continuous map $\theta \colon \mathbb{R} \times M \to M$ satisfying the following properties for all $s, t \in \mathbb{R}$ and $p \in M$:

$$\theta\big(t, \theta(s, p)\big) = \theta(t + s, p), \qquad \theta(0, p) = p. \tag{9.2}$$

Given a global flow θ on M, we define two collections of maps as follows:

- For each $t \in \mathbb{R}$, define a continuous map $\theta_t \colon M \to M$ by

$$\theta_t(p) = \theta(t, p).$$

The defining properties (9.2) are equivalent to the **group laws**

$$\theta_t \circ \theta_s = \theta_{t+s}, \qquad \theta_0 = \mathrm{Id}_M. \tag{9.3}$$

As is the case for any continuous group action, each map $\theta_t \colon M \to M$ is a homeomorphism, and if the flow is smooth, θ_t is a diffeomorphism.
- For each $p \in M$, define a curve $\theta^{(p)} \colon \mathbb{R} \to M$ by

$$\theta^{(p)}(t) = \theta(t, p).$$

The image of this curve is the orbit of p under the group action.

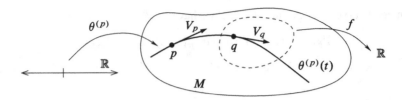

Fig. 9.4 The infinitesimal generator of a global flow

The next proposition shows that every smooth global flow is derived from the integral curves of some smooth vector field in precisely the way we described above. If $\theta\colon \mathbb{R} \times M \to M$ is a smooth global flow, for each $p \in M$ we define a tangent vector $V_p \in T_pM$ by

$$V_p = \theta^{(p)\prime}(0).$$

The assignment $p \mapsto V_p$ is a (rough) vector field on M, which is called the **infinitesimal generator of** θ, for reasons we will explain below.

Proposition 9.7. *Let $\theta\colon \mathbb{R} \times M \to M$ be a smooth global flow on a smooth manifold M. The infinitesimal generator V of θ is a smooth vector field on M, and each curve $\theta^{(p)}$ is an integral curve of V.*

Proof. To show that V is smooth, it suffices by Proposition 8.14 to show that Vf is smooth for every smooth real-valued function f defined on an open subset $U \subseteq M$. For any such f and any $p \in U$, just note that

$$Vf(p) = V_pf = \theta^{(p)\prime}(0)f = \frac{d}{dt}\bigg|_{t=0} f\big(\theta^{(p)}(t)\big) = \frac{\partial}{\partial t}\bigg|_{(0,p)} f\big(\theta(t,p)\big).$$

Because $f(\theta(t,p))$ is a smooth function of (t,p) by composition, so is its partial derivative with respect to t. Thus, $Vf(p)$ depends smoothly on p, so V is smooth.

Next we need to show that $\theta^{(p)}$ is an integral curve of V, which means that $\theta^{(p)\prime}(t) = V_{\theta^{(p)}(t)}$ for all $p \in M$ and all $t \in \mathbb{R}$. Let $t_0 \in \mathbb{R}$ be arbitrary, and set $q = \theta^{(p)}(t_0) = \theta_{t_0}(p)$, so what we have to show is $\theta^{(p)\prime}(t_0) = V_q$ (see Fig. 9.4). By the group law, for all t,

$$\theta^{(q)}(t) = \theta_t(q) = \theta_t\big(\theta_{t_0}(p)\big) = \theta_{t+t_0}(p) = \theta^{(p)}(t + t_0). \tag{9.4}$$

Therefore, for any smooth real-valued function f defined in a neighborhood of q,

$$V_qf = \theta^{(q)\prime}(0)f = \frac{d}{dt}\bigg|_{t=0} f\big(\theta^{(q)}(t)\big) = \frac{d}{dt}\bigg|_{t=0} f\big(\theta^{(p)}(t + t_0)\big)$$

$$= \theta^{(p)\prime}(t_0)f, \tag{9.5}$$

which was to be shown. $\square$

Example 9.8 (Global Flows). The two vector fields on the plane described in Example 9.1 both had integral curves defined for all $t \in \mathbb{R}$, so they generate global flows. Using the results of that example, we can write down the flows explicitly.

(a) The flow of $V = \partial/\partial x$ in $\mathbb{R}^2$ is the map $\tau \colon \mathbb{R} \times \mathbb{R}^2 \to \mathbb{R}^2$ given by

$$\tau_t(x, y) = (x + t, y).$$

For each nonzero $t \in \mathbb{R}$, τ_t translates the plane to the right ($t > 0$) or left ($t < 0$) by a distance $|t|$.

(b) The flow of $W = x\,\partial/\partial y - y\,\partial/\partial x$ is the map $\theta \colon \mathbb{R} \times \mathbb{R}^2 \to \mathbb{R}^2$ given by

$$\theta_t(x, y) = (x \cos t - y \sin t, \ x \sin t + y \cos t).$$

For each $t \in \mathbb{R}$, θ_t rotates the plane through an angle t about the origin.   //

The Fundamental Theorem on Flows

We have seen that every smooth global flow gives rise to a smooth vector field whose integral curves are precisely the curves defined by the flow. Conversely, we would like to be able to say that every smooth vector field is the infinitesimal generator of a smooth global flow. However, it is easy to see that this cannot be the case, because there are smooth vector fields whose integral curves are not defined for all $t \in \mathbb{R}$. Here are two examples.

Example 9.9. Let $M = \mathbb{R}^2 \smallsetminus \{0\}$ with standard coordinates (x, y), and let V be the vector field $\partial/\partial x$ on M. The unique integral curve of V starting at $(-1, 0) \in M$ is $\gamma(t) = (t - 1, 0)$. However, in this case, γ cannot be extended continuously past $t = 1$. This is intuitively evident because of the "hole" in M at the origin; to prove it rigorously, suppose $\tilde{\gamma}$ is any continuous extension of γ past $t = 1$. Then $\gamma(t) \to \tilde{\gamma}(1) \in \mathbb{R}^2 \smallsetminus \{0\}$ as $t \nearrow 1$. But we can also consider γ as a map into $\mathbb{R}^2$ by composing with the inclusion $M \hookrightarrow \mathbb{R}^2$, and it is obvious from the formula that $\gamma(t) \to (0, 0)$ as $t \nearrow 1$. Since limits in $\mathbb{R}^2$ are unique, this is a contradiction.   //

Example 9.10. For a more subtle example, let M be all of $\mathbb{R}^2$ and let $W = x^2 \partial/\partial x$. You can check easily that the unique integral curve of W starting at $(1, 0)$ is

$$\gamma(t) = \left(\frac{1}{1 - t}, 0 \right).$$

This curve also cannot be extended past $t = 1$, because its x-coordinate is unbounded as $t \nearrow 1$.   //

For this reason, we make the following definitions. If M is a manifold, a *flow domain* for M is an open subset $\mathcal{D} \subseteq \mathbb{R} \times M$ with the property that for each $p \in M$, the set $\mathcal{D}^{(p)} = \{t \in \mathbb{R} : (t, p) \in \mathcal{D}\}$ is an open interval containing 0 (Fig. 9.5). A *flow* on M is a continuous map $\theta \colon \mathcal{D} \to M$, where $\mathcal{D} \subseteq \mathbb{R} \times M$ is a flow domain, that satisfies the following group laws: for all $p \in M$,

$$\theta(0, p) = p, \tag{9.6}$$

and for all $s \in \mathcal{D}^{(p)}$ and $t \in \mathcal{D}^{(\theta(s, p))}$ such that $s + t \in \mathcal{D}^{(p)}$,

$$\theta\big(t, \theta(s, p)\big) = \theta(t + s, p). \tag{9.7}$$

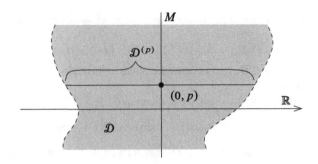

Fig. 9.5 A flow domain

We sometimes call θ a *local flow* to distinguish it from a global flow as defined earlier. The unwieldy term *local one-parameter group action* is also used.

If θ is a flow, we define $\theta_t(p) = \theta^{(p)}(t) = \theta(t, p)$ whenever $(t, p) \in \mathcal{D}$, just as for a global flow. For each $t \in \mathbb{R}$, we also define

$$M_t = \big\{ p \in M : (t, p) \in \mathcal{D} \big\}, \tag{9.8}$$

so that

$$p \in M_t \Leftrightarrow t \in \mathcal{D}^{(p)} \Leftrightarrow (t, p) \in \mathcal{D}.$$

If θ is smooth, the *infinitesimal generator of* θ is defined by $V_p = \theta^{(p)\prime}(0)$.

Proposition 9.11. *If $\theta \colon \mathcal{D} \to M$ is a smooth flow, then the infinitesimal generator V of θ is a smooth vector field, and each curve $\theta^{(p)}$ is an integral curve of V.*

Proof. The proof is essentially identical to the analogous proof for global flows, Proposition 9.7. In the proof that V is smooth, we need only note that for any $p_0 \in M$, $\theta(t, p)$ is defined and smooth for all (t, p) sufficiently close to $(0, p_0)$ because $\mathcal{D}$ is open. In the proof that $\theta^{(p)}$ is an integral curve, we need to verify that all of the expressions in (9.4) and (9.5) make sense. Suppose $t_0 \in \mathcal{D}^{(p)}$. Because both $\mathcal{D}^{(p)}$ and $\mathcal{D}^{(\theta_{t_0}(p))}$ are open intervals containing 0, there is a positive number ε such that $t + t_0 \in \mathcal{D}^{(p)}$ and $t \in \mathcal{D}^{(\theta_{t_0}(p))}$ whenever $|t| < \varepsilon$, and then $\theta_t(\theta_{t_0}(p)) = \theta_{t+t_0}(p)$ by definition of a flow. The rest of the proof goes through just as before. $\qquad\square$

The next theorem is the main result of this section. A *maximal integral curve* is one that cannot be extended to an integral curve on any larger open interval, and a *maximal flow* is a flow that admits no extension to a flow on a larger flow domain.

Theorem 9.12 (Fundamental Theorem on Flows). *Let V be a smooth vector field on a smooth manifold M. There is a unique smooth maximal flow $\theta \colon \mathcal{D} \to M$ whose infinitesimal generator is V. This flow has the following properties:*

(a) *For each $p \in M$, the curve $\theta^{(p)} \colon \mathcal{D}^{(p)} \to M$ is the unique maximal integral curve of V starting at p.*

(b) *If $s \in \mathcal{D}^{(p)}$, then $\mathcal{D}^{(\theta(s,p))}$ is the interval $\mathcal{D}^{(p)} - s = \big\{ t - s : t \in \mathcal{D}^{(p)} \big\}$.*

(c) *For each $t \in \mathbb{R}$, the set M_t is open in M, and $\theta_t \colon M_t \to M_{-t}$ is a diffeomorphism with inverse θ_{-t}.*

Proof. Proposition 9.2 shows that there exists an integral curve starting at each point $p \in M$. Suppose $\gamma, \tilde{\gamma} \colon J \to M$ are two integral curves of V defined on the same open interval J such that $\gamma(t_0) = \tilde{\gamma}(t_0)$ for some $t_0 \in J$. Let $\mathcal{S}$ be the set of $t \in J$ such that $\gamma(t) = \tilde{\gamma}(t)$. Clearly, $\mathcal{S} \neq \varnothing$, because $t_0 \in \mathcal{S}$ by hypothesis, and $\mathcal{S}$ is closed in J by continuity. On the other hand, suppose $t_1 \in \mathcal{S}$. Then in a smooth coordinate neighborhood around the point $p = \gamma(t_1)$, γ and $\tilde{\gamma}$ are both solutions to same ODE with the same initial condition $\gamma(t_1) = \tilde{\gamma}(t_1) = p$. By the uniqueness part of Theorem D.1, $\gamma \equiv \tilde{\gamma}$ on an interval containing t_1, which implies that $\mathcal{S}$ is open in J. Since J is connected, $\mathcal{S} = J$, which implies that $\gamma = \tilde{\gamma}$ on all of J. Thus, any two integral curves that agree at one point agree on their common domain.

For each $p \in M$, let $\mathcal{D}^{(p)}$ be the union of all open intervals $J \subseteq \mathbb{R}$ containing 0 on which an integral curve starting at p is defined. Define $\theta^{(p)} \colon \mathcal{D}^{(p)} \to M$ by letting $\theta^{(p)}(t) = \gamma(t)$, where γ is any integral curve starting at p and defined on an open interval containing 0 and t. Since all such integral curves agree at t by the argument above, $\theta^{(p)}$ is well defined, and is obviously the unique maximal integral curve starting at p.

Now let $\mathcal{D} = \{(t, p) \in \mathbb{R} \times M : t \in \mathcal{D}^{(p)}\}$, and define $\theta \colon \mathcal{D} \to M$ by $\theta(t, p) = \theta^{(p)}(t)$. As usual, we also write $\theta_t(p) = \theta(t, p)$. By definition, θ satisfies property (a) in the statement of the fundamental theorem: for each $p \in M$, $\theta^{(p)}$ is the unique maximal integral curve of V starting at p. To verify the group laws, fix any $p \in M$ and $s \in \mathcal{D}^{(p)}$, and write $q = \theta(s, p) = \theta^{(p)}(s)$. The curve $\gamma \colon \mathcal{D}^{(p)} - s \to M$ defined by $\gamma(t) = \theta^{(p)}(t + s)$ starts at q, and the translation lemma shows that γ is an integral curve of V. By uniqueness of ODE solutions, γ agrees with $\theta^{(q)}$ on their common domain, which is equivalent to the second group law (9.7), and the first group law (9.6) is immediate from the definition. By maximality of $\theta^{(q)}$, the domain of γ cannot be larger than $\mathcal{D}^{(q)}$, which means that $\mathcal{D}^{(p)} - s \subseteq \mathcal{D}^{(q)}$. Since $0 \in \mathcal{D}^{(p)}$, this implies that $-s \in \mathcal{D}^{(q)}$, and the group law implies that $\theta^{(q)}(-s) = p$. Applying the same argument with $(-s, q)$ in place of (s, p), we find that $\mathcal{D}^{(q)} + s \subseteq \mathcal{D}^{(p)}$, which is the same as $\mathcal{D}^{(q)} \subseteq \mathcal{D}^{(p)} - s$. This proves (b).

Next we show that $\mathcal{D}$ is open in $\mathbb{R} \times M$ (so it is a flow domain), and that $\theta \colon \mathcal{D} \to M$ is smooth. Define a subset $W \subseteq \mathcal{D}$ as the set of all $(t, p) \in \mathcal{D}$ such that θ is defined and smooth on a product neighborhood of (t, p) of the form $J \times U \subseteq \mathcal{D}$, where $J \subseteq \mathbb{R}$ is an open interval containing 0 and t and $U \subseteq M$ is a neighborhood of p. Then W is open in $\mathbb{R} \times M$, and the restriction of θ to W is smooth, so it suffices to show that $W = \mathcal{D}$. Suppose this is not the case. Then there exists some point $(\tau, p_0) \in \mathcal{D} \smallsetminus W$. For simplicity, assume $\tau > 0$; the argument for $\tau < 0$ is similar.

Let $t_0 = \inf\{t \in \mathbb{R} : (t, p_0) \notin W\}$ (Fig. 9.6). By the ODE theorem (applied in smooth coordinates around p_0), θ is defined and smooth in some product neighborhood of $(0, p_0)$, so $t_0 > 0$. Since $t_0 \leq \tau$ and $\mathcal{D}^{(p_0)}$ is an open interval containing 0 and τ, it follows that $t_0 \in \mathcal{D}^{(p_0)}$. Let $q_0 = \theta^{(p_0)}(t_0)$. By the ODE theorem again, there exist $\varepsilon > 0$ and a neighborhood U_0 of q_0 such that $(-\varepsilon, \varepsilon) \times U_0 \subseteq W$. We will

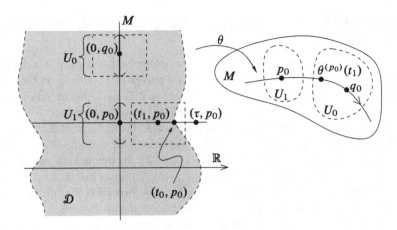

Fig. 9.6 Proof that $\mathcal{D}$ is open

use the group law to show that θ extends smoothly to a neighborhood of (t_0, p_0), which contradicts our choice of t_0.

Choose some $t_1 < t_0$ such that $t_1 + \varepsilon > t_0$ and $\theta^{(p_0)}(t_1) \in U_0$. Since $t_1 < t_0$, we have $(t_1, p_0) \in W$, and so there is a product neighborhood $(t_1 - \delta, t_1 + \delta) \times U_1 \subseteq W$. By definition of W, this implies that θ is defined and smooth on $[0, t_1 + \delta) \times U_1$. Because $\theta(t_1, p_0) \in U_0$, we can choose U_1 small enough that θ maps $\{t_1\} \times U_1$ into U_0. Define $\tilde{\theta} \colon [0, t_1 + \varepsilon) \times U_1 \to M$ by

$$\tilde{\theta}(t, p) = \begin{cases} \theta_t(p), & p \in U_1, \ 0 \le t < t_1, \\ \theta_{t-t_1} \circ \theta_{t_1}(p), & p \in U_1, \ t_1 - \varepsilon < t < t_1 + \varepsilon. \end{cases}$$

The group law for θ guarantees that these definitions agree where they overlap, and our choices of U_1, t_1, and ε ensure that this defines a smooth map. By the translation lemma, each map $t \mapsto \tilde{\theta}(t, p)$ is an integral curve of V, so $\tilde{\theta}$ is a smooth extension of θ to a neighborhood of (t_0, p_0), contradicting our choice of t_0. This completes the proof that $W = \mathcal{D}$.

Finally, we prove (c). The fact that M_t is open is an immediate consequence of the fact that $\mathcal{D}$ is open. From part (b) we deduce

$$p \in M_t \Rightarrow t \in \mathcal{D}^{(p)} \Rightarrow \mathcal{D}^{(\theta_t(p))} = \mathcal{D}^{(p)} - t$$

$$\Rightarrow -t \in \mathcal{D}^{(\theta_t(p))} \Rightarrow \theta_t(p) \in M_{-t},$$

which shows that θ_t maps M_t to M_{-t}. Moreover, the group laws then show that $\theta_{-t} \circ \theta_t$ is equal to the identity on M_t. Reversing the roles of t and $-t$ shows that $\theta_t \circ \theta_{-t}$ is the identity on M_{-t}, which completes the proof. $\qquad \square$

The flow whose existence and uniqueness are asserted in the fundamental theorem is called the *flow generated by* V, or just the *flow of* V. The term "infinitesimal generator" comes from the following picture: in a smooth chart, a good approximation to an integral curve can be obtained by composing many small straight-line motions, with the direction and length of each motion determined by the value of the vector field at the point arrived at in the previous step. Intuitively, one can think of a flow as a sequence of infinitely many infinitesimally small linear steps.

The naturality of integral curves (Proposition 9.6) translates into the following naturality statement for flows.

Proposition 9.13 (Naturality of Flows). *Suppose M and N are smooth manifolds, $F \colon M \to N$ is a smooth map, $X \in \mathfrak{X}(M)$, and $Y \in \mathfrak{X}(N)$. Let θ be the flow of X and η the flow of Y. If X and Y are F-related, then for each $t \in \mathbb{R}$, $F(M_t) \subseteq N_t$ and $\eta_t \circ F = F \circ \theta_t$ on M_t:*

$$
\begin{array}{ccc}
M_t & \xrightarrow{\ F\ } & N_t \\
\theta_t \big\downarrow & & \big\downarrow \eta_t \\
M_{-t} & \xrightarrow{\ F\ } & N_{-t}.
\end{array}
$$

Proof. By Proposition 9.6, for any $p \in M$, the curve $F \circ \theta^{(p)}$ is an integral curve of Y starting at $F \circ \theta^{(p)}(0) = F(p)$. By uniqueness of integral curves, therefore, the maximal integral curve $\eta^{(F(p))}$ must be defined at least on the interval $\mathcal{D}^{(p)}$, and $F \circ \theta^{(p)} = \eta^{(F(p))}$ on that interval. This means that

$$
p \in M_t \Rightarrow t \in \mathcal{D}^{(p)} \Rightarrow t \in \mathcal{D}^{(F(p))} \Rightarrow F(p) \in N_t,
$$

which is equivalent to $F(M_t) \subseteq N_t$, and

$$
F\big(\theta^{(p)}(t)\big) = \eta^{(F(p))}(t) \quad \text{for all } t \in \mathcal{D}^{(p)},
$$

which is equivalent to $\eta_t \circ F(p) = F \circ \theta_t(p)$ for all $p \in M_t$. $\square$

The next corollary is immediate.

Corollary 9.14 (Diffeomorphism Invariance of Flows). *Let $F \colon M \to N$ be a diffeomorphism. If $X \in \mathfrak{X}(M)$ and θ is the flow of X, then the flow of F_*X is $\eta_t = F \circ \theta_t \circ F^{-1}$, with domain $N_t = F(M_t)$ for each $t \in \mathbb{R}$.* $\square$

Complete Vector Fields

As we observed earlier in this chapter, not every smooth vector field generates a global flow. The ones that do are important enough to deserve a name. We say that a smooth vector field is **complete** if it generates a global flow, or equivalently if each of its maximal integral curves is defined for all $t \in \mathbb{R}$. For example, both of the vector fields on the plane whose flows we computed in Example 9.8 are complete, whereas those of Examples 9.9 and 9.10 are not.

It is not always easy to determine by looking at a vector field whether it is complete or not. If you can solve the ODE explicitly to find all of the integral curves, and they all exist for all time, then the vector field is complete. On the other hand, if you can find a single integral curve that cannot be extended to all of $\mathbb{R}$, as we did for the vector fields of Examples 9.9 and 9.10, then it is not complete. However, it is often impossible to solve the ODE explicitly, so it is useful to have some general criteria for determining when a vector field is complete.

We will show below that all compactly supported smooth vector fields, and therefore all smooth vector fields on a compact manifold, are complete. The proof will be based on the following lemma.

Lemma 9.15 (Uniform Time Lemma). *Let V be a smooth vector field on a smooth manifold M, and let θ be its flow. Suppose there is a positive number ε such that for every $p \in M$, the domain of $\theta^{(p)}$ contains $(-\varepsilon, \varepsilon)$. Then V is complete.*

Proof. Suppose for the sake of contradiction that for some $p \in M$, the domain $\mathcal{D}^{(p)}$ of $\theta^{(p)}$ is bounded above. (A similar proof works if it is bounded below.) Let $b = \sup \mathcal{D}^{(p)}$, let t_0 be a positive number such that $b - \varepsilon < t_0 < b$, and let $q = \theta^{(p)}(t_0)$. The hypothesis implies that $\theta^{(q)}(t)$ is defined at least for $t \in (-\varepsilon, \varepsilon)$. Define a curve $\gamma : (-\varepsilon, t_0 + \varepsilon) \to M$ by

$$
\gamma(t) = \begin{cases} \theta^{(p)}(t), & -\varepsilon < t < b, \\ \theta^{(q)}(t - t_0), & t_0 - \varepsilon < t < t_0 + \varepsilon. \end{cases}
$$

These two definitions agree where they overlap, because $\theta^{(q)}(t - t_0) = \theta_{t-t_0}(q) = \theta_{t-t_0} \circ \theta_{t_0}(p) = \theta_t(p) = \theta^{(p)}(t)$ by the group law for θ. By the translation lemma, γ is an integral curve starting at p. Since $t_0 + \varepsilon > b$, this is a contradiction. $\qquad\square$

Theorem 9.16. *Every compactly supported smooth vector field on a smooth manifold is complete.*

Proof. Suppose V is a compactly supported vector field on a smooth manifold M, and let $K = \operatorname{supp} V$. For each $p \in K$, there is a neighborhood U_p of p and a positive number ε_p such that the flow of V is defined at least on $(-\varepsilon_p, \varepsilon_p) \times U_p$. By compactness, finitely many such sets $U_{p_1}, \ldots, U_{p_k}$ cover K. With $\varepsilon = \min\{\varepsilon_{p_1}, \ldots, \varepsilon_{p_k}\}$, it follows that every maximal integral curve starting in K is defined at least on $(-\varepsilon, \varepsilon)$. Since $V \equiv 0$ outside of K, every integral curve starting in $M \smallsetminus K$ is constant and thus can be defined on all of $\mathbb{R}$. Thus the hypotheses of the uniform time lemma are satisfied, so V is complete. $\qquad\square$

Corollary 9.17. *On a compact smooth manifold, every smooth vector field is complete.* $\qquad\square$

Left-invariant vector fields on Lie groups form another class of vector fields that are always complete.

Theorem 9.18. *Every left-invariant vector field on a Lie group is complete.*

Proof. Let G be a Lie group, let $X \in \operatorname{Lie}(G)$, and let $\theta : \mathcal{D} \to G$ denote the flow of X. There is some $\varepsilon > 0$ such that $\theta^{(e)}$ is defined on $(-\varepsilon, \varepsilon)$.

Let $g \in G$ be arbitrary. Because X is L_g-related to itself, it follows from Proposition 9.6 that the curve $L_g \circ \theta^{(e)}$ is an integral curve of X starting at g and therefore

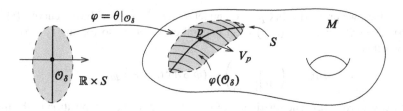

Fig. 9.7 A flowout

is equal to $\theta^{(g)}$. This shows that for each $g \in G$, the integral curve $\theta^{(g)}$ is defined at least on $(-\varepsilon, \varepsilon)$, so the uniform time lemma guarantees that X is complete. $\qquad\square$

Here is another useful property of integral curves.

Lemma 9.19 (Escape Lemma). *Suppose M is a smooth manifold and $V \in \mathfrak{X}(M)$. If $\gamma : J \to M$ is a maximal integral curve of V whose domain J has a finite least upper bound b, then for any $t_0 \in J$, $\gamma\big([t_0, b)\big)$ is not contained in any compact subset of M.*

Proof. Problem 9-6. $\qquad\square$

Flowouts

Flows provide the technical apparatus for many geometric constructions on manifolds. Most of those constructions are based on the following general theorem, which describes how flows behave in the vicinity of certain submanifolds.

Theorem 9.20 (Flowout Theorem). *Suppose M is a smooth manifold, $S \subseteq M$ is an embedded k-dimensional submanifold, and $V \in \mathfrak{X}(M)$ is a smooth vector field that is nowhere tangent to S. Let $\theta \colon \mathcal{D} \to M$ be the flow of V, let $\mathcal{O} = (\mathbb{R} \times S) \cap \mathcal{D}$, and let $\Phi = \theta|_{\mathcal{O}}$.*

(a) *$\Phi \colon \mathcal{O} \to M$ is an immersion.*
(b) *$\partial/\partial t \in \mathfrak{X}(\mathcal{O})$ is Φ-related to V.*
(c) *There exists a smooth positive function $\delta \colon S \to \mathbb{R}$ such that the restriction of Φ to $\mathcal{O}_\delta$ is injective, where $\mathcal{O}_\delta \subseteq \mathcal{O}$ is the flow domain*

$$\mathcal{O}_\delta = \big\{(t, p) \in \mathcal{O} : |t| < \delta(p)\big\}. \tag{9.9}$$

Thus, $\Phi(\mathcal{O}_\delta)$ is an immersed submanifold of M containing S, and V is tangent to this submanifold.
(d) *If S has codimension 1, then $\Phi|_{\mathcal{O}_\delta}$ a diffeomorphism onto an open submanifold of M.*

Remark. The submanifold $\Phi(\mathcal{O}_\delta) \subseteq M$ is called a **flowout from S along V** (see Fig. 9.7).

Proof. First we prove (b). Fix $p \in S$, and let $\sigma \colon \mathcal{D}^{(p)} \to \mathbb{R} \times S$ be the curve $\sigma(t) = (t, p)$. Then $\Phi \circ \sigma(t) = \theta(t, p)$ is an integral curve of V, so for any $t_0 \in \mathcal{D}^{(p)}$ it follows that

$$d\Phi_{(t_0, p)}\left(\frac{\partial}{\partial t}\bigg|_{(t_0, p)}\right) = (\Phi \circ \sigma)'(t_0) = V_{\Phi(t_0, p)}. \tag{9.10}$$

Next we prove (a). The restriction of Φ to $\{0\} \times S$ is the composition of the diffeomorphism $\{0\} \times S \approx S$ with the embedding $S \hookrightarrow M$, so it is an embedding. Thus, the restriction of $d\Phi_{(0, p)}$ to $T_p S$ (viewed as a subspace of $T_{(0, p)}\mathcal{O} \cong T_0\mathbb{R} \oplus T_p S$) is the inclusion $T_p S \hookrightarrow T_p M$. If $(E_1, \dots, E_k)$ is any basis for $T_p S$, it follows that $d\Phi_{(0, p)}$ maps the basis $(\partial/\partial t|_{(0, p)}, E_1, \dots, E_k)$ for $T_{(0, p)}\mathcal{O}$ to $(V_p, E_1, \dots, E_k)$. Since V_p is not tangent to S, this $(k + 1)$-tuple is linearly independent and thus $d\Phi_{(0, p)}$ is injective.

To show $d\Phi$ is injective at other points, we argue as in the proof of the equivariant rank theorem. Given $(t_0, p_0) \in \mathcal{O}$, let $\tau_{t_0} \colon \mathcal{O} \to \mathbb{R} \times S$ be the translation $\tau_{t_0}(t, p) = (t + t_0, p)$. By the group law for θ, the following diagram commutes (where the horizontal maps might be defined only in open subsets containing $(0, p_0)$ and p_0, respectively):

$$\begin{array}{ccc} \mathcal{O} & \xrightarrow{\tau_{t_0}} & \mathcal{O} \\ \Phi\downarrow & & \downarrow\Phi \\ M & \xrightarrow[\theta_{t_0}]{} & M. \end{array}$$

Both horizontal maps in the diagram above are local diffeomorphisms. Taking differentials, we obtain

$$\begin{array}{ccc} T_{(0, p_0)}\mathcal{O} & \xrightarrow{d(\tau_{t_0})_{(0, p_0)}} & T_{(t_0, p_0)}\mathcal{O} \\ d\Phi_{(0, p_0)}\bigg| & & \bigg|d\Phi_{(t_0, p_0)} \\ T_{p_0}M & \xrightarrow[d(\theta_{t_0})_{p_0}]{} & T_{\Phi(t_0, p_0)}M. \end{array}$$

Because the horizontal maps are isomorphisms, the two vertical maps have the same rank. Since we have already shown that $d\Phi_{(0, p_0)}$ has full rank, so does $d\Phi_{(t_0, p_0)}$. This completes the proof that Φ is an immersion.

Next we prove (c). Given a point $p_0 \in S$, choose a slice chart $(U, (x^i))$ for S in M centered at p_0, so that $U \cap S$ is the set where $x^{k+1} = \dots = x^n = 0$ (where $n = \dim M$). Because V is not tangent to S, one of the last $n - k$ components of V_{p_0}, say $V^j(p_0)$, must be nonzero. Shrinking U if necessary, we may assume that there is a constant $c > 0$ such that

$$|V^j(p)| \geq c \quad \text{for all } p \in U. \tag{9.11}$$

Since $\Phi^{-1}(U)$ is open in $\mathbb{R} \times S$, we may choose a number $\varepsilon_{p_0} > 0$ and a neighborhood W_{p_0} of p_0 in S such that $(-\varepsilon_{p_0}, \varepsilon_{p_0}) \times W_{p_0} \subseteq \mathcal{O}$ and

$\Phi\big((-\varepsilon_{p0},\varepsilon_{p0}) \times W_{p0}\big) \subseteq U$. Write the component functions of Φ in these local coordinates as

$$\Phi(t, p) = \big(\Phi^1(t, p), \ldots, \Phi^n(t, p)\big).$$

Because Φ is the restriction of the flow, the component function Φ^j satisfies

$$\frac{\partial \Phi^j}{\partial t}(t, p) = V^j\big(\Phi(t, p)\big), \qquad V^j(0, p) = 0.$$

By (9.11) and the fundamental theorem of calculus, $\big|\Phi^j(t, p)\big| \geq c\,|t|$, and thus for $(t, p) \in (-\varepsilon_{p0}, \varepsilon_{p0}) \times W_{p0}$ we conclude that $\Phi(t, p) \in S$ if and only if $t = 0$.

Choose a smooth partition of unity $\{\psi_p : p \in S\}$ subordinate to the open cover $\{W_p : p \in S\}$ of S, and define $f \colon S \to \mathbb{R}$ by

$$f(q) = \sum_{p \in S} \varepsilon_p \psi_p(q). \tag{9.12}$$

Then f is smooth and positive. For each $q \in S$, there are finitely many $p \in S$ such that $\psi_p(q) > 0$; if p_0 is one of these points such that ε_{p0} is maximum among all such ε_p, then

$$f(q) \leq \varepsilon_{p0} \sum_{p \in S} \psi_p(q) = \varepsilon_{p0}.$$

It follows that if $(t, q) \in \mathcal{O}$ such that $|t| < f(q)$, then $(t, q) \in (-\varepsilon_{p0}, \varepsilon_{p0}) \times W_{p0}$, so $\Phi(t, q) \in S$ if and only if $t = 0$.

Let $\delta = \frac{1}{2} f$. We will show that $\Phi|_{\mathcal{O}_\delta}$ is injective, where $\mathcal{O}_\delta$ is defined by (9.9). Suppose $\Phi(t, q) = \Phi(t', q')$ for some $(t, q), (t', q') \in \mathcal{O}_\delta$. By renaming the points if necessary, we may arrange that $f(q') \leq f(q)$. Our assumption means that $\theta_t(q) = \theta_{t'}(q')$, and the group law for θ then implies that $\theta_{t-t'}(q) = q' \in S$. The fact that (t, q) and (t', q') are in $\mathcal{O}_\delta$ implies that

$$|t - t'| \leq |t| + |t'| < \tfrac{1}{2} f(q) + \tfrac{1}{2} f(q') \leq f(q),$$

which forces $t = t'$ and thus $q = q'$.

Only (d) remains. If S has codimension 1, then $\Phi|_{\mathcal{O}_\delta}$ is an injective smooth immersion between manifolds of the same dimension, so it is an embedding (Proposition 4.22(d)) and a diffeomorphism onto an open submanifold (Proposition 5.1). $\square$

Regular Points and Singular Points

If V is a vector field on M, a point $p \in M$ is said to be a **singular point of V** if $V_p = 0$, and a **regular point** otherwise. The next proposition shows that the integral curves starting at regular and singular points behave very differently from each other.

Proposition 9.21. *Let V be a smooth vector field on a smooth manifold M, and let $\theta \colon \mathcal{D} \to M$ be the flow generated by V. If $p \in M$ is a singular point of V, then*

$\mathcal{D}^{(p)} = \mathbb{R}$ and $\theta^{(p)}$ is the constant curve $\theta^{(p)}(t) \equiv p$. If p is a regular point, then $\theta^{(p)} \colon \mathcal{D}^{(p)} \to M$ is a smooth immersion.

Proof. If $V_p = 0$, then the constant curve $\gamma \colon \mathbb{R} \to M$ given by $\gamma(t) \equiv p$ is clearly an integral curve of V, so by uniqueness and maximality it must be equal to $\theta^{(p)}$.

To verify the second statement, we prove its contrapositive: if $\theta^{(p)}$ is not an immersion, then p is a singular point. The assumption that $\theta^{(p)}$ is not an immersion means that $\theta^{(p)\prime}(s) = 0$ for some $s \in \mathcal{D}^{(p)}$. Write $q = \theta^{(p)}(s)$. Then the argument in the preceding paragraph implies that $\mathcal{D}^{(q)} = \mathbb{R}$ and $\theta^{(q)}(t) = q$ for all $t \in \mathbb{R}$. It follows from Theorem 9.12(b) that $\mathcal{D}^{(p)} = \mathbb{R}$ as well, and for all $t \in \mathbb{R}$ the group law gives

$$\theta^{(p)}(t) = \theta_t(p) = \theta_{t-s}\big(\theta_s(p)\big) = \theta_{t-s}(q) = q.$$

Setting $t = 0$ yields $p = q$, and thus $\theta^{(p)}(t) \equiv p$ and $V_p = \theta^{(p)\prime}(0) = 0$. $\square$

If $\theta \colon \mathcal{D} \to M$ is a flow, a point $p \in M$ is called an **equilibrium point of θ** if $\theta(t, p) = p$ for all $t \in \mathcal{D}^{(p)}$. Proposition 9.21 shows that the equilibrium points of a smooth flow are precisely the singular points of its infinitesimal generator.

The next theorem completely describes, up to diffeomorphism, exactly what a vector field looks like in a neighborhood of a regular point.

Theorem 9.22 (Canonical Form Near a Regular Point). *Let V be a smooth vector field on a smooth manifold M, and let $p \in M$ be a regular point of V. There exist smooth coordinates (s^i) on some neighborhood of p in which V has the coordinate representation $\partial/\partial s^1$. If $S \subseteq M$ is any embedded hypersurface with $p \in S$ and $V_p \notin T_p S$, then the coordinates can also be chosen so that s^1 is a local defining function for S.*

Proof. If no hypersurface S is given, choose any smooth coordinates $(U, (x^i))$ centered at p, and let $S \subseteq U$ be the hypersurface defined by $x^j = 0$, where j is chosen so that $V^j(p) \neq 0$. (Recall that p is a regular point of V.)

Regardless of whether S was given or was constructed as above, since $V_p \notin T_p S$, we can shrink S if necessary so that V is nowhere tangent to S. The flowout theorem then says that there is a flow domain $\mathcal{O}_\delta \subseteq \mathbb{R} \times S$ such that the flow of V restricts to a diffeomorphism Φ from $\mathcal{O}_\delta$ onto an open subset $W \subseteq M$ containing S. There is a product neighborhood $(-\varepsilon, \varepsilon) \times W_0$ of $(0, p)$ in $\mathcal{O}_\delta$. Choose a smooth local parametrization $X \colon \Omega \to S$ whose image is contained in W_0, where Ω is an open subset of $\mathbb{R}^{n-1}$ with coordinates denoted by $(s^2, \dots, s^n)$. It follows that the map $\Psi \colon (-\varepsilon, \varepsilon) \times \Omega \to M$ given by

$$\Psi\big(t, s^2, \dots, s^n\big) = \Phi\big(t, X\big(s^2, \dots, s^n\big)\big)$$

is a diffeomorphism onto a neighborhood of p in M. Because the diffeomorphism $(t, s^1, \dots, s^n) \mapsto (t, X(s^2, \dots, s^n))$ pushes $\partial/\partial t$ forward to itself and $\Phi_*(\partial/\partial t) = V$, it follows that $\Psi_*(\partial/\partial t) = V$. Thus Ψ^{-1} is a smooth coordinate chart in which V has the coordinate representation $\partial/\partial t$. Renaming t to s^1 completes the proof. $\square$

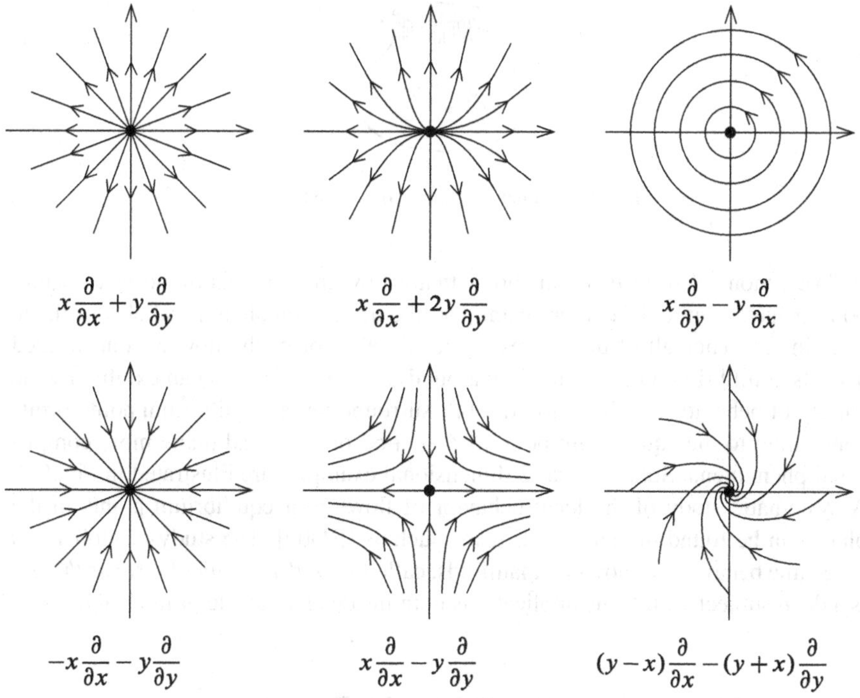
$x\frac{\partial}{\partial x}+y\frac{\partial}{\partial y}$
$x\frac{\partial}{\partial x}+2y\frac{\partial}{\partial y}$
$x\frac{\partial}{\partial y}-y\frac{\partial}{\partial x}$
$-x\frac{\partial}{\partial x}-y\frac{\partial}{\partial y}$
$x\frac{\partial}{\partial x}-y\frac{\partial}{\partial y}$
$(y-x)\frac{\partial}{\partial x}-(y+x)\frac{\partial}{\partial y}$

Fig. 9.8 Examples of flows near equilibrium points

The proof of the canonical form theorem actually provides a technique for finding coordinates that put a given vector field V in canonical form, at least when the corresponding system of ODEs can be explicitly solved: begin with a hypersurface S to which V is not tangent and a local parametrization $X\colon \Omega \to S$, and form the composite map $\Psi(t,s) = \theta_t\big(X(s)\big)$, where θ is the flow of V. The desired coordinate map is then the inverse of Ψ. The procedure is best illustrated by an example.

Example 9.23. Let $W = x\,\partial/\partial y - y\,\partial/\partial x$ on $\mathbb{R}^2$. We computed the flow of W in Example 9.8(b). The point $(1,0) \in \mathbb{R}^2$ is a regular point of W, because $W_{(1,0)} = \partial/\partial y|_{(1,0)} \neq 0$. Because W has nonzero y-coordinate there, we can take S to be the x-axis, parametrized by $X(s) = (s,0)$. We define $\Psi\colon \mathbb{R}^2 \to \mathbb{R}^2$ by

$$\Psi(t,s) = \theta_t(s,0) = (s\cos t, s\sin t),$$

and then solve locally for (t,s) in terms of (x,y) to obtain the following coordinate map in a neighborhood of $(1,0)$:

$$(t,s) = \Psi^{-1}(x,y) = \left(\tan^{-1}(y/x), \sqrt{x^2+y^2}\right). \tag{9.13}$$

It is easy to check that $W = \partial/\partial t$ in these coordinates. (They are, as you might have noticed, just polar coordinates with different names.) //

Fig. 9.9 A collar neighborhood of the boundary

The canonical form theorem shows that a flow in a neighborhood of a regular point behaves, up to diffeomorphism, just like translation along parallel coordinate lines in $\mathbb{R}^n$. Thus all of the interesting local behavior of the flow is concentrated near its equilibrium points. The flow around equilibrium points can exhibit a wide variety of behaviors, such as closed orbits surrounding the equilibrium point, orbits converging to the equilibrium point as $t \to +\infty$ or $-\infty$, and many more complicated phenomena. Some typical 2-dimensional examples are illustrated in Fig. 9.8. A systematic study of the local behavior of flows near equilibrium points in the plane can be found in many ODE texts, such as [BD09]. The study of global and long-time behaviors of flows on manifolds, called *smooth dynamical systems theory*, is a deep subject with many applications both inside and outside of mathematics.

Flows and Flowouts on Manifolds with Boundary

In general, a smooth vector field on a manifold with boundary need not generate a flow, because, for example, the integral curves starting at some boundary points might be defined only on half-open intervals. But there is a variant of the flowout theorem for manifolds with boundary, which has many important applications.

Suppose M is a smooth manifold with nonempty boundary. The next theorem describes a sort of "one-sided flowout" from ∂M, determined by a vector field that is inward pointing everywhere on ∂M.

Theorem 9.24 (Boundary Flowout Theorem). *Let M be a smooth manifold with nonempty boundary, and let N be a smooth vector field on M that is inward-pointing at each point of ∂M. There exist a smooth function $\delta: \partial M \to \mathbb{R}^+$ and a smooth embedding $\Phi: \mathcal{P}_\delta \to M$, where $\mathcal{P}_\delta = \{(t, p) : p \in \partial M, \ 0 \le t < \delta(p)\} \subseteq \mathbb{R} \times \partial M$, such that $\Phi(\mathcal{P}_\delta)$ is a neighborhood of ∂M, and for each $p \in \partial M$ the map $t \mapsto \Phi(t, p)$ is an integral curve of N starting at p.*

Proof. Problem 9-11. $\square$

Let M be a smooth manifold with boundary. A neighborhood of ∂M is called a *collar neighborhood* if it is the image of a smooth embedding $[0, 1) \times \partial M \to M$ that restricts to the obvious identification $\{0\} \times \partial M \to \partial M$. (See Fig. 9.9.)

Theorem 9.25 (Collar Neighborhood Theorem). *If M is a smooth manifold with nonempty boundary, then ∂M has a collar neighborhood.*

Proof. By the result of Problem 8-4, there exists a smooth vector field $N \in \mathfrak{X}(M)$ whose restriction to ∂M is everywhere inward-pointing. Let $\delta \colon M \to \mathbb{R}^+$ and $\Phi \colon \mathscr{P}_\delta \to M$ be as in Theorem 9.24, and define a map $\psi \colon [0, 1) \times \partial M \to \mathscr{P}_\delta$ by $\psi(t, p) = (t\delta(p), p)$. Then ψ is a diffeomorphism that restricts to the identity on $\{0\} \times \partial M$, and therefore the map $\Phi \circ \psi \colon [0, 1) \times \partial M \to M$ is a smooth embedding with open image that restricts to the usual identification $\{0\} \times \partial M \to \partial M$. The image of $\Phi \circ \psi$ is a collar neighborhood of ∂M. $\qquad\square$

Our first application of the collar neighborhood theorem shows (among other things) that every smooth manifold with boundary is homotopy equivalent to its interior.

Theorem 9.26. *Let M be a smooth manifold with nonempty boundary, and let $\iota \colon \operatorname{Int} M \hookrightarrow M$ denote inclusion. There exists a proper smooth embedding $R \colon M \to \operatorname{Int} M$ such that both $\iota \circ R \colon M \to M$ and $R \circ \iota \colon \operatorname{Int} M \to \operatorname{Int} M$ are smoothly homotopic to identity maps. Therefore, ι is a homotopy equivalence.*

Proof. Theorem 9.25 shows that ∂M has a collar neighborhood C in M, which is the image of a smooth embedding $E \colon [0, 1) \times \partial M \to M$ satisfying $E(0, x) = x$ for all $x \in \partial M$. To simplify notation, we will use this embedding to identify C with $[0, 1) \times \partial M$, and denote a point in C as an ordered pair (s, x), with $s \in [0, 1)$ and $x \in \partial M$; thus $(s, x) \in \partial M$ if and only if $s = 0$. For any $a \in (0, 1)$, let $C(a) = \{(s, x) \in C : 0 \le t < a\}$ and $M(a) = M \smallsetminus C(a)$, which is a regular domain in $\operatorname{Int} M$.

Let $\psi \colon [0, 1) \to \left[\frac{1}{3}, 1\right)$ be an increasing diffeomorphism that satisfies $\psi(s) = s$ for $\frac{2}{3} \le s < 1$, and define $R \colon M \to \operatorname{Int} M$ by

$$R(p) = \begin{cases} p, & p \in \operatorname{Int} M\left(\frac{2}{3}\right), \\ (\psi(s), x), & p = (s, x) \in C. \end{cases}$$

These definitions both give the identity map on the set $C \smallsetminus C\left(\frac{2}{3}\right)$ where they overlap, so R is smooth by the gluing lemma. It is a diffeomorphism onto the closed subset $M\left(\frac{1}{3}\right)$, so it is a proper smooth embedding of M into $\operatorname{Int} M$.

Define $H \colon M \times I \to M$ by

$$H(p, t) = \begin{cases} p, & p \in \operatorname{Int} M\left(\frac{2}{3}\right), \\ (ts + (1 - t)\psi(s), x), & p = (s, x) \in C. \end{cases}$$

As before, H is smooth, and a straightforward verification shows that it is a homotopy from $\iota \circ R$ to Id_M. If $p \in \operatorname{Int} M$, then $H(p, t) \in \operatorname{Int} M$ for all $t \in I$, so the restriction of H to $(\operatorname{Int} M) \times I$ is a smooth homotopy from $R \circ \iota$ to $\operatorname{Id}_{\operatorname{Int} M}$. $\qquad\square$

Theorem 9.26 is the main ingredient in the following generalization of the Whitney approximation theorem.

Theorem 9.27 (Whitney Approximation for Manifolds with Boundary). *If M and N are smooth manifolds with boundary, then every continuous map from M to N is homotopic to a smooth map.*

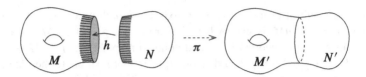

Fig. 9.10 Attaching manifolds along their boundaries

Proof. Theorem 6.26 takes care of the case in which $\partial N = \varnothing$, so we may assume that $\partial N \neq \varnothing$. Let $F \colon M \to N$ be a continuous map, let $\iota \colon \operatorname{Int} N \hookrightarrow N$ be inclusion, and let $R \colon N \to \operatorname{Int} N$ be the map constructed in Theorem 9.26, so that $\iota \circ R \colon N \to N$ is smoothly homotopic to Id_N. Theorem 6.26 shows that $R \circ F \colon M \to \operatorname{Int} N$ is homotopic to a smooth map G. It follows that $\iota \circ G \simeq \iota \circ R \circ F \simeq F$, so $\iota \circ G \colon M \to N$ is a smooth map homotopic to F. $\qquad \square$

The next theorem generalizes the main result of Theorem 6.29 to the case of maps into a manifold with boundary.

Theorem 9.28. *Suppose M and N are smooth manifolds with or without boundary. If $F, G \colon M \to N$ are homotopic smooth maps, then they are smoothly homotopic.*

Proof. Theorem 6.29 takes care of the case $\partial N = \varnothing$, so we may assume that N has nonempty boundary. Let $\iota \colon \operatorname{Int} N \hookrightarrow N$ and $R \colon N \to \operatorname{Int} N$ be as in Theorem 9.26. Then $R \circ F$ and $R \circ G$ are homotopic smooth maps from M to $\operatorname{Int} N$, so Theorem 6.29 shows that they are smoothly homotopic to each other. Thus we have smooth homotopies $F \simeq \iota \circ R \circ F \simeq \iota \circ R \circ G \simeq G$. By transitivity of smooth homotopy (Lemma 6.28), it follows that F is smoothly homotopic to G. $\qquad \square$

The following theorem is probably the most important application of the collar neighborhood theorem.

Theorem 9.29 (Attaching Smooth Manifolds Along Their Boundaries). *Let M and N be smooth n-manifolds with nonempty boundaries, and suppose $h \colon \partial N \to \partial M$ is a diffeomorphism (Fig. 9.10). Let $M \cup_h N$ be the adjunction space formed by identifying each $x \in \partial N$ with $h(x) \in \partial M$. Then $M \cup_h N$ is a topological manifold (without boundary), and has a smooth structure such that there are regular domains $M', N' \subseteq M \cup_h N$ diffeomorphic to M and N, respectively, and satisfying*

$$M' \cup N' = M \cup_h N, \qquad M' \cap N' = \partial M' = \partial N'. \tag{9.14}$$

If M and N are both compact, then $M \cup_h N$ is compact, and if they are both connected, then $M \cup_h N$ is connected.

Proof. For simplicity, let $X = M \cup_h N$ denote the quotient space and $\pi \colon M \amalg N \to X$ the quotient map. Let $V \subseteq M$ and $W \subseteq N$ be collar neighborhoods of ∂M and ∂N, respectively, and denote the corresponding diffeomorphisms by $\alpha \colon [0, 1) \times \partial M \to V$ and $\beta \colon [0, 1) \times \partial N \to W$. Define a continuous map

$\Phi\colon V \amalg W \to (-1,1) \times \partial M$ by

$$\Phi(x) = \begin{cases} (-t, p), & x = \alpha(t, p) \in V, \\ (t, h(q)), & x = \beta(t, q) \in W. \end{cases}$$

Then the restriction of Φ to V or W is a topological embedding with closed image, from which it follows easily that Φ is a closed map. Because Φ is constant on the fibers of π, it descends to a continuous map $\widetilde{\Phi}\colon \pi(V \amalg W) \to (-1,1) \times \partial M$. This map is bijective, and it is a homeomorphism because it too is a closed map: if $K \subseteq \pi(V \amalg W)$ is closed, then $\pi^{-1}(K)$ is closed in $V \amalg W$, and therefore $\widetilde{\Phi}(K) = \Phi(\pi^{-1}(K))$ is closed. Thus, $\pi(V \amalg W)$ is a topological n-manifold. On the other hand, the restriction of π to the saturated open subset $\operatorname{Int} M \amalg \operatorname{Int} N$ is an injective quotient map and thus a homeomorphism onto its image; this shows that X is locally Euclidean of dimension n. Since X is the union of the second-countable open subsets $\pi(\operatorname{Int} M \amalg \operatorname{Int} N)$ and $\pi(V \amalg W)$, it is second-countable. Any two fibers in $M \amalg N$ can be separated by saturated open subsets, so X is Hausdorff. Thus it is a topological n-manifold.

We define a collection of charts on X as follows:

$$\left(\pi(U), \; \varphi \circ \pi^{-1}|_{\pi(U)}\right), \qquad \text{for each smooth chart } (U, \varphi) \text{ for } \operatorname{Int} M \text{ or } \operatorname{Int} N;$$

$$\left(\widetilde{\Phi}^{-1}(U), \; \varphi \circ \widetilde{\Phi}|_{\widetilde{\Phi}^{-1}(U)}\right), \quad \text{for each smooth chart } (U, \varphi) \text{ for } (-1,1) \times \partial M.$$

These maps are compositions of homeomorphisms, so they define coordinate charts on X, and it is straightforward to check that they are all smoothly compatible and thus define a smooth structure on X. The restriction of π to M is continuous, closed, and injective, and thus it is a proper embedding. In terms of any of the smooth charts constructed above and corresponding charts on M, π has a coordinate representation that is either an identity map or an inclusion map, so it is a smooth embedding, and its image M' is therefore a regular domain in X. Similar considerations apply to N; and the relations (9.14) follow immediately from the definitions.

If M and N are compact, then X is the union of the compact sets M' and N', so it is compact; and if they are connected, then X is the union of the connected sets M' and N' with points of $\partial M' = \partial N'$ in common, so it is connected. $\qquad\square$

▶ **Exercise 9.30.** Suppose M and N are smooth n-manifolds with boundary, $A \subseteq \partial M$ and $B \subseteq \partial N$ are nonempty subsets that are unions of components of the respective boundaries, and $h\colon B \to A$ is a diffeomorphism. Verify that the proof of Theorem 9.29 goes through with only trivial changes to show that $M \cup_h N$ is a topological manifold with boundary, and can be given a smooth structure such that M and N are diffeomorphic to regular domains in $M \cup_h N$.

Example 9.31 (Connected Sums). Let M_1, M_2 be connected smooth manifolds of dimension n. For $i = 1, 2$, let U_i be a regular coordinate ball centered at some point $p_i \in M_i$, and let $M_i' = M_i \smallsetminus U_i$ (Fig. 9.11). Problem 5-8 shows that each M_i' is a smooth manifold with boundary whose boundary is diffeomorphic to $\mathbb{S}^{n-1}$. A *smooth connected sum of M_1 and M_2*, denoted by $M_1 \# M_2$, is a smooth manifold formed by choosing a diffeomorphism from ∂M_1 to ∂M_2 and attaching M_1' and M_2' along their boundaries. $\qquad$ //

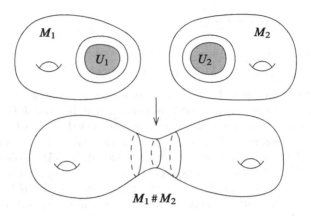

Fig. 9.11 A smooth connected sum

If M is any smooth manifold with boundary, Theorem 9.26 shows that M can be properly embedded into a smooth manifold without boundary (namely, a copy of Int M). The next example shows a different way that M can be so embedded; this construction has the advantage of embedding M into a compact manifold when M itself is compact.

Example 9.32 (The Double of a Smooth Manifold with Boundary). Let M be a smooth manifold with boundary. The *double of* M is the manifold $D(M) = M \cup_{\mathrm{Id}} M$, where Id: $\partial M \to \partial M$ is the identity map of ∂M; it is obtained from $M \amalg M$ by identifying each boundary point in one copy of M with the same boundary point in the other. It is a smooth manifold without boundary, and contains two regular domains diffeomorphic to M. It is easy to check that $D(M)$ is compact if and only if M is compact, and connected if and only if M is connected. (It is useful to extend the definition to manifolds without boundary by defining $D(M) = M \amalg M$ when $\partial M = \varnothing$.)   //

Although vector fields on manifolds with boundary do not always generate flows, there is one circumstance in which they do: when the vector field is everywhere tangent to the boundary. To prove this, we begin with the following special case.

Lemma 9.33. *Suppose M is a smooth manifold and $D \subseteq M$ is a regular domain. If V is a smooth vector field on M that is tangent to ∂D, then every integral curve of V that starts in D remains in D as long as it is defined.*

Proof. Suppose $\gamma: J \to M$ is an integral curve of V with $\gamma(0) \in D$. Define $\mathcal{T} \subseteq J$ by $\mathcal{T} = \{t \in J : \gamma(t) \in D\}$. We will show that $\mathcal{T}$ is both open and closed in J; since J is an interval, this implies $\mathcal{T} = J$ and proves the lemma.

Since D is closed in M (by definition of a regular domain), $\mathcal{T}$ is closed in J by continuity. To prove it is open, suppose $t_0 \in \mathcal{T}$. If $\gamma(t_0) \in \mathrm{Int}\, D$, then a neighborhood of t_0 is contained in $\mathcal{T}$ by continuity, so we can assume $\gamma(t_0) \in \partial D$. Because V is tangent to ∂D, Proposition 8.23 shows that there is a smooth vector field $W = V|_{\partial D}$ that is ι-related to V, where $\iota: \partial D \hookrightarrow M$ is inclusion. Let $\tilde{\gamma}: (t_0 - \varepsilon, t_0 + \varepsilon) \to$

∂D be an integral curve of W with $\widetilde{\gamma}(t_0) = \gamma(t_0)$. By naturality of integral curves (Proposition 9.6), $\iota \circ \widetilde{\gamma}$ is an integral curve of V with the same initial condition, so by uniqueness it must be equal to γ where both are defined. This shows that $\gamma(t) \in \partial D \subseteq D$ for t in some neighborhood of t_0, so $\mathcal{T}$ is open in J as claimed. $\square$

Theorem 9.34 (Flows on Manifolds with Boundary). *The conclusions of Theorem 9.12 remain true if M is a smooth manifold with boundary and V is a smooth vector field on M that is tangent to ∂M.*

Proof. Example 9.32 shows that we can consider M as a regular domain in its double $D(M)$. By the extension lemma for vector fields, we can extend V to a smooth vector field $\widetilde{V}$ on $D(M)$. Let $\widetilde{\theta} \colon \widetilde{\mathcal{D}} \to D(M)$ be the flow of $\widetilde{V}$, and let $\mathcal{D} = \widetilde{\mathcal{D}} \cap (\mathbb{R} \times M)$ and $\theta = \widetilde{\theta}|_{\mathcal{D}}$. Then Lemma 9.33 guarantees that θ maps $\mathcal{D}$ into M, and the rest of the conclusions follow from Theorem 9.12 applied to $\widetilde{V}$. $\square$

For manifolds with boundary, the canonical form theorem has the following variant.

Theorem 9.35 (Canonical Form Near a Regular Point on the Boundary). *Let M be a smooth manifold with boundary and let V be a smooth vector field on M that is tangent to ∂M. If $p \in \partial M$ is a regular point of V, there exist smooth boundary coordinates (s^i) on some neighborhood of p in which V has the coordinate representation $\partial/\partial s^1$.*

Proof. Problem 9-15. $\square$

Lie Derivatives

We know how to make sense of directional derivatives of real-valued functions on a manifold. Indeed, a tangent vector $v \in T_p M$ is by definition an operator that acts on a smooth function f to give a number vf that we interpret as a directional derivative of f at p. In Chapter 3 we showed that this number can be interpreted as the ordinary derivative of f along any curve whose initial velocity is v.

What about the directional derivative of a vector field? In Euclidean space, it makes perfectly good sense to define the directional derivative of a smooth vector field W in the direction of a vector $v \in T_p \mathbb{R}^n$. It is the vector

$$D_v W(p) = \frac{d}{dt}\bigg|_{t=0} W_{p+tv} = \lim_{t \to 0} \frac{W_{p+tv} - W_p}{t}. \tag{9.15}$$

An easy calculation shows that $D_v W(p)$ can be evaluated by applying D_v to each component of W separately:

$$D_v W(p) = D_v W^i(p) \left.\frac{\partial}{\partial x^i}\right|_p.$$

Unfortunately, this definition is heavily dependent upon the fact that $\mathbb{R}^n$ is a vector space, so that the tangent vectors W_{p+tv} and W_p can both be viewed as elements

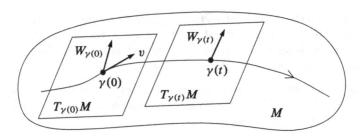

Fig. 9.12 The problem with directional derivatives of vector fields

of $\mathbb{R}^n$. If we search for a way to make invariant sense of (9.15) on a manifold, we see
very quickly what the problem is. To begin with, we can replace $p + tv$ by a curve
$\gamma(t)$ that starts at p and whose initial velocity is v. But even with this substitution,
the difference quotient still makes no sense because $W_{\gamma(t)}$ and $W_{\gamma(0)}$ are elements
of the two different vector spaces $T_{\gamma(t)}M$ and $T_{\gamma(0)}M$, respectively (see Fig. 9.12).
We got away with it in Euclidean space because there is a canonical identification
of each tangent space with $\mathbb{R}^n$ itself; but on a manifold there is no such identification. Thus there is no coordinate-independent way to make sense of the directional
derivative of W in the direction of a vector v.

This problem can be circumvented if we replace the vector $v \in T_pM$ with a
vector field $V \in \mathfrak{X}(M)$, so we can use the flow of V to push values of W back to
p and then differentiate. Thus we make the following definition. Suppose M is a
smooth manifold, V is a smooth vector field on M, and θ is the flow of V. For any
smooth vector field W on M, define a rough vector field on M, denoted by $\mathscr{L}_V W$
and called the **Lie derivative of W with respect to V**, by

$$
\begin{aligned}
(\mathscr{L}_V W)_p &= \frac{d}{dt}\bigg|_{t=0} d(\theta_{-t})_{\theta_t(p)}\big(W_{\theta_t(p)}\big) \\
&= \lim_{t \to 0} \frac{d(\theta_{-t})_{\theta_t(p)}\big(W_{\theta_t(p)}\big) - W_p}{t},
\end{aligned}
\tag{9.16}
$$

provided the derivative exists. For small $t \neq 0$, at least the difference quotient makes
sense: θ_t is defined in a neighborhood of p, and θ_{-t} is the inverse of θ_t, so both
$d(\theta_{-t})_{\theta_t(p)}\big(W_{\theta_t(p)}\big)$ and W_p are elements of T_pM (Fig. 9.13).

If M has nonempty boundary, this definition of $\mathscr{L}_V W$ makes sense as long as
V is tangent to ∂M so that its flow exists by Theorem 9.34. (We will define Lie
derivatives on more general manifolds with boundary below; see the remark after
the proof of Theorem 9.38.)

Lemma 9.36. *Suppose M is a smooth manifold with or without boundary, and
$V, W \in \mathfrak{X}(M)$. If $\partial M \neq \varnothing$, assume in addition that V is tangent to ∂M. Then
$(\mathscr{L}_V W)_p$ exists for every $p \in M$, and $\mathscr{L}_V W$ is a smooth vector field.*

Proof. Let θ be the flow of V. For arbitrary $p \in M$, let $\big(U, (x^i)\big)$ be a smooth chart
containing p. Choose an open interval J_0 containing 0 and an open subset $U_0 \subseteq U$

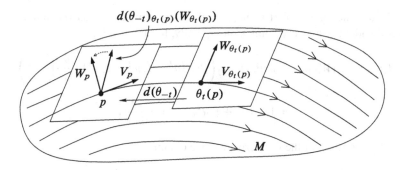

Fig. 9.13 The Lie derivative of a vector field

containing p such that θ maps $J_0 \times U_0$ into U. For $(t, x) \in J_0 \times U_0$, write the component functions of θ as $\left(\theta^1(t, x), \ldots, \theta^n(t, x)\right)$. Then for any $(t, x) \in J_0 \times U_0$, the matrix of $d(\theta_{-t})_{\theta_t(x)} \colon T_{\theta_t(x)}M \to T_x M$ is

$$\left(\frac{\partial \theta^i}{\partial x^j}(-t, \theta(t, x)) \right).$$

Therefore,

$$d(\theta_{-t})_{\theta_t(x)}\left(W_{\theta_t(x)}\right) = \frac{\partial \theta^i}{\partial x^j}(-t, \theta(t, x)) W^j\left(\theta(t, x)\right) \frac{\partial}{\partial x^i}\bigg|_x.$$

Because θ^i and W^j are smooth functions, the coefficient of $\partial/\partial x^i|_x$ depends smoothly on (t, x). It follows that $(\mathcal{L}_V W)_x$, which is obtained by taking the derivative of this expression with respect to t and setting $t = 0$, exists for each $x \in U_0$ and depends smoothly on x. $\qquad\square$

▶ **Exercise 9.37.** Suppose $v \in \mathbb{R}^n$ and W is a smooth vector field on an open subset of $\mathbb{R}^n$. Show that the directional derivative $D_v W(p)$ defined by (9.15) is equal to $(\mathcal{L}_V W)_p$, where V is the vector field $V = v^i \partial/\partial x^i$ with constant coefficients in standard coordinates.

The definition of $\mathcal{L}_V W$ is not very useful for computations, because typically the flow is difficult or impossible to write down explicitly. Fortunately, there is a simple formula for computing the Lie derivative without explicitly finding the flow.

Theorem 9.38. *If M is a smooth manifold and $V, W \in \mathfrak{X}(M)$, then $\mathcal{L}_V W = [V, W]$.*

Proof. Suppose $V, W \in \mathfrak{X}(M)$, and let $\mathcal{R}(V) \subseteq M$ be the set of regular points of V (the set of points $p \in M$ such that $V_p \neq 0$). Note that $\mathcal{R}(V)$ is open in M by continuity, and its closure is the support of V. We will show that $(\mathcal{L}_V W)_p = [V, W]_p$ for all $p \in M$, by considering three cases.

CASE 1: $p \in \mathcal{R}(V)$. In this case, we can choose smooth coordinates (u^i) on a neighborhood of p in which V has the coordinate representation $V = \partial/\partial u^1$ (Theorem 9.22). In these coordinates, the flow of V is $\theta_t(u) = (u^1 + t, u^2, \ldots, u^n)$. For each fixed t, the matrix of $d(\theta_{-t})_{\theta_t(x)}$ in these coordinates (the Jacobian matrix of θ_{-t}) is the identity at every point. Consequently, for any $u \in U$,

$$d(\theta_{-t})_{\theta_t(u)}(W_{\theta_t(u)}) = d(\theta_{-t})_{\theta_t(x)}\left(W^j\left(u^1 + t, u^2, \ldots, u^n\right) \frac{\partial}{\partial u^j}\bigg|_{\theta_t(u)} \right)$$

$$= W^j\left(u^1 + t, u^2, \ldots, u^n\right) \frac{\partial}{\partial u^j}\bigg|_u.$$

Using the definition of the Lie derivative, we obtain

$$(\mathscr{L}_V W)_u = \frac{d}{dt}\bigg|_{t=0} W^j\left(u^1 + t, u^2, \ldots, u^n\right) \frac{\partial}{\partial u^j}\bigg|_u = \frac{\partial W^j}{\partial u^1}\left(u^1, \ldots, u^n\right) \frac{\partial}{\partial u^j}\bigg|_u.$$

On the other hand, by virtue of formula (8.8) for the Lie bracket in coordinates, $[V, W]_u$ is easily seen to be equal to the same expression.

CASE 2: $p \in \operatorname{supp} V$. Because $\operatorname{supp} V$ is the closure of $\mathscr{R}(V)$, it follows by continuity from Case that $(\mathscr{L}_V W)_p = [V, W]_p$ for $p \in \operatorname{supp} V$.

CASE 3: $p \in M \smallsetminus \operatorname{supp} V$. In this case, $V \equiv 0$ on a neighborhood of p. On the one hand, this implies that θ_t is equal to the identity map in a neighborhood of p for all t, so $d(\theta_{-t})_{\theta_t(p)}(W_{\theta_t(p)}) = W_p$, which implies $(\mathscr{L}_V W)_p = 0$. On the other hand, $[V, W]_p = 0$ by formula (8.8).   □

This theorem allows us to extend the definition of the Lie derivative to arbitrary smooth vector fields on a smooth manifold M with boundary. Given $V, W \in \mathfrak{X}(M)$, we define $(\mathscr{L}_V W)_p$ for $p \in \partial M$ by embedding M in a smooth manifold $\widetilde{M}$ without boundary (such as the double of M), extending V and W to smooth vector fields on $\widetilde{M}$, and computing the Lie derivative there. By virtue of the preceding theorem, $(\mathscr{L}_V W)_p = [X, Y]_p$ is independent of the choice of extension.

Theorem 9.38 also gives us a geometric interpretation of the Lie bracket of two vector fields: it is the directional derivative of the second vector field along the flow of the first. A number of nonobvious properties of the Lie derivative follow immediately from things we already know about Lie brackets.

Corollary 9.39. *Suppose M is a smooth manifold with or without boundary, and $V, W, X \in \mathfrak{X}(M)$.*

(a) $\mathscr{L}_V W = -\mathscr{L}_W V$.
(b) $\mathscr{L}_V [W, X] = [\mathscr{L}_V W, X] + [W, \mathscr{L}_V X]$.
(c) $\mathscr{L}_{[V,W]} X = \mathscr{L}_V \mathscr{L}_W X - \mathscr{L}_W \mathscr{L}_V X$.
(d) *If $g \in C^\infty(M)$, then $\mathscr{L}_V(gW) = (Vg)W + g\mathscr{L}_V W$.*
(e) *If $F: M \to N$ is a diffeomorphism, then $F_*(\mathscr{L}_V X) = \mathscr{L}_{F_* V} F_* X$.*

▶ **Exercise 9.40.** Prove this corollary.

Part (d) of this corollary gives a meaning to the mysterious formula (8.11) for Lie brackets of vector fields multiplied by functions: because the Lie bracket $[fV, gW]$ can be thought of as the Lie derivative $\mathscr{L}_{fV}(gW)$, it satisfies a product rule in g

and W; and because it can also be thought of as $-\mathscr{L}_{gW}(fV)$, it satisfies a product rule in f and V as well. Expanding out these two product rules yields (8.11).

If V and W are vector fields on M and θ is the flow of V, the Lie derivative $(\mathscr{L}_V W)_p$, by definition, expresses the t-derivative of the time-dependent vector $d(\theta_{-t})_{\theta_t(p)}\left(W_{\theta_t(p)}\right) \in T_p M$ at $t = 0$. The next proposition shows how it can also be used to compute the derivative of this expression at other times. We will use this result in the proof of Theorem 9.42 below.

Proposition 9.41. *Suppose M is a smooth manifold with or without boundary and $V, W \in \mathfrak{X}(M)$. If $\partial M \neq \varnothing$, assume also that V is tangent to ∂M. Let θ be the flow of V. For any (t_0, p) in the domain of θ,*

$$\frac{d}{dt}\bigg|_{t=t_0} d(\theta_{-t})_{\theta_t(p)}\left(W_{\theta_t(p)}\right) = d\left(\theta_{-t_0}\right)\left((\mathscr{L}_V W)_{\theta_{t_0}(p)}\right). \qquad (9.17)$$

Proof. Let $p \in M$ be arbitrary, let $\mathcal{D}^{(p)} \subseteq \mathbb{R}$ denote the domain of the integral curve $\theta^{(p)}$, and consider the map $X: \mathcal{D}^{(p)} \to T_p M$ given by $X(t) = d(\theta_{-t})_{\theta_t(p)}\left(W_{\theta_t(p)}\right)$. The argument in the proof of Lemma 9.36 shows that X is a smooth curve in the vector space $T_p M$. Making the change of variables $t = t_0 + s$, we obtain

$$
\begin{aligned}
X'(t_0) &= \frac{d}{ds}\bigg|_{s=0} X(t_0 + s) = \frac{d}{ds}\bigg|_{s=0} d\left(\theta_{-t_0-s}\right)\left(W_{\theta_{s+t_0}(p)}\right) \\
&= \frac{d}{ds}\bigg|_{s=0} d\left(\theta_{-t_0}\right) \circ d(\theta_{-s})\left(W_{\theta_s(\theta_{t_0}(p))}\right) \\
&= d\left(\theta_{-t_0}\right)\left(\frac{d}{ds}\bigg|_{s=0} d(\theta_{-s})\left(W_{\theta_s(\theta_{t_0}(p))}\right)\right).
\end{aligned}
$$

(The last equality follows because $d\left(\theta_{-t_0}\right): T_{\theta_{t_0}(p)}M \to T_p M$ is a linear map that is independent of s. See Fig. 9.14.) By definition of the Lie derivative, this last expression is equal to the right-hand side of (9.17). $\qquad\square$

Commuting Vector Fields

Let M be a smooth manifold and $V, W \in \mathfrak{X}(M)$. We say that *V and W commute* if $VWf = WVf$ for every smooth function f, or equivalently if $[V, W] \equiv 0$. If θ is a smooth flow, a vector field W is said to be *invariant under θ* if W is θ_t-related to itself for each t; more precisely, this means that $W|_{M_t}$ is θ_t-related to $W|_{M_{-t}}$ for each t, or equivalently that $d(\theta_t)_p(W_p) = W_{\theta_t(p)}$ for all (t, p) in the domain of θ. The next proposition shows that these two concepts are intimately related.

Theorem 9.42. *For smooth vector fields V and W on a smooth manifold M, the following are equivalent:*

(a) *V and W commute.*

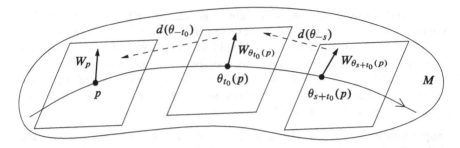

Fig. 9.14 Proof of Proposition 9.41

(b) W is invariant under the flow of V.
(c) V is invariant under the flow of W.

Proof. Suppose $V, W \in \mathfrak{X}(M)$, and let θ denote the flow of V. If (b) holds, then $W_{\theta_t(p)} = d(\theta_t)_p(W_p)$ whenever (t, p) is in the domain of θ. Applying $d(\theta_{-t})_{\theta_t(p)}$ to both sides, we conclude that $d(\theta_{-t})_{\theta_t(p)}(W_{\theta_t(p)}) = W_p$, which obviously implies $[V, W] = \mathscr{L}_V W = 0$ directly from the definition of the Lie derivative. The same argument shows that (c) implies (a).

To prove that (a) implies (b), assume that $[V, W] = \mathscr{L}_V W = 0$. Let $p \in M$ be arbitrary, and let $X(t) = d(\theta_{-t})_{\theta_t(p)}(W_{\theta_t(p)})$ for $t \in \mathcal{D}^{(p)}$. Proposition 9.41 shows that $X'(t) \equiv 0$. Since $X(0) = W_p$, this implies that $X(t) = W_p$ for all $t \in \mathcal{D}^{(p)}$, and applying $d(\theta_t)_p$ to both sides yields the identity that says W is invariant under θ. The same proof also shows that (a) implies (c). □

Corollary 9.43. *Every smooth vector field is invariant under its own flow.*

Proof. Use the preceding proposition together with the fact that $[V, V] \equiv 0$. □

The deepest characterization of commuting vector fields is in terms of the relationship between their respective flows. The next theorem says that two vector fields commute if and only if their flows commute. But before we state the theorem formally, we need to examine exactly what this means. Suppose V and W are smooth vector fields on M, and let θ and ψ denote their respective flows. If V and W are complete, it is clear what we should mean by saying their flows commute: simply that $\theta_t \circ \psi_s = \psi_s \circ \theta_t$ for all $s, t \in \mathbb{R}$. However, if either V or W is not complete, the most we can hope for is that this equation holds for all s and t such that both sides are defined. Unfortunately, even when the vector fields commute, their flows might not commute in this naive sense, because there are examples of commuting vector fields V and W and particular choices of t, s, and p for which both $\theta_t \circ \psi_s(p)$ and $\psi_s \circ \theta_t(p)$ are defined, but they are not equal (see Problem 9-19 for one such example). Here is the problem: if $\theta_t \circ \psi_s(p)$ is defined for $t = t_0$ and $s = s_0$, then by the properties of flow domains, it must be defined for all t in some open interval containing 0 and t_0, but the analogous statement need not be true of s—there might be values of s between 0 and s_0 for which the integral curve of V starting at $\psi_s(p)$ does not extend all the way to $t = t_0$.

Thus we make the following definition. If θ and ψ are flows on M, we say that *θ and ψ commute* if the following condition holds for every $p \in M$: whenever J and K are open intervals containing 0 such that one of the expressions $\theta_t \circ \psi_s(p)$ or $\psi_s \circ \theta_t(p)$ is defined for all $(s,t) \in J \times K$, both are defined and they are equal. For global flows, this is the same as saying that $\theta_t \circ \psi_s = \psi_s \circ \theta_t$ for all s and t.

Theorem 9.44. *Smooth vector fields commute if and only if their flows commute.*

Proof. Let V and W be smooth vector fields on a smooth manifold M, and let θ and ψ denote their respective flows. Assume first that V and W commute. Suppose that $p \in M$, and J and K are open intervals containing 0 such that $\psi_s \circ \theta_t(p)$ is defined for all $(s,t) \in J \times K$. (The same proof with V and W reversed works under the assumption that the other expression is defined on such a rectangle.) By Theorem 9.42, the hypothesis implies that V is invariant under ψ. Fix any $s \in J$, and consider the curve $\gamma\colon K \to M$ defined by $\gamma(t) = \psi_s \circ \theta_t(p) = \psi_s\big(\theta^{(p)}(t)\big)$. This curve satisfies $\gamma(0) = \psi_s(p)$, and its velocity at $t \in K$ is

$$\gamma'(t) = \frac{d}{dt}\Big(\psi_s\big(\theta^{(p)}(t)\big)\Big) = d(\psi_s)\big(\theta^{(p)'}(t)\big) = d(\psi_s)\big(V_{\theta^{(p)}(t)}\big) = V_{\gamma(t)}.$$

Thus, γ is an integral curve of V starting at $\psi_s(p)$. By uniqueness, therefore,

$$\gamma(t) = \theta^{\psi_s(p)}(t) = \theta_t\big(\psi_s(p)\big).$$

This proves that θ and ψ commute.

Conversely, assume that the flows commute, and let $p \in M$. If $\varepsilon > 0$ is chosen small enough that $\psi_s \circ \theta_t(p)$ is defined whenever $|s| < \varepsilon$ and $|t| < \varepsilon$, then the hypothesis guarantees that $\psi_s \circ \theta_t(p) = \theta_t \circ \psi_s(p)$ for all such s and t. This can be rewritten in the form

$$\psi^{\theta_t(p)}(s) = \theta_t\big(\psi^{(p)}(s)\big).$$

Differentiating this relation with respect to s, we get

$$W_{\theta_t(p)} = \frac{d}{ds}\bigg|_{s=0} \psi^{\theta_t(p)}(s) = \frac{d}{ds}\bigg|_{s=0} \theta_t\big(\psi^{(p)}(s)\big) = d(\theta_t)_p(W_p).$$

Applying $d(\theta_{-t})_{\theta_t(p)}$ to both sides, we conclude

$$d(\theta_{-t})_{\theta_t(p)}(W_{\theta_t(p)}) = W_p.$$

Differentiating with respect to t and applying the definition of the Lie derivative shows that $(\mathscr{L}_V W)_p = 0$. $\qquad\square$

Commuting Frames

Suppose M is a smooth n-manifold. Recall that a *local frame for M* is an n-tuple (E_i) of vector fields defined on an open subset $U \subseteq M$ such that $\big(E_i|_p\big)$ forms a basis for $T_p M$ at each $p \in U$. A smooth local frame (E_i) for M is called a *commuting frame* if $[E_i, E_j] = 0$ for all i and j. (Commuting frames are called *holonomic frames* by some authors.)

Example 9.45 (Commuting and Noncommuting Frames).

(a) The simplest examples of commuting frames are the coordinate frames. Given
    any smooth coordinate chart $(U, (x^i))$ for a smooth manifold M, (8.10) shows
    that the coordinate frame $(\partial/\partial x^i)$ is a commuting frame.
(b) The frame (E_1, E_2) for $\mathbb{R}^2$ over $\mathbb{R}^2 \smallsetminus \{0\}$ defined by (8.3) is not a commuting
    frame, because a straightforward computation shows that

$$[E_1, E_2] = \frac{y}{r^2}\frac{\partial}{\partial x} - \frac{x}{r^2}\frac{\partial}{\partial y} \neq 0. \qquad\qquad //$$

Because every coordinate frame is a commuting frame, and because Lie brackets
are invariantly defined, it follows that a necessary condition for a smooth frame to
be expressible as a coordinate frame in some smooth chart is that it be a commuting
frame. Thus, the computation above shows that (E_1, E_2) cannot be expressed as a
coordinate frame for $\mathbb{R}^2$ with respect to any choice of smooth local coordinates.

The next theorem shows that commuting is also a sufficient condition for a
smooth frame to be locally expressible as a coordinate frame.

Theorem 9.46 (Canonical Form for Commuting Vector Fields). *Let M be a
smooth n-manifold, and let $(V_1, \ldots, V_k)$ be a linearly independent k-tuple of smooth
commuting vector fields on an open subset $W \subseteq M$. For each $p \in W$, there exists a smooth coordinate chart $(U, (s^i))$ centered at p such that $V_i = \partial/\partial s^i$ for
$i = 1, \ldots, k$. If $S \subseteq W$ is an embedded codimension-k submanifold and p is a point
of S such that T_pS is complementary to the span of $(V_1|_p, \ldots, V_k|_p)$, then the coor-
dinates can also be chosen such that $S \cap U$ is the slice defined by $s^1 = \cdots = s^k = 0$.*

Proof. Let $p \in W$ be arbitrary. If no submanifold S is given, just let S be any
smooth embedded codimension-k submanifold S whose tangent space at p is com-
plementary to the span of $(V_1|_p, \ldots, V_k|_p)$ (e.g., an appropriate coordinate slice).
Let $(U, (x^i))$ be a slice chart for S centered at p, with $U \subseteq W$, and with $S \cap U$
equal to the slice $\{x \in U : x^1 = \cdots = x^k = 0\}$. Our assumptions ensure that the
vectors $\{V_1|_p, \ldots, V_k|_p, \partial/\partial x^{k+1}|_p, \ldots, \partial/\partial x^n|_p\}$ span T_pM. Since the theorem is
purely local, we may as well consider $V_1, \ldots, V_k$ as vector fields on $U \subseteq \mathbb{R}^n$, and
consider S to be the subset of U where the first k coordinates vanish. The basic idea
of this proof is similar to that of the flowout theorem, except that we have to do a bit
of extra work to make use of the hypothesis that the vector fields commute.

Let θ_i denote the flow of V_i for $i = 1, \ldots, k$. There exist $\varepsilon > 0$ and a neighborhood Y of p in U such that the composition $(\theta_1)_{t_1} \circ (\theta_2)_{t_2} \circ \cdots \circ (\theta_k)_{t_k}$ is defined
on Y and maps Y into U whenever $|t_1|, \ldots, |t_k|$ are all less than ε. (To see this,
just choose $\varepsilon_k > 0$ and $U_k \subseteq U$ such that θ_k maps $(-\varepsilon_k, \varepsilon_k) \times U_k$ into U, and then
inductively choose ε_i and U_i such that θ_i maps $(-\varepsilon_i, \varepsilon_i) \times U_i$ into U_{i+1}. Taking
$\varepsilon = \min\{\varepsilon_i\}$ and $Y = U_1$ does the trick.)

Define $\Omega \subseteq \mathbb{R}^{n-k}$ by

$$\Omega = \left\{ (s^{k+1}, \ldots, s^n) \in \mathbb{R}^{n-k} : (0, \ldots, 0, s^{k+1}, \ldots, s^n) \in Y \right\},$$

and define $\Phi \colon (-\varepsilon, \varepsilon)^k \times \Omega \to U$ by

$$\Phi\left(s^1, \ldots, s^k, s^{k+1}, \ldots, s^n\right) = (\theta_1)_{s^1} \circ \cdots \circ (\theta_k)_{s^k}\left(0, \ldots, 0, s^{k+1}, \ldots, s^n\right).$$

By construction, $\Phi(\{0\} \times \Omega) = S \cap Y$.

We show next that $\partial/\partial s^i$ is Φ-related to V_i for $i = 1, \ldots, k$. Because the flows θ_i commute, for any $i \in \{1, \ldots, k\}$ and any $s_0 \in (-\varepsilon, \varepsilon)^k \times \Omega$ we have

$$d\Phi_{s_0}\left(\left.\frac{\partial}{\partial s^i}\right|_{s_0}\right) f = \left.\frac{\partial}{\partial s^i}\right|_{s_0} f\left(\Phi\left(s^1, \ldots, s^n\right)\right)$$

$$= \left.\frac{\partial}{\partial s^i}\right|_{s_0} f\left((\theta_1)_{s^1} \circ \cdots \circ (\theta_k)_{s^k}\left(0, \ldots, 0, s^{k+1}, \ldots, s^n\right)\right)$$

$$= \left.\frac{\partial}{\partial s^i}\right|_{s_0} f\left((\theta_i)_{s^i} \circ (\theta_1)_{s^1} \circ \cdots \circ (\theta_{i-1})_{s^{i-1}} \circ (\theta_{i+1})_{s^{i+1}}\right.$$

$$\left. \circ \cdots \circ (\theta_k)_{s^k}\left(0, \ldots, 0, s^{k+1}, \ldots, s^n\right)\right).$$

For any $q \in M$, $t \mapsto (\theta_i)_t(q)$ is an integral curve of V_i, so this last expression is equal to $V_i|_{\Phi(s_0)} f$, which proves the claim.

Next we show that $d\Phi_0$ is invertible. The computation above shows that

$$d\Phi_0\left(\left.\frac{\partial}{\partial s^i}\right|_0\right) = V_i|_p, \quad i = 1, \ldots, k.$$

On the other hand, since $\Phi\left(0, \ldots, 0, s^{k+1}, \ldots, s^n\right) = \left(0, \ldots, 0, s^{k+1}, \ldots, s^n\right)$, it follows immediately that

$$d\Phi_0\left(\left.\frac{\partial}{\partial s^i}\right|_0\right) = \left.\frac{\partial}{\partial x^i}\right|_p, \quad i = k+1, \ldots, n.$$

It follows that $d\Phi_0$ takes the basis $\left(\partial/\partial s^1|_0, \ldots, \partial/\partial s^n|_0\right)$ for $T_0\mathbb{R}^n$ to the basis $\left(V_1|_p, \ldots, V_k|_p, \partial/\partial x^{k+1}|_p, \ldots, \partial/\partial x^n|_p\right)$ for T_pM. By the inverse function theorem, Φ is a diffeomorphism in a neighborhood of 0, and $\varphi = \Phi^{-1}$ is a smooth coordinate map that takes $\partial/\partial s^i$ to V_i for $i = 1, \ldots, k$, and takes S to the slice $s^1 = \cdots = s^k = 0$. $\qquad \square$

Just as in the case of a single vector field, the proof of Theorem 9.46 suggests a technique for finding explicit coordinates that put a set of commuting vector fields into canonical form, as long as their flows can be found explicitly. The method can be summarized as follows: Begin with an $(n-k)$-dimensional submanifold S whose tangent space at p is complementary to the span of $\left(V_1|_p, \ldots, V_k|_p\right)$. Then define Φ by starting at an arbitrary point in S and following the k flows successively for k arbitrary times. Because the flows commute, it does not matter in which order they are applied. An example will help to clarify the procedure.

Example 9.47. Consider the following two vector fields on $\mathbb{R}^2$:

$$V = x\frac{\partial}{\partial y} - y\frac{\partial}{\partial x}, \qquad W = x\frac{\partial}{\partial x} + y\frac{\partial}{\partial y}.$$

A computation shows that $[V, W] = 0$. Example 9.8 showed that the flow of V is

$$\theta_t(x, y) = (x\cos t - y\sin t, \ x\sin t + y\cos t),$$

and an easy verification shows that the flow of W is

$$\eta_t(x, y) = (e^t x, e^t y).$$

At $p = (1, 0)$, V_p and W_p are linearly independent. Because $k = n = 2$ in this case, we can take the subset S to be the single point $\{(1, 0)\}$, and define $\Phi: \mathbb{R}^2 \to \mathbb{R}^2$ by

$$\Phi(s, t) = \eta_t \circ \theta_s(1, 0) = (e^t \cos s, e^t \sin s).$$

In this case, we can solve for $(s, t) = \Phi^{-1}(x, y)$ explicitly in a neighborhood of $(1, 0)$ to obtain the coordinate map

$$(s, t) = \left(\tan^{-1}(y/x), \log\sqrt{x^2 + y^2}\right). \hspace{2cm} /\!/$$

Time-Dependent Vector Fields

All of the systems of differential equations we have encountered so far have been *autonomous* ones, meaning that when they are written in the form (9.1), the functions V^i on the right-hand sides do not depend explicitly on the independent variable t (see Appendix D). However, nonautonomous ODEs do arise in manifold theory, so it is worth exploring how the results of this chapter can be extended to cover this case. We will use this theory only in Chapter 22.

Let M be a smooth manifold. A *time-dependent vector field on M* is a continuous map $V: J \times M \to TM$, where $J \subseteq \mathbb{R}$ is an interval, such that $V(t, p) \in T_pM$ for each $(t, p) \in J \times M$. This means that for each $t \in J$, the map $V_t: M \to TM$ defined by $V_t(p) = V(t, p)$ is a vector field on M. If V is a time-dependent vector field on M, an *integral curve of V* is a differentiable curve $\gamma: J_0 \to M$, where J_0 is an interval contained in J, such that

$$\gamma'(t) = V(t, \gamma(t)) \quad \text{for all } t \in J_0.$$

Every ordinary vector field $X \in \mathfrak{X}(M)$ determines a time-dependent vector field defined on $\mathbb{R} \times M$, just by setting $V(t, p) = X_p$. (It is occasionally useful to consider time-dependent vector fields defined on more general open subsets of $\mathbb{R} \times M$; but for simplicity we restrict attention to a product set $J \times M$, and leave it to the interested reader to figure out how the results need to be modified for the more general case.)

A time-dependent vector field might not generate a flow, because two integral curves starting at the same point but at different times might follow different paths, whereas all integral curves of a flow through a given point have the same image. As a substitute for the fundamental theorem on flows, we have the following theorem.

Theorem 9.48 (Fundamental Theorem on Time-Dependent Flows). *Let M be a smooth manifold, let $J \subseteq \mathbb{R}$ be an open interval, and let $V \colon J \times M \to TM$ be a smooth time-dependent vector field on M. There exist an open subset $\mathcal{E} \subseteq J \times J \times M$ and a smooth map $\psi \colon \mathcal{E} \to M$ called the **time-dependent flow of V**, with the following properties:*

(a) *For each $t_0 \in J$ and $p \in M$, the set $\mathcal{E}^{(t_0,p)} = \{t \in J : (t, t_0, p) \in \mathcal{E}\}$ is an open interval containing t_0, and the smooth curve $\psi^{(t_0,p)} \colon \mathcal{E}^{(t_0,p)} \to M$ defined by $\psi^{(t_0,p)}(t) = \psi(t, t_0, p)$ is the unique maximal integral curve of V with initial condition $\psi^{(t_0,p)}(t_0) = p$.*

(b) *If $t_1 \in \mathcal{E}^{(t_0,p)}$ and $q = \psi^{(t_0,p)}(t_1)$, then $\mathcal{E}^{(t_1,q)} = \mathcal{E}^{(t_0,p)}$ and $\psi^{(t_1,q)} = \psi^{(t_0,p)}$.*

(c) *For each $(t_1, t_0) \in J \times J$, the set $M_{t_1,t_0} = \{p \in M : (t_1, t_0, p) \in \mathcal{E}\}$ is open in M, and the map $\psi_{t_1,t_0} \colon M_{t_1,t_0} \to M$ defined by $\psi_{t_1,t_0}(p) = \psi(t_1, t_0, p)$ is a diffeomorphism from M_{t_1,t_0} onto M_{t_0,t_1} with inverse ψ_{t_0,t_1}.*

(d) *If $p \in M_{t_1,t_0}$ and $\psi_{t_1,t_0}(p) \in M_{t_2,t_1}$, then $p \in M_{t_2,t_0}$ and*

$$\psi_{t_2,t_1} \circ \psi_{t_1,t_0}(p) = \psi_{t_2,t_0}(p). \tag{9.18}$$

Proof. This can be proved by following the outline of the proof of Theorem 9.12, using Theorem D.6 in place of Theorem D.1. However, it is much quicker to use the following trick to reduce it to the time-independent case.

Consider the smooth vector field $\widetilde{V}$ on $J \times M$ defined by

$$\widetilde{V}_{(s,p)} = \left(\left. \frac{\partial}{\partial s} \right|_s , \, V(s, p) \right),$$

where s is the standard coordinate on $J \subseteq \mathbb{R}$, and we identify $T_{(s,p)}(J \times M)$ with $T_s J \oplus T_p M$ as usual (see Proposition 3.14). Let $\widetilde{\theta} \colon \widetilde{\mathcal{D}} \to J \times M$ denote the flow of $\widetilde{V}$. If we write the component functions of $\widetilde{\theta}$ as

$$\widetilde{\theta}\big(t, (s, p)\big) = \big(\alpha\big(t, (s, p)\big), \beta\big(t, (s, p)\big)\big),$$

then $\alpha \colon \widetilde{\mathcal{D}} \to J$ and $\beta \colon \widetilde{\mathcal{D}} \to M$ satisfy

$$\frac{\partial \alpha}{\partial t}\big(t, (s, p)\big) = 1, \qquad\qquad \alpha\big(0, (s, p)\big) = s,$$

$$\frac{\partial \beta}{\partial t}\big(t, (s, p)\big) = V\big(\alpha\big(t, (s, p)\big), \beta\big(t, (s, p)\big)\big), \quad \beta\big(0, (s, p)\big) = p.$$

It follows immediately that $\alpha(t, (s, p)) = t + s$, and therefore β satisfies

$$\frac{\partial \beta}{\partial t}\big(t, (s, p)\big) = V\big(t + s, \beta\big(t, (s, p)\big)\big). \tag{9.19}$$

Let $\mathcal{E}$ be the subset of $\mathbb{R} \times J \times M$ defined by

$$\mathcal{E} = \big\{(t, t_0, p) : \big(t - t_0, (t_0, p)\big) \in \widetilde{\mathcal{D}}\big\}.$$

Clearly, $\mathcal{E}$ is open in $\mathbb{R} \times J \times M$ because $\widetilde{\mathcal{D}}$ is. Moreover, since α maps $\widetilde{\mathcal{D}}$ into J, if $(t, t_0, p) \in \mathcal{E}$, then $t = \alpha\big(t - t_0, (t_0, p)\big) \in J$, which implies that $\mathcal{E} \subseteq J \times J \times M$.

The fact that each set $M_{t_1,t_0} = \{p \in M : (t_1,t_0,p) \in \mathcal{E}\}$ is open in M follows immediately from the fact that $\mathcal{E}$ is open.

Now define $\psi \colon \mathcal{E} \to M$ by

$$\psi(t,t_0,p) = \beta\big(t - t_0, (t_0,p)\big).$$

Then ψ is smooth because β is, and it follows from (9.19) that $\psi^{(t_0,p)}(t) = \psi(t,t_0,p)$ is an integral curve of V with initial condition $\psi^{(t_0,p)}(t_0) = p$.

To prove uniqueness, suppose $t_0 \in J$ and $\gamma \colon J_0 \to M$ is any integral curve of V defined on some open interval $J_0 \subseteq J$ containing t_0 and satisfying $\gamma(t_0) = p$. Define a smooth curve $\tilde\gamma \colon J_0 \to J \times M$ by $\tilde\gamma(t) = \big(t,\gamma(t)\big)$. Then $\tilde\gamma$ is easily seen to be an integral curve of $\tilde V$ with initial condition $\tilde\gamma(t_0) = (t_0,p)$. By uniqueness and maximality of integral curves of $\tilde V$, we must have $\tilde\gamma(t) = \tilde\theta\big(t - t_0, (t_0,p)\big)$ on its whole domain, which implies that the domain of γ is contained in that of $\psi^{(t_0,p)}$, and $\gamma = \psi^{(t_0,p)}$ on that domain. It follows that $\psi^{(t_0,p)}$ is the unique maximal integral curve of V passing through p at $t = t_0$. This completes the proof of (a).

To prove (b), suppose $t_1 \in \mathcal{E}^{(t_0,p)}$ and $q = \psi^{(t_0,p)}(t_1)$. Then both $\psi^{(t_1,q)}$ and $\psi^{(t_0,p)}$ are integral curves of V that pass through q when $t = t_1$, so by uniqueness and maximality they must have the same domain and be equal on that domain.

Next, we prove (d). Suppose $p \in M_{t_1,t_0}$ and $\psi_{t_1,t_0}(p) \in M_{t_2,t_1}$, and set $q = \psi_{t_1,t_0}(p) = \psi^{(t_0,p)}(t_1)$. Then (b) implies that $\psi^{(t_1,q)}(t_2) = \psi^{(t_0,p)}(t_2)$. Unwinding the definitions yields (9.18).

Finally, we prove (c). Suppose $(t_1,t_0) \in J \times J$. We have already noted that M_{t_1,t_0} is open in M. To show that $\psi_{t_1,t_0}\big(M_{t_1,t_0}\big) \subseteq M_{t_0,t_1}$, let p be a point of M_{t_1,t_0}, and set $q = \psi_{t_1,t_0}(p)$. Part (b) implies that $\mathcal{E}^{(t_0,p)} = \mathcal{E}^{(t_1,q)}$, and thus $t_0 \in \mathcal{E}^{(t_0,p)} = \mathcal{E}^{(t_1,q)}$. This is equivalent to $(t_0,t_1,q) \in \mathcal{E}$, which in turn means $q \in M_{t_0,t_1}$ as claimed. To see that $\psi_{t_1,t_0} \colon M_{t_1,t_0} \to M_{t_0,t_1}$ is a diffeomorphism, just note that the same argument as above implies that $\psi_{t_0,t_1}\big(M_{t_0,t_1}\big) \subseteq M_{t_1,t_0}$, and then (d) implies that $\psi_{t_1,t_0} \circ \psi_{t_0,t_1}(p) = \psi_{t_1,t_1}(p) = p$ for all $p \in M_{t_0,t_1}$, and similarly that $\psi_{t_0,t_1} \circ \psi_{t_1,t_0}(q) = q$ for $q \in M_{t_1,t_0}$. $\qquad\square$

▶ **Exercise 9.49.** Let M be a smooth manifold. Suppose X is a (time-independent) smooth vector field on M, and $\theta \colon \mathcal{D} \to M$ is its flow. Let V be the time-dependent vector field defined by $V(t,p) = X_p$. Show that the time-dependent flow of V is given by $\psi(t,t_0,p) = \theta(t - t_0, p)$, with domain $\mathcal{E} = \{(t,t_0,p) : (t - t_0, p) \in \mathcal{D}\}$.

Example 9.50. Define a time-dependent vector field V on $\mathbb{R}^n$ by

$$V(t,x) = \frac{1}{t} x^i \frac{\partial}{\partial x^i}\bigg|_x, \qquad (t,x) \in (0,\infty) \times \mathbb{R}^n.$$

Suppose $t_0 \in (0,\infty)$ and $x_0 \in \mathbb{R}^n$ are arbitrary, and let $\gamma(t) = \big(x^1(t),\dots,x^n(t)\big)$ denote the integral curve of V with initial condition $\gamma(t_0) = x_0$. Then the components of γ satisfy the following nonautonomous system of differential equations:

$$\dot x^i(t) = \frac{1}{t} x^i(t),$$

$$x^i(t_0) = x_0^i.$$

The maximal solution to this system, as you can easily check, is $x^i(t) = tx_0^i/t_0$, defined for all $t > 0$. Therefore, the time-dependent flow of V is given by $\psi(t, t_0, x) = tx/t_0$ for $(t, t_0, x) \in (0, \infty) \times (0, \infty) \times \mathbb{R}^n$. //

First-Order Partial Differential Equations

One of the most powerful applications of the theory of flows is to partial differential equations. In its most general form, a *partial differential equation* (*PDE*) is any equation that relates an unknown function of two or more variables with its partial derivatives up to some order and with the independent variables. The *order* of the PDE is the highest-order derivative of the unknown function that appears.

The number of specialized techniques that have been developed to solve partial differential equations is staggering. (For an introduction to the general theory, you can consult one of the many excellent introductory books on the subject, such as [Eva98, Fol95, Joh91].) However, it is a remarkable fact that real-valued *first-order* PDEs can be reduced to *ordinary* differential equations by means of the theory of flows, and thus can be solved using only ODEs and a little differential-geometric insight but no specialized PDE theory. In this section, we describe how this is done for two special classes of first-order equations: first, linear equations; and then, somewhat more generally, quasilinear equations (which we define below). A PDE that is not quasilinear is said to be *fully nonlinear*; we will show how to treat fully nonlinear first-order equations in Chapter 22.

In coordinates, any first-order PDE for a single unknown function can be written

$$F\left(x^1, \ldots, x^n, u(x), \frac{\partial u}{\partial x^1}(x), \ldots, \frac{\partial u}{\partial x^n}(x)\right) = 0, \tag{9.20}$$

where u is an unknown function of n variables and F is a given smooth function of $2n + 1$ variables. (Smoothness is not strictly necessary, but we assume it throughout for simplicity.) The theory we will describe applies only when F and u are real-valued, so we assume that as well. (There is also a fascinating theory of complex-valued first-order PDEs, but it requires entirely different methods.)

Without further restrictions, most PDEs have a multitude of solutions—for example, the PDE $\partial u/\partial x = 0$ in the plane is solved by any smooth function u that depends on y alone—so in order to get a unique solution one generally stipulates that the solution should satisfy some extra conditions. For first-order equations, the appropriate condition is to specify "initial values" on a hypersurface: given a smooth hypersurface $S \subseteq \mathbb{R}^n$ and a smooth function $\varphi: S \to \mathbb{R}$, we seek a smooth function u that solves the PDE and also satisfies the initial condition

$$u|_S = \varphi. \tag{9.21}$$

The problem of finding a solution to (9.20) in a neighborhood of S subject to the initial condition (9.21) is called a *Cauchy problem*.

Not every Cauchy problem has a solution: for example, in $\mathbb{R}^2$, the equation $\partial u/\partial x = 1$ has no solution with $u = 0$ on the x-axis, because the equation and the initial condition contradict each other. To avoid such difficulties, one usually assumes that the Cauchy problem (9.20)–(9.21) is *noncharacteristic*, meaning that there is a certain geometric relationship between the equation and the initial data, which is sufficient to guarantee the existence of a solution near S. As we study the Cauchy problem in increasing generality, we will describe the noncharacteristic condition separately for each type of equation we treat. The most general form of the condition is given at the end of Chapter 22.

Linear Equations

The first type of equation we will treat is a first-order **linear PDE**, which is one that depends linearly or affinely on the unknown function and its derivatives. In coordinate form, the most general such equation can be written

$$a^1(x)\frac{\partial u}{\partial x^1}(x) + \cdots + a^n(x)\frac{\partial u}{\partial x^n}(x) + b(x)u(x) = f(x), \qquad (9.22)$$

where $a^1, \ldots, a^n$, b, and f are smooth, real-valued functions defined on some open subset $\Omega \subseteq \mathbb{R}^n$, and u is an unknown smooth function on Ω.

It should come as no surprise that flows of vector fields play a role in the solution of (9.22), because the first n terms on the left-hand side represent the action on u of a smooth vector field $A \in \mathfrak{X}(\Omega)$:

$$A_x = a^1(x) \left.\frac{\partial}{\partial x^1}\right|_x + \cdots + a^n(x) \left.\frac{\partial}{\partial x^n}\right|_x. \qquad (9.23)$$

In terms of A, we can rewrite (9.22) in the simple form $Au + bu = f$. In this form, it makes sense on any smooth manifold, and is no more difficult to solve in that generality, so we state our first theorem in that context. The Cauchy problem for $Au + bu = f$ with initial hypersurface S is said to be **noncharacteristic** if A is nowhere tangent to S.

Theorem 9.51 (The Linear First-Order Cauchy Problem). *Let M be a smooth manifold. Suppose we are given an embedded hypersurface $S \subseteq M$, a smooth vector field $A \in \mathfrak{X}(M)$ that is nowhere tangent to S, and functions $b, f \in C^\infty(M)$ and $\varphi \in C^\infty(S)$. Then for some neighborhood U of S in M, there exists a unique solution $u \in C^\infty(U)$ to the noncharacteristic Cauchy problem*

$$Au + bu = f, \qquad (9.24)$$

$$u|_S = \varphi. \qquad (9.25)$$

Proof. The flowout theorem gives us a neighborhood $\mathcal{O}_\delta$ of $\{0\} \times S$ in $\mathbb{R} \times S$, a neighborhood U of S in M, and a diffeomorphism $\Phi \colon \mathcal{O}_\delta \to U$ that satisfies $\Phi(0, p) = p$ for $p \in S$ and $\Phi_*(\partial/\partial t) = A$. Let us write $\hat{u} = u \circ \Phi$, $\hat{f} = f \circ \Phi$, and $\hat{b} = b \circ \Phi$. Proposition 8.16 shows that $\partial \hat{u}/\partial t = (Au) \circ \Phi$. Thus, $u \in C^\infty(U)$ satisfies (9.24)–(9.25) if and only if $\hat{u}$ satisfies

$$\frac{\partial \hat{u}}{\partial t}(t,p) = \hat{f}(t,p) - \hat{b}(t,p)\hat{u}(t,p), \quad (t,p) \in \mathcal{O}_\delta,$$

$$\hat{u}(0,p) = \varphi(p), \quad p \in S. \tag{9.26}$$

For each fixed $p \in S$, this is a linear first-order ODE initial value problem for $\hat{u}$ on the interval $-\delta(p) < t < \delta(p)$. As is shown in ODE texts, such a problem always has a unique solution on the whole interval, which can be written explicitly as

$$\hat{u}(t,p) = e^{-B(t,p)}\left(\varphi(p) + \int_0^t \hat{f}(\tau,p)e^{B(\tau,p)}\,d\tau\right), \quad \text{where}$$

$$B(\tau,p) = \int_0^\tau \hat{b}(\sigma,p)\,d\sigma.$$

This is a smooth function of (t,p) (as can be seen by choosing local coordinates for S and differentiating under the integral signs). Therefore, $u = \hat{u} \circ \Phi^{-1}$ is the unique solution on U to (9.24)–(9.25). □

This proof shows how to write down an explicit solution to the Cauchy problem, provided the flow of the vector field A can be found explicitly. The computations are usually easiest if we first choose a (local or global) parametrization $X \colon \Omega \to S$, and substitute $X(s)$ for p in (9.26). This amounts to using the canonical coordinates of Theorem 9.22 to transform the Cauchy problem to an ODE.

Example 9.52 (A Linear Cauchy Problem). Suppose we wish to solve the following Cauchy problem for a smooth function $u(x,y)$ in the plane:

$$x\frac{\partial u}{\partial y} - y\frac{\partial u}{\partial x} = x, \tag{9.27}$$

$$u(x,0) = x \quad \text{when } x > 0. \tag{9.28}$$

The vector field acting on u on the left-hand side of (9.27) is the vector field W of Example 9.23. The initial hypersurface S is the positive x-axis, and this problem is noncharacteristic because W is nowhere tangent to S. (Notice that this would not be the case if we took S to be the entire x-axis.) Using the computations of Example 9.23, we find that the transformation $(x,y) = \Psi(t,s) = (s\cos t, s\sin t)$ pushes $\partial/\partial t$ forward to W, and thus transforms (9.27)–(9.28) to the ODE initial value problem

$$\frac{\partial \hat{u}}{\partial t}(t,s) = s\cos t,$$

$$\hat{u}(0,s) = s.$$

This is solved by $\hat{u}(t,s) = s\sin t + s$. Substituting for (t,s) in terms of (x,y) using (9.13), we obtain the solution $u(x,y) = y + \sqrt{x^2 + y^2}$ to the original problem. //

Quasilinear Equations

The preceding results extend easily to certain nonlinear partial differential equations. A PDE is called **quasilinear** if it can be written as an affine equation in the highest-order derivatives of the unknown function, with coefficients that may depend on the function itself and its derivatives of lower order. Thus, in coordinates, a **quasilinear first-order PDE** is a differential equation of the form

$$a^1\left(x,u(x)\right)\frac{\partial u}{\partial x^1} + \cdots + a^n\left(x,u(x)\right)\frac{\partial u}{\partial x^n} = f\left(x,u(x)\right) \tag{9.29}$$

for an unknown real-valued function $u(x^1,\ldots,x^n)$, where $a^1,\ldots,a^n$ and f are smooth real-valued functions defined on some open subset $W \subseteq \mathbb{R}^{n+1}$. (For simplicity, this time we concentrate only on the local problem, and restrict our attention to open subsets of Euclidean space.)

We wish to solve a Cauchy problem for this equation with initial condition

$$u|_S = \varphi, \tag{9.30}$$

where $S \subseteq \mathbb{R}^n$ is a smooth, embedded hypersurface, and $\varphi\colon S \to \mathbb{R}$ is a smooth function whose graph is contained in W. A quasilinear Cauchy problem is said to be **noncharacteristic** if the vector field A^φ along S defined by

$$A^\varphi\big|_x = a^1\left(x,\varphi(x)\right)\frac{\partial}{\partial x^1}\bigg|_x + \cdots + a^n\left(x,\varphi(x)\right)\frac{\partial}{\partial x^n}\bigg|_x \tag{9.31}$$

is nowhere tangent to S. (Notice that in this case the noncharacteristic condition depends on the initial value φ, not just on the initial hypersurface.) We will show that a noncharacteristic Cauchy problem always has local solutions. (As we will see below, finding global solutions can be problematic because of the lack of uniqueness.)

Theorem 9.53 (The Quasilinear Cauchy Problem). *If the Cauchy problem* (9.29)–(9.30) *is noncharacteristic, then for each $p \in S$ there exists a neighborhood U of p in M on which there exists a unique solution u to* (9.29)–(9.30).

Proof. The key is to convert the dependent variable u to an additional independent variable. (This is a trick that is useful in many different contexts.) Define the **characteristic vector field** for (9.29) to be the vector field ξ on $W \subseteq \mathbb{R}^{n+1}$ given by

$$\xi_{(x,z)} = a^1(x,z)\frac{\partial}{\partial x^1}\bigg|_{(x,z)} + \cdots + a^n(x,z)\frac{\partial}{\partial x^n}\bigg|_{(x,z)} + f(x,z)\frac{\partial}{\partial z}\bigg|_{(x,z)}, \tag{9.32}$$

where we write $(x,z) = (x^1,\ldots,x^n,z)$. Suppose u is a smooth function defined on an open subset $V \subseteq \mathbb{R}^n$ whose graph $\Gamma(u) = \{(x,u(x)) : x \in V\}$ is contained in W. Then (9.29) holds if and only if $\xi(z - u(x)) = 0$ at all points of $\Gamma(u)$. Since $z - u(x)$ is a defining function for $\Gamma(u)$, it follows from Corollary 5.39 that u solves (9.29) if and only if ξ is tangent to $\Gamma(u)$. The idea is to construct the graph of u as the flowout by ξ from a suitable initial submanifold.

Let $\Gamma(\varphi) = \{(x, \varphi(x)) : x \in S\}$ denote the graph of φ; it is an $(n - 1)$-dimensional embedded submanifold of W. The projection $\pi \colon W \to \mathbb{R}^n$ onto the first n variables maps $\Gamma(\varphi)$ diffeomorphically onto S, so if ξ were tangent to $\Gamma(\varphi)$ at some point $(x, \varphi(x))$, then $d\pi(\xi_{(x,\varphi(x))})$ would be tangent to S at x. However, a direct computation using (9.32) and (9.31) shows that

$$d\pi\left(\xi_{(x,\varphi(x))}\right) = A^{\varphi}\big|_x,$$

so the noncharacteristic assumption guarantees that ξ is nowhere tangent to $\Gamma(\varphi)$.

We can apply the flowout theorem to the vector field ξ starting on $\Gamma(\varphi) \subseteq W$ to obtain an immersed n-dimensional submanifold $S \subseteq W$ containing $\Gamma(\varphi)$, such that ξ is everywhere tangent to S. If we can show that S is the graph of a smooth function u, at least locally near $\Gamma(\varphi)$, then u will be a solution to our problem.

Let $p \in S$ be arbitrary. At $(p, \varphi(p)) \in \Gamma(\varphi) \subseteq S$, the tangent space to S is spanned by the vector $\xi_{(p,\varphi(p))}$ together with $T_{(p,\varphi(p))}\Gamma(\varphi)$. The restriction of π to $\Gamma(\varphi)$ is a diffeomorphism onto S, so $d\pi$ maps $T_{(p,\varphi(p))}\Gamma(\varphi)$ isomorphically onto $T_p S$. On the other hand, as we noted above, $d\pi$ takes $\xi_{(p,\varphi(p))}$ to $A^{\varphi}|_p$. By the noncharacteristic assumption, $A^{\varphi}|_p \notin T_p S$, so $d\pi$ is injective on $T_{(p,\varphi(p))}S$, and thus for dimensional reasons $T_{(p,\varphi(p))}\mathbb{R}^{n+1} = T_{(p,\varphi(p))}S \oplus \operatorname{Ker} d\pi_{(p,\varphi(p))}$. Because $\operatorname{Ker} d\pi_{(p,\varphi(p))}$ is the tangent space to $\{p\} \times \mathbb{R}$, it follows that S intersects $\{p\} \times \mathbb{R}$ transversely at $(p, \varphi(p))$. By Corollary 6.33, there exist a neighborhood V of $(p, \varphi(p))$ in S and a neighborhood U of p in $\mathbb{R}^n$ such that V is the graph of a smooth function $u \colon U \to \mathbb{R}$. This function solves the Cauchy problem in U.

To prove uniqueness, we might need to shrink U. Because S is a flowout, it is the image of some open subset $\mathcal{O}_{\delta} \subseteq \mathbb{R} \times \Gamma(\varphi)$ under the flow of ξ. Choose V small enough that it is the image under the flow of a set of the form $(-\varepsilon, \varepsilon) \times Y \subseteq \mathcal{O}_{\delta}$, for some $\varepsilon > 0$ and some neighborhood Y of $(p, \varphi(p))$ in $\Gamma(\varphi)$. With this assumption, $\Gamma(u)$ is exactly the union of the images of the integral curves of ξ starting at points of Y and flowing for time $|t| < \varepsilon$. Suppose $\tilde{u}$ is any other solution to the same Cauchy problem on the same open subset U. As we noted above, this means that ξ is tangent to the graph of $\tilde{u}$, and the initial condition ensures that $Y = \Gamma(\varphi) \cap U \subseteq \Gamma(\tilde{u})$. Since the graph of $\tilde{u}$ is a properly embedded submanifold of $U \times \mathbb{R}$, Problem 9-2 shows that each integral curve of ξ in $U \times \mathbb{R}$ starting at a point of Y must lie entirely in the graph of $\tilde{u}$. Thus $\Gamma(\tilde{u}) \supseteq \Gamma(u)$. But then $\Gamma(\tilde{u})$ cannot contain any points that are not in $\Gamma(u)$ and still be the graph of a function, so $\tilde{u} = u$ on U. $\qquad\square$

To find an explicit solution to a quasilinear Cauchy problem, we begin by choosing a smooth local parametrization of S, written as $s \mapsto X(s)$ for $s = (s^2, \dots, s^n) \in \Omega \subseteq \mathbb{R}^{n-1}$. Then the map $\tilde{X} \colon \Omega \to \mathbb{R}^{n+1}$ given by $\tilde{X}(s) = (X(s), \varphi(X(s)))$ is a local parametrization of $\Gamma(\varphi)$, and a local parametrization of S is given by

$$\Psi(t, s) = \theta_t(\tilde{X}(s)),$$

where θ is the flow of ξ. To rewrite S as a graph, just invert the map $\pi \circ \Psi \colon \Omega \to \mathbb{R}^n$ locally by solving for $(x^1, \dots, x^n)$ in terms of $(t, s^2, \dots, s^n)$; then the z-component of Ψ, written as a function of $(x^1, \dots, x^n)$, is a solution to the Cauchy problem.

Example 9.54 (A Quasilinear Cauchy Problem). Suppose we wish to solve the following quasilinear Cauchy problem in the plane:

$$(u + 1)\frac{\partial u}{\partial x} + \frac{\partial u}{\partial y} = 0,$$

$$u(x, 0) = x.$$

The initial hypersurface S is the x-axis, and the initial value is $\varphi(x, 0) = x$. The vector field A^φ is $(x + 1)\partial/\partial x + \partial/\partial y$, which is nowhere tangent to the x-axis, so this problem is noncharacteristic.

The characteristic vector field is the following vector field on $\mathbb{R}^3$:

$$\xi = (z + 1)\frac{\partial}{\partial x} + \frac{\partial}{\partial y}.$$

We wish to find the flowout of ξ starting from $\Gamma(\varphi)$. We can parametrize S by $X(s) = (s, 0)$ for $s \in \mathbb{R}$, and then $\Gamma(\varphi)$ is parametrized by $\widetilde{X}(s) = (s, 0, s)$. Solving the system of ODEs associated to ξ with initial conditions $(x, y, z) = (s, 0, s)$, we find that the flowout of ξ starting from $\widetilde{X}(s)$ is parametrized by

$$\Psi(t, s) = (s + t + st, t, s).$$

The image of this map is the graph of our solution. To reparametrize the graph in terms of x and y, we invert the map $\pi \circ \Psi$; that is, we solve $(x, y) = (s + t + st, t)$ (locally) for t and s, yielding

$$(t, s) = \left(y, \frac{x - y}{1 + y}\right),$$

and therefore on the flowout manifold we have $z = s = (x - y)/(1 + y)$. The solution to our Cauchy problem is $u(x, y) = (x - y)/(1 + y)$. Note that it is defined only in a neighborhood of S (the set where $y > -1$), not on the whole plane.   //

The integral curves of ξ in $\mathbb{R}^{n+1}$ are called the **characteristic curves** (or **characteristics**) of the PDE (9.29). This solution technique, which boils down to constructing the graph of u as a union of characteristic curves, is called the **method of characteristics**. (For linear equations, the term *characteristic curves* is also sometimes applied to the integral curves of the vector field A defined by (9.23). Of course, the method described above for quasilinear equations can be applied to linear ones as well, but the technique of Example 9.52 is usually easier in the linear case.)

Theorems 9.51 and 9.53 only assert existence and uniqueness of a solution in a neighborhood of the initial submanifold. Cauchy problems do not always admit global solutions, and when global solutions do exist, they might not be unique (see Problem 9-23 for some examples). The basic problem is that the characteristic curves passing through the initial hypersurface might not reach all points of the manifold. Also, quasilinear problems have the added complication that the characteristic curves in $\mathbb{R}^{n+1}$ might cease to be transverse to the fibers of the projection

$\mathbb{R}^{n+1} \to \mathbb{R}^n$, or even if they are transverse, their projections into $\mathbb{R}^n$ starting at different points of $\Gamma(\varphi)$ might cross each other, even though the characteristic curves themselves do not. At any point in the image of two or more characteristic curves, the procedure above would produce two different values for u. Nevertheless, in specific cases, it is often possible to identify a neighborhood of S on which a unique solution exists by analyzing the behavior of the characteristics. For instance, consider the solution u that we produced on the set $U = \{(x, y) : y > -1\}$ in Example 9.54. Its graph contains the entire maximal integral curve starting at each point of $\Gamma(\varphi)$. Because of this, the argument we used to prove local uniqueness in Theorem 9.53 actually proves that the solution is globally unique in this case.

Problems

9-1. Suppose M is a smooth manifold, $X \in \mathfrak{X}(M)$, and γ is a maximal integral curve of X.
   (a) We say γ is **periodic** if there is a number $T > 0$ such that $\gamma(t + T) = \gamma(t)$ for all $t \in \mathbb{R}$. Show that exactly one of the following holds:

   - γ is constant.
   - γ is injective.
   - γ is periodic and nonconstant.

   (b) Show that if γ is periodic and nonconstant, then there exists a unique positive number T (called the **period of γ**) such that $\gamma(t) = \gamma(t')$ if and only if $t - t' = kT$ for some $k \in \mathbb{Z}$.
   (c) Show that the image of γ is an immersed submanifold of M, diffeomorphic to $\mathbb{R}$, $\mathbb{S}^1$, or $\mathbb{R}^0$.
   (*Used on pp. 398, 560.*)

9-2. Suppose M is a smooth manifold, $S \subseteq M$ is an immersed submanifold, and V is a smooth vector field on M that is tangent to S.
   (a) Show that for any integral curve γ of V such that $\gamma(t_0) \in S$, there exists $\varepsilon > 0$ such that $\gamma((t_0 - \varepsilon, t_0 + \varepsilon)) \subseteq S$.
   (b) Now assume S is properly embedded. Show that every integral curve that intersects S is contained in S.
   (c) Give a counterexample to (b) if S is not closed.
   (*Used on pp. 243, 491.*)

9-3. Compute the flow of each of the following vector fields on $\mathbb{R}^2$:
   (a) $V = y \dfrac{\partial}{\partial x} + \dfrac{\partial}{\partial y}$.
   (b) $W = x \dfrac{\partial}{\partial x} + 2y \dfrac{\partial}{\partial y}$.
   (c) $X = x \dfrac{\partial}{\partial x} - y \dfrac{\partial}{\partial y}$.
   (d) $Y = x \dfrac{\partial}{\partial y} + y \dfrac{\partial}{\partial x}$.

9-4. For any integer $n \geq 1$, define a flow on the odd-dimensional sphere $\mathbb{S}^{2n-1} \subseteq \mathbb{C}^n$ by $\theta(t, z) = e^{it} z$. Show that the infinitesimal generator of θ is a smooth nonvanishing vector field on $\mathbb{S}^{2n-1}$. [Remark: in the case $n = 2$, the integral curves of X are the curves γ_z of Problem 3-6, so this provides a simpler proof that each γ_z is smooth.] (*Used on p. 435.*)

9-5. Suppose M is a smooth, compact manifold that admits a nowhere vanishing smooth vector field. Show that there exists a smooth map $F: M \to M$ that is homotopic to the identity and has no fixed points.

9-6. Prove Lemma 9.19 (the escape lemma).

9-7. Let M be a connected smooth manifold. Show that the group of diffeomorphisms of M acts transitively on M: that is, for any $p, q \in M$, there is a diffeomorphism $F: M \to M$ such that $F(p) = q$. [Hint: first prove that if $p, q \in \mathbb{B}^n$ (the open unit ball in $\mathbb{R}^n$), there is a compactly supported smooth vector field on $\mathbb{B}^n$ whose flow θ satisfies $\theta_1(p) = q$.]

9-8. Let M be a smooth manifold and let $S \subseteq M$ be a compact embedded submanifold. Suppose $V \in \mathfrak{X}(M)$ is a smooth vector field that is nowhere tangent to S. Show that there exists $\varepsilon > 0$ such that the flow of V restricts to a smooth embedding $\Phi: (-\varepsilon, \varepsilon) \times S \to M$.

9-9. Suppose M is a smooth manifold and $S \subseteq M$ is an embedded hypersurface (not necessarily compact). Suppose further that there is a smooth vector field V defined on a neighborhood of S and nowhere tangent to S. Show that S has a neighborhood in M diffeomorphic to $(-1, 1) \times S$, under a diffeomorphism that restricts to the obvious identification $\{0\} \times S \approx S$. [Hint: using the notation of the flowout theorem, show that $\mathcal{O}_\delta \approx \mathcal{O}_1$.]

9-10. For each vector field in Problem 9-3, find smooth coordinates in a neighborhood of $(1, 0)$ for which the given vector field is a coordinate vector field.

9-11. Prove Theorem 9.24 (the boundary flowout theorem). [Hint: define Φ first in boundary coordinates and use uniqueness to glue together the local definitions. To obtain an embedding, make sure $\delta(p)$ is no more than half of the first time the integral curve starting at p hits the boundary (if it ever does).]

9-12. Suppose M_1 and M_2 are connected smooth n-manifolds and $M_1 \# M_2$ is their smooth connected sum (see Example 9.31). Show that the smooth structure on $M_1 \# M_2$ can be chosen in such a way that there are open subsets $\widetilde{M}_1, \widetilde{M}_2 \subseteq M_1 \# M_2$ that are diffeomorphic to $M_1 \smallsetminus \{p_1\}$ and $M_2 \smallsetminus \{p_2\}$, respectively, such that $\widetilde{M}_1 \cup \widetilde{M}_2 = M_1 \# M_2$ and $\widetilde{M}_1 \cap \widetilde{M}_2$ is diffeomorphic to $(-1, 1) \times \mathbb{S}^{n-1}$. (*Used on p. 465.*)

9-13. Prove that the conclusions of Theorems 5.29 and 5.53(b) (restricting the codomain of a smooth map to a submanifold or submanifold with boundary) remain true if M is allowed to be a smooth manifold with boundary.

9-14. Use the double to prove Theorem 6.18 (the Whitney immersion theorem) in the case that M has nonempty boundary.

9-15. Prove Theorem 9.35 (canonical form near a regular point on the boundary). [Hint: consider M as a regular domain in its double, and start with coordinates in which x^n is a local defining function for ∂M in $D(M)$.]

9-16. Give an example of smooth vector fields V, $\tilde{V}$, and W on $\mathbb{R}^2$ such that $V = \tilde{V} = \partial/\partial x$ along the x-axis but $\mathscr{L}_V W \neq \mathscr{L}_{\tilde{V}} W$ at the origin. [Remark: this shows that it is really necessary to know the vector field V to compute $(\mathscr{L}_V W)_p$; it is not sufficient just to know the vector V_p, or even to know the values of V along an integral curve of V.]

9-17. For each k-tuple of vector fields on $\mathbb{R}^3$ shown below, either find smooth coordinates (s^1, s^2, s^3) in a neighborhood of $(1, 0, 0)$ such that $V_i = \partial/\partial s^i$ for $i = 1, \ldots, k$, or explain why there are none.

(a) $k = 2$; $V_1 = \dfrac{\partial}{\partial x}$, $V_2 = \dfrac{\partial}{\partial x} + \dfrac{\partial}{\partial y}$.

(b) $k = 2$; $V_1 = (x + 1)\dfrac{\partial}{\partial x} - (y + 1)\dfrac{\partial}{\partial y}$, $V_2 = (x + 1)\dfrac{\partial}{\partial x} + (y + 1)\dfrac{\partial}{\partial y}$.

(c) $k = 3$; $V_1 = x\dfrac{\partial}{\partial y} - y\dfrac{\partial}{\partial x}$, $V_2 = y\dfrac{\partial}{\partial z} - z\dfrac{\partial}{\partial y}$, $V_3 = z\dfrac{\partial}{\partial x} - x\dfrac{\partial}{\partial z}$.

9-18. Define vector fields X and Y on the plane by

$$X = x\frac{\partial}{\partial x} - y\frac{\partial}{\partial y}, \qquad Y = x\frac{\partial}{\partial y} + y\frac{\partial}{\partial x}.$$

Compute the flows θ, ψ of X and Y, and verify that the flows do not commute by finding explicit open intervals J and K containing 0 such that $\theta_s \circ \psi_t$ and $\psi_t \circ \theta_s$ are both defined for all $(s, t) \in J \times K$, but they are unequal for some such (s, t).

9-19. Let M be $\mathbb{R}^3$ with the z-axis removed. Define $V, W \in \mathfrak{X}(M)$ by

$$V = \frac{\partial}{\partial x} - \frac{y}{x^2 + y^2}\frac{\partial}{\partial z}, \qquad W = \frac{\partial}{\partial y} + \frac{x}{x^2 + y^2}\frac{\partial}{\partial z},$$

and let θ and ψ be the flows of V and W, respectively. Prove that V and W commute, but there exist $p \in M$ and $s, t \in \mathbb{R}$ such that $\theta_t \circ \psi_s(p)$ and $\psi_s \circ \theta_t(p)$ are both defined but are not equal.

9-20. Suppose M is a compact smooth manifold and $V: J \times M \to TM$ is a smooth time-dependent vector field on M. Show that the domain of the time-dependent flow of V is all of $J \times J \times M$.

9-21. Let M be a smooth manifold. A **smooth isotopy of M** is a smooth map $H: M \times J \to M$, where $J \subseteq \mathbb{R}$ is an interval, such that for each $t \in J$, the map $H_t: M \to M$ defined by $H_t(p) = H(p, t)$ is a diffeomorphism. (In particular, if J is the unit interval, then H is a homotopy from H_0 to H_1 through diffeomorphisms.) This problem shows that smooth isotopies are closely related to time-dependent flows.

(a) Suppose $J \subseteq \mathbb{R}$ is an open interval and $H : M \times J \to M$ is a smooth isotopy. Show that the map $V : J \times M \to TM$ defined by

$$V(t, p) = \frac{\partial}{\partial t} H(p, t)$$

is a smooth time-dependent vector field on M, whose time-dependent flow is given by $\psi(t, t_0, p) = H_t \circ H_{t_0}^{-1}(p)$ with domain $J \times J \times M$.

(b) Conversely, suppose J is an open interval and $V : J \times M \to M$ is a smooth time-dependent vector field on M whose time-dependent flow is defined on $J \times J \times M$. For any $t_0 \in J$, show that the map $H : M \times J \to M$ defined by $H(t, p) = \psi(t, t_0, p)$ is a smooth isotopy of M.

9-22. Here are three Cauchy problems in $\mathbb{R}^2$. For each one, find an explicit solution $u(x, y)$ in a neighborhood of the initial submanifold:

(a) $y \dfrac{\partial u}{\partial x} + \dfrac{\partial u}{\partial y} = x,$ $u(x, 0) = \sin x.$

(b) $\dfrac{\partial u}{\partial x} + y \dfrac{\partial u}{\partial y} = 0,$ $u(x, 1) = e^{-x}.$

(c) $\dfrac{\partial u}{\partial x} + u \dfrac{\partial u}{\partial y} = y,$ $u(0, y) = 0.$

9-23. Consider again the Cauchy problems in Problem 9-22. Show that (a) has a unique global solution; (b) has a global solution, but it is not unique; and (c) has no global solutions. [Hint: consider which characteristic curves intersect the initial submanifold and which do not.]

9-24. Prove the converse to Euler's homogeneous function theorem (Problem 8-2): if $f \in C^\infty(\mathbb{R}^n \smallsetminus \{0\})$ satisfies $Vf = cf$, where V is the Euler vector field and $c \in \mathbb{R}$, then f is positively homogeneous of degree c.

Chapter 10
Vector Bundles

In Chapter 3, we saw that the tangent bundle of a smooth manifold has a natural structure as a smooth manifold in its own right. The natural coordinates we constructed on TM make it look, locally, like the Cartesian product of an open subset of M with $\mathbb{R}^n$. This kind of structure arises quite frequently—a collection of vector spaces, one for each point in M, glued together in a way that looks *locally* like the Cartesian product of M with $\mathbb{R}^n$, but globally may be "twisted." Such structures are called *vector bundles*, and are the main subject of this chapter.

The chapter begins with the definition of vector bundles and descriptions of a few examples. The most notable example, of course, is the tangent bundle of a smooth manifold. We then go on to discuss local and global sections of vector bundles (which correspond to vector fields in the case of the tangent bundle). The chapter continues with a discussion of the natural notions of maps between bundles, called *bundle homomorphisms*, and subsets of vector bundles that are themselves vector bundles, called *subbundles*. At the end of the chapter, we briefly introduce an important generalization of vector bundles, called *fiber bundles*.

There is a deep and extensive body of theory about vector bundles and fiber bundles on manifolds, which we cannot even touch. We introduce them primarily in order to have a convenient language for talking about the tangent bundle and structures like it; as you will see in the next few chapters, such structures exist in profusion on smooth manifolds.

Vector Bundles

Let M be a topological space. A *(real) vector bundle of rank k over M* is a topological space E together with a surjective continuous map $\pi: E \to M$ satisfying the following conditions:

(i) For each $p \in M$, the fiber $E_p = \pi^{-1}(p)$ over p is endowed with the structure of a k-dimensional real vector space.

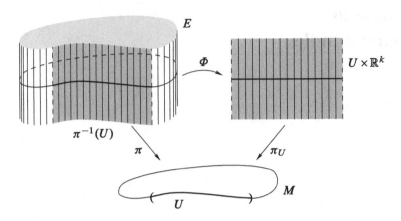

Fig. 10.1 A local trivialization of a vector bundle

(ii) For each $p \in M$, there exist a neighborhood U of p in M and a homeomor-
phism $\Phi \colon \pi^{-1}(U) \to U \times \mathbb{R}^k$ (called a *local trivialization of E over U*), satisfying the following conditions (Fig. 10.1):

- $\pi_U \circ \Phi = \pi$ (where $\pi_U \colon U \times \mathbb{R}^k \to U$ is the projection);
- for each $q \in U$, the restriction of Φ to E_q is a vector space isomorphism from
  E_q to $\{q\} \times \mathbb{R}^k \cong \mathbb{R}^k$.

If M and E are smooth manifolds with or without boundary, π is a smooth map,
and the local trivializations can be chosen to be diffeomorphisms, then E is called
a *smooth vector bundle*. In this case, we call any local trivialization that is a diffeomorphism onto its image a *smooth local trivialization*.

A rank-1 vector bundle is often called a *(real) line bundle*. *Complex vector bundles* are defined similarly, with "real vector space" replaced by "complex vector
space" and $\mathbb{R}^k$ replaced by $\mathbb{C}^k$ in the definition. We have no need to treat complex
vector bundles in this book, so all of our vector bundles are understood without
further comment to be real.

The space E is called the *total space of the bundle*, M is called its *base*, and
π is its *projection*. Depending on what we wish to emphasize, we sometimes omit
some of the ingredients from the notation, and write "E is a vector bundle over M,"
or "$E \to M$ is a vector bundle," or "$\pi \colon E \to M$ is a vector bundle."

▶ **Exercise 10.1.** Suppose E is a smooth vector bundle over M. Show that the projection map $\pi \colon E \to M$ is a surjective smooth submersion.

If there exists a local trivialization of E over all of M (called a *global trivialization of E*), then E is said to be a *trivial bundle*. In this case, E itself is homeomorphic to the product space $M \times \mathbb{R}^k$. If $E \to M$ is a smooth bundle that admits a
smooth global trivialization, then we say that E is *smoothly trivial*. In this case E
is *diffeomorphic* to $M \times \mathbb{R}^k$, not just homeomorphic. For brevity, when we say that
a smooth bundle is trivial, we always understand this to mean smoothly trivial, not
just trivial in the topological sense.

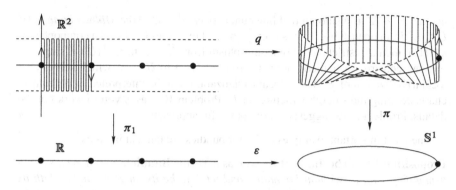

Fig. 10.2 Part of the Möbius bundle

Example 10.2 (Product Bundles). One particularly simple example of a rank-k vector bundle over any space M is the product space $E = M \times \mathbb{R}^k$ with $\pi = \pi_1 \colon M \times \mathbb{R}^k \to M$ as its projection. Any such bundle, called a ***product bundle***, is trivial (with the identity map as a global trivialization). If M is a smooth manifold with or without boundary, then $M \times \mathbb{R}^k$ is smoothly trivial. //

Although there are many vector bundles that are not trivial, the only one that is easy to visualize is the following.

Example 10.3 (The Möbius Bundle). Define an equivalence relation on $\mathbb{R}^2$ by declaring that $(x, y) \sim (x', y')$ if and only if $(x', y') = \big(x + n, (-1)^n y\big)$ for some $n \in \mathbb{Z}$. Let $E = \mathbb{R}^2/\!\!\sim$ denote the quotient space, and let $q \colon \mathbb{R}^2 \to E$ be the quotient map.

To visualize E, let S denote the strip $[0, 1] \times \mathbb{R} \subseteq \mathbb{R}^2$. The restriction of q to S is surjective and closed, so it is a quotient map. The only nontrivial identifications made by $q|_S$ are on the two boundary lines, so we can think of E as the space obtained from S by giving the right-hand edge a half-twist to turn it upside-down, and then pasting it to the left-hand edge (Fig. 10.2). For any $r > 0$, the image under the quotient map q of the rectangle $[0, 1] \times [-r, r]$ is a smooth compact manifold with boundary called a ***Möbius band***; you can make a paper model of this space by pasting the ends of a strip of paper together with a half-twist.

Consider the following commutative diagram:

$$
\begin{array}{ccc}
\mathbb{R}^2 & \xrightarrow{\ q\ } & E \\
{\scriptstyle \pi_1}\Big\downarrow & & \Big\downarrow{\scriptstyle \pi} \\
\mathbb{R} & \xrightarrow[\ \varepsilon\ ]{} & \mathbb{S}^1,
\end{array}
$$

where π_1 is the projection onto the first factor and $\varepsilon \colon \mathbb{R} \to \mathbb{S}^1$ is the smooth covering map $\varepsilon(x) = e^{2\pi i x}$. Because $\varepsilon \circ \pi_1$ is constant on each equivalence class, it descends to a continuous map $\pi \colon E \to \mathbb{S}^1$. A straightforward (if tedious) verification shows that E has a unique smooth manifold structure such that q is a smooth covering map and $\pi: E \to \mathbb{S}^1$ is a smooth real line bundle over $\mathbb{S}^1$, called the ***Möbius bundle***. (If $U \subseteq \mathbb{S}^1$ is an open subset that is evenly covered by ε, and $\tilde{U} \subseteq \mathbb{R}$ is a component of $\varepsilon^{-1}(U)$, then q restricts to a homeomorphism from $\tilde{U} \times \mathbb{R}$ to $\pi^{-1}(U)$. Using this, one can construct a homeomorphism from $\pi^{-1}(U)$ to $U \times \mathbb{R}$, which serves as a local trivialization of E. These local trivializations can be interpreted as coordinate charts defining the smooth structure on E. Problem 10-1 asks you to work out the details. Problem 21-9 suggests a more powerful approach.) $\qquad\qquad$ //

The most important examples of vector bundles are tangent bundles.

Proposition 10.4 (The Tangent Bundle as a Vector Bundle). *Let M be a smooth n-manifold with or without boundary, and let TM be its tangent bundle. With its standard projection map, its natural vector space structure on each fiber, and the topology and smooth structure constructed in Proposition 3.18, TM is a smooth vector bundle of rank n over M.*

Proof. Given any smooth chart (U, φ) for M with coordinate functions (x^i), define a map $\Phi: \pi^{-1}(U) \to U \times \mathbb{R}^n$ by

$$\Phi\left(v^i \frac{\partial}{\partial x^i}\bigg|_p\right) = \left(p, \left(v^1, \ldots, v^n\right)\right). \qquad (10.1)$$

This is linear on fibers and satisfies $\pi_1 \circ \Phi = \pi$. The composite map

$$\pi^{-1}(U) \xrightarrow{\Phi} U \times \mathbb{R}^n \xrightarrow{\varphi \times \mathrm{Id}_{\mathbb{R}^n}} \varphi(U) \times \mathbb{R}^n$$

is equal to the coordinate map $\tilde{\varphi}$ constructed in Proposition 3.18. Since both $\tilde{\varphi}$ and $\varphi \times \mathrm{Id}_{\mathbb{R}^n}$ are diffeomorphisms, so is Φ. Thus, Φ satisfies all the conditions for a smooth local trivialization. $\qquad\qquad \square$

Any bundle that is not trivial, of course, requires more than one local trivialization. The next lemma shows that the composition of two smooth local trivializations has a simple form where they overlap.

Lemma 10.5. *Let $\pi: E \to M$ be a smooth vector bundle of rank k over M. Suppose $\Phi: \pi^{-1}(U) \to U \times \mathbb{R}^k$ and $\Psi: \pi^{-1}(V) \to V \times \mathbb{R}^k$ are two smooth local trivializations of E with $U \cap V \neq \varnothing$. There exists a smooth map $\tau: U \cap V \to \mathrm{GL}(k, \mathbb{R})$ such that the composition $\Phi \circ \Psi^{-1}: (U \cap V) \times \mathbb{R}^k \to (U \cap V) \times \mathbb{R}^k$ has the form*

$$\Phi \circ \Psi^{-1}(p, v) = (p, \tau(p)v),$$

where $\tau(p)v$ denotes the usual action of the $k \times k$ matrix $\tau(p)$ on the vector $v \in \mathbb{R}^k$.

Proof. The following diagram commutes:

$$(U \cap V) \times \mathbb{R}^k \xleftarrow{\Psi} \pi^{-1}(U \cap V) \xrightarrow{\Phi} (U \cap V) \times \mathbb{R}^k$$

$$\pi_1 \qquad\qquad \pi \Big\downarrow \qquad\qquad \pi_1 \qquad\qquad (10.2)$$

$$U \cap V,$$

where the maps on top are to be interpreted as the restrictions of Ψ and Φ to $\pi^{-1}(U \cap V)$. It follows that $\pi_1 \circ (\Phi \circ \Psi^{-1}) = \pi_1$, which means that

$$\Phi \circ \Psi^{-1}(p, v) = (p, \sigma(p, v))$$

for some smooth map $\sigma \colon (U \cap V) \times \mathbb{R}^k \to \mathbb{R}^k$. Moreover, for each fixed $p \in U \cap V$, the map $v \mapsto \sigma(p, v)$ from $\mathbb{R}^k$ to itself is an invertible linear map, so there is a nonsingular $k \times k$ matrix $\tau(p)$ such that $\sigma(p, v) = \tau(p)v$. It remains only to show that the map $\tau \colon U \cap V \to \mathrm{GL}(k, \mathbb{R})$ is smooth. This is left to Problem 10-4. $\qquad\square$

The smooth map $\tau \colon U \cap V \to \mathrm{GL}(k, \mathbb{R})$ described in this lemma is called the *transition function* between the local trivializations Φ and Ψ. (This is one of the few situations in smooth manifold theory in which it is traditional to use the word "function" even though the codomain is not $\mathbb{R}$ or $\mathbb{R}^k$.) For example, if M is a smooth manifold and Φ and Ψ are the local trivializations of TM associated with two different smooth charts, then (3.12) shows that the transition function between them is the Jacobian matrix of the coordinate transition map.

Like the tangent bundle, vector bundles are often most easily described by giving a collection of vector spaces, one for each point of the base manifold. In order to make such a set into a smooth vector bundle, we would first have to construct a manifold topology and a smooth structure on the disjoint union of all the vector spaces, and then construct the local trivializations and show that they have the requisite properties. The next lemma provides a shortcut, by showing that it is sufficient to construct the local trivializations, as long as they overlap with smooth transition functions. (See also Problem 10-6 for a stronger form of this result.)

Lemma 10.6 (Vector Bundle Chart Lemma). *Let M be a smooth manifold with or without boundary, and suppose that for each $p \in M$ we are given a real vector space E_p of some fixed dimension k. Let $E = \coprod_{p \in M} E_p$, and let $\pi \colon E \to M$ be the map that takes each element of E_p to the point p. Suppose furthermore that we are given the following data:*

(i) *an open cover $\{U_\alpha\}_{\alpha \in A}$ of M*
(ii) *for each $\alpha \in A$, a bijective map $\Phi_\alpha \colon \pi^{-1}(U_\alpha) \to U_\alpha \times \mathbb{R}^k$ whose restriction to each E_p is a vector space isomorphism from E_p to $\{p\} \times \mathbb{R}^k \cong \mathbb{R}^k$*
(iii) *for each $\alpha, \beta \in A$ with $U_\alpha \cap U_\beta \neq \varnothing$, a smooth map $\tau_{\alpha\beta} \colon U_\alpha \cap U_\beta \to \mathrm{GL}(k, \mathbb{R})$ such that the map $\Phi_\alpha \circ \Phi_\beta^{-1}$ from $(U_\alpha \cap U_\beta) \times \mathbb{R}^k$ to itself has the form*

$$\Phi_\alpha \circ \Phi_\beta^{-1}(p, v) = (p, \tau_{\alpha\beta}(p)v) \tag{10.3}$$

Then E has a unique topology and smooth structure making it into a smooth manifold with or without boundary and a smooth rank-k vector bundle over M, with π as projection and $\{(U_\alpha, \Phi_\alpha)\}$ as smooth local trivializations.

Proof. For each point $p \in M$, choose some U_α containing p; choose a smooth chart (V_p, φ_p) for M such that $p \in V_p \subseteq U_\alpha$; and let $\widehat{V}_p = \varphi_p(V_p) \subseteq \mathbb{R}^n$ or $\mathbb{H}^n$ (where n is the dimension of M). Define a map $\widetilde{\varphi}_p \colon \pi^{-1}(V_p) \to \widehat{V}_p \times \mathbb{R}^k$ by $\widetilde{\varphi}_p = (\varphi_p \times \mathrm{Id}_{\mathbb{R}^k}) \circ$

Φ_α:

$$\pi^{-1}(V_p) \xrightarrow{\Phi_\alpha} V_p \times \mathbb{R}^k \xrightarrow{\varphi_p \times \mathrm{Id}_{\mathbb{R}^k}} \widehat{V}_p \times \mathbb{R}^k.$$

We will show that the collection of all such charts $\left\{\left(\pi^{-1}(V_p), \widetilde{\varphi}_p\right) : p \in M\right\}$ satisfies the conditions of the smooth manifold chart lemma (Lemma 1.35) or its counterpart for manifolds with boundary (Exercise 1.43), and therefore gives E the structure of a smooth manifold with or without boundary.

As a composition of bijective maps, $\widetilde{\varphi}_p$ is bijective onto an open subset of either $\mathbb{R}^n \times \mathbb{R}^k = \mathbb{R}^{n+k}$ or $\mathbb{H}^n \times \mathbb{R}^k \approx \mathbb{H}^{n+k}$. For any $p, q \in M$, it is easy to check that

$$\widetilde{\varphi}_p\left(\pi^{-1}(V_p) \cap \pi^{-1}(V_q)\right) = \varphi_p(V_p \cap V_q) \times \mathbb{R}^k,$$

which is open because φ_p is a homeomorphism onto an open subset of $\mathbb{R}^n$ or $\mathbb{H}^n$. Wherever two such charts overlap, we have

$$\widetilde{\varphi}_p \circ \widetilde{\varphi}_q^{-1} = \left(\varphi_p \times \mathrm{Id}_{\mathbb{R}^k}\right) \circ \Phi_\alpha \circ \Phi_\beta^{-1} \circ \left(\varphi_q \times \mathrm{Id}_{\mathbb{R}^k}\right)^{-1}.$$

Since $\varphi_p \times \mathrm{Id}_{\mathbb{R}^k}$, $\varphi_q \times \mathrm{Id}_{\mathbb{R}^k}$, and $\Phi_\alpha \circ \Phi_\beta^{-1}$ are diffeomorphisms, the composition is a diffeomorphism. Thus, conditions (i)–(iii) of Lemma 1.35 are satisfied. Because the open cover $\{V_p : p \in M\}$ has a countable subcover, (iv) is satisfied as well.

To check the Hausdorff condition (v), just note that any two points in the same space E_p lie in one of the charts we have constructed; while if $\xi \in E_p$ and $\eta \in E_q$ with $p \neq q$, we can choose V_p and V_q to be disjoint neighborhoods of p and q, so that the sets $\pi^{-1}(V_p)$ and $\pi^{-1}(V_q)$ are disjoint coordinate neighborhoods containing ξ and η, respectively. Thus we have given E the structure of a smooth manifold with or without boundary.

With respect to this structure, each of the maps Φ_α is a diffeomorphism, because in terms of the coordinate charts $\left(\pi^{-1}(V_p), \widetilde{\varphi}_p\right)$ for E and $\left(V_p \times \mathbb{R}^k, \varphi_p \times \mathrm{Id}_{\mathbb{R}^k}\right)$ for $V_p \times \mathbb{R}^k$, the coordinate representation of Φ_α is the identity map. The coordinate representation of π, with respect to the same chart for E and the chart (V_p, φ_p) for M, is $\pi(x, v) = x$, so π is smooth as well. Because each Φ_α maps E_p to $\{p\} \times \mathbb{R}^k$, it is immediate that $\pi_1 \circ \Phi_\alpha = \pi$, and Φ_α is linear on fibers by hypothesis. Thus, Φ_α satisfies all the conditions for a smooth local trivialization.

The fact that this is the unique such smooth structure follows easily from the requirement that the maps Φ_α be diffeomorphisms onto their images: any smooth structure satisfying the same conditions must include all of the charts we constructed, so it is equal to this one. $\qquad\square$

Here are some examples showing how the chart lemma can be used to construct new vector bundles from old ones.

Example 10.7 (Whitney Sums). Given a smooth manifold M and smooth vector bundles $E' \to M$ and $E'' \to M$ of ranks k' and k'', respectively, we will construct a new vector bundle over M called the **Whitney sum of E' and E''**, whose fiber at each $p \in M$ is the direct sum $E'_p \oplus E''_p$. The total space is defined as $E' \oplus E'' = \coprod_{p \in M} \left(E'_p \oplus E''_p\right)$, with the obvious projection $\pi : E' \oplus E'' \to M$. For each $p \in M$, choose a neighborhood U of p small enough that there exist local trivializations (U, Φ') of E' and (U, Φ'') of E'', and define $\Phi: \pi^{-1}(U) \to U \times \mathbb{R}^{k'+k''}$ by

$$\Phi(v', v'') = \left(\pi'(v'), \left(\pi_{\mathbb{R}^{k'}} \circ \Phi'(v'), \pi_{\mathbb{R}^{k''}} \circ \Phi''(v'') \right) \right).$$

Suppose we are given another such pair of local trivializations $(\widetilde{U}, \widetilde{\Phi}')$ and $(\widetilde{U}, \widetilde{\Phi}'')$. Let $\tau': (U \cap \widetilde{U}) \to \mathrm{GL}(k', \mathbb{R})$ and $\tau'': (U \cap \widetilde{U}) \to \mathrm{GL}(k'', \mathbb{R})$ be the corresponding transition functions. Then the transition function for $E' \oplus E''$ has the form

$$\widetilde{\Phi} \circ \Phi^{-1}\left(p, (v', v'') \right) = \left(p, \tau(p)(v', v'') \right),$$

where $\tau(p) = \tau'(p) \oplus \tau''(p) \in \mathrm{GL}(k' + k'', \mathbb{R})$ is the block diagonal matrix

$$\begin{pmatrix} \tau'(p) & 0 \\ 0 & \tau''(p) \end{pmatrix}.$$

Because this depends smoothly on p, it follows from the chart lemma that $E' \oplus E''$ is a smooth vector bundle over M. //

Example 10.8 (Restriction of a Vector Bundle). Suppose $\pi: E \to M$ is a rank-k vector bundle and $S \subseteq M$ is any subset. We define the *restriction of E to S* to be the set $E|_S = \bigcup_{p \in S} E_p$, with the projection $E|_S \to S$ obtained by restricting π. If $\Phi: \pi^{-1}(U) \to U \times \mathbb{R}^k$ is a local trivialization of E over $U \subseteq M$, it restricts to a bijective map $\Phi|_U: (\pi|_S)^{-1}(U \cap S) \to (U \cap S) \times \mathbb{R}^k$, and it is easy to check that these form local trivializations for a vector bundle structure on $E|_S$. If E is a smooth vector bundle and $S \subseteq M$ is an immersed or embedded submanifold, it follows easily from the chart lemma that $E|_S$ is a smooth vector bundle. In particular, if $S \subseteq M$ is a smooth (embedded or immersed) submanifold, then the restricted bundle $TM|_S$ is called the *ambient tangent bundle* over M. //

Local and Global Sections of Vector Bundles

Let $\pi: E \to M$ be a vector bundle. A *section of E* (sometimes called a *cross section*) is a section of the map π, that is, a continuous map $\sigma: M \to E$ satisfying $\pi \circ \sigma = \mathrm{Id}_M$. This means that $\sigma(p)$ is an element of the fiber E_p for each $p \in M$.

More generally, a *local section of E* is a continuous map $\sigma: U \to E$ defined on some open subset $U \subseteq M$ and satisfying $\pi \circ \sigma = \mathrm{Id}_U$ (see Fig. 10.3). To emphasize the distinction, a section defined on all of M is sometimes called a *global section*. Note that a local section of E over $U \subseteq M$ is the same as a global section of the restricted bundle $E|_U$. If M is a smooth manifold with or without boundary and E is a smooth vector bundle, a *smooth (local or global) section of E* is one that is a smooth map from its domain to E.

Just as with vector fields, for some purposes it is useful also to consider maps that would be sections except that they might not be continuous. Thus, we define a *rough (local or global) section of E* over a set $U \subseteq M$ to be a map $\sigma: U \to E$

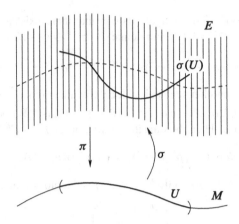

Fig. 10.3 A local section of a vector bundle

(not necessarily continuous) such that $\pi \circ \sigma = \mathrm{Id}_U$. A "section" without further qualification always means a continuous section.

The **zero section of E** is the global section $\zeta \colon M \to E$ defined by

$$\zeta(p) = 0 \in E_p \text{ for each } p \in M.$$

As in the case of vector fields, the **support** of a section σ is the closure of the set $\{p \in M : \sigma(p) \neq 0\}$.

▶ **Exercise 10.9.** Show that the zero section of every vector bundle is continuous, and the zero section of every smooth vector bundle is smooth. [Hint: consider $\Phi \circ \zeta$, where Φ is a local trivialization.]

Example 10.10 (Sections of Vector Bundles). Suppose M is a smooth manifold with or without boundary.

(a) Sections of TM are vector fields on M.
(b) Given an immersed submanifold $S \subseteq M$ with or without boundary, a section of the ambient tangent bundle $TM|_S \to S$ is called a **vector field along** S. It is a continuous map $X \colon S \to TM$ such that $X_p \in T_pM$ for each $p \in S$. This is different from a vector field *on* S, which satisfies $X_p \in T_pS$ at each point.
(c) If $E = M \times \mathbb{R}^k$ is a product bundle, there is a natural one-to-one correspondence between sections of E and continuous functions from M to $\mathbb{R}^k$: a continuous function $F \colon M \to \mathbb{R}^k$ determines a section $\widetilde{F} \colon M \to M \times \mathbb{R}^k$ by $\widetilde{F}(x) = (x, F(x))$, and vice versa. If M is a smooth manifold with or without boundary, then the section $\widetilde{F}$ is smooth if and only if F is.
(d) The correspondence in the preceding paragraph yields a natural identification between the space $C^\infty(M)$ and the space of smooth sections of the trivial line bundle $M \times \mathbb{R} \to M$. //

If $E \to M$ is a smooth vector bundle, the set of all smooth global sections of E is a vector space under pointwise addition and scalar multiplication:

$$(c_1\sigma_1 + c_2\sigma_2)(p) = c_1\sigma_1(p) + c_2\sigma_2(p).$$

This vector space is usually denoted by $\Gamma(E)$. (For particular vector bundles, we will often introduce specialized notations for their spaces of sections, such as the notation $\mathfrak{X}(M)$ introduced in Chapter 8 for the space of smooth sections of TM.)

Just like smooth vector fields, smooth sections of a smooth bundle $E \to M$ can be multiplied by smooth real-valued functions: if $f \in C^\infty(M)$ and $\sigma \in \Gamma(E)$, we obtain a new section $f\sigma$ defined by

$$(f\sigma)(p) = f(p)\sigma(p).$$

▶ **Exercise 10.11.** Let $E \to M$ be a smooth vector bundle.

(a) Show that if $\sigma, \tau \in \Gamma(E)$ and $f, g \in C^\infty(M)$, then $f\sigma + g\tau \in \Gamma(E)$.
(b) Show that $\Gamma(E)$ is a module over the ring $C^\infty(M)$.

Lemma 10.12 (Extension Lemma for Vector Bundles). *Let $\pi\colon E \to M$ be a smooth vector bundle over a smooth manifold M with or without boundary. Suppose A is a closed subset of M, and $\sigma\colon A \to E$ is a section of $E|_A$ that is smooth in the sense that σ extends to a smooth local section of E in a neighborhood of each point. For each open subset $U \subseteq M$ containing A, there exists a global smooth section $\tilde{\sigma} \in \Gamma(E)$ such that $\tilde{\sigma}|_A = \sigma$ and $\operatorname{supp} \tilde{\sigma} \subseteq U$.*

▶ **Exercise 10.13.** Prove the preceding lemma.

▶ **Exercise 10.14.** Let $\pi\colon E \to M$ be a smooth vector bundle. Show that each element of E is in the image of a smooth global section.

Local and Global Frames

The concept of local frames that we introduced in Chapter 8 extends readily to vector bundles. Let $E \to M$ be a vector bundle. If $U \subseteq M$ is an open subset, a k-tuple of local sections $(\sigma_1, \ldots, \sigma_k)$ of E over U is said to be *linearly independent* if their values $(\sigma_1(p), \ldots, \sigma_k(p))$ form a linearly independent k-tuple in E_p for each $p \in U$. Similarly, they are said to *span E* if their values span E_p for each $p \in U$. A *local frame for E over U* is an ordered k-tuple $(\sigma_1, \ldots, \sigma_k)$ of linearly independent local sections over U that span E; thus $(\sigma_1(p), \ldots, \sigma_k(p))$ is a basis for the fiber E_p for each $p \in U$. It is called a *global frame* if $U = M$. If $E \to M$ is a smooth vector bundle, a local or global frame is a *smooth frame* if each σ_i is a smooth section. We often denote a frame $(\sigma_1, \ldots, \sigma_k)$ by (σ_i).

The (local or global) frames for M that we defined in Chapter 8 are, in our new terminology, frames for the tangent bundle. We use both terms interchangeably depending on context: "frame for M" and "frame for TM" mean the same thing.

The next proposition is an analogue for vector bundles of Proposition 8.11.

Proposition 10.15 (Completion of Local Frames for Vector Bundles). *Suppose* $\pi\colon E \to M$ *is a smooth vector bundle of rank* k.

(a) *If* $(\sigma_1, \ldots, \sigma_m)$ *is a linearly independent m-tuple of smooth local sections of E over an open subset $U \subseteq M$, with $1 \le m < k$, then for each $p \in U$ there exist smooth sections $\sigma_{m+1}, \ldots, \sigma_k$ defined on some neighborhood V of p such that $(\sigma_1, \ldots, \sigma_k)$ is a smooth local frame for E over $U \cap V$.*

(b) *If* $(v_1, \ldots, v_m)$ *is a linearly independent m-tuple of elements of E_p for some $p \in M$, with $1 \le m \le k$, then there exists a smooth local frame (σ_i) for E over some neighborhood of p such that $\sigma_i(p) = v_i$ for $i = 1, \ldots, m$.*

(c) *If* $A \subseteq M$ *is a closed subset and $(\tau_1, \ldots, \tau_k)$ is a linearly independent k-tuple of sections of $E|_A$ that are smooth in the sense described in Lemma 10.12, then there exists a smooth local frame $(\sigma_1, \ldots, \sigma_k)$ for E over some neighborhood of A such that $\sigma_i|_A = \tau_i$ for $i = 1, \ldots, k$.*

▶ **Exercise 10.16.** Prove the preceding proposition.

Local frames for a vector bundle are intimately connected with local trivializations, as the next two examples show.

Example 10.17 (A Global Frame for a Product Bundle). If $E = M \times \mathbb{R}^k \to M$ is a product bundle, the standard basis $(e_1, \ldots, e_k)$ for $\mathbb{R}^k$ yields a global frame $(\tilde{e}_i)$ for E, defined by $\tilde{e}_i(p) = (p, e_i)$. If M is a smooth manifold with or without boundary, then this global frame is smooth.   //

Example 10.18 (Local Frames Associated with Local Trivializations). Suppose $\pi\colon E \to M$ is a smooth vector bundle. If $\Phi\colon \pi^{-1}(U) \to U \times \mathbb{R}^k$ is a smooth local trivialization of E, we can use the same idea as in the preceding example to construct a local frame for E over U. Define maps $\sigma_1, \ldots, \sigma_k\colon U \to E$ by $\sigma_i(p) = \Phi^{-1}(p, e_i) = \Phi^{-1} \circ \tilde{e}_i(p)$:

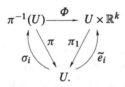

Then σ_i is smooth because Φ is a diffeomorphism, and the fact that $\pi_1 \circ \Phi = \pi$ implies that

$$\pi \circ \sigma_i(p) = \pi \circ \Phi^{-1}(p, e_i) = \pi_1(p, e_i) = p,$$

so σ_i is a section. To see that $(\sigma_i(p))$ forms a basis for E_p, just note that Φ restricts to an isomorphism from E_p to $\{p\} \times \mathbb{R}^k$, and $\Phi(\sigma_i(p)) = (p, e_i)$, so Φ takes $(\sigma_i(p))$ to the standard basis for $\{p\} \times \mathbb{R}^k \cong \mathbb{R}^k$. We say that this local frame (σ_i) is *associated with* Φ.   //

Proposition 10.19. *Every smooth local frame for a smooth vector bundle is associated with a smooth local trivialization as in Example 10.18.*

Proof. Suppose $E \to M$ is a smooth vector bundle and (σ_i) is a smooth local frame for E over an open subset $U \subseteq M$. We define a map $\Psi \colon U \times \mathbb{R}^k \to \pi^{-1}(U)$ by

$$\Psi\big(p,\big(v^1,\dots,v^k\big)\big) = v^i \sigma_i(p). \tag{10.4}$$

The fact that $(\sigma_i(p))$ forms a basis for E_p at each $p \in U$ implies that Ψ is bijective, and an easy computation shows that $\sigma_i = \Psi \circ \tilde{e}_i$. Thus, if we can show that Ψ is a diffeomorphism, then Ψ^{-1} will be a smooth local trivialization whose associated local frame is (σ_i).

Since Ψ is bijective, to show that it is a diffeomorphism it suffices to show that it is a local diffeomorphism. Given $q \in U$, we can choose a neighborhood V of q in M over which there exists a smooth local trivialization $\Phi \colon \pi^{-1}(V) \to V \times \mathbb{R}^k$, and by shrinking V if necessary we may assume that $V \subseteq U$. Since Φ is a diffeomorphism, if we can show that $\Phi \circ \Psi|_{V \times \mathbb{R}^k}$ is a diffeomorphism from $V \times \mathbb{R}^k$ to itself, it follows that Ψ restricts to a diffeomorphism from $V \times \mathbb{R}^k$ to $\pi^{-1}(V)$:

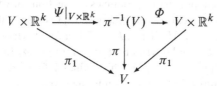

For each of our smooth sections σ_i, the composite map $\Phi \circ \sigma_i|_V \colon V \to V \times \mathbb{R}^k$ is smooth, and thus there are smooth functions $\sigma_i^1,\dots,\sigma_i^k \colon V \to \mathbb{R}$ such that

$$\Phi \circ \sigma_i(p) = \big(p,\big(\sigma_i^1(p),\dots,\sigma_i^k(p)\big)\big).$$

On $V \times \mathbb{R}^k$, therefore,

$$\Phi \circ \Psi\big(p,\big(v^1,\dots,v^k\big)\big) = \big(p,\big(v^i\sigma_i^1(p),\dots,v^i\sigma_i^k(p)\big)\big),$$

which is clearly smooth.

To show that $(\Phi \circ \Psi)^{-1}$ is smooth, note that the matrix $(\sigma_i^j(p))$ is invertible for each p, because $(\sigma_i(p))$ is a basis for E_p. Let $(\tau_i^j(p))$ denote the inverse matrix. Because matrix inversion is a smooth map from $\mathrm{GL}(k,\mathbb{R})$ to itself, the functions τ_i^j are smooth. It follows from the computations in the preceding paragraph that

$$(\Phi \circ \Psi)^{-1}\big(p,\big(w^1,\dots,w^k\big)\big) = \big(p,\big(w^i\tau_i^1(p),\dots,w^i\tau_i^k(p)\big)\big),$$

which is also smooth. $\square$

Corollary 10.20. *A smooth vector bundle is smoothly trivial if and only if it admits a smooth global frame.*

Proof. Example 10.18 and Proposition 10.19 show that there is a smooth local trivialization over an open subset $U \subseteq M$ if and only if there is a smooth local frame over U. The corollary is just the special case of this statement when $U = M$. $\square$

When applied to the tangent bundle of a smooth manifold M, this corollary says that TM is trivial if and only if M is parallelizable. (Recall that in Chapter 8 we defined a *parallelizable manifold* to be one that admits a smooth global frame for its tangent bundle.)

Corollary 10.21. *Let* $\pi\colon E \to M$ *be a smooth vector bundle of rank* k, *let* (V, φ) *be a smooth chart on* M *with coordinate functions* (x^i), *and suppose there exists a smooth local frame* (σ_i) *for* E *over* V. *Define* $\widetilde{\varphi}\colon \pi^{-1}(V) \to \varphi(V) \times \mathbb{R}^k$ *by*

$$\widetilde{\varphi}\big(v^i \sigma_i(p)\big) = \big(x^1(p), \dots, x^n(p), v^1, \dots, v^k\big).$$

Then $\big(\pi^{-1}(V), \widetilde{\varphi}\big)$ *is a smooth coordinate chart for* E.

Proof. Just check that $\widetilde{\varphi}$ is equal to the composition $(\varphi \times \mathrm{Id}_{\mathbb{R}^k}) \circ \Phi$, where Φ is the local trivialization associated with (σ_i). As a composition of diffeomorphisms, it is a diffeomorphism. $\qquad\square$

Just as smoothness of vector fields can be characterized in terms of their component functions in any smooth chart, smoothness of sections of vector bundles can be characterized in terms of local frames. Suppose (σ_i) is a smooth local frame for E over some open subset $U \subseteq M$. If $\tau\colon M \to E$ is a rough section, the value of τ at an arbitrary point $p \in U$ can be written $\tau(p) = \tau^i(p)\sigma_i(p)$ for some uniquely determined numbers $\big(\tau^1(p), \dots, \tau^n(p)\big)$. This defines k functions $\tau^i\colon U \to \mathbb{R}$, called the ***component functions of*** τ with respect to the given local frame.

Proposition 10.22 (Local Frame Criterion for Smoothness). *Let* $\pi\colon E \to M$ *be a smooth vector bundle, and let* $\tau\colon M \to E$ *be a rough section. If* (σ_i) *is a smooth local frame for* E *over an open subset* $U \subseteq M$, *then* τ *is smooth on* U *if and only if its component functions with respect to* (σ_i) *are smooth.*

Proof. Let $\Phi\colon \pi^{-1}(U) \to U \times \mathbb{R}^k$ be the local trivialization associated with the local frame (σ_i). Because Φ is a diffeomorphism, τ is smooth on U if and only if the composite map $\Phi \circ \tau$ is smooth on U. It is straightforward to check that $\Phi \circ \tau(p) = \big(p, \big(\tau^1(p), \dots, \tau^k(p)\big)\big)$, where (τ^i) are the component functions of τ with respect to (σ_i), so $\Phi \circ \tau$ is smooth if and only if the component functions τ^i are smooth. $\qquad\square$

▶ **Exercise 10.23.** Let $E \to M$ be a vector bundle. Show that a rough section of E is continuous if and only if its component functions in each local frame are continuous.

Proposition 10.22 applies equally well to local sections, since a local section of E over an open subset $V \subseteq M$ is a global section of the restricted bundle $E|_V$.

The correspondence between local frames and local trivializations leads to the following uniqueness result characterizing the smooth structure on the tangent bundle of a smooth manifold.

Proposition 10.24 (Uniqueness of the Smooth Structure on TM**).** *Let* M *be a smooth* n-*manifold with or without boundary. The topology and smooth structure on* TM *constructed in Proposition 3.18 are the unique ones with respect to which* $\pi\colon TM \to M$ *is a smooth vector bundle with the given vector space structure on the fibers, and such that all coordinate vector fields are smooth local sections.*

Proof. Suppose TM is endowed with some topology and smooth structure making it into a smooth vector bundle with the given properties. If (U, φ) is any smooth chart for M, the corresponding coordinate frame $(\partial/\partial x^i)$ is a smooth local frame over U, so by Proposition 10.19 there is a smooth local trivialization $\Phi \colon \pi^{-1}(U) \to U \times \mathbb{R}^n$ associated with this local frame. Referring back to the construction of Example 10.18, we see that this local trivialization is none other than the map Φ constructed in Proposition 10.4. It follows from Corollary 10.21 that the natural coordinate chart $\tilde{\varphi} = (\varphi \times \mathrm{Id}_{\mathbb{R}^n}) \circ \Phi$ belongs to the given smooth structure. Thus, the given smooth structure is equal to the one constructed in Proposition 3.18. $\qquad\square$

Bundle Homomorphisms

If $\pi \colon E \to M$ and $\pi' \colon E' \to M'$ are vector bundles, a continuous map $F \colon E \to E'$ is called a **bundle homomorphism** if there exists a map $f \colon M \to M'$ satisfying $\pi' \circ F = f \circ \pi$,

$$
\begin{array}{ccc}
E & \xrightarrow{\ F\ } & E' \\
\pi \downarrow & & \downarrow \pi' \\
M & \xrightarrow[f]{} & M',
\end{array}
$$

with the property that for each $p \in M$, the restricted map $F|_{E_p} \colon E_p \to E'_{f(p)}$ is linear. The relationship between F and f is expressed by saying that F **covers** f.

Proposition 10.25. *Suppose $\pi \colon E \to M$ and $\pi' \colon E \to M'$ are vector bundles and $F \colon E \to E'$ is a bundle homomorphism covering $f \colon M \to M'$. Then f is continuous and is uniquely determined by F. If the bundles and F are all smooth, then f is smooth as well.*

Proof. All of the conclusions follow from the easily verified fact that $f = \pi' \circ F \circ \zeta$, where $\zeta \colon M \to E$ is the zero section. $\qquad\square$

A bijective bundle homomorphism $F \colon E \to E'$ whose inverse is also a bundle homomorphism is called a **bundle isomorphism**; if F is also a diffeomorphism, it is called a **smooth bundle isomorphism**. If there exists a (smooth) bundle isomorphism between E and E', the two bundles are said to be **(smoothly) isomorphic**.

In the special case in which both E and E' are vector bundles over the same base space M, a slightly more restrictive notion of bundle homomorphism is usually more useful. A **bundle homomorphism over M** is a bundle homomorphism covering the identity map of M, or in other words, a continuous map $F \colon E \to E'$ such that $\pi' \circ F = \pi$,

$$
\begin{array}{ccc}
E & \xrightarrow{\ F\ } & E' \\
& {\scriptstyle \pi}\searrow \quad \swarrow{\scriptstyle \pi'} & \\
& M, &
\end{array}
$$

and whose restriction to each fiber is linear. If there exists a bundle homomorphism $F \colon E \to E'$ over M that is also a (smooth) bundle isomorphism, then we say that

E and *E'* are (**smoothly**) **isomorphic over** *M*. The next proposition shows that it is not necessary to check smoothness of the inverse.

Proposition 10.26. *Suppose E and E' are smooth vector bundles over a smooth manifold M with or without boundary, and F: E → E' is a bijective smooth bundle homomorphism over M. Then F is a smooth bundle isomorphism.*

Proof. Problem 10-11.   □

▶ **Exercise 10.27.** Show that a smooth rank-k vector bundle over M is smoothly trivial if and only if it is smoothly isomorphic over M to the product bundle $M \times \mathbb{R}^k$.

Example 10.28 (Bundle Homomorphisms).

(a) If $F: M \to N$ is a smooth map, the global differential $dF: TM \to TN$ is a smooth bundle homomorphism covering F.
(b) If $E \to M$ is a smooth vector bundle and $S \subseteq M$ is an immersed submanifold with or without boundary, then the inclusion map $E|_S \hookrightarrow E$ is a smooth bundle homomorphism covering the inclusion of S into M.   //

Suppose $E \to M$ and $E' \to M$ are smooth vector bundles over a smooth manifold M with or without boundary, and let $\Gamma(E)$, $\Gamma(E')$ denote their spaces of smooth global sections. If $F: E \to E'$ is a smooth bundle homomorphism over M, then composition with F induces a map $\widetilde{F}: \Gamma(E) \to \Gamma(E')$ as follows:

$$\widetilde{F}(\sigma)(p) = (F \circ \sigma)(p) = F\big(\sigma(p)\big). \tag{10.5}$$

It is easy to check that $\widetilde{F}(\sigma)$ is a section of E', and it is smooth by composition.

Because a bundle homomorphism is linear on fibers, the resulting map $\widetilde{F}$ on sections is linear over $\mathbb{R}$. In fact, it satisfies a stronger linearity property. A map $\mathcal{F}: \Gamma(E) \to \Gamma(E')$ is said to be **linear over** $C^\infty(M)$ if for any smooth functions $u_1, u_2 \in C^\infty(M)$ and smooth sections $\sigma_1, \sigma_2 \in \Gamma(E)$,

$$\mathcal{F}(u_1\sigma_1 + u_2\sigma_2) = u_1\mathcal{F}(\sigma_1) + u_2\mathcal{F}(\sigma_2).$$

It follows easily from the definition (10.5) that the map on sections induced by a smooth bundle homomorphism is linear over $C^\infty(M)$. The next lemma shows that the converse is true as well.

Lemma 10.29 (Bundle Homomorphism Characterization Lemma). *Let $\pi: E \to M$ and $\pi': E' \to M$ be smooth vector bundles over a smooth manifold M with or without boundary, and let $\Gamma(E)$, $\Gamma(E')$ denote their spaces of smooth sections. A map $\mathcal{F}: \Gamma(E) \to \Gamma(E')$ is linear over $C^\infty(M)$ if and only if there is a smooth bundle homomorphism $F: E \to E'$ over M such that $\mathcal{F}(\sigma) = F \circ \sigma$ for all $\sigma \in \Gamma(E)$.*

Proof. We noted above that the map on sections induced by a smooth bundle homomorphism is linear over $C^\infty(M)$. Conversely, suppose $\mathcal{F}: \Gamma(E) \to \Gamma(E')$ is linear over $C^\infty(M)$. First, we show that $\mathcal{F}$ acts locally: if $\sigma_1 \equiv \sigma_2$ in some open subset $U \subseteq M$, then $\mathcal{F}(\sigma_1) \equiv \mathcal{F}(\sigma_2)$ in U. Write $\tau = \sigma_1 - \sigma_2$; then by linearity of $\mathcal{F}$, it suffices to assume that τ vanishes in U and show that $\mathcal{F}(\tau)$ does too. Given $p \in U$, let $\psi \in C^\infty(M)$ be a smooth bump function supported in U and equal to 1 at p. Because $\psi\tau$ is identically zero on M, the fact that $\mathcal{F}$ is linear over $C^\infty(M)$ implies

$$0 = \mathcal{F}(\psi\tau) = \psi\mathcal{F}(\tau).$$

Evaluating at p shows that $\mathcal{F}(\tau)(p) = \psi(p)\mathcal{F}(\tau)(p) = 0$; since the same is true for every $p \in U$, the claim follows.

Next we show that $\mathcal{F}$ actually acts pointwise: if $\sigma_1(p) = \sigma_2(p)$, then $\mathcal{F}(\sigma_1)(p) = \mathcal{F}(\sigma_2)(p)$. Once again, it suffices to assume that $\tau(p) = 0$ and show that $\mathcal{F}(\tau)(p) = 0$. Let $(\sigma_1, \dots, \sigma_k)$ be a smooth local frame for E in some neighborhood U of p, and write τ in terms of this frame as $\tau = u^i\sigma_i$ for some smooth functions u^i defined in U. The fact that $\tau(p) = 0$ means that $u^1(p) = \cdots = u^k(p) = 0$. By the extension lemmas for vector bundles and for functions, there exist smooth global sections $\tilde\sigma_i \in \Gamma(E)$ that agree with σ_i in a neighborhood of p, and smooth functions $\tilde u^i \in C^\infty(M)$ that agree with u^i in some neighborhood of p. Then since $\tau = \tilde u^i\tilde\sigma_i$ on a neighborhood of p, we have

$$\mathcal{F}(\tau)(p) = \mathcal{F}(\tilde u^i\tilde\sigma_i)(p) = \tilde u^i(p)\mathcal{F}(\tilde\sigma_i)(p) = 0.$$

Define a bundle homomorphism $F: E \to E'$ as follows. For any $p \in M$ and $v \in E_p$, let $F(v) = \mathcal{F}(\tilde v)(p) \in E'_p$, where $\tilde v$ is any global smooth section of E such that $\tilde v(p) = v$. The discussion above shows that the resulting element of E'_p is independent of the choice of section. This map F clearly satisfies $\pi' \circ F = \pi$, and it is linear on each fiber because of the linearity of $\mathcal{F}$. It also satisfies $F \circ \sigma(p) = \mathcal{F}(\sigma)(p)$ for each $\sigma \in \Gamma(E)$ by definition. It remains only to show that F is smooth. It suffices to show that it is smooth in a neighborhood of each point.

Given $p \in M$, let (σ_i) be a smooth local frame for E on some neighborhood of p. By the extension lemma, there are global sections $\tilde\sigma_i$ that agree with σ_i in a (smaller) neighborhood U of p. Shrinking U further if necessary, we may also assume that there exists a smooth local frame (σ'_j) for E' over U. Because $\mathcal{F}$ maps smooth global sections of E to smooth global sections of E', there are smooth functions $A_i^j \in C^\infty(U)$ such that $\mathcal{F}(\tilde\sigma_i)|_U = A_i^j\sigma'_j$.

For any $q \in U$ and $v \in E_q$, we can write $v = v^i\sigma_i(q)$ for some real numbers $(v^1, \dots, v^k)$, and then

$$F(v^i\sigma_i(q)) = \mathcal{F}(v^i\tilde\sigma_i)(q) = v^i\mathcal{F}(\tilde\sigma_i)(q) = v^iA_i^j(q)\sigma'_j(q),$$

because $v^i\tilde\sigma_i$ is a global smooth section of E whose value at q is v. If Φ and Φ' denote the local trivializations of E and E' associated with the frames (σ_i) and (σ'_i), respectively, it follows that the composite map $\Phi' \circ F \circ \Phi^{-1}: U \times \mathbb{R}^k \to U \times \mathbb{R}^m$ has the form

$$\Phi' \circ F \circ \Phi^{-1}(q, (v^1, \dots, v^k)) = (q, (A_i^1(q)v^i, \dots, A_i^m(q)v^i)),$$

which is smooth. Because Φ and Φ' are diffeomorphisms, this shows that F is smooth on $\pi^{-1}(U)$. $\qquad\square$

Later, after we have developed more tools, we will see many examples of smooth bundle homomorphisms. For now, here are some elementary examples.

Example 10.30 (Bundle Homomorphisms Over Manifolds).

(a) If M is a smooth manifold and $f \in C^\infty(M)$, the map from $\mathfrak{X}(M)$ to itself defined by $X \mapsto fX$ is linear over $C^\infty(M)$ because $f(u_1 X_1 + u_2 X_2) = u_1 f X_1 + u_2 f X_2$, and thus defines a smooth bundle homomorphism over M from TM to itself.

(b) If Z is a smooth vector field on $\mathbb{R}^3$, the cross product with Z defines a map from $\mathfrak{X}(\mathbb{R}^3)$ to itself: $X \mapsto X \times Z$. Since it is linear over $C^\infty(\mathbb{R}^3)$ in X, it determines a smooth bundle homomorphism over $\mathbb{R}^3$ from $T\mathbb{R}^3$ to $T\mathbb{R}^3$.

(c) Given $Z \in \mathfrak{X}(\mathbb{R}^n)$, the Euclidean dot product defines a map $X \mapsto X \cdot Z$ from $\mathfrak{X}(\mathbb{R}^n)$ to $C^\infty(\mathbb{R}^n)$, which is linear over $C^\infty(\mathbb{R}^n)$ and thus determines a smooth bundle homomorphism over $\mathbb{R}^n$ from $T\mathbb{R}^n$ to the trivial line bundle $\mathbb{R}^n \times \mathbb{R}$.   //

Because of Lemma 10.29, we usually dispense with the notation $\widetilde{F}$ and use the same symbol for both a bundle homomorphism $F \colon E \to E'$ over M and the linear map $F \colon \Gamma(E) \to \Gamma(E')$ that it induces on sections, and we refer to a map of either of these types as a bundle homomorphism. Because the action on sections is obtained simply by applying the bundle homomorphism pointwise, this should cause no confusion. In fact, we have been doing the same thing all along in certain circumstances. For example, if $a \in \mathbb{R}$, we use the same notation $X \mapsto aX$ to denote both the operation of multiplying vectors in each tangent space $T_p M$ by a, and the operation of multiplying vector fields by a. Because multiplying by a is a bundle homomorphism from TM to itself, there is no ambiguity about what is meant.

It should be noted that most maps that involve differentiation are *not* bundle homomorphism. For example, if X is a smooth vector field on a smooth manifold M, the Lie derivative operator $\mathcal{L}_X \colon \mathfrak{X}(M) \to \mathfrak{X}(M)$ is not a bundle homomorphism from the tangent bundle to itself, because it is not linear over $C^\infty(M)$. As a rule of thumb, a linear map that takes smooth sections of one bundle to smooth sections of another is likely to be a bundle homomorphism if it acts pointwise, but not if it involves differentiation.

Subbundles

Given a vector bundle $\pi_E \colon E \to M$, a *subbundle of E* (see Fig. 10.4) is a vector bundle $\pi_D \colon D \to M$, in which D is a topological subspace of E and π_D is the restriction of π_E to D, such that for each $p \in M$, the subset $D_p = D \cap E_p$ is a linear subspace of E_p, and the vector space structure on D_p is the one inherited from E_p. Note that the condition that D be a vector bundle over M implies that all of the fibers D_p must be nonempty and have the same dimension. If $E \to M$ is a smooth bundle, then a subbundle of E is called a *smooth subbundle* if it is a smooth vector bundle and an embedded submanifold with or without boundary in E.

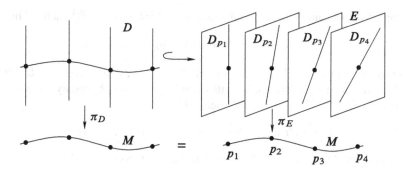

Fig. 10.4 A subbundle of a vector bundle

▶ **Exercise 10.31.** Given a smooth vector bundle $E \to M$ and a smooth subbundle $D \subseteq E$, show that the inclusion map $\iota\colon D \hookrightarrow E$ is a smooth bundle homomorphism over M.

The following lemma gives a convenient condition for checking that a union of subspaces $\{D_p \subseteq E_p : p \in M\}$ is a smooth subbundle.

Lemma 10.32 (Local Frame Criterion for Subbundles). *Let $\pi\colon E \to M$ be a smooth vector bundle, and suppose that for each $p \in M$ we are given an m-dimensional linear subspace $D_p \subseteq E_p$. Then $D = \bigcup_{p \in M} D_p \subseteq E$ is a smooth subbundle of E if and only if the following condition is satisfied:*

> *Each point of M has a neighborhood U on which there exist smooth local sections $\sigma_1, \ldots, \sigma_m\colon U \to E$ with the property   (10.6) that $\sigma_1(q), \ldots, \sigma_m(q)$ form a basis for D_q at each $q \in U$.*

Proof. If D is a smooth subbundle, then by definition each $p \in M$ has a neighborhood U over which there exists a smooth local trivialization of D, and Example 10.18 shows that there exists a smooth local frame for D over each such set U. Such a local frame is by definition a collection of smooth sections $\tau_1, \ldots, \tau_m\colon U \to D$ whose images form a basis for D_p at each point $p \in U$. The smooth sections of E that we seek are obtained by composing with the inclusion map $\iota\colon D \hookrightarrow E$: $\sigma_j = \iota \circ \tau_j$.

Conversely, suppose $E \to M$ is a smooth bundle of rank k, and $D \subseteq E$ satisfies (10.6). Each set $D \cap E_p$ is a linear subspace of E_p by hypothesis, so we need to show that D is an embedded submanifold with or without boundary in E and that the restriction of π makes it into a smooth vector bundle over M.

To prove that D is an embedded submanifold with or without boundary, it suffices to show that each $p \in M$ has a neighborhood U such that $D \cap \pi^{-1}(U)$ is an embedded submanifold (possibly with boundary) in $\pi^{-1}(U) \subseteq E$. Given $p \in M$, let $\sigma_1, \ldots, \sigma_m$ be smooth local sections of E satisfying (10.6) on a neighborhood of p. By Proposition 10.15, we can complete these to a smooth local frame $(\sigma_1, \ldots, \sigma_k)$ for E over some neighborhood U of p. By Proposition 10.19, this local frame is associated with a smooth local trivialization $\Phi \colon \pi^{-1}(U) \to U \times \mathbb{R}^k$, defined by

$$\Phi\left(s^1\sigma_1(q) + \cdots + s^k\sigma_k(q)\right) = \left(q, \left(s^1, \dots, s^k\right)\right).$$

This map Φ takes $D \cap \pi^{-1}(U)$ to the subset $\{(q, (s^1, \dots, s^m, 0, \dots, 0))\} \subseteq U \times \mathbb{R}^k$, which is an embedded submanifold (with boundary if U has a boundary). Moreover, the map $\Psi \colon D \cap \pi^{-1}(U) \to U \times \mathbb{R}^m$ defined by

$$\Psi\left(s^1\sigma_1(q) + \cdots + s^m\sigma_m(q)\right) = \left(q, \left(s^1, \dots, s^m\right)\right)$$

is a smooth local trivialization of D, so D is itself a smooth vector bundle.   □

Example 10.33 (Subbundles).

(a) If M is a smooth manifold and V is a nowhere-vanishing smooth vector field on M, then the set $D \subseteq TM$ whose fiber at each $p \in M$ is the linear span of V_p is a smooth 1-dimensional subbundle of TM.

(b) Suppose $E \to M$ is any trivial bundle, and let $(E_1, \dots, E_k)$ be a smooth global frame for E. If $0 \le m \le k$, the subset $D \subseteq E$ defined by $D_p = \text{span}\left(E_1|_p, \dots, E_m|_p\right)$ for each $p \in M$ is a smooth subbundle of E.

(c) Suppose M is a smooth manifold with or without boundary and $S \subseteq M$ is an immersed k-submanifold with or without boundary. Problem 10-14 asks you to prove that TS is a smooth rank-k subbundle of the ambient tangent bundle $TM|_S$.   //

The next theorem shows how to obtain many more subbundles. Suppose $E \to M$ and $E' \to M$ are vector bundles and $F \colon E \to E'$ is a bundle homomorphism over M. For each $p \in M$, the rank of the linear map $F|_{E_p}$ is called the *rank of F at p*. We say that F has *constant rank* if its rank is the same for all $p \in M$.

Theorem 10.34. *Let E and E' be smooth vector bundles over a smooth manifold M, and let $F \colon E \to E'$ be a smooth bundle homomorphism over M. Define subsets $\mathrm{Ker}\, F \subseteq E$ and $\mathrm{Im}\, F \subseteq E'$ by*

$$\mathrm{Ker}\, F = \bigcup_{p \in M} \mathrm{Ker}\left(F|_{E_p}\right), \qquad \mathrm{Im}\, F = \bigcup_{p \in M} \mathrm{Im}\left(F|_{E_p}\right).$$

Then $\mathrm{Ker}\, F$ and $\mathrm{Im}\, F$ are smooth subbundles of E and E', respectively, if and only if F has constant rank.

Proof. One direction is obvious: since the fibers of a bundle have the same dimension everywhere, the constant-rank condition is certainly necessary for $\mathrm{Ker}\, F$ and $\mathrm{Im}\, F$ to be subbundles. To prove sufficiency, suppose F has constant rank r, and let k and k' be the ranks of the bundles E and E', respectively. Let $p \in M$ be arbitrary, and choose a smooth local frame $(\sigma_1, \dots, \sigma_k)$ for E over a neighborhood U of p. For each i, the map $F \circ \sigma_i \colon U \to E'$ is a smooth local section of E', and these sections span $(\mathrm{Im}\, F)|_U$. After rearranging the indices if necessary, we can assume that the elements $\{F \circ \sigma_1(p), \dots, F \circ \sigma_r(p)\}$ form a basis for $\mathrm{Im}\left(F|_{E_p}\right)$, and by continuity they remain linearly independent in some neighborhood U_0 of p. Since F has

constant rank, this means that $(F \circ \sigma_1, \ldots, F \circ \sigma_r)$ forms a smooth local frame for
$\operatorname{Im} F$ over U_0. Since we can do the same in a neighborhood of each point, the local
frame criterion shows that $\operatorname{Im} F$ is a smooth subbundle of E'.

To prove that $\operatorname{Ker} F$ is also a smooth subbundle, let U_0 and (σ_i) be as above,
and let $V \subseteq E|_{U_0}$ be the smooth subbundle spanned by $\sigma_1, \ldots, \sigma_r$. The smooth
bundle homomorphism $F|_V : V \to (\operatorname{Im} F)|_{U_0}$ is bijective, and is thus a smooth
bundle isomorphism by Proposition 10.26. Define a smooth bundle homomorphism
$\Psi : E|_{U_0} \to E|_{U_0}$ by $\Psi(v) = v - (F|_V)^{-1} \circ F(v)$. If $v \in V$, then $F(v) = (F|_V)(v)$,
so $F(\Psi(v)) = F(v) - F \circ (F|_V)^{-1} \circ (F|_V)(v) = 0$. On the other hand, if $v \in$
$\operatorname{Ker} F$, then $\Psi(v) = v$, so again $F(\Psi(v)) = F(v) = 0$. Since V and $(\operatorname{Ker} F)|_{U_0}$
together span $E|_{U_0}$, it follows that Ψ takes its values in $(\operatorname{Ker} F)|_{U_0}$, and since it
restricts to the identity on $(\operatorname{Ker} F)|_{U_0}$, its image is exactly $(\operatorname{Ker} F)|_{U_0}$. Thus Ψ has
constant rank, and by the argument in the preceding paragraph, $(\operatorname{Ker} F)|_{U_0} = \operatorname{Im} \Psi$
is a smooth subbundle of $E|_{U_0}$. Since we can do the same thing in a neighborhood
of each point, $\operatorname{Ker} F$ is a smooth subbundle of E.   □

The next proposition illustrates another method for constructing interesting subbundles of the tangent bundle over submanifolds of $\mathbb{R}^n$.

Lemma 10.35 (Orthogonal Complement Bundles). *Let M be an immersed submanifold with or without boundary in $\mathbb{R}^n$, and D be a smooth rank-k subbundle of
$T\mathbb{R}^n|_M$. For each $p \in M$, let $D_p^\perp$ denote the orthogonal complement of D_p in $T_p\mathbb{R}^n$
with respect to the Euclidean dot product, and let $D^\perp \subseteq T\mathbb{R}^n|_M$ be the subset*

$$D^\perp = \{(p, v) \in T\mathbb{R}^n : p \in M, \ v \in D_p^\perp\}.$$

*Then $D^\perp$ is a smooth rank-$(n-k)$ subbundle of $T\mathbb{R}^n|_M$. For each $p \in M$, there is
a smooth orthonormal frame for $D^\perp$ on a neighborhood of p.*

Proof. Let $p \in M$ be arbitrary, and let $(X_1, \ldots, X_k)$ be a smooth local frame for
D over some neighborhood V of p in M. Because immersed submanifolds are
locally embedded, by shrinking V if necessary, we may assume that it is a single
slice in some coordinate ball or half-ball $U \subseteq \mathbb{R}^n$. Since V is closed in U, Proposition 8.11(c) shows that we can complete $(X_1, \ldots, X_k)$ to a smooth local frame
$(\widetilde{X}_1, \ldots, \widetilde{X}_n)$ for $T\mathbb{R}^n$ over U, and then Lemma 8.13 yields a smooth *orthonormal*
frame (E_j) over U such that $\operatorname{span}(E_1|_p, \ldots, E_k|_p) = \operatorname{span}(X_1|_p, \ldots, X_k|_p) = D_p$
for each $p \in U$. It follows that $(E_{k+1}, \ldots, E_n)$ restricts to a smooth orthonormal
frame for $D^\perp$ over V. Thus $D^\perp$ satisfies the local frame criterion, and is therefore
a smooth subbundle of $T\mathbb{R}^n|_M$.   □

Corollary 10.36 (The Normal Bundle to a Submanifold of $\mathbb{R}^n$). *If $M \subseteq \mathbb{R}^n$ is an
immersed m-dimensional submanifold with or without boundary, its normal bundle
NM is a smooth rank-$(n-m)$ subbundle of $T\mathbb{R}^n|_M$. For each $p \in M$, there exists
a smooth orthonormal frame for NM on a neighborhood of p.*

Proof. Apply Lemma 10.35 to the smooth subbundle $TM \subseteq T\mathbb{R}^n|_M$.   □

Fiber Bundles

We conclude this chapter by giving a brief introduction to an important generalization of vector bundles, in which the fibers are allowed to be arbitrary topological spaces instead of vector spaces. We can only touch on the subject here; but fiber bundles appear in many applications of manifold theory, so it is important to be at least familiar with the definitions.

Let M and F be topological spaces. A *fiber bundle over M with model fiber F* is a topological space E together with a surjective continuous map $\pi: E \to M$ with the property that for each $x \in M$, there exist a neighborhood U of x in M and a homeomorphism $\Phi: \pi^{-1}(U) \to U \times F$, called a *local trivialization of E over U*, such that the following diagram commutes:

$$
\begin{array}{ccc}
\pi^{-1}(U) & \xrightarrow{\ \Phi\ } & U \times F \\
& {\scriptstyle \pi}\searrow \quad \swarrow {\scriptstyle \pi_1} & \\
& U. &
\end{array}
$$

The space E is called the *total space of the bundle*, M is its *base*, and π is its *projection*. If E, M, and F are smooth manifolds with or without boundary, π is a smooth map, and the local trivializations can be chosen to be diffeomorphisms, then it is called a *smooth fiber bundle*.

A *trivial fiber bundle* is one that admits a local trivialization over the entire base space (a *global trivialization*). It is said to be *smoothly trivial* if it is a smooth bundle and the global trivialization is a diffeomorphism.

Example 10.37 (Fiber Bundles).

(a) Every product space $M \times F$ is a fiber bundle with projection $\pi_1: M \times F \to M$, called a *product fiber bundle*. It has a global trivialization given by the identity map $M \times F \to M \times F$, so every product bundle is trivial.

(b) Every rank-k vector bundle is a fiber bundle with model fiber $\mathbb{R}^k$.

(c) If $E \to \mathbb{S}^1$ is the Möbius bundle of Example 10.3, then the image of $\mathbb{R} \times [-1, 1]$ under the quotient map $q: \mathbb{R}^2 \to E$ is a fiber bundle over $\mathbb{S}^1$ with model fiber $[-1, 1]$. It is not a trivial bundle. (Can you prove it?)

(d) Every covering map $\pi: E \to M$ is a fiber bundle whose model fiber is discrete. To construct local trivializations, let S be a discrete space with the same cardinality as the fibers of π. For each evenly covered open subset $U \subseteq M$, define a map $\Phi: \pi^{-1}(U) \to U \times S$ by choosing a bijection between the set of components of $\pi^{-1}(U)$ and S, and letting $\Phi(x) = (\pi(x), c(x))$, where $c(x)$ is the element of S corresponding to the component containing x.   //

We will see a few more examples of fiber bundles as we go along.

Problems

10-1. Let E be the total space of the Möbius bundle constructed in Example 10.3.

(a) Show that E has a unique smooth structure such that the quotient map $q: \mathbb{R}^2 \to E$ is a smooth covering map.

(b) Show that $\pi: E \to \mathbb{S}^1$ is a smooth rank-1 vector bundle.

(c) Show that it is not a trivial bundle.

10-2. Let E be a vector bundle over a topological space M. Show that the projection map $\pi: E \to M$ is a homotopy equivalence.

10-3. Let VB denote the category whose objects are smooth vector bundles and whose morphisms are smooth bundle homomorphism, and let Diff denote the category whose objects are smooth manifolds and whose morphisms are smooth maps. Show that the assignment $M \mapsto TM$, $F \mapsto dF$ defines a covariant functor from Diff to VB, called the **tangent functor**. (*Used on p. 303.*)

10-4. Complete the proof of Lemma 10.5 by showing that $\tau: U \cap V \to \mathrm{GL}(k, \mathbb{R})$ is smooth. [Hint: use the same idea as in the proof of Proposition 7.37.]

10-5. Let $\pi: E \to M$ be a smooth vector bundle of rank k over a smooth manifold M with or without boundary. Suppose that $\{U_\alpha\}_{\alpha \in A}$ is an open cover of M, and for each $\alpha \in A$ we are given a smooth local trivialization $\Phi_\alpha: \pi^{-1}(U_\alpha) \to U_\alpha \times \mathbb{R}^k$ of E. For each $\alpha, \beta \in A$ such that $U_\alpha \cap U_\beta \neq \varnothing$, let $\tau_{\alpha\beta}: U_\alpha \cap U_\beta \to \mathrm{GL}(k, \mathbb{R})$ be the transition function defined by (10.3). Show that the following identity is satisfied for all $\alpha, \beta, \gamma \in A$:

$$\tau_{\alpha\beta}(p)\tau_{\beta\gamma}(p) = \tau_{\alpha\gamma}(p), \qquad p \in U_\alpha \cap U_\beta \cap U_\gamma. \tag{10.7}$$

(The juxtaposition on the left-hand side represents matrix multiplication.)

10-6. VECTOR BUNDLE CONSTRUCTION THEOREM: Let M be a smooth manifold with or without boundary, and let $\{U_\alpha\}_{\alpha \in A}$ be an open cover of M. Suppose for each $\alpha, \beta \in A$ we are given a smooth map $\tau_{\alpha\beta}: U_\alpha \cap U_\beta \to \mathrm{GL}(k, \mathbb{R})$ such that (10.7) is satisfied for all $\alpha, \beta, \gamma \in A$. Show that there is a smooth rank-k vector bundle $E \to M$ with smooth local trivializations $\Phi_\alpha: \pi^{-1}(U_\alpha) \to U_\alpha \times \mathbb{R}^k$ whose transition functions are the given maps $\tau_{\alpha\beta}$. [Hint: define an appropriate equivalence relation on $\coprod_{\alpha \in A}(U_\alpha \times \mathbb{R}^k)$, and use the vector bundle chart lemma.]

10-7. Compute the transition function for $T\mathbb{S}^2$ associated with the two local trivializations determined by stereographic coordinates (Problem 1-7).

10-8. Let Vec_1 be the category whose objects are finite-dimensional real vector spaces and whose morphisms are linear isomorphisms. If $\mathscr{F}$ is a covariant functor from Vec_1 to itself, for each finite-dimensional vector space V we get a map $\mathscr{F}: \mathrm{GL}(V) \to \mathrm{GL}(\mathscr{F}(V))$ sending each isomorphism $A: V \to V$ to the induced isomorphism $\mathscr{F}(A): \mathscr{F}(V) \to \mathscr{F}(V)$. We say $\mathscr{F}$ is a **smooth functor** if this map is smooth for every V. Given a smooth vector bundle $E \to M$ and a smooth functor $\mathscr{F}: \mathsf{Vec}_1 \to \mathsf{Vec}_1$, show that there is a smooth vector bundle $\mathscr{F}(E) \to M$ whose fiber at each point $p \in M$ is $\mathscr{F}(E_p)$. (*Used on p. 299.*)

10-9. EXTENSION LEMMA FOR SECTIONS OF RESTRICTED BUNDLES: Suppose M is a smooth manifold, $E \to M$ is a smooth vector bundle, and $S \subseteq M$ is an embedded submanifold with or without boundary. For any smooth section σ of the restricted bundle $E|_S \to S$, show that there exist a neighborhood U of S in M and a smooth section $\tilde{\sigma}$ of $E|_U$ such that $\sigma = \tilde{\sigma}|_S$. If E has positive rank, show that every smooth section of $E|_S$ extends smoothly to all of M if and only if S is properly embedded.

10-10. Suppose M is a compact smooth manifold and $E \to M$ is a smooth vector bundle of rank k. Use transversality to prove that E admits a smooth section σ with the following property: if $k > \dim M$, then σ is nowhere vanishing; while if $k \leq \dim M$, then the set of points where σ vanishes is a smooth compact codimension-k submanifold of M. Use this to show that M admits a smooth vector field with only finitely many singular points.

10-11. Prove Proposition 10.26 (a bijective bundle homomorphism is a bundle isomorphism).

10-12. Let $\pi \colon E \to M$ and $\tilde{\pi} \colon \tilde{E} \to M$ be two smooth rank-k vector bundles over a smooth manifold M with or without boundary. Suppose $\{U_\alpha\}_{\alpha \in A}$ is an open cover of M such that both E and $\tilde{E}$ admit smooth local trivializations over each U_α. Let $\{\tau_{\alpha\beta}\}$ and $\{\tilde{\tau}_{\alpha\beta}\}$ denote the transition functions determined by the given local trivializations of E and $\tilde{E}$, respectively. Show that E and $\tilde{E}$ are smoothly isomorphic over M if and only if for each $\alpha \in A$ there exists a smooth map $\sigma_\alpha \colon U_\alpha \to \mathrm{GL}(k, \mathbb{R})$ such that

$$\tilde{\tau}_{\alpha\beta}(p) = \sigma_\alpha(p)\tau_{\alpha\beta}(p)\sigma_\beta(p)^{-1}, \qquad p \in U_\alpha \cap U_\beta.$$

10-13. Let $U = \mathbb{S}^1 \smallsetminus \{1\}$ and $V = \mathbb{S}^1 \smallsetminus \{-1\}$, and define $\tau \colon U \cap V \to \mathrm{GL}(1, \mathbb{R})$ by

$$\tau(z) = \begin{cases} (1), & \operatorname{Im} z > 0, \\ (-1), & \operatorname{Im} z < 0. \end{cases}$$

By the result of Problem 10-6, there is a smooth real line bundle $F \to \mathbb{S}^1$ that is trivial over U and V, and has τ as transition function. Show that F is smoothly isomorphic over $\mathbb{S}^1$ to the Möbius bundle of Example 10.3.

10-14. Suppose M is a smooth manifold with or without boundary, and $S \subseteq M$ is an immersed submanifold with or without boundary. Identifying $T_p S$ as a subspace of $T_p M$ for each $p \in S$ in the usual way, show that TS is a smooth subbundle of $TM|_S$. (See Example 10.33.)

10-15. Let V be a finite-dimensional real vector space, and let $G_k(V)$ be the Grassmannian of k-dimensional subspaces of V (see Example 1.36). Let T be the subset of $G_k(V) \times V$ defined by

$$T = \{(S, v) \in G_k(V) \times V : v \in S\}.$$

Show that T is a smooth rank-k subbundle of the product bundle $G_k(V) \times V \to G_k(V)$, and is thus a smooth rank-k vector bundle over $G_k(V)$.

[Remark: T is called the **tautological vector bundle** over $G_k(V)$, because the fiber over each point $S \in G_k(V)$ is S itself.]

10-16. Show that the tautological vector bundle over $G_1(\mathbb{R}^2)$ is smoothly isomorphic to the Möbius bundle. (See Problems 10-1, 10-13, and 10-15.)

10-17. Suppose $M \subseteq \mathbb{R}^n$ is an immersed submanifold. Prove that the ambient tangent bundle $T\mathbb{R}^n|_M$ is isomorphic to the Whitney sum $TM \oplus NM$, where $NM \to M$ is the normal bundle.

10-18. Suppose S is a properly embedded codimension-k submanifold of $\mathbb{R}^n$. Show that the following are equivalent:

(a) There exists a smooth defining function for S on some neighborhood U of S in $\mathbb{R}^n$, that is, a smooth function $\Phi: U \to \mathbb{R}^k$ such that S is a regular level set of Φ.

(b) The normal bundle NS is a trivial vector bundle.

10-19. Suppose $\pi: E \to M$ is a fiber bundle with fiber F. Prove the following:

(a) π is an open quotient map.

(b) If the bundle is smooth, then π is a smooth submersion.

(c) π is a proper map if and only if F is compact.

(d) E is compact if and only if both M and F are compact.

(*Used on p. 560.*)

Chapter 11
The Cotangent Bundle

In this chapter we introduce a construction that is not typically seen in elementary calculus: *tangent covectors*, which are linear functionals on the tangent space at a point $p \in M$. The space of all covectors at p is a vector space called the *cotangent space* at p; in linear-algebraic terms, it is the dual space to $T_p M$. The union of all cotangent spaces at all points of M is a vector bundle called the *cotangent bundle*.

Whereas tangent vectors give us a coordinate-free interpretation of derivatives of curves, it turns out that derivatives of real-valued functions on a manifold are most naturally interpreted as tangent covectors. Thus we define the differential of a real-valued function as a covector field (a smooth section of the cotangent bundle); it is a coordinate-independent analogue of the gradient. We then explore the behavior of covector fields under smooth maps, and show that covector fields on the codomain of a smooth map always pull back to covector fields on the domain.

In the second half of the chapter we introduce line integrals of covector fields, which are the natural generalization of the line integrals of elementary calculus. Then we explore the relationships among three closely related types of covector fields: exact (those that are the differentials of functions), conservative (those whose line integrals around closed curves are zero), and closed (those that satisfy a certain differential equation in coordinates). This leads to a far-reaching generalization of the fundamental theorem of calculus to line integrals on manifolds.

Covectors

Let V be a finite-dimensional vector space. (As usual, all of our vector spaces are assumed to be real.) We define a ***covector on*** V to be a real-valued linear functional on V, that is, a linear map $\omega \colon V \to \mathbb{R}$. The space of all covectors on V is itself a real vector space under the obvious operations of pointwise addition and scalar multiplication. It is denoted by V^* and called the ***dual space of*** V.

The next proposition expresses the most important fact about V^* in the finite-dimensional case. Recall from Exercise B.13 that a linear map is uniquely determined by specifying its values on the elements of any basis.

Proposition 11.1. *Let V be a finite-dimensional vector space. Given any basis* $(E_1, \ldots, E_n)$ *for V, let $\varepsilon^1, \ldots, \varepsilon^n \in V^*$ be the covectors defined by*

$$\varepsilon^i(E_j) = \delta^i_j,$$

*where δ^i_j is the Kronecker delta symbol defined by (4.4). Then $(\varepsilon^1, \ldots, \varepsilon^n)$ is a basis for V^*, called the **dual basis to** (E_j). Therefore, $\dim V^* = \dim V$.*

▶ **Exercise 11.2.** Prove Proposition 11.1.

For example, we can apply this to the standard basis $(e_1, \ldots, e_n)$ for $\mathbb{R}^n$. The dual basis is denoted by $(e^1, \ldots, e^n)$ (note the upper indices), and is called the **standard dual basis**. These basis covectors are the linear functionals on $\mathbb{R}^n$ given by

$$e^i(v) = e^i(v^1, \ldots, v^n) = v^i.$$

In other words, e^i is the linear functional that picks out the ith component of a vector. In matrix notation, a linear map from $\mathbb{R}^n$ to $\mathbb{R}$ is represented by a $1 \times n$ matrix, called a **row matrix**. The basis covectors can therefore also be thought of as the linear functionals represented by the row matrices

$$e^1 = (1\ 0\ \ldots\ 0), \quad e^2 = (0\ 1\ 0\ \ldots\ 0), \quad \ldots, \quad e^n = (0\ \ldots\ 0\ 1).$$

In general, if (E_j) is a basis for V and (ε^i) is its dual basis, then for any vector $v = v^j E_j \in V$, we have (using the summation convention)

$$\varepsilon^i(v) = v^j \varepsilon^i(E_j) = v^j \delta^i_j = v^i.$$

Thus, just as in the case of $\mathbb{R}^n$, the ith basis covector ε^i picks out the ith component of a vector with respect to the basis (E_j). More generally, Proposition 11.1 shows that we can express an arbitrary covector $\omega \in V^*$ in terms of the dual basis as

$$\omega = \omega_i \varepsilon^i, \tag{11.1}$$

where the components are determined by $\omega_i = \omega(E_i)$. The action of ω on a vector $v = v^j E_j$ is

$$\omega(v) = \omega_i v^i. \tag{11.2}$$

We always write basis covectors with upper indices, and components of a covector with lower indices, because this helps to ensure that mathematically meaningful summations such as (11.1) and (11.2) always follow our index conventions.

Suppose V and W are vector spaces and $A \colon V \to W$ is a linear map. We define a linear map $A^* \colon W^* \to V^*$, called the **dual map** or **transpose of A**, by

$$(A^* \omega)(v) = \omega(Av) \quad \text{for } \omega \in W^*, \ v \in V.$$

▶ **Exercise 11.3.** Show that $A^* \omega$ is actually a linear functional on V, and that A^* is a linear map.

Proposition 11.4. *The dual map satisfies the following properties*:

(a) $(A \circ B)^* = B^* \circ A^*$.
(b) $(\mathrm{Id}_V)^* \colon V^* \to V^*$ *is the identity map of* V^*.

▶ **Exercise 11.5.** Prove the preceding proposition.

Corollary 11.6. *The assignment that sends a vector space to its dual space and a linear map to its dual map is a contravariant functor from the category of real vector spaces to itself.* □

Apart from the fact that the dimension of V^* is the same as that of V, the second most important fact about dual spaces is the following characterization of the *second dual space* $V^{**} = (V^*)^*$. For each vector space V there is a natural, basis-independent map $\xi \colon V \to V^{**}$, defined as follows. For each vector $v \in V$, define a linear functional $\xi(v) \colon V^* \to \mathbb{R}$ by

$$\xi(v)(\omega) = \omega(v) \quad \text{for } \omega \in V^*. \tag{11.3}$$

▶ **Exercise 11.7.** Let V be a vector space.

(a) For any $v \in V$, show that $\xi(v)(\omega)$ depends linearly on ω, so $\xi(v) \in V^{**}$.
(b) Show that the map $\xi \colon V \to V^{**}$ is linear.

Proposition 11.8. *For any finite-dimensional vector space* V*, the map* $\xi \colon V \to V^{**}$ *is an isomorphism.*

Proof. Because $\dim V = \dim V^{**}$, it suffices to verify that ξ is injective (see Exercise B.22(c)). Suppose $v \in V$ is not zero. Extend v to a basis $(v = E_1, \ldots, E_n)$ for V, and let $(\varepsilon^1, \ldots, \varepsilon^n)$ denote the dual basis for V^*. Then $\xi(v) \neq 0$ because

$$\xi(v)(\varepsilon^1) = \varepsilon^1(v) = \varepsilon^1(E_1) = 1. \qquad \square$$

The preceding proposition shows that when V is finite-dimensional, we can unambiguously identify V^{**} with V itself, because the map ξ is canonically defined, without reference to any basis. It is important to observe that although V^* is also isomorphic to V (for the simple reason that any two finite-dimensional vector spaces of the same dimension are isomorphic), there is no *canonical* isomorphism $V \cong V^*$. One way to make this statement precise is indicated in Problem 11-1. Note also that the conclusion of Proposition 11.8 is always false when V is infinite-dimensional (see Problem 11-2).

Because of Proposition 11.8, the real number $\omega(v)$ obtained by applying a covector ω to a vector v is sometimes denoted by either of the more symmetric-looking notations $\langle \omega, v \rangle$ and $\langle v, \omega \rangle$; both expressions can be thought of either as the action of the covector $\omega \in V^*$ on the vector $v \in V$, or as the action of the linear functional $\xi(v) \in V^{**}$ on the element $\omega \in V^*$. There should be no cause for confusion with the use of the same angle bracket notation for inner products: whenever one of the arguments is a vector and the other a covector, the notation $\langle \omega, v \rangle$ is always to be interpreted as the natural pairing between vectors and covectors, not as an inner product. We typically omit any mention of the map ξ, and think of $v \in V$ either as a vector or as a linear functional on V^*, depending on the context.

There is also a symmetry between bases and dual bases for a finite-dimensional vector space V: any basis for V determines a dual basis for V^*, and conversely, any basis for V^* determines a dual basis for $V^{**} = V$. If (ε^i) is the basis for V^* dual to a basis (E_j) for V, then (E_j) is the basis dual to (ε^i), because both statements are equivalent to the relation $\langle \varepsilon^i, E_j \rangle = \delta^i_j$.

Tangent Covectors on Manifolds

Now let M be a smooth manifold with or without boundary. For each $p \in M$, we define the *cotangent space at p*, denoted by $T^*_p M$, to be the dual space to $T_p M$:

$$T^*_p M = (T_p M)^*.$$

Elements of $T^*_p M$ are called *tangent covectors at p*, or just *covectors at p*.

Given smooth local coordinates (x^i) on an open subset $U \subseteq M$, for each $p \in U$ the coordinate basis $(\partial/\partial x^i |_p)$ gives rise to a dual basis for $T^*_p M$, which we denote for the moment by $(\lambda^i |_p)$. (In a short while, we will come up with a better notation.) Any covector $\omega \in T^*_p M$ can thus be written uniquely as $\omega = \omega_i \lambda^i |_p$, where

$$\omega_i = \omega \left(\frac{\partial}{\partial x^i} \bigg|_p \right).$$

Suppose now that $(\tilde{x}^j)$ is another set of smooth coordinates whose domain contains p, and let $(\tilde{\lambda}^j |_p)$ denote the basis for $T^*_p M$ dual to $(\partial/\partial \tilde{x}^j |_p)$. We can compute the components of the same covector ω with respect to the new coordinate system as follows. First observe that the computations in Chapter 3 show that the coordinate vector fields transform as follows:

$$\frac{\partial}{\partial x^i} \bigg|_p = \frac{\partial \tilde{x}^j}{\partial x^i}(p) \frac{\partial}{\partial \tilde{x}^j} \bigg|_p . \tag{11.4}$$

(Here we use the same notation p to denote either a point in M or its coordinate representation as appropriate.) Writing ω in both systems as $\omega = \omega_i \lambda^i |_p = \tilde{\omega}_j \tilde{\lambda}^j |_p$, we can use (11.4) to compute the components ω_i in terms of $\tilde{\omega}_j$:

$$\omega_i = \omega \left(\frac{\partial}{\partial x^i} \bigg|_p \right) = \omega \left(\frac{\partial \tilde{x}^j}{\partial x^i}(p) \frac{\partial}{\partial \tilde{x}^j} \bigg|_p \right) = \frac{\partial \tilde{x}^j}{\partial x^i}(p) \tilde{\omega}_j . \tag{11.5}$$

As we mentioned in Chapter 3, in the early days of smooth manifold theory, before most of the abstract coordinate-free definitions we are using were developed, mathematicians tended to think of a tangent vector at a point p as an assignment of an n-tuple of real numbers to each smooth coordinate system, with the property that the n-tuples $(v^1, \ldots, v^n)$ and $(\tilde{v}^1, \ldots, \tilde{v}^n)$ assigned to two different coordinate systems (x^i) and $(\tilde{x}^j)$ were related by the transformation law that we derived in

Chapter 3:

$$\widetilde{v}^j = \frac{\partial \widetilde{x}^j}{\partial x^i}(p)v^i. \tag{11.6}$$

Similarly, a tangent covector was thought of as an n-tuple $(\omega_1, \ldots, \omega_n)$ that transforms, by virtue of (11.5), according to the following slightly different rule:

$$\omega_i = \frac{\partial \widetilde{x}^j}{\partial x^i}(p)\widetilde{\omega}_j. \tag{11.7}$$

Since the transformation law (11.4) for the coordinate partial derivatives follows directly from the chain rule, it can be thought of as fundamental. Thus it became customary to call tangent covectors *covariant vectors* because their components transform in the same way as ("vary with") the coordinate partial derivatives, with the Jacobian matrix $(\partial \widetilde{x}^j / \partial x^i)$ multiplying the objects associated with the "new" coordinates $(\widetilde{x}^j)$ to obtain those associated with the "old" coordinates (x^i). Analogously, tangent vectors were called *contravariant vectors*, because their components transform in the opposite way. (Remember, it was the component n-tuples that were thought of as the objects of interest.) Admittedly, these terms do not make a lot of sense, but by now they are well entrenched, and we will see them again in Chapter 12. Note that this use of the terms covariant and contravariant has nothing to do with the covariant and contravariant functors of category theory!

Covector Fields

For any smooth manifold M with or without boundary, the disjoint union

$$T^*M = \coprod_{p \in M} T_p^* M$$

is called the *cotangent bundle of M*. It has a natural projection map $\pi \colon T^*M \to M$ sending $\omega \in T_p^*M$ to $p \in M$. As above, given any smooth local coordinates (x^i) on an open subset $U \subseteq M$, for each $p \in U$ we denote the basis for T_p^*M dual to $(\partial/\partial x^i|_p)$ by $(\lambda^i|_p)$. This defines n maps $\lambda^1, \ldots, \lambda^n \colon U \to T^*M$, called *coordinate covector fields*.

Proposition 11.9 (The Cotangent Bundle as a Vector Bundle). *Let M be a smooth n-manifold with or without boundary. With its standard projection map and the natural vector space structure on each fiber, the cotangent bundle T^*M has a unique topology and smooth structure making it into a smooth rank-n vector bundle over M for which all coordinate covector fields are smooth local sections.*

Proof. The proof is just like that of Theorem 10.4. Given a smooth chart (U, φ) on M, with coordinate functions (x^i), define $\Phi \colon \pi^{-1}(U) \to U \times \mathbb{R}^n$ by

$$\Phi\left(\xi_i \lambda^i\big|_p\right) = \left(p, (\xi_1, \ldots, \xi_n)\right),$$

where λ^i is the ith coordinate covector field associated with (x^i). Suppose $(\tilde{U}, \tilde{\varphi})$ is another smooth chart with coordinate functions $(\tilde{x}^j)$, and let $\tilde{\Phi} \colon \pi^{-1}(\tilde{U}) \to \tilde{U} \times \mathbb{R}^n$ be defined analogously. On $\pi^{-1}(U \cap \tilde{U})$, it follows from (11.5) that

$$\Phi \circ \tilde{\Phi}^{-1}\left(p, (\tilde{\xi}_1, \dots, \tilde{\xi}_n)\right) = \left(p, \left(\frac{\partial \tilde{x}^j}{\partial x^1}(p)\tilde{\xi}_j, \dots, \frac{\partial \tilde{x}^j}{\partial x^n}(p)\tilde{\xi}_j\right)\right).$$

The $\mathrm{GL}(n, \mathbb{R})$-valued function $(\partial \tilde{x}^j / \partial x^i)$ is smooth, so it follows from the vector bundle chart lemma that T^*M has a smooth structure making it into a smooth vector bundle for which the maps Φ are smooth local trivializations. Uniqueness follows as in the proof of Proposition 10.24.   □

▶ **Exercise 11.10.** Suppose M is a smooth manifold and $E \to M$ is a smooth vector bundle over M. Define the **dual bundle to** E to be the bundle $E^* \to M$ whose total space is the disjoint union $E^* = \coprod_{p \in M} E_p^*$, where E_p^* is the dual space to E_p, with the obvious projection. Show that $E^* \to M$ is a smooth vector bundle, whose transition functions are given by $\tau^*(p) = (\tau(p)^{-1})^T$ for any transition function $\tau \colon U \to \mathrm{GL}(k, \mathbb{R})$ of E.

As in the case of the tangent bundle, smooth local coordinates for M yield smooth local coordinates for its cotangent bundle. If (x^i) are smooth coordinates on an open subset $U \subseteq M$, Corollary 10.21 shows that the map from $\pi^{-1}(U)$ to $\mathbb{R}^{2n}$ given by

$$\xi_i \lambda^i \big|_p \mapsto (x^1(p), \dots, x^n(p), \xi_1, \dots, \xi_n)$$

is a smooth coordinate chart for T^*M. We call (x^i, ξ_i) the **natural coordinates for** T^*M associated with (x^i). (In this situation, we must forgo our insistence that coordinate functions have upper indices, because the fiber coordinates ξ_i are already required by our index conventions to have lower indices. Nonetheless, the convention still holds that each index to be summed over in a given term appears once as a superscript and once as a subscript.)

A (local or global) section of T^*M is called a **covector field** or a **(differential) 1-form**. (The reason for the latter terminology will become clear in Chapter 14, when we define differential k-forms for $k > 1$.) Like sections of other bundles, covector fields without further qualification are assumed to be merely continuous; when we make different assumptions, we use the terms **rough covector field** and **smooth covector field** with the obvious meanings. As we did with vector fields, we write the value of a covector field ω at a point $p \in M$ as ω_p instead of $\omega(p)$, to avoid conflict with the notation for the action of a covector on a vector. If ω itself has subscripts or superscripts, we usually use the notation $\omega|_p$ instead. In any smooth local coordinates on an open subset $U \subseteq M$, a (rough) covector field ω can be written in terms of the coordinate covector fields (λ^i) as $\omega = \omega_i \lambda^i$ for n functions $\omega_i \colon U \to \mathbb{R}$ called the **component functions of** ω. They are characterized by

$$\omega_i(p) = \omega_p\left(\frac{\partial}{\partial x^i}\bigg|_p\right).$$

If ω is a (rough) covector field and X is a vector field on M, then we can form a function $\omega(X)\colon M \to \mathbb{R}$ by

$$\omega(X)(p) = \omega_p(X_p), \quad p \in M.$$

If we write $\omega = \omega_i \lambda^i$ and $X = X^j \partial/\partial x^j$ in terms of local coordinates, then $\omega(X)$ has the local coordinate representation $\omega(X) = \omega_i X^i$.

Just as in the case of vector fields, there are several ways to check for smoothness of a covector field.

Proposition 11.11 (Smoothness Criteria for Covector Fields). *Let M be a smooth manifold with or without boundary, and let $\omega\colon M \to T^*M$ be a rough covector field. The following are equivalent:*

(a) *ω is smooth.*
(b) *In every smooth coordinate chart, the component functions of ω are smooth.*
(c) *Each point of M is contained in some coordinate chart in which ω has smooth component functions.*
(d) *For every smooth vector field $X \in \mathfrak{X}(M)$, the function $\omega(X)$ is smooth on M.*
(e) *For every open subset $U \subseteq M$ and every smooth vector field X on U, the function $\omega(X)\colon U \to \mathbb{R}$ is smooth on U.*

▶ **Exercise 11.12.** Prove this proposition. [Suggestion: try proving (a) $\Rightarrow$ (b) $\Rightarrow$ (c) $\Rightarrow$ (a), and (c) $\Rightarrow$ (d) $\Rightarrow$ (e) $\Rightarrow$ (b). The only tricky part is (d) $\Rightarrow$ (e); look at the proof of Proposition 8.14 for ideas.]

Of course, since any open subset of a smooth manifold (with boundary) is again a smooth manifold (with boundary), the preceding proposition applies equally well to covector fields defined only on some open subset of M.

Coframes

Let M be a smooth manifold with or without boundary, and let $U \subseteq M$ be an open subset. A *local coframe for M over U* is an ordered n-tuple of covector fields $(\varepsilon^1, \ldots, \varepsilon^n)$ defined on U such that $(\varepsilon^i|_p)$ forms a basis for T_p^*M at each point $p \in U$. If $U = M$, it is called a *global coframe*. (A local coframe for M is just a local frame for the vector bundle T^*M, in the terminology of Chapter 10.)

Example 11.13 (Coordinate Coframes). For any smooth chart $(U, (x^i))$, the coordinate covector fields (λ^i) defined above constitute a local coframe over U, called a *coordinate coframe*. By Proposition 11.11(c), every coordinate frame is smooth, because its component functions in the given chart are constants.   //

Given a local frame $(E_1, \ldots, E_n)$ for TM over an open subset U, there is a uniquely determined (rough) local coframe $(\varepsilon^1, \ldots, \varepsilon^n)$ over U such that $(\varepsilon^i|_p)$ is the dual basis to $(E_i|_p)$ for each $p \in U$, or equivalently $\varepsilon^i(E_j) = \delta_j^i$. This coframe is called the *coframe dual to (E_i)*. Conversely, if we start with a local coframe (ε^i)

over an open subset $U \subseteq M$, there is a uniquely determined local frame (E_i), called the *frame dual to* (ε^i), determined by $\varepsilon^i(E_j) = \delta^i_j$. For example, in a smooth chart, the coordinate frame $(\partial/\partial x^i)$ and the coordinate coframe (λ^i) are dual to each other.

Lemma 11.14. *Let M be a smooth manifold with or without boundary. If (E_i) is a rough local frame over an open subset $U \subseteq M$ and (ε^i) is its dual coframe, then (E_i) is smooth if and only if (ε^i) is smooth.*

Proof. It suffices to show that for each $p \in U$, the frame (E_i) is smooth in a neighborhood of p if and only if (ε^i) is. Given $p \in U$, let $(V, (x^i))$ be a smooth coordinate chart such that $p \in V \subseteq U$. In V, we can write

$$E_i = a_i^k \frac{\partial}{\partial x^k}, \qquad \varepsilon^j = b_l^j \lambda^l,$$

for some matrices of real-valued functions (a_i^k) and (b_l^j) defined on V. By virtue of Propositions 8.1 and 11.11, the vector fields E_i are smooth on V if and only if the functions a_i^k are smooth, and the covector fields ε^j are smooth on V if and only if the functions b_l^j are smooth. The fact that $\varepsilon^j(E_i) = \delta^j_i$ implies that the matrices (a_i^k) and (b_l^j) are inverses of each other. Because matrix inversion is a smooth map from $\mathrm{GL}(n, \mathbb{R})$ to itself, either one of these matrix-valued functions is smooth if and only if the other one is smooth. $\qquad\square$

Given a local coframe (ε^i) over an open subset $U \subseteq M$, every (rough) covector field ω on U can be expressed in terms of the coframe as $\omega = \omega_i \varepsilon^i$ for some functions $\omega_1, \ldots, \omega_n \colon U \to \mathbb{R}$, called the *component functions of ω with respect to the given coframe*. The component functions are determined by $\omega_i = \omega(E_i)$, where (E_i) is the frame dual to (ε^i). This leads to another way of characterizing smoothness of covector fields.

Proposition 11.15 (Coframe Criterion for Smoothness of Covector Fields). *Let M be a smooth manifold with or without boundary, and let ω be a rough covector field on M. If (ε^i) is a smooth coframe on an open subset $U \subseteq M$, then ω is smooth on U if and only if its component functions with respect to (ε^i) are smooth.*

▶ **Exercise 11.16.** Prove the preceding proposition.

We denote the real vector space of all smooth covector fields on M by $\mathfrak{X}^*(M)$. As smooth sections of a vector bundle, elements of $\mathfrak{X}^*(M)$ can be multiplied by smooth real-valued functions: if $f \in C^\infty(M)$ and $\omega \in \mathfrak{X}^*(M)$, the covector field $f\omega$ is defined by

$$(f\omega)_p = f(p)\omega_p. \tag{11.8}$$

Because it is the space of smooth sections of a vector bundle, $\mathfrak{X}^*(M)$ is a module over $C^\infty(M)$.

Geometrically, we think of a vector field on M as an arrow attached to each point of M. What kind of geometric picture can we form of a covector field? The key idea is that a nonzero linear functional $\omega_p \in T_p^* M$ is completely determined by

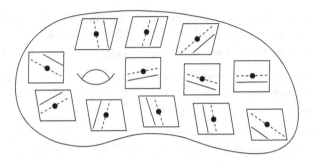

Fig. 11.1   A covector field

two pieces of data: its kernel, which is a linear hyperplane in T_pM (a codimension-1 linear subspace); and the set of vectors v for which $\omega_p(v) = 1$, which is an affine hyperplane parallel to the kernel (Fig. 11.1). (Actually, the set where $\omega_p(v) = 1$ alone suffices, but it is useful to visualize the two parallel hyperplanes.) The value of $\omega_p(v)$ for any other vector v is then obtained by linear interpolation or extrapolation.

Thus, you can visualize a covector field as defining a pair of hyperplanes in each tangent space, one through the origin and another parallel to it, and varying continuously from point to point. Where the covector field is small, one of the hyperplanes becomes very far from the kernel, eventually disappearing altogether at points where the covector field takes the value zero.

The Differential of a Function

In elementary calculus, the gradient of a smooth real-valued function f on $\mathbb{R}^n$ is defined as the vector field whose components are the partial derivatives of f. In our notation, this would read

$$\operatorname{grad} f = \sum_{i=1}^{n} \frac{\partial f}{\partial x^i} \frac{\partial}{\partial x^i}. \tag{11.9}$$

Unfortunately, in this form, the gradient does not make sense independently of coordinates. (The fact that it violates our index conventions is a strong clue.)

▶ **Exercise 11.17.** Let $f(x, y) = x^2$ on $\mathbb{R}^2$, and let X be the vector field

$$X = \operatorname{grad} f = 2x \frac{\partial}{\partial x}.$$

Compute the coordinate expression for X in polar coordinates (on some open subset on which they are defined) using (11.4) and show that it is *not* equal to

$$\frac{\partial f}{\partial r} \frac{\partial}{\partial r} + \frac{\partial f}{\partial \theta} \frac{\partial}{\partial \theta}.$$

Although the partial derivatives of a smooth function cannot be interpreted in a coordinate-independent way as the components of a vector field, it turns out that they *can* be interpreted as the components of a covector field. This is the most important application of covector fields.

Let f be a smooth real-valued function on a smooth manifold M with or without boundary. (As usual, all of this discussion applies to functions defined on an open subset $U \subseteq M$, simply by replacing M with U throughout.) We define a covector field df, called the **differential of** f, by

$$df_p(v) = vf \quad \text{for } v \in T_p M.$$

(We will discuss the relationship between this differential and the differential of a smooth map below; see the paragraph just after Exercise 11.24.)

Proposition 11.18. *The differential of a smooth function is a smooth covector field.*

Proof. It is straightforward to verify that at each point $p \in M$, $df_p(v)$ depends linearly on v, so that df_p is indeed a covector at p. To see that df is smooth, we use Proposition 11.11(d): for any smooth vector field X on M, the function $df(X)$ is smooth because it is equal to Xf. □

To see what df looks like more concretely, we need to compute its coordinate representation. Let (x^i) be smooth coordinates on an open subset $U \subseteq M$, and let (λ^i) be the corresponding coordinate coframe on U. Write df in coordinates as $df_p = A_i(p)\lambda^i|_p$ for some functions $A_i \colon U \to \mathbb{R}$; then the definition of df implies

$$A_i(p) = df_p\left(\frac{\partial}{\partial x^i}\bigg|_p\right) = \frac{\partial}{\partial x^i}\bigg|_p f = \frac{\partial f}{\partial x^i}(p).$$

This yields the following formula for the coordinate representation of df:

$$df_p = \frac{\partial f}{\partial x^i}(p)\lambda^i\big|_p. \tag{11.10}$$

Thus, the component functions of df in any smooth coordinate chart are the partial derivatives of f with respect to those coordinates. Because of this, we can think of df as an analogue of the classical gradient, reinterpreted in a way that makes coordinate-independent sense on a manifold.

If we apply (11.10) to the special case in which f is one of the coordinate functions $x^j \colon U \to \mathbb{R}$, we obtain

$$dx^j\big|_p = \frac{\partial x^j}{\partial x^i}(p)\lambda^i\big|_p = \delta_i^j \lambda^i\big|_p = \lambda^j\big|_p.$$

In other words, *the coordinate covector field λ^j is none other than the differential dx^j!* Therefore, the formula (11.10) for df_p can be rewritten as

$$df_p = \frac{\partial f}{\partial x^i}(p)dx^i\big|_p,$$

or as an equation between covector *fields* instead of covectors:

$$df = \frac{\partial f}{\partial x^i}dx^i. \tag{11.11}$$

In particular, in the 1-dimensional case, this reduces to

$$df = \frac{df}{dx}dx.$$

Thus, we have recovered the familiar classical expression for the differential of a function f in coordinates. Henceforth, we abandon the notation λ^i for the coordinate coframe, and use dx^i instead.

Example 11.19. If $f(x, y) = x^2 y \cos x$ on $\mathbb{R}^2$, then df is given by the formula

$$df = \frac{\partial \left(x^2 y \cos x\right)}{\partial x} dx + \frac{\partial \left(x^2 y \cos x\right)}{\partial y} dy$$

$$= \left(2xy \cos x - x^2 y \sin x\right) dx + x^2 \cos x\, dy. \qquad //$$

Proposition 11.20 (Properties of the Differential). *Let M be a smooth manifold with or without boundary, and let $f, g \in C^\infty(M)$.*

(a) *If a and b are constants, then $d(af + bg) = a\, df + b\, dg$.*
(b) *$d(fg) = f\, dg + g\, df$.*
(c) *$d(f/g) = (g\, df - f\, dg)/g^2$ on the set where $g \neq 0$.*
(d) *If $J \subseteq \mathbb{R}$ is an interval containing the image of f, and $h \colon J \to \mathbb{R}$ is a smooth function, then $d(h \circ f) = (h' \circ f)\, df$.*
(e) *If f is constant, then $df = 0$.*

▶ **Exercise 11.21.** Prove Proposition 11.20.

One very important property of the differential is the following characterization of smooth functions with vanishing differentials.

Proposition 11.22 (Functions with Vanishing Differentials). *If f is a smooth real-valued function on a smooth manifold M with or without boundary, then $df = 0$ if and only if f is constant on each component of M.*

Proof. It suffices to assume that M is connected and show that $df = 0$ if and only if f is constant. One direction is immediate: if f is constant, then $df = 0$ by Proposition 11.20(e). Conversely, suppose $df = 0$, let $p \in M$, and let $\mathcal{C} = \{q \in M : f(q) = f(p)\}$. If q is any point in $\mathcal{C}$, let U be a smooth coordinate ball (or half-ball, in case $q \in \partial M$) centered at q. From (11.11) we see that $\partial f/\partial x^i \equiv 0$ in U for each i, so by elementary calculus f is constant on U. This shows that $\mathcal{C}$ is open, and since it is closed by continuity, it must be all of M. Thus, f is everywhere equal to the constant $f(p)$. □

In elementary calculus, one thinks of df as an approximation for the small change in the value of f caused by small changes in the independent variables x^i. In our present context, df has the same meaning, provided we interpret everything appropriately. Suppose M is a smooth manifold and $f \in C^\infty(M)$, and let p be a point in M. By choosing smooth coordinates on a neighborhood of p, we can think of f as a function on an open subset $U \subseteq \mathbb{R}^n$. Recall that $dx^i|_p$ is the linear functional that picks out the ith component of a tangent vector at p. Writing

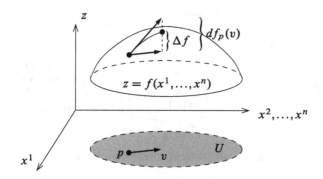

Fig. 11.2 The differential as an approximation to Δf

$\Delta f = f(p + v) - f(p)$ for $v \in \mathbb{R}^n$, Taylor's theorem shows that Δf is well approximated when v is small by

$$\Delta f \approx \frac{\partial f}{\partial x^i}(p)v^i = \frac{\partial f}{\partial x^i}(p)dx^i\big|_p(v) = df_p(v)$$

(where now we are considering v as an element of $T_p\mathbb{R}^n$ via our usual identification $T_p\mathbb{R}^n \leftrightarrow \mathbb{R}^n$). In other words, df_p is the linear functional that best approximates Δf near p (Fig. 11.2). The great power of the concept of the differential comes from the fact that we can define df invariantly on any manifold, without resorting to vague arguments involving infinitesimals.

The next result is an analogue of Proposition 3.24 for the differential.

Proposition 11.23 (Derivative of a Function Along a Curve). *Suppose M is a smooth manifold with or without boundary, $\gamma: J \to M$ is a smooth curve, and $f: M \to \mathbb{R}$ is a smooth function. Then the derivative of the real-valued function $f \circ \gamma: J \to \mathbb{R}$ is given by*

$$(f \circ \gamma)'(t) = df_{\gamma(t)}\big(\gamma'(t)\big). \tag{11.12}$$

Proof. See Fig. 11.3. Directly from the definitions, for any $t_0 \in J$,

$$
\begin{aligned}
df_{\gamma(t_0)}\big(\gamma'(t_0)\big) &= \gamma'(t_0)f && \text{(definition of } df) \\
&= d\gamma_{t_0}\left(\frac{d}{dt}\Big|_{t_0}\right)f && \text{(definition of } \gamma'(t)) \\
&= \frac{d}{dt}\Big|_{t_0}(f \circ \gamma) && \text{(definition of } d\gamma) \\
&= (f \circ \gamma)'(t_0) && \text{(definition of } d/dt|_{t_0}). \qquad \square
\end{aligned}
$$

▶ **Exercise 11.24.** For a smooth real-valued function $f: M \to \mathbb{R}$, show that $p \in M$ is a critical point of f if and only if $df_p = 0$.

You may have noticed that for a smooth real-valued function $f: M \to \mathbb{R}$, we now have two different definitions for the differential of f at a point $p \in M$. In

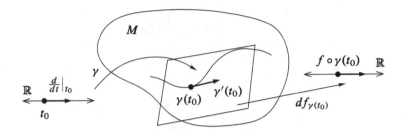

Fig. 11.3 Derivative of a function along a curve

Chapter 3, we defined df_p as a linear map from $T_p M$ to $T_{f(p)}\mathbb{R}$. In this chapter, we defined df_p as a covector at p, which is to say a linear map from $T_p M$ to $\mathbb{R}$. These are really the same object, once we take into account the canonical identification between $\mathbb{R}$ and $T_{f(p)}\mathbb{R}$; one easy way to see this is to note that both are represented in coordinates by the row matrix whose components are the partial derivatives of f.

Similarly, if γ is a smooth curve in M, we have two different meanings for the expression $(f \circ \gamma)'(t)$. On the one hand, $f \circ \gamma$ can be interpreted as a smooth curve in $\mathbb{R}$, and thus $(f \circ \gamma)'(t)$ is its velocity at the point $f \circ \gamma(t)$, which is an element of the tangent space $T_{f \circ \gamma(t)}\mathbb{R}$. Proposition 3.24 shows that this tangent vector is equal to $df_{\gamma(t)}(\gamma'(t))$, thought of as an element of $T_{f \circ \gamma(t)}\mathbb{R}$. On the other hand, $f \circ \gamma$ can also be considered simply as a real-valued function of one real variable, and then $(f \circ \gamma)'(t)$ is just its ordinary derivative. Proposition 11.23 shows that this derivative is equal to $df_{\gamma(t)}(\gamma'(t))$, thought of as a real number. Which of these interpretations we choose depends on the purpose we have in mind.

Pullbacks of Covector Fields

As we have seen, a smooth map yields a linear map on tangent vectors called the differential. Dualizing this leads to a linear map on covectors going in the opposite direction.

Let $F\colon M \to N$ be a smooth map between smooth manifolds with or without boundary, and let $p \in M$ be arbitrary. The differential $dF_p\colon T_p M \to T_{F(p)}N$ yields a dual linear map

$$dF_p^*\colon T_{F(p)}^* N \to T_p^* M,$$

called the **(pointwise) pullback by F at p**, or the **cotangent map of F**. Unraveling the definitions, we see that dF_p^* is characterized by

$$dF_p^*(\omega)(v) = \omega(dF_p(v)), \quad \text{for } \omega \in T_{F(p)}^* N,\ v \in T_p M.$$

Observe that the assignments $(M, p) \mapsto T_p^* M$ and $F \mapsto dF_p^*$ yield a contravariant functor from the category of pointed smooth manifolds to the category of real vector spaces. Because of this, the convention of calling elements of $T^* M$ "covariant vectors" is particularly unfortunate; but this terminology is so deeply entrenched that one has no choice but to go along with it.

When we discussed vector fields, we made a point of noting that pushforwards of vector fields under smooth maps are defined only in the special cases of diffeomorphisms or Lie group homomorphisms. The surprising thing about covectors is that covector fields *always* pull back to covector fields. Given a smooth map $F : M \to N$ and a covector field ω on N, define a rough covector field $F^*\omega$ on M, called the **pullback of ω by F**, by

$$(F^*\omega)_p = dF_p^* (\omega_{F(p)}). \tag{11.13}$$

It acts on a vector $v \in T_p M$ by

$$(F^*\omega)_p(v) = \omega_{F(p)} (dF_p(v)).$$

In contrast to the vector field case, there is no ambiguity here about what point to pull back from: the value of $F^*\omega$ at p is the pullback of ω at $F(p)$. We will prove in Proposition 11.26 below that $F^*\omega$ is continuous, and is smooth when ω is smooth. Before we do so, let us prove two important properties of the pullback.

Proposition 11.25. *Let $F : M \to N$ be a smooth map between smooth manifolds with or without boundary. Suppose u is a continuous real-valued function on N, and ω is a covector field on N. Then*

$$F^*(u\omega) = (u \circ F)F^*\omega. \tag{11.14}$$

If in addition u is smooth, then

$$F^*du = d(u \circ F). \tag{11.15}$$

Proof. To prove (11.14) we compute

$$
\begin{aligned}
\left(F^*(u\omega)\right)_p &= dF_p^* \left((u\omega)_{F(p)}\right) && \text{(by (11.13))}\\
&= dF_p^* \left(u\big(F(p)\big)\omega_{F(p)}\right) && \text{(by (11.8))}\\
&= u\big(F(p)\big)dF_p^* \left(\omega_{F(p)}\right) && \text{(by linearity of } dF_p^*)\\
&= u\big(F(p)\big)(F^*\omega)_p && \text{(by (11.13))}\\
&= \left((u \circ F)F^*\omega\right)_p && \text{(by (11.8))}.
\end{aligned}
$$

For (11.15), we let $v \in T_p M$ be arbitrary, and compute

$$
\begin{aligned}
(F^*du)_p(v) &= \left(dF_p^* \left(du_{F(p)}\right)\right)(v) && \text{(by (11.13))}\\
&= du_{F(p)} \left(dF_p(v)\right) && \text{(by definition of } dF_p^*)\\
&= dF_p(v)u && \text{(by definition of } du)\\
&= v(u \circ F) && \text{(by definition of } dF_p)\\
&= d(u \circ F)_p(v) && \text{(by definition of } d(u \circ F)). \qquad \square
\end{aligned}
$$

Proposition 11.26. *Suppose $F : M \to N$ is a smooth map between smooth manifolds with or without boundary, and let ω be a covector field on N. Then $F^*\omega$ is a (continuous) covector field on M. If ω is smooth, then so is $F^*\omega$.*

Proof. Let $p \in M$ be arbitrary, and choose smooth coordinates (y^j) for N in a neighborhood V of $F(p)$. Let $U = F^{-1}(V)$, which is a neighborhood of p. Writing ω in coordinates as $\omega = \omega_j \, dy^j$ for continuous functions ω_j on V and using Proposition 11.25 twice (applied to $F|_U$), we have the following computation in U:

$$F^*\omega = F^*\left(\omega_j dy^j\right) = (\omega_j \circ F)F^* dy^j = (\omega_j \circ F)d\left(y^j \circ F\right). \qquad (11.16)$$

This expression is continuous, and is smooth if ω is smooth.   □

Formula (11.16) for the pullback of a covector field can also be written in the following way:

$$F^*\omega = (\omega_j \circ F)\,d\left(y^j \circ F\right) = (\omega_j \circ F)\,dF^j, \qquad (11.17)$$

where F^j is the jth component function of F in these coordinates. Using either of these formulas, the computation of pullbacks in coordinates is exceedingly simple, as the next example shows.

Example 11.27. Let $F: \mathbb{R}^3 \to \mathbb{R}^2$ be the map given by

$$(u, v) = F(x, y, z) = \left(x^2 y, y \sin z\right),$$

and let $\omega \in \mathfrak{X}^*\left(\mathbb{R}^2\right)$ be the covector field

$$\omega = u \, dv + v \, du.$$

According to (11.16), the pullback $F^*\omega$ is given by

$$\begin{aligned}
F^*\omega &= (u \circ F)d(v \circ F) + (v \circ F)d(u \circ F) \\
&= \left(x^2 y\right) d(y \sin z) + (y \sin z)d\left(x^2 y\right) \\
&= x^2 y(\sin z \, dy + y \cos z \, dz) + y \sin z \left(2xy \, dx + x^2 \, dy\right) \\
&= 2xy^2 \sin z \, dx + 2x^2 y \sin z \, dy + x^2 y^2 \cos z \, dz.
\end{aligned}$$

//

In other words, to compute $F^*\omega$, all you need to do is substitute the component functions of F for the coordinate functions of N everywhere they appear in ω!

This also yields an easy way to remember the transformation law for a covector field under a change of coordinates. Again, an example will convey the idea better than a general formula.

Example 11.28. Let (r, θ) be polar coordinates on, say, the right half-plane $H = \{(x, y) : x > 0\}$. We can think of the change of coordinates $(x, y) = (r \cos \theta, r \sin \theta)$ as the coordinate expression for the identity map of H, but using (r, θ) as coordinates for the domain and (x, y) for the codomain. Then the pullback formula (11.17) tells us that we can compute the polar coordinate expression for a covector field simply by substituting $x = r \cos \theta$, $y = r \sin \theta$ and expanding. For example,

$$\begin{aligned}
x \, dy - y \, dx &= \mathrm{Id}^*(x \, dy - y \, dx) \\
&= (r \cos \theta)d(r \sin \theta) - (r \sin \theta)d(r \cos \theta)
\end{aligned}$$

$$= (r\cos\theta)(\sin\theta\,dr + r\cos\theta\,d\theta) - (r\sin\theta)(\cos\theta\,dr - r\sin\theta\,d\theta)$$

$$= (r\cos\theta\sin\theta - r\sin\theta\cos\theta)\,dr + \left(r^2\cos^2\theta + r^2\sin^2\theta\right)\,d\theta$$

$$= r^2\,d\theta. \hspace{6cm} /\!/$$

Restricting Covector Fields to Submanifolds

In Chapter 8, we considered the conditions under which a vector field restricts to a submanifold. The restriction of covector fields to submanifolds is much simpler.

Suppose M is a smooth manifold with or without boundary, $S \subseteq M$ is an immersed submanifold with or without boundary, and $\iota\colon S \hookrightarrow M$ is the inclusion map. If ω is any smooth covector field on M, the pullback by ι yields a smooth covector field $\iota^*\omega$ on S. To see what this means, let $v \in T_pS$ be arbitrary, and compute

$$(\iota^*\omega)_p(v) = \omega_p\left(d\iota_p(v)\right) = \omega_p(v),$$

since $d\iota_p\colon T_pS \to T_pM$ is just the inclusion map, under our usual identification of T_pS with a subspace of T_pM. Thus, $\iota^*\omega$ is just the restriction of ω to vectors tangent to S. For this reason, $\iota^*\omega$ is often called the **restriction of ω to S**. Be warned, however, that $\iota^*\omega$ might equal zero at a given point of S, even though *considered as a covector field on M*, ω might not vanish there. An example will help to clarify this distinction.

Example 11.29. Let $\omega = dy$ on $\mathbb{R}^2$, and let S be the x-axis, considered as an embedded submanifold of $\mathbb{R}^2$. As a covector field on $\mathbb{R}^2$, ω is nonzero everywhere, because one of its component functions is always 1. However, the restriction $\iota^*\omega$ is identically zero, because y vanishes identically on S:

$$\iota^*\omega = \iota^*\,dy = d(y \circ \iota) = 0. \hspace{4cm} /\!/$$

To distinguish the two ways in which we might interpret the statement "ω vanishes on S," one usually says that ω **vanishes along S** or **vanishes at points of S** if $\omega_p = 0$ for every point $p \in S$. The weaker condition that $\iota^*\omega = 0$ is expressed by saying that **the restriction of ω to S vanishes**, or **the pullback of ω to S vanishes**.

▶ **Exercise 11.30.** Suppose M is a smooth manifold with or without boundary and $S \subseteq M$ is an immersed submanifold with or without boundary. For any $f \in C^\infty(M)$, show that $d\left(f|_S\right) = \iota^*(df)$. Conclude that the pullback of df to S is zero if and only if f is constant on each component of S.

Line Integrals

Another important application of covector fields is to make coordinate-independent sense of the notion of a line integral.

We begin with the simplest case: an interval in the real line. Suppose $[a, b] \subseteq \mathbb{R}$ is a compact interval, and ω is a smooth covector field on $[a, b]$. (This means that the component function of ω admits a smooth extension to some neighborhood of $[a,b]$.) If we let t denote the standard coordinate on $\mathbb{R}$, then ω can be written $\omega_t = f(t)\,dt$ for some smooth function $f\colon [a,b] \to \mathbb{R}$. The similarity between this and the usual notation $\int f(t)\,dt$ for an integral suggests that there might be a connection between covector fields and integrals, and indeed there is. We define the **integral of ω over $[a,b]$** to be

$$\int_{[a,b]} \omega = \int_a^b f(t)\,dt.$$

The next proposition should convince you that this is more than just a trick of notation.

Proposition 11.31 (Diffeomorphism Invariance of the Integral). *Let ω be a smooth covector field on the compact interval $[a,b] \subseteq \mathbb{R}$. If $\varphi\colon [c,d] \to [a,b]$ is an increasing diffeomorphism (meaning that $t_1 < t_2$ implies $\varphi(t_1) < \varphi(t_2)$), then*

$$\int_{[c,d]} \varphi^* \omega = \int_{[a,b]} \omega.$$

Proof. If we let s denote the standard coordinate on $[c,d]$ and t that on $[a,b]$, then (11.17) shows that the pullback $\varphi^*\omega$ has the coordinate expression $(\varphi^*\omega)_s = f(\varphi(s))\varphi'(s)\,ds$. Inserting this into the definition of the line integral and using the change of variables formula for ordinary integrals, we obtain

$$\int_{[c,d]} \varphi^*\omega = \int_c^d f(\varphi(s))\varphi'(s)\,ds = \int_a^b f(t)\,dt = \int_{[a,b]} \omega. \qquad \square$$

▶ **Exercise 11.32.** Prove that if $\varphi\colon [c,d] \to [a,b]$ is a decreasing diffeomorphism, then

$$\int_{[c,d]} \varphi^*\omega = -\int_{[a,b]} \omega.$$

Now let M be a smooth manifold with or without boundary. By a **curve segment in M** we mean a continuous curve $\gamma\colon [a,b] \to M$ whose domain is a compact interval. It is a **smooth curve segment** if it is smooth when $[a,b]$ is considered as a manifold with boundary (or, equivalently, if γ has an extension to a smooth curve defined in a neighborhood of each endpoint). It is a **piecewise smooth curve segment** if there exists a finite partition $a = a_0 < a_1 < \cdots < a_k = b$ of $[a,b]$ such that $\gamma|_{[a_{i-1},a_i]}$ is smooth for each i (Fig. 11.4). Continuity of γ means that $\gamma(t)$ approaches the same value as t approaches any of the points a_i (other than a_0 or a_k) from the left or the right. Smoothness of γ on each subinterval means that γ has one-sided velocity vectors at each such a_i when approaching from the left or the right, but these one-sided velocities need not be equal.

Proposition 11.33. *If M is a connected smooth manifold with or without boundary, any two points of M can be joined by a piecewise smooth curve segment.*

Proof. Let p be an arbitrary point of M, and define a subset $\mathcal{C} \subseteq M$ by

$$\mathcal{C} = \{q \in M : \text{there is a piecewise smooth curve segment in } M \text{ from } p \text{ to } q\}.$$

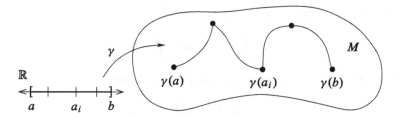

Fig. 11.4 A piecewise smooth curve segment

Clearly, $p \in \mathcal{C}$, so $\mathcal{C}$ is nonempty. To show that $\mathcal{C} = M$, we need to show that it is open and closed in M.

Let $q \in \mathcal{C}$ be arbitrary, which means that there is a piecewise smooth curve segment γ going from p to q. Let U be a smooth coordinate ball (or half-ball) centered at q. If q' is any point in U, then it is easy to construct a piecewise smooth curve segment from p to q' by first following γ from p to q, and then following a straight-line path in coordinates from q to q'. Thus $U \subseteq \mathcal{C}$, which shows that $\mathcal{C}$ is open in M. On the other hand, if $q \in \partial\mathcal{C}$, let U be a smooth coordinate ball or half-ball around q as above. The fact that q is a boundary point of $\mathcal{C}$ means that there is some point $q' \in \mathcal{C} \cap U$. In this case, we can construct a piecewise smooth curve from p to q by first following one from p to q' and then following a straight-line path in coordinates from q' to q. This shows that $q \in \mathcal{C}$, so $\mathcal{C}$ is also closed. $\qquad\square$

If $\gamma\colon [a,b] \to M$ is a smooth curve segment and ω is a smooth covector field on M, we define the *line integral of ω over γ* to be the real number

$$\int_\gamma \omega = \int_{[a,b]} \gamma^*\omega.$$

Because $\gamma^*\omega$ is a smooth covector field on $[a,b]$, this definition makes sense. More generally, if γ is piecewise smooth, we define

$$\int_\gamma \omega = \sum_{i=1}^{k} \int_{[a_{i-1},a_i]} \gamma^*\omega,$$

where $[a_{i-1}, a_i]$, $i = 1, \ldots, k$, are subintervals on which γ is smooth.

Proposition 11.34 (Properties of Line Integrals). *Let M be a smooth manifold with or without boundary. Suppose $\gamma\colon [a,b] \to M$ is a piecewise smooth curve segment, and $\omega, \omega_1, \omega_2 \in \mathfrak{X}^*(M)$.*

(a) *For any $c_1, c_2 \in \mathbb{R}$,*

$$\int_\gamma (c_1\omega_1 + c_2\omega_2) = c_1 \int_\gamma \omega_1 + c_2 \int_\gamma \omega_2.$$

(b) *If γ is a constant map, then $\int_\gamma \omega = 0$.*

(c) *If $\gamma_1 = \gamma|_{[a,c]}$ and $\gamma_2 = \gamma|_{[c,b]}$ with $a < c < b$, then*

$$\int_\gamma \omega = \int_{\gamma_1} \omega + \int_{\gamma_2} \omega.$$

(d) *If $F: M \to N$ is any smooth map and $\eta \in \mathfrak{X}^*(N)$, then*

$$\int_\gamma F^*\eta = \int_{F \circ \gamma} \eta.$$

▶ **Exercise 11.35.** Prove Proposition 11.34.

Example 11.36. Let $M = \mathbb{R}^2 \smallsetminus \{0\}$, let ω be the covector field on M given by

$$\omega = \frac{x\,dy - y\,dx}{x^2 + y^2},$$

and let $\gamma: [0, 2\pi] \to M$ be the curve segment defined by $\gamma(t) = (\cos t, \sin t)$. Since $\gamma^*\omega$ can be computed by substituting $x = \cos t$ and $y = \sin t$ everywhere in the formula for ω, we find that

$$\int_\gamma \omega = \int_{[0,2\pi]} \frac{\cos t(\cos t\,dt) - \sin t(-\sin t\,dt)}{\sin^2 t + \cos^2 t} = \int_0^{2\pi} dt = 2\pi. \qquad \text{//}$$

One of the most significant features of line integrals is that they are independent of parametrization, in a sense we now make precise. If $\gamma: [a,b] \to M$ and $\tilde{\gamma}: [c,d] \to M$ are piecewise smooth curve segments, we say that $\tilde{\gamma}$ is a ***reparametrization of*** γ if $\tilde{\gamma} = \gamma \circ \varphi$ for some diffeomorphism $\varphi: [c,d] \to [a,b]$. If φ is an increasing function, we say that $\tilde{\gamma}$ is a ***forward reparametrization***, and if φ is decreasing, it is a ***backward reparametrization***. (More generally, with obvious modifications one can allow φ to be piecewise smooth.)

Proposition 11.37 (Parameter Independence of Line Integrals). *Suppose M is a smooth manifold with or without boundary, $\omega \in \mathfrak{X}^*(M)$, and γ is a piecewise smooth curve segment in M. For any reparametrization $\tilde{\gamma}$ of γ, we have*

$$\int_{\tilde{\gamma}} \omega = \begin{cases} \displaystyle\int_\gamma \omega & \text{if } \tilde{\gamma} \text{ is a forward reparametrization,} \\[2mm] -\displaystyle\int_\gamma \omega & \text{if } \tilde{\gamma} \text{ is a backward reparametrization.} \end{cases}$$

Proof. First assume that $\gamma: [a,b] \to M$ is smooth, and suppose $\tilde{\gamma} = \gamma \circ \varphi$, where $\varphi: [c,d] \to [a,b]$ is an increasing diffeomorphism. Then Proposition 11.31 implies

$$\int_{\tilde{\gamma}} \omega = \int_{[c,d]} (\gamma \circ \varphi)^*\omega = \int_{[c,d]} \varphi^*\gamma^*\omega = \int_{[a,b]} \gamma^*\omega = \int_\gamma \omega.$$

When φ is decreasing, the analogous result follows from Exercise 11.32. If γ is only piecewise smooth, the result follows simply by applying the preceding argument on each subinterval where γ is smooth. $\qquad\square$

The next proposition gives a useful alternative expression for a line integral.

Proposition 11.38. *If* $\gamma\colon [a,b] \to M$ *is a piecewise smooth curve segment, the line integral of* ω *over* γ *can also be expressed as the ordinary integral*

$$\int_\gamma \omega = \int_a^b \omega_{\gamma(t)}\big(\gamma'(t)\big)\, dt. \tag{11.18}$$

Proof. First suppose that γ is smooth and that its image is contained in the domain of a single smooth chart. Writing the coordinate representations of γ and ω as $\big(\gamma^1(t),\dots,\gamma^n(t)\big)$ and $\omega_i\, dx^i$, respectively, we have

$$\omega_{\gamma(t)}\big(\gamma'(t)\big) = \omega_i\big(\gamma(t)\big)\, dx^i\big(\gamma'(t)\big) = \omega_i\big(\gamma(t)\big)\dot{\gamma}^i(t).$$

Combining this with the coordinate formula (11.17) for the pullback, we obtain

$$(\gamma^*\omega)_t = (\omega_i \circ \gamma)(t)\, d\big(\gamma^i\big)_t = \omega_i\big(\gamma(t)\big)\dot{\gamma}^i(t)\, dt = \omega_{\gamma(t)}\big(\gamma'(t)\big)\, dt.$$

Therefore, by definition of the line integral,

$$\int_\gamma \omega = \int_{[a,b]} \gamma^*\omega = \int_a^b \omega_{\gamma(t)}\big(\gamma'(t)\big)\, dt.$$

If γ is an arbitrary smooth curve segment, by compactness there exists a finite partition $a = a_0 < a_1 < \cdots < a_k = b$ of $[a,b]$ such that $\gamma\big([a_{i-1},a_i]\big)$ is contained in the domain of a single smooth chart for each $i = 1,\dots,k$, so we can apply the computation above on each such subinterval. Finally, if γ is only piecewise smooth, we simply apply the same argument on each subinterval on which γ is smooth. $\square$

There is one special case in which a line integral is trivial to compute: the line integral of a differential.

Theorem 11.39 (Fundamental Theorem for Line Integrals). *Let M be a smooth manifold with or without boundary. Suppose f is a smooth real-valued function on M and $\gamma\colon [a,b] \to M$ is a piecewise smooth curve segment in M. Then*

$$\int_\gamma df = f\big(\gamma(b)\big) - f\big(\gamma(a)\big).$$

Proof. Suppose first that γ is smooth. By Propositions 11.23 and 11.38,

$$\int_\gamma df = \int_a^b df_{\gamma(t)}\big(\gamma'(t)\big)\, dt = \int_a^b (f \circ \gamma)'(t)\, dt.$$

By the fundamental theorem of calculus, this is equal to $f \circ \gamma(b) - f \circ \gamma(a)$.

If γ is merely piecewise smooth, let $a = a_0 < \cdots < a_k = b$ be the endpoints of the subintervals on which γ is smooth. Applying the above argument on each subinterval and summing, we find that

$$\int_\gamma df = \sum_{i=1}^k \big(f\big(\gamma(a_i)\big) - f\big(\gamma(a_{i-1})\big)\big) = f\big(\gamma(b)\big) - f\big(\gamma(a)\big),$$

because the contributions from all the interior points cancel. $\square$

Conservative Covector Fields

Theorem 11.39 shows that the line integral of any covector field that can be written as the differential of a smooth function can be computed easily once the smooth function is known. For this reason, there is a special term for covector fields with this property. A smooth covector field ω on a smooth manifold M with or without boundary is said to be **exact** (or an **exact differential**) on M if there is a function $f \in C^\infty(M)$ such that $\omega = df$. In this case, the function f is called a **potential for ω**. The potential is not uniquely determined, but by Proposition 11.22, the difference between any two potentials for ω must be constant on each component of M.

Because exact differentials are so easy to integrate, it is important to develop criteria for deciding whether a covector field is exact. Theorem 11.39 provides an important clue. It shows that the line integral of an exact covector field depends only on the endpoints $p = \gamma(a)$ and $q = \gamma(b)$: any other curve segment from p to q would give the same value for the line integral. In particular, if γ is a **closed curve segment**, meaning that $\gamma(a) = \gamma(b)$, then the integral of df over γ is zero.

We say that a smooth covector field ω is **conservative** if the line integral of ω over every piecewise smooth closed curve segment is zero. This terminology comes from physics, where a force field is called conservative if the change in energy caused by the force acting along any closed path is zero ("energy is conserved"). (In elementary physics, force fields are usually thought of as vector fields rather than covector fields; see Problem 11-15 for the connection.)

Conservative covector fields can also be characterized by path independence.

Proposition 11.40. *A smooth covector field ω is conservative if and only if its line integrals are path-independent, in the sense that $\int_\gamma \omega = \int_{\tilde\gamma} \omega$ whenever γ and $\tilde\gamma$ are piecewise smooth curve segments with the same starting and ending points.*

▶ **Exercise 11.41.** Prove Proposition 11.40. [Remark: this would be harder to prove if we defined conservative fields in terms of smooth curves instead of piecewise smooth ones.]

Theorem 11.42. *Let M be a smooth manifold with or without boundary. A smooth covector field on M is conservative if and only if it is exact.*

Proof. If $\omega \in \mathfrak{X}^*(M)$ is exact, Theorem 11.39 shows that it is conservative, so we need only prove the converse. Suppose ω is conservative, and assume for the moment that M is connected. Because the line integrals of ω are path-independent, we can adopt the following notation: for any points $p, q \in M$, we use the notation $\int_p^q \omega$ to denote the value of any line integral of the form $\int_\gamma \omega$, where γ is a piecewise smooth curve segment from p to q. Because a backward reparametrization of a path from p to q is a path from q to p, Proposition 11.37 implies $\int_q^p \omega = -\int_p^q \omega$; and for any three points $p_1, p_2, p_3 \in M$, Proposition 11.34(c) implies that

$$\int_{p_1}^{p_2} \omega + \int_{p_2}^{p_3} \omega = \int_{p_1}^{p_3} \omega. \tag{11.19}$$

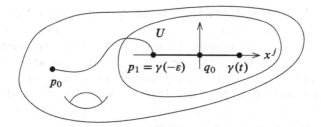

Fig. 11.5 Proof that a conservative covector field is exact

Now choose any base point $p_0 \in M$, and define a function $f : M \to \mathbb{R}$ by $f(q) = \int_{p_0}^{q} \omega$. We need to show that f is smooth and $df = \omega$. To accomplish this, let $q_0 \in M$ be arbitrary, let $(U, (x^i))$ be a smooth chart centered at q_0, and write the coordinate representation of ω in U as $\omega = \omega_i \, dx^i$. We need to show that

$$\frac{\partial f}{\partial x^j}(q_0) = \omega_j(q_0)$$

for $j = 1, \ldots, n$, which implies $df_{q_0} = \omega_{q_0}$.

First suppose $q_0 \in \operatorname{Int} M$. Fix j, and let $\gamma : [-\varepsilon, \varepsilon] \to U$ be the smooth curve segment defined in coordinates by $\gamma(t) = (0, \ldots, t, \ldots, 0)$, with t in the jth place, and with ε chosen small enough that $\gamma[-\varepsilon, \varepsilon] \subseteq U$ (Fig. 11.5). Let $p_1 = \gamma(-\varepsilon)$, and define a new function $\tilde{f} : M \to \mathbb{R}$ by $\tilde{f}(q) = \int_{p_1}^{q} \omega$. Note that (11.19) implies that for all $q \in M$,

$$f(q) - \tilde{f}(q) = \int_{p_0}^{q} \omega - \int_{p_1}^{q} \omega = \int_{p_0}^{q} \omega + \int_{q}^{p_1} \omega = \int_{p_0}^{p_1} \omega,$$

which does not depend on q. Thus $\tilde{f}$ and f differ by a constant, so it suffices to show that $\partial \tilde{f} / \partial x^j(q_0) = \omega_j(q_0)$.

Now $\gamma'(t) = \partial/\partial x^j|_{\gamma(t)}$ by construction, so

$$\omega_{\gamma(t)}\big(\gamma'(t)\big) = \omega_i\big(\gamma(t)\big) dx^i\left(\frac{\partial}{\partial x^j}\bigg|_{\gamma(t)}\right) = \omega_j\big(\gamma(t)\big).$$

Since the restriction of γ to $[-\varepsilon, t]$ is a smooth curve from p_1 to $\gamma(t)$, we have

$$\tilde{f} \circ \gamma(t) = \int_{p_1}^{\gamma(t)} \omega = \int_{-\varepsilon}^{t} \omega_{\gamma(s)}\big(\gamma'(s)\big) \, ds = \int_{-\varepsilon}^{t} \omega_j\big(\gamma(s)\big) \, ds.$$

Thus, by the fundamental theorem of calculus,

$$\frac{\partial \tilde{f}}{\partial x^j}(q_0) = \gamma'(0)\tilde{f} = \frac{d}{dt}\bigg|_{t=0} \tilde{f} \circ \gamma(t)$$

$$= \frac{d}{dt}\bigg|_{t=0} \int_{-\varepsilon}^{t} \omega_j\big(\gamma(s)\big) \, ds = \omega_j\big(\gamma(0)\big) = \omega_j(q_0).$$

This shows that $df_{q_0} = \omega_{q_0}$ when $q_0 \in \operatorname{Int} M$.

For $q_0 \in \partial M$, the chart $(U, (x^i))$ is a boundary chart centered at q_0. The proof above shows that $\partial f / \partial x^j (q_0) = \omega_j (q_0)$ except in the case $j = n = \dim M$; but that case requires a special argument because x^n takes on only nonnegative values in a boundary chart. In that case we simply set $p_1 = \gamma(0)$ instead of $p_1 = \gamma(-\varepsilon)$, and proceed as before. Then the same argument shows that $df_{q_0} = \omega_{q_0}$ in this case as well. This completes the proof that $df = \omega$ when M is connected. Since the component functions of ω are smooth and equal to the partial derivatives of f in coordinates, this also shows that f is smooth.

Finally, if M is not connected, let $\{M_i\}$ be the components of M. The argument above shows that for each i there is a function $f_i \in C^\infty(M_i)$ such that $df_i = \omega$ on M_i. Letting $f \colon M \to \mathbb{R}$ be the function that is equal to f_i on M_i for each i, we have $df = \omega$, thus completing the proof.   □

It would be nice if every smooth covector field were exact, for then the evaluation of any line integral would just be a matter of finding a potential function and evaluating it at the endpoints, a process analogous to evaluating an ordinary integral by finding an indefinite integral (also called a *primitive* or *antiderivative*). However, this is too much to hope for.

Example 11.43. The covector field ω of Example 11.36 cannot be exact on $\mathbb{R}^2 \smallsetminus \{0\}$, because it is not conservative: the computation in that example showed that $\int_\gamma \omega = 2\pi \neq 0$, where γ is the unit circle traversed counterclockwise.   //

Because exactness has such important consequences for the evaluation of line integrals, we would like to have an easy way to check whether a given covector field is exact. Fortunately, there is a very simple necessary condition, which follows from the fact that partial derivatives of smooth functions can be taken in any order.

To see what this condition is, suppose $\omega \in \mathfrak{X}^*(M)$ is exact. Let f be any potential function for ω, and let $(U, (x^i))$ be any smooth chart on M. Because f is smooth, it satisfies the following identity on U:

$$\frac{\partial^2 f}{\partial x^i \partial x^j} = \frac{\partial^2 f}{\partial x^j \partial x^i}. \tag{11.20}$$

Writing $\omega = \omega_i dx^i$ in coordinates, we see that $\omega = df$ is equivalent to $\omega_i = \partial f / \partial x^i$. Substituting this into (11.20), we find that the component functions of ω satisfy the following identity for each pair of indices i and j:

$$\frac{\partial \omega_j}{\partial x^i} = \frac{\partial \omega_i}{\partial x^j}. \tag{11.21}$$

We say that a smooth covector field ω is *closed* if its components in every smooth chart satisfy (11.21). The following proposition summarizes the computation above.

Proposition 11.44. *Every exact covector field is closed.*   □

One technical difficulty in checking directly from the definition that a covector field is closed is that it would require checking that (11.21) holds in *every* coordinate chart. The next proposition gives an alternative characterization of closed covector fields that is coordinate independent, and incidentally shows that it suffices to check (11.21) in *some* coordinate chart around each point.

Proposition 11.45. *Let ω be a smooth covector field on a smooth manifold M with or without boundary. The following are equivalent:*

(a) *ω is closed.*
(b) *ω satisfies (11.21) in some smooth chart around every point.*
(c) *For any open subset $U \subseteq M$ and smooth vector fields $X, Y \in \mathfrak{X}(U)$,*

$$X(\omega(Y)) - Y(\omega(X)) = \omega([X, Y]). \tag{11.22}$$

Proof. We will prove that (a) $\Rightarrow$ (b) $\Rightarrow$ (c) $\Rightarrow$ (a). The implication (a) $\Rightarrow$ (b) is immediate from the definition of closed covector fields.

To prove (b) $\Rightarrow$ (c), assume (b) holds, and suppose $U \subseteq M$ and $X, Y \in \mathfrak{X}(U)$ as in the statement of (c). It suffices to verify that (11.22) holds in a neighborhood of each point of U. In any coordinate chart $(V, (x^i))$ with $V \subseteq U$, we can write $\omega = \omega_i dx^i$, $X = X^j \partial/\partial x^j$, and $Y = Y^k \partial/\partial x^k$, and compute

$$X(\omega(Y)) = X(\omega_i Y^i) = Y^i X \omega_i + \omega_i X Y^i = Y^i X^j \frac{\partial \omega_i}{\partial x^j} + \omega_i X Y^i.$$

If we repeat the same computation with X and Y reversed and subtract, the terms involving derivatives of ω_i cancel by virtue of (11.21). Thus we get

$$X(\omega(Y)) - Y(\omega(X)) = \omega_i (X Y^i - Y X^i).$$

Formula (8.9) shows that this last expression is equal to $\omega([X, Y])$.

Finally, if ω satisfies (c), in any coordinate chart we can apply (11.22) with $X = \partial/\partial x^i$ and $Y = \partial/\partial x^j$, noting that $[X, Y] = 0$ in that case, to obtain (11.21). $\square$

One consequence of this proposition is that closedness can be easily checked using criterion (b), so many covector fields can be shown quickly not to be exact because they are not closed. Another is the following corollary.

Corollary 11.46. *Suppose $F: M \to N$ is a local diffeomorphism. Then the pullback $F^*: \mathfrak{X}^*(N) \to \mathfrak{X}^*(M)$ takes closed covector fields to closed covector fields, and exact ones to exact ones.*

Proof. The result for exact covector fields follows immediately from (11.15). For closed covector fields, if (U, φ) is any smooth chart for N, then $\varphi \circ F$ is a smooth chart for M in a neighborhood of each point of $F^{-1}(U)$. In these coordinates, the coordinate representation of F is the identity, so if ω satisfies (11.21) in U, then $F^*\omega$ satisfies (11.21) in $F^{-1}(U)$. $\square$

Example 11.47. Consider the following covector field on $\mathbb{R}^2$:

$$\omega = y \cos xy \, dx + x \cos xy \, dy.$$

It is easy to check that

$$\frac{\partial(y \cos xy)}{\partial y} = \frac{\partial(x \cos xy)}{\partial x} = \cos xy - xy \sin xy,$$

so ω is closed. In fact, you might guess that $\omega = d(\sin xy)$. On the other hand, the covector field

$$\eta = x \cos xy \, dx + y \cos xy \, dy$$

is not closed, because

$$\frac{\partial(x \cos xy)}{\partial y} = -x^2 \sin xy, \qquad \frac{\partial(y \cos xy)}{\partial x} = -y^2 \sin xy.$$

Thus η is not exact.   //

The question then naturally arises whether the converse of Proposition 11.44 is true: Is every closed covector field exact? The answer is *almost* yes, but there is an important restriction. It turns out that the answer to the question depends in a subtle way on the shape of the domain, as the next example illustrates.

Example 11.48. Look once again at the covector field ω of Example 11.36. A straightforward computation shows that ω is closed; but as we observed above, it is not exact on $\mathbb{R}^2 \smallsetminus \{0\}$. On the other hand, if we restrict the domain to the right half-plane $U = \{(x, y) : x > 0\}$, a computation shows that $\omega = d \left(\tan^{-1} y/x \right)$ there. This can be seen more clearly in polar coordinates, where $\omega = d\theta$. The problem, of course, is that there is no smooth (or even continuous) angle function on all of $\mathbb{R}^2 \smallsetminus \{0\}$, which is a consequence of the "hole" in the center.   //

This last example illustrates a key principle: the question of whether a particular closed covector field is exact is a global one, depending on the shape of the domain in question. This observation is the starting point for *de Rham cohomology*, which expresses a deep relationship between smooth structures and topology. We will define de Rham cohomology and study this relationship in Chapter 17, but for now we can prove the following result. If V is a finite-dimensional vector space, a subset $U \subseteq V$ is said to be *star-shaped* if there is a point $c \in U$ such that for every $x \in U$, the line segment from c to x is entirely contained in U (Fig. 11.6). For example, every convex subset is star-shaped.

Theorem 11.49 (Poincaré Lemma for Covector Fields). *If U is a star-shaped open subset of $\mathbb{R}^n$ or $\mathbb{H}^n$, then every closed covector field on U is exact.*

Proof. Suppose U is star-shaped with respect to $c \in U$, and let $\omega = \omega_i dx^i$ be a closed covector field on U. As in the proof of Theorem 11.42, we will construct a potential function for ω by integrating along smooth curve segments from c. However, in this case we do not know a priori that the line integrals are path-independent, so we must integrate along specific paths.

Because diffeomorphisms take closed forms to closed forms and exact ones to exact ones, we can apply a translation to U to arrange that $c = 0$. For any point $x \in U$, let $\gamma_x : [0, 1] \to U$ denote the line segment from 0 to x, parametrized as $\gamma_x(t) = tx$. The hypothesis guarantees that the image of γ_x lies entirely in U for each $x \in U$. Define a function $f : U \to \mathbb{R}$ by

$$f(x) = \int_{\gamma_x} \omega. \tag{11.23}$$

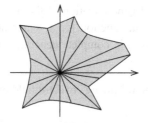

Fig. 11.6 A star-shaped subset of $\mathbb{R}^2$

We need to show that f is a potential for ω, or equivalently that $\partial f/\partial x^j = \omega_j$ for $j = 1, \ldots, n$. To begin, we compute

$$f(x) = \int_0^1 \omega_{\gamma_x(t)}\big(\gamma_x'(t)\big)\, dt = \int_0^1 \omega_i(tx)x^i\, dt. \tag{11.24}$$

(The summation convention is in effect.) To compute the partial derivatives of f, we note that the integrand is smooth in all variables, so it is permissible to differentiate under the integral sign to obtain

$$\frac{\partial f}{\partial x^j}(x) = \int_0^1 \left(t\frac{\partial \omega_i}{\partial x^j}(tx)x^i + \omega_j(tx)\right) dt.$$

Because ω is closed, this reduces to

$$\frac{\partial f}{\partial x^j}(x) = \int_0^1 \left(t\frac{\partial \omega_j}{\partial x^i}(tx)x^i + \omega_j(tx)\right) dt$$

$$= \int_0^1 \frac{d}{dt}\big(t\omega_j(tx)\big)\, dt = \Big[t\omega_j(tx)\Big]_{t=0}^{t=1} = \omega_j(x). \qquad \square$$

Corollary 11.50 (Local Exactness of Closed Covector Fields). *Let ω be a closed covector field on a smooth manifold M with or without boundary. Then every point of M has a neighborhood on which ω is exact.*

Proof. Let $p \in M$ be arbitrary. The hypothesis implies that ω satisfies (11.21) in some smooth coordinate ball or half-ball (U, φ) containing p. Because balls and half-balls are convex, we can apply Theorem 11.49 to the coordinate representation of ω and conclude that there is a function $f \in C^\infty(U)$ such that $\omega|_U = df$. $\qquad \square$

The key to constructing a potential function in Theorem 11.49 is that we can reach every point $x \in M$ by a definite path from c to x, chosen to vary smoothly as x varies. That is what fails in the case of the closed covector field ω on the punctured plane (Example 11.43): because of the hole, it is impossible to choose a smoothly varying family of paths starting at a fixed base point and reaching every point of the domain exactly once. In Chapter 16 we will generalize Theorem 11.49 to show that every closed covector field is exact on any simply connected manifold.

When you actually have to *compute* a potential function for a given covector field that is known to be exact, there is a much simpler procedure that almost always works. It is a straightforward generalization of the method introduced in calculus texts for computing a potential function for a vector field that is known to be a gradient. Rather than describe it in complete generality, we illustrate it with an example.

Example 11.51. Let ω be the following covector field on $\mathbb{R}^3$:

$$\omega = e^{y^2} dx + 2xye^{y^2} dy - 2z\, dz.$$

You can check that ω is closed. For f to be a potential for ω, it must satisfy

$$\frac{\partial f}{\partial x} = e^{y^2}, \quad \frac{\partial f}{\partial y} = 2xye^{y^2}, \quad \frac{\partial f}{\partial z} = -2z. \tag{11.25}$$

Holding y and z fixed and integrating the first equation with respect to x, we obtain

$$f(x, y, z) = \int e^{y^2} dx = xe^{y^2} + C_1(y, z),$$

where the "constant" of integration $C_1(y, z)$ may depend on the choice of (y, z). Now the second equation of (11.25) implies

$$2xye^{y^2} = \frac{\partial}{\partial y}\left(xe^{y^2} + C_1(y, z)\right) = 2xye^{y^2} + \frac{\partial C_1}{\partial y},$$

which forces $\partial C_1/\partial y = 0$, so C_1 is actually a function of z only. Finally, the third equation implies

$$-2z = \frac{\partial}{\partial z}\left(xe^{y^2} + C_1(z)\right) = \frac{dC_1}{dz},$$

from which we conclude that $C_1(z) = -z^2 + C$, where C is an arbitrary constant. Thus a potential function for ω is given by $f(x, y, z) = xe^{y^2} - z^2$. Any other potential differs from this one by a constant.   //

You should convince yourself that the formal procedure we followed in this example is equivalent to choosing an arbitrary base point $c \in \mathbb{R}^3$ and defining $f(x, y, z)$ by integrating ω along a path from c to (x, y, z) consisting of three straight line segments parallel to the axes. This works for any closed covector field on an open rectangle in $\mathbb{R}^n$ (which we know must be exact, because a rectangle is convex). In practice, once a formula is found for f on some open rectangle, the same formula typically works for the entire domain. (This is because most of the covector fields for which one can explicitly compute the integrals are real-analytic, and real-analytic functions are determined by their behavior in any open subset.)

Problems

11-1. (a) Suppose V and W are finite-dimensional vector spaces and $A \colon V \to W$ is any linear map. Show that the following diagram commutes:

$$\begin{CD} V @>A>> W \\ @V\xi_V VV @VV\xi_W V \\ V^{**} @>>(A^*)^*> W^{**}, \end{CD}$$

where ξ_V and ξ_W denote the isomorphisms defined by (11.3) for V and W, respectively.

(b) Show that there does not exist a way to assign to each finite-dimensional vector space V an isomorphism $\beta_V \colon V \to V^*$ such that for every linear map $A \colon V \to W$, the following diagram commutes:

$$\begin{CD} V @>A>> W \\ @V\beta_V VV @VV\beta_W V \\ V^* @<<A^*< W^*. \end{CD}$$

11-2. Suppose V is an infinite-dimensional real vector space, and let V^* denote the vector space of all linear functionals from V to $\mathbb{R}$. (This is often called the *algebraic dual space of* V to distinguish it from the space of continuous linear functionals when V is endowed with a norm or a topology.)

(a) Prove that there is an injective linear map from V to V^*. [Hint: see Exercise B.5.]

(b) Prove that there is no injective linear map from V^* to V. [Hint: if S is a basis for V, prove that there is a linearly independent subset of V^* with the same cardinality as the power set of S.]

(c) Use (a) and (b) to prove there is no isomorphism between V and V^{**}. (*Used on p. 489.*)

11-3. Let Vec_1 be the category of finite-dimensional vector spaces and linear isomorphisms as in Problem 10-8. Define a functor $\mathscr{F} \colon \mathsf{Vec}_1 \to \mathsf{Vec}_1$ by setting $\mathscr{F}(V) = V^*$ for a vector space V, and $\mathscr{F}(A) = (A^{-1})^*$ for an isomorphism A. Show that $\mathscr{F}$ is a smooth covariant functor, and that for every M, $\mathscr{F}(TM)$ and T^*M are canonically smoothly isomorphic vector bundles.

11-4. Let M be a smooth manifold with or without boundary and p be a point of M. Let $\mathcal{J}_p$ denote the subspace of $C^\infty(M)$ consisting of smooth functions that vanish at p, and let $\mathcal{J}_p^2$ be the subspace of $\mathcal{J}_p$ spanned by functions of the form fg for some $f, g \in \mathcal{J}_p$.

(a) Show that $f \in \mathcal{J}_p^2$ if and only if in any smooth local coordinates, its first-order Taylor polynomial at p is zero. (Because of this, a function in $\mathcal{J}_p^2$ is said to *vanish to second order*.)

(b) Define a map $\Phi: \mathcal{J}_p \to T_p^* M$ by setting $\Phi(f) = df_p$. Show that the restriction of Φ to $\mathcal{J}_p^2$ is zero, and that Φ descends to a vector space isomorphism from $\mathcal{J}_p/\mathcal{J}_p^2$ to $T_p^* M$.

[Remark: Problem 3-8 showed that tangent vectors at p can be viewed as equivalence classes of smooth curves, which are smooth maps from (subsets of) $\mathbb{R}$ to M. This problem shows that covectors at p can be viewed dually as equivalence classes of smooth functions from M to $\mathbb{R}$: a covector is an equivalence class of smooth functions that vanish at p, with two such functions considered equivalent if they differ by a function that vanishes to second order. In some treatments of smooth manifold theory, $T_p^* M$ is defined first in this way, and then $T_p M$ is defined as the dual space $(\mathcal{J}_p/\mathcal{J}_p^2)^*$.]

11-5. For any smooth manifold M, show that $T^* M$ is a trivial vector bundle if and only if TM is trivial.

11-6. Suppose M is a smooth n-manifold, $p \in M$, and $y^1, \ldots, y^k$ are smooth real-valued functions defined on a neighborhood of p in M. Prove the following statements.

(a) If $k = n$ and $(dy^1|_p, \ldots, dy^n|_p)$ is a basis for $T_p^* M$, then $(y^1, \ldots, y^n)$ are smooth coordinates for M in some neighborhood of p.

(b) If $(dy^1|_p, \ldots, dy^k|_p)$ is a linearly independent k-tuple of covectors and $k < n$, then there are smooth functions $y^{k+1}, \ldots, y^n$ such that $(y^1, \ldots, y^n)$ are smooth coordinates for M in a neighborhood of p.

(c) If $(dy^1|_p, \ldots, dy^k|_p)$ span $T_p^* M$, there are indices $i_1, \ldots, i_n$ such that $(y^{i_1}, \ldots, y^{i_n})$ are smooth coordinates for M in a neighborhood of p.

(*Used on p. 584.*)

11-7. In the following problems, M and N are smooth manifolds, $F: M \to N$ is a smooth map, and $\omega \in \mathfrak{X}^*(N)$. Compute $F^*\omega$ in each case.

(a) $M = N = \mathbb{R}^2$;
$F(s,t) = (st, e^t)$;
$\omega = x \, dy - y \, dx$.

(b) $M = \mathbb{R}^2$ and $N = \mathbb{R}^3$;
$F(\theta, \varphi) = ((\cos \varphi + 2) \cos \theta, \ (\cos \varphi + 2) \sin \theta, \ \sin \varphi)$;
$\omega = z^2 \, dx$.

(c) $M = \{(s,t) \in \mathbb{R}^2 : s^2 + t^2 < 1\}$ and $N = \mathbb{R}^3 \smallsetminus \{0\}$;
$F(s,t) = \left(s, t, \sqrt{1 - s^2 - t^2}\right)$;
$\omega = (1 - x^2 - y^2) \, dz$.

11-8. (a) Suppose $F: M \to N$ is a diffeomorphism, and let $dF^*: T^* N \to T^* M$ be the map whose restriction to each cotangent space $T_q^* N$ is equal to $dF^*_{F^{-1}(q)}$. Show that dF^* is a smooth bundle homomorphism.

(b) Let Diff_1 be the category whose objects are smooth manifolds, but whose only morphisms are diffeomorphisms; and let VB be the category whose objects are smooth vector bundles and whose morphisms are smooth bundle homomorphisms. Show that the assignment $M \mapsto$

T^*M, $F \mapsto dF^*$ defines a contravariant functor from Diff$_1$ to VB, called the *cotangent functor*.

(*Used on p. 592.*)

11-9. Let $f: \mathbb{R}^3 \to \mathbb{R}$ be the function $f(x, y, z) = x^2 + y^2 + z^2$, and let $F: \mathbb{R}^2 \to \mathbb{R}^3$ be the following map (the inverse of the stereographic projection):

$$F(u, v) = \left(\frac{2u}{u^2 + v^2 + 1}, \frac{2v}{u^2 + v^2 + 1}, \frac{u^2 + v^2 - 1}{u^2 + v^2 + 1} \right).$$

Compute F^*df and $d(f \circ F)$ separately, and verify that they are equal.

11-10. In each of the cases below, M is a smooth manifold and $f: M \to \mathbb{R}$ is a smooth function. Compute the coordinate representation for df, and determine the set of all points $p \in M$ at which $df_p = 0$.
   (a) $M = \{(x, y) \in \mathbb{R}^2 : x > 0\}$; $f(x, y) = x/(x^2 + y^2)$. Use standard coordinates (x, y).
   (b) M and f are as in (a); this time use polar coordinates (r, θ).
   (c) $M = S^2 \subseteq \mathbb{R}^3$; $f(p) = z(p)$ (the z-coordinate of p as a point in $\mathbb{R}^3$). Use north and south stereographic coordinates (Problem 1-7).
   (d) $M = \mathbb{R}^n$; $f(x) = |x|^2$. Use standard coordinates.

11-11. Let M be a smooth manifold, and $C \subseteq M$ be an embedded submanifold. Let $f \in C^\infty(M)$, and suppose $p \in C$ is a point at which f attains a local maximum or minimum value among points in C. Given a smooth local defining function $\Phi: U \to \mathbb{R}^k$ for C on a neighborhood U of p in M, show that there are real numbers $\lambda_1, \ldots, \lambda_k$ (called *Lagrange multipliers*) such that

$$df_p = \lambda_1 d\Phi^1\big|_p + \cdots + \lambda_k d\Phi^k\big|_p.$$

11-12. Show that any two points in a connected smooth manifold can be joined by a smooth curve segment.

11-13. The *length* of a smooth curve segment $\gamma: [a, b] \to \mathbb{R}^n$ is defined to be the value of the (ordinary) integral

$$L(\gamma) = \int_a^b |\gamma'(t)| \, dt.$$

Show that there is no smooth covector field $\omega \in \mathfrak{X}^*(\mathbb{R}^n)$ with the property that $\int_\gamma \omega = L(\gamma)$ for every smooth curve γ.

11-14. Consider the following two covector fields on $\mathbb{R}^3$:

$$\omega = -\frac{4z \, dx}{(x^2 + 1)^2} + \frac{2y \, dy}{y^2 + 1} + \frac{2x \, dz}{x^2 + 1},$$

$$\eta = -\frac{4xz \, dx}{(x^2 + 1)^2} + \frac{2y \, dy}{y^2 + 1} + \frac{2 \, dz}{x^2 + 1}.$$

(a) Set up and evaluate the line integral of each covector field along the straight line segment from $(0,0,0)$ to $(1,1,1)$.

(b) Determine whether either of these covector fields is exact.

(c) For each one that is exact, find a potential function and use it to recompute the line integral.

11-15. LINE INTEGRALS OF VECTOR FIELDS: Let X be a smooth vector field on an open subset $U \subseteq \mathbb{R}^n$. Given a piecewise smooth curve segment $\gamma: [a,b] \to U$, define the **line integral of X over γ**, denoted by $\int_\gamma X \cdot ds$, as

$$\int_\gamma X \cdot ds = \int_a^b X_{\gamma(t)} \cdot \gamma'(t) \, dt,$$

where the dot on the right-hand side denotes the Euclidean dot product between tangent vectors at $\gamma(t)$, identified with elements of $\mathbb{R}^n$. A **conservative vector field** is one whose line integral around every piecewise smooth closed curve is zero.

(a) Show that X is conservative if and only if there exists a smooth function $f \in C^\infty(U)$ such that $X = \operatorname{grad} f$. [Hint: consider the covector field ω defined by $\omega_x(v) = X_x \cdot v$.]

(b) Suppose $n = 3$. Show that if X is conservative, then $\operatorname{curl} X = 0$, where

$$\operatorname{curl} X = \left(\frac{\partial X^3}{\partial x^2} - \frac{\partial X^2}{\partial x^3} \right) \frac{\partial}{\partial x^1} + \left(\frac{\partial X^1}{\partial x^3} - \frac{\partial X^3}{\partial x^1} \right) \frac{\partial}{\partial x^2}$$

$$+ \left(\frac{\partial X^2}{\partial x^1} - \frac{\partial X^1}{\partial x^2} \right) \frac{\partial}{\partial x^3}. \tag{11.26}$$

(c) Show that if $U \subseteq \mathbb{R}^3$ is star-shaped, then X is conservative on U if and only if $\operatorname{curl} X = 0$.

11-16. Let M be a compact manifold of positive dimension. Show that every exact covector field on M vanishes at least at two points in each component of M.

11-17. Let $\mathbb{T}^n = \mathbb{S}^1 \times \cdots \times \mathbb{S}^1 \subseteq \mathbb{C}^n$ denote the n-torus. For each $j = 1,\ldots,n$, let $\gamma_j: [0,1] \to \mathbb{T}^n$ be the curve segment

$$\gamma_j(t) = \left(1,\ldots, e^{2\pi i t},\ldots, 1\right) \quad \text{(with } e^{2\pi i t} \text{ in the } j\text{th place).}$$

Show that a closed covector field ω on $\mathbb{T}^n$ is exact if and only if $\int_{\gamma_j} \omega = 0$ for $j = 1,\ldots,n$. [Hint: consider first $(\varepsilon^n)^* \omega$, where $\varepsilon^n: \mathbb{R}^n \to \mathbb{T}^n$ is the smooth covering map $\varepsilon^n\left(x^1,\ldots,x^n\right) = \left(e^{2\pi i x^1},\ldots, e^{2\pi i x^n}\right)$.]

11-18. This problem shows how to give a rigorous meaning to words like "natural" and "canonical" that are so often used informally in mathematics. Suppose C and D are categories, and $\mathscr{F}, \mathscr{G}$ are (covariant or contravariant) functors from C to D. A **natural transformation** λ from $\mathscr{F}$ to $\mathscr{G}$ is a rule that assigns to each object $X \in \operatorname{Ob}(\mathsf{C})$ a morphism $\lambda_X \in \operatorname{Hom}_{\mathsf{D}}\left(\mathscr{F}(X), \mathscr{G}(X)\right)$ in such a way that for every pair of objects $X, Y \in \mathrm{Ob}(C)$ and every morphism $f \in \mathrm{Hom}_C(X, Y)$, the following diagram commutes:

$$
\begin{array}{ccc}
\mathcal{F}(X) & \xrightarrow{\mathcal{F}(f)} & \mathcal{F}(Y) \\
\lambda_X \downarrow & & \downarrow \lambda_Y \\
\mathcal{G}(X) & \xrightarrow[\mathcal{G}(f)]{} & \mathcal{G}(Y).
\end{array}
$$

(If either $\mathcal{F}$ or $\mathcal{G}$ is contravariant, the corresponding horizontal arrow should be reversed.)

(a) Let $\mathsf{Vec}_\mathbb{R}$ denote the category of real vector spaces and linear maps, and let $\mathcal{D}$ be the contravariant functor from $\mathsf{Vec}_\mathbb{R}$ to itself that sends each vector space to its dual space and each linear map to its dual map. Show that the assignment $V \mapsto \xi_V$, where $\xi_V \colon V \to V^{**}$ is the map defined by $\xi_V(v)\omega = \omega(v)$, is a natural transformation from the identity functor of $\mathsf{Vec}_\mathbb{R}$ to $\mathcal{D} \circ \mathcal{D}$.

(b) Show that there does not exist a natural transformation from the identity functor of $\mathsf{Vec}_\mathbb{R}$ to $\mathcal{D}$.

(c) Let Diff_1 be the category of smooth manifolds and diffeomorphisms and VB the category of smooth vector bundles and smooth bundle homomorphisms, and let $T, T^* \colon \mathsf{Diff}_1 \to \mathsf{VB}$ be the tangent and cotangent functors, respectively (see Problems 10-3 and 11-8). Show that there does not exist a natural transformation from T to T^*.

(d) Let $\mathfrak{X} \colon \mathsf{Diff}_1 \to \mathsf{Vec}_\mathbb{R}$ be the covariant functor given by $M \mapsto \mathfrak{X}(M)$, $F \mapsto F_*$; and let $\mathfrak{X} \times \mathfrak{X} \colon \mathsf{Diff}_1 \to \mathsf{Vec}_\mathbb{R}$ be the covariant functor given by $M \mapsto \mathfrak{X}(M) \times \mathfrak{X}(M)$, $F \mapsto F_* \times F_*$. Show that the Lie bracket is a natural transformation from $\mathfrak{X} \times \mathfrak{X}$ to $\mathfrak{X}$.

(Used on pp. 189, 347, 376.)

Chapter 12
Tensors

Much of the technology of smooth manifold theory is designed to allow the concepts of linear algebra to be applied to smooth manifolds. Calculus tells us how to approximate smooth objects by linear ones, and the abstract definitions of manifold theory give a way to interpret these linear approximations in a coordinate-independent way.

In this chapter we carry this idea much further, by generalizing from linear maps to *multilinear* ones—those that take several vectors as input and depend linearly on each one separately. Although linear maps are paramount in differential geometry, there are many situations in which multilinear maps play important geometric roles. We will introduce a unified language for talking about multilinear maps: the language of tensors. This leads to the concepts of tensors and tensor fields on manifolds.

We begin with tensors on a vector space, which are multilinear generalizations of covectors; a covector is the special case of a tensor of rank one. We give two alternative definitions of tensors on a vector space: on the one hand, they are elements of the abstract "tensor product" of the dual vector space with itself; on the other hand, they are real-valued multilinear functions of several vectors. Each definition is useful in certain contexts. We deal primarily with covariant tensors, but we also give a brief introduction to contravariant tensors and tensors of mixed variance.

Next we introduce two special classes of tensors: the *symmetric tensors*, whose values are unchanged by permutations of their arguments, and the *alternating tensors*, whose values change sign when two argument are interchanged.

We then move to smooth manifolds, where we define tensor fields and tensor bundles. After describing the coordinate representations of tensor fields, we describe how they can be pulled back by smooth maps. We also show how the Lie derivative operator can be extended to tensors: the Lie derivative of a tensor field with respect to a vector field is a measure of the rate of change of the tensor field along the flow of the vector field.

Tensors will pervade the rest of the book, and we will see significant applications of them when we study Riemannian metrics, differential forms, orientations, integration, de Rham cohomology, foliations, and symplectic structures.

J.M. Lee, *Introduction to Smooth Manifolds*, Graduate Texts in Mathematics 218,
DOI 10.1007/978-1-4419-9982-5_12, © Springer Science+Business Media New York 2013

Multilinear Algebra

We have seen some of the important roles played in manifold theory by covectors, which are real-valued linear functions on a vector space. In their simplest form, tensors are just real-valued *multilinear* functions of one or more variables; simple examples include covectors, inner products, and determinants. To set the stage for our study of tensors, in this section we develop some of the basic properties of multilinear functions in a general setting.

Suppose $V_1, \ldots, V_k$, and W are vector spaces. A map $F: V_1 \times \cdots \times V_k \to W$ is said to be *multilinear* if it is linear as a function of each variable separately when the others are held fixed: for each i,

$$F(v_1, \ldots, a v_i + a' v_i', \ldots, v_k) = a F(v_1, \ldots, v_i, \ldots, v_k) + a' F(v_1, \ldots, v_i', \ldots, v_k).$$

(A multilinear function of one variable is just a linear function, and a multilinear function of two variables is generally called *bilinear*.) Let us write $L(V_1, \ldots, V_k; W)$ for the set of all multilinear maps from $V_1 \times \cdots \times V_k$ to W. It is a vector space under the usual operations of pointwise addition and scalar multiplication:

$$(F + F')(v_1, \ldots, v_k) = F(v_1, \ldots, v_k) + F'(v_1, \ldots, v_k),$$
$$(a F)(v_1, \ldots, v_k) = a\big(F(v_1, \ldots, v_k)\big).$$

Here are a few examples to keep in mind.

Example 12.1 (Some Familiar Multilinear Functions).

(a) *The dot product* in $\mathbb{R}^n$ is a scalar-valued bilinear function of two vectors, used to compute lengths of vectors and angles between them.
(b) *The cross product* in $\mathbb{R}^3$ is a vector-valued bilinear function of two vectors, used to compute areas of parallelograms and to find a third vector orthogonal to two given ones.
(c) *The determinant* is a real-valued multilinear function of n vectors in $\mathbb{R}^n$, used to detect linear independence and to compute the volume of the parallelepiped spanned by the vectors.
(d) *The bracket* in a Lie algebra $\mathfrak{g}$ is a $\mathfrak{g}$-valued bilinear function of two elements of $\mathfrak{g}$.   //

The next example is probably not as familiar, but it is extremely important.

Example 12.2 (Tensor Products of Covectors). Suppose V is a vector space, and $\omega, \eta \in V^*$. Define a function $\omega \otimes \eta: V \times V \to \mathbb{R}$ by

$$\omega \otimes \eta(v_1, v_2) = \omega(v_1)\eta(v_2),$$

where the product on the right is just ordinary multiplication of real numbers. The linearity of ω and η guarantees that $\omega \otimes \eta$ is a bilinear function of v_1 and v_2, so it is an element of $L(V, V; \mathbb{R})$. For example, if (e^1, e^2) denotes the standard dual basis for $(\mathbb{R}^2)^*$, then $e^1 \otimes e^2: \mathbb{R}^2 \times \mathbb{R}^2 \to \mathbb{R}$ is the bilinear function

$$e^1 \otimes e^2\big((w, x), (y, z)\big) = wz.$$ //

The last example can be generalized to arbitrary real-valued multilinear functions as follows: let $V_1, \ldots, V_k, W_1, \ldots, W_l$ be real vector spaces, and suppose $F \in L(V_1, \ldots, V_k; \mathbb{R})$ and $G \in L(W_1, \ldots, W_l; \mathbb{R})$. Define a function

$$F \otimes G : V_1 \times \cdots \times V_k \times W_1 \times \cdots \times W_l \to \mathbb{R}$$

by

$$F \otimes G(v_1, \ldots, v_k, w_1, \ldots, w_l) = F(v_1, \ldots, v_k)G(w_1, \ldots, w_l). \qquad (12.1)$$

It follows from the multilinearity of F and G that $F \otimes G(v_1, \ldots, v_k, w_1, \ldots, w_l)$ depends linearly on each argument v_i or w_j separately, so $F \otimes G$ is an element of $L(V_1, \ldots, V_k, W_1, \ldots, W_l; \mathbb{R})$, called the **tensor product of F and G**.

▶ **Exercise 12.3.** Show that the tensor product operation is bilinear and associative: $F \otimes G$ depends bilinearly on F and G, and $(F \otimes G) \otimes H = F \otimes (G \otimes H)$.

Because of the result of the preceding exercise, we can write tensor products of three or more multilinear functions unambiguously without parentheses. If $F_1, \ldots, F_l$ are multilinear functions depending on $k_1, \ldots, k_l$ variables, respectively, their tensor product $F_1 \otimes \cdots \otimes F_l$ is a multilinear function of $k = k_1 + \cdots + k_l$ variables, whose action on k vectors is given by inserting the first k_1 vectors into F_1, the next k_2 vectors into F_2, and so forth, and multiplying the results together. For example, if F and G are multilinear functions of two vectors and H is a multilinear function of three, then

$$F \otimes G \otimes H(v_1, \ldots, v_7) = F(v_1, v_2)G(v_3, v_4)H(v_5, v_6, v_7).$$

If $\omega^j \in V_j^*$ for $j = 1, \ldots, k$, then $\omega^1 \otimes \cdots \otimes \omega^k \in L(V_1, \ldots, V_k; \mathbb{R})$ is the multilinear function given by

$$\omega^1 \otimes \cdots \otimes \omega^k(v_1, \ldots, v_k) = \omega^1(v_1) \cdots \omega^k(v_k). \qquad (12.2)$$

The tensor product operation is important in part because of its role in the following proposition. The notation in this proposition is ugly because of the profusion of indices, but the underlying idea is simple: a basis for any space of multilinear functions can be formed by taking all possible tensor products of basis covectors.

Proposition 12.4 (A Basis for the Space of Multilinear Functions). *Let $V_1, \ldots, V_k$ be real vector spaces of dimensions $n_1, \ldots, n_k$, respectively. For each $j \in \{1, \ldots, k\}$, let $\left(E_1^{(j)}, \ldots, E_{n_j}^{(j)}\right)$ be a basis for V_j, and let $\left(\varepsilon_{(j)}^1, \ldots, \varepsilon_{(j)}^{n_j}\right)$ be the corresponding dual basis for V_j^*. Then the set*

$$\mathcal{B} = \left\{ \varepsilon_{(1)}^{i_1} \otimes \cdots \otimes \varepsilon_{(k)}^{i_k} : 1 \leq i_1 \leq n_1, \ldots, 1 \leq i_k \leq n_k \right\}$$

is a basis for $L(V_1, \ldots, V_k; \mathbb{R})$, which therefore has dimension equal to $n_1 \cdots n_k$.

Proof. We need to show that $\mathcal{B}$ is linearly independent and spans $L(V_1, \ldots, V_k; \mathbb{R})$. Suppose $F \in L(V_1, \ldots, V_k; \mathbb{R})$ is arbitrary. For each ordered k-tuple $(i_1, \ldots, i_k)$ of integers with $1 \leq i_j \leq n_j$, define a number $F_{i_1 \ldots i_k}$ by

$$F_{i_1 \ldots i_k} = F\left(E_{i_1}^{(1)}, \ldots, E_{i_k}^{(k)}\right). \tag{12.3}$$

We will show that

$$F = F_{i_1 \ldots i_k} \varepsilon_{(1)}^{i_1} \otimes \cdots \otimes \varepsilon_{(k)}^{i_k}$$

(with the summation convention in effect as usual), from which it follows that $\mathcal{B}$ spans $L(V_1, \ldots, V_k; \mathbb{R})$. For any k-tuple of vectors $(v_1, \ldots, v_k) \in V_1 \times \cdots \times V_k$, write $v_1 = v_1^{i_1} E_{i_1}^{(1)}, \ldots, v_k = v_k^{i_k} E_{i_k}^{(k)}$, and compute

$$F_{i_1 \ldots i_k} \varepsilon_{(1)}^{i_1} \otimes \cdots \otimes \varepsilon_{(k)}^{i_k}(v_1, \ldots, v_k) = F_{i_1 \ldots i_k} \varepsilon_{(1)}^{i_1}(v_1) \cdots \varepsilon_{(k)}^{i_k}(v_k)$$

$$= F_{i_1 \ldots i_k} v_1^{i_1} \cdots v_k^{i_k},$$

while $F(v_1, \ldots, v_k)$ is equal to the same thing by multilinearity. This proves the claim.

To show that $\mathcal{B}$ is linearly independent, suppose some linear combination equals zero:

$$F_{i_1 \ldots i_k} \varepsilon_{(1)}^{i_1} \otimes \cdots \otimes \varepsilon_{(k)}^{i_k} = 0.$$

Apply this to any ordered k-tuple of basis vectors, $\left(E_{j_1}^{(1)}, \ldots, E_{j_k}^{(k)}\right)$. By the same computation as above, this implies that each coefficient $F_{j_1 \ldots j_k}$ is zero. Thus, the only linear combination of elements of $\mathcal{B}$ that sums to zero is the trivial one. $\square$

This proof shows, by the way, that the components $F_{i_1 \ldots i_k}$ of a multilinear function F in terms of the basis elements in $\mathcal{B}$ are given by (12.3). Thus, F is completely determined by its action on all possible sequences of basis vectors.

Abstract Tensor Products of Vector Spaces

The results of the previous section showed that the vector space of multilinear functions $L(V_1, \ldots, V_k; \mathbb{R})$ can be viewed as the set of all linear combinations of objects of the form $\omega^1 \otimes \cdots \otimes \omega^k$, where $\omega^1, \ldots, \omega^k$ are covectors. In this section, we give a construction that makes sense of such linear combinations of tensor products in a more abstract setting. The construction is a bit involved, but the idea is simple: given finite-dimensional vector spaces $V_1, \ldots, V_k$, we will construct a new vector space $V_1 \otimes \cdots \otimes V_k$ whose dimension is the product of the dimensions of the V_i's, and which consists of "formal linear combinations" of objects of the form $v_1 \otimes \cdots \otimes v_k$ for $v_i \in V_i$, defined in such a way that $v_1 \otimes \cdots \otimes v_k$ depends linearly on each v_i separately. (Many of the concepts we introduce in this section—at least the parts that do not refer explicitly to finite bases—work equally well in the infinite-dimensional case; but we mostly restrict our attention to the finite-dimensional case in order to keep things simple.)

To begin, we need to make sense of "formal linear combinations." Let S be a set. Roughly speaking, a formal linear combination of elements of S is an expression of the form $\sum_{i=1}^{m} a_i x_i$, where $a_1, \ldots, a_m$ are real numbers and $x_1, \ldots, x_m$ are elements of S. Of course, since we are not assuming that S has any algebraic structure, we cannot literally add elements of S together or multiply them by numbers. But the essential feature of such an expression is that it is completely determined by which elements of S appear in the sum, and what coefficients appear with them. Thus, we make the following definition: for any set S, a *formal linear combination of elements of* S is a function $f : S \to \mathbb{R}$ such that $f(s) = 0$ for all but finitely many $s \in S$. The *free (real) vector space on* S, denoted by $\mathcal{F}(S)$, is the set of all formal linear combinations of elements of S. Under pointwise addition and scalar multiplication, $\mathcal{F}(S)$ becomes a vector space over $\mathbb{R}$.

For each element $x \in S$, there is a function $\delta_x \in \mathcal{F}(S)$ that takes the value 1 on x and zero on all other elements of S; typically we *identify* this function with x itself, and thus think of S as a subset of $\mathcal{F}(S)$. Every element $f \in \mathcal{F}(S)$ can then be written uniquely in the form $f = \sum_{i=1}^{m} a_i x_i$, where $x_1, \dots, x_m$ are the elements of S for which $f(x_i) \neq 0$, and $a_i = f(x_i)$. Thus, S is a basis for $\mathcal{F}(S)$, which is therefore finite-dimensional if and only if S is a finite set.

Proposition 12.5 (Characteristic Property of the Free Vector Space). *For any set S and any vector space W, every map $A: S \to W$ has a unique extension to a linear map $\bar{A}: \mathcal{F}(S) \to W$.*

▶ **Exercise 12.6.** Prove the preceding proposition.

Now let $V_1, \dots, V_k$ be real vector spaces. We begin by forming the free vector space $\mathcal{F}(V_1 \times \cdots \times V_k)$, which is the set of all finite formal linear combinations of k-tuples $(v_1, \dots, v_k)$ with $v_i \in V_i$ for $i = 1, \dots, k$. Let $\mathcal{R}$ be the subspace of $\mathcal{F}(V_1 \times \cdots \times V_k)$ spanned by all elements of the following forms:

$$(v_1, \dots, av_i, \dots, v_k) - a(v_1, \dots, v_i, \dots, v_k),$$
$$(v_1, \dots, v_i + v_i', \dots, v_k) - (v_1, \dots, v_i, \dots, v_k) - (v_1, \dots, v_i', \dots, v_k), \tag{12.4}$$

with $v_j, v_j' \in V_j$, $i \in \{1, \dots, k\}$, and $a \in \mathbb{R}$.

Define the *tensor product of the spaces $V_1, \dots, V_k$*, denoted by $V_1 \otimes \cdots \otimes V_k$, to be the following quotient vector space:

$$V_1 \otimes \cdots \otimes V_k = \mathcal{F}(V_1 \times \cdots \times V_k)/\mathcal{R},$$

and let $\Pi: \mathcal{F}(V_1 \times \cdots \times V_k) \to V_1 \otimes \cdots \otimes V_k$ be the natural projection. The equivalence class of an element $(v_1, \dots, v_k)$ in $V_1 \otimes \cdots \otimes V_k$ is denoted by

$$v_1 \otimes \cdots \otimes v_k = \Pi(v_1, \dots, v_k), \tag{12.5}$$

and is called the *(abstract) tensor product of $v_1, \dots, v_k$*. It follows from the definition that abstract tensor products satisfy

$$v_1 \otimes \cdots \otimes av_i \otimes \cdots \otimes v_k = a(v_1 \otimes \cdots \otimes v_i \otimes \cdots \otimes v_k),$$
$$v_1 \otimes \cdots \otimes (v_i + v_i') \otimes \cdots \otimes v_k = (v_1 \otimes \cdots \otimes v_i \otimes \cdots \otimes v_k)$$
$$+ (v_1 \otimes \cdots \otimes v_i' \otimes \cdots \otimes v_k).$$

Note that the definition implies that every element of $V_1 \otimes \cdots \otimes V_k$ can be expressed as a linear combination of elements of the form $v_1 \otimes \cdots \otimes v_k$ for $v_i \in V_i$; but it is not true in general that every element of the tensor product space is of the form $v_1 \otimes \cdots \otimes v_k$ (see Problem 12-1).

Proposition 12.7 (Characteristic Property of the Tensor Product Space). *Let $V_1, \ldots, V_k$ be finite-dimensional real vector spaces. If $A \colon V_1 \times \cdots \times V_k \to X$ is any multilinear map into a vector space X, then there is a unique linear map $\tilde{A} \colon V_1 \otimes \cdots \otimes V_k \to X$ such that the following diagram commutes:*

$$
\begin{array}{ccc}
V_1 \times \cdots \times V_k & \xrightarrow{\; A \;} & X, \\
\Big\downarrow{\scriptstyle \pi} & \nearrow{\scriptstyle \tilde{A}} & \\
V_1 \otimes \cdots \otimes V_k & &
\end{array}
\tag{12.6}
$$

where π is the map $\pi(v_1, \ldots, v_k) = v_1 \otimes \cdots \otimes v_k$.

Proof. First note that any map $A \colon V_1 \times \cdots \times V_k \to X$ extends uniquely to a linear map $\bar{A} \colon \mathcal{F}(V_1 \times \cdots \times V_k) \to X$ by the characteristic property of the free vector space. This map is characterized by the fact that $\bar{A}(v_1, \ldots, v_k) = A(v_1, \ldots, v_k)$ whenever $(v_1, \ldots, v_k) \in V_1 \times \cdots \times V_k \subseteq \mathcal{F}(V_1 \times \cdots \times V_k)$. The fact that A is multilinear means precisely that the subspace $\mathcal{R}$ is contained in the kernel of $\bar{A}$, because

$$
\bar{A}(v_1, \ldots, a v_i, \ldots, v_k) = A(v_1, \ldots, a v_i, \ldots, v_k) = a A(v_1, \ldots, v_i, \ldots, v_k)
$$
$$
= a \bar{A}(v_1, \ldots, v_i, \ldots, v_k) = \bar{A}\big(a(v_1, \ldots, v_i, \ldots, v_k)\big),
$$

with a similar computation for the other expression in (12.4). Therefore, $\bar{A}$ descends to a linear map $\tilde{A} \colon V_1 \otimes \cdots \otimes V_k = \mathcal{F}(V_1 \times \cdots \times V_k)/\mathcal{R} \to X$ satisfying $\tilde{A} \circ \Pi = \bar{A}$. Since π is equal to the inclusion $V_1 \times \cdots \times V_k \hookrightarrow \mathcal{F}(V_1 \times \cdots \times V_k)$ followed by Π, this implies $\tilde{A} \circ \pi = A$, which is (12.6). Uniqueness follows from the fact that every element of $V_1 \otimes \cdots \otimes V_k$ can be written as a linear combination of elements of the form $v_1 \otimes \cdots \otimes v_k$, and $\tilde{A}$ is uniquely determined on such elements by

$$
\tilde{A}(v_1 \otimes \cdots \otimes v_k) = \bar{A}(v_1, \ldots, v_k) = A(v_1, \ldots, v_k). \qquad \square
$$

The reason this is called the characteristic property is that it uniquely characterizes the tensor product up to isomorphism; see Problem 12-3.

The next result is an analogue of Proposition 12.4 for abstract tensor product spaces.

Proposition 12.8 (A Basis for the Tensor Product Space). *Suppose $V_1, \ldots, V_k$ are real vector spaces of dimensions $n_1, \ldots, n_k$, respectively. For each $j = 1, \ldots, k$, suppose $\big(E_1^{(j)}, \ldots, E_{n_j}^{(j)}\big)$ is a basis for V_j. Then the set*

$$
\mathcal{E} = \Big\{ E_{i_1}^{(1)} \otimes \cdots \otimes E_{i_k}^{(k)} : 1 \leq i_1 \leq n_1, \ldots, 1 \leq i_k \leq n_k \Big\}
$$

is a basis for $V_1 \otimes \cdots \otimes V_k$, which therefore has dimension equal to $n_1 \cdots n_k$.

Proof. Elements of the form $v_1 \otimes \cdots \otimes v_k$ span the tensor product space by definition; expanding each v_i in such an expression in terms of its basis representation shows that $\mathcal{C}$ spans $V_1 \otimes \cdots \otimes V_k$.

To prove that $\mathcal{C}$ is linearly independent, assume that some linear combination of elements of $\mathcal{C}$ is equal to zero:

$$a^{i_1 \ldots i_k} E_{i_1}^{(1)} \otimes \cdots \otimes E_{i_k}^{(k)} = 0.$$

For each ordered k-tuple of indices $(m_1, \ldots, m_k)$, define a multilinear function $\tau^{m_1 \ldots m_k} : V_1 \times \cdots \times V_k \to \mathbb{R}$ by

$$\tau^{m_1 \ldots m_k}(v_1, \ldots, v_k) = \varepsilon_{(1)}^{m_1}(v_1) \cdots \varepsilon_{(k)}^{m_k}(v_k),$$

where $\left(\varepsilon_{(j)}^i\right)$ is the basis for V_j^* dual to $\left(E_i^{(j)}\right)$. Because $\tau^{m_1 \ldots m_k}$ is multilinear, it descends to a linear function $\widetilde{\tau}^{m_1 \ldots m_k} : V_1 \otimes \cdots \otimes V_k \to \mathbb{R}$ by the characteristic property of the tensor product. It follows that

$$0 = \widetilde{\tau}^{m_1 \ldots m_k}\left(a^{i_1 \ldots i_k} E_{i_1}^{(1)} \otimes \cdots \otimes E_{i_k}^{(k)}\right)$$

$$= a^{i_1 \ldots i_k} \tau^{m_1 \ldots m_k}\left(E_{i_1}^{(1)}, \ldots, E_{i_k}^{(k)}\right) = a^{m_1 \ldots m_k},$$

which shows that $\mathcal{C}$ is linearly independent. $\qquad\qquad\qquad\qquad\qquad\square$

Proposition 12.9 (Associativity of Tensor Product Spaces). *Let V_1, V_2, V_3 be finite-dimensional real vector spaces. There are unique isomorphisms*

$$V_1 \otimes (V_2 \otimes V_3) \cong V_1 \otimes V_2 \otimes V_3 \cong (V_1 \otimes V_2) \otimes V_3,$$

under which elements of the forms $v_1 \otimes (v_2 \otimes v_3)$, $v_1 \otimes v_2 \otimes v_3$, and $(v_1 \otimes v_2) \otimes v_3$ all correspond.

Proof. We construct the isomorphism $V_1 \otimes V_2 \otimes V_3 \cong (V_1 \otimes V_2) \otimes V_3$; the other one is constructed similarly. The map $\alpha : V_1 \times V_2 \times V_3 \to (V_1 \otimes V_2) \otimes V_3$ defined by

$$\alpha(v_1, v_2, v_3) = (v_1 \otimes v_2) \otimes v_3$$

is obviously multilinear, and thus by the characteristic property of the tensor product it descends uniquely to a linear map $\widetilde{\alpha} : V_1 \otimes V_2 \otimes V_3 \to (V_1 \otimes V_2) \otimes V_3$ satisfying $\widetilde{\alpha}(v_1 \otimes v_2 \otimes v_3) = (v_1 \otimes v_2) \otimes v_3$ for all $v_1 \in V_1$, $v_2 \in V_2$, and $v_3 \in V_3$. Because $(V_1 \otimes V_2) \otimes V_3$ is spanned by elements of the form $(v_1 \otimes v_2) \otimes v_3$, $\widetilde{\alpha}$ is surjective, and therefore it is an isomorphism for dimensional reasons. It is clearly the unique such isomorphism, because any other would have to agree with $\widetilde{\alpha}$ on the set of all elements of the form $v_1 \otimes v_2 \otimes v_3$, which spans $V_1 \otimes V_2 \otimes V_3$. $\qquad\square$

The connection between tensor products in this abstract setting and the more concrete tensor products of multilinear functionals that we defined earlier is based on the following proposition.

Proposition 12.10 (Abstract vs. Concrete Tensor Products). *If $V_1, \ldots, V_k$ are finite-dimensional vector spaces, there is a canonical isomorphism*

$$V_1^* \otimes \cdots \otimes V_k^* \cong L(V_1, \ldots, V_k; \mathbb{R}),$$

under which the abstract tensor product defined by (12.5) corresponds to the tensor product of covectors defined by (12.2).

Proof. First, define a map $\Phi \colon V_1^* \times \cdots \times V_k^* \to L(V_1, \ldots, V_k; \mathbb{R})$ by

$$\Phi(\omega^1, \ldots, \omega^k)(v_1, \ldots, v_k) = \omega^1(v_1) \cdots \omega^k(v_k).$$

(There is no implied summation in this formula.) The expression on the right depends linearly on each v_i, so $\Phi(\omega^1, \ldots, \omega^k)$ is indeed an element of the space $L(V_1, \ldots, V_k; \mathbb{R})$. It is easy to check that Φ is multilinear as a function of $\omega^1, \ldots, \omega^k$, so by the characteristic property it descends uniquely to a linear map $\widetilde{\Phi}$ from $V_1^* \otimes \cdots \otimes V_k^*$ to $L(V_1, \ldots, V_k; \mathbb{R})$, which satisfies

$$\widetilde{\Phi}(\omega^1 \otimes \cdots \otimes \omega^k)(v_1, \ldots, v_k) = \omega^1(v_1) \cdots \omega^k(v_k).$$

It follows immediately from the definition that $\widetilde{\Phi}$ takes abstract tensor products to tensor products of covectors. It also takes the basis of $V_1^* \otimes \cdots \otimes V_k^*$ given by Proposition 12.8 to the basis for $L(V_1, \ldots, V_k; \mathbb{R})$ of Proposition 12.4, so it is an isomorphism. (Although we used bases to prove that $\widetilde{\Phi}$ is an isomorphism, $\widetilde{\Phi}$ itself is canonically defined without reference to any basis.) □

Using this canonical isomorphism, we henceforth use the notation $V_1^* \otimes \cdots \otimes V_k^*$ to denote either the abstract tensor product space or the space $L(V_1, \ldots, V_k; \mathbb{R})$, focusing on whichever interpretation is more convenient for the problem at hand. Since we are assuming the vector spaces are all finite-dimensional, we can also identify each V_j with its second dual space V_j^{**}, and thereby obtain another canonical identification

$$V_1 \otimes \cdots \otimes V_k \cong L(V_1^*, \ldots, V_k^*; \mathbb{R}).$$

Covariant and Contravariant Tensors on a Vector Space

Let V be a finite-dimensional real vector space. If k is a positive integer, a ***covariant k-tensor on V*** is an element of the k-fold tensor product $V^* \otimes \cdots \otimes V^*$, which we typically think of as a real-valued multilinear function of k elements of V:

$$\alpha \colon \underbrace{V \times \cdots \times V}_{k \text{ copies}} \to \mathbb{R}.$$

The number k is called the ***rank of*** α. A 0-tensor is, by convention, just a real number (a real-valued function depending multilinearly on no vectors!). We denote the vector space of all covariant k-tensors on V by the shorthand notation

$$T^k(V^*) = \underbrace{V^* \otimes \cdots \otimes V^*}_{k \text{ copies}}.$$

Let us look at some examples.

Example 12.11 (Covariant Tensors). Let V be a finite-dimensional vector space.

(a) Every linear functional $\omega\colon V \to \mathbb{R}$ is multilinear, so a covariant 1-tensor is just a covector. Thus, $T^1(V^*)$ is equal to V^*.
(b) A covariant 2-tensor on V is a real-valued bilinear function of two vectors, also called a **bilinear form**. One example is the dot product on $\mathbb{R}^n$. More generally, every inner product is a covariant 2-tensor.
(c) The determinant, thought of as a function of n vectors, is a covariant n-tensor on $\mathbb{R}^n$.   //

For some purposes, it is important to generalize the notion of covariant tensors as follows. For any finite-dimensional real vector space V, we define the space of **contravariant tensors on V of rank k** to be the vector space

$$T^k(V) = \underbrace{V \otimes \cdots \otimes V}_{k \text{ copies}}.$$

In particular, $T^1(V) = V$, and by convention $T^0(V) = \mathbb{R}$. Because we are assuming that V is finite-dimensional, it is possible to identify this space with the set of multilinear functionals of k covectors:

$$T^k(V) \cong \{\text{multilinear functions } \alpha\colon \underbrace{V^* \times \cdots \times V^*}_{k \text{ copies}} \to \mathbb{R}\}.$$

But for most purposes, it is easier to think of contravariant tensors simply as elements of the abstract tensor product space.

Even more generally, for any nonnegative integers k, l, we define the space of **mixed tensors on V of type (k, l)** as

$$T^{(k,l)}(V) = \underbrace{V \otimes \cdots \otimes V}_{k \text{ copies}} \otimes \underbrace{V^* \otimes \cdots \otimes V^*}_{l \text{ copies}}.$$

Some of these spaces are identical:

$$T^{(0,0)}(V) = T^0(V^*) = T^0(V) = \mathbb{R},$$
$$T^{(0,1)}(V) = T^1(V^*) = V^*,$$
$$T^{(1,0)}(V) = T^1(V) = V,$$
$$T^{(0,k)}(V) = T^k(V^*),$$
$$T^{(k,0)}(V) = T^k(V).$$

(Be aware that the notation $T^{(k,l)}(V)$ is not universal. Another notation that is in common use for this space is $T^k_l(V)$. To make matters worse, some books reverse the roles of k and l in either of these notations; for example, the previous edition of this text used the notation $T^l_k(V)$ for the space we are here denoting by $T^{(k,l)}(V)$.)

We have chosen the notation given here because it is common and nearly always means the same thing. The moral, as usual, is that when you read any differential geometry book, you need to make sure you understand the author's notational conventions.)

When V is finite-dimensional, any choice of basis for V automatically yields bases for all of the tensor spaces over V. The following corollary follows immediately from Proposition 12.8.

Corollary 12.12. *Let V be an n-dimensional real vector space. Suppose (E_i) is any basis for V and (ε^j) is the dual basis for V^*. Then the following sets constitute bases for the tensor spaces over V:*

$$\left\{ \varepsilon^{i_1} \otimes \cdots \otimes \varepsilon^{i_k} : 1 \leq i_1, \ldots, i_k \leq n \right\} \quad \text{for } T^k(V^*);$$

$$\left\{ E_{i_1} \otimes \cdots \otimes E_{i_k} : 1 \leq i_1, \ldots, i_k \leq n \right\} \quad \text{for } T^k(V);$$

$$\left\{ E_{i_1} \otimes \cdots \otimes E_{i_k} \otimes \varepsilon^{j_1} \otimes \cdots \otimes \varepsilon^{j_l} : 1 \leq i_1, \ldots, i_k, j_1, \ldots, j_l \leq n \right\} \quad \text{for } T^{(k,l)}(V).$$

Therefore, $\dim T^k(V^*) = \dim T^k(V) = n^k$ *and* $\dim T^{(k,l)}(V) = n^{k+l}$. $\qquad\square$

In particular, once a basis is chosen for V, every covariant k-tensor $\alpha \in T^k(V^*)$ can be written uniquely in the form

$$\alpha = \alpha_{i_1 \ldots i_k} \varepsilon^{i_1} \otimes \cdots \otimes \varepsilon^{i_k},$$

where the n^k coefficients $\alpha_{i_1 \ldots i_k}$ are determined by

$$\alpha_{i_1 \ldots i_k} = \alpha \left(E_{i_1}, \ldots, E_{i_k} \right).$$

For example, $T^2(V^*)$ is the space of bilinear forms on V, and every bilinear form can be written as $\beta = \beta_{ij} \varepsilon^i \otimes \varepsilon^j$ for some uniquely determined $n \times n$ matrix (β_{ij}).

In this book we are concerned primarily with covariant tensors. Thus tensors will always be understood to be covariant unless we explicitly specify otherwise. However, it is important to be aware that contravariant and mixed tensors play important roles in more advanced parts of differential geometry, especially Riemannian geometry.

Symmetric and Alternating Tensors

In general, rearranging the arguments of a covariant tensor need not have any predictable effect on its value. However, some special tensors—the dot product, for example—do not change their values when their arguments are rearranged. Others—notably the determinant—change sign whenever two arguments are interchanged. In this section, we describe two classes of tensors that change in a simple way when their arguments are rearranged: the symmetric ones and the alternating ones.

Symmetric Tensors

Let V be a finite-dimensional vector space. A covariant k-tensor α on V is said to be **symmetric** if its value is unchanged by interchanging any pair of arguments:

$$\alpha(v_1, \ldots, v_i, \ldots, v_j, \ldots, v_k) = \alpha(v_1, \ldots, v_j, \ldots, v_i, \ldots, v_k)$$

whenever $1 \leq i < j \leq k$.

▶ **Exercise 12.13.** Show that the following are equivalent for a covariant k-tensor α:

(a) α is symmetric.
(b) For any vectors $v_1, \ldots, v_k \in V$, the value of $\alpha(v_1, \ldots, v_k)$ is unchanged when $v_1, \ldots, v_k$ are rearranged in any order.
(c) The components $\alpha_{i_1 \ldots i_k}$ of α with respect to any basis are unchanged by any permutation of the indices.

The set of symmetric covariant k-tensors is a linear subspace of the space $T^k(V^*)$ of all covariant k-tensors on V; we denote this subspace by $\Sigma^k(V^*)$. There is a natural projection from $T^k(V^*)$ to $\Sigma^k(V^*)$ defined as follows. First, let S_k denote the **symmetric group on k elements**, that is, the group of permutations of the set $\{1, \ldots, k\}$. Given a k-tensor α and a permutation $\sigma \in S_k$, we define a new k-tensor ${}^\sigma\alpha$ by

$$ {}^\sigma\alpha(v_1, \ldots, v_k) = \alpha\left(v_{\sigma(1)}, \ldots, v_{\sigma(k)}\right). $$

Note that ${}^\tau({}^\sigma\alpha) = {}^{\tau\sigma}\alpha$, where $\tau\sigma$ represents the composition of τ and σ, that is, $\tau\sigma(i) = \tau(\sigma(i))$. (This is the reason for putting σ before α in the notation ${}^\sigma\alpha$, instead of after it.) We define a projection $\mathrm{Sym}\colon T^k(V^*) \to \Sigma^k(V^*)$ called **symmetrization** by

$$\mathrm{Sym}\,\alpha = \frac{1}{k!} \sum_{\sigma \in S_k} {}^\sigma\alpha.$$

More explicitly, this means that

$$(\mathrm{Sym}\,\alpha)(v_1, \ldots, v_k) = \frac{1}{k!} \sum_{\sigma \in S_k} \alpha\left(v_{\sigma(1)}, \ldots, v_{\sigma(k)}\right).$$

Proposition 12.14 (Properties of Symmetrization). *Let α be a covariant tensor on a finite-dimensional vector space.*

(a) $\mathrm{Sym}\,\alpha$ *is symmetric.*
(b) $\mathrm{Sym}\,\alpha = \alpha$ *if and only if α is symmetric.*

Proof. Suppose $\alpha \in T^k(V^*)$. If $\tau \in S_k$ is any permutation, then

$$(\mathrm{Sym}\,\alpha)\left(v_{\tau(1)}, \ldots, v_{\tau(k)}\right) = \frac{1}{k!} \sum_{\sigma \in S_k} {}^\sigma\alpha\left(v_{\tau(1)}, \ldots, v_{\tau(k)}\right)$$

$$= \frac{1}{k!} \sum_{\sigma \in S_k} {}^{\tau\sigma}\alpha(v_1, \ldots, v_k)$$

$$= \frac{1}{k!} \sum_{\eta \in S_k} {}^\eta\alpha(v_1, \ldots, v_k)$$

$$= (\mathrm{Sym}\ \alpha)(v_1, \ldots, v_k),$$

where we have substituted $\eta = \tau\sigma$ in the second-to-last line and used the fact that η runs over all of S_k as σ does. This shows that Sym α is symmetric.

If α is symmetric, then it follows from Exercise 12.13(b) that ${}^\sigma\alpha = \alpha$ for every $\sigma \in S_k$, so it follows immediately that Sym $\alpha = \alpha$. On the other hand, if Sym $\alpha = \alpha$, then α is symmetric because part (a) shows that Sym α is. $\qquad\square$

If α and β are symmetric tensors on V, then $\alpha \otimes \beta$ is not symmetric in general. However, using the symmetrization operator, it is possible to define a new product that takes a pair of symmetric tensors and yields another symmetric tensor. If $\alpha \in \Sigma^k(V^*)$ and $\beta \in \Sigma^l(V^*)$, we define their **symmetric product** to be the $(k + l)$-tensor $\alpha\beta$ (denoted by juxtaposition with no intervening product symbol) given by

$$\alpha\beta = \mathrm{Sym}(\alpha \otimes \beta).$$

More explicitly, the action of $\alpha\beta$ on vectors $v_1, \ldots, v_{k+l}$ is given by

$$\alpha\beta(v_1, \ldots, v_{k+l}) = \frac{1}{(k+l)!} \sum_{\sigma \in S_{k+l}} \alpha\left(v_{\sigma(1)}, \ldots, v_{\sigma(k)}\right) \beta\left(v_{\sigma(k+1)}, \ldots, v_{\sigma(k+l)}\right).$$

Proposition 12.15 (Properties of the Symmetric Product).

(a) *The symmetric product is symmetric and bilinear: for all symmetric tensors α, β, γ and all $a, b \in \mathbb{R}$,*

$$\alpha\beta = \beta\alpha,$$

$$(a\alpha + b\beta)\gamma = a\alpha\gamma + b\beta\gamma = \gamma(a\alpha + b\beta).$$

(b) *If α and β are covectors, then*

$$\alpha\beta = \tfrac{1}{2}(\alpha \otimes \beta + \beta \otimes \alpha).$$

▶ **Exercise 12.16.** Prove Proposition 12.15.

Alternating Tensors

We continue to assume that V is a finite-dimensional real vector space. A covariant k-tensor α on V is said to be **alternating** (or **antisymmetric** or **skew-symmetric**) if it changes sign whenever two of its arguments are interchanged. This means that for all vectors $v_1, \ldots, v_k \in V$ and every pair of distinct indices i, j it satisfies

$$\alpha(v_1, \ldots, v_i, \ldots, v_j, \ldots, v_k) = -\alpha(v_1, \ldots, v_j, \ldots, v_i, \ldots, v_k).$$

Alternating covariant k-tensors are also variously called **exterior forms**, **multi-covectors**, or **k-covectors**. The subspace of all alternating covariant k-tensors on V is denoted by $\Lambda^k(V^*) \subseteq T^k(V^*)$.

Recall that for any permutation $\sigma \in S_k$, the **sign of** σ, denoted by $\operatorname{sgn}\sigma$, is equal to $+1$ if σ is even (i.e., can be written as a composition of an even number of transpositions), and -1 if σ is odd (see Proposition B.26). The following exercise is an analogue of Exercise 12.13.

▶ **Exercise 12.17.** Show that the following are equivalent for a covariant k-tensor α:

(a) α is alternating.
(b) For any vectors $v_1, \dots, v_k$ and any permutation $\sigma \in S_k$,

$$\alpha(v_{\sigma(1)}, \dots, v_{\sigma(k)}) = (\operatorname{sgn}\sigma)\alpha(v_1, \dots, v_k).$$

(c) With respect to any basis, the components $\alpha_{i_1 \dots i_k}$ of α change sign whenever two indices are interchanged.

Every 0-tensor (which is just a real number) is both symmetric and alternating, because there are no arguments to interchange. Similarly, every 1-tensor is both symmetric and alternating. An alternating 2-tensor on V is a skew-symmetric bilinear form. It is interesting to note that every covariant 2-tensor β can be expressed as a sum of an alternating tensor and a symmetric one, because

$$\beta(v, w) = \tfrac{1}{2}\big(\beta(v, w) - \beta(w, v)\big) + \tfrac{1}{2}\big(\beta(v, w) + \beta(w, v)\big)$$
$$= \alpha(v, w) + \sigma(v, w),$$

where $\alpha(v, w) = \tfrac{1}{2}\big(\beta(v, w) - \beta(w, v)\big)$ is an alternating tensor, and $\sigma(v, w) = \tfrac{1}{2}\big(\beta(v, w) + \beta(w, v)\big)$ is symmetric. This is not true for tensors of higher rank, as Problem 12-7 shows.

There are analogues of symmetrization and symmetric products that apply to alternating tensors, but we will put off introducing them until Chapter 14, where we will study the properties of alternating tensors in much more detail.

Tensors and Tensor Fields on Manifolds

Now let M be a smooth manifold with or without boundary. We define the **bundle of covariant k-tensors on M** by

$$T^k T^* M = \coprod_{p \in M} T^k\big(T_p^* M\big).$$

Analogously, we define the **bundle of contravariant k-tensors** by

$$T^k TM = \coprod_{p \in M} T^k\big(T_p M\big),$$

and the **bundle of mixed tensors of type (k, l)** by

$$T^{(k,l)} TM = \coprod_{p \in M} T^{(k,l)}\big(T_p M\big).$$

There are natural identifications

$$T^{(0,0)}TM = T^0T^*M = T^0TM = M \times \mathbb{R},$$

$$T^{(0,1)}TM = T^1T^*M = T^*M,$$

$$T^{(1,0)}TM = T^1TM = TM,$$

$$T^{(0,k)}TM = T^kT^*M,$$

$$T^{(k,0)}TM = T^kTM.$$

▶ **Exercise 12.18.** Show that T^kT^*M, T^kTM, and $T^{(k,l)}TM$ have natural structures as smooth vector bundles over M, and determine their ranks.

Any one of these bundles is called a **tensor bundle over M**. (Thus, the tangent and cotangent bundles are special cases of tensor bundles.) A section of a tensor bundle is called a **(covariant, contravariant, or mixed) tensor field on M**. A **smooth tensor field** is a section that is smooth in the usual sense of smooth sections of vector bundles. Using the identifications above, we see that contravariant 1-tensor fields are the same as vector fields, and covariant 1-tensor fields are covector fields. Because a 0-tensor is just a real number, a 0-tensor field is the same as a continuous real-valued function.

The spaces of smooth sections of these tensor bundles, $\Gamma(T^kT^*M)$, $\Gamma(T^kTM)$, and $\Gamma(T^{(k,l)}TM)$, are infinite-dimensional vector spaces over $\mathbb{R}$, and modules over $C^\infty(M)$. In any smooth local coordinates (x^i), sections of these bundles can be written (using the summation convention) as

$$A = \begin{cases} A_{i_1 \dots i_k} \, dx^{i_1} \otimes \cdots \otimes dx^{i_k}, & A \in \Gamma(T^kT^*M); \\ A^{i_1 \dots i_k} \dfrac{\partial}{\partial x^{i_1}} \otimes \cdots \otimes \dfrac{\partial}{\partial x^{i_k}}, & A \in \Gamma(T^kTM); \\ A^{i_1 \dots i_k}_{j_1 \dots j_l} \dfrac{\partial}{\partial x^{i_1}} \otimes \cdots \otimes \dfrac{\partial}{\partial x^{i_k}} \otimes dx^{j_1} \otimes \cdots \otimes dx^{j_l}, & A \in \Gamma(T^{(k,l)}TM). \end{cases}$$

The functions $A_{i_1 \dots i_k}$, $A^{i_1 \dots i_k}$, or $A^{i_1 \dots i_k}_{j_1 \dots j_l}$ are called the **component functions of A** in the chosen coordinates. Because smooth covariant tensor fields occupy most of our attention, we adopt the following shorthand notation for the space of all smooth covariant k-tensor fields:

$$\mathcal{T}^k(M) = \Gamma(T^kT^*M).$$

Proposition 12.19 (Smoothness Criteria for Tensor Fields). *Let M be a smooth manifold with or without boundary, and let $A \colon M \to T^kT^*M$ be a rough section. The following are equivalent.*

(a) *A is smooth.*

(b) *In every smooth coordinate chart, the component functions of A are smooth.*

(c) *Each point of M is contained in some coordinate chart in which A has smooth component functions.*

(d) *If $X_1, \ldots, X_k \in \mathfrak{X}(M)$, then the function $A(X_1, \ldots, X_k)\colon M \to \mathbb{R}$, defined by*

$$A(X_1, \ldots, X_k)(p) = A_p\big(X_1|_p, \ldots, X_k|_p\big),$$

is smooth.

(e) *Whenever $X_1, \ldots, X_k$ are smooth vector fields defined on some open subset $U \subseteq M$, the function $A(X_1, \ldots, X_k)$ is smooth on U.*

▶ **Exercise 12.20.** Prove Proposition 12.19.

▶ **Exercise 12.21.** Formulate and prove smoothness criteria analogous to those of Proposition 12.19 for contravariant and mixed tensor fields.

Proposition 12.22. *Suppose M is a smooth manifold with or without boundary, $A \in \mathcal{T}^k(M)$, $B \in \mathcal{T}^l(M)$, and $f \in C^\infty(M)$. Then fA and $A \otimes B$ are also smooth tensor fields, whose components in any smooth local coordinate chart are*

$$(fA)_{i_1 \ldots i_k} = f A_{i_1 \ldots i_k},$$

$$(A \otimes B)_{i_1 \ldots i_{k+l}} = A_{i_1 \ldots i_k} B_{i_{k+1} \ldots i_{k+l}}.$$

▶ **Exercise 12.23.** Prove Proposition 12.22.

Proposition 12.19(d) shows that if A is a smooth covariant k-tensor field on M and $X_1, \ldots, X_k$ are smooth vector fields, then $A(X_1, \ldots, X_k)$ is a smooth real-valued function on M. Thus A induces a map

$$\underbrace{\mathfrak{X}(M) \times \cdots \times \mathfrak{X}(M)}_{k \text{ copies}} \to C^\infty(M).$$

It is easy to see that this map is multilinear over $\mathbb{R}$. In fact, more is true: it is **multilinear over $C^\infty(M)$**, which means that for $f, f' \in C^\infty(M)$ and $X_i, X_i' \in \mathfrak{X}(M)$, we have

$$A\big(X_1, \ldots, fX_i + f'X_i', \ldots, X_k\big)$$
$$= fA(X_1, \ldots, X_i, \ldots, X_k) + f'A\big(X_1, \ldots, X_i', \ldots, X_k\big).$$

This property turns out to be characteristic of smooth tensor fields, as the next lemma shows.

Lemma 12.24 (Tensor Characterization Lemma). *A map*

$$\mathcal{A}\colon \underbrace{\mathfrak{X}(M) \times \cdots \times \mathfrak{X}(M)}_{k \text{ copies}} \to C^\infty(M), \tag{12.7}$$

is induced by a smooth covariant k-tensor field as above if and only if it is multilinear over $C^\infty(M)$.

Proof. We already noted that if A is a smooth covariant k-tensor field, then the map $(X_1, \ldots, X_k) \mapsto A(X_1, \ldots, X_k)$ is multilinear over $C^\infty(M)$. To prove the converse, we proceed as in the proof of the bundle homomorphism characterization lemma (Lemma 10.29).

Suppose, therefore, that $\mathcal{A}$ is a map as in (12.7), and assume that $\mathcal{A}$ is multilinear over $C^\infty(M)$. We show first that $\mathcal{A}$ acts locally. If X_i is a smooth vector field that vanishes on a neighborhood U of p, we can choose a bump function ψ supported in U such that $\psi(p) = 1$; then because $\psi X_i \equiv 0$ we have

$$0 = \mathcal{A}(X_1, \ldots, \psi X_i, \ldots, X_k)(p) = \psi(p)\mathcal{A}(X_1, \ldots, X_i, \ldots, X_k)(p).$$

It follows as in the proof of Lemma 10.29 that the value of $\mathcal{A}(X_1, \ldots, X_k)$ at p depends only on the values of $X_1, \ldots, X_k$ in a neighborhood of p.

Next we show that $\mathcal{A}$ actually acts pointwise. If $X_i|_p = 0$, then in any coordinate chart centered at p we can write $X_i = X_i^j \partial/\partial x^j$, where the component functions X_i^j all vanish at p. By the extension lemma for vector fields, we can find global smooth vector fields E_j on M such that $E_j = \partial/\partial x^j$ in some neighborhood of p; and similarly the locally defined functions X_i^j can be extended to global smooth functions f_i^j on M that agree with X_i^j in a neighborhood of p. It follows from the multilinearity of $\mathcal{A}$ over $C^\infty(M)$ and the fact that $f_i^j E_j = X_i$ in a neighborhood of p that

$$\mathcal{A}(X_1, \ldots, X_i, \ldots, X_k)(p) = \mathcal{A}(X_1, \ldots, f_i^j E_j, \ldots, X_k)(p)$$

$$= f_i^j(p)\mathcal{A}(X_1, \ldots, E_j, \ldots, X_k)(p) = 0.$$

It follows by linearity that $\mathcal{A}(X_1, \ldots, X_k)$ depends only on the value of X_i at p.

Now we define a rough tensor field $A: M \to T^k T^*M$ by

$$A_p(v_1, \ldots, v_k) = \mathcal{A}(V_1, \ldots, V_k)(p)$$

for $p \in M$ and $v_1, \ldots, v_k \in T_pM$, where $V_1, \ldots, V_k$ are any extensions of $v_1, \ldots, v_k$ to smooth global vector fields on M. The discussion above shows that this is independent of the choices of extensions, and the resulting tensor field is smooth by Proposition 12.19(d). □

A *symmetric tensor field* on a manifold (with or without boundary) is simply a covariant tensor field whose value at each point is a symmetric tensor. The symmetric product of two or more tensor fields is defined pointwise, just like the tensor product. Thus, for example, if A and B are smooth covector fields, their symmetric product is the smooth 2-tensor field AB, which by Proposition 12.15(b) is given by

$$AB = \tfrac{1}{2}(A \otimes B + B \otimes A).$$

Alternating tensor fields are called *differential forms*; we will study them in depth beginning in Chapter 14.

Pullbacks of Tensor Fields

Just like covector fields, covariant tensor fields can be pulled back by a smooth map to yield tensor fields on the domain. (This construction works only for covariant tensor fields, which is one reason why we focus most of our attention on the covariant case.)

Suppose $F: M \to N$ is a smooth map. For any point $p \in M$ and any k-tensor $\alpha \in T^k\left(T^*_{F(p)}N\right)$, we define a tensor $dF^*_p(\alpha) \in T^k\left(T^*_p M\right)$, called the **pointwise pullback of α by F at p**, by

$$dF^*_p(\alpha)(v_1,\dots,v_k) = \alpha\big(dF_p(v_1),\dots,dF_p(v_k)\big)$$

for any $v_1,\dots,v_k \in T_p M$. If A is a covariant k-tensor field on N, we define a rough k-tensor field F^*A on M, called the **pullback of A by F**, by

$$(F^*A)_p = dF^*_p\big(A_{F(p)}\big).$$

This tensor field acts on vectors $v_1,\dots,v_k \in T_p M$ by

$$(F^*A)_p(v_1,\dots,v_k) = A_{F(p)}\big(dF_p(v_1),\dots,dF_p(v_k)\big).$$

Proposition 12.25 (Properties of Tensor Pullbacks). *Suppose $F: M \to N$ and $G: N \to P$ are smooth maps, A and B are covariant tensor fields on N, and f is a real-valued function on N.*

(a) $F^*(fB) = (f \circ F)F^*B$.
(b) $F^*(A \otimes B) = F^*A \otimes F^*B$.
(c) $F^*(A + B) = F^*A + F^*B$.
(d) F^*B *is a (continuous) tensor field, and is smooth if B is smooth.*
(e) $(G \circ F)^*B = F^*(G^*B)$.
(f) $(\mathrm{Id}_N)^*B = B$.

▶ **Exercise 12.26.** Prove Proposition 12.25.

If f is a continuous real-valued function (i.e., a 0-tensor field) and B is a k-tensor field, then it is consistent with our definitions to interpret $f \otimes B$ as fB, and F^*f as $f \circ F$. With these interpretations, property (a) of the preceding proposition is really just a special case of (b).

▶ **Exercise 12.27.** Suppose $F: M \to N$ is a smooth map and A, B are symmetric tensor fields on N. Show that F^*A and F^*B are symmetric, and $F^*(AB) = (F^*A)(F^*B)$.

The following corollary is an immediate consequence of Proposition 12.25.

Corollary 12.28. *Let $F: M \to N$ be smooth, and let B be a covariant k-tensor field on N. If $p \in M$ and $\left(y^i\right)$ are smooth coordinates for N on a neighborhood of $F(p)$, then F^*B has the following expression in a neighborhood of p:*

$$F^*\left(B_{i_1\dots i_k}\, dy^{i_1} \otimes \cdots \otimes dy^{i_k}\right)$$
$$= \left(B_{i_1\dots i_k} \circ F\right) d\left(y^{i_1} \circ F\right) \otimes \cdots \otimes d\left(y^{i_k} \circ F\right). \qquad \square$$

In words, this corollary just says that F^*B is computed by the same technique we described in Chapter 11 for computing the pullback of a covector field: wherever you see y^i in the expression for B, just substitute the ith component function of F and expand.

Example 12.29 (Pullback of a Tensor Field). Let $M = \{(r, \theta) : r > 0, |\theta| < \pi/2\}$ and $N = \{(x, y) : x > 0\}$, and let $F : M \to \mathbb{R}^2$ be the smooth map $F(r, \theta) = (r \cos \theta, r \sin \theta)$. The pullback of the tensor field $A = x^{-2} dy \otimes dy$ by F can be computed easily by substituting $x = r \cos \theta$, $y = r \sin \theta$ and simplifying:

$$F^* A = (r \cos \theta)^{-2} d(r \sin \theta) \otimes d(r \sin \theta)$$

$$= (r \cos \theta)^{-2} (\sin \theta \, dr + r \cos \theta \, d\theta) \otimes (\sin \theta \, dr + r \cos \theta \, d\theta)$$

$$= r^{-2} \tan^2 \theta \, dr \otimes dr + r^{-1} \tan \theta (d\theta \otimes dr + dr \otimes d\theta) + d\theta \otimes d\theta. \quad \text{//}$$

In general, there is neither a pushforward nor a pullback operation for mixed tensor fields. However, in the special case of a diffeomorphism, tensor fields of any variance can be pushed forward and pulled back at will (see Problem 12-10).

Lie Derivatives of Tensor Fields

The Lie derivative operation can be extended to tensor fields of arbitrary rank. As usual, we focus on covariant tensors; the analogous results for contravariant or mixed tensors require only minor modifications.

Suppose M is a smooth manifold, V is a smooth vector field on M, and θ is its flow. (For simplicity, we discuss only the case $\partial M = \varnothing$ here, but these definitions and results carry over essentially unchanged to manifolds with boundary as long as V is tangent to the boundary, so that its flow exists by Theorem 9.34.) For any $p \in M$, if t is sufficiently close to zero, then θ_t is a diffeomorphism from a neighborhood of p to a neighborhood of $\theta_t(p)$, so $d(\theta_t)_p^*$ pulls back tensors at $\theta_t(p)$ to ones at p by the formula

$$d(\theta_t)_p^* \big(A_{\theta_t(p)} \big)(v_1, \ldots, v_k) = A_{\theta_t(p)} \big(d(\theta_t)_p(v_1), \ldots, d(\theta_t)_p(v_k) \big).$$

Note that $d(\theta_t)_p^* \big(A_{\theta_t(p)} \big)$ is just the value of the pullback tensor field $\theta_t^* A$ at p.

Given a smooth covariant tensor field A on M, we define the **Lie derivative of A with respect to V**, denoted by $\mathcal{L}_V A$, by

$$(\mathcal{L}_V A)_p = \frac{d}{dt} \bigg|_{t=0} (\theta_t^* A)_p = \lim_{t \to 0} \frac{d(\theta_t)_p^* \big(A_{\theta_t(p)} \big) - A_p}{t}, \tag{12.8}$$

provided the derivative exists (Fig. 12.1). Because the expression being differentiated lies in $T^k \big(T_p^* M \big)$ for all t, $(\mathcal{L}_V A)_p$ makes sense as an element of $T^k (T_p^* M)$. The following lemma is an analogue of Lemma 9.36, and is proved in exactly the same way.

Lemma 12.30. *With M, V, and A as above, the derivative in* (12.8) *exists for every $p \in M$ and defines $\mathcal{L}_V A$ as a smooth tensor field on M.*

▶ **Exercise 12.31.** Prove the preceding lemma.

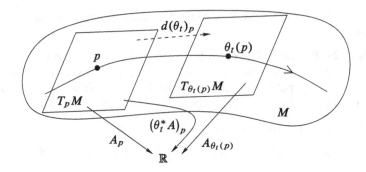

Fig. 12.1 The Lie derivative of a tensor field

Proposition 12.32. *Let M be a smooth manifold and let $V \in \mathfrak{X}(M)$. Suppose f is a smooth real-valued function (regarded as a 0-tensor field) on M, and A, B are smooth covariant tensor fields on M.*

(a) $\mathscr{L}_V f = V f$.
(b) $\mathscr{L}_V (fA) = (\mathscr{L}_V f) A + f \mathscr{L}_V A$.
(c) $\mathscr{L}_V (A \otimes B) = (\mathscr{L}_V A) \otimes B + A \otimes \mathscr{L}_V B$.
(d) *If $X_1, \dots, X_k$ are smooth vector fields and A is a smooth k-tensor field,*

$$\mathscr{L}_V \big(A(X_1, \dots, X_k) \big) = (\mathscr{L}_V A)(X_1, \dots, X_k) + A(\mathscr{L}_V X_1, \dots, X_k)$$

$$+ \cdots + A(X_1, \dots, \mathscr{L}_V X_k). \tag{12.9}$$

Proof. Let θ be the flow of V. For a real-valued function f, we can write

$$\theta_t^* f(p) = f \big(\theta_t(p) \big) = f \circ \theta^{(p)}(t).$$

Thus the definition of $\mathscr{L}_V f$ reduces to the ordinary derivative with respect to t of the composite function $f \circ \theta^{(p)}$. Because $\theta^{(p)}$ is an integral curve of V, it follows from Proposition 3.24 that

$$(\mathscr{L}_V f)(p) = \frac{d}{dt}\bigg|_{t=0} f \circ \theta^{(p)} = df_p\big(\theta^{(p)\prime}(0) \big) = df_p(V_p) = V f(p).$$

This proves (a).

The other assertions can be proved by the technique we used in Theorem 9.38: in a neighborhood of a regular point for V, if (u^i) are coordinates in which $V = \partial/\partial u^1$, then it follows immediately from the definition that $\mathscr{L}_V$ acts on a tensor field simply by taking the partial derivative of its coefficients with respect to u^1, and (b)–(d) all follow from the ordinary product rule. The same relations hold on the support of V by continuity, and on the complement of the support because the flow of V is trivial there.   □

One consequence of this proposition is the following formula expressing the Lie derivative of any smooth covariant tensor field in terms of Lie brackets and ordinary directional derivatives of functions, which allows us to compute Lie derivatives without first determining the flow.

Corollary 12.33. *If V is a smooth vector field and A is a smooth covariant k-tensor field, then for any smooth vector fields $X_1, \ldots, X_k$,*

$$(\mathcal{L}_V A)(X_1, \ldots, X_k) = V\big(A(X_1, \ldots, X_k)\big) - A\big([V, X_1], X_2, \ldots, X_k\big)$$
$$- \cdots - A\big(X_1, \ldots, X_{k-1}, [V, X_k]\big). \tag{12.10}$$

Proof. Just solve (12.9) for $(\mathcal{L}_V A)(X_1, \ldots, X_k)$, and replace $\mathcal{L}_V f$ by Vf and $\mathcal{L}_V X_i$ by $[V, X_i]$. $\qquad\square$

Corollary 12.34. *If $f \in C^\infty(M)$, then $\mathcal{L}_V(df) = d(\mathcal{L}_V f)$.*

Proof. Using (12.10), for any $X \in \mathfrak{X}(M)$ we compute

$$(\mathcal{L}_V df)(X) = V\big(df(X)\big) - df\big([V, X]\big) = VXf - [V, X]f$$
$$= VXf - (VXf - XVf) = XVf$$
$$= d(Vf)(X) = d(\mathcal{L}_V f)(X). \qquad\square$$

One drawback of formula (12.10) is that in order to calculate what $\mathcal{L}_V A$ does to vectors $v_1, \ldots, v_k$ at a point $p \in M$, one must first extend them to vector fields in a neighborhood of p. But Proposition 12.32 and Corollary 12.34 lead to an easy method for computing Lie derivatives of smooth tensor fields in coordinates that avoids this problem, since any tensor field can be written locally as a linear combination of functions multiplied by tensor products of exact 1-forms. The next example illustrates the technique.

Example 12.35. Suppose A is an arbitrary smooth covariant 2-tensor field, and V is a smooth vector field. We compute the Lie derivative $\mathcal{L}_V A$ in smooth local coordinates (x^i). First, we observe that $\mathcal{L}_V dx^i = d(\mathcal{L}_V x^i) = d(Vx^i) = dV^i$. Therefore,

$$\mathcal{L}_V A = \mathcal{L}_V\big(A_{ij} dx^i \otimes dx^j\big)$$
$$= \mathcal{L}_V(A_{ij}) dx^i \otimes dx^j + A_{ij}\big(\mathcal{L}_V dx^i\big) \otimes dx^j + A_{ij} dx^i \otimes \big(\mathcal{L}_V dx^j\big)$$
$$= VA_{ij}\, dx^i \otimes dx^j + A_{ij}\, dV^i \otimes dx^j + A_{ij}\, dx^i \otimes dV^j$$
$$= \left(VA_{ij} + A_{kj}\frac{\partial V^k}{\partial x^i} + A_{ik}\frac{\partial V^k}{\partial x^j}\right) dx^i \otimes dx^j. \qquad /\!/$$

Recall that the Lie derivative of a vector field W with respect to V is zero if and only if W is invariant under the flow of V (see Theorem 9.42). It turns out that the Lie derivative of a covariant tensor field has exactly the same interpretation. If A is a smooth tensor field on M and θ is a flow on M, we say that A **is invariant under** θ if for each t, the map θ_t pulls A back to itself wherever it is defined; more precisely, this means

$$d(\theta_t)^*_p\big(A_{\theta_t(p)}\big) = A_p \tag{12.11}$$

for all (t, p) in the domain of θ. If θ is a global flow, this is equivalent to $\theta_t^* A = A$ for all $t \in \mathbb{R}$.

In order to prove the connection between Lie derivatives and invariance under flows, we need the following proposition, which shows how the Lie derivative can be used to compute t-derivatives at times other than $t = 0$. It is a generalization to tensor fields of Proposition 9.41.

Proposition 12.36. *Suppose M is a smooth manifold with or without boundary and $V \in \mathfrak{X}(M)$. If $\partial M \neq \varnothing$, assume in addition that V is tangent to ∂M. Let θ be the flow of V. For any smooth covariant tensor field A and any (t_0, p) in the domain of θ,*

$$\left.\frac{d}{dt}\right|_{t=t_0} \left(\theta_t^* A\right)_p = \left(\theta_{t_0}^* (\mathscr{L}_V A)\right)_p . \tag{12.12}$$

Proof. After expanding the definitions of the pullbacks in (12.12), we see that we have to prove

$$\left.\frac{d}{dt}\right|_{t=t_0} d(\theta_t)_p^* \left(A_{\theta_t(p)}\right) = d(\theta_{t_0})_p^* \left((\mathscr{L}_V A)_{\theta_{t_0}(p)}\right).$$

Just as in the proof of Proposition 9.41, the change of variables $t = s + t_0$ yields

$$\left.\frac{d}{dt}\right|_{t=t_0} d(\theta_t)_p^* \left(A_{\theta_t(p)}\right) = \left.\frac{d}{ds}\right|_{s=0} d(\theta_{s+t_0})_p^* \left(A_{\theta_{s+t_0}(p)}\right)$$

$$= \left.\frac{d}{ds}\right|_{s=0} d(\theta_{t_0})_p^* d(\theta_s)_{\theta_{t_0}(p)}^* \left(A_{\theta_s(\theta_{t_0}(p))}\right)$$

$$= d(\theta_{t_0})_p^* \left.\frac{d}{ds}\right|_{s=0} d(\theta_s)_{\theta_{t_0}(p)}^* \left(A_{\theta_s(\theta_{t_0}(p))}\right)$$

$$= d(\theta_{t_0})_p^* \left((\mathscr{L}_V A)_{\theta_{t_0}(p)}\right). \qquad \square$$

Theorem 12.37. *Let M be a smooth manifold and let $V \in \mathfrak{X}(M)$. A smooth covariant tensor field A is invariant under the flow of V if and only if $\mathscr{L}_V A = 0$.*

▶ **Exercise 12.38.** Prove Theorem 12.37.

Problems

12-1. Give an example of finite-dimensional vector spaces V and W and a specific element $\alpha \in V \otimes W$ that cannot be expressed as $v \otimes w$ for $v \in V$ and $w \in W$.

12-2. For any finite-dimensional real vector space V, prove that there are canonical isomorphisms $\mathbb{R} \otimes V \cong V \cong V \otimes \mathbb{R}$.

12-3. Let V and W be finite-dimensional real vector spaces. Show that the tensor product space $V \otimes W$ is uniquely determined up to canonical isomorphism by its characteristic property (Proposition 12.7). More precisely, suppose $\tilde{\pi} : V \times W \to Z$ is a bilinear map into a vector space Z with the following property: for any bilinear map $B \colon V \times W \to Y$, there is a unique linear map $\widetilde{B} \colon Z \to Y$ such that the following diagram commutes:

Then there is a unique isomorphism $\Phi \colon V \otimes W \to Z$ such that $\widetilde{\pi} = \Phi \circ \pi$, where $\pi \colon V \times W \to V \otimes W$ is the canonical projection. [Remark: this shows that the details of the construction used to define the tensor product space are irrelevant, as long as the resulting space satisfies the characteristic property.]

12-4. Let $V_1, \ldots, V_k$ and W be finite-dimensional real vector spaces. Prove that there is a canonical (basis-independent) isomorphism

$$V_1^* \otimes \cdots \otimes V_k^* \otimes W \cong L(V_1, \ldots, V_k; W).$$

(In particular, this means that $V^* \otimes W$ is canonically isomorphic to the space $L(V; W)$ of linear maps from V to W.)

12-5. Let V be an n-dimensional real vector space. Show that

$$\dim \Sigma^k(V^*) = \binom{n+k-1}{k} = \frac{(n+k-1)!}{k!(n-1)!}.$$

12-6. (a) Let α be a covariant k-tensor on a finite-dimensional real vector space V. Show that $\operatorname{Sym} \alpha$ is the unique symmetric k-tensor satisfying

$$(\operatorname{Sym} \alpha)(v, \ldots, v) = \alpha(v, \ldots, v)$$

for all $v \in V$.

(b) Show that the symmetric product is associative: for all symmetric tensors α, β, γ,

$$(\alpha\beta)\gamma = \alpha(\beta\gamma).$$

[Hint: use part (a).]

(c) Let $\omega^1, \ldots, \omega^k$ be covectors on a finite-dimensional vector space. Show that their symmetric product satisfies

$$\omega^1 \cdots \omega^k = \frac{1}{k!} \sum_{\sigma \in S_k} \omega^{\sigma(1)} \otimes \cdots \otimes \omega^{\sigma(k)}.$$

12-7. Let (e^1, e^2, e^3) be the standard dual basis for $(\mathbb{R}^3)^*$. Show that $e^1 \otimes e^2 \otimes e^3$ is not equal to a sum of an alternating tensor and a symmetric tensor.

12-8. Let M be a smooth n-manifold, and let A be a smooth covariant k-tensor field on M. If $(U, (x^i))$ and $(\tilde{U}, (\tilde{x}^j))$ are overlapping smooth charts on M, we can write

$$A = A_{i_1 \ldots i_k} \, dx^{i_1} \otimes \cdots \otimes dx^{i_k} = \tilde{A}_{j_1 \ldots j_k} \, d\tilde{x}^{j_1} \otimes \cdots \otimes d\tilde{x}^{j_k}.$$

Compute a transformation law analogous to (11.7) expressing the component functions $A_{i_1 \ldots i_k}$ in terms of $\tilde{A}_{j_1 \ldots j_k}$.

12-9. Generalize the coordinate transformation law of Problem 12-8 to mixed tensor fields of any rank.

12-10. Show that for every pair of nonnegative integers k, l and every diffeomorphism $F : M \to N$, there are pushforward and pullback isomorphisms

$$F_* : \Gamma\left(T^{(k,l)} TM\right) \to \Gamma\left(T^{(k,l)} TN\right),$$

$$F^* : \Gamma\left(T^{(k,l)} TN\right) \to \Gamma\left(T^{(k,l)} TM\right),$$

such that F^* agrees with the usual pullback on covariant tensor fields, F_* agrees with the usual pushforward on contravariant 1-tensor fields (i.e., vector fields), and the following conditions are satisfied:
(a) $F_* = (F^*)^{-1}$.
(b) $F^*(A \otimes B) = F^* A \otimes F^* B$.
(c) $(F \circ G)_* = F_* \circ G_*$.
(d) $(F \circ G)^* = G^* \circ F^*$.
(e) $(\mathrm{Id}_M)^* = (\mathrm{Id}_M)_* = \mathrm{Id} : \Gamma\left(T^{(k,l)} TM\right) \to \Gamma\left(T^{(k,l)} TM\right)$.
(f) $F^*\left(A(X_1, \ldots, X_k)\right) = F^* A\left(F_*^{-1}(X_1), \ldots, F_*^{-1}(X_k)\right)$ for $A \in \mathcal{T}^k(N)$ and $X_1, \ldots, X_k \in \mathfrak{X}(N)$.

12-11. Suppose M is a smooth manifold, A is a smooth covariant tensor field on M, and $V, W \in \mathfrak{X}(M)$. Show that

$$\mathcal{L}_V \mathcal{L}_W A - \mathcal{L}_W \mathcal{L}_V A = \mathcal{L}_{[V,W]} A.$$

[Hint: use induction on the rank of A, beginning with Corollary 12.33 for 1-tensor fields.]

12-12. Let M be a smooth manifold and $V \in \mathfrak{X}(M)$. Show that the Lie derivative operators on covariant tensor fields, $\mathcal{L}_V : \mathcal{T}^k(M) \to \mathcal{T}^k(M)$ for $k \geq 0$, are uniquely characterized by the following properties:
(a) $\mathcal{L}_V$ is linear over $\mathbb{R}$.
(b) $\mathcal{L}_V f = Vf$ for $f \in \mathcal{T}^0(M) = C^\infty(M)$.
(c) $\mathcal{L}_V(A \otimes B) = \mathcal{L}_V A \otimes B + A \otimes \mathcal{L}_V B$ for $A \in \mathcal{T}^k(M)$, $B \in \mathcal{T}^l(M)$.
(d) $\mathcal{L}_V(\omega(X)) = (\mathcal{L}_V \omega)(X) + \omega([V, X])$ for $\omega \in \mathcal{T}^1(M)$, $X \in \mathfrak{X}(M)$.
[Remark: the Lie derivative operators on tensor fields are sometimes *defined* as the unique operators satisfying these properties. This definition has the virtue of making sense on a manifold with boundary, where the flow of V might not exist.]

Chapter 13
Riemannian Metrics

In this chapter, for the first time, we introduce *geometry* into smooth manifold theory. As is so much of this subject, our approach to geometry is modeled on the theory of finite-dimensional vector spaces. To define geometric concepts such as lengths and angles on a vector space, one uses an inner product. For manifolds, the appropriate structure is a *Riemannian metric*, which is essentially a choice of inner product on each tangent space, varying smoothly from point to point. A choice of Riemannian metric allows us to define geometric concepts such as lengths, angles, and distances on smooth manifolds.

Riemannian geometry is a deep subject in its own right. To develop all of the machinery needed for a complete treatment of it would require another whole book. (If you want to dig more deeply into it than we can here, you might start with [LeeRM], which gives a concise introduction to the subject, and for which you already have most of the necessary background.) But we can at least introduce the main definitions and some important examples.

After defining Riemannian metrics and the main constructions associated with them, we show how submanifolds of Riemannian manifolds inherit induced Riemannian metrics. Then we show how a Riemannian metric leads to a distance function, which allows us to consider connected Riemannian manifolds as metric spaces.

At the end of the chapter, we briefly describe a generalization of Riemannian metrics, called *pseudo-Riemannian metrics*, which play a central role in Einstein's general theory of relativity.

Riemannian Manifolds

The most important examples of symmetric tensors on a vector space are inner products. Any inner product allows us to define lengths of vectors and angles between them, and thus to do Euclidean geometry.

Transferring these ideas to manifolds, we obtain one of the most important applications of tensors to differential geometry. Let M be a smooth manifold with or without boundary. A ***Riemannian metric on M*** is a smooth symmetric covariant

2-tensor field on M that is positive definite at each point. A ***Riemannian manifold*** is a pair (M, g), where M is a smooth manifold and g is a Riemannian metric on M. One sometimes simply says "M is a Riemannian manifold" if M is understood to be endowed with a specific Riemannian metric. A ***Riemannian manifold with boundary*** is defined similarly.

Note that a Riemannian metric is not the same thing as a metric in the sense of metric spaces, although the two concepts are related, as we will see below. Because of this ambiguity, we usually use the term "distance function" for a metric in the metric space sense, and reserve "metric" for a Riemannian metric. In any event, which type of metric is being considered should always be clear from the context.

If g is a Riemannian metric on M, then for each $p \in M$, the 2-tensor g_p is an inner product on $T_p M$. Because of this, we often use the notation $\langle v, w \rangle_g$ to denote the real number $g_p(v, w)$ for $v, w \in T_p M$.

In any smooth local coordinates (x^i), a Riemannian metric can be written

$$g = g_{ij}\, dx^i \otimes dx^j,$$

where (g_{ij}) is a symmetric positive definite matrix of smooth functions. The symmetry of g allows us to write g also in terms of symmetric products as follows:

$$
\begin{aligned}
g &= g_{ij}\, dx^i \otimes dx^j \\
&= \tfrac{1}{2}\big(g_{ij}\, dx^i \otimes dx^j + g_{ji}\, dx^i \otimes dx^j\big) \quad \text{(since } g_{ij} = g_{ji}) \\
&= \tfrac{1}{2}\big(g_{ij}\, dx^i \otimes dx^j + g_{ij}\, dx^j \otimes dx^i\big) \quad \text{(switch } i \leftrightarrow j \text{ in the second term)} \\
&= g_{ij}\, dx^i\, dx^j \quad\quad\quad\quad\quad\quad\quad\quad\quad \text{(by Proposition 12.15(b)).}
\end{aligned}
$$

Example 13.1 (The Euclidean Metric). The simplest example of a Riemannian metric is the ***Euclidean metric*** $\bar{g}$ on $\mathbb{R}^n$, given in standard coordinates by

$$\bar{g} = \delta_{ij}\, dx^i\, dx^j,$$

where δ_{ij} is the Kronecker delta. It is common to abbreviate the symmetric product of a tensor α with itself by α^2, so the Euclidean metric can also be written

$$\bar{g} = \big(dx^1\big)^2 + \cdots + \big(dx^n\big)^2.$$

Applied to vectors $v, w \in T_p \mathbb{R}^n$, this yields

$$\bar{g}_p(v, w) = \delta_{ij} v^i w^j = \sum_{i=1}^{n} v^i w^i = v \cdot w.$$

In other words, $\bar{g}$ is the 2-tensor field whose value at each point is the Euclidean dot product. (As you may recall, we warned in Chapter 1 that expressions involving the Euclidean dot product are likely to violate our index conventions and therefore to require explicit summation signs. This can usually be avoided by writing the metric coefficients δ_{ij} explicitly, as in $\delta_{ij} v^i w^j$.)   //

Example 13.2 (Product Metrics). If (M, g) and $(\widetilde{M}, \widetilde{g})$ are Riemannian manifolds, we can define a Riemannian metric $\widehat{g} = g \oplus \widetilde{g}$ on the product manifold $M \times \widetilde{M}$, called the **product metric**, as follows:

$$\widehat{g}\big((v, \widetilde{v}), (w, \widetilde{w})\big) = g(v, w) + \widetilde{g}(\widetilde{v}, \widetilde{w}) \tag{13.1}$$

for any $(v, \widetilde{v}), (w, \widetilde{w}) \in T_p M \oplus T_q \widetilde{M} \cong T_{(p,q)}(M \times \widetilde{M})$. Given any local coordinates $(x^1, \ldots, x^n)$ for M and $(y^1, \ldots, y^m)$ for $\widetilde{M}$, we obtain local coordinates $(x^1, \ldots, x^n, y^1, \ldots, y^m)$ for $M \times \widetilde{M}$, and you can check that the product metric is represented locally by the block diagonal matrix

$$(\widehat{g}_{ij}) = \begin{pmatrix} g_{ij} & 0 \\ 0 & \widetilde{g}_{ij} \end{pmatrix}.$$

For example, it is easy to verify that the Euclidean metric on $\mathbb{R}^{n+m}$ is the same as the product metric determined by the Euclidean metrics on $\mathbb{R}^n$ and $\mathbb{R}^m$. //

One pleasant feature of Riemannian metrics is that they exist in great abundance. (For another approach, see Problem 13-18.)

Proposition 13.3 (Existence of Riemannian Metrics). *Every smooth manifold with or without boundary admits a Riemannian metric.*

Proof. Let M be a smooth manifold with or without boundary, and choose a covering of M by smooth coordinate charts $(U_\alpha, \varphi_\alpha)$. In each coordinate domain, there is a Riemannian metric $g_\alpha = \varphi_\alpha^* \overline{g}$, whose coordinate expression is $\delta_{ij}\, dx^i\, dx^j$. Let $\{\psi_\alpha\}$ be a smooth partition of unity subordinate to the cover $\{U_\alpha\}$, and define

$$g = \sum_\alpha \psi_\alpha g_\alpha,$$

with each term interpreted to be zero outside supp ψ_α. By local finiteness, there are only finitely many nonzero terms in a neighborhood of each point, so this expression defines a smooth tensor field. It is obviously symmetric, so only positivity needs to be checked. If $v \in T_p M$ is any nonzero vector, then

$$g_p(v, v) = \sum_\alpha \psi_\alpha(p) g_\alpha|_p(v, v).$$

This sum is nonnegative, because each term is nonnegative. At least one of the functions ψ_α is strictly positive at p (because they sum to 1). Because $g_\alpha|_p(v, v) > 0$, it follows that $g_p(v, v) > 0$. □

It is important to observe that there is an enormous amount of choice in the construction of a metric g for a given manifold, so there is nothing canonical about it. In particular, different metrics on the same manifold can have vastly different geometric properties. For example, Problem 13-20 describes four metrics on $\mathbb{R}^2$ that behave in strikingly different ways.

Below are just a few of the geometric constructions that can be defined on a Riemannian manifold (M, g) with or without boundary.

- The **length** or **norm** of a tangent vector $v \in T_pM$ is defined to be

$$|v|_g = \langle v, v \rangle_g^{1/2} = g_p(v, v)^{1/2}.$$

- The **angle** between two nonzero tangent vectors $v, w \in T_pM$ is the unique $\theta \in [0, \pi]$ satisfying

$$\cos \theta = \frac{\langle v, w \rangle_g}{|v|_g |w|_g}.$$

- Tangent vectors $v, w \in T_pM$ are said to be **orthogonal** if $\langle v, w \rangle_g = 0$. This means either one or both vectors are zero, or the angle between them is $\pi/2$.

One highly useful tool for the study of Riemannian manifolds is *orthonormal frames*. Let (M, g) be an n-dimensional Riemannian manifold with or without boundary. Just as we did for the case of $\mathbb{R}^n$ in Chapter 8, we say that a local frame $(E_1, \ldots, E_n)$ for M on an open subset $U \subseteq M$ is an **orthonormal frame** if the vectors $(E_1|_p, \ldots, E_n|_p)$ form an orthonormal basis for T_pM at each point $p \in U$, or equivalently if $\langle E_i, E_j \rangle_g = \delta_{ij}$.

Example 13.4. The coordinate frame $\left(\partial/\partial x^i \right)$ is a global orthonormal frame for $\mathbb{R}^n$ with the Euclidean metric. //

Example 13.5. The frame (E_1, E_2) on $\mathbb{R}^2 \smallsetminus \{0\}$ defined in Example 8.12 is a local orthonormal frame for $\mathbb{R}^2$. As we observed in Example 9.45, it is not a coordinate frame in any coordinates. //

The next proposition is proved in just the same way as Lemma 8.13, with the Euclidean dot product replaced by the inner product $\langle \cdot, \cdot \rangle_g$.

Proposition 13.6. *Suppose (M, g) is a Riemannian manifold with or without boundary, and (X_j) is a smooth local frame for M over an open subset $U \subseteq M$. Then there is a smooth orthonormal frame (E_j) over U such that* $\operatorname{span}\left(E_1|_p, \ldots, E_j|_p \right) = \operatorname{span}\left(X_1|_p, \ldots, X_j|_p \right)$ *for each $j = 1, \ldots, n$ and each $p \in U$.*

▶ **Exercise 13.7.** Prove the preceding proposition.

Corollary 13.8 (Existence of Local Orthonormal Frames). *Let (M, g) be a Riemannian manifold with or without boundary. For each $p \in M$, there is a smooth orthonormal frame on a neighborhood of p.*

Proof. Start with a smooth coordinate frame and apply Proposition 13.6. □

Observe that Corollary 13.8 does *not* show that there are smooth coordinates on a neighborhood of p for which the *coordinate frame* is orthonormal. In fact, this is rarely the case, as we will see below.

Pullback Metrics

Suppose M, N are smooth manifolds with or without boundary, g is a Riemannian metric on N, and $F: M \to N$ is smooth. The pullback F^*g is a smooth 2-tensor

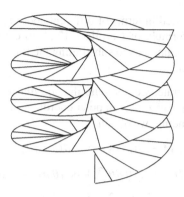

Fig. 13.1 A helicoid

field on M. If it is positive definite, it is a Riemannian metric on M, called the *pullback metric* determined by F. The next proposition shows when this is the case.

Proposition 13.9 (Pullback Metric Criterion). *Suppose $F: M \to N$ is a smooth map and g is a Riemannian metric on N. Then F^*g is a Riemannian metric on M if and only if F is a smooth immersion.*

▶ **Exercise 13.10.** Prove the preceding proposition.

If the coordinate representation for an immersion is known, then the pullback metric is easy to compute using the usual algorithm for computing pullbacks.

Example 13.11. Consider the smooth map $F: \mathbb{R}^2 \to \mathbb{R}^3$ given by

$$F(u, v) = (u \cos v, u \sin v, v).$$

It is a proper injective smooth immersion, and thus it is an embedding by Proposition 4.22. Its image is a surface called a *helicoid*; it looks like an infinitely wide spiral-shaped ramp (Fig. 13.1). The pullback metric $F^*\bar{g}$ can be computed by substituting the coordinate functions for F in place of x, y, z in the formula for $\bar{g}$:

$$
\begin{aligned}
F^*\bar{g} &= d(u \cos v)^2 + d(u \sin v)^2 + d(v)^2 \\
&= (\cos v \, du - u \sin v \, dv)^2 + (\sin v \, du + u \cos v \, dv)^2 + dv^2 \\
&= \cos^2 v \, du^2 - 2u \sin v \cos v \, du \, dv + u^2 \sin^2 v \, dv^2 \\
&\quad + \sin^2 v \, du^2 + 2u \sin v \cos v \, du \, dv + u^2 \cos^2 v \, dv^2 + dv^2 \\
&= du^2 + (u^2 + 1)dv^2.
\end{aligned}
$$

(By convention, when u is a real-valued function, the notation du^2 means the symmetric product $du \, du$, not $d(u^2)$.) //

To transform a Riemannian metric under a change of coordinates, we use the same technique as we used for covector fields: think of the change of coordinates as the identity map expressed in terms of different coordinates for the domain and codomain, and use the formula of Corollary 12.28 to compute the pullback. As before, in practice this just amounts to substituting the formulas for one set of coordinates in terms of the other.

Example 13.12. To illustrate, we compute the coordinate expression for the Euclidean metric $\bar{g} = dx^2 + dy^2$ on $\mathbb{R}^2$ in polar coordinates. Substituting $x = r \cos \theta$ and $y = r \sin \theta$ and expanding, we obtain

$$\bar{g} = dx^2 + dy^2 = d(r \cos \theta)^2 + d(r \sin \theta)^2$$
$$= (\cos \theta \, dr - r \sin \theta \, d\theta)^2 + (\sin \theta \, dr + r \cos \theta \, d\theta)^2$$
$$= \left(\cos^2 \theta + \sin^2 \theta\right) dr^2 + \left(r^2 \sin^2 \theta + r^2 \cos^2 \theta\right) d\theta^2$$
$$\quad + (-2r \cos \theta \sin \theta + 2r \sin \theta \cos \theta) dr \, d\theta$$
$$= dr^2 + r^2 \, d\theta^2. \qquad\qquad //$$

If (M, g) and $(\widetilde{M}, \widetilde{g})$ are both Riemannian manifolds, a smooth map $F \colon M \to \widetilde{M}$ is called a (**Riemannian**) **isometry** if it is a diffeomorphism that satisfies $F^* \widetilde{g} = g$. More generally, F is called a **local isometry** if every point $p \in M$ has a neighborhood U such that $F|_U$ is an isometry of U onto an open subset of $\widetilde{M}$; or equivalently, if F is a local diffeomorphism satisfying $F^* \widetilde{g} = g$.

If there exists a Riemannian isometry between (M, g) and $(\widetilde{M}, \widetilde{g})$, we say that they are **isometric** as Riemannian manifolds. If each point of M has a neighborhood that is isometric to an open subset of $(\widetilde{M}, \widetilde{g})$, then we say that (M, g) is **locally isometric** to $(\widetilde{M}, \widetilde{g})$. The study of properties of Riemannian manifolds that are invariant under (local or global) isometries is called **Riemannian geometry**.

One such property is *flatness*. A Riemannian n-manifold (M, g) is said to be a *flat Riemannian manifold*, and g is a *flat metric*, if (M, g) is locally isometric to $(\mathbb{R}^n, \bar{g})$.

▶ **Exercise 13.13.** Suppose (M, g) and $(\widetilde{M}, \widetilde{g})$ are isometric Riemannian manifolds. Show that g is flat if and only if $\widetilde{g}$ is flat.

The next theorem is the key to deciding whether a Riemannian metric is flat.

Theorem 13.14. *For a Riemannian manifold (M, g), the following are equivalent*:

(a) *g is flat.*
(b) *Each point of M is contained in the domain of a smooth coordinate chart in which g has the coordinate representation $g = \delta_{ij} \, dx^i \, dx^j$.*
(c) *Each point of M is contained in the domain of a smooth coordinate chart in which the coordinate frame is orthonormal.*
(d) *Each point of M is contained in the domain of a commuting orthonormal frame.*

Proof. The implications (a) $\Rightarrow$ (b) $\Rightarrow$ (c) $\Rightarrow$ (d) are easy consequences of the definitions, and are left to the reader. The remaining implication, (d) $\Rightarrow$ (a), follows from the canonical form theorem for commuting frames: if (E_i) is a commuting orthonormal frame for g on an open subset $U \subseteq M$, then Theorem 9.46 implies that each $p \in V$ is contained in the domain of a smooth chart (U, φ) in which the coordinate frame is equal to (E_i). This means $\varphi_* E_i = \partial/\partial x^i$, so the diffeomorphism $\varphi \colon U \to \varphi(U)$ satisfies

$$\varphi^* \overline{g}(E_i, E_j) = \overline{g}(\varphi_* E_i, \varphi_* E_j) = \overline{g}\left(\frac{\partial}{\partial x^i}, \frac{\partial}{\partial x^j}\right) = \delta_{ij} = g(E_i, E_j).$$

Bilinearity then shows that $\varphi^* \overline{g} = g$, so φ is an isometry between $(U, g|_U)$ and $\varphi(U)$ with the Euclidean metric. This shows that g is flat. $\qquad\square$

▶ **Exercise 13.15.** Complete the preceding proof by showing (a) $\Rightarrow$ (b) $\Rightarrow$ (c) $\Rightarrow$ (d).

It is not at all obvious from the definitions that there exist Riemannian metrics that are not flat. In fact, in the 1-dimensional case, every metric is flat, as Problem 13-6 shows. Later in this chapter, we will use Theorem 13.14 to show that most surfaces of revolution in $\mathbb{R}^3$, including $\mathbb{S}^2$, are not flat.

Riemannian Submanifolds

Pullback metrics are especially important for submanifolds. If (M, g) is a Riemannian manifold with or without boundary, every submanifold $S \subseteq M$ (immersed or embedded, with or without boundary) automatically inherits a pullback metric $\iota^* g$, where $\iota \colon S \hookrightarrow M$ is inclusion. In this setting, the pullback metric is also called the *induced metric* on S. By definition, this means for $v, w \in T_p S$ that

$$\left(\iota^* g\right)(v, w) = g\big(d\iota_p(v), d\iota_p(w)\big) = g(v, w),$$

because $d\iota_p \colon T_p S \to T_p M$ is our usual identification of $T_p S$ as a subspace of $T_p M$. Thus $\iota^* g$ is just the restriction of g to pairs of vectors tangent to S. With this metric, S is called a *Riemannian submanifold (with or without boundary) of M*.

Example 13.16. The metric $\mathring{g} = \iota^* \overline{g}$ induced on $\mathbb{S}^n$ by the usual inclusion $\iota \colon \mathbb{S}^n \hookrightarrow \mathbb{R}^{n+1}$ is called the *round metric* (or the *standard metric*) on the sphere. $\qquad\parallel$

If (M, g) is a Riemannian manifold and $\iota \colon S \hookrightarrow M$ is a Riemannian submanifold, it is usually easiest to compute the induced metric $\iota^* g$ in terms of a *local parametrization*; recall from Chapter 5 that this is an injective immersion X from an open subset $U \subseteq \mathbb{R}^k$ into M whose image is an open subset of S, and whose inverse is a smooth coordinate map for S. Since $\iota \circ X = X$, the coordinate representation of $\iota^* g$ is $X^* (\iota^* g) = X^* g$. The next two examples illustrate the procedure.

Example 13.17 (Induced Metrics in Graph Coordinates). Let $U \subseteq \mathbb{R}^n$ be an open subset, and let $S \subseteq \mathbb{R}^{n+1}$ be the graph of a smooth function $f \colon U \to \mathbb{R}$. The map $X \colon U \to \mathbb{R}^{n+1}$ given by $X(u^1, \ldots, u^n) = (u^1, \ldots, u^n, f(u))$ is a smooth global parametrization of S and the induced metric on S is given in graph coordinates by

$$X^*\bar{g} = X^* \left((dx^1)^2 + \cdots + (dx^{n+1})^2 \right) = (du^1)^2 + \cdots + (du^n)^2 + df^2.$$

For example, the upper hemisphere of $\mathbb{S}^2$ is parametrized by the map $X \colon \mathbb{B}^2 \to \mathbb{R}^3$ given by

$$X(u,v) = \left(u, v, \sqrt{1 - u^2 - v^2} \right).$$

In these coordinates, the round metric can be written

$$\mathring{g} = X^*\bar{g} = du^2 + dv^2 + \left(\frac{u\,du + v\,dv}{\sqrt{1 - u^2 - v^2}} \right)^2$$

$$= \frac{(1 - v^2)\,du^2 + (1 - u^2)\,dv^2 + 2uv\,du\,dv}{1 - u^2 - v^2}. \qquad \text{//}$$

Example 13.18 (Induced Metrics on Surfaces of Revolution). Let C be an embedded 1-dimensional submanifold of the half-plane $\{(r,z) : r > 0\}$, and let S_C be the surface of revolution generated by C as described in Example 5.17. To compute the induced metric on S_C, choose any smooth local parametrization $\gamma(t) = (a(t), b(t))$ for C, and note that the map $X(t, \theta) = (a(t)\cos\theta, a(t)\sin\theta, b(t))$ yields a smooth local parametrization of S_C, provided that (t, θ) is restricted to a sufficiently small open subset of the plane. Thus we can compute

$$X^*\bar{g} = d\left(a(t)\cos\theta \right)^2 + d\left(a(t)\sin\theta \right)^2 + d\left(b(t) \right)^2$$

$$= \left(a'(t)\cos\theta\,dt - a(t)\sin\theta\,d\theta \right)^2$$

$$+ \left(a'(t)\sin\theta\,dt + a(t)\cos\theta\,d\theta \right)^2 + \left(b'(t)\,dt \right)^2$$

$$= \left(a'(t)^2 + b'(t)^2 \right) dt^2 + a(t)^2\,d\theta^2.$$

In particular, if γ is a **unit-speed curve**, meaning that $|\gamma'(t)|^2 = a'(t)^2 + b'(t)^2 = 1$, this reduces to the simple formula $dt^2 + a(t)^2\,d\theta^2$.

Here are some familiar examples of surfaces of revolution.

(a) The embedded torus described in Example 5.17 is the surface of revolution generated by the circle $(r - 2)^2 + z^2 = 1$. Using the unit-speed parametrization $\gamma(t) = (2 + \cos t, \sin t)$ for the circle, we obtain the formula $dt^2 + (2 + \cos t)^2\,d\theta^2$ for the induced metric.

(b) The unit sphere (minus the north and south poles) is a surface of revolution whose generating curve is the semicircle parametrized by $\gamma(t) = (\sin t, \cos t)$ for $0 < t < \pi$. The induced metric is $dt^2 + \sin^2 t\,d\theta^2$.

(c) The unit cylinder $x^2 + y^2 = 1$ is a surface of revolution whose generating curve is the vertical line parametrized by $\gamma(t) = (1, t)$ for $t \in \mathbb{R}$. The induced metric is $dt^2 + d\theta^2$. //

Look again at the last example above. It shows that for each local parametrization of the cylinder given by $X(t, \theta) = (\cos \theta, \sin \theta, t)$, the induced metric $X^* \bar{g}$ is the Euclidean metric on the (t, θ)-plane. To put it another way, for any point p in the cylinder, a suitable restriction of X gives a Riemannian isometry between an open subset of $(\mathbb{R}^2, \bar{g})$ and a neighborhood of p in the cylinder with its induced metric. Thus the induced metric on the cylinder is *flat*. A two-dimensional being living in the cylinder would not be able to distinguish its surroundings from the Euclidean plane by local geometric measurements. This illustrates that the question of whether a metric is flat or not can sometimes have an unexpected answer.

To develop adequate machinery to determine systematically which metrics are flat and which are not would require techniques that are beyond the scope of this book. Just as proving two topological spaces are not homeomorphic requires finding topological invariants that distinguish them, in order to prove two Riemannian manifolds are not locally isometric, one must introduce local invariants that are preserved by Riemannian isometries, and show that different metrics have different invariants. The fundamental invariant of a Riemannian metric is called its *curvature*; this is a quantitative measure of how far the metric deviates from flatness. See, for example, [LeeRM] for an account of the theory of Riemannian curvature.

For the present, we have to content ourselves with the next proposition, which answers the question for surfaces of revolution using a rather ad hoc method.

Proposition 13.19 (Flatness Criterion for Surfaces of Revolution). *Let $C \subseteq H$ be a connected embedded 1-dimensional submanifold of the half-plane $H = \{(r, z) : r > 0\}$, and let S_C be the surface of revolution generated by C. The induced metric on S_C is flat if and only if C is part of a straight line.*

Proof. First assume C is part of a straight line. Then it has a parametrization of the form $\gamma(t) = (Pt + K, Qt + L)$ for some constants P, Q, K, L with P and Q not both zero. By rescaling the t variable, we may assume that γ is unit-speed. If $Q = 0$, then S_C is an open subset of the plane $z = L$ and is therefore flat. If $P = 0$, then S_C is part of the cylinder $x^2 + y^2 = K^2$, which can be shown to be flat in the same way as we did for the unit cylinder in Example 13.18(c). On the other hand, if neither P nor Q is zero, then S_C is part of a cone, and Example 13.18 shows that the induced metric is $dt^2 + (Pt + K)^2 \, d\theta^2$. In a neighborhood of any point, the change of coordinates $(u, v) = \big((t + K/P) \cos P\theta, (t + K/P) \sin P\theta\big)$ pulls the Euclidean metric $du^2 + dv^2$ back to $dt^2 + (Pt + K)^2 \, d\theta^2$, so this metric is flat. (Think of slitting a paper cone along one side and flattening it out.)

Conversely, assuming that S_C is flat, we will show that C is part of a straight line. Let $\gamma(t) = \big(a(t), b(t)\big)$ be a local parametrization of C. Using the result of Problem 13-5, we may assume that γ is unit-speed, so that $a'(t)^2 + b'(t)^2 = 1$. As in Example 13.18, the induced metric is $dt^2 + a^2 d\theta^2$. Thus the local frame

(E_1, E_2) given by

$$E_1 = \frac{\partial}{\partial t}, \qquad E_2 = \frac{1}{a}\frac{\partial}{\partial \theta},$$

is orthonormal. Any other orthonormal frame $(\tilde{E}_1, \tilde{E}_2)$ can be written in the form

$$\tilde{E}_1 = u E_1 + v E_2 = u\frac{\partial}{\partial t} + \frac{v}{a}\frac{\partial}{\partial \theta},$$

$$\pm\tilde{E}_2 = v E_1 - u E_2 = v\frac{\partial}{\partial t} - \frac{u}{a}\frac{\partial}{\partial \theta},$$

for some functions u and v depending smoothly on (t, θ) and satisfying $u^2 + v^2 = 1$. Because the metric is flat, it is possible to choose u and v such that $(\tilde{E}_1, \tilde{E}_2)$ is a commuting orthonormal frame (Theorem 13.14). Using formula (8.8) for Lie brackets in coordinates, this implies

$$0 = \pm[\tilde{E}_1, \tilde{E}_2] = \left(u\frac{\partial v}{\partial t} + \frac{v}{a}\frac{\partial v}{\partial \theta}\right)\frac{\partial}{\partial t} - \left(u\frac{\partial}{\partial t}\left(\frac{u}{a}\right) + \frac{v}{a}\frac{\partial}{\partial \theta}\left(\frac{u}{a}\right)\right)\frac{\partial}{\partial \theta}$$

$$- \left(v\frac{\partial u}{\partial t} - \frac{u}{a}\frac{\partial u}{\partial \theta}\right)\frac{\partial}{\partial t} - \left(v\frac{\partial}{\partial t}\left(\frac{v}{a}\right) - \frac{u}{a}\frac{\partial}{\partial \theta}\left(\frac{v}{a}\right)\right)\frac{\partial}{\partial \theta}.$$

To simplify this expression, we use the shorthand notations $f_\theta = \partial f/\partial \theta$ and $f_t = \partial f/\partial t$ for any function f. Note that $u^2 + v^2 = 1$ implies $uu_\theta + vv_\theta = uu_t + vv_t = 0$, and the fact that a depends only on t implies $a_\theta = 0$ and $a_t = a'$. Inserting these relations into the formula above and simplifying, we obtain

$$0 = (uv_t - vu_t)\frac{\partial}{\partial t} + \left(\frac{a' - vu_\theta + uv_\theta}{a^2}\right)\frac{\partial}{\partial \theta},$$

which implies

$$uv_t - vu_t = 0, \tag{13.2}$$

$$vu_\theta - uv_\theta = a'. \tag{13.3}$$

Because $u^2 + v^2 \equiv 1$, each point has a neighborhood on which either u or v is nonzero. On any open subset where $v \neq 0$, (13.2) implies that the t-derivative of u/v is zero. Thus we can write $u = fv$, where f is some function of θ alone. Then $u^2 + v^2 = 1$ implies that $v^2(f^2 + 1) = 1$, so $v = \pm 1/\sqrt{f^2 + 1}$ is also a function of θ alone, and so is $u = \pm\sqrt{1 - v^2}$. A similar argument applies where $u \neq 0$. But then (13.3) implies that a' is independent of t, so it is constant, and consequently so is $b' = \pm\sqrt{1 - (a')^2}$. It follows that a and b are affine functions of t, so each point of C has a neighborhood contained in a straight line. Since we are assuming C is connected, it follows that all of C is contained in a single straight line. $\qquad\square$

Corollary 13.20. *The round metric on* $\mathbb{S}^2$ *is not flat.* $\qquad\square$

The Normal Bundle

Suppose (M, g) is an n-dimensional Riemannian manifold with or without boundary, and $S \subseteq M$ is a k-dimensional Riemannian submanifold (also with or without boundary). Just as we did for submanifolds of $\mathbb{R}^n$, for any $p \in S$ we say that a vector $v \in T_p M$ is **normal to** S if v is orthogonal to every vector in $T_p S$ with respect to the inner product $\langle \cdot, \cdot \rangle_g$. The **normal space to** S **at** p is the subspace $N_p S \subseteq T_p M$ consisting of all vectors that are normal to S at p, and the **normal bundle of** S is the subset $NS \subseteq TM$ consisting of the union of all the normal spaces at points of S. The projection $\pi_{NS} \colon NS \to S$ is defined as the restriction to NS of $\pi \colon TM \to M$. The following proposition is proved in the same way as Corollary 10.36.

Proposition 13.21 (The Normal Bundle to a Riemannian Submanifold). *Let* (M, g) *be a Riemannian n-manifold with or without boundary. For any immersed k-dimensional submanifold $S \subseteq M$ with or without boundary, the normal bundle NS is a smooth rank-$(n - k)$ subbundle of $TM|_S$. For each $p \in S$, there is a smooth frame for NS on a neighborhood of p that is orthonormal with respect to g.*

▶ **Exercise 13.22.** Prove the preceding proposition.

The Riemannian Distance Function

One of the most important tools that a Riemannian metric gives us is the ability to define lengths of curves. Suppose (M, g) is a Riemannian manifold with or without boundary. If $\gamma \colon [a, b] \to M$ is a piecewise smooth curve segment, the **length of** γ is

$$L_g(\gamma) = \int_a^b |\gamma'(t)|_g \, dt.$$

Because $|\gamma'(t)|_g$ is continuous at all but finitely many values of t, and has well-defined limits from the left and right at those points, the integral is well defined.

▶ **Exercise 13.23.** Suppose $\gamma \colon [a, b] \to M$ is a piecewise smooth curve segment and $a < c < b$. Show that

$$L_g(\gamma) = L_g(\gamma|_{[a,c]}) + L_g(\gamma|_{[c,b]}).$$

▶ **Exercise 13.24.** Show that lengths of curves are local isometry invariants of Riemannian manifolds. More precisely, suppose (M, g) and $(\widetilde{M}, \widetilde{g})$ are Riemannian manifolds with or without boundary, and $F \colon M \to \widetilde{M}$ is a local isometry. Show that $L_{\widetilde{g}}(F \circ \gamma) = L_g(\gamma)$ for every piecewise smooth curve segment γ in M.

It is an extremely important fact that length is independent of parametrization in the following sense. In Chapter 11 we defined a *reparametrization* of a piecewise smooth curve segment $\gamma \colon [a, b] \to M$ to be a curve segment of the form $\widetilde{\gamma} = \gamma \circ \varphi$, where $\varphi \colon [c, d] \to [a, b]$ is a diffeomorphism.

Proposition 13.25 (Parameter Independence of Length). *Let (M, g) be a Riemannian manifold with or without boundary, and let $\gamma: [a, b] \to M$ be a piecewise smooth curve segment. If $\widetilde{\gamma}$ is a reparametrization of γ, then $L_g(\widetilde{\gamma}) = L_g(\gamma)$.*

Proof. First suppose that γ is smooth, and $\varphi: [c, d] \to [a, b]$ is a diffeomorphism such that $\widetilde{\gamma} = \gamma \circ \varphi$. The fact that φ is a diffeomorphism implies that either $\varphi' > 0$ or $\varphi' < 0$ everywhere. Let us assume first that $\varphi' > 0$. We have

$$L_g(\widetilde{\gamma}) = \int_c^d \left| \widetilde{\gamma}'(t) \right|_g dt = \int_c^d \left| \frac{d}{dt}(\gamma \circ \varphi)(t) \right|_g dt$$

$$= \int_c^d \left| \varphi'(t) \gamma'(\varphi(t)) \right|_g dt = \int_c^d \left| \gamma'(\varphi(t)) \right|_g \varphi'(t) \, dt$$

$$= \int_a^b \left| \gamma'(s) \right|_g ds = L_g(\gamma),$$

where the next-to-last equality uses the change of variables formula for integrals.

In the case $\varphi' < 0$, we just need to introduce two sign changes into the above calculation. The sign changes once when $\varphi'(t)$ is moved outside the absolute value signs, because $|\varphi'(t)| = -\varphi'(t)$. Then it changes again when we change variables, because φ reverses the direction of the integral. Since the two sign changes cancel each other, the result is the same.

If γ is only piecewise smooth, we just apply the same argument on each subinterval on which it is smooth.   □

Using curve segments as "measuring tapes," we can define distances between points on a Riemannian manifold. Suppose (M, g) is a connected Riemannian manifold. (The theory is most straightforward when $\partial M = \varnothing$, so we assume that for the rest of this section.) For any $p, q \in M$, the **(Riemannian) distance from p to q**, denoted by $d_g(p, q)$, is defined to be the infimum of $L_g(\gamma)$ over all piecewise smooth curve segments γ from p to q. Because any pair of points in M can be joined by a piecewise smooth curve segment (Proposition 11.33), this is well defined.

Example 13.26. In $(\mathbb{R}^n, \bar{g})$, Problem 13-10 shows that any straight line segment is the shortest piecewise smooth curve segment between its endpoints. Therefore, the distance function $d_{\bar{g}}$ is equal to the usual Euclidean distance:

$$d_{\bar{g}}(x, y) = |x - y|. \qquad //$$

▶ **Exercise 13.27.** Suppose (M, g) and $(\widetilde{M}, \widetilde{g})$ are connected Riemannian manifolds and $F: M \to \widetilde{M}$ is a Riemannian isometry. Show that $d_{\widetilde{g}}(F(p), F(q)) = d_g(p, q)$ for all $p, q \in M$.

We will see below that the Riemannian distance function turns M into a metric space whose topology is the same as the given manifold topology. The key is the following technical lemma, which shows that every Riemannian metric is locally comparable to the Euclidean metric in coordinates.

Lemma 13.28. *Let g be a Riemannian metric on an open subset $U \subseteq \mathbb{R}^n$. Given a compact subset $K \subseteq U$, there exist positive constants c, C such that for all $x \in K$ and all $v \in T_x\mathbb{R}^n$,*

$$c|v|_{\bar{g}} \le |v|_g \le C|v|_{\bar{g}}. \tag{13.4}$$

Proof. For any compact subset $K \subseteq U$, let $L \subseteq T\mathbb{R}^n$ be the set

$$L = \left\{ (x,v) \in T\mathbb{R}^n : x \in K, \ |v|_{\bar{g}} = 1 \right\}.$$

Under the canonical identification of $T\mathbb{R}^n$ with $\mathbb{R}^n \times \mathbb{R}^n$, L is just the product set $K \times \mathbb{S}^{n-1}$ and therefore is compact. Because the norm $|v|_g$ is continuous and strictly positive on L, there are positive constants c, C such that $c \le |v|_g \le C$ whenever $(x,v) \in L$. If $x \in K$ and v is a nonzero vector in $T_x\mathbb{R}^n$, let $\lambda = |v|_{\bar{g}}$. Then $(x, \lambda^{-1}v) \in L$, so by homogeneity of the norm,

$$|v|_g = \lambda \left| \lambda^{-1}v \right|_g \le \lambda C = C|v|_{\bar{g}}.$$

A similar computation shows that $|v|_g \ge c|v|_{\bar{g}}$. The same inequalities are trivially true when $v = 0$. $\qquad\square$

Theorem 13.29 (Riemannian Manifolds as Metric Spaces). *Let (M, g) be a connected Riemannian manifold. With the Riemannian distance function, M is a metric space whose metric topology is the same as the original manifold topology.*

Proof. It is immediate from the definition that $d_g(p,q) \ge 0$. Because every constant curve segment has length zero, it follows that $d_g(p,p) = 0$; and $d_g(p,q) = d_g(q,p)$ follows from the fact that any curve segment from p to q can be reparametrized to go from q to p. Suppose γ_1 and γ_2 are piecewise smooth curve segments from p to q and q to r, respectively (Fig. 13.2), and let γ be a piecewise smooth curve segment that first follows γ_1 and then follows γ_2 (reparametrized if necessary). Then

$$d_g(p,r) \le L_g(\gamma) = L_g(\gamma_1) + L_g(\gamma_2).$$

Taking the infimum over all such γ_1 and γ_2, we find that $d_g(p,r) \le d_g(p,q) + d_g(q,r)$. (This is one reason why it is important to define the distance function using piecewise smooth curves instead of just smooth ones.)

To complete the proof that (M, d_g) is a metric space, we need only show that $d_g(p,q) > 0$ if $p \ne q$. For this purpose, let $p, q \in M$ be distinct points, and let U be a smooth coordinate domain containing p but not q. Use the coordinate map as usual to identify U with an open subset in $\mathbb{R}^n$, and let $\bar{g}$ denote the Euclidean metric in these coordinates. If V is a regular coordinate ball of radius ε centered at p such that $\bar{V} \subseteq U$, Lemma 13.28 shows that there are positive constants c, C such that (13.4) is satisfied whenever $x \in \bar{V}$ and $v \in T_x M$. Then for any piecewise smooth curve segment γ lying entirely in $\bar{V}$, it follows that

$$cL_{\bar{g}}(\gamma) \le L_g(\gamma) \le CL_{\bar{g}}(\gamma).$$

Suppose $\gamma : [a,b] \to M$ is a piecewise smooth curve segment from p to q. Let t_0 be the infimum of all $t \in [a,b]$ such that $\gamma(t) \notin \bar{V}$ (Fig. 13.3). It follows that

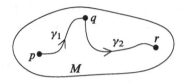
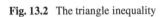
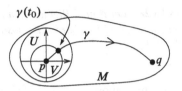

Fig. 13.2 The triangle inequality   **Fig. 13.3** Positivity of d_g

$\gamma(t_0) \in \partial V$ by continuity, and $\gamma(t) \in \bar{V}$ for $a \leq t \leq t_0$. Thus,

$$\mathrm{L}_g(\gamma) \geq \mathrm{L}_g\left(\gamma|_{[a,t_0]}\right) \geq c\mathrm{L}_{\bar{g}}\left(\gamma|_{[a,t_0]}\right) \geq cd_{\bar{g}}\left(p, \gamma(t_0)\right) = c\varepsilon.$$

Taking the infimum over all such γ, we conclude that $d_g(p,q) \geq c\varepsilon > 0$.

Finally, to show that the metric topology generated by d_g is the same as the given manifold topology on M, we need to show that the open subsets in the manifold topology are open in the metric topology, and vice versa. Suppose, first, that $U \subseteq M$ is open in the manifold topology. Let $p \in U$, and let V be a regular coordinate ball of radius ε around p such that $\bar{V} \subseteq U$ as above. The argument in the previous paragraph shows that $d_g(p,q) \geq c\varepsilon$ whenever $q \notin \bar{V}$. The contrapositive of this statement is that $d_g(p,q) < c\varepsilon$ implies $q \in \bar{V} \subseteq U$, or in other words, the metric ball of radius $c\varepsilon$ around p is contained in U. This shows that U is open in the metric topology.

Conversely, suppose that W is open in the metric topology, and let $p \in W$. Let V be a regular coordinate ball of radius r around p, let $\bar{g}$ be the Euclidean metric on $\bar{V}$ determined by the given coordinates, and let c, C be positive constants such that (13.4) is satisfied for $v \in T_q M$, $q \in \bar{V}$. Let $\varepsilon < r$ be a positive number small enough that the metric ball around p of radius $C\varepsilon$ is contained in W, and let V_ε be the set of points in $\bar{V}$ whose Euclidean distance from p is less than ε. If $q \in V_\varepsilon$, let γ be the straight-line segment in coordinates from p to q. Using Lemma 13.28 as above, we conclude that

$$d_g(p,q) \leq \mathrm{L}_g(\gamma) \leq C\mathrm{L}_{\bar{g}}(\gamma) < C\varepsilon.$$

This shows that V_ε is contained in the metric ball of radius $C\varepsilon$ around p, and therefore in W. Since V_ε is a neighborhood of p in the manifold topology, this shows that W is open in the manifold topology as well.   □

As a consequence of this theorem, all of the terminology of metric spaces can be carried over to connected Riemannian manifolds. Thus, a connected Riemannian manifold (M, g) is said to be **complete**, and g is said to be a **complete Riemannian metric**, if (M, d_g) is a complete metric space (i.e., if every Cauchy sequence in M converges to a point in M); and a subset $B \subseteq M$ is said to be **bounded** if there exists a constant K such that $d_g(x, y) \leq K$ for all $x, y \in B$. Problems 13-17 and 13-18 outline two different proofs that every connected smooth manifold admits a complete Riemannian metric.

Recall that a topological space is said to be **metrizable** if it admits a distance function whose metric topology is the same as the given topology.

Corollary 13.30. *Every smooth manifold with or without boundary is metrizable.*

Proof. First suppose M is a smooth manifold without boundary, and choose any Riemannian metric g on M. If M is connected, Theorem 13.29 shows that M is metrizable. More generally, let $\{M_i\}$ be the connected components of M, and choose a point $p_i \in M_i$ for each i. For $x \in M_i$ and $y \in M_j$, define $d_g(x, y)$ as in Theorem 13.29 when $i = j$, and otherwise

$$d_g(x, y) = d_g(x, p_i) + 1 + d_g(p_j, y).$$

(Think of building a "bridge" of length 1 between each pair of chosen points p_i, p_j in different components, so to get from x to y, you have to go to p_i, cross the bridge to p_j, and then go from p_j to y.) It is straightforward to check that this is a distance function that induces the given topology on M. Finally, if M has nonempty boundary, just embed M into its double (Example 9.32), and note that a subspace of a metrizable topological space is always metrizable. $\square$

The Tangent–Cotangent Isomorphism

Another convenient feature of every Riemannian metric is that it provides a natural isomorphism between the tangent and cotangent bundles. Given a Riemannian metric g on a smooth manifold M with or without boundary, we define a bundle homomorphism $\hat{g} \colon TM \to T^*M$ as follows. For each $p \in M$ and each $v \in T_pM$, we let $\hat{g}(v) \in T_p^*M$ be the covector defined by

$$\hat{g}(v)(w) = g_p(v, w) \quad \text{for all } w \in T_pM.$$

To see that this is a smooth bundle homomorphism, it is easiest to consider its action on smooth vector fields:

$$\hat{g}(X)(Y) = g(X, Y) \quad \text{for } X, Y \in \mathfrak{X}(M).$$

Because $\hat{g}(X)(Y)$ is linear over $C^\infty(M)$ as a function of Y, it follows from the tensor characterization lemma (Lemma 12.24) that $\hat{g}(X)$ is a smooth covector field; and because $\hat{g}(X)$ is linear over $C^\infty(M)$ as a function of X, this defines $\hat{g}$ as a smooth bundle homomorphism by the bundle homomorphism characterization lemma (Lemma 10.29). As usual, we use the same symbol for both the pointwise bundle homomorphism $\hat{g} \colon TM \to T^*M$ and the linear map on sections $\hat{g} \colon \mathfrak{X}(M) \to \mathfrak{X}^*(M)$.

Note that $\hat{g}$ is injective at each point, because $\hat{g}(v) = 0$ for some $v \in T_pM$ implies

$$0 = \hat{g}(v)(v) = \langle v, v \rangle_g,$$

which in turn implies $v = 0$. For dimensional reasons, therefore, $\hat{g}$ is bijective, so it is a bundle isomorphism (see Proposition 10.26).

In any smooth coordinates (x^i), we can write $g = g_{ij} \, dx^i \, dx^j$. Thus, if X and Y are smooth vector fields, we have

$$\hat{g}(X)(Y) = g_{ij} X^i Y^j,$$

which implies that the covector field $\hat{g}(X)$ has the coordinate expression

$$\hat{g}(X) = g_{ij} X^i \, dx^j.$$

In other words, $\hat{g}$ is the bundle homomorphism whose matrix with respect to coordinate frames for TM and T^*M is the same as the matrix of g itself. (Actually, it is the transpose of the matrix of g, but because (g_{ij}) is symmetric, these are the same.)

It is customary to denote the components of the covector field $\hat{g}(X)$ by

$$\hat{g}(X) = X_j \, dx^j, \quad \text{where } X_j = g_{ij} X^i.$$

Because of this, one says that $\hat{g}(X)$ is obtained from X by **lowering an index**. The notation $X^\flat$ is frequently used for $\hat{g}(X)$, because the symbol $\flat$ ("flat") is used in musical notation to indicate that a tone is to be lowered.

The matrix of the inverse map $\hat{g}^{-1}\colon T_p^* M \to T_p M$ is thus the inverse of (g_{ij}). (Because (g_{ij}) is the matrix of the isomorphism $\hat{g}$, it is invertible at each point.) We let (g^{ij}) denote the matrix-valued function whose value at $p \in M$ is the inverse of the matrix $(g_{ij}(p))$, so that

$$g^{ij} g_{jk} = g_{kj} g^{ji} = \delta_k^i.$$

Because g_{ij} is a symmetric matrix, so is g^{ij}, as you can easily check. Thus for a covector field $\omega \in \mathfrak{X}^*(M)$, the vector field $\hat{g}^{-1}(\omega)$ has the coordinate representation

$$\hat{g}^{-1}(\omega) = \omega^i \frac{\partial}{\partial x^i}, \quad \text{where } \omega^i = g^{ij} \omega_j.$$

We use the notation $\omega^\sharp$ ("ω-sharp") for $\hat{g}^{-1}(\omega)$, and say that $\omega^\sharp$ is obtained from ω by **raising an index**. Because the symbols $\flat$ and $\sharp$ are borrowed from musical notation, these two inverse isomorphisms are frequently called the **musical isomorphisms**. A handy mnemonic device for keeping the flat and sharp operations straight is to remember that the value of $\omega^\sharp$ at each point is a vector, which we visualize as a (sharp) arrow; while the value of $X^\flat$ is a covector, which we visualize by means of its (flat) level sets.

The most important use of the sharp operation is to reinstate the gradient as a vector field on Riemannian manifolds. For any smooth real-valued function f on a Riemannian manifold (M, g) with or without boundary, we define a vector field called the **gradient of f** by

$$\text{grad} \, f = (df)^\sharp = \hat{g}^{-1}(df).$$

Unraveling the definitions, we see that for any $X \in \mathfrak{X}(M)$, the gradient satisfies

$$\langle \text{grad} \, f, X \rangle_g = \hat{g}(\text{grad} \, f)(X) = df(X) = Xf.$$

Thus grad f is the unique vector field that satisfies

$$\langle \operatorname{grad} f, X \rangle_g = Xf \quad \text{for every vector field } X,$$

or equivalently,

$$\langle \operatorname{grad} f, \cdot \rangle_g = df.$$

In smooth coordinates, grad f has the expression

$$\operatorname{grad} f = g^{ij} \frac{\partial f}{\partial x^i} \frac{\partial}{\partial x^j}.$$

In particular, this shows that grad f is smooth. On $\mathbb{R}^n$ with the Euclidean metric, this reduces to

$$\operatorname{grad} f = \delta^{ij} \frac{\partial f}{\partial x^i} \frac{\partial}{\partial x^j} = \sum_{i=1}^{n} \frac{\partial f}{\partial x^i} \frac{\partial}{\partial x^i}.$$

Thus our new definition of the gradient in this case coincides with the gradient from elementary calculus. In other coordinates, however, the gradient does not generally have the same form.

Example 13.31. Let us compute the gradient of a function $f \in C^\infty(\mathbb{R}^2)$ with respect to the Euclidean metric in polar coordinates. From Example 13.12 we see that the matrix of $\bar{g}$ in polar coordinates is $\begin{pmatrix} 1 & 0 \\ 0 & r^2 \end{pmatrix}$, so its inverse matrix is $\begin{pmatrix} 1 & 0 \\ 0 & 1/r^2 \end{pmatrix}$. Inserting this into the formula for the gradient, we obtain

$$\operatorname{grad} f = \frac{\partial f}{\partial r} \frac{\partial}{\partial r} + \frac{1}{r^2} \frac{\partial f}{\partial \theta} \frac{\partial}{\partial \theta}. \qquad\qquad /\!/$$

Problem 13-21 shows that the gradient of a function f on a Riemannian manifold has the same geometric interpretation as it has in Euclidean space: its direction is the direction in which f is increasing fastest, and is orthogonal to the level sets of f; and its length is the maximum directional derivative of f in any direction.

Pseudo-Riemannian Metrics

An important generalization of Riemannian metrics is obtained by relaxing the positivity requirement. A symmetric 2-tensor g on a vector space V is said to be **nondegenerate** if the linear map $\hat{g} \colon V \to V^*$ defined by $\hat{g}(v)(w) = g(v, w)$ is an isomorphism, or equivalently if for every nonzero $v \in V$ there exists $w \in V$ such that $g(v, w) \neq 0$. Just as any inner product can be transformed to the Euclidean one by switching to an orthonormal basis, every nondegenerate symmetric 2-tensor can be transformed by a change of basis to one whose matrix is diagonal with all entries equal to ± 1 (the proof is an adaptation of the Gram–Schmidt algorithm). The numbers r and s of positive and negative diagonal entries, respectively, are independent of the choice of basis (a fact known as *Sylvester's law of inertia*; see [FIS03] for a proof). Thus the ordered pair (r, s), called the **signature of g**, is an invariant of g.

A *pseudo-Riemannian metric* on a smooth manifold M is a smooth symmetric 2-tensor field whose value is nondegenerate at each point, with the same signature everywhere on M. Pseudo-Riemannian metrics with signature $(n-1, 1)$ (or $(1, n-1)$, depending on the convention used) are called *Lorentz metrics*; they play a central role in physics, where they are used to model gravitation in Einstein's general theory of relativity.

We do not pursue the subject of pseudo-Riemannian metrics any further, except to note that the proof of the existence of Riemannian metrics does not carry over to the pseudo-Riemannian case, since it is not generally true that a linear combination of nondegenerate 2-tensors with positive coefficients is necessarily nondegenerate. Indeed, not every manifold admits a Lorentz metric (cf. [HE73, p. 39]).

Problems

13-1. If (M, g) is a Riemannian n-manifold with or without boundary, let $UM \subseteq TM$ be the subset $UM = \{(x, v) \in TM : |v|_g = 1\}$, called the *unit tangent bundle of M*. Show that UM is a smooth fiber bundle over M with model fiber $\mathbb{S}^{n-1}$.

13-2. In the proof of Proposition 13.3 we used a partition of unity to patch together locally defined Riemannian metrics to obtain a global one. A crucial part of the proof was verifying that the global tensor field so obtained was positive definite. The key to the success of this argument is the fact that the set of inner products on a given tangent space is a convex subset of the vector space of all symmetric 2-tensors. This problem outlines a generalization of this construction to arbitrary vector bundles. Suppose that E is a smooth vector bundle over a smooth manifold M with or without boundary, and $V \subseteq E$ is an open subset with the property that for each $p \in M$, the intersection of V with the fiber E_p is convex and nonempty. By a "section of V," we mean a (local or global) section of E whose image lies in V.

   (a) Show that there exists a smooth global section of V.
   (b) Suppose $\sigma : A \to V$ is a smooth section of V defined on a closed subset $A \subseteq M$. (This means that σ extends to a smooth section of V in a neighborhood of each point of A.) Show that there exists a smooth global section $\tilde{\sigma}$ of V whose restriction to A is equal to σ. Show that if V contains the image of the zero section of E, then $\tilde{\sigma}$ can be chosen to be supported in any predetermined neighborhood of A.

   (*Used on pp. 381, 430.*)

13-3. Let M be a smooth manifold. Prove the following statements.
   (a) If there exists a global nonvanishing vector field on M, then there exists a global *smooth* nonvanishing vector field. [Hint: imitate the proof of Theorem 6.21, with the constants $F(x_i)$ replaced by constant-coefficient vector fields in coordinates, and with absolute values replaced by norms in some Riemannian metric.]

(b) If there exists a linearly independent k-tuple of vector fields on M, then there exists such a k-tuple of *smooth* vector fields.

13-4. Let $\overset{\circ}{g}$ denote the round metric on $\mathbb{S}^n$. Compute the coordinate representation of $\overset{\circ}{g}$ in stereographic coordinates (see Problem 1-7).

13-5. Suppose (M, g) is a Riemannian manifold. A smooth curve $\gamma: J \to M$ is said to be a **unit-speed curve** if $|\gamma'(t)|_g \equiv 1$. Prove that every smooth curve with nowhere-vanishing velocity has a unit-speed reparametrization. (*Used on p. 335.*)

13-6. Prove that every Riemannian 1-manifold is flat. [Hint: use Problem 13-5. Note that this implies the round metric on $\mathbb{S}^1$ is flat!]

13-7. Show that a product of flat metrics is flat.

13-8. Let $\mathbb{T}^n = \mathbb{S}^1 \times \cdots \times \mathbb{S}^1 \subseteq \mathbb{C}^n$, and let g be the metric on $\mathbb{T}^n$ induced from the Euclidean metric on $\mathbb{C}^n$ (identified with $\mathbb{R}^{2n}$). Show that g is flat.

13-9. Let $H \subseteq \mathbb{R}^3$ be the helicoid (the image of the embedding $F: \mathbb{R}^2 \to \mathbb{R}^3$ of Example 13.11), and let $C \subseteq \mathbb{R}^3$ be the **catenoid**, which is the surface of revolution generated by the curve $\gamma(t) = (\cosh t, t)$. Show that H is locally isometric to C but not globally isometric.

13-10. Show that the shortest path between two points in Euclidean space is a straight line segment. More precisely, for $x, y \in \mathbb{R}^n$, let $\gamma: [0, 1] \to \mathbb{R}^n$ be the curve segment $\gamma(t) = x + t(y - x)$, and show that any other piecewise smooth curve segment $\widetilde{\gamma}$ from x to y satisfies $L_{\bar{g}}(\widetilde{\gamma}) > L_{\bar{g}}(\gamma)$ unless $\widetilde{\gamma}$ is a reparametrization of γ. [Hint: first, consider the case in which both x and y lie on the x^1-axis.] (*Used on p. 338.*)

13-11. Let $M = \mathbb{R}^2 \smallsetminus \{0\}$, and let g be the restriction to M of the Euclidean metric $\bar{g}$. Show that there are points $p, q \in M$ for which there is no piecewise smooth curve segment γ from p to q in M with $L_g(\gamma) = d_g(p, q)$.

13-12. Consider $\mathbb{R}^n$ as a Riemannian manifold with the Euclidean metric $\bar{g}$.
    (a) Suppose $U, V \subseteq \mathbb{R}^n$ are connected open sets, $\varphi, \psi: U \to V$ are Riemannian isometries, and for some $p \in U$ they satisfy $\varphi(p) = \psi(p)$ and $d\varphi_p = d\psi_p$. Show that $\varphi = \psi$. [Hint: first, use Problem 13-10 to show that φ and ψ take lines to lines.]
    (b) Show that the set of maps from $\mathbb{R}^n$ to itself given by the action of $E(n)$ on $\mathbb{R}^n$ described in Example 7.32 is the full group of Riemannian isometries of $(\mathbb{R}^n, \bar{g})$.

13-13. Let (M, g) be a Riemannian manifold. A smooth vector field V on M is called a **Killing vector field for g** (named after the late nineteenth/early twentieth-century German mathematician Wilhelm Killing) if the flow of V acts by isometries of g.
    (a) Show that the set of all Killing vector fields on M constitutes a Lie subalgebra of $\mathfrak{X}(M)$. [Hint: see Corollary 9.39(c).]

(b) Show that a smooth vector field V on M is a Killing vector field if and only if it satisfies the following equation in each smooth local coordinate chart:

$$V^k \frac{\partial g_{ij}}{\partial x^k} + g_{jk} \frac{\partial V^k}{\partial x^i} + g_{ik} \frac{\partial V^k}{\partial x^j} = 0.$$

13-14. Let $K \subseteq \mathfrak{X}(\mathbb{R}^n)$ denote the Lie algebra of Killing vector fields with respect to the Euclidean metric (see Problem 13-13), and let $K_0 \subseteq K$ denote the subspace consisting of fields that vanish at the origin.

(a) Show that the map

$$V \mapsto \left(\frac{\partial V^i}{\partial x^j}(0) \right)$$

is an injective linear map from K_0 to $\mathfrak{o}(n)$. [Hint: If V is in the kernel of this map and θ is its flow, show that the linear map $d(\theta_t)_0 \colon T_0\mathbb{R}^n \to T_0\mathbb{R}^n$ is independent of t, and use the result of Problem 13-12(a).]

(b) Show that the following vector fields form a basis for K:

$$\frac{\partial}{\partial x^i}, \quad 1 \le i \le n; \qquad x^i \frac{\partial}{\partial x^j} - x^j \frac{\partial}{\partial x^i}, \quad 1 \le i < j \le n.$$

13-15. Let (M, g) be a Riemannian manifold, and let $\widehat{g}$ be the product metric on $M \times \mathbb{R}$ determined by g and the Euclidean metric on $\mathbb{R}$. Let $X = 0 \oplus d/dt$ be the product vector field on $M \times \mathbb{R}$ determined by the zero vector field on M and the standard coordinate vector field d/dt on $\mathbb{R}$ (see Problem 8-17). Show that X is a Killing vector field for $(M \times \mathbb{R}, \widehat{g})$.

13-16. Suppose $g = f(t)dt^2$ is a Riemannian metric on $\mathbb{R}$. Show that g is complete if and only if both of the following improper integrals diverge:

$$\int_0^\infty \sqrt{f(t)}\, dt, \qquad \int_{-\infty}^0 \sqrt{f(t)}\, dt.$$

13-17. Let M be a connected noncompact smooth manifold and let g be a Riemannian metric on M. Prove that there exists a positive function $h \in C^\infty(M)$ such that the Riemannian metric $\widetilde{g} = hg$ is complete. Use this to prove that every connected smooth manifold admits a complete Riemannian metric. [Hint: let $f \colon M \to \mathbb{R}$ be an exhaustion function, and show that h can be chosen so that f is bounded on $\widetilde{g}$-bounded sets.]

13-18. Suppose (M, g) is a connected Riemannian manifold, $S \subseteq M$ is a connected embedded submanifold, and $\widetilde{g}$ is the induced Riemannian metric on S.

(a) Prove that $d_{\widetilde{g}}(p, q) \ge d_g(p, q)$ for $p, q \in S$.

(b) Prove that if (M, g) is complete and S is properly embedded, then $(S, \widetilde{g})$ is complete.

(c) Use (b) together with the Whitney embedding theorem to prove (without quoting Proposition 13.3 or Problem 13-17) that every connected smooth manifold admits a complete Riemannian metric.

13-19. The following example shows that the converse of Problem 13-18(b) does not hold. Define $F: \mathbb{R} \to \mathbb{R}^2$ by $F(t) = ((e^t + 1)\cos t, (e^t + 1)\sin t)$. Show that F is an embedding that is not proper, yet $\mathbb{R}$ is complete in the metric induced from the Euclidean metric on $\mathbb{R}^2$.

13-20. Consider the embeddings $F_1, F_2, F_3, F_4: \mathbb{R}^2 \to \mathbb{R}^3$ defined as follows:

$$F_1(u, v) = (u, v, 0);$$

$$F_2(u, v) = (u, e^v, 0);$$

$$F_3(u, v) = (u, v, u^2 + v^2);$$

$$F_4(u, v) = \left(\frac{2u}{u^2 + v^2 + 1}, \frac{2v}{u^2 + v^2 + 1}, \frac{u^2 + v^2 - 1}{u^2 + v^2 + 1} \right).$$

For each i, let $g_i = F_i^* \bar{g}$. For each of the Riemannian manifolds $(\mathbb{R}^2, g_1)$ through $(\mathbb{R}^2, g_4)$, answer the following questions: Is it bounded? Is it complete? Is it flat? Prove your answers correct. [Hint: if you have trouble analyzing g_4, look at Problem 1-7.]

13-21. Let (M, g) be a Riemannian manifold, let $f \in C^\infty(M)$, and let $p \in M$ be a regular point of f.
   (a) Show that among all unit vectors $v \in T_p M$, the directional derivative vf is greatest when v points in the same direction as grad $f|_p$, and the length of grad $f|_p$ is equal to the value of the directional derivative in that direction.
   (b) Show that grad $f|_p$ is normal to the level set of f through p.
   (*Used on p. 391.*)

13-22. For any smooth manifold M with or without boundary, show that the vector bundles TM and T^*M are smoothly isomorphic over M. [Remark: Problem 11-18 shows that this isomorphism cannot be *natural*, in the sense that there does not exist a rule that assigns to every smooth manifold M a bundle isomorphism $\lambda_M: TM \to T^*M$ in such a way that for every diffeomorphism $F: M \to N$, the two bundle isomorphisms λ_M and λ_N are related by $\lambda_M = dF^* \circ \lambda_N \circ dF$.]

13-23. Is there a smooth covector field on $\mathbb{S}^2$ that vanishes at exactly one point?

13-24. Let M be a compact smooth n-manifold, and suppose f is a smooth real-valued function on M that has only finitely many critical points $\{p_1, \dots, p_k\}$, with corresponding critical values $\{c_1, \dots, c_k\}$ labeled so that $c_1 \leq \cdots \leq c_k$. For any $a < b \in \mathbb{R}$, define $M_a = f^{-1}(a)$, $M_{[a,b]} = f^{-1}([a,b])$, and $M_{(a,b)} = f^{-1}((a,b))$. If a and b are regular values, note that M_a and M_b are embedded hypersurfaces in M, $M_{(a,b)}$ is an open submanifold, and $M_{[a,b]}$ is a regular domain by Proposition 5.47 (see Fig. 13.4).
   (a) Choose a Riemannian metric g on M, let X be the vector field $X = $ grad $f / |$ grad $f|_g^2$ on $M \setminus \{p_1, \dots, p_k\}$, and let θ denote the flow of X. Show that $f(\theta_t(p)) = f(p) + t$ whenever $\theta_t(p)$ is defined.

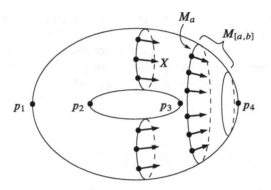

Fig. 13.4 The setup for Problem 13-24

(b) Let $[a, b] \subseteq \mathbb{R}$ be a compact interval containing no critical values of f. Show that θ restricts to a diffeomorphism from $[0, b - a] \times M_a$ to $M_{[a,b]}$.

[Remark: this result shows that M can be decomposed as a union of simpler "building blocks"—the product sets $M_{[c_i + \varepsilon, c_{i+1} - \varepsilon]} \approx I \times M_{c_i + \varepsilon}$, and the neighborhoods $M_{(c_i - \varepsilon, c_i + \varepsilon)}$ of the critical points. This is the starting point of **Morse theory**, which is one of the deepest applications of differential geometry to topology. The next step would be to analyze the behavior of f near each critical point, and use this analysis to determine exactly how the level sets change topologically when crossing a critical level. See [Mil63] for an excellent introduction.]

13-25. Suppose M is a smooth manifold that admits a proper smooth function $f : M \to \mathbb{R}$ with no critical points. Show that M is diffeomorphic to $N \times \mathbb{R}$ for some compact smooth manifold N. [Hint: let $X = \operatorname{grad} f / |\operatorname{grad} f|_g^2$, defined with respect to some Riemannian metric on M. Show that X is complete, and use its flowout to define the diffeomorphism.]

Chapter 14
Differential Forms

When we introduced tensors in Chapter 12, we observed that there are two special classes of tensors whose values change predictably when their arguments are rearranged: *symmetric tensors* and *alternating tensors*. We saw a significant application of symmetric tensors in Chapter 13, in the form of Riemannian metrics. In this chapter, we begin to develop the theory of alternating tensors, and especially *differential forms*, which are alternating tensor fields on manifolds. It might come as a surprise, but these innocent-sounding objects turn out to be considerably more important in smooth manifold theory than symmetric tensor fields.

Much of the theory of differential forms can be viewed as a generalization of the theory of covector fields—which are, after all, the simplest examples of differential forms. Most fundamentally, covector fields are objects that can be integrated over curves in a coordinate invariant way; similarly, it turns out that differential forms are objects that can be integrated over higher-dimensional submanifolds. We will develop the theory of integration of differential forms in Chapter 16; the first section of that chapter gives a heuristic explanation of why alternating tensor fields make sense as objects to integrate.

In addition to their role in integration, differential forms provide a framework for generalizing such diverse concepts from multivariable calculus as the cross product, curl, divergence, and Jacobian determinant. They are also essential to the theories of orientations (Chapter 15), de Rham cohomology (Chapters 17–18), foliations (Chapter 19), and symplectic manifolds (Chapter 22).

Because of the many uses of differential forms, differential geometers have developed an array of technical tools for manipulating them. The purpose of this chapter is to describe those tools. If you have never been exposed to differential forms before, you might find much of the theory in this chapter unmotivated the first time through; but rest assured that your effort will be repaid many times over in the many applications of differential forms throughout the rest of the book.

We begin the chapter by examining the algebra of alternating tensors on a finite-dimensional vector space. After exploring the computational properties of these tensors in a linear-algebraic setting, we transfer everything to smooth manifolds, and begin to explore the properties of differential forms.

J.M. Lee, *Introduction to Smooth Manifolds*, Graduate Texts in Mathematics 218,
DOI 10.1007/978-1-4419-9982-5_14, © Springer Science+Business Media New York 2013

The heart of the chapter is the introduction of the most important operation on differential forms, called the *exterior derivative*, which generalizes the differential of a smooth function that we introduced in Chapter 11, as well as the gradient, divergence, and curl operators of multivariable calculus. It acts on a differential form and yields another differential form of one higher degree. It is remarkable in that it is one of the very few differential operators that are naturally defined on every smooth manifold without any arbitrary choices.

At the end of the chapter, we will see how the exterior derivative can be used to simplify the computation of Lie derivatives of differential forms.

The Algebra of Alternating Tensors

Let V be a finite-dimensional (real) vector space. Recall that a covariant k-tensor on V is said to be *alternating* if its value changes sign whenever two arguments are interchanged, or equivalently if any permutation of the arguments causes its value to be multiplied by the sign of the permutation (see Exercise 12.17). Alternating covariant k-tensors are also called *exterior forms*, *multicovectors*, or *k-covectors*. The vector space of all k-covectors on V is denoted by $\Lambda^k(V^*)$. As we noted in Chapter 12, all 0-tensors and 1-tensors are alternating.

The next lemma gives two more useful characterizations of alternating tensors.

Lemma 14.1. *Let α be a covariant k-tensor on a finite-dimensional vector space V. The following are equivalent:*

(a) α *is alternating.*
(b) $\alpha(v_1,\dots,v_k) = 0$ *whenever the k-tuple $(v_1,\dots,v_k)$ is linearly dependent.*
(c) α *gives the value zero whenever two of its arguments are equal:*

$$\alpha(v_1,\dots,w,\dots,w,\dots,v_k) = 0.$$

Proof. The implications (a) $\Rightarrow$ (c) and (b) $\Rightarrow$ (c) are immediate. We complete the proof by showing that (c) implies both (a) and (b).

Assume that α satisfies (c). For any vectors $v_1,\dots,v_k$, the hypothesis implies

$$
\begin{aligned}
0 &= \alpha(v_1,\dots,v_i + v_j,\dots,v_i + v_j,\dots,v_k) \\
&= \alpha(v_1,\dots,v_i,\dots,v_i,\dots,v_k) + \alpha(v_1,\dots,v_i,\dots,v_j,\dots,v_k) \\
&\quad + \alpha(v_1,\dots,v_j,\dots,v_i,\dots,v_k) + \alpha(v_1,\dots,v_j,\dots,v_j,\dots,v_k) \\
&= \alpha(v_1,\dots,v_i,\dots,v_j,\dots,v_k) + \alpha(v_1,\dots,v_j,\dots,v_i,\dots,v_k).
\end{aligned}
$$

Thus α is alternating. On the other hand, if $(v_1,\dots,v_k)$ is a linearly dependent k-tuple, then one of the v_i's can be written as a linear combination of the others. For simplicity, let us assume that $v_k = \sum_{j=1}^{k-1} a^j v_j$. Then multilinearity of α implies

$$\alpha(v_1, \ldots, v_k) = \sum_{j=1}^{k-1} a^j \alpha(v_1, \ldots, v_{k-1}, v_j).$$

In each of these terms, α has two identical arguments, so every term is zero.   □

In Chapter 12, we defined a projection Sym: $T^k(V^*) \to \Sigma^k(V^*)$ called *symmetrization*. We define a similar projection Alt: $T^k(V^*) \to \Lambda^k(V^*)$, called **alternation**, as follows:

$$\text{Alt } \alpha = \frac{1}{k!} \sum_{\sigma \in S_k} (\text{sgn } \sigma)(^\sigma \alpha),$$

where S_k is the symmetric group on k elements. More explicitly, this means

$$(\text{Alt } \alpha)(v_1, \ldots, v_k) = \frac{1}{k!} \sum_{\sigma \in S_k} (\text{sgn } \sigma) \alpha \left(v_{\sigma(1)}, \ldots, v_{\sigma(k)} \right).$$

Example 14.2. If α is any 1-tensor, then Alt $\alpha = \alpha$. If β is a 2-tensor, then

$$(\text{Alt } \beta)(v, w) = \tfrac{1}{2} \big(\beta(v, w) - \beta(w, v) \big).$$

For a 3-tensor γ,

$$(\text{Alt } \gamma)(v, w, x) = \tfrac{1}{6} \big(\gamma(v, w, x) + \gamma(w, x, v) + \gamma(x, v, w)$$
$$- \gamma(w, v, x) - \gamma(v, x, w) - \gamma(x, w, v) \big). \qquad //$$

The next proposition is the analogue for alternating tensors of Proposition 12.14.

Proposition 14.3 (Properties of Alternation). *Let α be a covariant tensor on a finite-dimensional vector space.*

(a) Alt α *is alternating.*
(b) Alt $\alpha = \alpha$ *if and only if α is alternating.*

▶ **Exercise 14.4.** Prove Proposition 14.3.

Elementary Alternating Tensors

For computations with alternating tensors, the following notation is exceedingly useful. Given a positive integer k, an ordered k-tuple $I = (i_1, \ldots, i_k)$ of positive integers is called a **multi-index of length k**. If I is such a multi-index and $\sigma \in S_k$ is a permutation of $\{1, \ldots, k\}$, we write I_σ for the following multi-index:

$$I_\sigma = \big(i_{\sigma(1)}, \ldots, i_{\sigma(k)} \big).$$

Note that $I_{\sigma\tau} = (I_\sigma)_\tau$ for $\sigma, \tau \in S_k$.

Let V be an n-dimensional vector space, and suppose $(\varepsilon^1, \ldots, \varepsilon^n)$ is any basis for V^*. We now define a collection of k-covectors on V that generalize the determinant function on $\mathbb{R}^n$. For each multi-index $I = (i_1, \ldots, i_k)$ of length k such that $1 \leq i_1, \ldots, i_k \leq n$, define a covariant k-tensor $\varepsilon^I = \varepsilon^{i_1 \cdots i_k}$ by

$$\varepsilon^I(v_1, \ldots, v_k) = \det \begin{pmatrix} \varepsilon^{i_1}(v_1) & \ldots & \varepsilon^{i_1}(v_k) \\ \vdots & \ddots & \vdots \\ \varepsilon^{i_k}(v_1) & \ldots & \varepsilon^{i_k}(v_k) \end{pmatrix} = \det \begin{pmatrix} v_1^{i_1} & \ldots & v_k^{i_1} \\ \vdots & \ddots & \vdots \\ v_1^{i_k} & \ldots & v_k^{i_k} \end{pmatrix}. \tag{14.1}$$

In other words, if $\mathbf{v}$ denotes the $n \times k$ matrix whose columns are the components of the vectors $v_1, \ldots, v_k$ with respect to the basis (E_i) dual to (ε^i), then $\varepsilon^I(v_1, \ldots, v_k)$ is the determinant of the $k \times k$ submatrix consisting of rows $i_1, \ldots, i_k$ of $\mathbf{v}$. Because the determinant changes sign whenever two columns are interchanged, it is clear that ε^I is an alternating k-tensor. We call ε^I an **elementary alternating tensor** or **elementary k-covector**.

Example 14.5. In terms of the standard dual basis (e^1, e^2, e^3) for $(\mathbb{R}^3)^*$, we have

$$e^{13}(v, w) = v^1 w^3 - w^1 v^3;$$

$$e^{123}(v, w, x) = \det(v, w, x). \qquad\qquad /\!/$$

In order to streamline computations with the elementary k-covectors, it is useful to extend the Kronecker delta notation in the following way. If I and J are multi-indices of length k, we define

$$\delta_J^I = \det \begin{pmatrix} \delta_{j_1}^{i_1} & \ldots & \delta_{j_k}^{i_1} \\ \vdots & \ddots & \vdots \\ \delta_{j_1}^{i_k} & \ldots & \delta_{j_k}^{i_k} \end{pmatrix}.$$

▶ **Exercise 14.6.** Show that

$$\delta_J^I = \begin{cases} \operatorname{sgn}\sigma & \text{if neither } I \text{ nor } J \text{ has a repeated index and } J = I_\sigma \text{ for some } \sigma \in S_k, \\ 0 & \text{if } I \text{ or } J \text{ has a repeated index or } J \text{ is not a permutation of } I. \end{cases}$$

Lemma 14.7 (Properties of Elementary k-Covectors). *Let (E_i) be a basis for V, let (ε^i) be the dual basis for V^*, and let ε^I be as defined above.*

(a) *If I has a repeated index, then $\varepsilon^I = 0$.*
(b) *If $J = I_\sigma$ for some $\sigma \in S_k$, then $\varepsilon^I = (\operatorname{sgn}\sigma)\varepsilon^J$.*
(c) *The result of evaluating ε^I on a sequence of basis vectors is*

$$\varepsilon^I(E_{j_1}, \ldots, E_{j_k}) = \delta_J^I.$$

Proof. If I has a repeated index, then for any vectors $v_1, \ldots, v_k$, the determinant in (14.1) has two identical rows and thus is equal to zero, which proves (a). On the other hand, if J is obtained from I by interchanging two indices, then the corresponding determinants have opposite signs; this implies (b). Finally, (c) follows immediately from the definition of ε^I. $\qquad\qquad\qquad\qquad\qquad\qquad\qquad\qquad\qquad\qquad\qquad\qquad\square$

The significance of the elementary k-covectors is that they provide a convenient basis for $\Lambda^k(V^*)$. Of course, the ε^I's are not all linearly independent, because some of them are zero and the ones corresponding to different permutations of the same multi-index are scalar multiples of each other. But as the next proposition shows, we can get a basis by restricting attention to an appropriate subset of multi-indices. A multi-index $I = (i_1, \ldots, i_k)$ is said to be **increasing** if $i_1 < \cdots < i_k$. It is useful to use a primed summation sign to denote a sum over only increasing multi-indices, so that, for example,

$$\sideset{}{'}\sum_I \alpha_I \varepsilon^I = \sum_{\{I : i_1 < \cdots < i_k\}} \alpha_I \varepsilon^I.$$

Proposition 14.8 (A Basis for $\Lambda^k(V^*)$). *Let V be an n-dimensional vector space. If (ε^i) is any basis for V^*, then for each positive integer $k \leq n$, the collection of k-covectors*

$$\mathcal{E} = \{\varepsilon^I : I \text{ is an increasing multi-index of length } k\}$$

is a basis for $\Lambda^k(V^*)$. Therefore,

$$\dim \Lambda^k(V^*) = \binom{n}{k} = \frac{n!}{k!(n-k)!}.$$

If $k > n$, then $\dim \Lambda^k(V^*) = 0$.

Proof. The fact that $\Lambda^k(V^*)$ is the trivial vector space when $k > n$ follows immediately from Lemma 14.1(b), since every k-tuple of vectors is linearly dependent in that case. For the case $k \leq n$, we need to show that the set $\mathcal{E}$ spans $\Lambda^k(V^*)$ and is linearly independent. Let (E_i) be the basis for V dual to (ε^i).

To show that $\mathcal{E}$ spans $\Lambda^k(V^*)$, let $\alpha \in \Lambda^k(V^*)$ be arbitrary. For each multi-index $I = (i_1, \ldots, i_k)$ (not necessarily increasing), define a real number α_I by

$$\alpha_I = \alpha(E_{i_1}, \ldots, E_{i_k}).$$

The fact that α is alternating implies that $\alpha_I = 0$ if I contains a repeated index, and $\alpha_J = (\operatorname{sgn}\sigma)\alpha_I$ if $J = I_\sigma$ for $\sigma \in S_k$. For any multi-index J, therefore, Lemma 14.7 gives

$$\sideset{}{'}\sum_I \alpha_I \varepsilon^I(E_{j_1}, \ldots, E_{j_k}) = \sideset{}{'}\sum_I \alpha_I \delta_J^I = \alpha_J = \alpha(E_{j_1}, \ldots, E_{j_k}).$$

Thus $\sum_I' \alpha_I \varepsilon^I = \alpha$, so $\mathcal{E}$ spans $\Lambda^k(V^*)$.

To show that $\mathcal{E}$ is a linearly independent set, suppose the identity $\sum_I' \alpha_I \varepsilon^I = 0$ holds for some coefficients α_I. Let J be any increasing multi-index. Applying both sides of the identity to the vectors $(E_{j_1}, \dots, E_{j_k})$ and using Lemma 14.7, we get

$$0 = \sum_I{}' \alpha_I \varepsilon^I (E_{j_1}, \dots, E_{j_k}) = \alpha_J.$$

Thus each coefficient α_J is zero. $\square$

In particular, for an n-dimensional vector space V, this proposition implies that $\Lambda^n(V^*)$ is 1-dimensional and is spanned by $\varepsilon^{1 \cdots n}$. By definition, this elementary n-covector acts on vectors $(v_1, \dots, v_n)$ by taking the determinant of their component matrix $\mathbf{v} = (v_j^i)$. For example, on $\mathbb{R}^n$ with the standard basis, $e^{1 \cdots n}$ is precisely the determinant function.

One consequence of this is the following useful description of the behavior of an n-covector on an n-dimensional space under linear maps. Recall that if $T \colon V \to V$ is a linear map, the determinant of T is defined to be the determinant of the matrix representation of T with respect to any basis (see Appendix B, p. 633).

Proposition 14.9. *Suppose V is an n-dimensional vector space and $\omega \in \Lambda^n(V^*)$. If $T \colon V \to V$ is any linear map and $v_1, \dots, v_n$ are arbitrary vectors in V, then*

$$\omega(Tv_1, \dots, Tv_n) = (\det T)\omega(v_1, \dots, v_n). \tag{14.2}$$

Proof. Let (E_i) be any basis for V, and let (ε^i) be the dual basis. Let (T_i^j) denote the matrix of T with respect to this basis, and let $T_i = TE_i = T_i^j E_j$. By Proposition 14.8, we can write $\omega = c \varepsilon^{1 \cdots n}$ for some real number c.

Since both sides of (14.2) are multilinear functions of $v_1, \dots, v_n$, it suffices to verify the identity when the v_i's are basis vectors. Furthermore, since both sides are alternating, we only need to check the case $(v_1, \dots, v_n) = (E_1, \dots, E_n)$. In this case, the right-hand side of (14.2) is

$$(\det T) c \varepsilon^{1 \cdots n} (E_1, \dots, E_n) = c \det T.$$

On the other hand, the left-hand side reduces to

$$\omega(TE_1, \dots, TE_n) = c \varepsilon^{1 \cdots n} (T_1, \dots, T_n) = c \det \left(\varepsilon^j (T_i) \right) = c \det \left(T_i^j \right),$$

which is equal to the right-hand side. $\square$

The Wedge Product

In Chapter 12, we defined the symmetric product, which takes a pair of symmetric tensors α, β and yields another symmetric tensor $\alpha\beta = \mathrm{Sym}(\alpha \otimes \beta)$ whose rank is the sum of the ranks of the original ones.

In this section we define a similar product operation for alternating tensors. One way to define it would be to mimic what we did in the symmetric case and define the product of alternating tensors ω and η to be $\mathrm{Alt}(\omega \otimes \eta)$. However, we will use a different definition that looks more complicated at first but turns out to be much better suited to computation.

We continue with the assumption that V is a finite-dimensional real vector space. Given $\omega \in \Lambda^k(V^*)$ and $\eta \in \Lambda^l(V^*)$, we define their **wedge product** or **exterior product** to be the following $(k+l)$-covector:

$$\omega \wedge \eta = \frac{(k+l)!}{k!\,l!} \operatorname{Alt}(\omega \otimes \eta). \tag{14.3}$$

The mysterious coefficient is motivated by the simplicity of the statement of the following lemma.

Lemma 14.10. *Let V be an n-dimensional vector space and let $(\varepsilon^1, \ldots, \varepsilon^n)$ be a basis for V^*. For any multi-indices $I = (i_1, \ldots, i_k)$ and $J = (j_1, \ldots, j_l)$,*

$$\varepsilon^I \wedge \varepsilon^J = \varepsilon^{IJ}, \tag{14.4}$$

where $IJ = (i_1, \ldots, i_k, j_1, \ldots, j_l)$ is obtained by concatenating I and J.

Proof. By multilinearity, it suffices to show that

$$\varepsilon^I \wedge \varepsilon^J \left(E_{p_1}, \ldots, E_{p_{k+l}}\right) = \varepsilon^{IJ} \left(E_{p_1}, \ldots, E_{p_{k+l}}\right) \tag{14.5}$$

for any sequence $\left(E_{p_1}, \ldots, E_{p_{k+l}}\right)$ of basis vectors. We consider several cases.

CASE 1: $P = (p_1, \ldots, p_{k+l})$ *has a repeated index.* In this case, both sides of (14.5) are zero by Lemma 14.1(c).

CASE 2: P *contains an index that does not appear in either I or J.* In this case, the right-hand side is zero by Lemma 14.7(c). Similarly, each term in the expansion of the left-hand side involves either ε^I or ε^J evaluated on a sequence of basis vectors that is not a permutation of I or J, respectively, so the left-hand side is also zero.

CASE 3: $P = IJ$ *and P has no repeated indices.* In this case, the right-hand side of (14.5) is equal to 1 by Lemma 14.7(c), so we need to show that the left-hand side is also equal to 1. By definition,

$$\varepsilon^I \wedge \varepsilon^J \left(E_{p_1}, \ldots, E_{p_{k+l}}\right)$$
$$= \frac{(k+l)!}{k!\,l!} \operatorname{Alt}\left(\varepsilon^I \otimes \varepsilon^J\right)\left(E_{p_1}, \ldots, E_{p_{k+l}}\right)$$
$$= \frac{1}{k!\,l!} \sum_{\sigma \in S_{k+l}} (\operatorname{sgn}\sigma) \varepsilon^I \left(E_{p_{\sigma(1)}}, \ldots, E_{p_{\sigma(k)}}\right) \varepsilon^J \left(E_{p_{\sigma(k+1)}}, \ldots, E_{p_{\sigma(k+l)}}\right).$$

By Lemma 14.7 again, the only terms in the sum above that give nonzero values are those in which σ permutes the first k indices and the last l indices of P separately. In other words, σ must be of the form $\sigma = \tau\eta$, where $\tau \in S_k$ acts by permuting $\{1, \ldots, k\}$ and $\eta \in S_l$ acts by permuting $\{k+1, \ldots, k+l\}$. Since

$\mathrm{sgn}(\tau\eta) = (\mathrm{sgn}\,\tau)(\mathrm{sgn}\,\eta)$, we have

$$\varepsilon^I \wedge \varepsilon^J \left(E_{p_1}, \dots, E_{p_{k+l}} \right)$$

$$= \frac{1}{k!\,l!} \sum_{\substack{\tau \in S_k \\ \eta \in S_l}} (\mathrm{sgn}\,\tau)(\mathrm{sgn}\,\eta)\varepsilon^I \left(E_{p_{\tau(1)}}, \dots, E_{p_{\tau(k)}} \right) \varepsilon^J \left(E_{p_{k+\eta(1)}}, \dots, E_{p_{k+\eta(l)}} \right)$$

$$= \left(\frac{1}{k!} \sum_{\tau \in S_k} (\mathrm{sgn}\,\tau)\varepsilon^I \left(E_{p_{\tau(1)}}, \dots, E_{p_{\tau(k)}} \right) \right)$$

$$\times \left(\frac{1}{l!} \sum_{\eta \in S_l} (\mathrm{sgn}\,\eta)\varepsilon^J \left(E_{p_{k+\eta(1)}}, \dots, E_{p_{k+\eta(l)}} \right) \right)$$

$$= \left(\mathrm{Alt}\,\varepsilon^I \right) \left(E_{p_1}, \dots, E_{p_k} \right) \left(\mathrm{Alt}\,\varepsilon^J \right) \left(E_{p_{k+1}}, \dots, E_{p_{k+l}} \right)$$

$$= \varepsilon^I \left(E_{p_1}, \dots, E_{p_k} \right) \varepsilon^J \left(E_{p_{k+1}}, \dots, E_{p_{k+l}} \right) = 1.$$

CASE 4: *P is a permutation of IJ and has no repeated indices.* In this case, applying a permutation to P brings us back to Case . Since the effect of the permutation is to multiply both sides of (14.5) by the same sign, the result holds in this case as well.   □

Proposition 14.11 (Properties of the Wedge Product). *Suppose ω, ω', η, η', and ξ are multicovectors on a finite-dimensional vector space V.*

(a) BILINEARITY: *For $a, a' \in \mathbb{R}$,*

$$(a\omega + a'\omega') \wedge \eta = a(\omega \wedge \eta) + a'(\omega' \wedge \eta),$$

$$\eta \wedge (a\omega + a'\omega') = a(\eta \wedge \omega) + a'(\eta \wedge \omega').$$

(b) ASSOCIATIVITY:

$$\omega \wedge (\eta \wedge \xi) = (\omega \wedge \eta) \wedge \xi.$$

(c) ANTICOMMUTATIVITY: *For $\omega \in \Lambda^k(V^*)$ and $\eta \in \Lambda^l(V^*)$,*

$$\omega \wedge \eta = (-1)^{kl} \eta \wedge \omega. \tag{14.6}$$

(d) *If $\left(\varepsilon^i \right)$ is any basis for V^* and $I = (i_1, \dots, i_k)$ is any multi-index, then*

$$\varepsilon^{i_1} \wedge \cdots \wedge \varepsilon^{i_k} = \varepsilon^I. \tag{14.7}$$

(e) *For any covectors $\omega^1, \dots, \omega^k$ and vectors $v_1, \dots, v_k$,*

$$\omega^1 \wedge \cdots \wedge \omega^k (v_1, \dots, v_k) = \det \left(\omega^j (v_i) \right). \tag{14.8}$$

Proof. Bilinearity follows immediately from the definition, because the tensor product is bilinear and Alt is linear. To prove associativity, note that Lemma 14.10 gives

$$\left(\varepsilon^I \wedge \varepsilon^J \right) \wedge \varepsilon^K = \varepsilon^{IJ} \wedge \varepsilon^K = \varepsilon^{IJK} = \varepsilon^I \wedge \varepsilon^{JK} = \varepsilon^I \wedge \left(\varepsilon^J \wedge \varepsilon^K \right).$$

The general case follows from bilinearity. Similarly, using Lemma 14.10 again, we get

$$\varepsilon^I \wedge \varepsilon^J = \varepsilon^{IJ} = (\operatorname{sgn}\tau)\varepsilon^{JI} = (\operatorname{sgn}\tau)\varepsilon^J \wedge \varepsilon^I,$$

where τ is the permutation that sends IJ to JI. It is easy to check that $\operatorname{sgn}\tau = (-1)^{kl}$, because τ can be decomposed as a composition of kl transpositions (each index of I must be moved past each of the indices of J). Anticommutativity then follows from bilinearity.

Part (d) is an immediate consequence of Lemma 14.10 and induction. To prove (e), we note that the special case in which each ω^j is one of the basis covectors ε^{i_j} reduces to (14.7). Since both sides of (14.8) are multilinear in $(\omega^1, \ldots, \omega^k)$, this suffices. $\qquad\qquad\qquad\Box$

Because of part (d) of this lemma, henceforth we generally use the notations ε^I and $\varepsilon^{i_1} \wedge \cdots \wedge \varepsilon^{i_k}$ interchangeably.

A k-covector η is said to be **decomposable** if it can be expressed in the form $\eta = \omega^1 \wedge \cdots \wedge \omega^k$, where $\omega^1, \ldots, \omega^k$ are covectors. It is important to be aware that not every k-covector is decomposable when $k > 1$ (see Problem 14-2); however, it follows from Propositions 14.8 and 14.11(d) that every k-covector can be written as a linear combination of decomposable ones.

The definition and computational properties of the wedge product can seem daunting at first sight. However, the only properties that you need to remember for most practical purposes are the ones expressed in the preceding proposition. In fact, these properties are more than enough to determine the wedge product uniquely, as the following exercise shows.

▶ **Exercise 14.12.** Show that the wedge product is the unique associative, bilinear, and anticommutative map $\Lambda^k(V^*) \times \Lambda^l(V^*) \to \Lambda^{k+l}(V^*)$ satisfying (14.7).

For any n-dimensional vector space V, define a vector space $\Lambda(V^*)$ by

$$\Lambda(V^*) = \bigoplus_{k=0}^{n} \Lambda^k(V^*).$$

It follows from Proposition 14.8 that $\dim \Lambda(V^*) = 2^n$. Proposition 14.11 shows that the wedge product turns $\Lambda(V^*)$ into an associative algebra, called the **exterior algebra** (or **Grassmann algebra**) **of** V. This algebra is not commutative, but it has a closely related property. An algebra A is said to be **graded** if it has a direct sum decomposition $A = \bigoplus_{k \in \mathbb{Z}} A^k$ such that the product satisfies $(A^k)(A^l) \subseteq A^{k+l}$ for each k and l. A graded algebra is **anticommutative** if the product satisfies $ab = (-1)^{kl}ba$ for $a \in A^k$, $b \in A^l$. Proposition 14.11(c) shows that $\Lambda(V^*)$ is an anticommutative graded algebra (where we interpret $A^k = \Lambda^k(V^*)$ for $0 \leq k \leq n$, and $A^k = \{0\}$ otherwise).

As we observed at the beginning of this section, one could also define the wedge product without the unwieldy coefficient of (14.3). Some authors choose this alternative definition of the wedge product. To avoid confusion, we denote it by $\bar{\wedge}$:

$$\omega \,\bar{\wedge}\, \eta = \operatorname{Alt}(\omega \otimes \eta). \qquad\qquad (14.9)$$

With this definition, (14.4) is replaced by

$$\varepsilon^I \overline{\wedge} \varepsilon^J = \frac{k!l!}{(k+l)!} \varepsilon^{IJ},$$

and (14.8) is replaced by

$$\omega^1 \overline{\wedge} \cdots \overline{\wedge} \omega^k (v_1, \ldots, v_k) = \frac{1}{k!} \det \left(\omega^i (v_j) \right) \qquad (14.10)$$

whenever $\omega^1, \ldots, \omega^k$ are covectors, as you can check.

Because of (14.8), we call definition (14.3) the **determinant convention** for the wedge product, and (14.9) the **Alt convention**. The choice of which definition to use is largely a matter of taste. Although the definition of the Alt convention is perhaps a bit more natural, the computational advantages of the determinant convention make it preferable for most applications, and we use it exclusively in this book. (But see Problem 14-3 for an argument in favor of the Alt convention.) The determinant convention is most common in introductory differential geometry texts, and is used, for example, in [Boo86, Cha06, dC92, LeeJeff09, Pet06, Spi99]. The Alt convention is used in [KN69] and is more common in complex differential geometry.

Interior Multiplication

There is an important operation that relates vectors with alternating tensors. Let V be a finite-dimensional vector space. For each $v \in V$, we define a linear map $i_v \colon \Lambda^k(V^*) \to \Lambda^{k-1}(V^*)$, called **interior multiplication by v**, as follows:

$$i_v \omega(w_1, \ldots, w_{k-1}) = \omega(v, w_1, \ldots, w_{k-1}).$$

In other words, $i_v \omega$ is obtained from ω by inserting v into the first slot. By convention, we interpret $i_v \omega$ to be zero when ω is a 0-covector (i.e., a number). Another common notation is

$$v \lrcorner \omega = i_v \omega.$$

This is often read "v into ω."

Lemma 14.13. *Let V be a finite-dimensional vector space and $v \in V$.*

(a) $i_v \circ i_v = 0$.
(b) *If $\omega \in \Lambda^k(V^*)$ and $\eta \in \Lambda^l(V^*)$,*

$$i_v(\omega \wedge \eta) = (i_v \omega) \wedge \eta + (-1)^k \omega \wedge (i_v \eta). \qquad (14.11)$$

Proof. On k-covectors for $k \geq 2$, part (a) is immediate from the definition, because any alternating tensor gives zero when two of its arguments are identical. On 1-covectors and 0-covectors, it follows from the fact that $i_v \equiv 0$ on 0-covectors.

To prove (b), it suffices to consider the case in which both ω and η are decomposable, since every alternating tensor of positive rank can be written as a linear combination of decomposable ones. It is straightforward to verify that (b) follows in this special case from the following general formula for covectors $\omega^1, \ldots, \omega^k$:

$$v \lrcorner \left(\omega^1 \wedge \cdots \wedge \omega^k\right) = \sum_{i=1}^{k} (-1)^{i-1} \omega^i(v) \omega^1 \wedge \cdots \wedge \widehat{\omega^i} \wedge \cdots \wedge \omega^k, \qquad (14.12)$$

where the hat indicates that ω^i is omitted.

To prove (14.12), let us write $v_1 = v$ and apply both sides to an arbitrary $(k-1)$-tuple of vectors $(v_2, \ldots, v_k)$; then what we have to prove is

$$\left(\omega^1 \wedge \cdots \wedge \omega^k\right)(v_1, \ldots, v_k)$$

$$= \sum_{i=1}^{k} (-1)^{i-1} \omega^i(v_1) \left(\omega^1 \wedge \cdots \wedge \widehat{\omega^i} \wedge \cdots \wedge \omega^k\right)(v_2, \ldots, v_k). \qquad (14.13)$$

The left-hand side of (14.13) is the determinant of the matrix $\mathbf{v}$ whose (i, j)-entry is $\omega^i(v_j)$. To simplify the right-hand side, let $\mathbf{v}^i_j$ denote the $(k-1) \times (k-1)$ submatrix of $\mathbf{v}$ obtained by deleting the ith row and jth column. Then the right-hand side of (14.13) is

$$\sum_{i=1}^{k} (-1)^{i-1} \omega^i(v_1) \det \mathbf{v}^i_1.$$

This is just the expansion of $\det \mathbf{v}$ by minors along the first column, and therefore is equal to $\det \mathbf{v}$. □

When the wedge product is defined using the Alt convention, interior multiplication of a vector with a k-form has to be defined with an extra factor of k:

$$\bar{\iota}_v \omega(w_1, \ldots, w_{k-1}) = k\omega(v, w_1, \ldots, w_{k-1}).$$

This definition ensures that interior multiplication $\bar{\iota}_v$ still satisfies (14.11)—the factor of k compensates for the difference between the factors of $1/k!$ and $1/(k-1)!$ that occur when the left-hand and right-hand sides of (14.13) are evaluated using the Alt convention.

Differential Forms on Manifolds

Now we turn our attention to an n-dimensional smooth manifold M (with or without boundary). Recall that $T^k T^* M$ is the bundle of covariant k-tensors on M. The subset of $T^k T^* M$ consisting of alternating tensors is denoted by $\Lambda^k T^* M$:

$$\Lambda^k T^* M = \coprod_{p \in M} \Lambda^k(T_p^* M).$$

▶ **Exercise 14.14.** Show that $\Lambda^k T^* M$ is a smooth subbundle of $T^k T^* M$, and therefore is a smooth vector bundle of rank $\binom{n}{k}$ over M.

A section of $\Lambda^k T^* M$ is called a ***differential k-form***, or just a ***k-form***; this is a (continuous) tensor field whose value at each point is an alternating tensor. The integer k is called the ***degree*** of the form. We denote the vector space of smooth k-forms by

$$\Omega^k(M) = \Gamma(\Lambda^k T^* M).$$

The wedge product of two differential forms is defined pointwise: $(\omega \wedge \eta)_p = \omega_p \wedge \eta_p$. Thus, the wedge product of a k-form with an l-form is a $(k + l)$-form. If f is a 0-form and η is a k-form, we interpret the wedge product $f \wedge \eta$ to mean the ordinary product $f\eta$. If we define

$$\Omega^*(M) = \bigoplus_{k=0}^{n} \Omega^k(M), \tag{14.14}$$

then the wedge product turns $\Omega^*(M)$ into an associative, anticommutative graded algebra.

In any smooth chart, a k-form ω can be written locally as

$$\omega = \sum_I{}' \omega_I \, dx^{i_1} \wedge \cdots \wedge dx^{i_k} = \sum_I{}' \omega_I \, dx^I,$$

where the coefficients ω_I are continuous functions defined on the coordinate domain, and we use dx^I as an abbreviation for $dx^{i_1} \wedge \cdots \wedge dx^{i_k}$ (not to be mistaken for the differential of a real-valued function x^I). Proposition 10.22 shows that ω is smooth on U if and only if the component functions ω_I are smooth. In terms of differential forms, the result of Lemma 14.7(c) translates to

$$dx^{i_1} \wedge \cdots \wedge dx^{i_k}\left(\frac{\partial}{\partial x^{j_1}}, \dots, \frac{\partial}{\partial x^{j_k}}\right) = \delta_J^I.$$

Thus the component functions ω_I of ω are determined by

$$\omega_I = \omega\left(\frac{\partial}{\partial x^{i_1}}, \dots, \frac{\partial}{\partial x^{i_k}}\right).$$

Example 14.15. A 0-form is just a continuous real-valued function, and a 1-form is a covector field. On $\mathbb{R}^3$, some examples of smooth 2-forms are given by

$$\omega = (\sin xy) \, dy \wedge dz;$$

$$\eta = dx \wedge dy + dx \wedge dz + dy \wedge dz.$$

Every 3-form on $\mathbb{R}^3$ is a continuous real-valued function times $dx \wedge dy \wedge dz$.   //

If $F: M \to N$ is a smooth map and ω is a differential form on N, the pullback $F^*\omega$ is a differential form on M, defined as for any covariant tensor field:

$$(F^*\omega)_p(v_1, \dots, v_k) = \omega_{F(p)}(dF_p(v_1), \dots, dF_p(v_k)).$$

Lemma 14.16. *Suppose* $F \colon M \to N$ *is smooth.*

(a) $F^* \colon \Omega^k(N) \to \Omega^k(M)$ *is linear over* $\mathbb{R}$.
(b) $F^*(\omega \wedge \eta) = (F^*\omega) \wedge (F^*\eta)$.
(c) *In any smooth chart,*

$$F^*\left(\sideset{}{'}\sum_I \omega_I \, dy^{i_1} \wedge \cdots \wedge dy^{i_k}\right) = \sideset{}{'}\sum_I (\omega_I \circ F) \, d\left(y^{i_1} \circ F\right) \wedge \cdots \wedge d\left(y^{i_k} \circ F\right).$$

▶ **Exercise 14.17.** Prove this lemma.

This lemma gives a computational rule for pullbacks of differential forms similar to the ones we developed for covector fields and arbitrary tensor fields earlier.

Example 14.18. Define $F \colon \mathbb{R}^2 \to \mathbb{R}^3$ by $F(u, v) = (u, v, u^2 - v^2)$, and let ω be the 2-form $y \, dx \wedge dz + x \, dy \wedge dz$ on $\mathbb{R}^3$. The pullback $F^*\omega$ is computed as follows:

$$
\begin{aligned}
F^*(y \, dx \wedge dz + x \, dy \wedge dz) &= v \, du \wedge d\left(u^2 - v^2\right) + u \, dv \wedge d\left(u^2 - v^2\right) \\
&= v \, du \wedge (2u \, du - 2v \, dv) + u \, dv \wedge (2u \, du - 2v \, dv) \\
&= -2v^2 \, du \wedge dv + 2u^2 \, dv \wedge du,
\end{aligned}
$$

where we have used the fact $du \wedge du = dv \wedge dv = 0$ by anticommutativity. Because $dv \wedge du = -du \wedge dv$, this simplifies to

$$F^*\omega = -2\left(u^2 + v^2\right) du \wedge dv. \qquad\qquad \text{//}$$

The same technique can also be used to compute the expression for a differential form in another smooth chart.

Example 14.19. Let $\omega = dx \wedge dy$ on $\mathbb{R}^2$. Thinking of the transformation to polar coordinates $x = r \cos\theta$, $y = r \sin\theta$ as an expression for the identity map with respect to different coordinates on the domain and codomain, we obtain

$$
\begin{aligned}
dx \wedge dy &= d(r \cos\theta) \wedge d(r \sin\theta) \\
&= (\cos\theta \, dr - r \sin\theta \, d\theta) \wedge (\sin\theta \, dr + r \cos\theta \, d\theta) \\
&= r \, dr \wedge d\theta. \qquad\qquad \text{//}
\end{aligned}
$$

The similarity between this formula and the formula for changing a double integral from Cartesian to polar coordinates is striking. The following proposition generalizes this.

Proposition 14.20 (Pullback Formula for Top-Degree Forms). *Let* $F \colon M \to N$ *be a smooth map between* n-*manifolds with or without boundary. If* (x^i) *and* (y^j) *are smooth coordinates on open subsets* $U \subseteq M$ *and* $V \subseteq N$, *respectively, and* u *is a continuous real-valued function on* V, *then the following holds on* $U \cap F^{-1}(V)$:

$$F^*\left(u \, dy^1 \wedge \cdots \wedge dy^n\right) = (u \circ F)(\det DF) \, dx^1 \wedge \cdots \wedge dx^n, \qquad (14.15)$$

where DF *represents the Jacobian matrix of* F *in these coordinates.*

Proof. Because the fiber of $\Lambda^n T^* M$ is spanned by $dx^1 \wedge \cdots \wedge dx^n$ at each point, it suffices to show that both sides of (14.15) give the same result when evaluated on $(\partial/\partial x^1, \ldots, \partial/\partial x^n)$. From Lemma 14.16,

$$F^* \left(u \, dy^1 \wedge \cdots \wedge dy^n \right) = (u \circ F) dF^1 \wedge \cdots \wedge dF^n.$$

Proposition 14.11(e) shows that

$$dF^1 \wedge \cdots \wedge dF^n \left(\frac{\partial}{\partial x^1}, \ldots, \frac{\partial}{\partial x^n} \right) = \det \left(dF^j \left(\frac{\partial}{\partial x^i} \right) \right) = \det \left(\frac{\partial F^j}{\partial x^i} \right).$$

Therefore, the left-hand side of (14.15) gives $(u \circ F) \det DF$ when applied to $(\partial/\partial x^1, \ldots, \partial/\partial x^n)$. On the other hand, the right-hand side gives the same thing, because $dx^1 \wedge \cdots \wedge dx^n (\partial/\partial x^1, \ldots, \partial/\partial x^n) = 1$.   □

Corollary 14.21. *If* $\left(U, (x^i) \right)$ *and* $\left(\tilde{U}, (\tilde{x}^j) \right)$ *are overlapping smooth coordinate charts on* M, *then the following identity holds on* $U \cap \tilde{U}$:

$$d\tilde{x}^1 \wedge \cdots \wedge d\tilde{x}^n = \det \left(\frac{\partial \tilde{x}^j}{\partial x^i} \right) dx^1 \wedge \cdots \wedge dx^n. \qquad (14.16)$$

Proof. Apply the previous proposition with F equal to the identity map of $U \cap \tilde{U}$, but using coordinates (x^i) in the domain and $(\tilde{x}^j)$ in the codomain.   □

Interior multiplication also extends naturally to vector fields and differential forms, simply by letting it act pointwise: if $X \in \mathfrak{X}(M)$ and $\omega \in \Omega^k(M)$, define a $(k-1)$-form $X \lrcorner \omega = i_X \omega$ by

$$(X \lrcorner \omega)_p = X_p \lrcorner \omega_p.$$

▶ **Exercise 14.22.** Let X be a smooth vector field on M.

(a) Show that if ω is a smooth differential form, then $i_X \omega$ is smooth.
(b) Verify that $i_X : \Omega^k(M) \to \Omega^{k-1}(M)$ is linear over $C^\infty(M)$ and therefore corresponds to a smooth bundle homomorphism $i_X : \Lambda^k T^* M \to \Lambda^{k-1} T^* M$.

Exterior Derivatives

In this section we define a natural differential operator on smooth forms, called the *exterior derivative*. It is a generalization of the differential of a function.

To give some idea of where the motivation for the exterior derivative comes from, let us look back at a question we addressed in Chapter 11. Recall that not all 1-forms are differentials of functions: given a smooth 1-form ω, a necessary condition for the existence of a smooth function f such that $\omega = df$ is that ω be *closed*, which means that it satisfies

$$\frac{\partial \omega_j}{\partial x^i} - \frac{\partial \omega_i}{\partial x^j} = 0 \qquad (14.17)$$

in every smooth coordinate chart. Since this is a coordinate-independent property by Proposition 11.45, one might hope that the expression on the left side of (14.17) would have a meaning of its own. The key is that it is antisymmetric in the indices i and j, so it can be interpreted as the ij-component of an alternating tensor field, i.e., a 2-form. We can define a 2-form $d\omega$ locally in each smooth chart by

$$d\omega = \sum_{i<j} \left(\frac{\partial \omega_j}{\partial x^i} - \frac{\partial \omega_i}{\partial x^j} \right) dx^i \wedge dx^j, \tag{14.18}$$

so it follows that ω is closed if and only if $d\omega = 0$ in each chart.

It turns out that $d\omega$ is actually well defined globally, independently of the choice of coordinate chart, and this definition has a significant generalization to differential forms of all degrees. For each smooth manifold M with or without boundary, we will show that there is a differential operator $d: \Omega^k(M) \to \Omega^{k+1}(M)$ satisfying $d(d\omega) = 0$ for all ω. Thus, it will follow that a necessary condition for a smooth k-form ω to be equal to $d\eta$ for some $(k-1)$-form η is that $d\omega = 0$.

The definition of d on Euclidean space is straightforward: if $\omega = \sum'_J \omega_J \, dx^J$ is a smooth k-form on an open subset $U \subseteq \mathbb{R}^n$ or $\mathbb{H}^n$, we define its *exterior derivative* $d\omega$ to be the following $(k+1)$-form:

$$d\left(\sum_J' \omega_J \, dx^J \right) = \sum_J' d\omega_J \wedge dx^J, \tag{14.19}$$

where $d\omega_J$ is the differential of the function ω_J. In somewhat more detail, this is

$$d\left(\sum_J' \omega_J \, dx^{j_1} \wedge \cdots \wedge dx^{j_k} \right) = \sum_J' \sum_i \frac{\partial \omega_J}{\partial x^i} dx^i \wedge dx^{j_1} \wedge \cdots \wedge dx^{j_k}. \tag{14.20}$$

Observe that when ω is a 1-form, this becomes

$$d\left(\omega_j \, dx^j \right) = \sum_{i,j} \frac{\partial \omega_j}{\partial x^i} dx^i \wedge dx^j$$

$$= \sum_{i<j} \frac{\partial \omega_j}{\partial x^i} dx^i \wedge dx^j + \sum_{i>j} \frac{\partial \omega_j}{\partial x^i} dx^i \wedge dx^j$$

$$= \sum_{i<j} \left(\frac{\partial \omega_j}{\partial x^i} - \frac{\partial \omega_i}{\partial x^j} \right) dx^i \wedge dx^j$$

after we interchange i and j in the second sum and use the fact that $dx^j \wedge dx^i = -dx^i \wedge dx^j$, so this is consistent with our earlier definition. For a smooth 0-form f (a real-valued function), (14.20) reduces to

$$df = \frac{\partial f}{\partial x^i} dx^i,$$

which is just the differential of f.

In order to transfer this definition to manifolds, we need to check that it satisfies the following properties.

Proposition 14.23 (Properties of the Exterior Derivative on $\mathbb{R}^n$).

(a) d *is linear over* $\mathbb{R}$.

(b) *If* ω *is a smooth* k*-form and* η *is a smooth* l*-form on an open subset* $U \subseteq \mathbb{R}^n$ *or* $\mathbb{H}^n$, *then*

$$d(\omega \wedge \eta) = d\omega \wedge \eta + (-1)^k \omega \wedge d\eta.$$

(c) $d \circ d \equiv 0$.

(d) d *commutes with pullbacks: if* U *is an open subset of* $\mathbb{R}^n$ *or* $\mathbb{H}^n$, V *is an open subset of* $\mathbb{R}^m$ *or* $\mathbb{H}^m$, $F: U \to V$ *is a smooth map, and* $\omega \in \Omega^k(V)$, *then*

$$F^*(d\omega) = d(F^*\omega). \tag{14.21}$$

Proof. Linearity of d is an immediate consequence of the definition. To prove (b), by linearity it suffices to consider terms of the form $\omega = u\,dx^I \in \Omega^k(U)$ and $\eta = v\,dx^J \in \Omega^l(U)$ for smooth real-valued functions u and v. First, though, we need to know that d satisfies $d(u\,dx^I) = du \wedge dx^I$ for *any* multi-index I, not just increasing ones. If I has repeated indices, then clearly $d(u\,dx^I) = 0 = du \wedge dx^I$. If not, let σ be the permutation sending I to an increasing multi-index J. Then

$$d(u\,dx^I) = (\operatorname{sgn}\sigma)d(u\,dx^J) = (\operatorname{sgn}\sigma)\,du \wedge dx^J = du \wedge dx^I.$$

Using this, we compute

$$\begin{aligned}
d(\omega \wedge \eta) &= d\left((u\,dx^I) \wedge (v\,dx^J)\right) \\
&= d(uv\,dx^I \wedge dx^J) \\
&= (v\,du + u\,dv) \wedge dx^I \wedge dx^J \\
&= (du \wedge dx^I) \wedge (v\,dx^J) + (-1)^k (u\,dx^I) \wedge (dv \wedge dx^J) \\
&= d\omega \wedge \eta + (-1)^k \omega \wedge d\eta,
\end{aligned}$$

where the $(-1)^k$ comes from the fact that $dv \wedge dx^I = (-1)^k dx^I \wedge dv$ because dv is a 1-form and dx^I is a k-form.

We prove (c) first for the special case of a 0-form, which is just a real-valued function. In this case,

$$\begin{aligned}
d(du) &= d\left(\frac{\partial u}{\partial x^j}\,dx^j\right) = \frac{\partial^2 u}{\partial x^i \partial x^j}\,dx^i \wedge dx^j \\
&= \sum_{i<j}\left(\frac{\partial^2 u}{\partial x^i \partial x^j} - \frac{\partial^2 u}{\partial x^j \partial x^i}\right) dx^i \wedge dx^j = 0.
\end{aligned}$$

For the general case, we use the $k = 0$ case together with (b) to compute

$$d(d\omega) = d\left(\sum_J{}' d\omega_J \wedge dx^{j_1} \wedge \cdots \wedge dx^{j_k}\right)$$

$$= \sum_J{}' d(d\omega_J) \wedge dx^{j_1} \wedge \cdots \wedge dx^{j_k}$$

$$+ \sum_J{}' \sum_{i=1}^{k} (-1)^i d\omega_J \wedge dx^{j_1} \wedge \cdots \wedge d\left(dx^{j_i}\right) \wedge \cdots \wedge dx^{j_k} = 0.$$

Finally, to prove (d), again it suffices to consider $\omega = u\, dx^{i_1} \wedge \cdots \wedge dx^{i_k}$. For such a form, the left-hand side of (14.21) is

$$F^*\left(d\left(u\, dx^{i_1} \wedge \cdots \wedge dx^{i_k}\right)\right) = F^*\left(du \wedge dx^{i_1} \wedge \cdots \wedge dx^{i_k}\right)$$

$$= d(u \circ F) \wedge d\left(x^{i_1} \circ F\right) \wedge \cdots \wedge d\left(x^{i_k} \circ F\right),$$

and the right-hand side is

$$d\left(F^*\left(u\, dx^{i_1} \wedge \cdots \wedge dx^{i_k}\right)\right) = d\left((u \circ F)\, d\left(x^{i_1} \circ F\right) \wedge \cdots \wedge d\left(x^{i_k} \circ F\right)\right)$$

$$= d(u \circ F) \wedge d\left(x^{i_1} \circ F\right) \wedge \cdots \wedge d\left(x^{i_k} \circ F\right),$$

so they are equal. $\square$

These results allow us to transplant the definition of the exterior derivative to manifolds.

Theorem 14.24 (Existence and Uniqueness of Exterior Differentiation). *Suppose M is a smooth manifold with or without boundary. There are unique operators $d\colon \Omega^k(M) \to \Omega^{k+1}(M)$ for all k, called **exterior differentiation**, satisfying the following four properties:*

(i) *d is linear over $\mathbb{R}$.*
(ii) *If $\omega \in \Omega^k(M)$ and $\eta \in \Omega^l(M)$, then*

$$d(\omega \wedge \eta) = d\omega \wedge \eta + (-1)^k \omega \wedge d\eta.$$

(iii) *$d \circ d \equiv 0$.*
(iv) *For $f \in \Omega^0(M) = C^\infty(M)$, df is the differential of f, given by $df(X) = Xf$.*

In any smooth coordinate chart, d is given by (14.19).

Proof. First, we prove existence. Suppose $\omega \in \Omega^k(M)$. We wish to define $d\omega$ by means of the coordinate formula (14.19) in each chart; more precisely, this means that for each smooth chart (U, φ) for M, we wish to set

$$d\omega = \varphi^* d\left(\varphi^{-1*}\omega\right). \tag{14.22}$$

To see that this is well defined, we just note that for any other smooth chart (V, ψ), the map $\varphi \circ \psi^{-1}$ is a diffeomorphism between open subsets of $\mathbb{R}^n$ or $\mathbb{H}^n$, so Proposition 14.23(d) implies

$$\left(\varphi \circ \psi^{-1}\right)^* d\left(\varphi^{-1*}\omega\right) = d\left(\left(\varphi \circ \psi^{-1}\right)^* \varphi^{-1*}\omega\right).$$

Together with the fact that $\left(\varphi \circ \psi^{-1}\right)^* = \psi^{-1*} \circ \varphi^*$, this implies $\varphi^* d\left(\varphi^{-1*}\omega\right) = \psi^* d\left(\psi^{-1*}\omega\right)$, so $d\omega$ is well defined. It satisfies (i)–(iv) by virtue of Proposition 14.23.

To prove uniqueness, suppose that d is any operator satisfying (i)–(iv). First we need to show that $d\omega$ is determined locally: if ω_1 and ω_2 are k-forms that agree on an open subset $U \subseteq M$, then $d\omega_1 = d\omega_2$ on U. To see this, let $p \in U$ be arbitrary, let $\eta = \omega_1 - \omega_2$, and let $\psi \in C^\infty(M)$ be a bump function that is identically 1 on some neighborhood of p and supported in U. Then $\psi\eta$ is identically zero, so (i)–(iv) imply $0 = d(\psi\eta) = d\psi \wedge \eta + \psi \, d\eta$. Evaluating this at p and using the facts that $\psi(p) = 1$ and $d\psi_p = 0$, we conclude that $d\omega_1|_p - d\omega_2|_p = d\eta_p = 0$.

Now let $\omega \in \Omega^k(M)$ be arbitrary, and let (U, φ) be any smooth coordinate chart on M. We can write ω in coordinates as $\sum_I' \omega_I \, dx^I$ on U. For any $p \in U$, by means of a bump function we can construct global smooth functions $\widetilde\omega_I$ and $\widetilde x^i$ on M that agree with ω_I and x^i in a neighborhood of p. By virtue of (i)–(iv) together with the observation in the preceding paragraph, it follows that (14.19) holds at p. Since p was arbitrary, this d must be equal to the one we defined above. $\square$

If $A = \bigoplus_k A^k$ is a graded algebra, a linear map $T \colon A \to A$ is said to be a **map of degree m** if $T\left(A^k\right) \subseteq A^{k+m}$ for each k. It is said to be an **antiderivation** if it satisfies $T(xy) = (Tx)y + (-1)^k x(Ty)$ whenever $x \in A^k$ and $y \in A^l$. The preceding theorem can be summarized by saying that the differential on functions extends uniquely to an antiderivation of $\Omega^*(M)$ of degree $+1$ whose square is zero.

▶ **Exercise 14.25.** Suppose M is a smooth manifold and $X \in \mathfrak{X}(M)$. Show that interior multiplication $i_X \colon \Omega^*(M) \to \Omega^*(M)$ is an antiderivation of degree -1 whose square is zero.

Another important feature of the exterior derivative is that it commutes with all pullbacks.

Proposition 14.26 (Naturality of the Exterior Derivative). *If $F \colon M \to N$ is a smooth map, then for each k the pullback map $F^* \colon \Omega^k(N) \to \Omega^k(M)$ commutes with d: for all $\omega \in \Omega^k(N)$,*

$$F^*(d\omega) = d(F^*\omega). \tag{14.23}$$

Proof. If (U, φ) and (V, ψ) are smooth charts for M and N, respectively, we can apply Proposition 14.23(d) to the coordinate representation $\psi \circ F \circ \varphi^{-1}$. Using (14.22) twice, we compute as follows on $U \cap F^{-1}(V)$:

$$F^*(d\omega) = F^*\psi^* d\left(\psi^{-1*}\omega\right)$$
$$= \varphi^* \circ \left(\psi \circ F \circ \varphi^{-1}\right)^* d\left(\psi^{-1*}\omega\right)$$

$$= \varphi^* d\left(\left(\psi \circ F \circ \varphi^{-1}\right)^* \psi^{-1*}\omega\right)$$
$$= \varphi^* d\left(\varphi^{-1*} F^*\omega\right)$$
$$= d(F^*\omega). \qquad \qquad \square$$

Extending the terminology we introduced for covector fields in Chapter 11, we say that a smooth differential form $\omega \in \Omega^k(M)$ is **closed** if $d\omega = 0$, and **exact** if there exists a smooth $(k-1)$-form η on M such that $\omega = d\eta$. Because the exterior derivative of a 1-form satisfies (14.18), this definition of closed 1-forms agrees with the one we gave in Chapter 11.

The fact that $d \circ d = 0$ implies that every exact form is closed. In Chapter 11, we saw that the converse might not be true: the 1-form ω of Example 11.36 is closed but not exact on $\mathbb{R}^2 \smallsetminus \{0\}$. On the other hand, we showed there that every closed 1-form is locally exact. We will return to these questions in Chapter 17, where we will show that this behavior is typical: closed forms are always locally exact but not necessarily globally, so the question of whether a given closed form is exact depends on global properties of the manifold.

Exterior Derivatives and Vector Calculus in $\mathbb{R}^3$

Example 14.27. Let us work out the exterior derivatives of arbitrary 1-forms and 2-forms on $\mathbb{R}^3$. Any smooth 1-form can be written

$$\omega = P\,dx + Q\,dy + R\,dz$$

for some smooth functions P, Q, R. Using (14.19) and the fact that the wedge product of any 1-form with itself is zero, we compute

$$d\omega = dP \wedge dx + dQ \wedge dy + dR \wedge dz$$
$$= \left(\frac{\partial P}{\partial x}dx + \frac{\partial P}{\partial y}dy + \frac{\partial P}{\partial z}dz\right) \wedge dx + \left(\frac{\partial Q}{\partial x}dx + \frac{\partial Q}{\partial y}dy + \frac{\partial Q}{\partial z}dz\right) \wedge dy$$
$$+ \left(\frac{\partial R}{\partial x}dx + \frac{\partial R}{\partial y}dy + \frac{\partial R}{\partial z}dz\right) \wedge dz$$
$$= \left(\frac{\partial Q}{\partial x} - \frac{\partial P}{\partial y}\right)dx \wedge dy + \left(\frac{\partial R}{\partial x} - \frac{\partial P}{\partial z}\right)dx \wedge dz$$
$$+ \left(\frac{\partial R}{\partial y} - \frac{\partial Q}{\partial z}\right)dy \wedge dz.$$

An arbitrary 2-form on $\mathbb{R}^3$ can be written

$$\eta = u\,dx \wedge dy + v\,dx \wedge dz + w\,dy \wedge dz.$$

A similar computation shows that

$$d\eta = \left(\frac{\partial u}{\partial z} - \frac{\partial v}{\partial y} + \frac{\partial w}{\partial x}\right)dx \wedge dy \wedge dz. \qquad \qquad /\!/$$

Recall the classical vector calculus operators on $\mathbb{R}^n$: the (Euclidean) gradient of a function $f \in C^\infty (\mathbb{R}^n)$ and the **divergence** of a vector field $X \in \mathfrak{X} (\mathbb{R}^n)$ are defined by

$$\operatorname{grad} f = \sum_{i=1}^{n} \frac{\partial f}{\partial x^i} \frac{\partial}{\partial x^i}, \qquad \operatorname{div} X = \sum_{i=1}^{n} \frac{\partial X^i}{\partial x^i}. \tag{14.24}$$

In addition, in the case $n = 3$, the curl of a vector field $X \in \mathfrak{X} (\mathbb{R}^3)$ is defined by (11.26). It is interesting to note that the components of the 2-form $d\omega$ in the preceding example are exactly the components of the curl of the vector field with components (P, Q, R) (except perhaps in a different order and with different signs). Similarly, but for signs and ordering of terms, there is a strong analogy between the formula for $d\eta$ and the divergence of a vector field. These analogies can be made precise in the following way.

The Euclidean metric on $\mathbb{R}^3$ yields an index-lowering isomorphism $\flat \colon \mathfrak{X} (\mathbb{R}^3) \to \Omega^1 (\mathbb{R}^3)$. Interior multiplication yields another map $\beta \colon \mathfrak{X} (\mathbb{R}^3) \to \Omega^2 (\mathbb{R}^3)$ as follows:

$$\beta(X) = X \lrcorner (dx \wedge dy \wedge dz). \tag{14.25}$$

It is easy to check that β is linear over $C^\infty (\mathbb{R}^3)$, so it corresponds to a smooth bundle homomorphism from TM to $\Lambda^2 T^* \mathbb{R}^3$. It is a bundle isomorphism because it is injective and both TM and $\Lambda^2 T^* \mathbb{R}^3$ are bundles of rank 3. Similarly, we define a smooth bundle isomorphism $* \colon C^\infty (\mathbb{R}^3) \to \Omega^3 (\mathbb{R}^3)$ by

$$*(f) = f \, dx \wedge dy \wedge dz. \tag{14.26}$$

The relationships among all of these operators are summarized in the following diagram:

$$\begin{array}{ccccccc}
C^\infty (\mathbb{R}^3) & \xrightarrow{\operatorname{grad}} & \mathfrak{X} (\mathbb{R}^3) & \xrightarrow{\operatorname{curl}} & \mathfrak{X} (\mathbb{R}^3) & \xrightarrow{\operatorname{div}} & C^\infty (\mathbb{R}^3) \\
\downarrow{\scriptstyle \mathrm{Id}} & & \downarrow{\scriptstyle \flat} & & \downarrow{\scriptstyle \beta} & & \downarrow{\scriptstyle *} \\
\Omega^0 (\mathbb{R}^3) & \xrightarrow[d]{} & \Omega^1 (\mathbb{R}^3) & \xrightarrow[d]{} & \Omega^2 (\mathbb{R}^3) & \xrightarrow[d]{} & \Omega^3 (\mathbb{R}^3).
\end{array} \tag{14.27}$$

▶ **Exercise 14.28.** Prove that diagram (14.27) commutes, and use it to give a quick proof that curl ∘ grad ≡ 0 and div ∘ curl ≡ 0 on $\mathbb{R}^3$. Prove also that the analogues of the left-hand and right-hand squares commute when $\mathbb{R}^3$ is replaced by $\mathbb{R}^n$ for any n.

The desire to generalize these vector calculus operators from $\mathbb{R}^3$ to higher dimensions was one of the main motivations for developing the theory of differential forms. The curl, in particular, makes sense as an operator on vector fields only in dimension 3, whereas the exterior derivative expresses the same information but makes sense in all dimensions.

An Invariant Formula for the Exterior Derivative

In addition to the coordinate formula (14.19) that we used in the definition of d, there is another formula for d that is often useful, not least because it is manifestly coordinate-independent. The formula for 1-forms is by far the most important, and is the easiest to state and prove, so we begin with that. Note the similarity between this and the formula of Proposition 11.45.

Proposition 14.29 (Exterior Derivative of a 1-Form). *For any smooth 1-form ω and smooth vector fields X and Y,*

$$d\omega(X,Y) = X\big(\omega(Y)\big) - Y\big(\omega(X)\big) - \omega\big([X,Y]\big). \tag{14.28}$$

Proof. Since any smooth 1-form can be expressed locally as a sum of terms of the form $u\, dv$ for smooth functions u and v, it suffices to consider that case. Suppose $\omega = u\, dv$, and X, Y are smooth vector fields. Then the left-hand side of (14.28) is

$$d(u\, dv)(X,Y) = du \wedge dv(X,Y) = du(X)\, dv(Y) - dv(X)\, du(Y)$$

$$= Xu\, Yv - Xv\, Yu.$$

The right-hand side is

$$X\big(u\, dv(Y)\big) - Y\big(u\, dv(X)\big) - u\, dv\big([X,Y]\big)$$

$$= X(u\, Yv) - Y(u\, Xv) - u\,[X,Y]v$$

$$= (Xu\, Yv + u\, XYv) - (Yu\, Xv + u\, YXv) - u(XYv - YXv).$$

After the two $u\, XYv$ terms and the two $u\, YXv$ terms are canceled, this is equal to the left-hand side. $\square$

We will see some applications of (14.28) in later chapters. Here is our first one. It shows that the exterior derivative is in a certain sense dual to the Lie bracket. In particular, it shows that if we know all the Lie brackets of basis vector fields in a smooth local frame, we can compute the exterior derivatives of the dual covector fields, and vice versa.

Proposition 14.30. *Let M be a smooth n-manifold with or without boundary, let (E_i) be a smooth local frame for M, and let (ε^i) be the dual coframe. For each i, let b^i_{jk} denote the component functions of the exterior derivative of ε^i in this frame, and for each j, k, let c^i_{jk} be the component functions of the Lie bracket $[E_j, E_k]$:*

$$d\varepsilon^i = \sum_{j<k} b^i_{jk}\varepsilon^j \wedge \varepsilon^k; \qquad [E_j, E_k] = c^i_{jk} E_i.$$

Then $b^i_{jk} = -c^i_{jk}$.

▶ **Exercise 14.31.** Use (14.28) to prove the preceding proposition.

The generalization of (14.28) to higher-degree forms is more complicated.

Proposition 14.32 (Invariant Formula for the Exterior Derivative). *Let M be a smooth manifold with or without boundary, and $\omega \in \Omega^k(M)$. For any smooth vector fields $X_1, \ldots, X_{k+1}$ on M,*

$$
\begin{aligned}
d\omega(X_1, \ldots, X_{k+1}) \\
= \sum_{1 \le i \le k+1} (-1)^{i-1} X_i \big(\omega(X_1, \ldots, \widehat{X_i}, \ldots, X_{k+1}) \big) \\
+ \sum_{1 \le i < j \le k+1} (-1)^{i+j} \omega \big([X_i, X_j], X_1, \ldots, \widehat{X_i}, \ldots, \widehat{X_j}, \ldots, X_{k+1} \big), \quad (14.29)
\end{aligned}
$$

where the hats indicate omitted arguments.

Proof. For this proof, let us denote the entire expression on the right-hand side of (14.29) by $D\omega(X_1, \ldots, X_{k+1})$, and the two sums on the right-hand side by $I(X_1, \ldots, X_{k+1})$ and $II(X_1, \ldots, X_{k+1})$, respectively. Note that $D\omega$ is obviously multilinear over $\mathbb{R}$. We begin by showing that, like $d\omega$, it is actually multilinear over $C^\infty(M)$, which is to say that for $1 \le p \le k+1$ and $f \in C^\infty(M)$,

$$
D\omega(X_1, \ldots, fX_p, \ldots, X_{k+1}) = f D\omega(X_1, \ldots, X_p, \ldots, X_{k+1}).
$$

In the expansion of $I(X_1, \ldots, fX_p, \ldots, X_{k+1})$, f obviously factors out of the $i = p$ term. The other terms expand as follows:

$$
\begin{aligned}
\sum_{i \ne p} (-1)^{i-1} X_i \big(f\omega(X_1, \ldots, \widehat{X_i}, \ldots, X_{k+1}) \big) \\
= \sum_{i \ne p} (-1)^{i-1} \Big(f X_i \big(\omega(X_1, \ldots, \widehat{X_i}, \ldots, X_{k+1}) \big) \\
+ (X_i f) \omega \big(X_1, \ldots, \widehat{X_i}, \ldots, X_{k+1} \big) \Big).
\end{aligned}
$$

Therefore,

$$
\begin{aligned}
I(X_1, \ldots, fX_p, \ldots, X_{k+1}) \\
= f I(X_1, \ldots, X_p, \ldots, X_{k+1}) + \sum_{i \ne p} (-1)^{i-1} (X_i f) \omega \big(X_1, \ldots, \widehat{X_i}, \ldots, X_{k+1} \big).
\end{aligned}
$$

$$(14.30)$$

Consider next the expansion of II. Again, f factors out of all the terms in which $i \ne p$ and $j \ne p$. To expand the other terms, we use (8.11), which implies

$$
\begin{aligned}
[fX_p, X_j] = f[X_p, X_j] - (X_j f) X_p, \\
[X_i, fX_p] = f[X_i, X_p] + (X_i f) X_p.
\end{aligned}
$$

Inserting these formulas into the $i = p$ and $j = p$ terms, we obtain

$$\begin{aligned} II&(X_1,\ldots,fX_p,\ldots,X_{k+1}) \\ &= f II(X_1,\ldots,X_p,\ldots,X_{k+1}) \\ &\quad + \sum_{p<j}(-1)^{p+j+1}(X_j f)\omega\big(X_p,X_1,\ldots,\widehat{X_p},\ldots,\widehat{X_j},\ldots,X_{k+1}\big) \\ &\quad + \sum_{i<p}(-1)^{i+p}(X_i f)\omega\big(X_p,X_1,\ldots,\widehat{X_i},\ldots,\widehat{X_p},\ldots,X_{k+1}\big). \end{aligned}$$

Rearranging the arguments in these two sums so as to put X_p into its original position, we see that they exactly cancel the sum in (14.30). This completes the proof that $D\omega$ is multilinear over $C^\infty(M)$, so it defines a smooth $(k+1)$-tensor field.

Since both $D\omega$ and $d\omega$ are smooth tensor fields, we can verify the equation $D\omega = d\omega$ in any frame that is convenient. By multilinearity, it suffices to show that both sides give the same result when applied to an arbitrary sequence of basis vectors in some chosen local frame in a neighborhood of each point. The computations are greatly simplified by working in a coordinate frame, for which all the Lie brackets vanish. Thus, let $\big(U,(x^i)\big)$ be an arbitrary smooth chart on M. Because both $d\omega$ and $D\omega$ depend linearly on ω, we may assume that $\omega = u\,dx^I$ for some smooth function u and some increasing multi-index $I = (i_1,\ldots,i_k)$, so

$$d\omega = du \wedge dx^I = \sum_m \frac{\partial u}{\partial x^m}\,dx^m \wedge dx^I.$$

If $J = (j_1,\ldots,j_{k+1})$ is any multi-index of length $k+1$, it follows that

$$d\omega\left(\frac{\partial}{\partial x^{j_1}},\ldots,\frac{\partial}{\partial x^{j_{k+1}}}\right) = \sum_m \frac{\partial u}{\partial x^m}\delta_J^{mI}.$$

The only terms in this sum that can possibly be nonzero are those for which m is equal to one of the indices in J, say $m = j_p$. In this case, it is easy to check that $\delta_J^{mI} = (-1)^{p-1}\delta_{\widehat{J}_p}^{I}$, where $\widehat{J}_p = \big(j_1,\ldots,\widehat{j}_p,\ldots,j_{k+1}\big)$, so

$$d\omega\left(\frac{\partial}{\partial x^{j_1}},\ldots,\frac{\partial}{\partial x^{j_{k+1}}}\right) = \sum_{1\le p\le k+1}(-1)^{p-1}\frac{\partial u}{\partial x^{j_p}}\delta_{\widehat{J}_p}^{I}. \qquad (14.31)$$

On the other hand, because all the Lie brackets are zero, we have

$$\begin{aligned} D\omega&\left(\frac{\partial}{\partial x^{j_1}},\ldots,\frac{\partial}{\partial x^{j_{k+1}}}\right) \\ &= \sum_{1\le p\le k+1}(-1)^{p-1}\frac{\partial}{\partial x^{j_p}}\left(u\,dx^I\left(\frac{\partial}{\partial x^{j_1}},\ldots,\widehat{\frac{\partial}{\partial x^{j_p}}},\ldots,\frac{\partial}{\partial x^{j_{k+1}}}\right)\right) \\ &= \sum_{1\le p\le k+1}(-1)^{p-1}\frac{\partial u}{\partial x^{j_p}}\delta_{\widehat{J}_p}^{I}, \end{aligned}$$

which agrees with (14.31). $\qquad\qquad\qquad\qquad\qquad\qquad\qquad\qquad\qquad\square$

It is worth remarking that formula (14.29) can be used to give an invariant definition of d, as well as an alternative proof of Theorem 14.24 on the existence, uniqueness, and properties of d. As the proof of Proposition 14.32 showed, the right-hand side of (14.29) is multilinear over $C^\infty(M)$ as a function of $(X_1, \ldots, X_{k+1})$. By the tensor characterization lemma (Lemma 12.24), therefore, it defines a smooth covariant $(k + 1)$-tensor field. A straightforward (if slightly tedious) verification shows that it changes sign whenever two of its arguments are interchanged, so in fact it defines a smooth $(k + 1)$-form, which we could have used as a definition of $d\omega$. The rest of the proof of Proposition 14.32 then shows that $d\omega$ is actually given locally by the coordinate formula (14.19), and so the properties asserted in Theorem 14.24 follow just as before. We have chosen to define d by means of its coordinate formula because that formula is generally much easier to remember and to work with. Except in the $k = 1$ case, the invariant formula (14.29) is too complicated to be of much use for computation; in addition, it has the serious flaw that in order to compute the action of $d\omega$ on vectors $(v_1, \ldots, v_k)$ at a point $p \in M$, one must first extend them to vector *fields* in a neighborhood of p. Nonetheless, it does have some important theoretical consequences, so it is useful to know that it exists.

Lie Derivatives of Differential Forms

In Chapter 12, we derived some formulas for computing Lie derivatives of smooth tensor fields (see Corollary 12.33 and Example 12.35), which apply equally well to differential forms. However, in the case of differential forms, the exterior derivative yields a much more powerful formula for computing Lie derivatives, which also has significant theoretical consequences. As we did in Chapter 12, we restrict attention to the case of manifolds without boundary for simplicity; but these results extend easily to manifolds with boundary and vector fields tangent to the boundary.

First, we note a simple fact that will be useful in both proofs and computations: Lie differentiation satisfies a product rule with respect to wedge products.

Proposition 14.33. *Suppose M is a smooth manifold, $V \in \mathfrak{X}(M)$, and $\omega, \eta \in \Omega^*(M)$. Then*

$$\mathscr{L}_V(\omega \wedge \eta) = (\mathscr{L}_V \omega) \wedge \eta + \omega \wedge (\mathscr{L}_V \eta).$$

▶ **Exercise 14.34.** Prove the preceding proposition.

The next theorem is the main result of this section. It gives a remarkable formula for Lie derivatives of differential forms, which dates back to Élie Cartan (1869–1951), the French mathematician who invented the theory of differential forms.

Theorem 14.35 (Cartan's Magic Formula). *On a smooth manifold M, for any smooth vector field V and any smooth differential form ω,*

$$\mathscr{L}_V \omega = V \lrcorner (d\omega) + d(V \lrcorner \omega). \tag{14.32}$$

Proof. We prove that (14.32) holds for smooth k-forms by induction on k. We begin with a smooth 0-form f, in which case

$$V \lrcorner (df) + d(V \lrcorner f) = V \lrcorner df = df(V) = Vf = \mathcal{L}_V f,$$

which is (14.32).

Now let $k \geq 1$, and suppose (14.32) has been proved for forms of degree less than k. Let ω be an arbitrary smooth k-form, written in smooth local coordinates as

$$\omega = {\sum_I}' \omega_I \, dx^{i_1} \wedge \cdots \wedge dx^{i_k}.$$

Writing $u = x^{i_1}$ and $\beta = \omega_I \, dx^{i_2} \wedge \cdots \wedge dx^{i_k}$, we see that each term in this sum can be written in the form $du \wedge \beta$, where u is a smooth function and β is a smooth $(k-1)$-form. Corollary 12.34 showed that $\mathcal{L}_V du = d(\mathcal{L}_V u) = d(Vu)$. Thus Proposition 14.33 and the induction hypothesis imply

$$\mathcal{L}_V(du \wedge \beta) = (\mathcal{L}_V \, du) \wedge \beta + du \wedge (\mathcal{L}_V \beta)$$

$$= d(Vu) \wedge \beta + du \wedge \big(V \lrcorner d\beta + d(V \lrcorner \beta)\big). \tag{14.33}$$

On the other hand, using the facts that both d and interior multiplication by V are antiderivations, and $V \lrcorner du = du(V) = Vu$, we compute

$$V \lrcorner d(du \wedge \beta) + d\big(V \lrcorner (du \wedge \beta)\big)$$

$$= V \lrcorner (-du \wedge d\beta) + d\big((Vu)\beta - du \wedge (V \lrcorner \beta)\big)$$

$$= -(Vu)d\beta + du \wedge (V \lrcorner d\beta) + d(Vu) \wedge \beta + (Vu) \, d\beta + du \wedge d(V \lrcorner \beta).$$

After the $(Vu) \, d\beta$ terms are canceled, this is equal to (14.33). $\qquad\square$

Corollary 14.36 (The Lie Derivative Commutes with d). *If V is a smooth vector field and ω is a smooth differential form, then*

$$\mathcal{L}_V(d\omega) = d(\mathcal{L}_V \omega).$$

Proof. This follows from Cartan's formula and the fact that $d \circ d = 0$:

$$\mathcal{L}_V d\omega = V \lrcorner d(d\omega) + d(V \lrcorner d\omega) = d(V \lrcorner d\omega);$$

$$d\mathcal{L}_V \omega = d(V \lrcorner d\omega) + d\big(d(V \lrcorner \omega)\big) = d(V \lrcorner d\omega). \qquad\square$$

Problems

14-1. Show that covectors $\omega^1, \ldots, \omega^k$ on a finite-dimensional vector space are linearly dependent if and only if $\omega^1 \wedge \cdots \wedge \omega^k = 0$.

14-2. For what values of k and n is it true that every k-covector on $\mathbb{R}^n$ is decomposable? [Suggestion: first do the cases $n = 3$ and $n = 4$, and then see if you can figure out how to generalize your results to other dimensions.]

14-3. We have two ways to think about covariant k-tensors on a finite-dimensional vector space V: concretely, as k-multilinear functionals on V, and abstractly, as elements of the abstract tensor product space $V^* \otimes \cdots \otimes V^*$. However, we have defined alternating and symmetric tensors only in terms of the concrete definition. This problem outlines an abstract approach to alternating tensors. (Symmetric tensors can be handled similarly.) Suppose V is a finite-dimensional real vector space. Let $\mathcal{A}$ denote the subspace of the k-fold abstract tensor product $V^* \otimes \cdots \otimes V^*$ spanned by all elements of the form $\omega^1 \otimes \cdots \otimes \omega^k$ with $\omega^i = \omega^j$ for some $i \neq j$. (Thus $\mathcal{A}$ is the trivial subspace if $k < 2$.) Let $\mathrm{A}^k(V^*)$ denote the quotient vector space $(V^* \otimes \cdots \otimes V^*)/\mathcal{A}$.

(a) Show that there is a unique isomorphism $F \colon \mathrm{A}^k(V^*) \to \Lambda^k(V^*)$ such that the following diagram commutes:

$$
\begin{array}{ccc}
V^* \otimes \cdots \otimes V^* & \xrightarrow{\ \cong\ } & T^k(V^*) \\[4pt]
{\scriptstyle \pi}\big\downarrow & & \big\downarrow{\scriptstyle \mathrm{Alt}} \\[4pt]
\mathrm{A}^k(V^*) & \dashrightarrow[F] & \Lambda^k(V^*),
\end{array}
$$

where $\pi \colon V^* \otimes \cdots \otimes V^* \to \mathrm{A}^k(V^*)$ is the projection.

(b) Define a wedge product on $\bigoplus_k \mathrm{A}^k(V^*)$ by $\omega \wedge \eta = \pi(\widetilde{\omega} \otimes \widetilde{\eta})$, where $\widetilde{\omega}, \widetilde{\eta}$ are arbitrary tensors such that $\pi(\widetilde{\omega}) = \omega$, $\pi(\widetilde{\eta}) = \eta$. Show that this wedge product is well defined, and that F takes this wedge product to the Alt convention wedge product on $\Lambda(V^*)$.

[Remark: this is one reason why some authors consider the Alt convention for the wedge product to be more natural than the determinant convention.]

14-4. This chapter focused on alternating covariant tensors because of their many important applications in differential geometry. Alternating *contravariant* tensors have a few applications as well (one is described in Problem 21-14). If V is a finite-dimensional vector space, we can define alternating contravariant k-tensors either as multilinear functionals from $V^* \times \cdots \times V^*$ to $\mathbb{R}$ that change sign whenever two arguments are interchanged, or as elements of a quotient space $(V \otimes \cdots \otimes V)/\mathcal{A}$ analogous to the one defined in Problem 14-3. In the first case, the wedge product is defined just as in (14.3), but with the roles of vectors and covectors interchanged. In the second case, the wedge product is defined as the image of the tensor product in the quotient space as in Problem 14-3. Whichever definition is used, alternating contravariant k-tensors are called **multivectors** or **k-vectors**. For this problem, choose whichever of these definitions you prefer.

(a) Show that an ordered k-tuple $(v_1, \ldots, v_k)$ of elements of V is linearly dependent if and only if $v_1 \wedge \cdots \wedge v_k = 0$.

(b) Show that two linearly independent ordered k-tuples $(v_1, \ldots, v_k)$ and $(w_1, \ldots, w_k)$ have the same span if and only if

$$v_1 \wedge \cdots \wedge v_k = c \, w_1 \wedge \cdots \wedge w_k$$

for some nonzero real number c.

14-5. CARTAN'S LEMMA: Let M be a smooth n-manifold with or without boundary, and let $(\omega^1, \ldots, \omega^k)$ be an ordered k-tuple of smooth 1-forms on an open subset $U \subseteq M$ such that $(\omega^1|_p, \ldots, \omega^k|_p)$ is linearly independent for each $p \in U$. Given smooth 1-forms $\alpha^1, \ldots, \alpha^k$ on U such that

$$\sum_{i=1}^{k} \alpha^i \wedge \omega^i = 0,$$

show that each α^i can be written as a linear combination of $\omega^1, \ldots, \omega^k$ with smooth coefficients.

14-6. Define a 2-form ω on $\mathbb{R}^3$ by

$$\omega = x \, dy \wedge dz + y \, dz \wedge dx + z \, dx \wedge dy.$$

(a) Compute ω in spherical coordinates (ρ, φ, θ) defined by $(x, y, z) = (\rho \sin \varphi \cos \theta, \rho \sin \varphi \sin \theta, \rho \cos \varphi)$.

(b) Compute $d\omega$ in both Cartesian and spherical coordinates and verify that both expressions represent the same 3-form.

(c) Compute the pullback $\iota_{\mathbb{S}^2}^* \omega$ to $\mathbb{S}^2$, using coordinates (φ, θ) on the open subset where these coordinates are defined.

(d) Show that $\iota_{\mathbb{S}^2}^* \omega$ is nowhere zero.

14-7. In each of the following cases, M and N are smooth manifolds; $F \colon M \to N$ is a smooth map; and ω is a smooth differential form on N. In each case, compute $d\omega$ and $F^*\omega$, and verify by direct computation that $F^*(d\omega) = d(F^*\omega)$. (Cf. Problem 11-7.)

(a) $M = N = \mathbb{R}^2$;
$F(s, t) = (st, e^t)$;
$\omega = x \, dy$.

(b) $M = \mathbb{R}^2$ and $N = \mathbb{R}^3$;
$F(\theta, \varphi) = ((\cos \varphi + 2) \cos \theta, \ (\cos \varphi + 2) \sin \theta, \ \sin \varphi)$;
$\omega = y \, dz \wedge dx$.

(c) $M = \{(u, v) \in \mathbb{R}^2 : u^2 + v^2 < 1\}$ and $N = \mathbb{R}^3 \smallsetminus \{0\}$;
$F(u, v) = \left(u, v, \sqrt{1 - u^2 - v^2} \right)$;
$$\omega = \frac{x \, dy \wedge dz + y \, dz \wedge dx + z \, dx \wedge dy}{(x^2 + y^2 + z^2)^{3/2}}.$$

14-8. For each nonnegative integer k, show that there is a contravariant functor $\Omega^k \colon \text{Diff} \to \text{Vec}_{\mathbb{R}}$, which to each smooth manifold M assigns the vector space $\Omega^k(M)$ and to each smooth map F the pullback F^*. Show that the exterior derivative is a natural transformation from Ω^k to Ω^{k+1}. (See Problem 11-18.)

14-9. Let M, N be smooth manifolds, and suppose $\pi \colon M \to N$ is a surjective smooth submersion with connected fibers. We say that a tangent vector $v \in T_p M$ is **vertical** if $d\pi_p(v) = 0$. Suppose $\omega \in \Omega^k(M)$. Show that there exists $\eta \in \Omega^k(N)$ such that $\omega = \pi^*\eta$ if and only if $v \lrcorner \omega_p = 0$ and $v \lrcorner d\omega_p = 0$ for every $p \in M$ and every vertical vector $v \in T_p M$. [Hint: first, do the case in which $\pi \colon \mathbb{R}^{n+m} \to \mathbb{R}^n$ is projection onto the first n coordinates.]

Chapter 15
Orientations

The purpose of this chapter is to introduce a subtle but important property of smooth manifolds called *orientation*. This word stems from the Latin *oriens* ("east"), and originally meant "turning toward the east" or more generally "positioning with respect to one's surroundings." Thus, an orientation of a line or curve is a simply a choice of direction along it. As we saw in Chapter 11, the sign of a line integral depends on a choice of preferred direction along the curve.

Mathematicians have extended the sense of the word "orientation" to higher-dimensional manifolds, as a choice between two inequivalent ways in which objects can be situated with respect to their surroundings. For 2-dimensional manifolds, an orientation is essentially a choice of which rotational direction should be considered "clockwise" and which "counterclockwise." For 3-dimensional ones, it is a choice between "left-handedness" and "right-handedness." The general definition of an orientation is an adaptation of these everyday concepts to arbitrary dimensions.

As we will see in this chapter, a vector space always has exactly two choices of orientation. In $\mathbb{R}^n$, there is a standard orientation that we can all agree on; but in other vector spaces, an arbitrary choice has to be made. For manifolds, the situation is much more complicated. On a sphere, it is possible to decide unambiguously which rotational direction is counterclockwise, by looking at the surface from the outside (Fig. 15.1). On the other hand, a Möbius band (Fig. 15.2) has the curious property that a figure moving around on the surface can come back to its starting point transformed into its mirror image, so it is impossible to decide consistently which of the two possible rotational directions on the surface to call "clockwise" and which "counterclockwise," or which is the "front" side and which is the "back" side. The analogous phenomenon in a 3-manifold would be a right-handed person who takes a long trip and comes back left-handed. Manifolds like the sphere, in which it is possible to choose a consistent orientation, are said to be *orientable*; those like the Möbius band in which it is not possible are said to be *nonorientable*.

In this chapter we develop the theory of orientations of smooth manifolds. They have numerous applications, most notably in the theory of integration on manifolds, which we will study in Chapter 16.

J.M. Lee, *Introduction to Smooth Manifolds*, Graduate Texts in Mathematics 218, 377
DOI 10.1007/978-1-4419-9982-5_15, © Springer Science+Business Media New York 2013

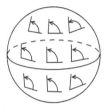

Fig. 15.1 A sphere is orientable   **Fig. 15.2** A Möbius band is not orientable

We begin the chapter with an introduction to orientations of vector spaces, and then show how this theory can be carried over to manifolds. Next, we explore the ways in which orientations can be induced on hypersurfaces and on boundaries of manifolds with boundary. Then we treat the special case of orientations on Riemannian manifolds and Riemannian hypersurfaces. At the end of the chapter, we explore the close relationship between orientability and covering maps.

Orientations of Vector Spaces

We begin with orientations of vector spaces. We are all familiar with certain informal rules for singling out preferred ordered bases of $\mathbb{R}^1$, $\mathbb{R}^2$, and $\mathbb{R}^3$ (see Fig. 15.3). We usually choose a basis for $\mathbb{R}^1$ that points to the right (i.e., in the positive direction). A natural family of preferred ordered bases for $\mathbb{R}^2$ consists of those for which the rotation from the first vector to the second is in the counterclockwise direction. And every student of vector calculus encounters "right-handed" bases in $\mathbb{R}^3$: these are the ordered bases (E_1, E_2, E_3) with the property that when the fingers of your right hand curl from E_1 to E_2, your thumb points in the direction of E_3.

Although "to the right," "counterclockwise," and "right-handed" are not mathematical terms, it is easy to translate the rules for selecting preferred bases of $\mathbb{R}^1$, $\mathbb{R}^2$, and $\mathbb{R}^3$ into rigorous mathematical language: you can check that in all three cases, the preferred bases are the ones whose transition matrices from the standard basis have positive determinants.

In an abstract vector space for which there is no canonical basis, we no longer have any way to determine which bases are "correctly oriented." For example, if V is the vector space of polynomials in one real variable of degree at most 2, who is to say which of the ordered bases $(1, x, x^2)$ and $(x^2, x, 1)$ is "right-handed"? All we can say in general is what it means for two bases to have the "same orientation."

Thus we are led to introduce the following definition. Let V be a real vector space of dimension $n \geq 1$. We say that two ordered bases $(E_1, \dots, E_n)$ and $(\widetilde{E}_1, \dots, \widetilde{E}_n)$ for V are **consistently oriented** if the transition matrix (B_i^j), defined by

$$E_i = B_i^j \, \widetilde{E}_j, \tag{15.1}$$

has positive determinant.

▶ **Exercise 15.1.** Show that being consistently oriented is an equivalence relation on the set of all ordered bases for V, and show that there are exactly two equivalence classes.

If $\dim V = n \geq 1$, we define an **orientation for** V as an equivalence class of ordered bases. If $(E_1, \ldots, E_n)$ is any ordered basis for V, we denote the orientation that it determines by $[E_1, \ldots, E_n]$, and the opposite orientation by $-[E_1, \ldots, E_n]$. A vector space together with a choice of orientation is called an **oriented vector space**. If V is oriented, then any ordered basis $(E_1, \ldots, E_n)$ that is in the given orientation is said to be **oriented** or **positively oriented**. Any basis that is not in the given orientation is said to be **negatively oriented**.

For the special case of a zero-dimensional vector space V, we define an orientation of V to be simply a choice of one of the numbers ± 1.

Example 15.2. The orientation $[e_1, \ldots, e_n]$ of $\mathbb{R}^n$ determined by the standard basis is called the **standard orientation**. You should convince yourself that, in our usual way of representing the axes graphically, an oriented basis for $\mathbb{R}$ is one that points to the right; an oriented basis for $\mathbb{R}^2$ is one for which the rotation from the first basis vector to the second is counterclockwise; and an oriented basis for $\mathbb{R}^3$ is a right-handed one. (These can be taken as mathematical definitions for the words "right," "counterclockwise," and "right-handed.") The standard orientation for $\mathbb{R}^0$ is defined to be $+1$. //

There is an important connection between orientations and alternating tensors, expressed in the following proposition.

Proposition 15.3. *Let V be a vector space of dimension n. Each nonzero element $\omega \in \Lambda^n(V^*)$ determines an orientation $\mathcal{O}_\omega$ of V as follows: if $n \geq 1$, then $\mathcal{O}_\omega$ is the set of ordered bases $(E_1, \ldots, E_n)$ such that $\omega(E_1, \ldots, E_n) > 0$; while if $n = 0$, then $\mathcal{O}_\omega$ is $+1$ if $\omega > 0$, and -1 if $\omega < 0$. Two nonzero n-covectors determine the same orientation if and only if each is a positive multiple of the other.*

Proof. The 0-dimensional case is immediate, since a nonzero element of $\Lambda^0(V^*)$ is just a nonzero real number. Thus we may assume $n \geq 1$. Let ω be a nonzero element of $\Lambda^n(V^*)$, and let $\mathcal{O}_\omega$ denote the set of ordered bases on which ω gives positive values. We need to show that $\mathcal{O}_\omega$ is exactly one equivalence class.

Suppose (E_i) and $(\widetilde{E}_j)$ are any two ordered bases for V, and let $B: V \to V$ be the linear map sending E_j to $\widetilde{E}_j$. This means that $\widetilde{E}_j = BE_j = B_j^i E_i$, so B is the transition matrix between the two bases. By Proposition 14.9,

$$\omega(\widetilde{E}_1, \ldots, \widetilde{E}_n) = \omega(BE_1, \ldots, BE_n) = (\det B)\omega(E_1, \ldots, E_n).$$

It follows that the basis $(\widetilde{E}_j)$ is consistently oriented with (E_i) if and only if $\omega(E_1, \ldots, E_n)$ and $\omega(\widetilde{E}_1, \ldots, \widetilde{E}_n)$ have the same sign, which is the same as saying that $\mathcal{O}_\omega$ is one equivalence class. The last statement then follows easily. $\square$

If V is an oriented n-dimensional vector space and ω is an n-covector that determines the orientation of V as described in this proposition, we say that ω is a

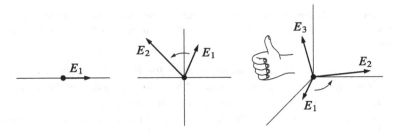

Fig. 15.3 Oriented bases for $\mathbb{R}^1$, $\mathbb{R}^2$, and $\mathbb{R}^3$

(*positively*) *oriented n-covector*. For example, the n-covector $e^1 \wedge \cdots \wedge e^n$ is positively oriented for the standard orientation on $\mathbb{R}^n$.

For any n-dimensional vector space V, the space $\Lambda^n(V^*)$ is 1-dimensional.
Proposition 15.3 shows that choosing an orientation for V is equivalent to choosing one of the two components of $\Lambda^n(V^*) \smallsetminus \{0\}$. This formulation also works for
0-dimensional vector spaces, and explains why we have defined an orientation of a
0-dimensional space in the way we did.

Orientations of Manifolds

Let M be a smooth manifold with or without boundary. We define a *pointwise
orientation* on M to be a choice of orientation of each tangent space. By itself, this
is not a very useful concept, because the orientations of nearby points may have
no relation to each other. For example, a pointwise orientation on $\mathbb{R}^n$ might switch
randomly from point to point between the standard orientation and its opposite. In
order for orientations to have some relationship with the smooth structure, we
need an extra condition to ensure that the orientations of nearby tangent spaces are
consistent with each other.

Let M be a smooth n-manifold with or without boundary, endowed with a pointwise orientation. If (E_i) is a local frame for TM, we say that (E_i) is (*positively*)
oriented if $(E_1|_p, \ldots, E_n|_p)$ is a positively oriented basis for $T_p M$ at each point
$p \in U$. A *negatively oriented* frame is defined analogously.

A pointwise orientation is said to be *continuous* if every point of M is in the
domain of an oriented local frame. (Recall that by definition the vector fields that
make up a local frame are continuous.) An *orientation of M* is a continuous pointwise orientation. We say that M is *orientable* if there exists an orientation for it, and
nonorientable if not. An *oriented manifold* is an ordered pair $(M, \mathcal{O})$, where M is
an orientable smooth manifold and $\mathcal{O}$ is a choice of orientation for M; an *oriented
manifold with boundary* is defined similarly. For each $p \in M$, the orientation of
$T_p M$ determined by $\mathcal{O}$ is denoted by $\mathcal{O}_p$. When it is not important to name the orientation explicitly, we use the usual shorthand expression "M is an oriented smooth
manifold" (or "manifold with boundary").

If M is zero-dimensional, this definition just means that an orientation of M is a
choice of ± 1 attached to each of its points. The continuity condition is vacuous in

this case, and the notion of oriented frames is not useful. Clearly, every 0-manifold is orientable.

▶ **Exercise 15.4.** Suppose M is an oriented smooth n-manifold with or without boundary, and $n \geq 1$. Show that every local frame with connected domain is either positively oriented or negatively oriented. Show that the connectedness assumption is necessary.

The next two propositions give ways of specifying orientations on manifolds that are more practical to use than the definition.

Proposition 15.5 (The Orientation Determined by an n-Form). *Let M be a smooth n-manifold with or without boundary. Any nonvanishing n-form ω on M determines a unique orientation of M for which ω is positively oriented at each point. Conversely, if M is given an orientation, then there is a smooth nonvanishing n-form on M that is positively oriented at each point.*

Remark. Because of this proposition, if M is a smooth n-manifold with or without boundary, any nonvanishing n-form on M is called an ***orientation form***. If M is oriented and ω is an orientation form determining the given orientation, we also say that ω is ***(positively) oriented***. It is easy to check that if ω and $\widetilde{\omega}$ are two positively oriented smooth forms on M, then $\widetilde{\omega} = f\omega$ for some strictly positive smooth real-valued function f. If M is a 0-manifold, a nonvanishing 0-form (i.e., real-valued function) assigns the orientation $+1$ to points where it is positive and -1 to points where it is negative.

Proof. Let ω be a nonvanishing n-form on M. Then ω defines a pointwise orientation by Proposition 15.3, so all we need to check is that it is continuous. This is trivially true when $n = 0$, so assume $n \geq 1$. Given $p \in M$, let (E_i) be any local frame on a connected neighborhood U of p, and let (ε^i) be the dual coframe. On U, the expression for ω in this frame is $\omega = f\varepsilon^1 \wedge \cdots \wedge \varepsilon^n$ for some continuous function f. The fact that ω is nonvanishing means that f is nonvanishing, and therefore

$$\omega(E_1, \ldots, E_n) = f \neq 0$$

at all points of U. Since U is connected, it follows that this expression is either always positive or always negative on U, and therefore the given frame is either positively oriented or negatively oriented. If negatively, we can replace E_1 by $-E_1$ to obtain a new frame that is positively oriented. Thus, the pointwise orientation determined by ω is continuous.

Conversely, suppose M is oriented, and let $\Lambda^n_+ T^*M \subseteq \Lambda^n T^*M$ be the open subset consisting of positively oriented n-covectors at all points of M. At any point $p \in M$, the intersection of $\Lambda^n_+ T^*M$ with the fiber $\Lambda^n(T^*_p M)$ is an open half-line, and therefore convex. By the usual partition-of-unity argument (see Problem 13-2), there exists a smooth global section of $\Lambda^n_+ T^*M$ (i.e., a positively oriented smooth global n-form). □

A smooth coordinate chart on an oriented smooth manifold with or without boundary is said to be ***(positively) oriented*** if the coordinate frame $(\partial/\partial x^i)$ is positively oriented, and **negatively oriented** if the coordinate frame is negatively oriented. A smooth atlas $\{(U_\alpha, \varphi_\alpha)\}$ is said to be **consistently oriented** if for each α, β, the transition map $\varphi_\beta \circ \varphi_\alpha^{-1}$ has positive Jacobian determinant everywhere on $\varphi_\alpha(U_\alpha \cap U_\beta)$.

Proposition 15.6 (The Orientation Determined by a Coordinate Atlas). *Let M be a smooth positive-dimensional manifold with or without boundary. Given any consistently oriented smooth atlas for M, there is a unique orientation for M with the property that each chart in the given atlas is positively oriented. Conversely, if M is oriented and either $\partial M = \varnothing$ or $\dim M > 1$, then the collection of all oriented smooth charts is a consistently oriented atlas for M.*

Proof. First, suppose M has a consistently oriented smooth atlas. Each chart in the atlas determines a pointwise orientation at each point of its domain. Wherever two of the charts overlap, the transition matrix between their respective coordinate frames is the Jacobian matrix of the transition map, which has positive determinant by hypothesis, so they determine the same pointwise orientation at each point. The orientation thus determined is continuous because each point is in the domain of an oriented coordinate frame.

Conversely, assume M is oriented and either $\partial M = \varnothing$ or $\dim M > 1$. Each point is in the domain of a smooth chart, and if the chart is negatively oriented, we can replace x^1 by $-x^1$ to obtain a new chart that is positively oriented. The fact that these charts all are positively oriented guarantees that their transition maps have positive Jacobian determinants, so they form a consistently oriented atlas. (This does not work for boundary charts when $\dim M = 1$ because of our convention that the last coordinate is nonnegative in a boundary chart.) $\square$

Proposition 15.7 (Product Orientations). *Suppose $M_1, \ldots, M_k$ are orientable smooth manifolds. There is a unique orientation on $M_1 \times \cdots \times M_k$, called the **product orientation**, with the following property: if for each $i = 1, \ldots, k$, ω_i is an orientation form for the given orientation on M_i, then $\pi_1^* \omega_1 \wedge \cdots \wedge \pi_k^* \omega_k$ is an orientation form for the product orientation.*

▶ **Exercise 15.8.** Prove the preceding proposition.

Proposition 15.9. *Let M be a connected, orientable, smooth manifold with or without boundary. Then M has exactly two orientations. If two orientations of M agree at one point, they are equal.*

▶ **Exercise 15.10.** Prove the preceding proposition.

Proposition 15.11 (Orientations of Codimension-0 Submanifolds). *Suppose M is an oriented smooth manifold with or without boundary, and $D \subseteq M$ is a smooth codimension-0 submanifold with or without boundary. Then the orientation of M restricts to an orientation of D. If ω is an orientation form for M, then $\iota_D^* \omega$ is an orientation form for D.*

▶ **Exercise 15.12.** Prove the preceding proposition.

Let M and N be oriented smooth manifolds with or without boundary, and suppose $F: M \to N$ is a local diffeomorphism. If M and N are positive-dimensional, we say that F is **orientation-preserving** if for each $p \in M$, the isomorphism dF_p takes oriented bases of $T_p M$ to oriented bases of $T_{F(p)} N$, and **orientation-reversing** if it takes oriented bases of $T_p M$ to negatively oriented bases of $T_{F(p)} N$. If M and N are 0-manifolds, then F is orientation-preserving if for every $p \in M$, the points p and $F(p)$ have the same orientation; and it is orientation-reversing if they have the opposite orientation.

▶ **Exercise 15.13.** Suppose M and N are oriented positive-dimensional smooth manifolds with or without boundary, and $F: M \to N$ is a local diffeomorphism. Show that the following are equivalent.

(a) F is orientation-preserving.
(b) With respect to any oriented smooth charts for M and N, the Jacobian matrix of F has positive determinant.
(c) For any positively oriented orientation form ω for N, the form $F^*\omega$ is positively oriented for M.

▶ **Exercise 15.14.** Show that a composition of orientation-preserving maps is orientation-preserving.

Here is another important method for constructing orientations.

Proposition 15.15 (The Pullback Orientation). *Suppose M and N are smooth manifolds with or without boundary. If $F: M \to N$ is a local diffeomorphism and N is oriented, then M has a unique orientation, called the **pullback orientation induced by F**, such that F is orientation-preserving.*

Proof. For each $p \in M$, there is a unique orientation on $T_p M$ that makes the isomorphism $dF_p: T_p M \to T_{F(p)} N$ orientation-preserving. This defines a pointwise orientation on M, and provided it is continuous, it is the unique orientation on M with respect to which F is orientation-preserving. To see that it is continuous, just choose a smooth orientation form ω for N and note that $F^*\omega$ is a smooth orientation form for M. $\square$

In the situation of the preceding proposition, if $\mathcal{O}$ denotes the given orientation on N, the pullback orientation on M is denoted by $F^*\mathcal{O}$.

▶ **Exercise 15.16.** Suppose $F: M \to N$ and $G: N \to P$ are local diffeomorphisms and $\mathcal{O}$ is an orientation on P. Show that $(G \circ F)^*\mathcal{O} = F^*(G^*\mathcal{O})$.

Recall that a smooth manifold is said to be *parallelizable* if it admits a smooth global frame.

Proposition 15.17. *Every parallelizable smooth manifold is orientable.*

Proof. Suppose M is parallelizable, and let $(E_1, \dots, E_n)$ be a global smooth frame for M. Define a pointwise orientation on M by declaring the basis $(E_1|_p, \dots, E_n|_p)$ to be positively oriented at each $p \in M$. This pointwise orientation is continuous, because every point of M is in the domain of the (global) oriented frame (E_i). $\square$

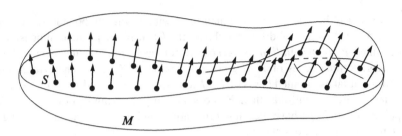

Fig. 15.4   A vector field along a submanifold

Example 15.18. The preceding proposition implies that Euclidean space $\mathbb{R}^n$, the n-torus $\mathbb{T}^n$, the spheres $\mathbb{S}^1$, $\mathbb{S}^3$, and $\mathbb{S}^7$, and products of them are all orientable, because they are all parallelizable. Therefore, any codimension-0 submanifold of one of these manifolds is also orientable. Likewise, every Lie group is orientable because it is parallelizable.   //

In the case of Lie groups, we can say more. If G is a Lie group, an orientation of G is said to be **left-invariant** if L_g is orientation-preserving for every $g \in G$.

Proposition 15.19. *Every Lie group has precisely two left-invariant orientations, corresponding to the two orientations of its Lie algebra.*

▶ **Exercise 15.20.** Prove the preceding proposition.

Orientations of Hypersurfaces

If M is an oriented smooth manifold and S is a smooth submanifold of M (with or without boundary), S might not inherit an orientation from M, even if S is embedded. Clearly, it is not sufficient to restrict an orientation form from M to S, since the restriction of an n-form to a manifold of lower dimension must necessarily be zero. A useful example to consider is the Möbius band, which is not orientable (see Example 15.38 below), even though it can be embedded in $\mathbb{R}^3$.

In this section we focus our attention on immersed or embedded hypersurfaces (codimension-1 submanifolds). With one extra piece of information (a vector field that is nowhere tangent to the hypersurface), we can use an orientation on M to induce an orientation on a hypersurface in M.

Suppose M is a smooth manifold with or without boundary, and $S \subseteq M$ is a smooth submanifold (immersed or embedded, with or without boundary). Recall (Example 10.10) that a *vector field along* S is a section of the ambient tangent bundle $TM|_S$, i.e., a continuous map $N : S \to TM$ with the property that $N_p \in T_p M$ for each $p \in S$ (Fig. 15.4). For example, any vector field on M restricts to a vector field along S, but in general, not every vector field along S is of this form (see Problem 10-9).

Proposition 15.21. *Suppose M is an oriented smooth n-manifold with or without boundary, S is an immersed hypersurface with or without boundary in M, and N is*

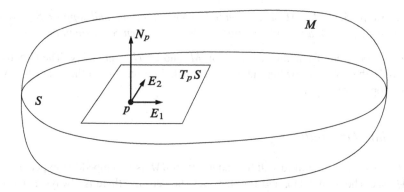

Fig. 15.5 The orientation induced by a nowhere tangent vector field

a vector field along S that is nowhere tangent to S. Then S has a unique orientation such that for each $p \in S$, $(E_1, \ldots, E_{n-1})$ is an oriented basis for $T_p S$ if and only if $(N_p, E_1, \ldots, E_{n-1})$ is an oriented basis for $T_p M$. If ω is an orientation form for M, then $\iota_S^*(N \lrcorner\, \omega)$ is an orientation form for S with respect to this orientation, where $\iota_S \colon S \hookrightarrow M$ is inclusion.

Remark. See Fig. 15.5 for an illustration of the $n = 3$ case. When $n = 1$, since S is a 0-manifold, this proposition should be interpreted as follows: at each point $p \in S$, we assign the orientation $+1$ to p if N_p is an oriented basis for $T_p M$, and -1 if N_p is negatively oriented. With this understanding, the proof below goes through in the $n = 1$ case without modification.

Proof. Let ω be an orientation form for M. Then $\sigma = \iota_S^*(N \lrcorner\, \omega)$ is an $(n-1)$-form on S. (Recall that the pullback ι_S^* is really just restriction to vectors tangent to S.) It will follow that σ is an orientation form for S if we can show that it never vanishes. Given any basis $(E_1, \ldots, E_{n-1})$ for $T_p S$, the fact that N is nowhere tangent to S implies that $(N_p, E_1, \ldots, E_{n-1})$ is a basis for $T_p M$. The fact that ω is nonvanishing implies that

$$\sigma_p(E_1, \ldots, E_{n-1}) = \omega_p(N_p, E_1, \ldots, E_{n-1}) \neq 0.$$

Since $\sigma_p(E_1, \ldots, E_{n-1}) > 0$ if and only if $\omega_p(N_p, E_1, \ldots, E_{n-1}) > 0$, the orientation determined by σ is the one defined in the statement of the proposition. $\square$

Example 15.22. The sphere $\mathbb{S}^n$ is a hypersurface in $\mathbb{R}^{n+1}$, to which the vector field $N = x^i \, \partial/\partial x^i$ is nowhere tangent, so this vector field induces an orientation on $\mathbb{S}^n$. This shows that all spheres are orientable. We define the **standard orientation of $\mathbb{S}^n$** to be the orientation determined by N. Unless otherwise specified, we always use this orientation. (The standard orientation on $\mathbb{S}^0$ is the one that assigns the orientation $+1$ to the point $+1 \in \mathbb{S}^0$ and -1 to $-1 \in \mathbb{S}^0$.) //

Not every hypersurface admits a nowhere tangent vector field. (See Problem 15-6.) However, the following proposition gives a sufficient condition that holds in many cases.

Proposition 15.23. *Let M be an oriented smooth manifold, and suppose $S \subseteq M$ is a regular level set of a smooth function $f : M \to \mathbb{R}$. Then S is orientable.*

Proof. Choose any Riemannian metric on M, and let $N = \text{grad } f|_S$. The hypotheses imply that N is a nowhere tangent vector field along S, so the result follows from Proposition 15.21. $\qquad\square$

Boundary Orientations

If M is a smooth manifold with boundary, then ∂M is an embedded hypersurface in M (see Theorem 5.11), and Problem 8-4 showed that there is always a smooth outward-pointing vector field along ∂M. Because an outward-pointing vector field is nowhere tangent to ∂M, it determines an orientation on ∂M.

Proposition 15.24 (The Induced Orientation on a Boundary). *Let M be an oriented smooth n-manifold with boundary, $n \geq 1$. Then ∂M is orientable, and all outward-pointing vector fields along ∂M determine the same orientation on ∂M.*

Remark. The orientation on ∂M determined by any outward-pointing vector field is called the ***induced orientation*** or the ***Stokes orientation*** on ∂M. (The second term is chosen because of the role this orientation plays in Stokes's theorem, to be discussed in Chapter 16.)

Proof. Let $n = \dim M$, let ω be an orientation form for M, and let N be a smooth outward-pointing vector field along ∂M. The $(n-1)$-form $\iota^*_{\partial M}(N \lrcorner \omega)$ is an orientation form for ∂M by Proposition 15.21, so ∂M is orientable.

To show that this orientation is independent of the choice of N, let $p \in \partial M$ be arbitrary, and let (x^i) be smooth boundary coordinates for M on a neighborhood of p. If N and $\tilde{N}$ are two different outward-pointing vector fields along ∂M, Proposition 5.41 shows that the last components $N^n(p)$ and $\tilde{N}^n(p)$ are both negative. Both $(N_p, \partial/\partial x^1|_p, \dots, \partial/\partial x^{n-1}|_p)$ and $(\tilde{N}_p, \partial/\partial x^1|_p, \dots, \partial/\partial x^{n-1}|_p)$ are bases for $T_p M$, and the transition matrix between them has determinant equal to $N^n(p)/\tilde{N}^n(p) > 0$. Thus, both bases determine the same orientation for $T_p M$, so N and $\tilde{N}$ determine the same orientation for $T_p \partial M$. (When $n = 1$, the bases in question are just (N_p) and $(\tilde{N}_p)$, which determine the same orientation because they are both negative multiples of $\partial/\partial x^1|_p$.) $\qquad\square$

Example 15.25. This proposition gives a simpler proof that $\mathbb{S}^n$ is orientable, because it is the boundary of the closed unit ball. The orientation thus induced on $\mathbb{S}^n$ is the standard one, as you can check. $\qquad\mathbin{/\mkern-5mu/}$

Example 15.26. Let us determine the induced orientation on $\partial \mathbb{H}^n$ when $\mathbb{H}^n$ itself has the standard orientation inherited from $\mathbb{R}^n$. We can identify $\partial \mathbb{H}^n$ with $\mathbb{R}^{n-1}$ under the correspondence $(x^1, \dots, x^{n-1}, 0) \leftrightarrow (x^1, \dots, x^{n-1})$. Since the vector field $-\partial/\partial x^n$ is outward-pointing along $\partial \mathbb{H}^n$, the standard coordinate frame for $\mathbb{R}^{n-1}$ is positively oriented for $\partial \mathbb{H}^n$ if and only if $[-\partial/\partial x^n, \partial/\partial x^1, \dots, \partial/\partial x^{n-1}]$ is the

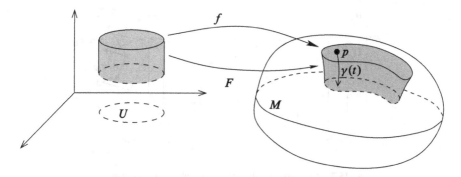

Fig. 15.6 Orientation criterion for a boundary parametrization

standard orientation for $\mathbb{R}^n$. This orientation satisfies

$$\left[-\partial/\partial x^n, \partial/\partial x^1, \ldots, \partial/\partial x^{n-1}\right] = -\left[\partial/\partial x^n, \partial/\partial x^1, \ldots, \partial/\partial x^{n-1}\right]$$
$$= (-1)^n \left[\partial/\partial x^1, \ldots, \partial/\partial x^{n-1}, \partial/\partial x^n\right].$$

Thus, the induced orientation on $\partial \mathbb{H}^n$ is equal to the standard orientation on $\mathbb{R}^{n-1}$ when n is even, but it is *opposite* to the standard orientation when n is odd. In particular, the standard coordinates on $\partial \mathbb{H}^n \approx \mathbb{R}^{n-1}$ are positively oriented if and only if n is even. (This fact will be important in the proof of Stokes's theorem in Chapter 16.) //

For many purposes, the most useful way of describing submanifolds is by means of local parametrizations. The next lemma gives a useful criterion for checking whether a local parametrization of a boundary is orientation-preserving.

Lemma 15.27. *Let M be an oriented smooth n-manifold with boundary. Suppose $U \subseteq \mathbb{R}^{n-1}$ is open, a, b are real numbers with $a < b$, and $F \colon (a, b] \times U \to M$ is a smooth embedding that restricts to an embedding of $\{b\} \times U$ into ∂M. Then the parametrization $f \colon U \to \partial M$ given by $f(x) = F(b, x)$ is orientation-preserving for ∂M if and only if F is orientation-preserving for M.*

Proof. Let x be an arbitrary point of U, and let $p = f(x) = F(b, x) \in \partial M$ (Fig. 15.6). The hypothesis that F is an embedding means that the linear map $dF_{(b,x)} \colon \left(T_b \mathbb{R} \oplus T_x \mathbb{R}^{n-1}\right) \to T_p M$ is bijective. Since the restriction of $dF_{(b,x)}$ to $T_x \mathbb{R}^{n-1}$ is equal to $df_x \colon T_x \mathbb{R}^{n-1} \to T_p \partial M$, which is already injective, it follows that $dF\left(\partial/\partial s|_{(b,x)}\right) \notin T_p \partial M$ (where s denotes the coordinate on $(a, b]$).

Define a smooth curve $\gamma \colon [0, \varepsilon) \to M$ by

$$\gamma(t) = F(b - t, x).$$

This curve satisfies $\gamma(0) = p$ and $\gamma'(0) = -dF\left(\partial/\partial s|_{(b,x)}\right) \notin T_p \partial M$. It follows that $-dF\left(\partial/\partial s|_{(b,x)}\right)$ is inward-pointing, and therefore $dF\left(\partial/\partial s|_{(b,x)}\right)$ is outward-pointing.

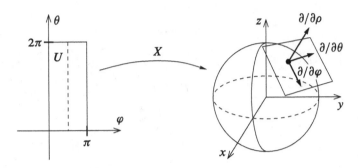

Fig. 15.7 Parametrizing the sphere via spherical coordinates

The definition of the induced orientation yields the following equivalences:

F is orientation-preserving for M

$\Leftrightarrow$ $\left(dF(\partial/\partial s), dF(\partial/\partial x^1), \ldots, dF(\partial/\partial x^{n-1})\right)$ is oriented for TM

$\Leftrightarrow$ $\left(dF(\partial/\partial x^1), \ldots, dF(\partial/\partial x^{n-1})\right)$ is oriented for $T\partial M$

$\Leftrightarrow$ $\left(df(\partial/\partial x^1), \ldots, df(\partial/\partial x^{n-1})\right)$ is oriented for $T\partial M$

$\Leftrightarrow$ f is orientation-preserving for ∂M. $\square$

Here is an illustration of how the lemma can be used.

Example 15.28. Spherical coordinates (Example C.38) yield a smooth local parametrization of $\mathbb{S}^2$ as follows. Let U be the open rectangle $(0, \pi) \times (0, 2\pi) \subseteq \mathbb{R}^2$, and let $X : U \to \mathbb{R}^3$ be the following map:

$$X(\varphi, \theta) = (\sin\varphi \cos\theta, \sin\varphi \sin\theta, \cos\varphi)$$

(Fig. 15.7).We can check whether X preserves or reverses orientation by using the fact that it is the restriction of the 3-dimensional spherical coordinate parametrization $F : (0, 1] \times U \to \overline{\mathbb{B}}^3$ defined by

$$F(\rho, \varphi, \theta) = (\rho \sin\varphi \cos\theta, \rho \sin\varphi \sin\theta, \rho \cos\varphi).$$

Because $F(1, \varphi, \theta) = X(\varphi, \theta)$, the hypotheses of Lemma 15.27 are satisfied. By direct computation, the Jacobian determinant of F is $\rho^2 \sin\varphi$, which is positive on $(0, 1] \times U$. By virtue of Lemma 15.27, X is orientation-preserving.   //

The Riemannian Volume Form

Let (M, g) be an oriented Riemannian manifold of positive dimension. We know from Proposition 13.6 that there is a smooth orthonormal frame $(E_1, \ldots, E_n)$ in a neighborhood of each point of M. By replacing E_1 by $-E_1$ if necessary, we can find an *oriented* orthonormal frame in a neighborhood of each point.

Proposition 15.29. *Suppose (M, g) is an oriented Riemannian n-manifold with or without boundary, and $n \geq 1$. There is a unique smooth orientation form $\omega_g \in \Omega^n(M)$, called the* **Riemannian volume form***, that satisfies*

$$\omega_g(E_1, \ldots, E_n) = 1 \tag{15.2}$$

for every local oriented orthonormal frame (E_i) for M.

Proof. Suppose first that such a form ω_g exists. If $(E_1, \ldots, E_n)$ is any local oriented orthonormal frame on an open subset $U \subseteq M$ and $(\varepsilon^1, \ldots, \varepsilon^n)$ is the dual coframe, we can write $\omega_g = f \, \varepsilon^1 \wedge \cdots \wedge \varepsilon^n$ on U. The condition (15.2) then reduces to $f = 1$, so

$$\omega_g = \varepsilon^1 \wedge \cdots \wedge \varepsilon^n. \tag{15.3}$$

This proves that such a form is uniquely determined.

To prove existence, we would like to *define* ω_g in a neighborhood of each point by (15.3), so we need to check that this definition is independent of the choice of oriented orthonormal frame. If $(\widetilde{E}_1, \ldots, \widetilde{E}_n)$ is another oriented orthonormal frame, with dual coframe $(\widetilde{\varepsilon}^1, \ldots, \widetilde{\varepsilon}^n)$, let

$$\widetilde{\omega}_g = \widetilde{\varepsilon}^1 \wedge \cdots \wedge \widetilde{\varepsilon}^n.$$

We can write

$$\widetilde{E}_i = A_i^j E_j$$

for some matrix (A_i^j) of smooth functions. The fact that both frames are orthonormal means that $(A_i^j(p)) \in O(n)$ for each p, so $\det(A_i^j) = \pm 1$, and the fact that the two frames are consistently oriented forces the positive sign. We compute

$$\omega_g(\widetilde{E}_1, \ldots, \widetilde{E}_n) = \det(\varepsilon^j(\widetilde{E}_i)) = \det(A_i^j) = 1 = \widetilde{\omega}_g(\widetilde{E}_1, \ldots, \widetilde{E}_n).$$

Thus $\omega_g = \widetilde{\omega}_g$, so defining ω_g in a neighborhood of each point by (15.3) with respect to some smooth oriented orthonormal frame yields a global n-form. The resulting form is clearly smooth and satisfies (15.2) for every oriented orthonormal frame. $\qquad\square$

▶ **Exercise 15.30.** Suppose (M, g) and $(\widetilde{M}, \widetilde{g})$ are positive-dimensional Riemannian manifolds with or without boundary, and $F \colon M \to \widetilde{M}$ is a local isometry. Show that $F^*\omega_{\widetilde{g}} = \omega_g$.

Although the expression for the Riemannian volume form with respect to an oriented orthonormal frame is particularly simple, it is also useful to have an expression for it in coordinates.

Proposition 15.31. *Let (M, g) be an oriented Riemannian n-manifold with or without boundary, $n \geq 1$. In any oriented smooth coordinates (x^i), the Riemannian volume form has the local coordinate expression*

$$\omega_g = \sqrt{\det(g_{ij})} \, dx^1 \wedge \cdots \wedge dx^n,$$

where g_{ij} are the components of g in these coordinates.

Proof. Let $(U, (x^i))$ be an oriented smooth chart, and let $p \in M$. In these coordinates, $\omega_g = f\, dx^1 \wedge \cdots \wedge dx^n$ for some positive coefficient function f. To compute f, let (E_i) be any smooth oriented orthonormal frame defined on a neighborhood of p, and let (ε^i) be the dual coframe. If we write the coordinate frame in terms of the orthonormal frame as

$$\frac{\partial}{\partial x^i} = A_i^j E_j,$$

then we can compute

$$f = \omega_g \left(\frac{\partial}{\partial x^1}, \ldots, \frac{\partial}{\partial x^n} \right) = \varepsilon^1 \wedge \cdots \wedge \varepsilon^n \left(\frac{\partial}{\partial x^1}, \ldots, \frac{\partial}{\partial x^n} \right)$$

$$= \det \left(\varepsilon^j \left(\frac{\partial}{\partial x^i} \right) \right) = \det(A_i^j).$$

On the other hand, observe that

$$g_{ij} = \left\langle \frac{\partial}{\partial x^i}, \frac{\partial}{\partial x^j} \right\rangle_g = \left\langle A_i^k E_k, A_j^l E_l \right\rangle_g = A_i^k A_j^l \langle E_k, E_l \rangle_g = \sum_k A_i^k A_j^k.$$

This last expression is the (i, j)-entry of the matrix product $A^T A$, where $A = (A_i^j)$. Thus,

$$\det(g_{ij}) = \det(A^T A) = \det A^T \det A = (\det A)^2,$$

from which it follows that $f = \det A = \pm\sqrt{\det(g_{ij})}$. Since both frames $(\partial/\partial x^i)$ and (E_j) are oriented, the sign must be positive. $\square$

Hypersurfaces in Riemannian Manifolds

Let (M, g) be an oriented Riemannian manifold with or without boundary, and suppose $S \subseteq M$ is an immersed hypersurface with or without boundary. Any unit normal vector field along S is nowhere tangent to S, so it determines an orientation of S by Proposition 15.21. The next proposition gives a simple formula for the volume form of the induced metric on S with respect to this orientation.

Proposition 15.32. *Let (M, g) be an oriented Riemannian manifold with or without boundary, let $S \subseteq M$ be an immersed hypersurface with or without boundary, and let $\tilde{g}$ denote the induced metric on S. Suppose N is a smooth unit normal vector field along S. With respect to the orientation of S determined by N, the volume form of $(S, \tilde{g})$ is given by*

$$\omega_{\tilde{g}} = \iota_S^*(N \lrcorner \omega_g).$$

Proof. By Proposition 15.21, the $(n-1)$-form $\iota_S^*(N \lrcorner \omega_g)$ is an orientation form for S. To prove that it is the volume form for the induced Riemannian metric, we

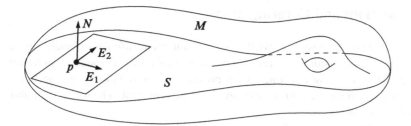

Fig. 15.8 A hypersurface in a Riemannian manifold

need only show that it gives the value 1 whenever it is applied to an oriented orthonormal frame for S. Thus, let $(E_1, \ldots, E_{n-1})$ be such a frame. At each point $p \in S$, the basis $(N_p, E_1|_p, \ldots, E_{n-1}|_p)$ is orthonormal (Fig. 15.8), and is oriented for $T_p M$ (this is the definition of the orientation determined by N). Thus

$$\iota_S^*(N \lrcorner \, \omega_g)(E_1, \ldots, E_{n-1}) = \omega_g(N, E_1, \ldots, E_{n-1}) = 1,$$

which proves the result. □

The result of Proposition 15.32 takes on particular importance in the case of a Riemannian manifold with boundary, because of the following proposition.

Proposition 15.33. *Suppose M is any Riemannian manifold with boundary. There is a unique smooth outward-pointing unit normal vector field N along ∂M.*

Proof. First, we prove uniqueness. At any point $p \in \partial M$, the subspace $(T_p \partial M)^\perp \subseteq T_p M$ is 1-dimensional, so there are exactly two unit vectors at p that are normal to ∂M. Since any unit normal vector N is nowhere tangent to ∂M, it must have nonzero x^n-component in any smooth boundary chart. Thus, exactly one of the two choices of unit normal has negative x^n-component, which is equivalent to being outward-pointing.

To prove existence, let $f : M \to \mathbb{R}$ be a boundary defining function (Proposition 5.43), and let N be the restriction to ∂M of the unit vector field $-\operatorname{grad} f / |\operatorname{grad} f|_g$. Because $df \neq 0$ at points of ∂M, N is well-defined and smooth on ∂M. Then N is normal to ∂M by Problem 13-21, and outward pointing by Exercise 5.44, because $Nf = -\langle \operatorname{grad} f, \operatorname{grad} f \rangle_g / |\operatorname{grad} f|_g = -|\operatorname{grad} f|_g < 0$. □

The next corollary is immediate.

Corollary 15.34. *If (M, g) is an oriented Riemannian manifold with boundary and $\tilde{g}$ is the induced Riemannian metric on ∂M, then the volume form of $\tilde{g}$ is*

$$\omega_{\tilde{g}} = \iota_{\partial M}^*(N \lrcorner \, \omega_g),$$

where N is the outward unit normal vector field along ∂M. □

Orientations and Covering Maps

Although it is often easy to prove that a given smooth manifold is orientable by constructing an orientation for it, proving that a manifold is *not* orientable can be much trickier. The theory of covering spaces provides one of the most useful techniques for doing so. In this section, we explore the close relationship between orientability and covering maps.

Our first result is a simple application of pullback orientations.

Proposition 15.35. *If $\pi: E \to M$ is a smooth covering map and M is orientable, then E is also orientable.*

Proof. Because a covering map is a local diffeomorphism, this follows immediately from Proposition 15.15. $\qquad\square$

The next theorem is more interesting. If G is a Lie group acting smoothly on a smooth manifold E (on the left, say), we say the action is an ***orientation-preserving action*** if for each $g \in G$, the diffeomorphism $x \mapsto g \cdot x$ is orientation-preserving.

Theorem 15.36. *Suppose E is a connected, oriented, smooth manifold with or without boundary, and $\pi: E \to M$ is a smooth normal covering map. Then M is orientable if and only if the action of $\mathrm{Aut}_\pi(E)$ on E is orientation-preserving.*

Proof. Let $\mathcal{O}_E$ denote the given orientation on E. First suppose M is orientable, and let q be an arbitrary point in E. Because M is connected, it has exactly two orientations, and one of them has the property that $d\pi_q: T_q E \to T_{\pi(q)}M$ is orientation-preserving. Call that orientation $\mathcal{O}_M$. The pullback orientation $\pi^*\mathcal{O}_M$ agrees with the given orientation at q, so it must be equal to $\mathcal{O}_E$ by Proposition 15.9. Suppose $\varphi \in \mathrm{Aut}_\pi(E)$. The fact that $\pi \circ \varphi = \pi$ implies that

$$\varphi^*\mathcal{O}_E = \varphi^*(\pi^*\mathcal{O}_M) = (\pi \circ \varphi)^*\mathcal{O}_M = \pi^*\mathcal{O}_M = \mathcal{O}_E.$$

Thus, φ is orientation-preserving.

Conversely, suppose the action of $\mathrm{Aut}_\pi(E)$ is orientation-preserving, and let $p \in M$. If $U \subseteq M$ is any evenly covered neighborhood of p, there is a smooth section $\sigma: U \to E$, which induces an orientation $\sigma^*\mathcal{O}_E$ on U. Suppose $\sigma_1: U \to E$ is any other smooth local section over U. Because π is a normal covering, $\mathrm{Aut}_\pi(E)$ acts transitively on each fiber of π, so there is a covering automorphism φ such that $\sigma_1(p) = \varphi(\sigma(p))$. Then $\varphi \circ \sigma$ is a local section of π that agrees with σ_1 at p, and thus $\sigma_1 = \varphi \circ \sigma$ on all of U. Because φ is orientation-preserving, $\sigma_1^*\mathcal{O}_E = \sigma^*\varphi^*\mathcal{O}_E = \sigma^*\mathcal{O}_E$, so the orientations induced by σ and σ_1 are equal. Thus, we can define a global orientation $\mathcal{O}_M$ on M by defining it on each evenly covered open subset to be the pullback orientation induced by any local section; the argument above shows that the orientations so defined agree where they overlap. $\qquad\square$

Here are two applications of the preceding theorem.

Example 15.37 (Orientability of Projective Spaces). For $n \geq 1$, consider the smooth covering map $q : \mathbb{S}^n \to \mathbb{RP}^n$ of Example 4.35. The only nontrivial covering automorphism of q is the antipodal map $\alpha(x) = -x$. Problem 15-3 shows that α is orientation-preserving if and only if n is odd, so it follows that $\mathbb{RP}^n$ is orientable if and only if n is odd. //

Example 15.38 (The Möbius Bundle and the Möbius Band). Let E be the total space of the Möbius bundle (Example 10.3). The quotient map $q : \mathbb{R}^2 \to E$ used to define E is a smooth normal covering map, and the covering automorphism group is isomorphic to $\mathbb{Z}$, acting on $\mathbb{R}^2$ by $n \cdot (x, y) = (x + n, (-1)^n y)$. (You can check this directly from the definitions, or you can accept this for now and wait until we have developed more machinery in Chapter 21, where a simpler proof is available; see Problem 21-9.) For n odd, the diffeomorphism $(x, y) \mapsto n \cdot (x, y)$ of $\mathbb{R}^2$ pulls back the orientation form $dx \wedge dy$ to $-dx \wedge dy$, so the action of $\mathrm{Aut}_\pi(E)$ is not orientation-preserving. Thus, Theorem 15.36 shows that E is not orientable.

For each $r > 0$, the image under q of the rectangle $[0, 1] \times [-r, r]$ is a Möbius band M_r. Because q restricts to a smooth covering map from $\mathbb{R} \times [-r, r]$ to M_r, the same argument shows that a Möbius band is not orientable either. //

The Orientation Covering

Next we show that every nonorientable smooth manifold M has an orientable two-sheeted covering manifold. The fiber over a point $p \in M$ will correspond to the two orientations of $T_p M$.

In order to handle the orientable and nonorientable cases in a uniform way, it is useful to expand our definition of covering maps slightly, by allowing "covering spaces" that are not connected. If N and M are topological spaces, let us say that a map $\pi : N \to M$ is a **generalized covering map** if it satisfies all of the requirements for a covering map except that N might not be connected: this means that N is locally path-connected, π is surjective and continuous, and each point $p \in M$ has a neighborhood that is evenly covered by π. If in addition N and M are smooth manifolds with or without boundary and π is a local diffeomorphism, we say it is a **generalized smooth covering map**.

Lemma 15.39. *Suppose N and M are topological spaces and $\pi : N \to M$ is a generalized covering map. If M is connected, then the restriction of π to each component of N is a covering map.*

Proof. Suppose W is a component of N. If U is any open subset of M that is evenly covered by π, then each component of $\pi^{-1}(U)$ is connected and therefore contained in a single component of N. It follows that $\left(\pi|_W\right)^{-1}(U) = \pi^{-1}(U) \cap W$ is either the empty set or a nonempty disjoint union of components of $\pi^{-1}(U)$, each of which is mapped homeomorphically onto U by $\pi|_W$. In particular, this means that each point in $\pi(W)$ has a neighborhood that is evenly covered by $\pi|_W$.

To complete the proof, we just need to show that $\pi|_W$ is surjective. Because π is a local homeomorphism, $\pi(W)$ is an open subset of M. On the other hand, if

$p \in M \smallsetminus \pi(W)$, and U is a neighborhood of p that is evenly covered by π, then the discussion in the preceding paragraph shows that $\left(\pi|_W\right)^{-1}(U) = \varnothing$, which implies that $U \subseteq M \smallsetminus \pi(W)$. Therefore, $\pi(W)$ is closed in M. Because W is not empty, $\pi(W)$ is all of M. ☐

Let M be a connected, smooth, positive-dimensional manifold with or without boundary, and let $\widehat{M}$ denote the set of orientations of all tangent spaces to M:

$$\widehat{M} = \left\{(p, \mathcal{O}_p) : p \in M \text{ and } \mathcal{O}_p \text{ is an orientation of } T_p M\right\}.$$

Define the projection $\widehat{\pi} \colon \widehat{M} \to M$ by sending an orientation of $T_p M$ to the point p itself: $\widehat{\pi}(p, \mathcal{O}_p) = p$. Since each tangent space has exactly two orientations, each fiber of this map has cardinality 2. The map $\widehat{\pi} \colon \widehat{M} \to M$ is called the **orientation covering of** M.

Proposition 15.40 (Properties of the Orientation Covering). *Suppose M is a connected, smooth, positive-dimensional manifold with or without boundary, and let $\widehat{\pi} \colon \widehat{M} \to M$ be its orientation covering. Then $\widehat{M}$ can be given the structure of a smooth, oriented manifold with or without boundary, with the following properties:*

(a) *$\widehat{\pi} \colon \widehat{M} \to M$ is a generalized smooth covering map.*
(b) *A connected open subset $U \subseteq M$ is evenly covered by $\widehat{\pi}$ if and only if U is orientable.*
(c) *If $U \subseteq M$ is an evenly covered open subset, then every orientation of U is the pullback orientation induced by a local section of $\widehat{\pi}$ over U.*

Proof. We first topologize $\widehat{M}$ by defining a basis for it. For each pair $(U, \mathcal{O})$, where U is an open subset of M and $\mathcal{O}$ is an orientation on U, define a subset $\widehat{U}_{\mathcal{O}} \subseteq \widehat{M}$ as follows:

$$\widehat{U}_{\mathcal{O}} = \left\{(p, \mathcal{O}_p) \in \widehat{M} : p \in U \text{ and } \mathcal{O}_p \text{ is the orientation of } T_p M \text{ determined by } \mathcal{O}\right\}.$$

We will show that the collection of all subsets of the form $\widehat{U}_{\mathcal{O}}$ is a basis for a topology on $\widehat{M}$. Given an arbitrary point $(p, \mathcal{O}_p) \in \widehat{M}$, let U be an orientable neighborhood of p in M, and let $\mathcal{O}$ be an orientation on it. After replacing $\mathcal{O}$ by $-\mathcal{O}$ if necessary, we may assume that the given orientation $\mathcal{O}_p$ is same as the orientation of $T_p M$ determined by $\mathcal{O}$. It follows that $(p, \mathcal{O}_p) \in \widehat{U}_{\mathcal{O}}$, so the collection of all sets of the form $\widehat{U}_{\mathcal{O}}$ covers $\widehat{M}$. If $\widehat{U}_{\mathcal{O}}$ and $\widehat{U}'_{\mathcal{O}'}$ are two such sets and $(p, \mathcal{O}_p)$ is a point in their intersection, then $\mathcal{O}_p$ is the orientation of $T_p M$ determined by both $\mathcal{O}$ and $\mathcal{O}'$. If V is the component of $U \cap U'$ containing p, then the restricted orientations $\mathcal{O}|_V$ and $\mathcal{O}'|_V$ agree at p and therefore are identical by Proposition 15.9, so it follows that $\widehat{U}_{\mathcal{O}} \cap \widehat{U}'_{\mathcal{O}'}$ contains the basis set $\widehat{V}_{\mathcal{O}|_V}$. Thus, we have defined a topology on $\widehat{M}$. Note that for each orientable open subset $U \subseteq M$ and each orientation $\mathcal{O}$ of U, $\widehat{\pi}$ maps the basis set $\widehat{U}_{\mathcal{O}}$ bijectively onto U. Because the orientable open subsets form a basis for the topology of M, this implies that $\widehat{\pi}$ restricts to a homeomorphism from $\widehat{U}_{\mathcal{O}}$ to U. In particular, $\widehat{\pi}$ is a local homeomorphism.

Next we show that with this topology, $\widehat{\pi}$ is a generalized covering map. Suppose $U \subseteq M$ is an orientable connected open subset and $\mathcal{O}$ is an orientation for U.

Then $\hat{\pi}^{-1}(U)$ is the disjoint union of open subsets $\hat{U}_{\mathcal{O}}$ and $\hat{U}_{-\mathcal{O}}$, and $\hat{\pi}$ restricts to a homeomorphism from each of these sets to U. Thus, each such set U is evenly covered, and it follows that $\hat{\pi}$ is a generalized covering map. By Lemma 15.39, $\hat{\pi}$ restricts to an ordinary covering map on each component of $\hat{M}$, and so Proposition 4.40 shows that each such component is a topological n-manifold with or without boundary and has a unique smooth structure making $\hat{\pi}$ into a smooth covering map. These smooth structures combine to give a smooth structure on all of $\hat{M}$. This completes the proof of (a).

Next we give $\hat{M}$ an orientation. Let $\hat{p} = (p, \mathcal{O}_p)$ be a point in $\hat{M}$. By definition, $\mathcal{O}_p$ is an orientation of $T_p M$, so we can give $T_{\hat{p}}\hat{M}$ the unique orientation $\mathcal{O}_{\hat{p}}$ such that $d\hat{\pi}_{\hat{p}} \colon T_{\hat{p}}\hat{M} \to T_p M$ is orientation-preserving. This defines a pointwise orientation $\hat{\mathcal{O}}$ on $\hat{M}$. On each basis open subset $\hat{U}_{\mathcal{O}}$, the orientation $\hat{\mathcal{O}}$ agrees with the pullback orientation induced from $(U, \mathcal{O})$ by (the restriction of) $\hat{\pi}$, so it is continuous.

Next we prove (b). We showed earlier that every orientable connected open subset of M is evenly covered by $\hat{\pi}$. Conversely, if $U \subseteq M$ is any evenly covered open subset, then there is a smooth local section $\sigma \colon U \to \hat{M}$ of $\hat{\pi}$ by Proposition 4.36, which pulls $\hat{\mathcal{O}}$ back to an orientation on U by Proposition 15.15.

Finally, to prove (c), assume $U \subseteq M$ is evenly covered and therefore orientable. Given any orientation $\mathcal{O}$ of U, define a section $\sigma \colon U \to \hat{M}$ by setting $\sigma(p) = (p, \mathcal{O}_p)$. To see that σ is continuous, suppose $\hat{U}'_{\mathcal{O}'}$ is any basis open subset of $\hat{M}$. Then for each component V of $U \cap U'$, the restricted orientations $\mathcal{O}|_V$ and $\mathcal{O}'|_V$ must either agree or disagree on all of V, so $\sigma^{-1}(\hat{U}'_{\mathcal{O}'})$ is a union of such components and therefore open. $\qquad \square$

Theorem 15.41 (Orientation Covering Theorem). *Suppose M is a connected smooth manifold with or without boundary, and let $\hat{\pi} \colon \hat{M} \to M$ be its orientation covering.*

(a) *If M is orientable, then $\hat{M}$ has exactly two components, and the restriction of $\hat{\pi}$ to each component is a diffeomorphism onto M.*

(b) *If M is nonorientable, then $\hat{M}$ is connected, and $\hat{\pi}$ is a two-sheeted smooth covering map.*

Proof. If M is orientable, then Proposition 15.40(b) shows that M is evenly covered by $\hat{\pi}$, which means that $\hat{M}$ has two components, each mapped diffeomorphically onto M.

Now assume M is nonorientable. We show first that $\hat{M}$ is connected. Let W be a component of $\hat{M}$. Lemma 15.39 shows that $\hat{\pi}|_W$ is a covering map, so its fibers all have the same cardinality. Because the fibers of $\hat{\pi}$ have cardinality 2 and W is not empty, the fibers of $\hat{\pi}|_W$ must have cardinality 1 or 2. If the cardinality were 1, then $\hat{\pi}|_W$ would be an injective smooth covering map and thus a diffeomorphism, and its inverse would be a smooth section of $\hat{\pi}$, which would induce an orientation on M. Thus, the cardinality must be 2, which implies that $W = \hat{M}$. Because $\hat{M}$ is connected, $\hat{\pi}$ is a covering map by Lemma 15.39, and because it is a local diffeomorphism it is a smooth covering map. $\qquad \square$

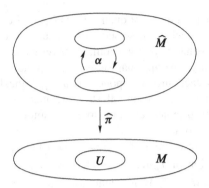

Fig. 15.9 The nontrivial covering automorphism of $\widehat{M}$

The orientation covering is sometimes called the *oriented double covering of M*. There are other ways of constructing it besides the one we have given here, but as the next theorem shows, the specific details of the construction do not matter, because they all yield isomorphic covering manifolds.

Theorem 15.42 (Uniqueness of the Orientation Covering). *Let M be a nonorientable connected smooth manifold with or without boundary, and let $\widehat{\pi} \colon \widehat{M} \to M$ be its orientation covering. If $\widetilde{M}$ is an oriented smooth manifold with or without boundary that admits a two-sheeted smooth covering map $\widetilde{\pi} \colon \widetilde{M} \to M$, then there exists a unique orientation-preserving diffeomorphism $\varphi \colon \widetilde{M} \to \widehat{M}$ such that $\widehat{\pi} \circ \varphi = \widetilde{\pi}$.*

Proof. See Problem 15-11.   □

By invoking a little more covering space theory, we obtain the following sufficient topological condition for orientability. If G is a group and $H \subseteq G$ is a subgroup, the *index of H in G* is the cardinality of the set of left cosets of H in G. (If H is a normal subgroup, it is just the cardinality of the quotient group G/H.)

Theorem 15.43. *Let M be a connected smooth manifold with or without boundary, and suppose the fundamental group of M has no subgroup of index 2. Then M is orientable. In particular, if M is simply connected then it is orientable.*

Proof. Suppose M is not orientable, and let $\widehat{\pi} \colon \widehat{M} \to M$ be its orientation covering, which is an honest covering map in this case. Choose any point $q \in \widehat{M}$, and let $p = \widehat{\pi}(q) \in M$. Let $\alpha \colon \widehat{M} \to \widehat{M}$ be the map that interchanges the two points in each fiber of $\widehat{\pi}$ (Fig. 15.9). To prove that α is smooth, suppose $U \subseteq M$ is any evenly covered open subset and $U_0, U_1 \subseteq \widehat{M}$ are the two components of $\widehat{\pi}^{-1}(U)$. Since $\widehat{\pi}$ restricts to a diffeomorphism from each component onto U, we can write $\alpha|_{U_0} = \left(\widehat{\pi}|_{U_1}\right)^{-1} \circ \left(\widehat{\pi}|_{U_0}\right)$, which is smooth. Similarly, $\alpha|_{U_1}$ is also smooth. Since the collection of all such sets U_0, U_1 is an open covering of $\widehat{M}$, it follows that α is smooth, and it is a covering automorphism because it satisfies $\widehat{\pi} \circ \alpha = \widehat{\pi}$. In fact, since a covering automorphism is determined by what it does to one point, α is the unique nontrivial element of the automorphism group $\mathrm{Aut}_{\widehat{\pi}}(\widehat{M})$, which is therefore equal to the two-element group $\{\mathrm{Id}_{\widehat{M}}, \alpha\}$. Because the automorphism group acts transitively on fibers, $\widehat{\pi}$ is a normal covering map. Let H denote the subgroup $\widehat{\pi}_*\big(\pi_1(\widehat{M}, q)\big)$ of $\pi_1(M, p)$. A fundamental result in the theory of covering spaces (see, e.g., [LeeTM, Chap. 12]) is that the quotient group $\pi_1(M, p)/H$ is isomorphic to $\mathrm{Aut}_{\widehat{\pi}}(\widehat{M})$. Therefore, H has index 2 in $\pi_1(M, p)$. $\qquad\square$

Problems

15-1. Suppose M is a smooth manifold that is the union of two orientable open submanifolds with connected intersection. Show that M is orientable. Use this to give another proof that $\mathbb{S}^n$ is orientable.

15-2. Suppose M and N are oriented smooth manifolds with or without boundary, and $F\colon M \to N$ is a local diffeomorphism. Show that if M is connected, then F is either orientation-preserving or orientation-reversing.

15-3. Suppose $n \ge 1$, and let $\alpha\colon \mathbb{S}^n \to \mathbb{S}^n$ be the antipodal map: $\alpha(x) = -x$. Show that α is orientation-preserving if and only if n is odd. [Hint: consider the map $F\colon \overline{\mathbb{B}}^n \to \overline{\mathbb{B}}^n$ given by $F(x) = -x$, and use Corollary 15.34.] (*Used on pp. 393, 435.*)

15-4. Let θ be a smooth flow on an oriented smooth manifold with or without boundary. Show that for each $t \in \mathbb{R}$, θ_t is orientation-preserving wherever it is defined. (*Used on p. 425.*)

15-5. Let M be a smooth manifold with or without boundary. Show that the total spaces of TM and T^*M are orientable.

15-6. Let $U \subseteq \mathbb{R}^3$ be the open subset $\left\{(x, y, z) : \left(\sqrt{x^2 + y^2} - 2\right)^2 + z^2 < 1\right\}$ (the solid torus bounded by the torus of revolution of Example 5.17). Define a map $F\colon \mathbb{R}^2 \to U$ by

$$F(u, v) = \big(\cos 2\pi u (2 + \tanh v \cos \pi u),$$

$$\sin 2\pi u (2 + \tanh v \cos \pi u), \tanh v \sin \pi u\big).$$

(a) Show that F descends to a smooth embedding of E into U, where E is the total space of the Möbius bundle of Example 10.3.

(b) Let S be the image of F. Show that S is a properly embedded smooth submanifold of U.

(c) Show that there is no unit normal vector field along S.

(d) Show that S has no global defining function in U.

15-7. Suppose M is an oriented Riemannian manifold with or without boundary, and $S \subseteq M$ is an oriented smooth hypersurface with or without boundary. Show that there is a unique smooth unit normal vector field along S that determines the given orientation of S.

15-8. Suppose M is an orientable Riemannian manifold, and $S \subseteq M$ is an immersed or embedded submanifold with or without boundary. Prove the following statements.
(a) If S has trivial normal bundle, then S is orientable.
(b) If S is an orientable hypersurface, then S has trivial normal bundle.

15-9. Let S be an oriented, embedded, 2-dimensional submanifold with boundary in $\mathbb{R}^3$, and let $C = \partial S$ with the induced orientation. By Problem 15-7, there is a unique smooth unit normal vector field N on S that determines the orientation. Let T be the oriented unit tangent vector field on C, and let V be the unique unit vector field tangent to S along C that is orthogonal to T and inward-pointing. Show that (T_p, V_p, N_p) is an oriented orthonormal basis for $\mathbb{R}^3$ at each $p \in C$.

15-10. CHARACTERISTIC PROPERTY OF THE ORIENTATION COVERING: Let M be a connected nonorientable smooth manifold with or without boundary, and let $\widehat{\pi} : \widehat{M} \to M$ be its orientation covering. Prove that if X is any oriented smooth manifold with or without boundary, and $F : X \to M$ is any local diffeomorphism, then there exists a unique orientation-preserving local diffeomorphism $\widehat{F} : X \to \widehat{M}$ such that $\widehat{\pi} \circ \widehat{F} = F$:

15-11. Prove Theorem 15.42 (uniqueness of the orientation covering). [Hint: use Problem 15-10.]

15-12. Show that every orientation-reversing diffeomorphism of $\mathbb{R}$ has a fixed point.

15-13. CLASSIFICATION OF SMOOTH 1-MANIFOLDS: Let M be a connected smooth 1-manifold. Show that M is diffeomorphic to either $\mathbb{R}$ or $\mathbb{S}^1$, as follows:
(a) First, do the case in which M is orientable by showing that M admits a nonvanishing smooth vector field and using Problem 9-1.
(b) Now let M be arbitrary, and prove that M is orientable by showing that its universal covering manifold is diffeomorphic to $\mathbb{R}$ and using the result of Problem 15-12.
Conclude that the smooth structures on both $\mathbb{R}$ and $\mathbb{S}^1$ are unique up to diffeomorphism.

15-14. CLASSIFICATION OF SMOOTH 1-MANIFOLDS WITH BOUNDARY: Show that every connected smooth 1-manifold with nonempty boundary is diffeomorphic to either $[0, 1]$ or $[0, \infty)$. [Hint: use the double.]

15-15. Let M be a nonorientable embedded hypersurface in $\mathbb{R}^n$, and let NM be its normal bundle with projection $\pi_{NM} : NM \to M$. Show that the set

$$W = \{(x, v) \in NM : |v| = 1\}$$

is an embedded submanifold of NM, and the restriction of π_{NM} to W is a smooth covering map isomorphic to the orientation covering of M. [Hint: consider the orientation determined by $v \lrcorner \left(dx^1 \wedge \cdots \wedge dx^n\right)$.]

15-16. Let E be the total space of the Möbius bundle as in Example 15.38. Show that the orientation covering of E is diffeomorphic to the cylinder $\mathbb{S}^1 \times \mathbb{R}$.

Chapter 16
Integration on Manifolds

In Chapter 11, we introduced line integrals of covector fields, which generalize ordinary integrals to the setting of curves in manifolds. It is also useful to generalize *multiple* integrals to manifolds. In this chapter, we carry out that generalization.

As we show in the beginning of this chapter, there is no way to define the integral of a *function* in a coordinate-independent way on a smooth manifold. On the other hand, differential forms turn out to have just the right properties for defining integrals intrinsically.

We begin the chapter with a heuristic discussion of the measurement of volume, to motivate the central role played by alternating tensors in integration theory. We will see that a k-covector on a vector space can be interpreted as "signed k-dimensional volume meter." This suggests that a k-form on a smooth manifold might be thought of as a way of assigning "signed volumes" to k-dimensional submanifolds. The purpose of this chapter is to make this rigorous.

First, we define the integral of a differential form over a domain in Euclidean space, and then we show how to use diffeomorphism invariance and partitions of unity to extend this definition to n-forms on oriented n-manifolds. The key feature of the definition is that it is invariant under orientation-preserving diffeomorphisms.

After developing the general theory of integration of differential forms, we prove one of the most important theorems in differential geometry: *Stokes's theorem*. It is a generalization of the fundamental theorem of calculus and of the fundamental theorem for line integrals, as well as of the three great classical theorems of vector analysis: Green's theorem for vector fields in the plane; the divergence theorem for vector fields in space; and (the classical version of) Stokes's theorem for surface integrals in $\mathbb{R}^3$. Then we extend the theorem to manifolds with corners, which will be useful in our treatment of de Rham cohomology in Chapters 17 and 18.

Next, we show how these ideas play out on a Riemannian manifold. We prove Riemannian versions of the divergence theorem and of Stokes's theorem for surface integrals, of which the classical theorems are special cases.

At the end of the chapter, we show how to extend the theory of integration to nonorientable manifolds by introducing *densities*, which are fields that can be integrated on any manifold, not just oriented ones.

J.M. Lee, *Introduction to Smooth Manifolds*, Graduate Texts in Mathematics 218,
DOI 10.1007/978-1-4419-9982-5_16, © Springer Science+Business Media New York 2013

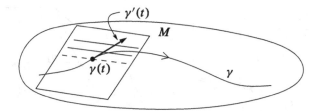

Fig. 16.1 A covector field as a "signed length meter"

The Geometry of Volume Measurement

How might we make coordinate-independent sense of multiple integrals? First, observe that there is no way to integrate real-valued *functions* in a coordinate-independent way on a manifold, at least without adding further structure such as a Riemannian metric. It is easy to see why, even in the simplest case: suppose $C \subseteq \mathbb{R}^n$ is a closed ball, and $f : C \to \mathbb{R}$ is the constant function $f(x) \equiv 1$. Then

$$\int_C f \, dV = \operatorname{Vol}(C),$$

which is clearly not invariant under coordinate transformations, even if we just restrict attention to linear ones.

On the other hand, in Chapter 11 we showed that covector fields could be integrated in a natural way along curves. Let us think a bit more geometrically about why this is so. A covector field on a manifold M assigns a number to each tangent vector, in such a way that multiplying the tangent vector by a constant has the effect of multiplying the resulting number by the same constant. Thus, a covector field can be thought of as assigning a "signed length meter" to each 1-dimensional subspace of each tangent space (Fig. 16.1), and it does so in a coordinate-independent way. Computing the line integral of a covector field, in effect, assigns a "length" to a curve by using this varying measuring scale along the points of the curve.

Now we wish to seek a kind of "field" that can be integrated in a coordinate-independent way over submanifolds of dimension $k > 1$. Its value at each point should be something that we can interpret as a "signed volume meter" on k-dimensional subspaces of the tangent space, a machine ω that accepts any k tangent vectors $(v_1, \dots, v_k)$ at a point and returns a number $\omega(v_1, \dots, v_k)$ that we might think of as the "signed volume" of the parallelepiped spanned by those vectors, measured according to a scale determined by ω.

The most obvious example of such a machine is the determinant in $\mathbb{R}^n$. For example, it is shown in most linear algebra texts that for any two vectors $v_1, v_2 \in \mathbb{R}^2$, $\det(v_1, v_2)$ is, up to a sign, the area of the parallelogram spanned by v_1, v_2. It is not hard to show (see Problem 16-1) that the analogous fact is true in all dimensions. The determinant, remember, is an example of an alternating tensor.

Let us consider what properties we might expect a general "signed k-dimensional volume meter" ω to have. To be consistent with our intuition about volume, multi-

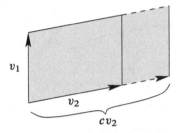

Fig. 16.2 Scaling by a constant

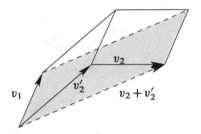

Fig. 16.3 Sum of two vectors

plying any one of the vectors by a constant should scale the volume by that same constant (Fig. 16.2),

and the volume of a k-dimensional parallelepiped formed by adding together two vectors in the i th place should be the sum of the volumes of the two parallelepipeds with the original vectors in the i th place (Fig. 16.3):

$$\omega(v_1, \ldots, cv_i, \ldots, v_k) = c\omega(v_1, \ldots, v_i, \ldots, v_k),$$

$$\omega(v_1, \ldots, v_i + v_i', \ldots, v_k) = \omega(v_1, \ldots, v_i, \ldots, v_k) + \omega(v_1, \ldots, v_i', \ldots, v_k).$$

(Note that the vectors in Fig. 16.3 are all assumed to lie in one plane.) This suggests that ω should be multilinear, and thus should be a covariant k-tensor.

There is one more property that we should expect: since a linearly dependent k-tuple of vectors spans a parallelepiped of zero k-dimensional volume, ω should give zero whenever it is applied to a such a k-tuple. By Lemma 14.1, this forces ω to be alternating. Thus, alternating tensor fields are promising objects for integrating in a coordinate-independent way. In this chapter, we show how this is done.

Integration of Differential Forms

Just as we began our treatment of line integrals by first defining integrals of 1-forms over intervals in $\mathbb{R}$, we begin here by defining integrals of n-forms over suitable subsets of $\mathbb{R}^n$. For the time being, let us restrict attention to the case $n \geq 1$. You should make sure that you are familiar with the basic properties of multiple integrals in $\mathbb{R}^n$, as summarized in Appendix C.

Recall that a *domain of integration* in $\mathbb{R}^n$ is a bounded subset whose boundary has measure zero. Let $D \subseteq \mathbb{R}^n$ be a domain of integration, and let ω be a (continuous) n-form on $\overline{D}$. Any such form can be written as $\omega = f\, dx^1 \wedge \cdots \wedge dx^n$ for some continuous function $f: \overline{D} \to \mathbb{R}$. We define the *integral of ω over D* to be

$$\int_D \omega = \int_D f\, dV.$$

This can be written more suggestively as

$$\int_D f\, dx^1 \wedge \cdots \wedge dx^n = \int_D f\, dx^1 \cdots dx^n.$$

In simple terms, to compute the integral of a form such as $f \, dx^1 \wedge \cdots \wedge dx^n$, just "erase the wedges"!

Somewhat more generally, let U be an open subset of $\mathbb{R}^n$ or $\mathbb{H}^n$, and suppose ω is a compactly supported n-form on U. We define

$$\int_U \omega = \int_D \omega,$$

where $D \subseteq \mathbb{R}^n$ or $\mathbb{H}^n$ is any domain of integration (such as a rectangle) containing supp ω, and ω is extended to be zero on the complement of its support. It is easy to check that this definition does not depend on what domain D is chosen.

Like the definition of the integral of a 1-form over an interval, our definition of the integral of an n-form might look like a trick of notation. The next proposition shows why it is natural.

Proposition 16.1. *Suppose D and E are open domains of integration in $\mathbb{R}^n$ or $\mathbb{H}^n$, and $G \colon \bar{D} \to \bar{E}$ is a smooth map that restricts to an orientation-preserving or orientation-reversing diffeomorphism from D to E. If ω is an n-form on $\bar{E}$, then*

$$\int_D G^*\omega = \begin{cases} \displaystyle\int_E \omega & \text{if } G \text{ is orientation-preserving,} \\[2ex] -\displaystyle\int_E \omega & \text{if } G \text{ is orientation-reversing.} \end{cases}$$

Proof. Let us use $(y^1, \ldots, y^n)$ to denote standard coordinates on E, and $(x^1, \ldots, x^n)$ to denote those on D. Suppose first that G is orientation-preserving. With $\omega = f \, dy^1 \wedge \cdots \wedge dy^n$, the change of variables formula (Theorem C.26) together with formula (14.15) for pullbacks of n-forms yields

$$\int_E \omega = \int_E f \, dV = \int_D (f \circ G) \, |\det DG| \, dV = \int_D (f \circ G)(\det DG) \, dV$$

$$= \int_D (f \circ G)(\det DG) \, dx^1 \wedge \cdots \wedge dx^n = \int_D G^*\omega.$$

If G is orientation-reversing, the same computation holds except that a negative sign is introduced when the absolute value signs are removed. □

We would like to extend this theorem to compactly supported n-forms defined on open subsets. However, since we cannot guarantee that arbitrary open subsets or arbitrary compact subsets are domains of integration, we need the following lemma.

Lemma 16.2. *Suppose U is an open subset of $\mathbb{R}^n$ or $\mathbb{H}^n$, and K is a compact subset of U. Then there is an open domain of integration D such that $K \subseteq D \subseteq \bar{D} \subseteq U$.*

Proof. For each $p \in K$, there is an open ball or half-ball containing p whose closure is contained in U. By compactness, finitely many such sets $B_1, \ldots, B_m$ cover K (Fig. 16.4). Since the boundary of an open ball is a codimension-1 submanifold, and the boundary of an open half-ball is contained in a union of two such

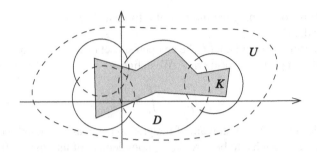

Fig. 16.4 A domain of integration containing a compact set

submanifolds, the boundary of each has measure zero by Corollary 6.12. The set $D = B_1 \cup \cdots \cup B_m$ is the required domain of integration. $\qquad\square$

Proposition 16.3. *Suppose U, V are open subsets of $\mathbb{R}^n$ or $\mathbb{H}^n$, and $G \colon U \to V$ is an orientation-preserving or orientation-reversing diffeomorphism. If ω is a compactly supported n-form on V, then*

$$\int_V \omega = \pm \int_U G^* \omega,$$

with the positive sign if G is orientation-preserving, and the negative sign otherwise.

Proof. Let E be an open domain of integration such that supp $\omega \subseteq E \subseteq \overline{E} \subseteq V$ (Fig. 16.5). Since diffeomorphisms take interiors to interiors, boundaries to boundaries, and sets of measure zero to sets of measure zero, $D = G^{-1}(E) \subseteq U$ is an open domain of integration containing supp $G^* \omega$. The result follows from Proposition 16.1. $\qquad\square$

Integration on Manifolds

Using the results of the previous section, we can now make sense of the integral of a differential form over an oriented manifold. Let M be an oriented smooth n-manifold with or without boundary, and let ω be an n-form on M. Suppose first that ω is compactly supported in the domain of a single smooth chart (U, φ) that is either positively or negatively oriented. We define the ***integral of ω over M*** to be

$$\int_M \omega = \pm \int_{\varphi(U)} \left(\varphi^{-1} \right)^* \omega, \tag{16.1}$$

with the positive sign for a positively oriented chart, and the negative sign otherwise. (See Fig. 16.6.) Since $(\varphi^{-1})^* \omega$ is a compactly supported n-form on the open subset $\varphi(U) \subseteq \mathbb{R}^n$ or $\mathbb{H}^n$, its integral is defined as discussed above.

Proposition 16.4. *With ω as above, $\int_M \omega$ does not depend on the choice of smooth chart whose domain contains supp ω.*

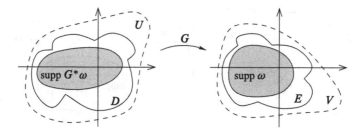

Fig. 16.5 Diffeomorphism invariance of the integral of a form on an open subset

Proof. Suppose (U, φ) and $(\tilde{U}, \tilde{\varphi})$ are two smooth charts such that supp $\omega \subseteq U \cap \tilde{U}$ (Fig. 16.7). If both charts are positively oriented or both are negatively oriented, then $\tilde{\varphi} \circ \varphi^{-1}$ is an orientation-preserving diffeomorphism from $\varphi(U \cap \tilde{U})$ to $\tilde{\varphi}(U \cap \tilde{U})$, so Proposition 16.3 implies that

$$\int_{\tilde{\varphi}(\tilde{U})} (\tilde{\varphi}^{-1})^* \omega = \int_{\tilde{\varphi}(U \cap \tilde{U})} (\tilde{\varphi}^{-1})^* \omega = \int_{\varphi(U \cap \tilde{U})} (\tilde{\varphi} \circ \varphi^{-1})^* (\tilde{\varphi}^{-1})^* \omega$$

$$= \int_{\varphi(U \cap \tilde{U})} (\varphi^{-1})^* (\tilde{\varphi})^* (\tilde{\varphi}^{-1})^* \omega = \int_{\varphi(U)} (\varphi^{-1})^* \omega.$$

If the charts are oppositely oriented, then the two definitions given by (16.1) have opposite signs, but this is compensated by the fact that $\tilde{\varphi} \circ \varphi^{-1}$ is orientation-reversing, so Proposition 16.3 introduces an extra negative sign into the computation above. In either case, the two definitions of $\int_M \omega$ agree. □

To integrate over an entire manifold, we combine this definition with a partition of unity. Suppose M is an oriented smooth n-manifold with or without boundary, and ω is a compactly supported n-form on M. Let $\{U_i\}$ be a finite open cover of supp ω by domains of positively or negatively oriented smooth charts, and let $\{\psi_i\}$ be a subordinate smooth partition of unity. Define the **integral of ω over M** to be

$$\int_M \omega = \sum_i \int_M \psi_i \omega. \tag{16.2}$$

(The reason we allow for negatively oriented charts is that it may not be possible to find positively oriented boundary charts on a 1-manifold with boundary, as noted in the proof of Proposition 15.6.) Since for each i, the n-form $\psi_i \omega$ is compactly supported in U_i, each of the terms in this sum is well defined according to our discussion above. To show that the integral is well defined, we need only examine the dependence on the open cover and the partition of unity.

Proposition 16.5. *The definition of $\int_M \omega$ given above does not depend on the choice of open cover or partition of unity.*

Proof. Suppose $\{\tilde{U}_j\}$ is another finite open cover of supp ω by domains of positively or negatively oriented smooth charts, and $\{\tilde{\psi}_j\}$ is a subordinate smooth parti-

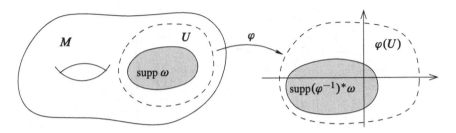

Fig. 16.6 The integral of a form over a manifold

tion of unity. For each i, we compute

$$\int_M \psi_i \omega = \int_M \left(\sum_j \tilde{\psi}_j \right) \psi_i \omega = \sum_j \int_M \tilde{\psi}_j \psi_i \omega.$$

Summing over i, we obtain

$$\sum_i \int_M \psi_i \omega = \sum_{i,j} \int_M \tilde{\psi}_j \psi_i \omega.$$

Observe that each term in this last sum is the integral of a form that is compactly supported in a single smooth chart (e.g., in U_i), so by Proposition 16.4 each term is well defined, regardless of which coordinate map we use to compute it. The same argument, starting with $\int_M \tilde{\psi}_j \omega$, shows that

$$\sum_j \int_M \tilde{\psi}_j \omega = \sum_{i,j} \int_M \tilde{\psi}_j \psi_i \omega.$$

Thus, both definitions yield the same value for $\int_M \omega$.   □

As usual, we have a special definition in the zero-dimensional case. The integral of a compactly supported 0-form (i.e., a real-valued function) f over an oriented 0-manifold M is defined to be the sum

$$\int_M f = \sum_{p \in M} \pm f(p),$$

where we take the positive sign at points where the orientation is positive and the negative sign at points where it is negative. The assumption that f is compactly supported implies that there are only finitely many nonzero terms in this sum.

If $S \subseteq M$ is an oriented immersed k-dimensional submanifold (with or without boundary), and ω is a k-form on M whose restriction to S is compactly supported, we interpret $\int_S \omega$ to mean $\int_S \iota_S^* \omega$, where $\iota_S \colon S \hookrightarrow M$ is inclusion. In particular, if M is a compact, oriented, smooth n-manifold with boundary and ω is an $(n-1)$-form on M, we can interpret $\int_{\partial M} \omega$ unambiguously as the integral of $\iota_{\partial M}^* \omega$ over ∂M, where ∂M is always understood to have the induced orientation.

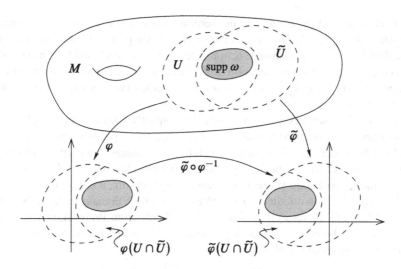

Fig. 16.7 Coordinate independence of the integral

It is worth remarking that it is possible to extend the definition of the integral to some noncompactly supported forms, and such integrals are important in many applications. However, in such cases the resulting multiple integrals are improper, so one must pay close attention to convergence issues. For the purposes we have in mind, the cases we have described here are quite sufficient.

Proposition 16.6 (Properties of Integrals of Forms). *Suppose M and N are non-empty oriented smooth n-manifolds with or without boundary, and ω, η are compactly supported n-forms on M.*

(a) LINEARITY: *If $a, b \in \mathbb{R}$, then*

$$\int_M a\omega + b\eta = a \int_M \omega + b \int_M \eta.$$

(b) ORIENTATION REVERSAL: *If $-M$ denotes M with the opposite orientation, then*

$$\int_{-M} \omega = - \int_M \omega.$$

(c) POSITIVITY: *If ω is a positively oriented orientation form, then $\int_M \omega > 0$.*

(d) DIFFEOMORPHISM INVARIANCE: *If $F : N \to M$ is an orientation-preserving or orientation-reversing diffeomorphism, then*

$$\int_M \omega = \begin{cases} \displaystyle\int_N F^*\omega & \text{if } F \text{ is orientation-preserving,} \\[2mm] -\displaystyle\int_N F^*\omega & \text{if } F \text{ is orientation-reversing.} \end{cases}$$

Proof. Parts (a) and (b) are left as an exercise. Suppose ω is a positively oriented orientation form for M. This means that if (U, φ) is a positively oriented smooth chart, then $(\varphi^{-1})^* \omega$ is a positive function times $dx^1 \wedge \cdots \wedge dx^n$, and for a negatively oriented chart it is a negative function times the same form. Therefore, each term in the sum (16.2) defining $\int_M \omega$ is nonnegative, with at least one strictly positive term, thus proving (c).

To prove (d), it suffices to assume that ω is compactly supported in a single positively or negatively oriented smooth chart, because any compactly supported n-form on M can be written as a finite sum of such forms by means of a partition of unity. Thus, suppose (U, φ) is a positively oriented smooth chart on M whose domain contains the support of ω. When F is orientation-preserving, it is easy to check that $(F^{-1}(U), \varphi \circ F)$ is an oriented smooth chart on N whose domain contains the support of $F^* \omega$, and the result then follows immediately from Proposition 16.3. The cases in which the chart is negatively oriented or F is orientation-reversing then follow from this result together with (b). $\square$

▶ **Exercise 16.7.** Prove parts (a) and (b) of the preceding proposition.

Although the definition of the integral of a form based on partitions of unity is very convenient for theoretical purposes, it is useless for doing actual computations. It is generally quite difficult to write down a smooth partition of unity explicitly, and even when one can be written down, one would have to be exceptionally lucky to be able to compute the resulting integrals (think of trying to integrate $e^{-1/x}$).

For computational purposes, it is much more convenient to "chop up" the manifold into a finite number of pieces whose boundaries are sets of measure zero, and compute the integral on each piece separately by means of local parametrizations. One way to do this is described below.

Proposition 16.8 (Integration Over Parametrizations). *Let M be an oriented smooth n-manifold with or without boundary, and let ω be a compactly supported n-form on M. Suppose $D_1, \ldots, D_k$ are open domains of integration in $\mathbb{R}^n$, and for $i = 1, \ldots, k$, we are given smooth maps $F_i : \bar{D}_i \to M$ satisfying*

(i) *F_i restricts to an orientation-preserving diffeomorphism from D_i onto an open subset $W_i \subseteq M$;*
(ii) *$W_i \cap W_j = \varnothing$ when $i \neq j$;*
(iii) *$\operatorname{supp} \omega \subseteq \bar{W}_1 \cup \cdots \cup \bar{W}_k$.*

Then

$$\int_M \omega = \sum_{i=1}^k \int_{D_i} F_i^* \omega. \tag{16.3}$$

Proof. As in the preceding proof, it suffices to assume that ω is supported in the domain of a single oriented smooth chart (U, φ). In fact, by restricting to sufficiently nice charts, we may assume that U is precompact, $Y = \varphi(U)$ is a domain of integration in $\mathbb{R}^n$ or $\mathbb{H}^n$, and φ extends to a diffeomorphism from $\bar{U}$ to $\bar{Y}$.

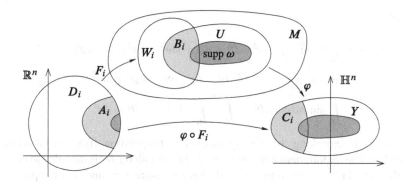

Fig. 16.8 Integrating over parametrizations

For each i, define open subsets $A_i \subseteq D_i$, $B_i \subseteq W_i$, and $C_i \subseteq Y$ (Fig. 16.8) by

$$A_i = F_i^{-1}(U \cap W_i), \quad B_i = U \cap W_i = F_i(A_i), \quad C_i = \varphi(B_i) = \varphi_i \circ F_i(A_i).$$

Because $\overline{D}_i$ is compact, it is straightforward to check that $\partial W_i \subseteq F_i(\partial D_i)$, and therefore ∂W_i has measure zero in M, and $\partial C_i = \varphi(\partial B_i)$ has measure zero in $\mathbb{R}^n$.

The support of $(\varphi^{-1})^* \omega$ is contained in $\overline{C}_1 \cup \cdots \cup \overline{C}_k$, and any two of these sets intersect only on their boundaries, which have measure zero. Thus by Proposition C.23,

$$\int_M \omega = \int_Y (\varphi^{-1})^* \omega = \sum_{i=1}^k \int_{C_i} (\varphi^{-1})^* \omega.$$

The proof is completed by applying Proposition 16.1 to each term above, using the diffeomorphism $\varphi \circ F_i : A_i \to C_i$:

$$\int_{C_i} (\varphi^{-1})^* \omega = \int_{A_i} (\varphi \circ F_i)^* (\varphi^{-1})^* \omega = \int_{A_i} F_i^* \omega = \int_{D_i} F_i^* \omega.$$

Summing over i, we obtain (16.3).   □

Example 16.9. Let us use this technique to compute the integral of a 2-form over $\mathbb{S}^2$, oriented as the boundary of $\overline{\mathbb{B}}^3$. Let ω be the following 2-form on $\mathbb{R}^3$:

$$\omega = x\,dy \wedge dz + y\,dz \wedge dx + z\,dx \wedge dy.$$

Let D be the open rectangle $(0, \pi) \times (0, 2\pi)$, and let $F: \overline{D} \to \mathbb{S}^2$ be the spherical coordinate parametrization $F(\varphi, \theta) = (\sin \varphi \cos \theta, \sin \varphi \sin \theta, \cos \varphi)$. Example 15.28 showed that $F|_D$ is orientation-preserving, so it satisfies the hypotheses of Proposition 16.8. Note that

$$F^* dx = \cos \varphi \cos \theta \, d\varphi - \sin \varphi \sin \theta \, d\theta,$$
$$F^* dy = \cos \varphi \sin \theta \, d\varphi + \sin \varphi \cos \theta \, d\theta,$$
$$F^* dz = -\sin \varphi \, d\varphi.$$

Therefore,

$$\int_{\mathbb{S}^2} \omega = \int_D \left(-\sin^3 \varphi \cos^2 \theta \, d\theta \wedge d\varphi + \sin^3 \varphi \sin^2 \theta \, d\varphi \wedge d\theta \right.$$
$$\left. + \cos^2 \varphi \sin \varphi \cos^2 \theta \, d\varphi \wedge d\theta - \cos^2 \varphi \sin \varphi \sin^2 \theta \, d\theta \wedge d\varphi \right)$$
$$= \int_D \sin \varphi \, d\varphi \wedge d\theta = \int_0^{2\pi} \int_0^{\pi} \sin \varphi \, d\varphi \, d\theta = 4\pi. \qquad\qquad //$$

It is worth remarking that the hypotheses of Proposition 16.8 can be relaxed somewhat. The requirement that each map F_i be smooth on $\bar{D}_i$ is included to ensure that the boundaries of the image sets W_i have measure zero and that the pullback forms $F_i^* \omega$ are continuous on $\bar{D}_i$. Provided the open subsets W_i together fill up all of M except for a set of measure zero, we can allow maps F_i that do not extend smoothly to the boundary, by interpreting the resulting integrals of unbounded forms either as improper Riemann integrals or as Lebesgue integrals. For example, if the closed upper hemisphere of $\mathbb{S}^2$ is parametrized by the map $F \colon \bar{\mathbb{B}}^2 \to \mathbb{S}^2$ given by $F(u, v) = \left(u, v, \sqrt{1 - u^2 - v^2} \right)$, then F is continuous but not smooth up to the boundary, but the conclusion of the proposition still holds. We leave it to the interested reader to work out reasonable conditions under which such a generalization of Proposition 16.8 holds.

Integration on Lie Groups

Let G be a Lie group. A covariant tensor field A on G is said to be **left-invariant** if $L_g^* A = A$ for all $g \in G$.

Proposition 16.10. *Let G be a compact Lie group endowed with a left-invariant orientation. Then G has a unique positively oriented left-invariant n-form ω_G with the property that $\int_G \omega_G = 1$.*

Proof. If $\dim G = 0$, we just let ω_G be the constant function $1/k$, where k is the cardinality of G. Otherwise, let $E_1, \dots, E_n$ be a left-invariant global frame on G (i.e., a basis for the Lie algebra of G). By replacing E_1 with $-E_1$ if necessary, we may assume that this frame is positively oriented. Let $\varepsilon^1, \dots, \varepsilon^n$ be the dual coframe. Left invariance of E_j implies that

$$\left(L_g^* \varepsilon^i \right) (E_j) = \varepsilon^i (L_{g*} E_j) = \varepsilon^i (E_j) = \delta_j^i,$$

which shows that $L_g^* \varepsilon^i = \varepsilon^i$, so ε^i is left-invariant.
Let $\omega_G = \varepsilon^1 \wedge \cdots \wedge \varepsilon^n$. Then

$$L_g^* (\omega_G) = L_g^* \varepsilon^1 \wedge \cdots \wedge L_g^* \varepsilon^n = \varepsilon^1 \wedge \cdots \wedge \varepsilon^n = \omega_G,$$

so ω_G is left-invariant as well. Because $\omega_G(E_1, \dots, E_n) = 1 > 0$, ω_G is an orientation form for the given orientation. Clearly, any positive constant multiple of ω_G is also a left-invariant orientation form. Conversely, if $\tilde{\omega}_G$ is any other left-invariant

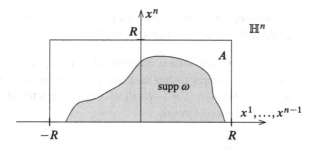

Fig. 16.9 Proof of Stokes's theorem

orientation form, we can write $\tilde{\omega}_G|_e = c\omega_G|_e$ for some positive number c. Using left-invariance, we find that

$$\tilde{\omega}_G|_g = L^*_{g^{-1}}\tilde{\omega}_G|_e = cL^*_{g^{-1}}\omega_G|_e = c\omega_G|_g,$$

which proves that $\tilde{\omega}_G$ is a positive constant multiple of ω_G.

Since G is compact and oriented, $\int_G \omega_G$ is a positive real number, so we can define $\tilde{\omega}_G = \left(\int_G \omega_G\right)^{-1}\omega_G$. Clearly, $\tilde{\omega}_G$ is the unique positively oriented left-invariant orientation form with integral 1. $\qquad\square$

Remark. The orientation form whose existence is asserted in this proposition is called the **Haar volume form on G**. Similarly, the map $f \mapsto \int_G f\,\omega_G$ is called the **Haar integral**. Observe that the proof above did not use the fact that G was compact until the last paragraph; thus every Lie group has a left-invariant orientation form that is uniquely defined up to a constant multiple. It is only in the compact case, however, that we can use the volume normalization to single out a unique one.

Stokes's Theorem

In this section we state and prove the central result in the theory of integration on manifolds, Stokes's theorem. It is a far-reaching generalization of the fundamental theorem of calculus and of the classical theorems of vector calculus.

Theorem 16.11 (Stokes's Theorem). *Let M be an oriented smooth n-manifold with boundary, and let ω be a compactly supported smooth $(n-1)$-form on M. Then*

$$\int_M d\omega = \int_{\partial M} \omega. \tag{16.4}$$

Remark. The statement of this theorem is concise and elegant, but it requires a bit of interpretation. First, as usual, ∂M is understood to have the induced (Stokes) orientation, and the ω on the right-hand side is to be interpreted as $\iota^*_{\partial M}\omega$. If $\partial M = \varnothing$, then the right-hand side is to be interpreted as zero. When M is 1-dimensional, the right-hand integral is really just a finite sum.

With these understandings, we proceed with the proof of the theorem. You should check that it works correctly when $n = 1$ and when $\partial M = \varnothing$.

Proof. We begin with a very special case: suppose M is the upper half-space $\mathbb{H}^n$ itself. Then because ω has compact support, there is a number $R > 0$ such that supp ω is contained in the rectangle $A = [-R, R] \times \cdots \times [-R, R] \times [0, R]$ (Fig. 16.9). We can write ω in standard coordinates as

$$\omega = \sum_{i=1}^{n} \omega_i \, dx^1 \wedge \cdots \wedge \widehat{dx^i} \wedge \cdots \wedge dx^n,$$

where the hat means that dx^i is omitted. Therefore,

$$
\begin{aligned}
d\omega &= \sum_{i=1}^{n} d\omega_i \wedge dx^1 \wedge \cdots \wedge \widehat{dx^i} \wedge \cdots \wedge dx^n \\
&= \sum_{i,j=1}^{n} \frac{\partial \omega_i}{\partial x^j} \, dx^j \wedge dx^1 \wedge \cdots \wedge \widehat{dx^i} \wedge \cdots \wedge dx^n \\
&= \sum_{i=1}^{n} (-1)^{i-1} \frac{\partial \omega_i}{\partial x^i} \, dx^1 \wedge \cdots \wedge dx^n.
\end{aligned}
$$

Thus we compute

$$
\begin{aligned}
\int_{\mathbb{H}^n} d\omega &= \sum_{i=1}^{n} (-1)^{i-1} \int_A \frac{\partial \omega_i}{\partial x^i} \, dx^1 \wedge \cdots \wedge dx^n \\
&= \sum_{i=1}^{n} (-1)^{i-1} \int_0^R \int_{-R}^R \cdots \int_{-R}^R \frac{\partial \omega_i}{\partial x^i}(x) \, dx^1 \cdots dx^n.
\end{aligned}
$$

We can change the order of integration in each term so as to do the x^i integration first. By the fundamental theorem of calculus, the terms for which $i \neq n$ reduce to

$$
\begin{aligned}
&\sum_{i=1}^{n-1} (-1)^{i-1} \int_0^R \int_{-R}^R \cdots \int_{-R}^R \frac{\partial \omega_i}{\partial x^i}(x) \, dx^1 \cdots dx^n \\
&= \sum_{i=1}^{n-1} (-1)^{i-1} \int_0^R \int_{-R}^R \cdots \int_{-R}^R \frac{\partial \omega_i}{\partial x^i}(x) \, dx^i \, dx^1 \cdots \widehat{dx^i} \cdots dx^n \\
&= \sum_{i=1}^{n-1} (-1)^{i-1} \int_0^R \int_{-R}^R \cdots \int_{-R}^R \Big[\omega_i(x) \Big]_{x^i = -R}^{x^i = R} dx^1 \cdots \widehat{dx^i} \cdots dx^n = 0,
\end{aligned}
$$

because we have chosen R large enough that $\omega = 0$ when $x^i = \pm R$. The only term that might not be zero is the one for which $i = n$. For that term we have

$$\int_{\mathbb{H}^n} d\omega = (-1)^{n-1} \int_{-R}^{R} \cdots \int_{-R}^{R} \int_{0}^{R} \frac{\partial \omega_n}{\partial x^n}(x) dx^n \, dx^1 \cdots dx^{n-1}$$

$$= (-1)^{n-1} \int_{-R}^{R} \cdots \int_{-R}^{R} \left[\omega_n(x)\right]_{x^n=0}^{x^n=R} dx^1 \cdots dx^{n-1}$$

$$= (-1)^{n} \int_{-R}^{R} \cdots \int_{-R}^{R} \omega_n\left(x^1,\ldots,x^{n-1},0\right) dx^1 \cdots dx^{n-1}, \qquad (16.5)$$

because $\omega_n = 0$ when $x^n = R$.

To compare this to the other side of (16.4), we compute as follows:

$$\int_{\partial \mathbb{H}^n} \omega = \sum_i \int_{A \cap \partial \mathbb{H}^n} \omega_i\left(x^1,\ldots,x^{n-1},0\right) dx^1 \wedge \cdots \wedge \widehat{dx^i} \wedge \cdots \wedge dx^n.$$

Because x^n vanishes on $\partial \mathbb{H}^n$, the pullback of dx^n to the boundary is identically zero (see Exercise 11.30). Thus, the only term above that is nonzero is the one for which $i = n$, which becomes

$$\int_{\partial \mathbb{H}^n} \omega = \int_{A \cap \partial \mathbb{H}^n} \omega_n\left(x^1,\ldots,x^{n-1},0\right) dx^1 \wedge \cdots \wedge dx^{n-1}.$$

Taking into account the fact that the coordinates $\left(x^1,\ldots,x^{n-1}\right)$ are positively oriented for $\partial \mathbb{H}^n$ when n is even and negatively oriented when n is odd (Example 15.26), we find that this is equal to (16.5).

Next we consider another special case: $M = \mathbb{R}^n$. In this case, the support of ω is contained in a cube of the form $A = [-R, R]^n$. Exactly the same computation goes through, except that in this case the $i = n$ term vanishes like all the others, so the left-hand side of (16.4) is zero. Since M has empty boundary in this case, the right-hand side is zero as well.

Now let M be an arbitrary smooth manifold with boundary, but consider an $(n-1)$-form ω that is compactly supported in the domain of a single positively or negatively oriented smooth chart (U, φ). Assuming that φ is a positively oriented boundary chart, the definition yields

$$\int_M d\omega = \int_{\mathbb{H}^n} \left(\varphi^{-1}\right)^* d\omega = \int_{\mathbb{H}^n} d\left(\left(\varphi^{-1}\right)^* \omega\right).$$

By the computation above, this is equal to

$$\int_{\partial \mathbb{H}^n} \left(\varphi^{-1}\right)^* \omega, \qquad (16.6)$$

where $\partial \mathbb{H}^n$ is given the induced orientation. Since $d\varphi$ takes outward-pointing vectors on ∂M to outward-pointing vectors on $\mathbb{H}^n$ (by Proposition 5.41), it follows that $\varphi|_{U \cap \partial M}$ is an orientation-preserving diffeomorphism onto $\varphi(U) \cap \partial \mathbb{H}^n$, and thus (16.6) is equal to $\int_{\partial M} \omega$. For a negatively oriented smooth boundary chart, the same argument applies with an additional negative sign on each side of the equation. For an interior chart, we get the same computations with $\mathbb{H}^n$ replaced by $\mathbb{R}^n$. This proves the theorem in this case.

Finally, let ω be an arbitrary compactly supported smooth $(n-1)$-form. Choosing a cover of supp ω by finitely many domains of positively or negatively oriented smooth charts $\{U_i\}$, and choosing a subordinate smooth partition of unity $\{\psi_i\}$, we can apply the preceding argument to $\psi_i \omega$ for each i and obtain

$$\int_{\partial M} \omega = \sum_i \int_{\partial M} \psi_i \omega = \sum_i \int_M d(\psi_i \omega) = \sum_i \int_M d\psi_i \wedge \omega + \psi_i \, d\omega$$

$$= \int_M d\left(\sum_i \psi_i\right) \wedge \omega + \int_M \left(\sum_i \psi_i\right) d\omega = 0 + \int_M d\omega,$$

because $\sum_i \psi_i \equiv 1$. $\square$

Example 16.12. Let M be a smooth manifold and suppose $\gamma: [a,b] \to M$ is a smooth embedding, so that $S = \gamma([a,b])$ is an embedded 1-submanifold with boundary in M. If we give S the orientation such that γ is orientation-preserving, then for any smooth function $f \in C^\infty(M)$, Stokes's theorem says that

$$\int_\gamma df = \int_{[a,b]} \gamma^* df = \int_S df = \int_{\partial S} f = f(\gamma(b)) - f(\gamma(a)).$$

Thus Stokes's theorem reduces to the fundamental theorem for line integrals (Theorem 11.39) in this case. In particular, when $\gamma: [a,b] \to \mathbb{R}$ is the inclusion map, then Stokes's theorem is just the ordinary fundamental theorem of calculus.   //

Two special cases of Stokes's theorem arise so frequently that they are worthy of special note. The proofs are immediate.

Corollary 16.13 (Integrals of Exact Forms). *If M is a compact oriented smooth manifold without boundary, then the integral of every exact form over M is zero:*

$$\int_M d\omega = 0 \quad \text{if } \partial M = \varnothing.$$ $\square$

Corollary 16.14 (Integrals of Closed Forms over Boundaries). *Suppose M is a compact oriented smooth manifold with boundary. If ω is a closed form on M, then the integral of ω over ∂M is zero:*

$$\int_{\partial M} \omega = 0 \quad \text{if } d\omega = 0 \text{ on } M.$$ $\square$

These results have the following extremely useful applications to submanifolds.

Corollary 16.15. *Suppose M is a smooth manifold with or without boundary, $S \subseteq M$ is an oriented compact smooth k-dimensional submanifold (without boundary), and ω is a closed k-form on M. If $\int_S \omega \neq 0$, then both of the following are true:*

(a) ω *is not exact on* M.

(b) S *is not the boundary of an oriented compact smooth submanifold with boundary in* M. □

Example 16.16. It follows from the computation of Example 11.36 that the closed 1-form $\omega = (x\,dy - y\,dx)/(x^2 + y^2)$ has nonzero integral over $\mathbb{S}^1$. We already observed that ω is not exact on $\mathbb{R}^2 \smallsetminus \{0\}$. The preceding corollary tells us in addition that $\mathbb{S}^1$ is not the boundary of a compact regular domain in $\mathbb{R}^2 \smallsetminus \{0\}$. //

The following classical result is an easy application of Stokes's theorem.

Theorem 16.17 (Green's Theorem). *Suppose D is a compact regular domain in $\mathbb{R}^2$, and P, Q are smooth real-valued functions on D. Then*

$$\int_D \left(\frac{\partial Q}{\partial x} - \frac{\partial P}{\partial y} \right) dx\,dy = \int_{\partial D} P\,dx + Q\,dy.$$

Proof. This is just Stokes's theorem applied to the 1-form $P\,dx + Q\,dy$. □

Manifolds with Corners

In many applications of Stokes's theorem it is necessary to deal with geometric objects such as triangles, squares, or cubes that are topological manifolds with boundary, but are not smooth manifolds with boundary because they have "corners." It is easy to generalize Stokes's theorem to this setting, and we do so in this section.

Let $\overline{\mathbb{R}}^n_+$ denote the subset of $\mathbb{R}^n$ where all of the coordinates are nonnegative:

$$\overline{\mathbb{R}}^n_+ = \left\{ (x^1, \dots, x^n) \in \mathbb{R}^n : x^1 \geq 0, \dots, x^n \geq 0 \right\}.$$

This space is the model for the type of corners we are concerned with.

▶ **Exercise 16.18.** Prove that $\overline{\mathbb{R}}^n_+$ is homeomorphic to the upper half-space $\mathbb{H}^n$.

Suppose M is a topological n-manifold with boundary. A **chart with corners** for M is a pair (U, φ), where $U \subseteq M$ is open and φ is a homeomorphism from U to a (relatively) open subset $\hat{U} \subseteq \overline{\mathbb{R}}^n_+$ (Fig. 16.10). Two charts with corners (U, φ), (V, ψ) are smoothly compatible if the composite map $\varphi \circ \psi^{-1} \colon \psi(U \cap V) \to \varphi(U \cap V)$ is smooth. (As usual, this means that it admits a smooth extension in an open neighborhood of each point.)

A **smooth structure with corners** on a topological manifold with boundary is a maximal collection of smoothly compatible interior charts and charts with corners whose domains cover M. A topological manifold with boundary together with a smooth structure with corners is called a **smooth manifold with corners**. Any chart with corners in the given smooth structure with corners is called a **smooth chart with corners** for M.

Example 16.19. Any closed rectangle in $\mathbb{R}^n$ is a smooth n-manifold with corners. //

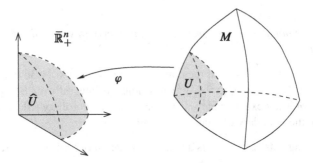

Fig. 16.10 A chart with corners

Because of the result of Exercise 16.18, charts with corners are topologically indistinguishable from boundary charts. Thus, from the topological point of view there is no difference between manifolds with boundary and manifolds with corners. The difference is in the smooth structure, because in dimensions greater than 1, the compatibility condition for charts with corners is different from that for boundary charts. In the case $n = 1$, though, $\bar{\mathbb{R}}^1_+$ is actually equal to $\mathbb{H}^1$, so smooth 1-manifolds with corners are no different from smooth manifolds with boundary.

The boundary of $\bar{\mathbb{R}}^n_+$ in $\mathbb{R}^n$ is the set of points at which at least one coordinate vanishes. The points in $\bar{\mathbb{R}}^n_+$ at which more than one coordinate vanishes are called its **corner points**. For example, the corner points of $\bar{\mathbb{R}}^3_+$ are the origin together with all the points on the positive x-, y-, and z-axes.

Proposition 16.20 (Invariance of Corner Points). *Let M be a smooth n-manifold with corners, $n \geq 2$, and let $p \in M$. If $\varphi(p)$ is a corner point for some smooth chart with corners (U, φ), then the same is true for every such chart whose domain contains p.*

Proof. Suppose (U, φ) and (V, ψ) are two smooth charts with corners such that $\varphi(p)$ is a corner point but $\psi(p)$ is not (Fig. 16.11). To simplify notation, let us assume without loss of generality that $\varphi(p)$ has coordinates $(x^1, \dots, x^k, 0, \dots, 0)$ with $k \leq n - 2$. Then $\psi(V)$ contains an open subset of some $(n-1)$-dimensional linear subspace $S \subseteq \mathbb{R}^n$, with $\psi(p) \in S$. (If $\psi(p) \in \partial\bar{\mathbb{R}}^n_+$, take S to be the unique subspace defined by an equation of the form $x^i = 0$ that contains $\psi(p)$. If $\psi(p)$ is an interior point, any $(n-1)$-dimensional subspace containing $\psi(p)$ will do.)

Let $S' = S \cap \psi(U \cap V)$, and let $\alpha \colon S' \to \mathbb{R}^n$ be the restriction of $\varphi \circ \psi^{-1}$ to S'. Because $\varphi \circ \psi^{-1}$ is a diffeomorphism from $\psi(U \cap V)$ to $\varphi(U \cap V)$, it follows that $\psi \circ \varphi^{-1} \circ \alpha$ is the identity of S', and therefore $d\alpha_{\psi(p)}$ is an injective linear map. Let $T = d\alpha_{\psi(p)}(T_{\psi(p)}S) \subseteq \mathbb{R}^n$. Because T is $(n-1)$-dimensional, it must contain a vector v such that one of the last two components, v^{n-1} or v^n, is nonzero (otherwise, T would be contained in a codimension-2 subspace). Renumbering the coordinates and replacing v by $-v$ if necessary, we may assume that $v^n < 0$.

Now let $\gamma \colon (-\varepsilon, \varepsilon) \to S$ be a smooth curve such that $\gamma(0) = p$ and $d\alpha(\gamma'(0)) = v$. Then $\alpha \circ \gamma(t)$ has negative x^n coordinate for small $t > 0$, which contradicts the fact that α takes its values in $\bar{\mathbb{R}}^n_+$. $\square$

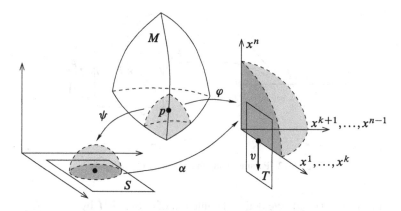

Fig. 16.11 Invariance of corner points

If M is a smooth manifold with corners, a point $p \in M$ is called a **corner point** if $\varphi(p)$ is a corner point in $\overline{\mathbb{R}}_+^n$ with respect to some (and hence every) smooth chart with corners (U, φ). Similarly, p is called a **boundary point** if $\varphi(p) \in \partial \overline{\mathbb{R}}_+^n$ with respect to some (hence every) such chart. For example, the set of corner points of the unit cube $[0, 1]^3 \subseteq \mathbb{R}^3$ is the union of its eight vertices and twelve edges.

Every smooth manifold with or without boundary is also a smooth manifold with corners (but with no corner points). Conversely, a smooth manifold with corners is a smooth manifold with boundary if and only if it has no corner points. The boundary of a smooth manifold with corners, however, is in general not a smooth manifold with corners (e.g., think of the boundary of a cube). In fact, even the boundary of $\overline{\mathbb{R}}_+^n$ itself is not a smooth manifold with corners. It is, however, a union of finitely many such: $\partial \overline{\mathbb{R}}_+^n = H_1 \cup \cdots \cup H_n$, where

$$H_i = \left\{ \left(x^1, \ldots, x^n \right) \in \overline{\mathbb{R}}_+^n : x^i = 0 \right\} \tag{16.7}$$

is an $(n-1)$-dimensional smooth manifold with corners contained in the subspace defined by $x^i = 0$.

The usual flora and fauna of smooth manifolds—smooth maps, partitions of unity, tangent vectors, covectors, tensors, differential forms, orientations, and integrals of differential forms—can be defined on smooth manifolds with corners in exactly the same way as we have done for smooth manifolds and smooth manifolds with boundary, using smooth charts with corners in place of smooth boundary charts. The details are left to the reader.

In addition, for Stokes's theorem we need to integrate a differential form over the boundary of a smooth manifold with corners. Since the boundary is not itself a smooth manifold with corners, this requires a separate (albeit routine) definition. Let M be an oriented smooth n-manifold with corners, and suppose ω is an $(n-1)$-form on ∂M that is compactly supported in the domain of a single oriented smooth

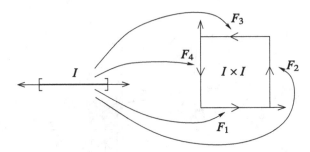

Fig. 16.12 Parametrizing the boundary of the square

chart with corners (U, φ). We define the integral of ω over ∂M by

$$\int_{\partial M} \omega = \sum_{i=1}^{n} \int_{H_i} \left(\varphi^{-1}\right)^* \omega,$$

where H_i, defined by (16.7), is given the induced orientation as part of the boundary of the set where $x^i \geq 0$. In other words, we simply integrate ω in coordinates over the codimension-1 portion of the boundary. Finally, if ω is an arbitrary compactly supported $(n-1)$-form on M, we define the integral of ω over ∂M by piecing together with a partition of unity just as in the case of a manifold with boundary.

In practice, of course, one does not evaluate such integrals by using partitions of unity. Instead, one "chops up" the boundary into pieces that can be parametrized by domains of integration, just as for ordinary manifolds with or without boundary. The following proposition is an analogue of Proposition 16.8.

Proposition 16.21. *The statement of Proposition 16.8 is true if M is replaced by the boundary of a compact, oriented, smooth n-manifold with corners.*

▶ **Exercise 16.22.** Show how the proof of Proposition 16.8 needs to be adapted to prove Proposition 16.21.

Example 16.23. Let $I \times I = [0,1] \times [0,1]$ be the unit square in $\mathbb{R}^2$, and suppose ω is a 1-form on $\partial(I \times I)$. Then it is not hard to check that the maps $F_i \colon I \to I \times I$ given by

$$F_1(t) = (t,0), \qquad F_2(t) = (1,t),$$
$$F_3(t) = (1-t,1), \qquad F_4(t) = (0,1-t), \tag{16.8}$$

satisfy the hypotheses of Proposition 16.21. (These four curve segments in sequence traverse the boundary of $I \times I$ in the counterclockwise direction; see Fig. 16.12.) Therefore,

$$\int_{\partial(I \times I)} \omega = \int_{F_1} \omega + \int_{F_2} \omega + \int_{F_3} \omega + \int_{F_4} \omega. \qquad \textit{//}$$

▶ **Exercise 16.24.** Verify the claims of the preceding example.

The next theorem is the main result of this section.

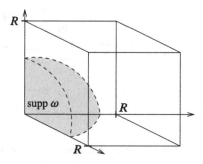

Fig. 16.13 Stokes's theorem for manifolds with corners

Theorem 16.25 (Stokes's Theorem on Manifolds with Corners). *Let M be an oriented smooth n-manifold with corners, and let ω be a compactly supported smooth $(n-1)$-form on M. Then*

$$\int_M d\omega = \int_{\partial M} \omega.$$

Proof. The proof is nearly identical to the proof of Stokes's theorem proper, so we just indicate where changes need to be made. By means of smooth charts and a partition of unity, we may reduce the theorem to the case in which either $M = \mathbb{R}^n$ or $M = \bar{\mathbb{R}}^n_+$. The $\mathbb{R}^n$ case yields zero on both sides of the equation, just as before. In the case of a chart with corners, ω is supported in some cube $[0, R]^n$ (Fig. 16.13), and we calculate exactly as in the proof of Theorem 16.11:

$$\int_{\bar{\mathbb{R}}^n_+} d\omega = \sum_{i=1}^n (-1)^{i-1} \int_0^R \cdots \int_0^R \frac{\partial \omega_i}{\partial x^i}(x)\, dx^1 \cdots dx^n$$

$$= \sum_{i=1}^n (-1)^{i-1} \int_0^R \cdots \int_0^R \frac{\partial \omega_i}{\partial x^i}(x)\, dx^i\, dx^1 \cdots \widehat{dx^i} \cdots dx^n$$

$$= \sum_{i=1}^n (-1)^{i-1} \int_0^R \cdots \int_0^R \left[\omega_i(x)\right]_{x^i=0}^{x^i=R} dx^1 \cdots \widehat{dx^i} \cdots dx^n$$

$$= \sum_{i=1}^n (-1)^{i} \int_0^R \cdots \int_0^R \omega_i\left(x^1,\dots,0,\dots,x^n\right) dx^1 \cdots \widehat{dx^i} \cdots dx^n$$

$$= \sum_{i=1}^n \int_{H_i} \omega = \int_{\partial \bar{\mathbb{R}}^n_+} \omega.$$

(The factor $(-1)^i$ disappeared because the induced orientation on H_i is $(-1)^i$ times that of the standard coordinates $\left(x^1,\dots,\widehat{x^i},\dots,x^n\right)$.) This completes the proof. $\square$

The preceding theorem has the following important application.

Theorem 16.26. *Suppose M is a smooth manifold and $\gamma_0, \gamma_1 \colon [a,b] \to M$ are path-homotopic piecewise smooth curve segments. For every closed 1-form ω on M,*

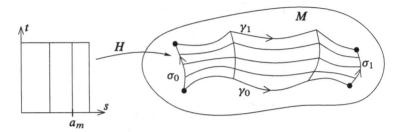

Fig. 16.14 Homotopic piecewise smooth curve segments

$$\int_{\gamma_0} \omega = \int_{\gamma_1} \omega.$$

Proof. By means of an affine reparametrization, we may as well assume for simplicity that $[a,b] = [0,1]$. Assume first that γ_0 and γ_1 are smooth. By Theorem 6.29, γ_0 and γ_1 are smoothly homotopic relative to $\{0,1\}$. Let $H: I \times I \to M$ be such a smooth homotopy. Since ω is closed, we have

$$\int_{I \times I} d(H^*\omega) = \int_{I \times I} H^* d\omega = 0.$$

On the other hand, $I \times I$ is a smooth manifold with corners, so Stokes's theorem implies

$$0 = \int_{I \times I} d(H^*\omega) = \int_{\partial(I \times I)} H^*\omega.$$

Using the parametrization of $\partial(I \times I)$ given in Example 16.23 together with Proposition 11.34(d), we obtain

$$0 = \int_{\partial(I \times I)} H^*\omega = \int_{F_1} H^*\omega + \int_{F_2} H^*\omega + \int_{F_3} H^*\omega + \int_{F_4} H^*\omega$$
$$= \int_{H \circ F_1} \omega + \int_{H \circ F_2} \omega + \int_{H \circ F_3} \omega + \int_{H \circ F_4} \omega,$$

where F_1, F_2, F_3, F_4 are defined by (16.8). The fact that H is a homotopy relative to $\{0,1\}$ means that $H \circ F_2$ and $H \circ F_4$ are constant maps, and therefore the second and fourth terms above are zero. The theorem then follows from the facts that $H \circ F_1 = \gamma_0$ and $H \circ F_3$ is a backward reparametrization of γ_1.

Next we consider the general case of piecewise smooth curves. We cannot simply apply the preceding result on each subinterval where γ_0 and γ_1 are smooth, because the restricted curves may not start and end at the same points. Instead, we prove the following more general claim: *Let $\gamma_0, \gamma_1 : I \to M$ be piecewise smooth curve segments (not necessarily with the same endpoints), and suppose $H: I \times I \to M$ is any homotopy between them (Fig. 16.14). Define curve segments $\sigma_0, \sigma_1 : I \to M$ by*

$$\sigma_0(t) = H(0,t), \qquad \sigma_1(t) = H(1,t),$$

and let $\tilde{\sigma}_0, \tilde{\sigma}_1$ be any smooth curve segments that are path-homotopic to σ_0, σ_1 respectively. Then

$$\int_{\gamma_1} \omega - \int_{\gamma_0} \omega = \int_{\tilde{\sigma}_1} \omega - \int_{\tilde{\sigma}_0} \omega. \tag{16.9}$$

When specialized to the case in which γ_0 and γ_1 are path-homotopic, this implies the theorem, because σ_0 and σ_1 are constant maps in that case.

Since γ_0 and γ_1 are piecewise smooth, there are only finitely many points $\{a_1, \ldots, a_m\}$ in $(0, 1)$ at which either γ_0 or γ_1 is not smooth. We prove the claim by induction on the number m of such points. When $m = 0$, both curves are smooth, and by Theorem 6.29 we may replace the given homotopy H by a smooth homotopy $\tilde{H}$. Recall from the proof of Theorem 6.29 that the smooth homotopy $\tilde{H}$ can actually be taken to be homotopic to H relative to $I \times \{0\} \cup I \times \{1\}$. Thus, for $i = 0, 1$, the curve $\tilde{\sigma}_i(t) = \tilde{H}(i, t)$ is a smooth curve segment that is path-homotopic to σ_i. In this setting, (16.9) just reduces to the integration formula of Example 16.23. Note that the integrals over $\tilde{\sigma}_0$ and $\tilde{\sigma}_1$ do not depend on which smooth curves path-homotopic to σ_0 and σ_1 are chosen, by the smooth case proved above.

Now let γ_0, γ_1 be homotopic piecewise smooth curves with m nonsmooth points $\{a_1, \ldots, a_m\}$, and suppose the claim is true for curves with fewer than m such points. For $i = 0, 1$, let γ_i' be the restriction of γ_i to $[0, a_m]$, and let γ_i'' be its restriction to $[a_m, 1]$. Let $\sigma: I \to M$ be the curve segment $\sigma(t) = H(a_m, t)$, and let $\tilde{\sigma}$ by any smooth curve segment that is path-homotopic to σ. Then, since γ_i' and γ_i'' have fewer than m nonsmooth points, the inductive hypothesis implies

$$\int_{\gamma_1} \omega - \int_{\gamma_0} \omega = \left(\int_{\gamma_1'} \omega - \int_{\gamma_0'} \omega \right) + \left(\int_{\gamma_1''} \omega - \int_{\gamma_0''} \omega \right)$$

$$= \left(\int_{\tilde{\sigma}} \omega - \int_{\tilde{\sigma}_0} \omega \right) + \left(\int_{\tilde{\sigma}_1} \omega - \int_{\tilde{\sigma}} \omega \right)$$

$$= \int_{\tilde{\sigma}_1} \omega - \int_{\tilde{\sigma}_0} \omega. \qquad \square$$

Corollary 16.27. *On a simply connected smooth manifold, every closed 1-form is exact.*

Proof. Suppose M is simply connected and ω is a closed 1-form on M. Since every piecewise smooth closed curve segment in M is path-homotopic to a constant curve, the preceding theorem shows that the integral of ω over every such curve is equal to 0. Thus, ω is conservative and therefore exact. $\qquad \square$

Integration on Riemannian Manifolds

In this section we explore what happens when the theory of integration and Stokes's theorem are specialized to Riemannian manifolds.

Integration of Functions on Riemannian Manifolds

We noted at the beginning of the chapter that real-valued functions cannot be integrated in a coordinate-independent way on an arbitrary manifold. However, with the additional structures of a Riemannian metric and an orientation, we can recover the notion of the integral of a real-valued function.

Suppose (M, g) is an oriented Riemannian manifold with or without boundary, and let ω_g denote its Riemannian volume form. If f is a compactly supported continuous real-valued function on M, then $f\omega_g$ is a compactly supported n-form, so we can define the **integral of f over M** to be $\int_M f\omega_g$. If M itself is compact, we define the **volume of M** by $\mathrm{Vol}(M) = \int_M \omega_g$.

Because of these definitions, the Riemannian volume form is often denoted by dV_g (or dA_g or ds_g in the 2-dimensional or 1-dimensional case, respectively). Then the integral of f over M is written $\int_M f\, dV_g$, and the volume of M as $\int_M dV_g$. Be warned, however, that this notation is *not* meant to imply that the volume form is the exterior derivative of an $(n-1)$-form; in fact, as we will see when we study de Rham cohomology, this is never the case on a compact manifold. You should just interpret dV_g as a notational convenience.

Proposition 16.28. *Let (M, g) be a nonempty oriented Riemannian manifold with or without boundary, and suppose f is a compactly supported continuous real-valued function on M satisfying $f \geq 0$. Then $\int_M f\, dV_g \geq 0$, with equality if and only if $f \equiv 0$.*

Proof. If f is supported in the domain of a single oriented smooth chart (U, φ), then Proposition 15.31 shows that

$$\int_M f\, dV_g = \int_{\varphi(U)} f(x)\sqrt{\det(g_{ij})}\, dx^1 \cdots dx^n \geq 0.$$

The same inequality holds in a negatively oriented chart because the negative sign from the chart cancels the negative sign in the expression for dV_g. The general case follows from this one, because $\int_M f\, dV_g$ is equal to a sum of terms like $\int_M \psi_i f\, dV_g$, where each integrand $\psi_i f$ is nonnegative and supported in a single smooth chart. If in addition f is positive somewhere, then it is positive on a nonempty open subset by continuity, so at least one of the integrals in this sum is positive. On the other hand, if f is identically zero, then clearly $\int_M f\, dV_g = 0$. $\square$

▶ **Exercise 16.29.** Suppose (M, g) is an oriented Riemannian manifold and $f : M \to \mathbb{R}$ is continuous and compactly supported. Prove that $\left| \int_M f\, dV_g \right| \leq \int_M |f|\, dV_g$.

The Divergence Theorem

Let (M, g) be an oriented Riemannian n-manifold (with or without boundary). We can generalize the classical divergence operator to this setting as follows. Multiplication by the Riemannian volume form defines a smooth bundle isomorphism

$*: C^\infty(M) \to \Omega^n(M)$:

$$* f = f\, dV_g. \tag{16.10}$$

In addition, as we did in Chapter 14 in the case of $\mathbb{R}^3$, we define a smooth bundle isomorphism $\beta: \mathfrak{X}(M) \to \Omega^{n-1}(M)$ as follows:

$$\beta(X) = X \lrcorner\, dV_g. \tag{16.11}$$

We need the following technical lemma.

Lemma 16.30. *Let (M, g) be an oriented Riemannian manifold with or without boundary. Suppose $S \subseteq M$ is an immersed hypersurface with the orientation determined by a unit normal vector field N, and $\tilde{g}$ is the induced metric on S. If X is any vector field along S, then*

$$\iota_S^*(\beta(X)) = \langle X, N \rangle_g\, dV_{\tilde{g}}. \tag{16.12}$$

Proof. Define two vector fields $X^\top$ and $X^\perp$ along S by

$$X^\perp = \langle X, N \rangle_g N,$$
$$X^\top = X - X^\perp.$$

Then $X = X^\perp + X^\top$, where $X^\perp$ is normal to S and $X^\top$ is tangent to it. Using this decomposition,

$$\beta(X) = X^\perp \lrcorner\, dV_g + X^\top \lrcorner\, dV_g.$$

Now pull back to S. Proposition 15.32 shows that the first term simplifies to

$$\iota_S^*(X^\perp \lrcorner\, dV_g) = \langle X, N \rangle_g \iota_S^*(N \lrcorner\, dV_g) = \langle X, N \rangle_g\, dV_{\tilde{g}}.$$

Thus (16.12) will be proved if we can show that $\iota_S^*(X^\top \lrcorner\, dV_g) = 0$. If $X_1, \ldots, X_{n-1}$ are any vectors tangent to S, then

$$(X^\top \lrcorner\, dV_g)(X_1, \ldots, X_{n-1}) = dV_g(X^\top, X_1, \ldots, X_{n-1}) = 0,$$

because any n-tuple of vectors in an $(n-1)$-dimensional vector space is linearly dependent. $\qquad\square$

Define the *divergence operator* div: $\mathfrak{X}(M) \to C^\infty(M)$ by

$$\operatorname{div} X = *^{-1} d(\beta(X)),$$

or equivalently,

$$d(X \lrcorner\, dV_g) = (\operatorname{div} X) dV_g.$$

▶ **Exercise 16.31.** Show that divergence operator on an oriented Riemannian manifold does not depend on the choice of orientation, and conclude that it is invariantly defined on all Riemannian manifolds.

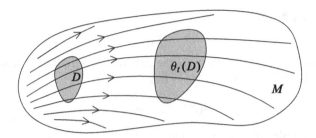

Fig. 16.15 Geometric interpretation of the divergence

The next theorem is a fundamental result about vector fields on Riemannian manifolds. In the special case of a compact regular domain in $\mathbb{R}^3$, it is often referred to as *Gauss's theorem*. (Later in the chapter, we will show that this theorem holds on nonorientable manifolds as well; see Theorem 16.48.)

Theorem 16.32 (The Divergence Theorem). *Let (M, g) be an oriented Riemannian manifold with boundary. For any compactly supported smooth vector field X on M,*

$$\int_M (\operatorname{div} X)\, dV_g = \int_{\partial M} \langle X, N\rangle_g\, dV_{\widetilde{g}},$$

where N is the outward-pointing unit normal vector field along ∂M and $\widetilde{g}$ is the induced Riemannian metric on ∂M.

Proof. By Stokes's theorem,

$$\int_M (\operatorname{div} X)\, dV_g = \int_M d\big(\beta(X)\big) = \int_{\partial M} \iota_S^* \beta(X).$$

The divergence theorem then follows from Lemma 16.30. □

The term "divergence" is used because of the following geometric interpretation. A smooth flow θ on M is said to be *volume-preserving* if for every compact regular domain D, we have $\operatorname{Vol}\big(\theta_t(D)\big) = \operatorname{Vol}(D)$ whenever the domain of θ_t contains D. It is called *volume-increasing*, *volume-decreasing*, *volume-nonincreasing*, or *volume-nondecreasing* if for every such D, $\operatorname{Vol}\big(\theta_t(D)\big)$ is strictly increasing, strictly decreasing, nonincreasing, or nondecreasing, respectively, as a function of t. Note that the properties of flow domains ensure that if D is contained in the domain of θ_t for some t, then the same is true for all times between 0 and t.

The next proposition shows that the divergence of a vector field can be interpreted as a measure of the tendency of its flow to "spread out," or diverge (see Fig. 16.15).

Proposition 16.33 (Geometric Interpretation of the Divergence). *Let M be an oriented Riemannian manifold, let $X \in \mathfrak{X}(M)$, and let θ be the flow of X. Then θ is*

(a) *volume-preserving if and only if $\operatorname{div} X = 0$ everywhere on M.*
(b) *volume-nondecreasing if and only if $\operatorname{div} X \geq 0$ everywhere on M.*

(c) *volume-nonincreasing if and only if* div $X \leq 0$ *everywhere on* M.

(d) *volume-increasing if and only if* div $X > 0$ *on a dense subset of* M.

(e) *volume-decreasing if and only if* div $X < 0$ *on a dense subset of* M.

Proof. First we establish some preliminary results. For each $t \in \mathbb{R}$, let M_t be the domain of θ_t. If D is a compact regular domain contained in M_t, then θ_t is an orientation-preserving diffeomorphism from D to $\theta_t(D)$ by the result of Problem 15-4, so

$$\text{Vol}\left(\theta_t(D)\right) = \int_{\theta_t(D)} dV_g = \int_D \theta_t^* \, dV_g.$$

Because the integrand on the right depends smoothly on (t, p) in the domain of θ, we can differentiate this expression with respect to t by differentiating under the integral sign. (Strictly speaking, we should use a partition of unity to express the integral as a sum of integrals over domains in $\mathbb{R}^n$, and then differentiate under the integral signs there; but the result is the same. The details are left to you.)

Using Cartan's magic formula for the Lie derivative of the Riemannian volume form, we obtain

$$\mathcal{L}_X dV_g = X \lrcorner d(dV_g) + d(X \lrcorner dV_g) = (\text{div } X) dV_g,$$

because $d(dV_g)$ is an $(n + 1)$-form on an n-manifold. Then Proposition 12.36 implies

$$\frac{d}{dt}\bigg|_{t=t_0} \text{Vol}\left(\theta_t(D)\right) = \int_D \frac{\partial}{\partial t}\bigg|_{t=t_0} (\theta_t^* \, dV_g) = \int_D \theta_{t_0}^* (\mathcal{L}_X dV_g)$$

$$= \int_D \theta_{t_0}^*\left((\text{div } X) dV_g\right) = \int_{\theta_{t_0}(D)} (\text{div } X) \, dV_g. \quad (16.13)$$

Now we can prove the "if" parts of all five equivalences. If div $X \equiv 0$, then it follows from (16.13) that $\text{Vol}\left(\theta_t(D)\right)$ is a constant function of t for every D, and thus θ is volume-preserving. Similarly, an inequality of the form div $X \geq 0$ or div $X \leq 0$ implies that $\text{Vol}\left(\theta_t(D)\right)$ is nondecreasing or nonincreasing, respectively. For part (d), suppose that div $X > 0$ on a dense subset of M, and let D be a compact regular domain in M. Then div $X \geq 0$ everywhere by continuity, so $\text{Vol}\left(\theta_t(D)\right)$ is nondecreasing by the argument above. Because Int D is an open subset of M (by Proposition 5.1), Int $\theta_t(D)$ is open for each t such that $D \subseteq M_t$, and therefore by density there is a point in Int $\theta_t(D)$ where div $X > 0$. Proposition 16.28 then shows that $\int_{\theta_t(D)} (\text{div } X) dV_g > 0$, and thus $\text{Vol}\left(\theta_t(D)\right)$ is strictly increasing by (16.13). A similar argument proves (e).

To prove the converses, we prove their contrapositives. We begin with (b). If there is a point where div $X < 0$, then by continuity there is an open subset $U \subseteq M$ on which div $X < 0$. The argument in the first part of the proof shows that X generates a volume-decreasing flow on U. In particular, for any regular coordinate ball B such that $\overline{B} \subseteq U$ and any $t > 0$ small enough to ensure that $\theta_t\left(\overline{B}\right) \subseteq U$, we have $\text{Vol}\left(\theta_t\left(\overline{B}\right)\right) < \text{Vol}\left(\overline{B}\right)$, which implies that θ is not volume-nondecreasing. The same argument with inequalities reversed proves (c). If div X is not identically zero, then there is an open subset on which it is either strictly positive or strictly negative, and then the argument above shows that it is not volume-preserving on that set, thus proving (a).

Next, consider (d). If the subset of M where div $X > 0$ is not dense, there is an open subset $U \subseteq M$ on which div $X \leq 0$. Then (c) shows that θ is volume-nonincreasing on U, so it cannot be volume-increasing on M. The argument for (e) is similar. $\square$

Surface Integrals

The original theorem that bears the name of Stokes concerned "surface integrals" of vector fields over surfaces in $\mathbb{R}^3$. Using the version of Stokes's theorem that we have proved, we cam generalize this to surfaces in Riemannian 3-manifolds.

Let (M, g) be an oriented Riemannian 3-manifold. Define the **curl operator**, denoted by curl: $\mathfrak{X}(M) \to \mathfrak{X}(M)$, by

$$\operatorname{curl} X = \beta^{-1} d \left(X^\flat \right),$$

where $\beta \colon \mathfrak{X}(M) \to \Omega^2(M)$ is defined in (16.11). Unwinding the definitions, we see that this is equivalent to

$$(\operatorname{curl} X) \lrcorner\, dV_g = d\left(X^\flat\right). \tag{16.14}$$

The operators div, grad, and curl on an oriented Riemannian 3-manifold M are related by the following commutative diagram analogous to (14.27):

$$\begin{array}{ccccccc}
C^\infty(M) & \xrightarrow{\operatorname{grad}} & \mathfrak{X}(M) & \xrightarrow{\operatorname{curl}} & \mathfrak{X}(M) & \xrightarrow{\operatorname{div}} & C^\infty(M) \\
\downarrow{\scriptstyle \operatorname{Id}} & & \downarrow{\scriptstyle \flat} & & \downarrow{\scriptstyle \beta} & & \downarrow{\scriptstyle *} \\
\Omega^0(M) & \xrightarrow[d]{} & \Omega^1(M) & \xrightarrow[d]{} & \Omega^2(M) & \xrightarrow[d]{} & \Omega^3(M).
\end{array} \tag{16.15}$$

The identities curl $\circ$ grad $\equiv 0$ and div $\circ$ curl $\equiv 0$ follow from $d \circ d \equiv 0$ just as they do in the Euclidean case. The curl operator is defined only in dimension 3 because it is only in that case that $\Lambda^2 T^* M$ is isomorphic to TM (via the map $\beta \colon X \mapsto X \lrcorner\, dV_g$).

Now suppose $S \subseteq M$ is a compact 2-dimensional submanifold with or without boundary, and N is a smooth unit normal vector field along S. Let dA denote the Riemannian volume form on S with respect to the induced metric $\iota_S^* g$ and the orientation determined by N, so that $dA = \iota_S^*(N \lrcorner\, dV_g)$ by Proposition 15.32. (See Fig. 16.16.) For any smooth vector field X defined on M, the **surface integral of X over S** (with respect to the given choice of unit normal field) is defined as

$$\int_S \langle X, N \rangle_g \, dA.$$

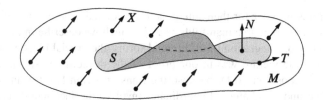

Fig. 16.16 The setup for a surface integral

The next result, in the special case in which $M = \mathbb{R}^3$, is the theorem usually referred to as *Stokes's theorem* in multivariable calculus texts.

Theorem 16.34 (Stokes's Theorem for Surface Integrals). *Suppose M is an oriented Riemannian 3-manifold with or without boundary, and S is a compact oriented 2-dimensional smooth submanifold with boundary in M. For any smooth vector field X on M,*

$$\int_S \langle \operatorname{curl} X, N \rangle_g \, dA = \int_{\partial S} \langle X, T \rangle_g \, ds,$$

where N is the smooth unit normal vector field along S that determines its orientation, ds is the Riemannian volume form for ∂S (with respect to the metric and orientation induced from S), and T is the unique positively oriented unit tangent vector field on ∂S.

Proof. The general version of Stokes's theorem applied to the 1-form $X^\flat$ yields

$$\int_S d(X^\flat) = \int_{\partial S} X^\flat.$$

Thus the theorem follows from the following two identities:

$$\iota_S^* d(X^\flat) = \langle \operatorname{curl} X, N \rangle_g \, dA, \tag{16.16}$$

$$\iota_{\partial S}^* X^\flat = \langle X, T \rangle_g \, ds. \tag{16.17}$$

Equation (16.16) is just the defining equation (16.14) for the curl combined with the result of Lemma 16.30. To prove (16.17), we note that $\iota_{\partial S}^* X^\flat$ is a smooth 1-form on a 1-manifold, and thus must be equal to $f \, ds$ for some smooth function f on ∂S. To evaluate f, we note that $ds(T) = 1$, and so the definition of $X^\flat$ yields

$$f = f \, ds(T) = X^\flat(T) = \langle X, T \rangle_g.$$

This proves (16.17) and thus the theorem. □

Densities

Although differential forms are natural objects to integrate on manifolds, and are essential for use in Stokes's theorem, they have the disadvantage of requiring oriented manifolds in order for their integrals to be defined. There is a way to define integration on nonorientable manifolds as well, which we describe in this section.

In the theory of integration of differential forms, the crucial place where orientations entered the picture was in our proof of the diffeomorphism-invariance of the integral (Proposition 16.1), because the transformation law for an n-form on an n-manifold under a change of coordinates involves the Jacobian determinant of the transition map, while transformation law for integrals involves the absolute value of the determinant. We had to restrict attention to orientation-preserving diffeomorphisms so that we could freely remove the absolute value signs. In this section we define objects whose transformation law involves the absolute value of the determinant, so that we no longer have this sign problem.

We begin, as always, in the linear-algebraic setting. Let V be an n-dimensional vector space. A *density on V* is a function

$$\mu \colon \underbrace{V \times \cdots \times V}_{n \text{ copies}} \to \mathbb{R}$$

satisfying the following condition: if $T \colon V \to V$ is any linear map, then

$$\mu(Tv_1, \ldots, Tv_n) = |\det T| \mu(v_1, \ldots, v_n). \tag{16.18}$$

(Compare this with the corresponding formula (14.2) for n-forms.) Observe that a density is *not* a tensor, because it is not linear over $\mathbb{R}$ in any of its arguments. Let $\mathcal{D}(V)$ denote the set of all densities on V.

Proposition 16.35 (Properties of Densities). *Let V be a vector space of dimension $n \geq 1$.*

(a) $\mathcal{D}(V)$ *is a vector space under the obvious vector operations:*

$$(c_1\mu_1 + c_2\mu_2)(v_1, \ldots, v_n) = c_1\mu_1(v_1, \ldots, v_n) + c_2\mu_2(v_1, \ldots, v_n).$$

(b) *If $\mu_1, \mu_2 \in \mathcal{D}(V)$ and $\mu_1(E_1, \ldots, E_n) = \mu_2(E_1, \ldots, E_n)$ for some basis (E_i) of V, then $\mu_1 = \mu_2$.*

(c) *If $\omega \in \Lambda^n(V^*)$, the map $|\omega| \colon V \times \cdots \times V \to \mathbb{R}$ defined by*

$$|\omega|(v_1, \ldots, v_n) = |\omega(v_1, \ldots, v_n)|$$

is a density.

(d) $\mathcal{D}(V)$ *is 1-dimensional, spanned by $|\omega|$ for any nonzero $\omega \in \Lambda^n(V^*)$.*

Proof. Part (a) is immediate from the definition. For part (b), suppose μ_1 and μ_2 give the same value when applied to $(E_1, \ldots, E_n)$. If $v_1, \ldots, v_n$ are arbitrary vectors in V, let $T \colon V \to V$ be the unique linear map that takes E_i to v_i for $i = 1, \ldots, n$. It follows that

$$\mu_1(v_1, \ldots, v_n) = \mu_1(TE_1, \ldots, TE_n)$$
$$= |\det T| \mu_1(E_1, \ldots, E_n)$$
$$= |\det T| \mu_2(E_1, \ldots, E_n)$$

$$= \mu_2(TE_1, \ldots, TE_n)$$
$$= \mu_2(v_1, \ldots, v_n).$$

Part (c) follows from Proposition 14.9:

$$|\omega|(Tv_1, \ldots, Tv_n) = |\omega(Tv_1, \ldots, Tv_n)|$$
$$= |(\det T)\omega(v_1, \ldots, v_n)|$$
$$= |\det T| \, |\omega| \, (v_1, \ldots, v_n).$$

Finally, to prove (d), suppose ω is any nonzero element of $\Lambda^n(V^*)$. If μ is an arbitrary element of $\mathcal{D}(V)$, it suffices to show that $\mu = c \, |\omega|$ for some $c \in \mathbb{R}$. Let (E_i) be a basis for V, and define $a, b \in \mathbb{R}$ by

$$a = |\omega| \, (E_1, \ldots, E_n) = |\omega(E_1, \ldots, E_n)|,$$
$$b = \mu(E_1, \ldots, E_n).$$

Because $\omega \neq 0$, it follows that $a \neq 0$. Thus, μ and $(b/a)|\omega|$ give the same result when applied to $(E_1, \ldots, E_n)$, so they are equal by part (b). □

A *positive density on* V is a density μ satisfying $\mu(v_1, \ldots, v_n) > 0$ whenever $(v_1, \ldots, v_n)$ is a linearly independent n-tuple. A *negative density* is defined similarly. If ω is a nonzero element of $\Lambda^n(V^*)$, then it is clear that $|\omega|$ is a positive density; more generally, a density $c \, |\omega|$ is positive, negative, or zero if and only if c has the same property. Thus, each density on V is either positive, negative, or zero, and the set of positive densities is a convex subset of $\mathcal{D}(V)$ (namely, a half-line).

Now let M be a smooth manifold with or without boundary. The set

$$\mathcal{D}M = \coprod_{p \in M} \mathcal{D}(T_pM)$$

is called the *density bundle of* M. Let $\pi \colon \mathcal{D}M \to M$ be the natural projection map taking each element of $\mathcal{D}(T_pM)$ to p.

Proposition 16.36. *If M is a smooth manifold with or without boundary, its density bundle is a smooth line bundle over M.*

Proof. We will construct local trivializations and use the vector bundle chart lemma (Lemma 10.6). Let $(U, (x^i))$ be any smooth coordinate chart on M, and let $\omega = dx^1 \wedge \cdots \wedge dx^n$. Proposition 16.35 shows that $|\omega_p|$ is a basis for $\mathcal{D}(T_pM)$ at each point $p \in U$. Therefore, the map $\Phi \colon \pi^{-1}(U) \to U \times \mathbb{R}$ given by

$$\Phi(c|\omega_p|) = (p, c)$$

is a bijection.

Now suppose $(\widetilde{U}, (\widetilde{x}^j))$ is another smooth chart with $U \cap \widetilde{U} \neq \varnothing$. Let $\widetilde{\omega} = d\widetilde{x}^1 \wedge \cdots \wedge d\widetilde{x}^n$, and define $\widetilde{\Phi} \colon \pi^{-1}(\widetilde{U}) \to \widetilde{U} \times \mathbb{R}$ correspondingly:

$$\widetilde{\Phi}(c|\widetilde{\omega}_p|) = (p, c).$$

It follows from the transformation law (14.16) for n-forms under changes of coordinates that

$$\Phi \circ \tilde{\Phi}^{-1}(p,c) = \Phi\big(c|\tilde{\omega}_p|\big) = \Phi\left(c\left|\det\left(\frac{\partial \tilde{x}^j}{\partial x^i}\right)\right||\omega_p|\right)$$

$$= \left(p, c\left|\det\left(\frac{\partial \tilde{x}^j}{\partial x^i}\right)\right|\right).$$

Thus, the hypotheses of Lemma 10.6 are satisfied, with the transition functions equal to $\left|\det\left(\partial \tilde{x}^j/\partial x^i\right)\right|$.   □

If M is a smooth n-manifold with or without boundary, a section of $\mathcal{D}M$ is called a **density on M**. (One might choose to call such a section a "density field" to distinguish it from a density on a vector space, but we do not do so.) If μ is a density and f is a continuous real-valued function, then $f\mu$ is again a density, which is smooth if both f and μ are. A density on M is said to be positive or negative if its value at each point has that property. Any nonvanishing n-form ω determines a positive density $|\omega|$, defined by $|\omega|_p = |\omega_p|$ for each $p \in M$. If ω is a nonvanishing n-form on an open subset $U \subseteq M$, then any density μ on U can be written $\mu = f|\omega|$ for some real-valued function f.

One important fact about densities is that every smooth manifold admits a global smooth positive density, without any orientability assumptions.

Proposition 16.37. *If M is a smooth manifold with or without boundary, there exists a smooth positive density on M.*

Proof. Because the set of positive elements of $\mathcal{D}M$ is an open subset whose intersection with each fiber is convex, the usual partition of unity argument (Problem 13-2) allows us to piece together local positive densities to obtain a global smooth positive density.   □

It is important to understand that this proposition works because positivity of a density is a well-defined property, independent of any choices of coordinates or orientations. There is no corresponding existence result for orientation forms because without a choice of orientation, there is no way to decide which n-forms are positive.

Under smooth maps, densities pull back in the same way as differential forms. If $F: M \to N$ is a smooth map between n-manifolds (with or without boundary) and μ is a density on N, we define a density $F^*\mu$ on M by

$$(F^*\mu)_p(v_1, \ldots, v_n) = \mu_{F(p)}\big(dF_p(v_1), \ldots, dF_p(v_n)\big).$$

Proposition 16.38. *Let $G: P \to M$ and $F: M \to N$ be smooth maps between n-manifolds with or without boundary, and let μ be a density on N.*

(a) *For any $f \in C^\infty(N)$, $F^*(f\mu) = (f \circ F)F^*\mu$.*
(b) *If ω is an n-form on N, then $F^*|\omega| = |F^*\omega|$.*
(c) *If μ is smooth, then $F^*\mu$ is a smooth density on M.*
(d) *$(F \circ G)^*\mu = G^*(F^*\mu)$.*

▶ **Exercise 16.39.** Prove the preceding proposition.

The next result shows how to compute the pullback of a density in coordinates. It is an analogue for densities of Proposition 14.20.

Proposition 16.40. *Suppose* $F: M \to N$ *is a smooth map between n-manifolds with or without boundary. If* (x^i) *and* (y^j) *are smooth coordinates on open subsets* $U \subseteq M$ *and* $V \subseteq N$, *respectively, and* u *is a continuous real-valued function on* V, *then the following holds on* $U \cap F^{-1}(V)$:

$$F^* \left(u \left| dy^1 \wedge \cdots \wedge dy^n \right| \right) = (u \circ F) \left| \det DF \right| \left| dx^1 \wedge \cdots \wedge dx^n \right|, \qquad (16.19)$$

where DF *represents the matrix of partial derivatives of* F *in these coordinates.*

Proof. Using Propositions 14.20 and 16.38, we obtain

$$
\begin{aligned}
F^* \left(u \left| dy^1 \wedge \cdots \wedge dy^n \right| \right) &= (u \circ F) F^* \left| dy^1 \wedge \cdots \wedge dy^n \right| \\
&= (u \circ F) \left| F^* (dy^1 \wedge \cdots \wedge dy^n) \right| \\
&= (u \circ F) \left| (\det DF) \, dx^1 \wedge \cdots \wedge dx^n \right| \\
&= (u \circ F) \left| \det DF \right| \left| dx^1 \wedge \cdots \wedge dx^n \right|. \qquad \square
\end{aligned}
$$

Now we turn to integration. As we did with forms, we begin by defining integrals of densities on subsets of $\mathbb{R}^n$. If $D \subseteq \mathbb{R}^n$ is a domain of integration and μ is a density on $\bar{D}$, we can write $\mu = f \left| dx^1 \wedge \cdots \wedge dx^n \right|$ for some uniquely determined continuous function $f: \bar{D} \to \mathbb{R}$. We define the ***integral of*** μ ***over*** D by

$$\int_D \mu = \int_D f \, dV,$$

or more suggestively,

$$\int_D f \left| dx^1 \wedge \cdots \wedge dx^n \right| = \int_D f \, dx^1 \cdots dx^n.$$

Similarly, if U is an open subset of $\mathbb{R}^n$ or $\mathbb{H}^n$ and μ is compactly supported in U, we define

$$\int_U \mu = \int_D \mu,$$

where D is any domain of integration containing the support of μ. The key fact is that this is diffeomorphism-invariant.

Proposition 16.41. *Suppose* U *and* V *are open subsets of* $\mathbb{R}^n$ *or* $\mathbb{H}^n$, *and* $G: U \to V$ *is a diffeomorphism. If* μ *is a compactly supported density on* V, *then*

$$\int_V \mu = \int_U G^* \mu.$$

Proof. The proof is essentially identical to that of Proposition 16.3, using (16.19) instead of (14.15). $\square$

Now let M be a smooth n-manifold (with or without boundary). If μ is a density on M whose support is contained in the domain of a single smooth chart (U, φ), the *integral of μ over M* is defined as

$$\int_M \mu = \int_{\varphi(U)} \left(\varphi^{-1} \right)^* \mu.$$

This is extended to arbitrary densities μ by setting

$$\int_M \mu = \sum_i \int_M \psi_i \mu,$$

where $\{\psi_i\}$ is a smooth partition of unity subordinate to an open cover of M by smooth charts. The fact that this is independent of the choices of coordinates or partition of unity follows just as in the case of forms.

The following proposition is proved in the same way as Proposition 16.6.

Proposition 16.42 (Properties of Integrals of Densities). *Suppose M and N are smooth n-manifolds with or without boundary, and μ, η are compactly supported densities on M.*

(a) LINEARITY: *If $a, b \in \mathbb{R}$, then*

$$\int_M a\mu + b\eta = a \int_M \mu + b \int_M \eta.$$

(b) POSITIVITY: *If μ is a positive density, then $\int_M \mu > 0$.*
(c) DIFFEOMORPHISM INVARIANCE: *If $F \colon N \to M$ is a diffeomorphism, then $\int_M \mu = \int_N F^* \mu$.*

▶ **Exercise 16.43.** Prove Proposition 16.42.

Just as for forms, integrals of densities are usually computed by cutting the manifold into pieces and parametrizing each piece, just as in Proposition 16.8. The details are left to the reader.

▶ **Exercise 16.44.** Formulate and prove an analogue of Proposition 16.8 for densities.

The Riemannian Density

Densities are particularly useful on Riemannian manifolds.

Proposition 16.45 (The Riemannian Density). *Let (M, g) be a Riemannian manifold with or without boundary. There is a unique smooth positive density μ_g on M, called the **Riemannian density**, with the property that*

$$\mu_g(E_1, \ldots, E_n) = 1 \qquad\qquad (16.20)$$

for any local orthonormal frame (E_i).

Proof. Uniqueness is immediate, because any two densities that agree on a basis must be equal. Given any point $p \in M$, let U be a connected smooth coordinate neighborhood of p. Since U is diffeomorphic to an open subset of Euclidean space, it is orientable. Any choice of orientation of U uniquely determines a Riemannian volume form ω_g on U, with the property that $\omega_g(E_1, \ldots, E_n) = 1$ for any oriented orthonormal frame. If we put $\mu_g = |\omega_g|$, it follows easily that μ_g is a smooth positive density on U satisfying (16.20). If U and V are two overlapping smooth coordinate neighborhoods, the two definitions of μ_g agree where they overlap by uniqueness, so this defines μ_g globally. $\square$

▶ **Exercise 16.46.** Let (M, g) be an oriented Riemannian manifold with or without boundary and let ω_g be its Riemannian volume form.

(a) Show that the Riemannian density of M is given by $\mu_g = |\omega_g|$.
(b) For any compactly supported continuous function $f: M \to \mathbb{R}$, show that

$$\int_M f\mu_g = \int_M f\omega_g.$$

▶ **Exercise 16.47.** Suppose (M, g) and $(\widetilde{M}, \widetilde{g})$ are Riemannian manifolds with or without boundary, and $F: M \to \widetilde{M}$ is a local isometry. Show that $F^*\mu_{\widetilde{g}} = \mu_g$.

Because of Exercise 16.46(b), it is customary to denote the Riemannian density simply by dV_g, and to specify when necessary whether the notation refers to a density or a form. If $f: M \to \mathbb{R}$ is a compactly supported continuous function, the *integral of f over M* is defined to be $\int_M f \, dV_g$. Exercise 16.46 shows that when M is oriented, it does not matter whether we interpret dV_g as the Riemannian volume form or the Riemannian density. (If the orientation of M is changed, then both the integral and dV_g change signs, so the result is the same.) When M is not orientable, however, we have no choice but to interpret it as a density.

One of the most useful applications of densities is that they enable us to generalize the divergence theorem to nonorientable manifolds. If X is a smooth vector field on M, Exercise 16.31 shows that the divergence of X can be defined even when M is not orientable. The next theorem shows that the divergence theorem holds in that case as well.

Theorem 16.48 (The Divergence Theorem in the Nonorientable Case). *Suppose (M, g) is a nonorientable Riemannian manifold with boundary. For any compactly supported smooth vector field X on M,*

$$\int_M (\operatorname{div} X)\mu_g = \int_{\partial M} \langle X, N \rangle_g \, \mu_{\widetilde{g}}, \qquad (16.21)$$

where N is the outward-pointing unit normal vector field along ∂M, $\widetilde{g}$ is the induced Riemannian metric on ∂M, and μ_g, $\mu_{\widetilde{g}}$ are the Riemannian densities of g and $\widetilde{g}$, respectively.

Proof. Let $\widehat{\pi}: \widehat{M} \to M$ be the orientation covering of M. Problem 5-12 shows that $\widehat{\pi}$ restricts to a smooth covering map from each component of $\partial\widehat{M}$ to a component of ∂M, so in the terminology of Chapter 15, $\hat{\pi} \colon \partial \widehat{M} \to \partial M$ is a generalized covering map.

Define metrics $\hat{g} = \hat{\pi}^* g$ on $\widehat{M}$ and $\bar{g} = \hat{\pi}^* \tilde{g}$ on $\partial \widehat{M}$. Denote the Riemannian volume forms of $\hat{g}$ and $\bar{g}$ by $\omega_{\hat{g}}$ and $\omega_{\bar{g}}$, respectively, and their Riemannian densities by $\mu_{\hat{g}}$ and $\mu_{\bar{g}}$. Because $\hat{\pi}$ is a local isometry, it is easy to check that the outward unit normal $\widehat{N}$ along $\partial \widehat{M}$ is $\hat{\pi}$-related to N. Moreover, it follows from Problem 8-18(a) that there is a unique smooth vector field $\widehat{X}$ on $\widehat{M}$ that is $\hat{\pi}$-related to X.

Since $\widehat{M}$ is an oriented smooth Riemannian manifold with boundary, we can apply the usual divergence theorem to it to obtain

$$2 \int_M (\operatorname{div} X)\mu_g = \int_{\widehat{M}} \hat{\pi}^* \left((\operatorname{div} X)\mu_g \right) \qquad \text{(by Problem 16-3)}$$

$$= \int_{\widehat{M}} (\operatorname{div} \widehat{X})\mu_{\hat{g}} \qquad (\hat{\pi} \text{ is a local isometry})$$

$$= \int_{\widehat{M}} (\operatorname{div} \widehat{X})\, \omega_{\hat{g}} \qquad \text{(by Exercise 16.46(b))}$$

$$= \int_{\partial \widehat{M}} \langle \widehat{X}, \widehat{N} \rangle_{\hat{g}}\, \omega_{\bar{g}} \qquad \text{(divergence theorem on } \widehat{M})$$

$$= \int_{\partial \widehat{M}} \langle \widehat{X}, \widehat{N} \rangle_{\hat{g}}\, \mu_{\bar{g}} \qquad \text{(by Exercise 16.46(b))}$$

$$= \int_{\partial \widehat{M}} (\hat{\pi}|_{\partial \widehat{M}})^* \left(\langle X, N \rangle_g \mu_{\tilde{g}} \right) \qquad (\hat{\pi}|_{\partial \widehat{M}} \text{ is a local isometry})$$

$$= 2 \int_{\partial M} \langle X, N \rangle_g\, \mu_{\tilde{g}} \qquad \text{(by Problem 16-3).}$$

Dividing both sides by 2 yields (16.21). $\square$

Problems

16-1. Let $v_1, \dots, v_n$ be any n linearly independent vectors in $\mathbb{R}^n$, and let P be the n-dimensional parallelepiped they span:

$$P = \{t_1 v_1 + \dots + t_n v_n : 0 \le t_i \le 1\}.$$

Show that $\operatorname{Vol}(P) = |\det(v_1, \dots, v_n)|$. (*Used on p. 401.*)

16-2. Let $\mathbb{T}^2 = \mathbb{S}^1 \times \mathbb{S}^1 \subseteq \mathbb{R}^4$ denote the 2-torus, defined as the set of points (w, x, y, z) such that $w^2 + x^2 = y^2 + z^2 = 1$, with the product orientation determined by the standard orientation on $\mathbb{S}^1$. Compute $\int_{\mathbb{T}^2} \omega$, where ω is the following 2-form on $\mathbb{R}^4$:

$$\omega = xyz\, dw \wedge dy.$$

16-3. Suppose E and M are smooth n-manifolds with or without boundary, and $\pi \colon E \to M$ is a smooth k-sheeted covering map or generalized covering map.

(a) Show that if E and M are oriented and π is orientation-preserving, then $\int_E \pi^* \omega = k \int_M \omega$ for any compactly supported n-form ω on M.
(b) Show that $\int_E \pi^* \mu = k \int_M \mu$ whenever μ is a compactly supported density on M.

16-4. Suppose M is an oriented compact smooth manifold with boundary. Show that there does not exist a retraction of M onto its boundary. [Hint: if the retraction is smooth, consider an orientation form on ∂M.]

16-5. Suppose M and N are oriented, compact, connected, smooth manifolds, and $F, G: M \to N$ are homotopic diffeomorphisms. Show that F and G are either both orientation-preserving or both orientation-reversing. [Hint: use Theorem 6.29 and Stokes's theorem on $M \times I$.]

16-6. THE HAIRY BALL THEOREM: *There exists a nowhere-vanishing vector field on $\mathbb{S}^n$ if and only if n is odd.* ("You cannot comb the hair on a ball.") Prove this by showing that the following are equivalent:
(a) There exists a nowhere-vanishing vector field on $\mathbb{S}^n$.
(b) There exists a continuous map $V: \mathbb{S}^n \to \mathbb{S}^n$ satisfying $V(x) \perp x$ (with respect to the Euclidean dot product on $\mathbb{R}^{n+1}$) for all $x \in \mathbb{S}^n$.
(c) The antipodal map $\alpha: \mathbb{S}^n \to \mathbb{S}^n$ is homotopic to $\mathrm{Id}_{\mathbb{S}^n}$.
(d) The antipodal map $\alpha: \mathbb{S}^n \to \mathbb{S}^n$ is orientation-preserving.
(e) n is odd.
[Hint: use Problems 9-4, 15-3, and 16-5.]

16-7. Show that any finite product $M_1 \times \cdots \times M_k$ of smooth manifolds with corners is again a smooth manifold with corners. Give a counterexample to show that a finite product of smooth manifolds with boundary need not be a smooth manifold with boundary.

16-8. Suppose M is a smooth manifold with corners, and let $\mathcal{C}$ denote the set of corner points of M. Show that $M \smallsetminus \mathcal{C}$ is a smooth manifold with boundary.

16-9. Let ω be the $(n-1)$-form on $\mathbb{R}^n \smallsetminus \{0\}$ defined by

$$\omega = |x|^{-n} \sum_{i=1}^n (-1)^{i-1} x^i \, dx^1 \wedge \cdots \wedge \widehat{dx^i} \wedge \cdots \wedge dx^n. \tag{16.22}$$

(a) Show that $\iota^*_{\mathbb{S}^{n-1}} \omega$ is the Riemannian volume form of $\mathbb{S}^{n-1}$ with respect to the round metric and the standard orientation.
(b) Show that ω is closed but not exact on $\mathbb{R}^n \smallsetminus \{0\}$.

16-10. Let D denote the torus of revolution in $\mathbb{R}^3$ obtained by revolving the circle $(r-2)^2 + z^2 = 1$ around the z-axis (Example 5.17), with its induced Riemannian metric and with the orientation determined by the outward unit normal.
(a) Compute the surface area of D.
(b) Compute the integral over D of the function $f(x, y, z) = z^2 + 1$.
(c) Compute the integral over D of the 2-form $\omega = z \, dx \wedge dy$.

16-11. Let (M, g) be a Riemannian n-manifold with or without boundary. In any smooth local coordinates (x^i), show that

$$\operatorname{div}\left(X^i \frac{\partial}{\partial x^i}\right) = \frac{1}{\sqrt{\det g}} \frac{\partial}{\partial x^i}\left(X^i \sqrt{\det g}\right),$$

where $\det g = \det(g_{kl})$ is the determinant of the component matrix of g in these coordinates.

16-12. Let (M, g) be a compact Riemannian manifold with boundary, let $\tilde{g}$ denote the induced Riemannian metric on ∂M, and let N be the outward unit normal vector field along ∂M.
   (a) Show that the divergence operator satisfies the following product rule for $f \in C^\infty(M)$, $X \in \mathfrak{X}(M)$:

$$\operatorname{div}(fX) = f \operatorname{div} X + \langle \operatorname{grad} f, X\rangle_g.$$

   (b) Prove the following "integration by parts" formula:

$$\int_M \langle \operatorname{grad} f, X\rangle_g \, dV_g = \int_{\partial M} f \langle X, N\rangle_g \, dV_{\tilde{g}} - \int_M (f \operatorname{div} X) \, dV_g.$$

   (c) Explain what this has to do with integration by parts.

16-13. Let (M, g) be a Riemannian n-manifold with or without boundary. The linear operator $\Delta \colon C^\infty(M) \to C^\infty(M)$ defined by $\Delta u = -\operatorname{div}(\operatorname{grad} u)$ is called the (geometric) *Laplacian*. Show that the Laplacian is given in any smooth local coordinates by

$$\Delta u = -\frac{1}{\sqrt{\det g}} \frac{\partial}{\partial x^i}\left(g^{ij} \sqrt{\det g} \frac{\partial u}{\partial x^j}\right).$$

Conclude that on $\mathbb{R}^n$ with the Euclidean metric and standard coordinates,

$$\Delta u = -\sum_{i=1}^n \frac{\partial^2 u}{(\partial x^i)^2}.$$

[Remark: there is no general agreement about the sign convention for the Laplacian on a Riemannian manifold, and many authors define Δ to be the negative of the operator we have defined. Although the geometric Laplacian defined here is the opposite of the traditional Laplacian on $\mathbb{R}^n$, it has two distinct advantages: our Laplacian has nonnegative eigenvalues (see Problem 16-15), and it agrees with the Laplace–Beltrami operator defined on differential forms (see Problems 17-2 and 17-3). When reading any book or article that mentions the Laplacian, you have to be careful to determine which sign convention the author is using.] (*Used on p. 465.*)

16-14. Let (M, g) be a Riemannian manifold with or without boundary. A function $u \in C^\infty(M)$ is said to be *harmonic* if $\Delta u = 0$ (see Problem 16-13).

(a) Suppose M is compact, and prove **Green's identities**:

$$\int_M u\Delta v\, dV_g = \int_M \langle \operatorname{grad} u, \operatorname{grad} v\rangle_g\, dV_g - \int_{\partial M} uNv\, dV_{\tilde g},$$

$$\int_M (u\Delta v - v\Delta u)\, dV_g = \int_{\partial M} (vNu - uNv)\, dV_{\tilde g},$$

where N and $\tilde g$ are as in Problem 16-12.

(b) Show that if M is compact and connected and $\partial M = \varnothing$, the only harmonic functions on M are the constants.

(c) Show that if M is compact and connected, $\partial M \neq \varnothing$, and u, v are harmonic functions on M whose restrictions to ∂M agree, then $u \equiv v$.

16-15. Let (M, g) be a compact connected Riemannian manifold without boundary, and let Δ be its geometric Laplacian. A real number λ is called an **eigenvalue of** Δ if there exists a smooth real-valued function u on M, not identically zero, such that $\Delta u = \lambda u$. In this case, u is called an **eigenfunction** corresponding to λ.

(a) Prove that 0 is an eigenvalue of Δ, and that all other eigenvalues are strictly positive.

(b) Prove that if u and v are eigenfunctions corresponding to distinct eigenvalues, then $\int_M uv\, dV_g = 0$.

16-16. Let M be a compact connected Riemannian n-manifold with nonempty boundary. A number $\lambda \in \mathbb{R}$ is called a **Dirichlet eigenvalue for M** if there exists a smooth real-valued function u on M, not identically zero, such that $\Delta u = \lambda u$ and $u|_{\partial M} = 0$. Similarly, λ is called a **Neumann eigenvalue** if there exists such a u satisfying $\Delta u = \lambda u$ and $Nu|_{\partial M} = 0$, where N is the outward unit normal.

(a) Show that every Dirichlet eigenvalue is strictly positive.

(b) Show that 0 is a Neumann eigenvalue, and all other Neumann eigenvalues are strictly positive.

16-17. DIRICHLET'S PRINCIPLE: Suppose M is a compact connected Riemannian n-manifold with nonempty boundary. Prove that a function $u \in C^\infty(M)$ is harmonic if and only if it minimizes $\int_M |\operatorname{grad} u|_g^2\, dV_g$ among all smooth functions with the same boundary values. [Hint: for any function $f \in C^\infty(M)$ that vanishes on ∂M, expand $\int_M |\operatorname{grad}(u + \varepsilon f)|_g^2\, dV_g$ and use Problem 16-12.]

16-18. Let (M, g) be an oriented Riemannian n-manifold. This problem outlines an important generalization of the operator $*\colon C^\infty(M) \to \Omega^n(M)$ defined in this chapter.

(a) For each $k = 1, \ldots, n$, show that g determines a unique inner product on $\Lambda^k(T_p^*M)$ (denoted by $\langle \cdot, \cdot\rangle_g$, just like the inner product on T_pM) satisfying

$$\langle \omega^1 \wedge \cdots \wedge \omega^k, \eta^1 \wedge \cdots \wedge \eta^k\rangle_g = \det\big(\langle (\omega^i)^\#, (\eta^j)^\#\rangle_g\big)$$

whenever $\omega^1, \ldots, \omega^k, \eta^1, \ldots, \eta^k$ are covectors at p. [Hint: define the inner product locally by declaring $\{\varepsilon^I|_p : I \text{ is increasing}\}$ to be an orthonormal basis for $\Lambda^k(T_p^*M)$ whenever (ε^i) is the coframe dual to a local orthonormal frame, and then prove that the resulting inner product is independent of the choice of frame.]

(b) Show that the Riemannian volume form dV_g is the unique positively oriented n-form that has unit norm with respect to this inner product.

(c) For each $k = 0, \ldots, n$, show that there is a unique smooth bundle homomorphism $*: \Lambda^k T^*M \to \Lambda^{n-k} T^*M$ satisfying

$$\omega \wedge *\eta = \langle \omega, \eta \rangle_g \, dV_g$$

for all smooth k-forms ω, η. (For $k = 0$, interpret the inner product as ordinary multiplication.) This map is called the **Hodge star operator**. [Hint: first prove uniqueness, and then define $*$ locally by setting

$$*\left(\varepsilon^{i_1} \wedge \cdots \wedge \varepsilon^{i_k}\right) = \pm \varepsilon^{j_1} \wedge \cdots \wedge \varepsilon^{j_{n-k}}$$

in terms of an orthonormal coframe (ε^i), where the indices $j_1, \ldots, j_{n-k}$ are chosen so that $(i_1, \ldots, i_k, j_1, \ldots, j_{n-k})$ is some permutation of $(1, \ldots, n)$.]

(d) Show that $*: \Lambda^0 T^*M \to \Lambda^n T^*M$ is given by $* f = f \, dV_g$.

(e) Show that $* * \omega = (-1)^{k(n-k)} \omega$ if ω is a k-form.

16-19. Consider $\mathbb{R}^n$ as a Riemannian manifold with the Euclidean metric and the standard orientation.

(a) Calculate $* dx^i$ for $i = 1, \ldots, n$.

(b) Calculate $*\left(dx^i \wedge dx^j\right)$ in the case $n = 4$.

16-20. Let M be an oriented Riemannian 4-manifold. A 2-form ω on M is said to be **self-dual** if $*\omega = \omega$, and **anti-self-dual** if $*\omega = -\omega$.

(a) Show that every 2-form ω on M can be written uniquely as a sum of a self-dual form and an anti-self-dual form.

(b) On $M = \mathbb{R}^4$ with the Euclidean metric, determine the self-dual and anti-self-dual forms in standard coordinates.

16-21. Let (M, g) be an oriented Riemannian manifold and $X \in \mathfrak{X}(M)$. Show that

$$X \lrcorner \, dV_g = * X^\flat,$$

$$\operatorname{div} X = * d * X^\flat,$$

and, when $\dim M = 3$,

$$\operatorname{curl} X = \left(* dX^\flat\right)^\sharp.$$

16-22. Let (M, g) be a compact, oriented Riemannian n-manifold. For $1 \le k \le n$, define a map $d^*: \Omega^k(M) \to \Omega^{k-1}(M)$ by $d^*\omega = (-1)^{n(k+1)+1} * d * \omega$, where $*$ is the Hodge star operator defined in Problem 16-18. Extend this definition to 0-forms by defining $d^*\omega = 0$ for $\omega \in \Omega^0(M)$.

(a) Show that $d^* \circ d^* = 0$.

(b) Show that the formula

$$(\omega, \eta) = \int_M \langle \omega, \eta \rangle_g \, dV_g$$

defines an inner product on $\Omega^k(M)$ for each k, where $\langle \cdot, \cdot \rangle_g$ is the pointwise inner product on forms defined in Problem 16-18.

(c) Show that $(d^*\omega, \eta) = (\omega, d\eta)$ for all $\omega \in \Omega^k(M)$ and $\eta \in \Omega^{k-1}(M)$. (*Used on p. 464.*)

16-23. This problem illustrates another approach to proving that certain Riemannian metrics are not flat. Let $\mathbb{B}^2$ be the unit disk in $\mathbb{R}^2$, and let g be the Riemannian metric on $\mathbb{B}^2$ given by

$$g = \frac{dx^2 + dy^2}{1 - x^2 - y^2}.$$

(a) Show that if g is flat, then for sufficiently small $r > 0$, the volume of the g-metric ball $\bar{B}_r^g(0)$ satisfies $\mathrm{Vol}_g\left(\bar{B}_r^g(0)\right) = \pi r^2$.

(b) For any $v \in \mathbb{B}^2$, by computing the g-length of the straight line from 0 to v, show that $d_g(0, v) \le \tanh^{-1}|v|$ (where $\tanh^{-1}$ denotes the inverse hyperbolic tangent function). Conclude that for any $r > 0$, the g-metric ball $\bar{B}_r^g(0)$ contains the Euclidean ball $\bar{B}_{\tanh r}(0) = \{v : |v| \le \tanh r\}$.

(c) Show that $\mathrm{Vol}_g\left(\bar{B}_r^g(0)\right) \ge \pi \sinh^2 r > \pi r^2$, and therefore g is not flat.

Chapter 17
De Rham Cohomology

In Chapter 14 we defined closed and exact forms: a smooth differential form ω is *closed* if $d\omega = 0$, and *exact* if it can be written $\omega = d\eta$. Because $d \circ d = 0$, every exact form is closed. In this chapter, we explore the implications of the converse question: Is every closed form exact? The answer, in general, is no: in Example 11.48, for instance, we saw a 1-form on the punctured plane that is closed but not exact; the failure of exactness seemed to be a consequence of the "hole" in the center of the domain. For higher-degree forms, the question of which closed forms are exact depends on subtle topological properties of the manifold, connected with the existence of "holes" of higher dimensions. Making this dependence quantitative leads to a new set of invariants of smooth manifolds, called the *de Rham cohomology groups*, which are the subject of this chapter.

Knowledge of which closed forms are exact has many important consequences. For example, Stokes's theorem implies that if ω is exact, then the integral of ω over any compact submanifold without boundary is zero. Proposition 11.42 showed that a smooth 1-form is conservative if and only if it is exact.

We begin by defining the de Rham cohomology groups and proving some of their basic properties, including diffeomorphism invariance. Then we prove an important generalization of this fact: the de Rham groups are in fact *homotopy invariants*, which implies in particular that they are topological invariants. Next, after computing the de Rham groups in some simple cases, we state a general theorem, called the *Mayer–Vietoris theorem*, that expresses the de Rham groups of a manifold in terms of those of its open subsets. Using this, we compute all of the de Rham groups of spheres and the top-degree groups of compact manifolds. Then we give an important application of these ideas to topology: there is a homotopically invariant integer associated with any continuous map between connected, compact, oriented, smooth manifolds of the same dimension, called the *degree* of the map.

At the end of the chapter, we prove the Mayer–Vietoris theorem.

J.M. Lee, *Introduction to Smooth Manifolds*, Graduate Texts in Mathematics 218, DOI 10.1007/978-1-4419-9982-5_17, © Springer Science+Business Media New York 2013

The de Rham Cohomology Groups

In Chapter 11, we studied the closed 1-form

$$\omega = \frac{x \, dy - y \, dx}{x^2 + y^2}, \tag{17.1}$$

and showed that it is not exact on $\mathbb{R}^2 \smallsetminus \{0\}$, but it is exact on some smaller domains such as the right half-plane $H = \{(x, y) : x > 0\}$, where it is equal to $d\theta$ (see Example 11.48).

As we will see in this chapter, that behavior is typical: closed forms are always *locally* exact, so whether a given closed form is exact depends on the global shape of the domain. To capture this dependence, we make the following definitions.

Let M be a smooth manifold with or without boundary, and let p be a nonnegative integer. Because $d : \Omega^p(M) \to \Omega^{p+1}(M)$ is linear, its kernel and image are linear subspaces. We define

$$\mathcal{Z}^p(M) = \operatorname{Ker}\left(d : \Omega^p(M) \to \Omega^{p+1}(M)\right) = \{\text{closed } p\text{-forms on } M\},$$

$$\mathcal{B}^p(M) = \operatorname{Im}\left(d : \Omega^{p-1}(M) \to \Omega^p(M)\right) = \{\text{exact } p\text{-forms on } M\}.$$

By convention, we consider $\Omega^p(M)$ to be the zero vector space when $p < 0$ or $p > n = \dim M$, so that, for example, $\mathcal{B}^0(M) = 0$ and $\mathcal{Z}^n(M) = \Omega^n(M)$.

The fact that every exact form is closed implies that $\mathcal{B}^p(M) \subseteq \mathcal{Z}^p(M)$. Thus, it makes sense to define the **de Rham cohomology group in degree** p (or the **pth de Rham group**) **of** M to be the quotient vector space

$$H_{dR}^p(M) = \frac{\mathcal{Z}^p(M)}{\mathcal{B}^p(M)}.$$

(It is a real vector space, and thus in particular a group under vector addition. Perhaps "de Rham cohomology space" would be a more appropriate term, but because most other cohomology theories produce only groups it is traditional to use the term *group* in this context as well, bearing in mind that these "groups" are actually real vector spaces.) It is clear that $H_{dR}^p(M) = 0$ for $p < 0$ or $p > \dim M$, because $\Omega^p(M) = 0$ in those cases. For $0 \le p \le n$, the definition implies that $H_{dR}^p(M) = 0$ if and only if every closed p-form on M is exact.

Example 17.1. The fact that there is a closed 1-form on $\mathbb{R}^2 \smallsetminus \{0\}$ that is not exact means that $H_{dR}^1\left(\mathbb{R}^2 \smallsetminus \{0\}\right) \ne 0$ (see Example 11.48). On the other hand, the Poincaré lemma for 1-forms (Theorem 11.49) implies that $H_{dR}^1(U) = 0$ for any star-shaped open subset $U \subseteq \mathbb{R}^n$. //

The first order of business is to show that the de Rham groups are diffeomorphism invariants. For any closed p-form ω on M, we let $[\omega]$ denote the equivalence class of ω in $H_{dR}^p(M)$, called the **cohomology class of** ω. If $[\omega] = [\omega']$ (that is, if ω and ω' differ by an exact form), we say that ω and ω' are **cohomologous**.

Proposition 17.2 (Induced Cohomology Maps). *For any smooth map $F: M \to$ N between smooth manifolds with or without boundary, the pullback $F^*: \Omega^p(N) \to$ $\Omega^p(M)$ carries $Z^p(N)$ into $Z^p(M)$ and $B^p(N)$ into $B^p(M)$. It thus descends to a linear map, still denoted by F^*, from $H_{dR}^p(N)$ to $H_{dR}^p(M)$, called the **induced cohomology map**. It has the following properties:*

(a) *If $G: N \to P$ is another smooth map, then*

$$(G \circ F)^* = F^* \circ G^*: H_{dR}^p(P) \to H_{dR}^p(M).$$

(b) *If Id denotes the identity map of M, then Id^* is the identity map of $H_{dR}^p(M)$.*

Proof. If ω is closed, then $d(F^*\omega) = F^*(d\omega) = 0$, so $F^*\omega$ is also closed. If $\omega = d\eta$ is exact, then $F^*\omega = F^*(d\eta) = d(F^*\eta)$, which is also exact. Therefore, F^* maps $Z^p(N)$ into $Z^p(M)$ and $B^p(N)$ into $B^p(M)$. The induced cohomology map $F^*: H_{dR}^p(N) \to H_{dR}^p(M)$ is defined in the obvious way: for a closed p-form ω, let

$$F^*[\omega] = [F^*\omega].$$

If $\omega' = \omega + d\eta$, then $[F^*\omega'] = [F^*\omega + d(F^*\eta)] = [F^*\omega]$, so this map is well defined. Properties (a) and (b) follow immediately from the analogous properties of the pullback map on forms. $\square$

The next two corollaries are immediate.

Corollary 17.3 (Functoriality). *For any integer p, the assignment $M \mapsto H_{dR}^p(M)$, $F \mapsto F^*$ is a contravariant functor from the category of smooth manifolds with boundary to the category of real vector spaces.* $\square$

Corollary 17.4 (Diffeomorphism Invariance of de Rham Cohomology). *Diffeomorphic smooth manifolds (with or without boundary) have isomorphic de Rham cohomology groups.* $\square$

Elementary Computations

The direct computation of the de Rham groups is not easy in general. However, there are a number of special cases that can be easily computed by various techniques. In this section, we describe a few of those cases. We begin with disjoint unions.

Proposition 17.5 (Cohomology of Disjoint Unions). *Let $\{M_j\}$ be a countable collection of smooth n-manifolds with or without boundary, and let $M = \coprod_j M_j$. For each p, the inclusion maps $\iota_j: M_j \hookrightarrow M$ induce an isomorphism from $H_{dR}^p(M)$ to the direct product space $\prod_j H_{dR}^p(M_j)$.*

Proof. The pullback maps $\iota_j^*: \Omega^p(M) \to \Omega^p(M_j)$ already induce an isomorphism from $\Omega^p(M)$ to $\prod_j \Omega^p(M_j)$, namely

$$\omega \mapsto (\iota_1^*\omega, \iota_2^*\omega, \dots) = (\omega|_{M_1}, \omega|_{M_2}, \dots).$$

This map is injective because any smooth p-form whose restriction to each M_j is zero must itself be zero, and it is surjective because giving an arbitrary smooth p-form on each M_j defines one on M. □

Because of this proposition, each de Rham group of a disconnected manifold is just the direct product of the corresponding groups of its components. Thus, we can concentrate henceforth on computing the de Rham groups of connected manifolds.

Next we give an explicit characterization of de Rham cohomology in degree zero.

Proposition 17.6 (Cohomology in Degree Zero). *If M is a connected smooth manifold with or without boundary, then $H^0_{dR}(M)$ is equal to the space of constant functions and is therefore 1-dimensional.*

Proof. Because there are no (-1)-forms, $\mathcal{B}^0(M) = 0$. A closed 0-form is a smooth real-valued function f such that $df = 0$, and since M is connected, this is true if and only if f is constant. Therefore, $H^0_{dR}(M) = \mathcal{Z}^0(M) = \{\text{constants}\}$. □

Corollary 17.7 (Cohomology of Zero-Manifolds). *Suppose M is a manifold of dimension 0. Then $H^0_{dR}(M)$ is a direct product of 1-dimensional vector spaces, one for each point of M, and all other de Rham cohomology groups of M are zero.*

Proof. The statement about $H^0_{dR}(M)$ follows from Propositions 17.5 and 17.6, and the cohomology groups in nonzero degrees vanish for dimensional reasons. □

Homotopy Invariance

In this section we present a profound generalization of Corollary 17.4, one surprising consequence of which is that the de Rham cohomology groups are actually *topological* invariants. In fact, they are something much more: they are **homotopy invariants**, which means that homotopy equivalent manifolds have isomorphic de Rham groups. (See p. 614 for the definition of homotopy equivalence.)

The underlying fact that allows us to prove the homotopy invariance of de Rham cohomology is that homotopic smooth maps induce the same cohomology map. To motivate the proof, suppose $F, G: M \to N$ are smooth maps, and let us think about what it means to prove that $F^* = G^*$. Given a closed p-form ω on N, we need somehow to produce a $(p-1)$-form η on M such that

$$G^*\omega - F^*\omega = d\eta,$$

from which it follows that $G^*[\omega] - F^*[\omega] = [d\eta] = 0$. One might hope to construct η in a systematic way, resulting in a map h from closed p-forms on N to $(p-1)$-forms on M that satisfies

$$d(h\omega) = G^*\omega - F^*\omega. \tag{17.2}$$

Instead of defining $h\omega$ only when ω is closed, it turns out to be far simpler to define a map h from the space of *all* smooth p-forms on N to the space of smooth

$(p-1)$-forms on M. Such a map cannot satisfy (17.2), but instead we will find a family of such maps, one for each p, satisfying

$$d(h\omega) + h(d\omega) = G^*\omega - F^*\omega. \tag{17.3}$$

This implies (17.2) when ω is closed.

In general, if $F, G \colon M \to N$ are smooth maps, a collection of linear maps $h \colon \Omega^p(N) \to \Omega^{p-1}(M)$ such that (17.3) is satisfied for all ω is called a **homotopy operator between F^* and G^***. (The term *cochain homotopy* is used frequently in the algebraic topology literature.) The next proposition follows immediately from the argument in the preceding paragraph.

Proposition 17.8. *Suppose M and N are smooth manifolds with or without boundary. If $F, G \colon M \to N$ are smooth maps and there exists a homotopy operator between the pullback maps F^* and G^*, then the induced cohomology maps $F^*, G^* \colon H^p_{\mathrm{dR}}(N) \to H^p_{\mathrm{dR}}(M)$ are equal.* $\qquad\square$

The key to our proof of homotopy invariance is to construct a homotopy operator first in the following special case. Let M be a smooth manifold with or without boundary, and for each $t \in I$, let $i_t \colon M \to M \times I$ be the map

$$i_t(x) = (x, t).$$

If M has empty boundary, then $M \times I$ is a smooth manifold with boundary, and all of the results above apply to it. But if $\partial M \neq \varnothing$, then $M \times I$ has to be considered as a smooth manifold with corners. It is straightforward to check that the definitions of the de Rham groups and induced homomorphisms make perfectly good sense on manifolds with corners, and Proposition 17.2 is valid in that context as well.

Lemma 17.9 (Existence of a Homotopy Operator). *For any smooth manifold M with or without boundary, there exists a homotopy operator between the two maps $i_0^*, i_1^* \colon \Omega^*(M \times I) \to \Omega^*(M)$.*

Proof. For each p, we need to define a linear map $h \colon \Omega^p(M \times I) \to \Omega^{p-1}(M)$ such that

$$h(d\omega) + d(h\omega) = i_1^*\omega - i_0^*\omega. \tag{17.4}$$

Let s denote the standard coordinate on $\mathbb{R}$, and let S be the vector field on $M \times \mathbb{R}$ given by $S_{(q,s)} = \left(0, \partial/\partial s|_s\right)$ under the usual identification $T_{(q,s)} M \leftrightarrow T_q M \times T_s \mathbb{R}$. Given a smooth p-form ω on $M \times I$, define $h\omega \in \Omega^{p-1}(M)$ by

$$h\omega = \int_0^1 i_t^*(S \lrcorner \omega)\, dt.$$

More specifically, for any $q \in M$, this means

$$(h\omega)_q = \int_0^1 i_t^*\left((S \lrcorner \omega)_{(q,t)}\right) dt,$$

where the integrand is interpreted as a function of t with values in the vector space $\Lambda^{p-1}(T_q^* M)$. On any smooth coordinate domain $U \subseteq M$, the components of the integrand are smooth functions of $(q,t) \in U \times I$, so the integral defines a smooth $(p-1)$-form on M. We can compute $d(h\omega)$ at any point by differentiating under the integral sign in local coordinates, which yields

$$d(h\omega) = \int_0^1 d\left(i_t^*(S \lrcorner \omega)\right) dt.$$

Therefore, using Cartan's magic formula, we obtain

$$h(d\omega) + d(h\omega) = \int_0^1 \left(i_t^*(S \lrcorner d\omega) + d\left(i_t^*(S \lrcorner \omega)\right)\right) dt$$

$$= \int_0^1 \left(i_t^*(S \lrcorner d\omega) + i_t^* d(S \lrcorner \omega)\right) dt$$

$$= \int_0^1 i_t^*(\mathcal{L}_S \omega)\, dt. \tag{17.5}$$

To simplify this last expression, we use the flow of S on $M \times \mathbb{R}$. (If M has nonempty boundary, note that S is tangent to $\partial(M \times \mathbb{R}) = \partial M \times \mathbb{R}$, so Theorem 9.34 applies.) The flow is given explicitly by $\theta_t(q,s) = (q, t+s)$, so S is complete. It follows that we can write $i_t = \theta_t \circ i_0$, and therefore by Proposition 12.36,

$$i_t^*(\mathcal{L}_S \omega) = i_0^*\left(\theta_t^*(\mathcal{L}_S \omega)\right) = i_0^*\left(\frac{d}{dt}\left(\theta_t^* \omega\right)\right) = \frac{d}{dt} i_0^*\left(\theta_t^* \omega\right) = \frac{d}{dt} i_t^* \omega.$$

Inserting this into (17.5) and applying the fundamental theorem of calculus, we obtain (17.4). $\qquad \square$

Proposition 17.10. *Suppose M and N are smooth manifolds with or without boundary, and $F, G \colon M \to N$ are homotopic smooth maps. For every p, the induced cohomology maps $F^*, G^* \colon H_{dR}^p(N) \to H_{dR}^p(M)$ are equal.*

Proof. The preceding lemma implies that the two cohomology maps i_0^* and i_1^* from $H_{dR}^p(M \times I)$ to $H_{dR}^p(M)$ are equal. By Theorem 9.28, there is a smooth homotopy $H \colon M \times I \to N$ from F to G. Because $F = H \circ i_0$ and $G = H \circ i_1$ (see Fig. 17.1), Proposition 17.2 implies

$$F^* = (H \circ i_0)^* = i_0^* \circ H^* = i_1^* \circ H^* = (H \circ i_1)^* = G^*. \qquad \square$$

The next theorem is the main result of this section.

Theorem 17.11 (Homotopy Invariance of de Rham Cohomology). *If M and N are homotopy equivalent smooth manifolds with or without boundary, then $H_{dR}^p(M) \cong H_{dR}^p(N)$ for each p. The isomorphisms are induced by any smooth homotopy equivalence $F \colon M \to N$.*

Proof. Suppose $F \colon M \to N$ is a homotopy equivalence, with homotopy inverse $G \colon N \to M$. By the Whitney approximation theorem (Theorem 6.26 or 9.27), there are smooth maps $\widetilde{F} \colon M \to N$ homotopic to F and $\widetilde{G} \colon N \to M$ homotopic to G.

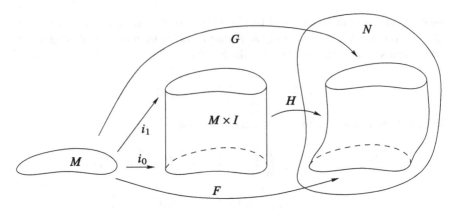

Fig. 17.1 Homotopic maps

Because homotopy is preserved by composition, it follows that $\widetilde{F} \circ \widetilde{G} \simeq F \circ G \simeq$ Id_N and $\widetilde{G} \circ \widetilde{F} \simeq G \circ F \simeq \mathrm{Id}_M$, so $\widetilde{F}$ and $\widetilde{G}$ are homotopy inverses of each other. Proposition 17.10 shows that, on cohomology,

$$\widetilde{F}^* \circ \widetilde{G}^* = \left(\widetilde{G} \circ \widetilde{F}\right)^* = (\mathrm{Id}_M)^* = \mathrm{Id}_{H^p_{\mathrm{dR}}(M)}.$$

The same argument shows that $\widetilde{G}^* \circ \widetilde{F}^*$ is also the identity, so $\widetilde{F}^* \colon H^p_{\mathrm{dR}}(N) \to$ $H^p_{\mathrm{dR}}(M)$ is an isomorphism.   □

Because every homeomorphism is a homotopy equivalence, the next corollary is immediate.

Corollary 17.12 (Topological Invariance of de Rham Cohomology). *The de Rham cohomology groups are topological invariants: if M and N are homeomorphic smooth manifolds with or without boundary, then their de Rham cohomology groups are isomorphic.*   □

This result is remarkable, because the definition of the de Rham groups of M is intimately tied up with its smooth structure, and we had no reason to expect that different differentiable structures on the same topological manifold should give rise to the same de Rham groups.

Computations Using Homotopy Invariance

We can use homotopy invariance to compute a number of de Rham groups. We begin with the simplest case of homotopy equivalence. A topological space X is said to be **contractible** if the identity map of X is homotopic to a constant map.

Theorem 17.13 (Cohomology of Contractible Manifolds). *If M is a contractible smooth manifold with or without boundary, then $H^p_{\mathrm{dR}}(M) = 0$ for $p \geq 1$.*

Proof. The assumption means there is some point $q \in M$ such that the identity map of M is homotopic to the constant map $c_q \colon M \to M$ sending all of M to q.

If $\iota_q \colon \{q\} \hookrightarrow M$ denotes the inclusion map, it follows that $c_q \circ \iota_q = \mathrm{Id}_{\{q\}}$ and $\iota_q \circ c_q \simeq \mathrm{Id}_M$, so ι_q is a homotopy equivalence. The result then follows from the homotopy invariance of H_{dR}^p together with the obvious fact that $H_{\mathrm{dR}}^p(\{q\}) = 0$ for $p \geq 1$ because $\{q\}$ is a 0-manifold. $\square$

In Theorem 11.49, we showed that every closed 1-form on a star-shaped open subset of $\mathbb{R}^n$ is exact. (Recall that a subset $U \subseteq \mathbb{R}^n$ is said to be *star-shaped* if there is a point $c \in U$ such that for every $x \in U$, the line segment from c to x is entirely contained in U.) The next theorem is a generalization of that result to forms of all degrees. Despite the apparent specialness of star-shaped domains, this theorem is one of the most important facts about de Rham cohomology.

Theorem 17.14 (The Poincaré Lemma). *If U is a star-shaped open subset of $\mathbb{R}^n$ or $\mathbb{H}^n$, then $H_{\mathrm{dR}}^p(U) = 0$ for $p \geq 1$.*

Proof. If U is star-shaped with respect to c, then it is contractible by the following straight-line homotopy:

$$H(x,t) = c + t(x - c).$$ $\square$

Corollary 17.15 (Local Exactness of Closed Forms). *Let M be a smooth manifold with or without boundary. Each point of M has a neighborhood on which every closed form is exact.*

Proof. Every point of M has a neighborhood diffeomorphic to an open ball in $\mathbb{R}^n$ or an open half-ball in $\mathbb{H}^n$, each of which is star-shaped. The result follows from the Poincaré lemma and the diffeomorphism invariance of de Rham cohomology. $\square$

Corollary 17.16 (Cohomology of Euclidean Spaces and Half-Spaces). *For any integers $n \geq 0$ and $p \geq 1$, $H_{\mathrm{dR}}^p(\mathbb{R}^n) = 0$ and $H_{\mathrm{dR}}^p(\mathbb{H}^n) = 0$.*

Proof. Both $\mathbb{R}^n$ and $\mathbb{H}^n$ are star-shaped. $\square$

Another case in which we can say quite a lot about de Rham cohomology is in degree 1. Suppose M is a connected smooth manifold and q is any point in M. Let $\mathrm{Hom}\big(\pi_1(M,q), \mathbb{R}\big)$ denote the set of group homomorphisms from $\pi_1(M,q)$ to the additive group $\mathbb{R}$; it is a vector space under pointwise addition of homomorphisms and multiplication by constants. We define a linear map $\Phi \colon H_{\mathrm{dR}}^1(M) \to \mathrm{Hom}\big(\pi_1(M,q), \mathbb{R}\big)$ as follows: given a cohomology class $[\omega] \in H_{\mathrm{dR}}^1(M)$, define $\Phi[\omega] \colon \pi_1(M,q) \to \mathbb{R}$ by

$$\Phi[\omega][\gamma] = \int_{\widetilde{\gamma}} \omega,$$

where $[\gamma]$ is any path homotopy class in $\pi_1(M,q)$, and $\widetilde{\gamma}$ is any piecewise smooth curve representing the same path class.

Theorem 17.17 (First Cohomology and the Fundamental Group). *Suppose M is a connected smooth manifold. For each $q \in M$, the linear map $\Phi \colon H_{\mathrm{dR}}^1(M) \to \mathrm{Hom}\big(\pi_1(M,q), \mathbb{R}\big)$ is well defined and injective.*

Remark. It is actually the case that Φ is an isomorphism, but we do not quite have the tools to prove this. See Problem 18-2.

Proof. Given $[\gamma] \in \pi_1(M, q)$, it follows from the Whitney approximation theorem that there is some smooth closed curve segment $\widetilde{\gamma}$ in the same path class as γ, and from Theorem 16.26 that $\int_{\widetilde{\gamma}} \omega$ gives the same result for every piecewise smooth curve $\widetilde{\gamma}$ in the given class. Moreover, if $\widetilde{\omega}$ is another smooth 1-form in the same cohomology class as ω, then $\widetilde{\omega} - \omega = df$ for some smooth function f, which implies

$$\int_{\widetilde{\gamma}} \widetilde{\omega} - \int_{\widetilde{\gamma}} \omega = \int_{\widetilde{\gamma}} df = f(q) - f(q) = 0.$$

Thus Φ is well defined. It follows from Proposition 11.34(c) that $\Phi[\omega]$ is a group homomorphism from $\pi_1(M, q)$ to $\mathbb{R}$, and from linearity of the line integral that Φ itself is a linear map.

To see that Φ is injective, suppose $\Phi[\omega]$ is the zero homomorphism. This means that $\int_{\widetilde{\gamma}} \omega = 0$ for every piecewise smooth closed curve $\widetilde{\gamma}$ starting at q. If σ is a piecewise smooth closed curve starting at some other point $q' \in M$, we can choose a piecewise smooth curve α from q to q', so that the path product $\alpha \cdot \sigma \cdot \overline{\alpha}$ is a closed curve based at q, where $\overline{\alpha}$ is a backward reparametrization of α. It then follows that

$$0 = \int_{\alpha \cdot \sigma \cdot \overline{\alpha}} \omega = \int_{\alpha} \omega + \int_{\sigma} \omega - \int_{\alpha} \omega = \int_{\sigma} \omega.$$

Thus, ω is conservative and therefore exact.   □

It follows from Corollary 16.27 that $H^1_{dR}(M) = 0$ when M is simply connected. The next corollary generalizes that result.

Corollary 17.18. *If M is a connected smooth manifold with finite fundamental group, then $H^1_{dR}(M) = 0$.*

Proof. There are no nontrivial homomorphisms from a finite group to $\mathbb{R}$.   □

▶ **Exercise 17.19.** A group Γ is called a ***torsion group*** if for each $g \in \Gamma$ there exists an integer k such that $g^k = 1$. Show that if M is a connected smooth manifold whose fundamental group is a torsion group, then $H^1_{dR}(M) = 0$.

The Mayer–Vietoris Theorem

In this section we state a general theorem that can be used to compute the de Rham cohomology groups of many manifolds, by expressing them as unions of open submanifolds with simpler cohomology. We use the theorem here to compute all of the de Rham cohomology groups of spheres and of punctured Euclidean spaces, and the top-degree cohomology groups of compact manifolds. In the next chapter, we will use it again as an essential ingredient in the proof of the de Rham theorem. Because the proof of the Mayer–Vietoris theorem is fairly technical, we defer it to the end of the chapter.

Here is the setup for the theorem. Suppose M is a smooth manifold with or without boundary, and U, V are open subsets of M such that $M = U \cup V$. We have

four inclusions,

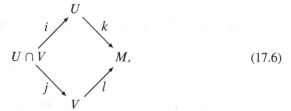

$$(17.6)$$

which induce pullback maps on differential forms,

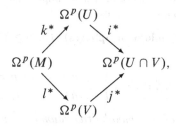

as well as corresponding induced cohomology maps. Note that these pullback maps are really just restrictions: for example, $k^*\omega = \omega|_U$. Consider the following sequence of maps:

$$0 \to \Omega^p(M) \xrightarrow{k^* \oplus l^*} \Omega^p(U) \oplus \Omega^p(V) \xrightarrow{i^* - j^*} \Omega^p(U \cap V) \to 0, \qquad (17.7)$$

where

$$(k^* \oplus l^*)\omega = (k^*\omega, l^*\omega),$$
$$(i^* - j^*)(\omega, \eta) = i^*\omega - j^*\eta. \qquad (17.8)$$

Because pullbacks commute with d, these maps descend to linear maps on the corresponding de Rham cohomology groups.

In the statement of the Mayer–Vietoris theorem, we will use the following standard algebraic terminology. Suppose we are given a sequence of vector spaces and linear maps:

$$\cdots \to V^{p-1} \xrightarrow{F_{p-1}} V^p \xrightarrow{F_p} V^{p+1} \xrightarrow{F_{p+1}} V^{p+2} \to \cdots . \qquad (17.9)$$

Such a sequence is said to be **exact** if the image of each map is equal to the kernel of the next: for each p,

$$\operatorname{Im} F_{p-1} = \operatorname{Ker} F_p.$$

Theorem 17.20 (Mayer–Vietoris). *Let M be a smooth manifold with or without boundary, and let U, V be open subsets of M whose union is M. For each p, there is a linear map $\delta \colon H^p_{\mathrm{dR}}(U \cap V) \to H^{p+1}_{\mathrm{dR}}(M)$ such that the following sequence, called the Mayer–Vietoris sequence for the open cover $\{U, V\}$, is exact:*

$$\cdots \xrightarrow{\delta} H^p_{dR}(M) \xrightarrow{k^* \oplus l^*} H^p_{dR}(U) \oplus H^p_{dR}(V) \xrightarrow{i^* - j^*} H^p_{dR}(U \cap V)$$

$$\xrightarrow{\delta} H^{p+1}_{dR}(M) \xrightarrow{k^* \oplus l^*} \cdots . \quad (17.10)$$

Computations Using the Mayer–Vietoris Theorem

Using the Mayer–Vietoris theorem, it is a simple matter to compute all of the de Rham cohomology groups of spheres.

Theorem 17.21 (Cohomology of Spheres). *For $n \geq 1$, the de Rham cohomology groups of $\mathbb{S}^n$ are*

$$H^p_{dR}(\mathbb{S}^n) \cong \begin{cases} \mathbb{R} & \text{if } p = 0 \text{ or } p = n, \\ 0 & \text{if } 0 < p < n. \end{cases} \quad (17.11)$$

The cohomology class of any smooth orientation form is a basis for $H^n_{dR}(\mathbb{S}^n)$.

Proof. Proposition 17.6 shows that $H^0_{dR}(\mathbb{S}^n) \cong \mathbb{R}$, so we need only prove (17.11) for $p \geq 1$. We do so by induction on n. For $n = 1$, note first that any orientation form on $\mathbb{S}^1$ has nonzero integral, so it is not exact by Corollary 16.13; thus $\dim H^1_{dR}(\mathbb{S}^1) \geq 1$. On the other hand, Theorem 17.17 implies that there is an injective linear map from $H^1_{dR}(\mathbb{S}^n)$ into $\mathrm{Hom}(\pi_1(\mathbb{S}^1, 1), \mathbb{R})$, which is 1-dimensional. Thus, $H^1_{dR}(\mathbb{S}^1)$ has dimension exactly 1, and is spanned by the cohomology class of any orientation form.

Next, suppose $n \geq 2$ and assume by induction that the theorem is true for $\mathbb{S}^{n-1}$. Because $\mathbb{S}^n$ is simply connected, $H^1_{dR}(\mathbb{S}^n) = 0$ by Corollary 17.18. For $p > 1$, we use the Mayer–Vietoris theorem as follows. Let N and S be the north and south poles in $\mathbb{S}^n$, respectively, and let $U = \mathbb{S}^n \smallsetminus \{S\}$, $V = \mathbb{S}^n \smallsetminus \{N\}$. By stereographic projection (Problem 1-7), both U and V are diffeomorphic to $\mathbb{R}^n$ (Fig. 17.2), and thus $U \cap V$ is diffeomorphic to $\mathbb{R}^n \smallsetminus \{0\}$.

Part of the Mayer–Vietoris sequence for $\{U, V\}$ reads

$$H^{p-1}_{dR}(U) \oplus H^{p-1}_{dR}(V) \to H^{p-1}_{dR}(U \cap V) \to H^p_{dR}(\mathbb{S}^n) \to H^p_{dR}(U) \oplus H^p_{dR}(V).$$

Because U and V are diffeomorphic to $\mathbb{R}^n$, the groups on both ends are trivial when $p > 1$, which implies that $H^p_{dR}(\mathbb{S}^n) \cong H^{p-1}_{dR}(U \cap V)$. Moreover, $U \cap V$ is diffeomorphic to $\mathbb{R}^n \smallsetminus \{0\}$ and therefore homotopy equivalent to $\mathbb{S}^{n-1}$, so in the end we conclude that $H^p_{dR}(\mathbb{S}^n) \cong H^{p-1}_{dR}(\mathbb{S}^{n-1})$ for $p > 1$, and (17.11) follows by induction. As in the $n = 1$ case, any smooth orientation form on $\mathbb{S}^n$ determines a nonzero cohomology class, which therefore spans $H^n_{dR}(\mathbb{S}^n)$. $\square$

▶ **Exercise 17.22.** Show that $\eta \in \Omega^n(\mathbb{S}^n)$ is exact if and only if $\int_{\mathbb{S}^n} \eta = 0$.

Corollary 17.23 (Cohomology of Punctured Euclidean Space). *Suppose $n \geq 2$ and $x \in \mathbb{R}^n$, and let $M = \mathbb{R}^n \smallsetminus \{x\}$. The only nontrivial de Rham groups of M are*

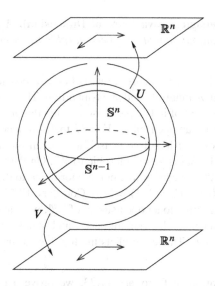

Fig. 17.2 Computing the de Rham cohomology of $\mathbb{S}^n$

$H_{dR}^0(M)$ and $H_{dR}^{n-1}(M)$, *both of which are* 1-*dimensional. A closed* $(n-1)$-*form* η *on* M *is exact if and only if* $\int_S \eta = 0$ *for some* (*and hence every*) $(n-1)$-*dimensional sphere* $S \subseteq M$ *centered at* x.

Proof. Let $S \subseteq M$ be any $(n-1)$-dimensional sphere centered at x. Because inclusion $\iota\colon S \hookrightarrow M$ is a homotopy equivalence, $\iota^*\colon H_{dR}^p(M) \to H_{dR}^p(S)$ is an isomorphism for each p, so the assertion about the dimension of $H_{dR}^p(M)$ follows from Theorem 17.21. If η is a closed $(n-1)$-form on M, it follows that η is exact if and only if $\iota^*\eta$ is exact on S, which in turn is true if and only if $\int_S \eta = \int_S \iota^*\eta = 0$ by Exercise 17.22. $\qquad\square$

▶ **Exercise 17.24.** Check that the statement and proof of Corollary 17.23 remain true if $\mathbb{R}^n \smallsetminus \{x\}$ is replaced by $\mathbb{R}^n \smallsetminus \bar{B}$ for some closed ball $\bar{B} \subseteq \mathbb{R}^n$.

Corollary 17.25. *Suppose* $n \geq 2$, $U \subseteq \mathbb{R}^n$ *is any open subset, and* $x \in U$. *Then* $H_{dR}^{n-1}(U \smallsetminus \{x\}) \neq 0$.

Proof. Because U is open, there is an $(n-1)$-dimensional sphere S centered at x such that $S \subseteq U \smallsetminus \{x\}$. Let $\iota\colon S \hookrightarrow U \smallsetminus \{x\}$ be inclusion and $r\colon U \smallsetminus \{x\} \to S$ be the radial projection onto S. Then r and ι are smooth with $r \circ \iota = \mathrm{Id}_S$. This implies $\iota^* \circ r^* = \mathrm{Id}_{H_{dR}^{n-1}(S)}$, and therefore $r^*\colon H_{dR}^{n-1}(S) \to H_{dR}^{n-1}(U \smallsetminus \{x\})$ is injective. Since $H_{dR}^{n-1}(S) \neq 0$ by Theorem 17.21, the result follows. $\qquad\square$

Here is an important application of the topological invariance of the de Rham cohomology groups. Recall the theorem on invariance of dimension (Theorem 1.2); it is a surprising fact that this purely topological theorem can be proved using de Rham cohomology. Before proving the theorem, we restate it here for convenience.

Theorem 17.26 (Topological Invariance of Dimension). *A nonempty n-dimensional topological manifold cannot be homeomorphic to an m-dimensional manifold unless m = n.*

Proof. If M is a topological n-manifold that is homeomorphic to an m-manifold, then M is itself both an n-manifold and an m-manifold. The case in which m or n is zero was already taken care of in Chapter 1, so assume that $m > n \geq 1$. Because M is an m-manifold, there is an open subset $V \subseteq M$ that is homeomorphic to $\mathbb{R}^m$. Because an open subset of an n-manifold is itself an n-manifold, any point $x \in V$ has a neighborhood $U \subseteq V$ that is homeomorphic to $\mathbb{R}^n$. On the one hand, because U is homeomorphic to $\mathbb{R}^n$, we can use the homeomorphism to define a smooth structure on U, and then $H_{\mathrm{dR}}^{m-1}(U \smallsetminus \{x\}) = 0$ by Corollary 17.23. On the other hand, because U is homeomorphic to an open subset of $\mathbb{R}^m$, we can use that homeomorphism to define another smooth structure on U, and then Corollary 17.25 implies that $H_{\mathrm{dR}}^{m-1}(U \smallsetminus \{x\}) \neq 0$. This contradicts the topological invariance of de Rham cohomology. $\qquad\square$

As another application of Corollary 17.23, we prove a generalization of the Poincaré lemma for compactly supported forms. We will use it below to compute top-degree cohomology groups.

Lemma 17.27 (Poincaré Lemma with Compact Support). *Let $n \geq p \geq 1$, and suppose ω is a compactly supported closed p-form on $\mathbb{R}^n$. If $p = n$, suppose in addition that $\int_{\mathbb{R}^n} \omega = 0$. Then there exists a compactly supported smooth $(p-1)$-form η on $\mathbb{R}^n$ such that $d\eta = \omega$.*

Remark. Of course, we know that ω is exact by the Poincaré lemma, so the novelty here is the claim that it is the exterior derivative of a *compactly supported* form.

Proof. When $n = p = 1$, we can write $\omega = f\,dx$ for some smooth, compactly supported function $f \in C^\infty(\mathbb{R})$. Define $F \colon \mathbb{R} \to \mathbb{R}$ by

$$F(x) = \int_{-\infty}^{x} f(t)\,dt.$$

By the fundamental theorem of calculus, $dF = F'\,dx = f\,dx = \omega$. Choose $R > 0$ such that $\operatorname{supp} f \subseteq [-R, R]$. When $x < -R$, $F(x) = 0$ by our choice of R. When $x > R$, the fact that $\int_{\mathbb{R}} \omega = 0$ translates to

$$F(x) = \int_{-\infty}^{x} f(t)\,dt = \int_{-\infty}^{\infty} f(t)\,dt = 0,$$

so, in fact, $\operatorname{supp} F \subseteq [-R, R]$. This completes the proof for the case $n = p = 1$.

Now assume $n \geq 2$, and let $B, B' \subseteq \mathbb{R}^n$ be open balls centered at the origin such that $\operatorname{supp}\omega \subseteq B \subseteq \bar{B} \subseteq B'$. By the ordinary Poincaré lemma, there exists a smooth (but not necessarily compactly supported) $(p-1)$-form η_0 on $\mathbb{R}^n$ such that $d\eta_0 = \omega$. This implies, in particular, that $d\eta_0 = 0$ on $\mathbb{R}^n \smallsetminus \bar{B}$. To complete the proof, we consider three cases.

CASE 1: $p = 1$. In this case η_0 is a smooth function. Because $\mathbb{R}^n \smallsetminus \bar{B}$ is connected when $n \geq 2$, it follows that η_0 is equal to a constant c there. Letting $\eta = \eta_0 - c$, we find that η is compactly supported and satisfies $d\eta = \omega$ as claimed.

CASE 2: $1 < p < n$. Now the restriction of η_0 to $\mathbb{R}^n \smallsetminus \bar{B}$ is a closed $(p-1)$-form. Because $H_{\mathrm{dR}}^{p-1}\left(\mathbb{R}^n \smallsetminus \bar{B}\right) = 0$ by Exercise 17.24, there is a smooth $(p-2)$-form γ on $\mathbb{R}^n \smallsetminus \bar{B}$ such that $d\gamma = \eta_0$ there. If we let ψ be a smooth bump function that is supported in $\mathbb{R}^n \smallsetminus \bar{B}$ and equal to 1 on $\mathbb{R}^n \smallsetminus \bar{B}'$, then $\eta = \eta_0 - d(\psi\gamma)$ is smooth on all of $\mathbb{R}^n$ and satisfies $d\eta = d\eta_0 = \omega$. Because $d(\psi\gamma) = d\gamma = \eta_0$ on $\mathbb{R}^n \smallsetminus \bar{B}'$, η is compactly supported.

CASE 3: $p = n$. In this case, we cannot use the same argument as in Case 2 because $H_{\mathrm{dR}}^{n-1}\left(\mathbb{R}^n \smallsetminus \bar{B}\right) \neq 0$. However, it follows from Corollary 17.23 and Exercise 17.24 that the restriction of η_0 to $\mathbb{R}^n \smallsetminus \bar{B}$ is exact provided its integral is zero over some sphere centered at the origin and contained in $\mathbb{R}^n \smallsetminus \bar{B}$. Stokes's theorem implies that

$$0 = \int_{\mathbb{R}^n} \omega = \int_{\bar{B}} \omega = \int_{\bar{B}'} d\eta_0 = \int_{\partial B'} \eta_0.$$

Thus η_0 is exact on $\mathbb{R}^n \smallsetminus \bar{B}$, and the proof proceeds exactly as in Case 2. $\qquad\square$

For some purposes it is useful to define a generalization of the de Rham cohomology groups using only compactly supported forms. Let M be a smooth manifold with or without boundary and let $\Omega_c^p(M)$ denote the vector space of compactly supported smooth p-forms on M. The **pth compactly supported de Rham cohomology group of M** is the quotient space

$$H_c^p(M) = \frac{\mathrm{Ker}\left(d : \Omega_c^p(M) \to \Omega_c^{p+1}(M)\right)}{\mathrm{Im}\left(d : \Omega_c^{p-1}(M) \to \Omega_c^p(M)\right)}.$$

Of course, when M is compact, this just reduces to ordinary de Rham cohomology. But for noncompact manifolds the two groups can be different, as the next theorem illustrates.

Theorem 17.28 (Compactly Supported Cohomology of $\mathbb{R}^n$). *For $n \geq 1$, the compactly supported de Rham cohomology groups of $\mathbb{R}^n$ are*

$$H_c^p\left(\mathbb{R}^n\right) \cong \begin{cases} 0 & \text{if } 0 \leq p < n, \\ \mathbb{R} & \text{if } p = n. \end{cases}$$

▶ **Exercise 17.29.** Prove this theorem.

In general, a smooth map need not pull back compactly supported forms to compactly supported ones, so it does not induce a map on compactly supported cohomology. However, a *proper* map does pull back compactly supported forms to compactly supported ones, so for a proper smooth map $F : M \to N$ there is an induced cohomology map $F^* : H_c^p(N) \to H_c^p(M)$ for each p.

Compactly supported cohomology has a number of important applications in algebraic topology. One important application is the Poincaré duality theorem, which is outlined in Problem 18-7.

Another application is to facilitate the computation of de Rham cohomology in the top degree. Suppose first that M is an oriented smooth n-manifold. There is a natural linear map $I\colon \Omega_c^n(M) \to \mathbb{R}$ given by integration over M:

$$I(\omega) = \int_M \omega.$$

Because the integral of the exterior derivative of a compactly supported $(n-1)$-form is zero, I descends to a linear map, still denoted by the same symbol, from $H_c^n(M)$ to $\mathbb{R}$. (Note that every smooth n-form on an n-manifold is closed.)

Theorem 17.30 (Top Cohomology, Orientable Compact Support Case). *If M is a connected oriented smooth n-manifold, then the integration map $I\colon H_c^n(M) \to \mathbb{R}$ is an isomorphism, so $H_c^n(M)$ is 1-dimensional.*

Proof. Because a connected 0-manifold is a single point, the 0-dimensional case is an immediate consequence of Corollary 17.7, so we may assume $n \geq 1$. Let $\bigl(U, (x^i)\bigr)$ be an oriented smooth coordinate chart on M, and let f be a smooth bump function with compact support in U. Then the n-form defined by $\theta_0 = f\, dx^1 \wedge \cdots \wedge dx^n$ in U and 0 outside U is smooth and compactly supported on M, and satisfies $I(\theta_0) > 0$. Thus, I is surjective, so we need only show that it is injective. In other words, we have to show the following: if ω is a smooth, compactly supported n-form on M satisfying $\int_M \omega = 0$, then there is a smooth, compactly supported $(n-1)$-form η such that $\omega = d\eta$.

Let $\{U_i\}$ be a countable cover of M by open subsets that are diffeomorphic to $\mathbb{R}^n$, and let $M_k = U_1 \cup \cdots \cup U_k$ for each k. Because M is connected, by renumbering the sequence if necessary, we can arrange the $M_k \cap U_{k+1} \neq \varnothing$ for each k. Since every compactly supported n-form is supported in M_k for some finite k, it suffices to prove that if $\omega \in \Omega_c^n(M_k)$ has zero integral, then $\omega = d\eta$ for some $\eta \in \Omega_c^{n-1}(M_k)$, and then we can extend η by zero to a compactly supported form on all of M. We will prove this claim by induction on k.

For $k = 1$, since $M_1 = U_1$ is diffeomorphic to $\mathbb{R}^n$, the claim reduces to Lemma 17.27. So assume that the claim is true for some $k \geq 1$, and suppose ω is a compactly supported smooth n-form on $M_{k+1} = M_k \cup U_{k+1}$ that satisfies $\int_{M_{k+1}} \omega = 0$.

Let $\theta \in \Omega_c^n(M_{k+1})$ be an auxiliary form that is supported in $M_k \cap U_{k+1}$ and satisfies $\int_{M_{k+1}} \theta = 1$. (Such a form is easily constructed by using a bump function in coordinates as above.) Let $\{\varphi, \psi\}$ be a smooth partition of unity for M_{k+1} subordinate to the cover $\{M_k, U_{k+1}\}$ (Fig. 17.3), and let $c = \int_{M_{k+1}} \varphi\omega$. Observe that $\varphi\omega - c\theta$ is compactly supported in M_k, and its integral is equal to zero by our choice of c. Therefore, by the induction hypothesis, there is a compactly supported smooth $(n-1)$-form α on M_k such that $d\alpha = \varphi\omega - c\theta$. Similarly, $\psi\omega + c\theta$ is compactly

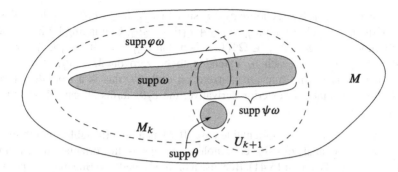

Fig. 17.3 Computing the top-degree cohomology

supported in U_{k+1}, and its integral is

$$\int_{U_{k+1}} (\psi\omega + c\theta) = \int_{M_{k+1}} (1-\varphi)\omega + c\int_{M_{k+1}} \theta$$

$$= \int_{M_{k+1}} \omega - \int_{M_{k+1}} \varphi\omega + c = 0.$$

Thus by Lemma 17.27, there exists another smooth $(n-1)$-form β, compactly supported in U_{k+1}, such that $d\beta = \psi\omega + c\theta$. Both α and β can be extended by zero to smooth compactly supported forms on M_{k+1}. We compute

$$d(\alpha + \beta) = (\varphi\omega - c\theta) + (\psi\omega + c\theta) = (\varphi + \psi)\omega = \omega,$$

which completes the inductive step. $\qquad\square$

Theorem 17.31 (Top Cohomology, Orientable Compact Case). *If M is a compact connected orientable smooth n-manifold, then $H_{dR}^n(M)$ is 1-dimensional, and is spanned by the cohomology class of any smooth orientation form.*

Proof. This follows from the preceding theorem, because $H_{dR}^p(M) = H_c^p(M)$ in that case, and the integral of any orientation form is nonzero. $\qquad\square$

Theorem 17.32 (Top Cohomology, Orientable Noncompact Case). *If M is a noncompact connected orientable smooth n-manifold, then $H_{dR}^n(M) = 0$.*

Proof. Choose an orientation on M. Let $f \in C^\infty(M)$ be a smooth exhaustion function. By adding a constant, we can arrange that $\inf_M f = 0$, and then connectedness and noncompactness of M imply that $f(M) = [0, \infty)$. For each positive integer i, let $V_i = f^{-1}\big((i-2, i)\big)$. Thus, $\{V_i\}_{i=1}^\infty$ is a cover of M by nonempty precompact open sets, with $V_i \cap V_j \neq \varnothing$ if and only if $j = i-1$, i, or $i+1$. Let $\{\psi_i\}$ be a smooth partition of unity subordinate to this cover, and for each i, let $\theta_i \in \Omega_c^n(M)$ be a smooth n-form compactly supported in $V_i \cap V_{i+1}$ with $\int_M \theta_i = 1$.

Suppose ω is any smooth n-form on M, and let $\omega_i = \psi_i\omega$ for each i, so $\omega_i \in \Omega_c^n(V_i)$. Let $c_1 = \int_{V_1} \omega_1$, so that $\omega_1 - c_1\theta_1$ is compactly supported in V_1 and has zero integral. It follows from Theorem 17.30 that there exists $\eta_1 \in \Omega_c^n(V_1)$ such that $d\eta_1 = \omega_1 - c_1\theta_1$. Next, choose $c_2 \in \mathbb{R}$ such that $\int_{V_2}(\omega_2 + c_1\theta_1 - c_2\theta_2) = 0$, so there exists $\eta_2 \in \Omega_c^n(V_2)$ with $d\eta_2 = \omega_2 + c_1\theta_1 - c_2\theta_2$. Continuing by induction, we can choose $c_j \in \mathbb{R}$ and $\eta_j \in \Omega_c^n(V_j)$ such that $d\eta_j = \omega_j + c_{j-1}\theta_{j-1} - c_j\theta_j$. Set $\eta = \sum_{j=1}^\infty \eta_j$, with each η_j extended to be zero on $M \smallsetminus V_j$. Because at most three terms in this sum are nonzero on each V_i, this is a smooth n-form on M. When we take its exterior derivative, the $c_j\theta_j$ terms all cancel, so $d\eta = \sum_j \omega_j = \omega$. $\qquad\square$

Next we consider the nonorientable case. If M is a nonorientable smooth manifold, the key to analyzing its cohomology groups is the orientation covering $\hat{\pi}\colon \widehat{M} \to M$ (see Theorem 15.41). Because a finite-sheeted covering map is a proper map by Exercise A.75, $\hat{\pi}$ induces cohomology maps on both compactly supported and ordinary de Rham cohomology. The next lemma shows that these maps are all injective.

Lemma 17.33. *Suppose M is a connected nonorientable smooth manifold and $\hat{\pi}\colon \widehat{M} \to M$ is its orientation covering. For each p, the induced cohomology maps $\hat{\pi}^*\colon H_{dR}^p(M) \to H_{dR}^p(\widehat{M})$ and $\hat{\pi}^*\colon H_c^p(M) \to H_c^p(\widehat{M})$ are injective.*

Proof. First, we prove the lemma for compactly supported cohomology. Suppose ω is a closed, compactly supported p-form on M such that $\hat{\pi}^*[\omega] = 0 \in H_c^p(\widehat{M})$. Then there exists $\eta \in \Omega_c^p(\widehat{M})$ such that $d\eta = \hat{\pi}^*\omega$. Let $\alpha\colon \widehat{M} \to \widehat{M}$ be the unique nontrivial covering automorphism of $\widehat{M}$ (see Fig. 15.9), and let $\tilde{\eta} = \frac{1}{2}(\eta + \alpha^*\eta)$, which is also compactly supported. Using the fact that $\alpha \circ \alpha = \mathrm{Id}_{\widehat{M}}$, we compute

$$\alpha^*\tilde{\eta} = \frac{1}{2}(\alpha^*\eta + (\alpha \circ \alpha)^*\eta) = \tilde{\eta}.$$

Because $\hat{\pi} \circ \alpha = \hat{\pi}$, this implies

$$d\tilde{\eta} = \frac{1}{2}(d\eta + d\alpha^*\eta) = \frac{1}{2}(d\eta + \alpha^* d\eta) = \frac{1}{2}(\hat{\pi}^*\omega + \alpha^*\hat{\pi}^*\omega) = \hat{\pi}^*\omega.$$

Let $U \subseteq M$ be any evenly covered open subset. There are exactly two smooth local sections $\sigma_1, \sigma_2\colon U \to \widehat{M}$ over U, which are related by $\sigma_2 = \alpha \circ \sigma_1$. Observe that

$$\sigma_2^*\tilde{\eta} = (\alpha \circ \sigma_1)^*\tilde{\eta} = \sigma_1^*\alpha^*\tilde{\eta} = \sigma_1^*\tilde{\eta}.$$

Therefore, we can define a smooth global $(p-1)$-form β on M by setting $\beta|_U = \sigma^*\tilde{\eta}$ for any smooth local section $\sigma\colon U \to \widehat{M}$; the argument above guarantees that the various definitions agree where they overlap. Because $\operatorname{supp}\beta = \hat{\pi}(\operatorname{supp}\tilde{\eta})$, it follows that β is compactly supported. To determine the exterior derivative of β, given $p \in M$, choose a smooth local section σ defined on a neighborhood U of p, and compute

$$d\beta = d\sigma^*\tilde{\eta} = \sigma^* d\tilde{\eta} = \sigma^*\hat{\pi}^*\omega = (\hat{\pi} \circ \sigma)^*\omega = \omega,$$

because $\hat{\pi} \circ \sigma = \mathrm{Id}_U$.

The argument for ordinary de Rham cohomology is the same, but with all references to compact support deleted. $\qquad\square$

Theorem 17.34 (Top Cohomology, Nonorientable Case). *If M is a connected nonorientable smooth n-manifold, then $H_c^n(M) = 0$ and $H_{dR}^n(M) = 0$.*

Proof. First consider the case of compactly supported cohomology. By the preceding lemma, it suffices to show that $\widehat{\pi}^*\colon H_c^n(M) \to H_c^n\big(\widehat{M}\big)$ is the zero map, where $\widehat{\pi}\colon \widehat{M} \to M$ is the orientation covering of M. Let $\alpha\colon \widehat{M} \to \widehat{M}$ be the nontrivial covering automorphism as in the preceding proof. Now, α cannot be orientation-preserving: if it were, the entire covering automorphism group $\{\mathrm{Id}_{\widehat{M}}, \alpha\}$ would be orientation-preserving, and then M would be orientable by Theorem 15.36. By connectedness of $\widehat{M}$ and the fact that α is a diffeomorphism, it follows that α is orientation-reversing.

Suppose ω is any compactly supported smooth n-form on M, and let $\widehat{\omega} = \widehat{\pi}^*\omega$. Because $\widehat{\pi}$ is proper, $\widehat{\omega}$ is compactly supported, and $\widehat{\pi} \circ \alpha = \widehat{\pi}$ implies

$$\alpha^*\widehat{\omega} = \alpha^*\widehat{\pi}^*\omega = (\widehat{\pi} \circ \alpha)^*\omega = \widehat{\pi}^*\omega = \widehat{\omega}.$$

Because α is orientation-reversing, we conclude from Proposition 16.6(d) that

$$\int_{\widehat{M}} \widehat{\omega} = -\int_{\widehat{M}} \alpha^*\widehat{\omega} = -\int_{\widehat{M}} \widehat{\omega}.$$

This implies that $\int_{\widehat{M}} \widehat{\omega} = 0$, so $[\widehat{\omega}] = 0 \in H_c^n\big(\widehat{M}\big)$ by Theorem 17.31. This completes the proof that $H_c^n(M) = 0$.

It remains only to handle ordinary cohomology. If M is compact, it follows from the argument above that $H_{dR}^n(M) = H_c^n(M) = 0$. On the other hand, if M is noncompact, then so is $\widehat{M}$, and Theorem 17.32 shows that $H_{dR}^n\big(\widehat{M}\big) = 0$. It follows from Lemma 17.33 that $H_{dR}^n(M) = 0$ as well. $\qquad\square$

Degree Theory

Now that we know the top-degree cohomology groups of all compact smooth manifolds, we can use them to draw a number of significant conclusions about smooth maps between certain compact manifolds of the same dimension. They all follow from the fact that we can associate an integer to each such map, called its *degree*, in such a way that homotopic maps have the same degree.

Theorem 17.35 (Degree of a Smooth Map). *Suppose M and N are compact, connected, oriented, smooth manifolds of dimension n, and $F\colon M \to N$ is a smooth map. There exists a unique integer k, called the **degree of F**, that satisfies both of the following conditions.*

(a) *For every smooth n-form ω on N,*

$$\int_M F^*\omega = k \int_N \omega.$$

(b) *If $q \in N$ is a regular value of F, then*

$$k = \sum_{x \in F^{-1}(q)} \mathrm{sgn}(x),$$

where $\mathrm{sgn}(x) = +1$ if dF_x is orientation-preserving, and -1 if it is orientation-reversing.

Proof. By Theorem 17.31, two smooth n-forms on either M or N are cohomologous if and only if they have the same integral. Let θ be any smooth n-form on N such that $\int_N \theta = 1$, and let $k = \int_M F^*\theta$. If $\omega \in \Omega^n(N)$ is arbitrary, then ω is cohomologous to $a\theta$, where $a = \int_N \omega$, and therefore $F^*\omega$ is cohomologous to $aF^*\theta$. It follows that

$$\int_M F^*\omega = a \int_M F^*\theta = ak = k \int_N \omega.$$

Thus k satisfies (a), and is clearly the only number that does so.

Next we show that k also has the characterization given in part (b), from which it follows that it is an integer. Let $q \in N$ be an arbitrary regular value of F. Because $F^{-1}(q)$ is a properly embedded 0-dimensional submanifold of M, it is finite. Suppose first that $F^{-1}(q)$ is not empty—say, $F^{-1}(q) = \{x_1, \ldots, x_m\}$. By the inverse function theorem, for each i there is a neighborhood U_i of x_i such that F is a diffeomorphism from U_i to a neighborhood W_i of q, and by shrinking the U_i's if necessary, we may assume that they are pairwise disjoint. Then $K = M \smallsetminus (U_1 \cup \cdots \cup U_m)$ is closed in M and thus compact, so $F(K)$ is closed in N and disjoint from q. Let W be the connected component of $W_1 \cap \cdots \cap W_m \cap \big(N \smallsetminus F(K)\big)$ containing q, and let $V_i = F^{-1}(W) \cap U_i$. It follows that W is a connected neighborhood of q whose preimage under F is the disjoint union $V_1 \amalg \cdots \amalg V_m$, and F restricts to a diffeomorphism from each V_i to W. Since each V_i is connected, the restriction of F to V_i must be either orientation-preserving or orientation-reversing.

Let ω be a smooth n-form on N that is compactly supported in W and satisfies $\int_N \omega = \int_W \omega = 1$. It follows from part (a) that $\int_M F^*\omega = k$. Since $F^*\omega$ is compactly supported in $F^{-1}(W)$, we have $\int_M F^*\omega = \sum_{i=1}^m \int_{V_i} F^*\omega$. From Proposition 16.6(d) we conclude that for each i, $\int_{V_i} F^*\omega = \pm \int_W \omega = \pm 1$, with the positive sign if F is orientation-preserving on V_i and the negative sign otherwise. This proves (b) when $F^{-1}(q) \neq \varnothing$.

On the other hand, suppose $F^{-1}(q) = \varnothing$. Then q has a neighborhood W contained in $N \smallsetminus F(M)$ (because $F(M)$ is compact and thus closed). If ω is any smooth n-form on N that is compactly supported in W, then $\int_M F^*\omega = 0$, so $k = 0$. This proves (b). $\qquad\square$

Much of the power of degree theory arises from the fact that the two different characterizations of the degree can be played off against each other. For example, it is often easy to compute the degree of a particular map simply by counting the points in the preimage of a regular value, with appropriate signs. On the other hand, the characterization in terms of differential forms makes it easy to prove many important properties, such as the ones given in the next proposition.

Proposition 17.36 (Properties of the Degree). *Suppose M, N, and P are compact, connected, oriented, smooth n-manifolds.*

(a) *If $F: M \to N$ and $G: N \to P$ are both smooth maps, then $\deg(G \circ F) = (\deg G)(\deg F)$.*

(b) *If $F: M \to N$ is a diffeomorphism, then $\deg F = +1$ if F is orientation-preserving and -1 if it is orientation-reversing.*

(c) *If two smooth maps $F_0, F_1: M \to N$ are homotopic, then they have the same degree.*

▶ **Exercise 17.37.** Prove the preceding proposition.

This proposition allows us to define the ***degree of a continuous map*** $F: M \to N$ between compact, connected, oriented, smooth n-manifolds, by letting $\deg F$ be the degree of any smooth map that is homotopic to F. The Whitney approximation theorem guarantees that there is such a map, and the preceding proposition guarantees that the degree is the same for every map homotopic to F.

Here are some applications of degree theory.

Theorem 17.38. *Suppose N is a compact, connected, oriented, smooth n-manifold, and X is a compact, oriented, smooth $(n + 1)$-manifold with connected boundary. If $f: \partial X \to N$ is a continuous map that has a continuous extension to X, then $\deg f = 0$.*

Proof. Suppose f has an extension to a continuous map $F: X \to N$. By the Whitney approximation theorem, there is a smooth map $\widetilde{F}: X \to N$ that is homotopic to F. Replacing F by $\widetilde{F}$ and f by $\widetilde{F}|_{\partial X}$, we may assume that both f and F are smooth.

Let ω be any smooth n-form on N. Then $d\omega = 0$ because it is an $(n + 1)$-form on an n-manifold. From Stokes's theorem, we obtain

$$\int_{\partial X} f^*\omega = \int_{\partial X} F^*\omega = \int_X d(F^*\omega) = \int_X F^* d\omega = 0.$$

It follows from Theorem 17.35 that f has degree zero. $\qquad\square$

Theorem 17.39 (Brouwer Fixed-Point Theorem). *Every continuous map from $\overline{\mathbb{B}}^n$ to itself has a fixed point.*

Proof. Suppose for the sake of contradiction that $F: \overline{\mathbb{B}}^n \to \overline{\mathbb{B}}^n$ is continuous and has no fixed points. We can define a continuous map $G: \overline{\mathbb{B}}^n \to \mathbb{S}^{n-1}$ by

$$G(x) = \frac{x - F(x)}{|x - F(x)|},$$

and let $g = G|_{\mathbb{S}^{n-1}}: \mathbb{S}^{n-1} \to \mathbb{S}^{n-1}$. On the one hand, the previous theorem implies that g has degree zero. On the other hand, consider the map $H: \mathbb{S}^{n-1} \times I \to \mathbb{S}^{n-1}$ defined by

$$H(x,t) = \frac{x - tF(x)}{|x - tF(x)|}.$$

The denominator never vanishes when $t = 1$ because F has no fixed points, and when $t < 1$ it cannot vanish because $|x| = 1$ while $|tF(x)| \leq t < 1$. Thus H is continuous, so it is a homotopy from the identity to g. It follows from Proposition 17.36 that g has degree 1, which is a contradiction.   □

With a little more machinery from algebraic topology, it is possible to give many more applications of degree theory. For example, it turns out that continuous maps from $\mathbb{S}^n$ to itself ($n \geq 1$) are classified up to homotopy by degree (see [Hat02, Cor. 4.25]). This is not true for other compact orientable manifolds, however; Problem 17-13 describes a counterexample.

Proof of the Mayer–Vietoris Theorem

In this section we give the proof of the Mayer–Vietoris theorem. For this purpose we need to introduce some simple algebraic concepts. More details about the ideas introduced here can be found in [LeeTM, Chap. 13] or in any textbook on algebraic topology.

Let $\mathcal{R}$ be a commutative ring, and suppose we are given a sequence of $\mathcal{R}$-modules and $\mathcal{R}$-linear maps:

$$\cdots \to A^{p-1} \xrightarrow{d} A^p \xrightarrow{d} A^{p+1} \to \cdots . \tag{17.12}$$

(In all of our applications, the ring will be either $\mathbb{Z}$, in which case we are looking at abelian groups and homomorphisms, or $\mathbb{R}$, in which case we have vector spaces and linear maps. The terminology of modules is just a convenient way to combine the two cases.) Such a sequence is said to be a *complex* if the composition of any two successive applications of d is the zero map:

$$d \circ d = 0 \colon A^p \to A^{p+2} \quad \text{for each } p.$$

Just as in the case of vector spaces, such a sequence of modules is called an *exact sequence* if the image of each d is equal to the kernel of the next. Clearly, every exact sequence is a complex, but the converse need not be true.

Let us denote the sequence (17.12) by A^*. If it is a complex, then the image of each map d is contained in the kernel of the next, so we define the *pth cohomology group of A^** to be the quotient module

$$H^p(A^*) = \frac{\text{Ker}(d \colon A^p \to A^{p+1})}{\text{Im}(d \colon A^{p-1} \to A^p)}.$$

It can be thought of as a quantitative measure of the failure of exactness at A^p. The obvious example is the *de Rham complex* of a smooth n-manifold M:

$$0 \to \Omega^0(M) \xrightarrow{d} \cdots \xrightarrow{d} \Omega^p(M) \xrightarrow{d} \Omega^{p+1}(M) \xrightarrow{d} \cdots \xrightarrow{d} \Omega^n(M) \to 0,$$

whose cohomology groups are the de Rham groups of M. (In algebraic topology, a complex as we have defined it is usually called a *cochain complex*, while a *chain*

complex is defined similarly except that the maps go in the direction of decreasing indices:

$$\cdots \to A_{p+1} \xrightarrow{\partial} A_p \xrightarrow{\partial} A_{p-1} \to \cdots.$$

In that case, the term **homology** is used in place of cohomology.)

If A^* and B^* are complexes, a **cochain map from A^* to B^***, denoted by $F: A^* \to B^*$, is a collection of linear maps $F: A^p \to B^p$ (it is easiest to use the same symbol for all of the maps) such that the following diagram commutes for each p:

$$
\begin{array}{ccc}
\cdots \longrightarrow A^p & \xrightarrow{\ d\ } & A^{p+1} \longrightarrow \cdots \\
F \downarrow & & \downarrow F \\
\cdots \longrightarrow B^p & \xrightarrow[\ d\ ]{} & B^{p+1} \longrightarrow \cdots.
\end{array}
$$

The fact that $F \circ d = d \circ F$ means that any cochain map induces a linear map on cohomology $F^*: H^p(A^*) \to H^p(B^*)$ for each p, just as in the case of de Rham cohomology. (A map between chain complexes satisfying the analogous relations is called a **chain map**; the same argument shows that a chain map induces a linear map on homology.)

A **short exact sequence of complexes** consists of three complexes A^*, B^*, C^*, together with cochain maps

$$0 \to A^* \xrightarrow{F} B^* \xrightarrow{G} C^* \to 0$$

such that each sequence

$$0 \to A^p \xrightarrow{F} B^p \xrightarrow{G} C^p \to 0$$

is exact. This means that F is injective, G is surjective, and $\operatorname{Im} F = \operatorname{Ker} G$.

Lemma 17.40 (The Zigzag Lemma). *Given a short exact sequence of complexes as above, for each p there is a linear map*

$$\delta: H^p(C^*) \to H^{p+1}(A^*),$$

*called the **connecting homomorphism**, such that the following sequence is exact:*

$$\cdots \xrightarrow{\delta} H^p(A^*) \xrightarrow{F^*} H^p(B^*) \xrightarrow{G^*} H^p(C^*) \xrightarrow{\delta} H^{p+1}(A^*) \xrightarrow{F^*} \cdots. \tag{17.13}$$

Proof. We sketch only the main idea; you can either carry out the details yourself or look them up.

The hypothesis means that the following diagram commutes and has exact horizontal rows:

$$
\begin{array}{ccccccccc}
0 & \longrightarrow & A^p & \xrightarrow{F} & B^p & \xrightarrow{G} & C^p & \longrightarrow & 0 \\
 & & \downarrow{d} & & \downarrow{d} & & \downarrow{d} & & \\
0 & \longrightarrow & A^{p+1} & \xrightarrow{F} & B^{p+1} & \xrightarrow{G} & C^{p+1} & \longrightarrow & 0 \\
 & & \downarrow{d} & & \downarrow{d} & & \downarrow{d} & & \\
0 & \longrightarrow & A^{p+2} & \xrightarrow{F} & B^{p+2} & \xrightarrow{G} & C^{p+2} & \longrightarrow & 0.
\end{array}
$$

Suppose $c^p \in C^p$ represents a cohomology class; this means that $dc^p = 0$. Since $G \colon B^p \to C^p$ is surjective, there is some element $b^p \in B^p$ such that $Gb^p = c^p$. Because the diagram commutes, $Gdb^p = dGb^p = dc^p = 0$, and therefore $db^p \in \operatorname{Ker} G = \operatorname{Im} F$. Thus, there exists $a^{p+1} \in A^{p+1}$ satisfying $Fa^{p+1} = db^p$. By commutativity of the diagram again, $Fda^{p+1} = dFa^{p+1} = ddb^p = 0$. Since F is injective, this implies $da^{p+1} = 0$, so a^{p+1} represents a cohomology class in $H^{p+1}(A^*)$. The connecting homomorphism δ is defined by setting $\delta[c^p] = [a^{p+1}]$ for any such $a^{p+1} \in A^{p+1}$, that is, provided there exists $b^p \in B^p$ such that

$$
Gb^p = c^p, \qquad Fa^{p+1} = db^p.
$$

A number of facts have to be verified: that the cohomology class $[a^{p+1}]$ is well defined, independently of the choices made along the way; that the resulting map δ is linear; and that the resulting sequence (17.13) is exact. Each of these verifications is a routine "diagram chase" like the one we used to define δ; the details are left as an exercise. $\qquad\qquad\qquad\qquad\qquad\qquad\qquad\qquad\qquad\qquad\qquad\qquad\qquad\square$

▶ **Exercise 17.41.** Complete (or look up) the proof of the zigzag lemma.

Proof of the Mayer–Vietoris Theorem. Suppose M is a smooth manifold with or without boundary, and U, V are open subsets of M whose union is M. The heart of the proof is to show that the sequence (17.7) is exact for each p. Because pullback maps commute with the exterior derivative, (17.7) therefore defines a short exact sequence of cochain maps, and the Mayer–Vietoris theorem follows immediately from the zigzag lemma.

We begin by proving exactness at $\Omega^p(M)$, which just means showing that $k^* \oplus l^*$ is injective. Suppose that $\sigma \in \Omega^p(M)$ satisfies $(k^* \oplus l^*)\sigma = (\sigma|_U, \sigma|_V) = (0,0)$. This means that the restrictions of σ to U and V are both zero. Since $\{U, V\}$ is an open cover of M, this implies that σ is zero.

To prove exactness at $\Omega^p(U) \oplus \Omega^p(V)$, first observe that

$$
(i^* - j^*) \circ (k^* \oplus l^*)(\sigma) = (i^* - j^*)(\sigma|_U, \sigma|_V) = \sigma|_{U \cap V} - \sigma|_{U \cap V} = 0,
$$

which shows that $\operatorname{Im}(k^* \oplus l^*) \subseteq \operatorname{Ker}(i^* - j^*)$. Conversely, suppose we are given $(\eta, \eta') \in \Omega^p(U) \oplus \Omega^p(V)$ such that $(i^* - j^*)(\eta, \eta') = 0$. This means that $\eta|_{U \cap V} =$

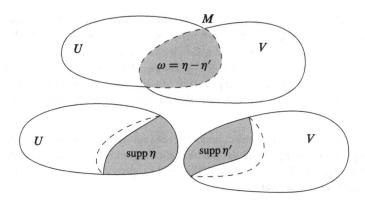

Fig. 17.4 Surjectivity of $i^* - j^*$

$\eta'|_{U \cap V}$, so there is a global smooth p-form σ on M defined by

$$\sigma = \begin{cases} \eta & \text{on } U, \\ \eta' & \text{on } V. \end{cases}$$

Clearly, $(\eta, \eta') = (k^* \oplus l^*)\sigma$, so $\mathrm{Ker}(i^* - j^*) \subseteq \mathrm{Im}(k^* \oplus l^*)$.

Exactness at $\Omega^p(U \cap V)$ means that $i^* - j^*$ is surjective. This is the only non-trivial part of the proof, and the only part that really uses any properties of smooth manifolds and differential forms.

Let $\omega \in \Omega^p(U \cap V)$ be arbitrary. We need to show that there exist $\eta \in \Omega^p(U)$ and $\eta' \in \Omega^p(V)$ such that

$$\omega = (i^* - j^*)(\eta, \eta') = i^*\eta - j^*\eta' = \eta|_{U \cap V} - \eta'|_{U \cap V}.$$

(See Fig. 17.4.) Let $\{\varphi, \psi\}$ be a smooth partition of unity subordinate to the open cover $\{U, V\}$ of M, and define $\eta \in \Omega^p(U)$ by

$$\eta = \begin{cases} \psi\omega & \text{on } U \cap V, \\ 0 & \text{on } U \smallsetminus \mathrm{supp}\, \psi. \end{cases} \tag{17.14}$$

On the set $(U \cap V) \smallsetminus \mathrm{supp}\, \psi$ where these definitions overlap, they both give zero, so this defines η as a smooth p-form on U. Similarly, define $\eta' \in \Omega^p(V)$ by

$$\eta' = \begin{cases} -\varphi\omega & \text{on } U \cap V, \\ 0 & \text{on } V \smallsetminus \mathrm{supp}\, \varphi. \end{cases} \tag{17.15}$$

Then we have

$$\eta|_{U \cap V} - \eta'|_{U \cap V} = \psi\omega - (-\varphi\omega) = (\psi + \varphi)\omega = \omega,$$

which was to be proved. $\qquad\qquad\qquad\qquad\qquad\qquad\qquad\qquad\qquad\qquad\square$

For use in the next chapter, we record the following corollary to the proof, which explicitly characterizes the connecting homomorphism δ.

Corollary 17.42. *The connecting homomorphism in the Mayer–Vietoris sequence,* $\delta\colon H^p_{dR}(U \cap V) \to H^{p+1}_{dR}(M)$, *is defined as follows. For each* $\omega \in Z^p(U \cap V)$, *there are p-forms* $\eta \in \Omega^p(U)$ *and* $\eta' \in \Omega^p(V)$ *such that* $\omega = \eta|_{U \cap V} - \eta'|_{U \cap V}$; *and then* $\delta[\omega] = [\sigma]$, *where* σ *is the* $(p+1)$*-form on* M *that is equal to* $d\eta$ *on* U *and to* $d\eta'$ *on* V. *If* $\{\varphi, \psi\}$ *is a smooth partition of unity subordinate to* $\{U, V\}$, *we can take* $\eta = \psi\omega$ *and* $\eta' = -\varphi\omega$, *both extended by zero outside the supports of* ψ *and* φ.

Proof. A characterization of the connecting homomorphism was given in the proof of the zigzag lemma. Specializing this characterization to the situation of the short exact sequence (17.7), we find that $\delta[\omega] = [\sigma]$, provided there exists $(\eta, \eta') \in \Omega^p(U) \oplus \Omega^p(V)$ such that

$$i^*\eta - j^*\eta' = \omega, \qquad (k^*\sigma, l^*\sigma) = (d\eta, d\eta'). \qquad (17.16)$$

Just as in the proof of the Mayer–Vietoris theorem, if $\{\varphi, \psi\}$ is a smooth partition of unity subordinate to $\{U, V\}$, then formulas (17.14) and (17.15) define smooth forms $\eta \in \Omega^p(U)$ and $\eta' \in \Omega^p(V)$ satisfying the first equation of (17.16). Given such forms η, η', the fact that ω is closed implies that $d\eta = d\eta'$ on $U \cap V$. Thus there is a smooth $(p+1)$-form σ on M that is equal to $d\eta$ on U and $d\eta'$ on V, and it satisfies the second equation of (17.16). $\qquad\qquad\square$

Problems

17-1. Let M be a smooth manifold with or without boundary, and let $\omega \in \Omega^p(M)$, $\eta \in \Omega^q(M)$ be closed forms. Show that the de Rham cohomology class of $\omega \wedge \eta$ depends only on the cohomology classes of ω and η, and thus there is a well-defined bilinear map $\cup\colon H^p_{dR}(M) \times H^q_{dR}(M) \to H^{p+q}_{dR}(M)$, called the **cup product**, given by $[\omega] \cup [\eta] = [\omega \wedge \eta]$.

17-2. Let (M, g) be an oriented compact Riemannian n-manifold. For each $0 \le p \le n$, the **Laplace–Beltrami operator** $\Delta\colon \Omega^p(M) \to \Omega^p(M)$ is the linear map defined by

$$\Delta\omega = d\,d^*\omega + d^*\,d\omega,$$

where d^* is the operator defined in Problem 16-22. A smooth form $\omega \in \Omega^p(M)$ is said to be **harmonic** if $\Delta\omega = 0$. Show that the following are equivalent for any $\omega \in \Omega^p(M)$.

(a) ω is harmonic.

(b) $d\omega = 0$ and $d^*\omega = 0$.

(c) $d\omega = 0$ and ω is the unique smooth p-form in its de Rham cohomology class with minimum norm $\|\omega\| = (\omega, \omega)^{1/2}$. (Here $(\cdot, \cdot)$ is the inner product on $\Omega^p(M)$ defined in Problem 16-22.)

[Hint: for (c), consider $f(t) = \|\omega + d(t\,d^*\omega)\|^2$.] [Remark: there is a deep theorem called the *Hodge theorem*, which says that on every compact, oriented Riemannian manifold, there is a unique harmonic form in every de Rham cohomology class. See [Gil95] or [War83] for a proof.]

17-3. Let (M, g) be an oriented Riemannian manifold, and let $\Delta = dd^* + d^*d$ be the Laplace–Beltrami operator on p-forms as in Problem 17-2. When $p = 0$, show that Δ agrees with the geometric Laplacian $\Delta u = -\operatorname{div}(\operatorname{grad} u)$ defined on real-valued functions in Problem 16-13.

17-4. Suppose $U \subseteq \mathbb{R}^n$ is open and star-shaped with respect to 0, and $\omega = \sum' \omega_I dx^I$ is a closed p-form on U. Show either directly or by tracing through the proof of the Poincaré lemma that the $(p-1)$-form η given explicitly by the formula

$$\eta = \sum_I{}' \sum_{q=1}^{p} (-1)^{q-1} \left(\int_0^1 t^{p-1} \omega_I(tx)\, dt \right) x^{i_q}\, dx^{i_1} \wedge \cdots \wedge \widehat{dx^{i_q}} \wedge \cdots \wedge dx^{i_p}$$

satisfies $d\eta = \omega$. In the case that ω is a smooth closed 1-form, show that η is equal to the potential function f defined in Theorem 11.49.

17-5. For each $n \geq 1$, compute the de Rham cohomology groups of $\mathbb{R}^n \smallsetminus \{e_1, -e_1\}$; and for each nonzero cohomology group, give specific differential forms whose cohomology classes form a basis.

17-6. Let M be a connected smooth manifold of dimension $n \geq 3$. For any $x \in M$ and $0 \leq p \leq n - 2$, prove that the map $H_{dR}^p(M) \to H_{dR}^p(M \smallsetminus \{x\})$ induced by inclusion $M \smallsetminus \{x\} \hookrightarrow M$ is an isomorphism. Prove that the same is true for $p = n - 1$ if M is compact and orientable. [Hint: use the Mayer–Vietoris theorem. The cases $p = 0$, $p = 1$, and $p = n - 1$ require special handling.]

17-7. Let M_1, M_2 be connected smooth manifolds of dimension $n \geq 3$, and let $M_1 \# M_2$ denote their smooth connected sum (Example 9.31). Prove that $H_{dR}^p(M_1 \# M_2) \cong H_{dR}^p(M_1) \oplus H_{dR}^p(M_2)$ for $0 < p < n - 1$. Prove that the same is true for $p = n - 1$ if M_1 and M_2 are both compact and orientable. [Hint: use Problems 9-12 and 17-6.]

17-8. Suppose M is a compact, connected, orientable, smooth n-manifold.
 (a) Show that there is a one-to-one correspondence between orientations of M and orientations of the vector space $H_{dR}^n(M)$, under which the cohomology class of a smooth orientation form is an oriented basis for $H_{dR}^n(M)$.
 (b) Now suppose M and N are smooth n-manifolds with given orientations. Show that a diffeomorphism $F: M \to N$ is orientation preserving if and only if $F^*: H_{dR}^n(N) \to H_{dR}^n(M)$ is orientation preserving.

17-9. Prove Theorem 1.37 (topological invariance of the boundary).

17-10. Let p be a nonzero polynomial in one variable with complex coefficients, and let $\tilde{p}: \mathbb{CP}^1 \to \mathbb{CP}^1$ be the smooth map defined in Problem 2-9. Prove that the degree of $\tilde{p}$ (as a smooth map between manifolds) is equal to the degree of the polynomial p in the usual sense.

17-11. This problem shows that some parts of degree theory can be extended to proper maps between noncompact manifolds. Suppose M and N are noncompact, connected, oriented, smooth n-manifolds.

(a) Suppose $F: M \to N$ is a proper smooth map. Prove that there is a unique integer k called the **degree of** F such that for each smooth, compactly supported n-form ω on N,

$$\int_M F^*\omega = k \int_N \omega,$$

and for each regular value q of F,

$$k = \sum_{x \in F^{-1}(q)} \text{sgn}(dF_x),$$

where $\text{sgn}(dF_x)$ is defined in Theorem 17.35.

(b) By considering the maps $F, G: \mathbb{C} \to \mathbb{C}$ given by $F(z) = z$ and $G(z) = z^2$, show that the degree of a proper map is not a homotopy invariant.

17-12. Suppose M and N are compact, connected, oriented, smooth n-manifolds, and $F: M \to N$ is a smooth map. Prove that if $\int_M F^*\eta \neq 0$ for some $\eta \in \Omega^n(N)$, then F is surjective. Give an example to show that F can be surjective even if $\int_M F^*\eta = 0$ for every $\eta \in \Omega^n(N)$.

17-13. Let $\mathbb{T}^2 = \mathbb{S}^1 \times \mathbb{S}^1$ be the 2-torus. Consider the two maps $f, g: \mathbb{T}^2 \to \mathbb{T}^2$ given by $f(w, z) = (w, z)$ and $g(w, z) = (z, \bar{w})$. Show that f and g have the same degree, but are not homotopic. [Suggestion: consider the induced homomorphisms on the first cohomology group or the fundamental group.]

Chapter 18
The de Rham Theorem

The topological invariance of the de Rham groups suggests that there should be some purely topological way of computing them. There is indeed, and the connection between the de Rham groups and topology was first proved by Georges de Rham himself in the 1930s. The theorem that bears his name is a major landmark in the development of smooth manifold theory. The purpose of this chapter is to give a proof of this theorem.

In the category of topological spaces, there are a number of functorial ways of associating to each space an algebraic object such as a group or a vector space, so that homeomorphic spaces have isomorphic objects. Most of these measure, in a certain sense, the existence of "holes" in different dimensions. You are already familiar with the simplest such functor: the fundamental group. In the beginning of this chapter, we describe the next most straightforward ones, called the *singular homology groups* and *singular cohomology groups*. Because a complete treatment of singular theory would be far beyond the scope of this book, we can only summarize the basic ideas here. For more details, you can consult a standard textbook on algebraic topology, such as [Hat02], [Bre93], or [Mun84]. (See also [LeeTM, Chap. 13] for a more concise treatment.) After introducing the basic definitions, we prove that singular homology can be computed by restricting attention only to smooth simplices.

At the end of the chapter we turn our attention to the de Rham theorem, which shows that integration of differential forms over smooth simplices induces isomorphisms between the de Rham groups and the singular cohomology groups.

Singular Homology

We begin with a brief summary of singular homology theory. Suppose $v_0, \dots, v_p$ are any $p + 1$ points in some Euclidean space $\mathbb{R}^n$. They are said to be **affinely independent** (or **in general position**) if they are not contained in any $(p - 1)$-dimensional

J.M. Lee, *Introduction to Smooth Manifolds*, Graduate Texts in Mathematics 218,   467
DOI 10.1007/978-1-4419-9982-5_18, © Springer Science+Business Media New York 2013

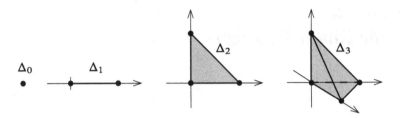

Fig. 18.1 Standard p-simplices for $p = 0, 1, 2, 3$

affine subspace. A *geometric p-simplex* is a subset of $\mathbb{R}^n$ of the form

$$\left\{\sum_{i=0}^{p} t_i v_i : 0 \le t_i \le 1 \text{ and } \sum_{i=0}^{p} t_i = 1\right\},$$

for some $p + 1$ affinely independent points $\{v_0, \ldots, v_p\}$. The integer p (one less than the number of vertices) is called the *dimension* of the simplex. The points $v_0, \ldots, v_p$ are called its *vertices*, and the geometric simplex with these vertices is denoted by $[v_0, \ldots, v_p]$. It is a compact convex set, in fact the smallest convex set containing $\{v_0, \ldots, v_p\}$. The simplices whose vertices are nonempty subsets of $\{v_0, \ldots, v_p\}$ are called the *faces* of the simplex. The $(p - 1)$-dimensional faces are called its *boundary faces*. There are precisely $p + 1$ boundary faces, obtained by omitting each of the vertices in turn; the i th boundary face $[v_0, \ldots, \widehat{v_i}, \ldots, v_p]$ (with v_i omitted) is denoted by $\partial_i[v_0, \ldots, v_p]$, and is called the *face opposite v_i*.

▶ **Exercise 18.1.** Show that a geometric p-simplex is a p-dimensional smooth manifold with corners smoothly embedded in $\mathbb{R}^n$.

The *standard p-simplex* is the simplex $\Delta_p = [e_0, e_1, \ldots, e_p] \subseteq \mathbb{R}^p$, where $e_0 = 0$ and e_i is the i th standard basis vector. For example, $\Delta_0 = \{0\}$, $\Delta_1 = [0, 1]$, Δ_2 is the triangle with vertices $(0, 0)$, $(1, 0)$, and $(0, 1)$ together with its interior, and Δ_3 is a solid tetrahedron (Fig. 18.1).

Let M be a topological space. A continuous map $\sigma \colon \Delta_p \to M$ is called a *singular p-simplex in M*. The *singular chain group of M in degree p*, denoted by $C_p(M)$, is the free abelian group generated by all singular p-simplices in M. An element of this group, called a *singular p-chain*, is a finite formal linear combination of singular p-simplices in M with integer coefficients.

One special case that arises frequently is that in which the space M is a convex subset of some Euclidean space $\mathbb{R}^m$. In that case, for any ordered $(p + 1)$-tuple of points $(w_0, \ldots, w_p)$ in M, not necessarily affinely independent, there is a unique affine map from $\mathbb{R}^p$ to $\mathbb{R}^m$ that takes e_i to w_i for $i = 0, \ldots, p$. (The map is easily constructed by first finding a linear map that takes e_i to $w_i - w_0$ for $i = 1, \ldots, p$, and then translating by w_0.) The restriction of this affine map to Δ_p is denoted by $A(w_0, \ldots, w_p)$, and is called an *affine singular simplex in M*.

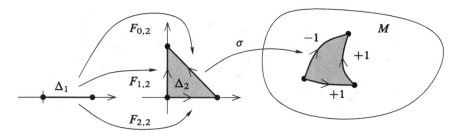

Fig. 18.2 The singular boundary operator

For each $i = 0, \ldots, p$, we define the *i th face map in Δ_p* to be the affine singular $(p-1)$-simplex $F_{i,p} \colon \Delta_{p-1} \to \Delta_p$ defined by

$$F_{i,p} = A\left(e_0, \ldots, \widehat{e_i}, \ldots, e_p\right).$$

It maps Δ_{p-1} homeomorphically onto the boundary face $\partial_i \Delta_p$. Explicitly, it is the unique affine map sending $e_0 \mapsto e_0, \ldots, e_{i-1} \mapsto e_{i-1}, e_i \mapsto e_{i+1}, \ldots, e_{p-1} \mapsto e_p$.

The *boundary* of a singular p-simplex $\sigma \colon \Delta_p \to M$ is the singular $(p-1)$-chain $\partial \sigma$ defined by

$$\partial \sigma = \sum_{i=0}^{p} (-1)^i \sigma \circ F_{i,p}.$$

For example, if σ is a singular 2-simplex, its boundary is a formal sum of three singular 1-simplices with coefficients ± 1, as indicated in Fig. 18.2. This extends uniquely to a group homomorphism $\partial \colon C_p(M) \to C_{p-1}(M)$, called the *singular boundary operator*. The basic fact about the boundary operator is the next lemma.

Lemma 18.2. *If c is any singular chain, then $\partial(\partial c) = 0$.*

Sketch of Proof. The starting point is the fact that

$$F_{i,p} \circ F_{j,p-1} = F_{j,p} \circ F_{i-1,p-1} \tag{18.1}$$

when $i > j$, which can be verified by following what both compositions do to each of the vertices of Δ_{p-2}. Using this, the proof of the lemma is just a straightforward computation. $\square$

A singular p-chain c is called a *cycle* if $\partial c = 0$, and a *boundary* if $c = \partial b$ for some singular $(p+1)$-chain b. Let $Z_p(M)$ denote the set of singular p-cycles in M, and $B_p(M)$ the set of singular p-boundaries. Because ∂ is a homomorphism, $Z_p(M)$ and $B_p(M)$ are subgroups of $C_p(M)$, and because $\partial \circ \partial = 0$, they satisfy $B_p(M) \subseteq Z_p(M)$. The *pth singular homology group* of M is the quotient group

$$H_p(M) = \frac{Z_p(M)}{B_p(M)}.$$

To put it another way, the sequence of abelian groups and homomorphisms

$$\cdots \to C_{p+1}(M) \xrightarrow{\partial} C_p(M) \xrightarrow{\partial} C_{p-1}(M) \to \cdots$$

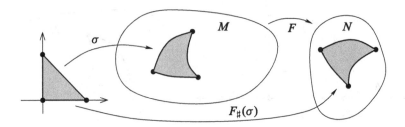

Fig. 18.3 The homology homomorphism induced by a continuous map

is a complex, called the ***singular chain complex***, and $H_p(M)$ is the pth homology group of this complex. The equivalence class in $H_p(M)$ of a singular p-cycle c is called its ***homology class***, and is denoted by $[c]$. We say that two p-cycles are ***homologous*** if they differ by a boundary.

A continuous map $F: M \to N$ induces a homomorphism $F_\sharp: C_p(M) \to C_p(N)$ on each singular chain group, defined by $F_\sharp(\sigma) = F \circ \sigma$ for any singular simplex σ (Fig. 18.3) and extended linearly to chains. An easy computation shows that $F_\sharp \circ \partial = \partial \circ F_\sharp$, so $F_\sharp$ is a chain map, and therefore induces a homomorphism on the singular homology groups, denoted by $F_*: H_p(M) \to H_p(N)$. It is immediate that $(G \circ F)_* = G_* \circ F_*$ and $(\mathrm{Id}_M)_* = \mathrm{Id}_{H_p(M)}$, so pth singular homology defines a covariant functor from the category of topological spaces and continuous maps to the category of abelian groups and homomorphisms. In particular, homeomorphic spaces have isomorphic singular homology groups.

Intuitively, you will not go too far astray if you visualize a singular p-chain in M as representing something like a compact p-dimensional submanifold of M with boundary (although, because there is no requirement that singular chains be smooth, or topological embeddings, or even injective, a chain might not *look* at all like a submanifold; hence the designation "singular"). A closed p-chain, then, is like a compact submanifold without boundary, and it represents the trivial homology class if and only if it is the boundary of a $(p + 1)$-chain. Thus a nontrivial element of $H_p(M)$ is rather like a compact p-dimensional submanifold of M that does not bound a compact $(p + 1)$-dimensional submanifold, and so must represent some kind of p-dimensional "hole" in M. (See Problem 18-3, which introduces *smooth triangulations* as a way of giving this intuition more substance.)

Proposition 18.3 (Properties of Singular Homology Groups).

(a) *For any one-point space $\{q\}$, $H_0(\{q\})$ is the infinite cyclic group generated by the homology class of the unique singular 0-simplex mapping Δ_0 to q, and $H_p(\{q\}) = 0$ for all $p \neq 0$.*

(b) *Let $\{M_j\}$ be any collection of topological spaces, and let $M = \coprod_j M_j$. The inclusion maps $\iota_j: M_j \hookrightarrow M$ induce an isomorphism $\bigoplus_j H_p(M_j) \cong H_p(M)$.*

(c) *Homotopy equivalent spaces have isomorphic singular homology groups.*

Sketch of Proof. In a one-point space $\{q\}$, there is exactly one singular p-simplex for each p, namely the constant map. The result of part (a) follows from an analysis of the boundary maps. Part (b) is immediate because the maps ι_j already induce an isomorphism on the chain level: $\bigoplus_j C_p(M_j) \cong C_p(M)$.

The main step in the proof of homotopy invariance is the construction for any space M of a linear map $h \colon C_p(M) \to C_{p+1}(M \times I)$ satisfying

$$h \circ \partial + \partial \circ h = (i_1)_\# - (i_0)_\#, \qquad (18.2)$$

where $i_k \colon M \to M \times I$ is the injection $i_k(x) = (x,k)$. From this it follows just as in the proof of Proposition 17.10 that homotopic maps induce the same homology homomorphism, and then in turn that homotopy equivalent spaces have isomorphic singular homology groups.   □

In addition to the properties above, singular homology satisfies the following version of the Mayer–Vietoris theorem. Suppose M is a topological space and $U, V \subseteq M$ are open subsets whose union is M. The usual diagram (17.6) of inclusions induces homology homomorphisms:

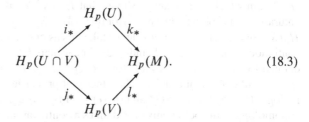

$$(18.3)$$

Theorem 18.4 (Mayer–Vietoris for Singular Homology). *Let M be a topological space and let U, V be open subsets of M whose union is M. For each p there is a connecting homomorphism $\partial_* \colon H_p(M) \to H_{p-1}(U \cap V)$ such that the following sequence is exact:*

$$\cdots \xrightarrow{\partial_*} H_p(U \cap V) \xrightarrow{\alpha} H_p(U) \oplus H_p(V) \xrightarrow{\beta} H_p(M)$$

$$\xrightarrow{\partial_*} H_{p-1}(U \cap V) \xrightarrow{\alpha} \cdots, \qquad (18.4)$$

where

$$\alpha[c] = \big(i_*[c], -j_*[c]\big), \qquad \beta\big([c],[c']\big) = k_*[c] + l_*[c'],$$

and $\partial_*[e] = [c]$, provided there exist $f \in C_p(U)$ and $f' \in C_p(V)$ such that $k_\# f + l_\# f'$ is homologous to e and $(i_\# c, -j_\# c) = (\partial f, \partial f')$.

Sketch of Proof. The basic idea, of course, is to construct a short exact sequence of complexes and use the zigzag lemma. The hardest part of the proof is showing that every homology class $[e] \in H_p(M)$ can be represented in the form $\beta([c],[c'])$, where c is a singular chain in U and c' is a singular chain in V. This is accomplished by systematically "subdividing" each chain into smaller ones, each of which maps only into U or V, and keeping careful track of the boundary operators.   □

Note that the maps α and β in this Mayer–Vietoris sequence can be replaced by

$$\tilde{\alpha}[c] = \big(i_*[c], j_*[c]\big), \qquad \tilde{\beta}\big([c], [c']\big) = k_*[c] - l_*[c'],$$

and the same proof goes through. If you consult various algebraic topology texts, you will find both definitions in use. We are using the definition given in the statement of the theorem because it leads to a cohomology exact sequence that is compatible with the Mayer–Vietoris sequence for de Rham cohomology; see the proof of the de Rham theorem below.

Singular Cohomology

In addition to the singular homology groups, for any topological space M and any abelian group G one can define a closely related sequence of groups $H^p(M; G)$ called the *singular cohomology groups with coefficients in* G. The precise definition is unimportant for our purposes; we are only concerned with the special case $G = \mathbb{R}$, in which case it can be shown that $H^p(M; \mathbb{R})$ is a real vector space that is naturally isomorphic to the space $\mathrm{Hom}\big(H_p(M), \mathbb{R}\big)$ of group homomorphisms from $H_p(M)$ into $\mathbb{R}$. (For simplicity, let us take this as our definition of $H^p(M, \mathbb{R})$.) Any continuous map $F \colon M \to N$ induces a linear map $F^* \colon H^p(N; \mathbb{R}) \to H^p(M; \mathbb{R})$, defined by $(F^*\gamma)[c] = \gamma(F_*[c])$ for each $\gamma \in H^p(N; \mathbb{R}) \cong \mathrm{Hom}\big(H_p(N), \mathbb{R}\big)$ and each singular p-chain c in M. The functorial properties of F_* carry over to cohomology: $(G \circ F)^* = F^* \circ G^*$ and $(\mathrm{Id}_M)^* = \mathrm{Id}_{H^p(M;\mathbb{R})}$. It follows that pth singular cohomology with coefficients in $\mathbb{R}$ defines a contravariant functor from the topological category to the category of real vector spaces and linear maps.

There is an important theorem of algebraic topology called the *universal coefficient theorem*, which shows how the singular cohomology groups with coefficients in an arbitrary group can be recovered from the singular homology groups. Thus, the cohomology groups do not contain any new information that is not already encoded in the homology groups; but they organize it in a different way that is more convenient for many purposes. In particular, the fact that the singular cohomology groups, like the de Rham cohomology groups, define *contravariant* functors makes it much easier to compare the two.

Proposition 18.5 (Properties of Singular Cohomology).

(a) *For any one-point space* $\{q\}$, $H^p(\{q\}; \mathbb{R})$ *is trivial except when* $p = 0$, *in which case it is* 1-*dimensional.*

(b) *If* $\{M_j\}$ *is any collection of topological spaces and* $M = \coprod_j M_j$, *then the inclusion maps* $\iota_j \colon M_j \hookrightarrow M$ *induce an isomorphism from* $H^p(M; \mathbb{R})$ *to* $\prod_j H^p(M_j; \mathbb{R})$.

(c) *Homotopy equivalent spaces have isomorphic singular cohomology groups.*

Sketch of Proof. These properties follow easily from the definitions and Proposition 18.3.   □

The key fact about the singular cohomology groups that we need is that they, too, satisfy a Mayer–Vietoris theorem.

Theorem 18.6 (Mayer–Vietoris for Singular Cohomology). *Suppose M, U, and V satisfy the hypotheses of Theorem 18.4. The following sequence is exact:*

$$\cdots \xrightarrow{\partial^*} H^p(M;\mathbb{R}) \xrightarrow{k^*\oplus l^*} H^p(U;\mathbb{R}) \oplus H^p(V;\mathbb{R}) \xrightarrow{i^*-j^*} H^p(U \cap V;\mathbb{R})$$

$$\xrightarrow{\partial^*} H^{p+1}(M;\mathbb{R}) \xrightarrow{k^*\oplus l^*} \cdots, \quad (18.5)$$

where the maps $k^* \oplus l^*$ and $i^* - j^*$ are defined as in (17.8), and ∂^* is defined by $\partial^*(\gamma) = \gamma \circ \partial_*$, with ∂_* as in Theorem 18.4.

Sketch of Proof. For any homomorphism $F \colon A \to B$ between abelian groups, there is a *dual homomorphism* $F^* \colon \mathrm{Hom}(B,\mathbb{R}) \to \mathrm{Hom}(A,\mathbb{R})$ given by $F^*(\gamma) = \gamma \circ F$. Applying this to the Mayer–Vietoris sequence (18.4) for singular homology, we obtain the cohomology sequence (18.5). Exactness of (18.5) is a consequence of the fact that the assignments $A \mapsto \mathrm{Hom}(A,\mathbb{R})$ and $F \mapsto F^*$ define an *exact functor*, meaning that it takes exact sequences to exact sequences. This in turn follows from the fact that $\mathbb{R}$ is an *injective group*: this means that whenever H is a subgroup of an abelian group G, every homomorphism from H into $\mathbb{R}$ extends to all of G. $\quad\square$

Smooth Singular Homology

The connection between the singular and de Rham cohomology groups will be established by integrating differential forms over singular chains. More precisely, given a singular p-simplex σ in a manifold M and a p-form ω on M, we would like to pull ω back by σ and integrate the resulting form over Δ_p. However, there is an immediate problem with this approach, because forms can be pulled back only by *smooth* maps, while singular simplices are in general only continuous. (Actually, since only first derivatives of the map appear in the formula for the pullback, it would be sufficient to consider C^1 maps, but merely continuous ones definitely will not do.) In this section we overcome this problem by showing that singular homology can be computed equally well with smooth simplices.

If M is a smooth manifold, a *smooth p-simplex in M* is a map $\sigma \colon \Delta_p \to M$ that is smooth in the sense that it has a smooth extension to a neighborhood of each point. The subgroup of $C_p(M)$ generated by smooth simplices is denoted by $C_p^\infty(M)$ and called the *smooth chain group in degree p*. Elements of this group, which are finite formal linear combinations of smooth simplices, are called *smooth chains*. Because the boundary of a smooth simplex is a smooth chain, we can define the *pth smooth singular homology group of M* to be the quotient group

$$H_p^\infty(M) = \frac{\mathrm{Ker}\big(\partial \colon C_p^\infty(M) \to C_{p-1}^\infty(M)\big)}{\mathrm{Im}\big(\partial \colon C_{p+1}^\infty(M) \to C_p^\infty(M)\big)}.$$

The inclusion map $\iota \colon C_p^\infty(M) \hookrightarrow C_p(M)$ commutes with the boundary operator, and so induces a map on homology: $\iota_* \colon H_p^\infty(M) \to H_p(M)$ by $\iota_*[c] = [\iota(c)]$.

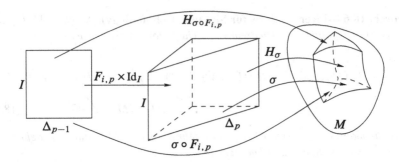

Fig. 18.4 The homotopy H_σ

Theorem 18.7 (Smooth Singular vs. Singular Homology). *For any smooth manifold M, the map $\iota_*\colon H_p^\infty(M) \to H_p(M)$ induced by inclusion is an isomorphism.*

The basic idea of the proof is to construct, with the help of the Whitney approximation theorem, two operators: first, a smoothing operator $s\colon C_p(M) \to C_p^\infty(M)$ such that $s \circ \partial = \partial \circ s$ and $s \circ \iota$ is the identity on $C_p^\infty(M)$; and second, a homotopy operator that shows that $\iota \circ s$ induces the identity map on $H_p(M)$. The details are highly technical, so unless algebraic topology is your primary interest, you may wish to skim the rest of this section on first reading.

The key to the proof is a systematic construction of a homotopy from each continuous simplex to a smooth one, in a way that respects the restriction to each boundary face of Δ_p. This is summarized in the following lemma.

Lemma 18.8. *Let M be a smooth manifold. For each integer $p \geq 0$ and each singular p-simplex $\sigma\colon \Delta_p \to M$, there exists a continuous map $H_\sigma\colon \Delta_p \times I \to M$ such that the following properties hold:*

(i) *H_σ is a homotopy from $\sigma(x) = H_\sigma(x,0)$ to a smooth p-simplex $\tilde{\sigma}(x) = H_\sigma(x,1)$.*

(ii) *For each face map $F_{i,p}\colon \Delta_{p-1} \to \Delta_p$,*

$$H_{\sigma \circ F_{i,p}} = H_\sigma \circ \left(F_{i,p} \times \mathrm{Id}_I\right), \qquad (18.6)$$

or more explicitly,

$$H_{\sigma \circ F_{i,p}}(x,t) = H_\sigma\left(F_{i,p}(x),t\right), \quad (x,t) \in \Delta_{p-1} \times I. \qquad (18.7)$$

(iii) *If σ is a smooth p-simplex, then H_σ is the constant homotopy $H_\sigma(x,t) = \sigma(x)$.*

Proof. We will construct the homotopies H_σ (see Fig. 18.4) by induction on the dimension of σ. To get started, for each 0-simplex $\sigma\colon \Delta_0 \to M$, we just define $H_\sigma(x,t) = \sigma(x)$. Since each 0-simplex is smooth and there are no face maps, conditions (i)–(iii) are automatically satisfied.

Now suppose by induction that for each $p' < p$ and for each p'-simplex σ' we have defined $H_{\sigma'}$ in such a way that the primed analogues of (i)–(iii) are satisfied.

Let $\sigma \colon \Delta_p \to M$ be an arbitrary singular p-simplex in M. If σ is smooth, we just let $H_\sigma(x,t) = \sigma(x)$, and (i)–(iii) are easily verified (using the fact that the restriction of σ to each boundary face is also smooth).

Assume that σ is not smooth, and let S be the subset

$$S = \left(\Delta_p \times \{0\}\right) \cup \left(\partial\Delta_p \times I\right) \subseteq \Delta_p \times I$$

(the bottom and side faces of the "prism" $\Delta_p \times I$). Recall that $\partial\Delta_p$ is the union of the boundary faces $\partial_i \Delta_p$ for $i = 0, \ldots, p$, and for each i, the face map $F_{i,p}$ is a homeomorphism from Δ_{p-1} onto $\partial_i \Delta_p$. Define $H_0 \colon S \to M$ by

$$H_0(x,t) = \begin{cases} \sigma(x), & x \in \Delta_p, \ t = 0; \\ H_{\sigma \circ F_{i,p}}\left(F_{i,p}^{-1}(x),t\right), & x \in \partial_i \Delta_p, \ t \in I. \end{cases}$$

We need to check that the various definitions agree where they overlap, which implies that H_0 is continuous by the gluing lemma.

When $t = 0$, the inductive hypothesis (i) applied to the singular $(p-1)$-simplex $\sigma \circ F_{i,p}$ implies that $H_{\sigma \circ F_{i,p}}(x,0) = \sigma \circ F_{i,p}(x)$. It follows that

$$H_{\sigma \circ F_{i,p}}\left(F_{i,p}^{-1}(x),0\right) = \sigma(x),$$

so the different definitions of H_0 agree at points where $t = 0$.

Suppose now that x is a point in the intersection of two boundary faces $\partial_i \Delta_p$ and $\partial_j \Delta_p$, and assume without loss of generality that $i > j$. Since $F_{i,p} \circ F_{j,p-1}$ is a homeomorphism from Δ_{p-2} onto $\partial_i \Delta_p \cap \partial_j \Delta_p$, we can write $x = F_{i,p} \circ F_{j,p-1}(y)$ for some point $y \in \Delta_{p-2}$. Then (18.7) applied with $\sigma \circ F_{i,p}$ in place of σ and $F_{j-1,p}$ in place of $F_{i,p}$ implies that

$$H_{\sigma \circ F_{i,p}}\left(F_{i,p}^{-1}(x),t\right) = H_{\sigma \circ F_{i,p}}\left(F_{j,p-1}(y),t\right) = H_{\sigma \circ F_{i,p} \circ F_{j,p-1}}(y,t).$$

On the other hand, thanks to (18.1), we can also write $x = F_{j,p} \circ F_{i-1,p-1}(y)$, and then the same argument applied to $\sigma \circ F_{j,p}$ yields

$$H_{\sigma \circ F_{j,p}}\left(F_{j,p}^{-1}(x),t\right) = H_{\sigma \circ F_{j,p}}\left(F_{i-1,p-1}(y),t\right) = H_{\sigma \circ F_{j,p} \circ F_{i-1,p-1}}(y,t).$$

Because of (18.1), this shows that the two definitions of $H_0(x,t)$ agree.

To extend H_0 to all of $\Delta_p \times I$, we use the fact that there is a retraction from $\Delta_p \times I$ onto S. For example, if q_0 is any point in the interior of Δ_p, then the map $R \colon \Delta_p \times I \to S$ obtained by radially projecting from the point $(q_0, 2) \in \mathbb{R}^p \times \mathbb{R}$ is such a retraction (see Fig. 18.5). Extend H_0 to a continuous map $H \colon \Delta_p \times I \to M$ by setting $H(x,t) = H_0(R(x,t))$. Because H agrees with H_0 on S, it is a homotopy from σ to some other (continuous) singular simplex $\sigma'(x) = H(x,1)$, and it satisfies (18.7) by construction. Our only remaining task is to modify H so that it becomes a homotopy from σ to a *smooth* simplex.

Before we do so, we need to observe that the restriction of H to each boundary face $\partial_i \Delta_p \times \{1\}$ is smooth: since these faces lie in S, H agrees with H_0 on each of these sets, and hypothesis (i) applied to $\sigma \circ F_{i,p}$ shows that H_0 is smooth there. By virtue of Lemma 18.9 below, this implies that the restriction of H to the entire set

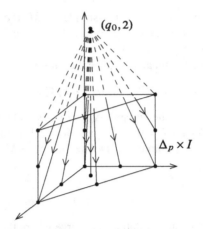

Fig. 18.5 A retraction from $\Delta_p \times I$ onto S

$\partial\Delta_p \times \{1\}$ is smooth. Let σ'' be any continuous extension of σ' to an open subset $U \subseteq \mathbb{R}^p$ containing Δ_p. (For example, σ'' could be defined by projecting points outside Δ_p to $\partial\Delta_p$ along radial lines from some point in the interior of Δ_p, and then applying σ'.) By the Whitney approximation theorem, σ'' is homotopic relative to $\partial\Delta_p$ to a smooth map, and restricting the homotopy to $\Delta_p \times I$ we obtain a homotopy $G\colon \sigma' \simeq \tilde\sigma$ from σ' to some smooth singular p-simplex $\tilde\sigma$, again relative to $\partial\Delta_p$.

Now let $u\colon \Delta_p \to \mathbb{R}$ be any continuous function that is equal to 1 on $\partial\Delta_p$ and satisfies $0 < u(x) < 1$ for $u \in \operatorname{Int}\Delta_p$. (For example, we could take $u\left(\sum_{0\le i\le p} t_i e_i\right) = 1 - t_0 t_1 \cdots t_p$, where $\sum_{0\le i\le p} t_i = 1$ and $e_0, \ldots, e_p$ are the vertices of Δ_p.) We combine the two homotopies H and G into a single homotopy $H_\sigma\colon \Delta_p \times I \to M$ by

$$H_\sigma(x,t) = \begin{cases} H\left(x, \dfrac{t}{u(x)}\right), & x \in \Delta_p,\ 0 \le t \le u(x), \\[2mm] G\left(x, \dfrac{t - u(x)}{1 - u(x)}\right), & x \in \operatorname{Int}\Delta_p,\ u(x) \le t \le 1. \end{cases}$$

Because $H(x,1) = \sigma'(x) = G(x,0)$, the gluing lemma shows that H_σ is continuous in $\operatorname{Int}\Delta_p \times I$. Also, $H_\sigma(x,t) = H\big(x, t/u(x)\big)$ in a neighborhood of $\partial\Delta_p \times [0,1)$, and thus is continuous there. It remains only to show that H_σ is continuous on $\partial\Delta_p \times \{1\}$. Let $x_0 \in \partial\Delta_p$ be arbitrary, and let $U \subseteq M$ be any neighborhood of $H_\sigma(x_0, 1) = H(x_0, 1)$. By continuity of H and u, there exists $\delta_1 > 0$ such that $H\big(x, t/u(x)\big) \in U$ whenever $|(x,t) - (x_0, 1)| < \delta_1$ and $0 \le t \le u(x)$. Since $G(x_0, t) = G(x_0, 0) = H(x_0, 1) = H_\sigma(x_0, 1) \in U$ for all $t \in I$, a simple compactness argument shows that there exists $\delta_2 > 0$ such that $|x - x_0| < \delta_2$ implies $G(x,t) \in U$ for all $t \in I$. Thus, if $|(x,t) - (x_0, 1)| < \min(\delta_1, \delta_2)$, we have $H_\sigma(x,t) \in U$ in both cases, showing that $(x_0, 1)$ has a neighborhood mapped into U by H_σ. Thus H_σ is continuous.

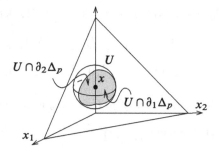

Fig. 18.6 Showing that f is smooth on $\partial \Delta_p$

It follows from the definition that $H_\sigma = H$ on $\partial \Delta_p \times I$, so (ii) is satisfied. For any $x \in \Delta_p$, $H_\sigma(x, 0) = H(x, 0) = \sigma(x)$. Moreover, when $x \in \text{Int}\, \Delta_p$, $H_\sigma(x, 1) = G(x, 1) = \tilde{\sigma}(x)$, and when $x \in \partial \Delta_p$, $H_\sigma(x, 1) = H(x, 1) = \sigma'(x) = \tilde{\sigma}(x)$ (because G is a homotopy relative to $\partial \Delta_p$). Thus, H_σ is a homotopy from σ to the smooth simplex $\tilde{\sigma}$, and (i) is satisfied as well. $\qquad\square$

Here is the lemma used in the preceding proof.

Lemma 18.9. *Let M be a smooth manifold, let Δ be a geometric p-simplex in $\mathbb{R}^n$, and let $f \colon \partial \Delta \to M$ be a continuous map whose restriction to each individual boundary face of Δ is smooth. Then f is smooth when considered as a map from the entire boundary $\partial \Delta$ to M.*

Proof. Let $(v_0, \dots, v_p)$ denote the vertices of Δ in some order, and for each $i = 0, \dots, p$, let $\partial_i \Delta = [v_0, \dots, \widehat{v_i}, \dots, v_p]$ be the boundary face opposite v_i. The hypothesis means that for each i and each $x \in \partial_i \Delta$, there exist an open subset $U_x \subseteq \mathbb{R}^n$ and a smooth map $\tilde{f} \colon U_x \to M$ whose restriction to $U_x \cap \partial_i \Delta$ agrees with f. We need to show that a single smooth extension can be chosen simultaneously for all the boundary faces containing x.

Suppose $x \in \partial \Delta$. Note that x is in one or more boundary faces of Δ, but cannot be in all of them. By reordering the vertices, we may assume that $x \in \partial_1 \Delta \cap \cdots \cap \partial_k \Delta$ for some $1 \le k \le p$, but $x \notin \partial_0 \Delta$. After composing with an affine diffeomorphism that takes v_i to e_i for $i = 0, \dots, p$, we may assume without loss of generality that $\Delta = \Delta_p$ and $x \notin \partial_0 \Delta_p$. Then the boundary faces containing x are precisely the intersections with Δ_p of the coordinate hyperplanes $x^1 = 0, \dots, x^k = 0$. For each i, there are a neighborhood U_i of x in $\mathbb{R}^n$ (which can be chosen disjoint from $\partial_0 \Delta_p$) and a smooth map $\tilde{f}_i \colon U_i \to M$ whose restriction to $U_i \cap \partial_i \Delta_p$ agrees with f.

Let $U = U_1 \cap \cdots \cap U_k$ (see Fig. 18.6). We show by induction on k that there is a smooth map $\tilde{f} \colon U \to M$ whose restriction to $U \cap \partial_i \Delta_p$ agrees with f for $i = 1, \dots, k$. Because the argument is local from this point on, after shrinking U if necessary we may replace M with a coordinate neighborhood of $f(x)$ that is diffeomorphic to $\mathbb{R}^m$; thus we henceforth identify M with $\mathbb{R}^m$.

For $k = 1$ there is nothing to prove, because $\tilde{f}_1$ is already such an extension. So suppose $k \ge 2$, and we have shown that there is a smooth map $\tilde{f}_0 \colon U \to M$ whose restriction to $U \cap \partial_i \Delta_p$ agrees with f for $i = 1, \ldots, k-1$. Define $\tilde{f}: U \to M$ by

$$\tilde{f}(x^1, \ldots, x^n) = \tilde{f}_0(x^1, \ldots, x^n) - \tilde{f}_0(x^1, \ldots, x^{k-1}, 0, x^{k+1}, \ldots, x^n)$$
$$+ \tilde{f}_k(x^1, \ldots, x^{k-1}, 0, x^{k+1}, \ldots, x^n).$$

For $i = 1, \ldots, k-1$, the restriction of $\tilde{f}$ to $U \cap \partial_i \Delta_p$ is given by

$$\tilde{f}(x^1, \ldots, x^{i-1}, 0, x^{i+1}, \ldots, x^n)$$
$$= \tilde{f}_0(x^1, \ldots, x^{i-1}, 0, x^{i+1}, \ldots, x^n)$$
$$- \tilde{f}_0(x^1, \ldots, x^{i-1}, 0, x^{i+1}, \ldots, x^{k-1}, 0, x^{k+1}, \ldots, x^n)$$
$$+ \tilde{f}_k(x^1, \ldots, x^{i-1}, 0, x^{i+1}, \ldots, x^{k-1}, 0, x^{k+1}, \ldots, x^n)$$
$$= f(x^1, \ldots, x^{i-1}, 0, x^{i+1}, \ldots, x^n),$$

since $\tilde{f}_0$ agrees with f when $x \in \Delta_p$ and $x^i = 0$, as does $\tilde{f}_k$ when $x \in \Delta_p$ and $x^k = 0$. Similarly, the restriction to $U \cap \partial_k \Delta_p$ is

$$\tilde{f}(x^1, \ldots, x^{k-1}, 0, x^{k+1}, \ldots, x^n) = \tilde{f}_0(x^1, \ldots, x^{k-1}, 0, x^{k+1}, \ldots, x^n)$$
$$- \tilde{f}_0(x^1, \ldots, x^{k-1}, 0, x^{k+1}, \ldots, x^n)$$
$$+ \tilde{f}_k(x^1, \ldots, x^{k-1}, 0, x^{k+1}, \ldots, x^n)$$
$$= f(x^1, \ldots, x^{k-1}, 0, x^{k+1}, \ldots, x^n).$$

This completes the inductive step and thus the proof.   □

Proof of Theorem 18.7. Let $i_0, i_1: \Delta_p \to \Delta_p \times I$ be the smooth embeddings $i_0(x) = (x, 0)$, $i_1(x) = (x, 1)$. Define a homomorphism $s: C_p(M) \to C_p^\infty(M)$ by setting

$$s\sigma = H_\sigma \circ i_1$$

for each singular p-simplex σ (where H_σ is the homotopy whose existence is proved in Lemma 18.8) and extending linearly to p-chains. Because of property (i) in Lemma 18.8, $s\sigma$ is a smooth p-simplex homotopic to σ.

Using (18.6), we can verify that s is a chain map: for each singular p-simplex σ,

$$s\partial\sigma = s \sum_{i=0}^{p} (-1)^i \sigma \circ F_{i,p} = \sum_{i=0}^{p} (-1)^i H_{\sigma \circ F_{i,p}} \circ i_1$$

$$= \sum_{i=0}^{p} (-1)^i H_\sigma \circ (F_{i,p} \times \mathrm{Id}_I) \circ i_1 = \sum_{i=0}^{p} (-1)^i H_\sigma \circ i_1 \circ F_{i,p}$$

$$= \partial(H_\sigma \circ i_1) = \partial s\sigma.$$

(In the fourth equality we used the fact that $(F_{i,p} \times \mathrm{Id}_I) \circ i_1(x) = \big(F_{i,p}(x), 1\big) = i_1 \circ F_{i,p}(x)$.) Therefore, s descends to a homomorphism $s_* \colon H_p(M) \to H_p^\infty(M)$. We will show that s_* is an inverse for $\iota_* \colon H_p^\infty(M) \to H_p(M)$.

First, observe that condition (iii) in Lemma 18.8 guarantees that $s \circ \iota$ is the identity map of $C_p^\infty(M)$, so clearly $s_* \circ \iota_*$ is the identity on $H_p^\infty(M)$. To show that $\iota_* \circ s_*$ is also the identity, we construct for each $p \geq 0$ a homotopy operator $h \colon C_p(M) \to C_{p+1}(M)$ satisfying

$$\partial \circ h + h \circ \partial = \iota \circ s - \mathrm{Id}_{C_p(M)}. \tag{18.8}$$

Once the existence of such an operator is known, it follows just as in the proof of Proposition 17.10 that $\iota_* \circ s_* = \mathrm{Id}_{H_p(M)}$: for any cycle $c \in C_p(M)$,

$$\iota_* \circ s_*[c] - [c] = [\iota \circ s(c) - c] = \big[\partial(hc) + h(\partial c)\big] = 0,$$

because $\partial c = 0$ and $\partial(hc)$ is a boundary.

To define the homotopy operator h, we need to introduce a family of affine singular simplices in the convex set $\Delta_p \times I \subseteq \mathbb{R}^p \times \mathbb{R}$. For each $i = 0, \dots, p$, let $E_i = (e_i, 0) \in \mathbb{R}^p \times \mathbb{R}$ and $E_i' = (e_i, 1) \in \mathbb{R}^p \times \mathbb{R}$, so that $E_0, \dots, E_p$ are the vertices of the geometric p-simplex $\Delta_p \times \{0\}$, and $E_0' \dots, E_p'$ are those of $\Delta_p \times \{1\}$. For each $i = 0, \dots, p$, let $G_{i,p} \colon \Delta_{p+1} \to \Delta_p \times I$ be the affine singular $(p+1)$-simplex

$$G_{i,p} = A\big(E_0, \dots, E_i, E_i', \dots, E_p'\big).$$

Thus, $G_{i,p}$ is the unique affine map that sends $e_0 \mapsto E_0, \dots, e_i \mapsto E_i, e_{i+1} \mapsto E_i', \dots,$ and $e_{p+1} \mapsto E_p'$. A routine computation shows that these maps compose with the face maps as follows:

$$G_{j,p} \circ F_{j,p+1} = G_{j-1,p} \circ F_{j,p+1} = A\big(E_0, \dots, E_{j-1}, E_j', \dots, E_p'\big). \tag{18.9}$$

In particular, this implies that

$$G_{p,p} \circ F_{p+1,p+1} = A(E_0, \dots, E_p) = i_0, \tag{18.10}$$

$$G_{0,p} \circ F_{0,p+1} = A(E_0', \dots, E_p') = i_1. \tag{18.11}$$

A similar computation shows that

$$(F_{j,p} \times \mathrm{Id}_I) \circ G_{i,p-1} = \begin{cases} G_{i+1,p} \circ F_{j,p+1}, & i \geq j, \\ G_{i,p} \circ F_{j+1,p+1}, & i < j. \end{cases} \tag{18.12}$$

We define $h \colon C_p(M) \to C_{p+1}(M)$ as follows:

$$h\sigma = \sum_{i=0}^{p} (-1)^i H_\sigma \circ G_{i,p}.$$

The proof that it satisfies the homotopy formula (18.8) is just a laborious computation using (18.7), (18.9), and (18.12):

$$
h(\partial\sigma) = h\sum_{j=0}^{p}(-1)^j \sigma \circ F_{j,p} = \sum_{i=0}^{p-1}\sum_{j=0}^{p}(-1)^{i+j} H_{\sigma \circ F_{j,p}} \circ G_{i,p-1}
$$

$$
= \sum_{i=0}^{p-1}\sum_{j=0}^{p}(-1)^{i+j} H_\sigma \circ (F_{j,p} \times \mathrm{Id}_I) \circ G_{i,p-1}
$$

$$
= \sum_{0 \le j \le i \le p-1}(-1)^{i+j} H_\sigma \circ G_{i+1,p} \circ F_{j,p+1}
$$

$$
+ \sum_{0 \le i < j \le p}(-1)^{i+j} H_\sigma \circ G_{i,p} \circ F_{j+1,p+1}, \tag{18.13}
$$

while

$$
\partial(h\sigma) = \partial\sum_{i=0}^{p}(-1)^i H_\sigma \circ G_{i,p} = \sum_{j=0}^{p+1}\sum_{i=0}^{p}(-1)^{i+j} H_\sigma \circ G_{i,p} \circ F_{j,p+1}.
$$

Writing separately the terms in $\partial(h\sigma)$ for which $i < j-1$, $i = j-1$, $i = j$, and $i > j$, we get

$$
\partial(h\sigma) = \sum_{\substack{0 \le i < j-1 \\ j \le p+1}}(-1)^{i+j} H_\sigma \circ G_{i,p} \circ F_{j,p+1} - \sum_{1 \le j \le p+1} H_\sigma \circ G_{j-1,p} \circ F_{j,p+1}
$$

$$
+ \sum_{0 \le j \le p} H_\sigma \circ G_{j,p} \circ F_{j,p+1} + \sum_{0 \le j < i \le p}(-1)^{i+j} H_\sigma \circ G_{i,p} \circ F_{j,p+1}.
$$

After substituting $j = j' + 1$ in the first of these four sums and $i = i' + 1$ in the last, we see that the first and last sums exactly cancel the two sums in the expression (18.13) for $h(\partial\sigma)$. Using (18.9), all the terms in the middle two sums cancel each other except those in which $j = 0$ and $j = p + 1$. Thanks to (18.10) and (18.11), these two terms simplify to

$$
h(\partial\sigma) + \partial(h\sigma) = -H_\sigma \circ G_{p,p} \circ F_{p+1,p+1} + H_\sigma \circ G_{0,p} \circ F_{0,p+1}
$$

$$
= -H_\sigma \circ i_0 + H_\sigma \circ i_1 = -\sigma + s\sigma.
$$

Since ι is an inclusion map, $s\sigma = \iota \circ s\sigma$ for any singular p-simplex σ, so this completes the proof. $\square$

The de Rham Theorem

In this section we state and prove the de Rham theorem. Before getting to the theorem itself, we need one more algebraic lemma. Its proof is another diagram chase like the proof of the zigzag lemma.

Lemma 18.10 (The Five Lemma). *Consider the following commutative diagram of modules and linear maps:*

$$
\begin{array}{ccccccccc}
A_1 & \xrightarrow{\alpha_1} & A_2 & \xrightarrow{\alpha_2} & A_3 & \xrightarrow{\alpha_3} & A_4 & \xrightarrow{\alpha_4} & A_5 \\
\downarrow{f_1} & & \downarrow{f_2} & & \downarrow{f_3} & & \downarrow{f_4} & & \downarrow{f_5} \\
B_1 & \xrightarrow{\beta_1} & B_2 & \xrightarrow{\beta_2} & B_3 & \xrightarrow{\beta_3} & B_4 & \xrightarrow{\beta_4} & B_5.
\end{array}
$$

If the horizontal rows are exact and f_1, f_2, f_4, and f_5 are isomorphisms, then f_3 is also an isomorphism.

▶ **Exercise 18.11.** Prove (or look up) the five lemma.

Suppose M is a smooth manifold, ω is a closed p-form on M, and σ is a smooth p-simplex in M. We define the ***integral of ω over σ*** to be

$$
\int_\sigma \omega = \int_{\Delta_p} \sigma^* \omega.
$$

This makes sense because Δ_p is a smooth p-submanifold with corners embedded in $\mathbb{R}^p$, and it inherits the orientation of $\mathbb{R}^p$. (Or we could just consider Δ_p as a domain of integration in $\mathbb{R}^p$.) Observe that when $p = 1$, this is the same as the line integral of ω over the smooth curve segment $\sigma : [0, 1] \to M$. If $c = \sum_{i=1}^k c_i \sigma_i$ is a smooth p-chain, the integral of ω over c is defined as

$$
\int_c \omega = \sum_{i=1}^k c_i \int_{\sigma_i} \omega.
$$

Theorem 18.12 (Stokes's Theorem for Chains). *If c is a smooth p-chain in a smooth manifold M, and ω is a smooth $(p-1)$-form on M, then*

$$
\int_{\partial c} \omega = \int_c d\omega.
$$

Proof. It suffices to prove the theorem when c is just a smooth simplex σ. Since Δ_p is a manifold with corners, Stokes's theorem says that

$$
\int_\sigma d\omega = \int_{\Delta_p} \sigma^* d\omega = \int_{\Delta_p} d\sigma^* \omega = \int_{\partial \Delta_p} \sigma^* \omega.
$$

The maps $\{F_{i,p} : 0 = 1, \ldots, p\}$ are parametrizations of the boundary faces of Δ_p satisfying the conditions of Proposition 16.21, except possibly that they might not be orientation-preserving. To check the orientations, note that $F_{i,p}$ is the restriction to $\Delta_p \cap \partial \mathbb{H}^p$ of the affine diffeomorphism sending the simplex $[e_0, \ldots, e_p]$ to $[e_0, \ldots, \widehat{e_i}, \ldots, e_p, e_i]$. This is easily seen to be orientation-preserving if and only if $(e_0, \ldots, \widehat{e_i}, \ldots, e_p, e_i)$ is an even permutation of $(e_0, \ldots, e_p)$, which is the case if and only if $p - i$ is even. Since the standard coordinates on $\partial \mathbb{H}^p$ are positively oriented if and only if p is even, the upshot is that $F_{i,p}$ is orientation-preserving for $\partial\Delta_p$ if and only if i is even. Thus, by Proposition 16.21,

$$\int_{\partial\Delta_p} \sigma^*\omega = \sum_{i=0}^{p}(-1)^i \int_{\Delta_{p-1}} F_{i,p}^*\sigma^*\omega = \sum_{i=0}^{p}(-1)^i \int_{\Delta_{p-1}} (\sigma \circ F_{i,p})^*\omega$$

$$= \sum_{i=0}^{p}(-1)^i \int_{\sigma\circ F_{i,p}} \omega.$$

By definition of the singular boundary operator, this is equal to $\int_{\partial\sigma}\omega$. $\qquad\square$

Using this theorem, we define a natural linear map $\mathcal{I}\colon H_{\mathrm{dR}}^p(M) \to H^p(M;\mathbb{R})$, called the *de Rham homomorphism*, as follows. For any $[\omega] \in H_{\mathrm{dR}}^p(M)$ and $[c] \in H_p(M) \cong H_p^\infty(M)$, we define

$$\mathcal{I}[\omega][c] = \int_{\widetilde{c}} \omega, \qquad (18.14)$$

where $\widetilde{c}$ is any smooth p-cycle representing the homology class $[c]$. This is well defined, because if $\widetilde{c}$, $\widetilde{c}'$ are smooth cycles representing the same homology class, then Theorem 18.7 guarantees that $\widetilde{c} - \widetilde{c}' = \partial\widetilde{b}$ for some smooth $(p+1)$-chain $\widetilde{b}$, which implies

$$\int_{\widetilde{c}} \omega - \int_{\widetilde{c}'} \omega = \int_{\partial\widetilde{b}} \omega = \int_{\widetilde{b}} d\omega = 0,$$

while if $\omega = d\eta$ is exact, then

$$\int_{\widetilde{c}} \omega = \int_{\widetilde{c}} d\eta = \int_{\partial\widetilde{c}} \eta = 0.$$

(Note that $\partial\widetilde{c} = 0$ because $\widetilde{c}$ represents a homology class, and $d\omega = 0$ because ω represents a cohomology class.) Clearly, $\mathcal{I}[\omega][c+c'] = \mathcal{I}[\omega][c] + \mathcal{I}[\omega][c']$, and the resulting homomorphism $\mathcal{I}[\omega]\colon H_p(M) \to \mathbb{R}$ depends linearly on ω. Thus, $\mathcal{I}[\omega]$ is a well-defined element of $\mathrm{Hom}(H_p(M),\mathbb{R}) \cong H^p(M;\mathbb{R})$.

Proposition 18.13 (Naturality of the de Rham Homomorphism). *For a smooth manifold M and nonnegative integer p, let $\mathcal{I}\colon H_{\mathrm{dR}}^p(M) \to H^p(M;\mathbb{R})$ denote the de Rham homomorphism.*

(a) *If $F\colon M \to N$ is a smooth map, then the following diagram commutes:*

$$
\begin{array}{ccc}
H_{\mathrm{dR}}^p(N) & \xrightarrow{\ F^*\ } & H_{\mathrm{dR}}^p(M) \\[4pt]
\mathcal{I}\big\downarrow & & \big\downarrow\mathcal{I} \\[4pt]
H^p(N;\mathbb{R}) & \xrightarrow[\ F^*\ ]{} & H^p(M;\mathbb{R}).
\end{array}
$$

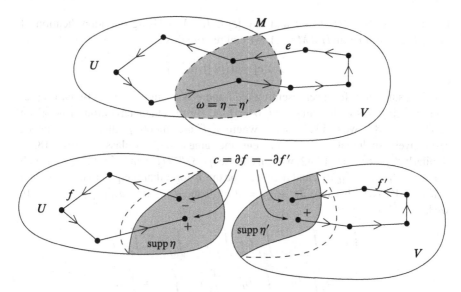

Fig. 18.7 Naturality of $\mathcal{I}$ with respect to connecting homomorphisms

(b) *If M is a smooth manifold and U, V are open subsets of M whose union is M, then the following diagram commutes:*

$$
\begin{array}{ccc}
H_{\mathrm{dR}}^{p-1}(U \cap V) & \xrightarrow{\ \delta\ } & H_{\mathrm{dR}}^{p}(M) \\
\Big\downarrow{\scriptstyle \mathcal{I}} & & \Big\downarrow{\scriptstyle \mathcal{I}} \\
H^{p-1}(U \cap V; \mathbb{R}) & \xrightarrow[\partial^{*}]{} & H^{p}(M; \mathbb{R}),
\end{array}
\tag{18.15}
$$

where δ and ∂^{*} are the connecting homomorphisms of the Mayer–Vietoris sequences for de Rham and singular cohomology, respectively.

Proof. Directly from the definitions, if σ is a smooth p-simplex in M and ω is a smooth p-form on N,

$$
\int_{\sigma} F^{*}\omega = \int_{\Delta_{p}} \sigma^{*}F^{*}\omega = \int_{\Delta_{p}} (F \circ \sigma)^{*}\omega = \int_{F \circ \sigma} \omega.
$$

This implies

$$
\mathcal{I}\big(F^{*}[\omega]\big)[\sigma] = \mathcal{I}[\omega][F \circ \sigma] = \mathcal{I}[\omega]\big(F_{*}[\sigma]\big) = F^{*}\big(\mathcal{I}[\omega]\big)[\sigma],
$$

which proves (a).

Now consider (b). Commutativity of this diagram means

$$
\mathcal{I}\big(\delta[\omega]\big)[e] = \big(\partial^{*}\mathcal{I}[\omega]\big)[e]
$$

for any $[\omega] \in H_{\mathrm{dR}}^{p-1}(U \cap V)$ and any $[e] \in H_p(M)$. Using our identification of $H^p(M; \mathbb{R})$ with $\mathrm{Hom}(H_p(M), \mathbb{R})$, we can rewrite this as

$$\mathcal{I}\big(\delta[\omega]\big)[e] = \mathcal{I}\big([\omega]\big)\big(\partial_*[e]\big).$$

If σ is a smooth p-form representing $\delta[\omega]$ and c is a smooth $(p-1)$-chain representing $\partial_*[e]$, this is the same as $\int_e \sigma = \int_c \omega$. By the characterization of ∂_* given in Theorem 18.4, we can let $c = \partial f$, where f, f' are smooth p-chains in U and V, respectively, such that $f + f'$ represents the same homology class as e (Fig. 18.7). Similarly, by Corollary 17.42, we can choose $\eta \in \Omega^{p-1}(U)$ and $\eta' \in \Omega^{p-1}(V)$ such that $\omega = \eta|_{U \cap V} - \eta'|_{U \cap V}$, and then let σ be the p-form that is equal to $d\eta$ on U and to $d\eta'$ on V. Then, because $\partial f + \partial f' = \partial e = 0$ and $d\eta|_{U \cap V} - d\eta'|_{U \cap V} = d\omega = 0$, we have

$$\int_c \omega = \int_{\partial f} \omega = \int_{\partial f} \eta - \int_{\partial f} \eta' = \int_{\partial f} \eta + \int_{\partial f'} \eta'$$

$$= \int_f d\eta + \int_{f'} d\eta' = \int_f \sigma + \int_{f'} \sigma = \int_e \sigma.$$

Thus the diagram commutes. $\square$

Theorem 18.14 (de Rham). *For every smooth manifold M and nonnegative integer p, the de Rham homomorphism $\mathcal{I}\colon H_{\mathrm{dR}}^p(M) \to H^p(M; \mathbb{R})$ is an isomorphism.*

Proof. Let us say that a smooth manifold M is a *de Rham manifold* if the homomorphism $\mathcal{I}\colon H_{\mathrm{dR}}^p(M) \to H^p(M; \mathbb{R})$ is an isomorphism for each p. Since $\mathcal{I}$ commutes with the cohomology maps induced by smooth maps (Proposition 18.13), any manifold that is diffeomorphic to a de Rham manifold is also de Rham. The theorem will be proved once we show that every smooth manifold is de Rham.

If M is any smooth manifold, let us call an open cover $\{U_i\}$ of M a *de Rham cover* if each subset U_i is a de Rham manifold, and every finite intersection $U_{i_1} \cap \cdots \cap U_{i_k}$ is de Rham. A de Rham cover that is also a basis for the topology of M is called a *de Rham basis* for M.

STEP 1: *If $\{M_j\}$ is any countable collection of de Rham manifolds, then their disjoint union is de Rham.* By Propositions 17.5 and 18.5(b), for both de Rham and singular cohomology the inclusions $\iota_j\colon M_j \hookrightarrow \coprod_j M_j$ induce isomorphisms between the cohomology groups of the disjoint union and the direct product of the cohomology groups of the manifolds M_j. By Proposition 18.13, $\mathcal{I}$ commutes with these isomorphisms.

STEP 2: *Every convex open subset of $\mathbb{R}^n$ is de Rham.* Let U be such a subset. By the Poincaré lemma, $H_{\mathrm{dR}}^p(U)$ is trivial when $p \neq 0$. Since U is homotopy equivalent to a one-point space, Proposition 18.5 implies that the singular cohomology groups of U are also trivial for $p \neq 0$. In the $p = 0$ case, $H_{\mathrm{dR}}^0(U)$ is the 1-dimensional space consisting of the constant functions, and $H^0(U; \mathbb{R}) = \mathrm{Hom}\big(H_0(U), \mathbb{R}\big)$ is also 1-dimensional because $H_0(U)$ is generated by any singular 0-simplex. If $\sigma\colon \Delta_0 \to M$ is a singular 0-simplex (which is smooth because any map from a 0-manifold is smooth), and f is the constant function equal to 1, then

$$\mathcal{I}[f][\sigma] = \int_{\Delta_0} \sigma^* f = (f \circ \sigma)(0) = 1.$$

Thus $\mathcal{I} \colon H^0_{dR}(U) \to H^0(U; \mathbb{R})$ is not the zero map, so it is an isomorphism.

STEP 3: *If M has a finite de Rham cover, then M is de Rham.* This is the heart of the proof. Suppose $M = U_1 \cup \cdots \cup U_k$, where the open subsets U_i and their finite intersections are de Rham. We prove the result by induction on k. For $k = 1$, the result is obvious. Suppose next that M has a de Rham cover consisting of two sets $\{U, V\}$. Putting together the Mayer–Vietoris sequences for de Rham and singular cohomology, we obtain the following commutative diagram, in which the horizontal rows are exact and the vertical maps are all de Rham homomorphisms:

$$
\begin{array}{ccccc}
H^{p-1}_{dR}(U) \oplus H^{p-1}_{dR}(V) & \longrightarrow & H^{p-1}_{dR}(U \cap V) & \longrightarrow & H^p_{dR}(M) \longrightarrow \\
\downarrow & & \downarrow & & \downarrow \\
H^{p-1}(U; \mathbb{R}) \oplus H^{p-1}(V; \mathbb{R}) & \longrightarrow & H^{p-1}(U \cap V; \mathbb{R}) & \longrightarrow & H^p(M; \mathbb{R}) \longrightarrow
\end{array}
$$

$$
\begin{array}{ccc}
H^p_{dR}(U) \oplus H^p_{dR}(V) & \longrightarrow & H^p_{dR}(U \cap V) \\
\downarrow & & \downarrow \\
H^p(U; \mathbb{R}) \oplus H^p(V; \mathbb{R}) & \longrightarrow & H^p(U \cap V; \mathbb{R}).
\end{array}
$$

The commutativity of the diagram is an immediate consequence of Proposition 18.13. By hypothesis the first, second, fourth, and fifth vertical maps are all isomorphisms, so by the five lemma the middle map is an isomorphism, which proves that M is de Rham.

Now assume the claim is true for smooth manifolds admitting a de Rham cover with $k \geq 2$ sets, and suppose $\{U_1, \ldots, U_{k+1}\}$ is a de Rham cover of M. Define $U = U_1 \cup \cdots \cup U_k$ and $V = U_{k+1}$. The hypothesis implies that U and V are de Rham, and $U \cap V$ is also de Rham because it has a k-fold de Rham cover given by $\{U_1 \cap U_{k+1}, \ldots, U_k \cap U_{k+1}\}$. Therefore, $M = U \cup V$ is also de Rham by the argument above.

STEP 4: *If M has a de Rham basis, then M is de Rham.* Suppose $\{U_\alpha\}$ is a de Rham basis for M. Let $f \colon M \to \mathbb{R}$ be an exhaustion function (see Proposition 2.28). For each integer m, define subsets A_m and A'_m of M by

$$A_m = \{q \in M : m \leq f(q) \leq m + 1\},$$
$$A'_m = \{q \in M : m - \tfrac{1}{2} < f(q) < m + \tfrac{3}{2}\}.$$

(See Fig. 18.8.) For each point $q \in A_m$, there is a basis open subset containing q and contained in A'_m. The collection of all such basis sets is an open cover of A_m. Since f is an exhaustion function, A_m is compact, and therefore it is covered by finitely many of these basis sets. Let B_m be the union of this finite collection of sets. This is a finite de Rham cover of B_m, so by Step 3, B_m is de Rham.

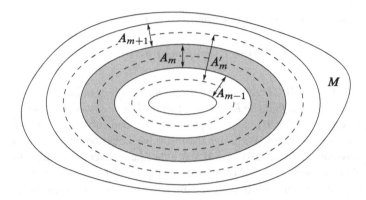

Fig. 18.8 Proof of the de Rham theorem, Step 4

Observe that $B_m \subseteq A'_m$, so B_m can have nonempty intersection with $B_{\widetilde{m}}$ only when $\widetilde{m} = m - 1, m$, or $m + 1$. Therefore, if we define

$$U = \bigcup_{m \text{ odd}} B_m, \qquad V = \bigcup_{m \text{ even}} B_m,$$

then U and V are disjoint unions of de Rham manifolds, and so they are both de Rham by Step 1. Finally, $U \cap V$ is de Rham because it is the disjoint union of the sets $B_m \cap B_{m+1}$ for $m \in \mathbb{Z}$, each of which has a finite de Rham cover consisting of sets of the form $U_\alpha \cap U_\beta$, where U_α and U_β are basis sets used to define B_m and B_{m+1}, respectively. Thus $M = U \cup V$ is de Rham by Step 3.

STEP 5: *Every open subset of $\mathbb{R}^n$ is de Rham.* If $U \subseteq \mathbb{R}^n$ is such a subset, then U has a basis consisting of Euclidean balls. Because each ball is convex, it is de Rham, and because any finite intersection of balls is again convex, finite intersections are also de Rham. Thus, U has a de Rham basis, so it is de Rham by Step 4.

STEP 6: *Every smooth manifold is de Rham.* Any smooth manifold has a basis of smooth coordinate domains. Since every smooth coordinate domain is diffeomorphic to an open subset of $\mathbb{R}^n$, as are their finite intersections, this is a de Rham basis. The claim therefore follows from Step 4. □

This result expresses a deep connection between the topological and analytic properties of a smooth manifold, and plays a central role in differential geometry. If one has some information about the topology of a manifold M, the de Rham theorem can be used to draw conclusions about solutions to differential equations such as $d\eta = \omega$ on M. Conversely, if one can prove that such solutions do or do not exist, then one can draw conclusions about the topology.

As befits so fundamental a theorem, the de Rham theorem has many and varied proofs. The elegant proof given here is due to Glen E. Bredon [Bre93]. Another common approach is via the theory of sheaves; for example, a proof using this technique can be found in [War83]. The sheaf-theoretic proof is extremely powerful and lends itself to countless generalizations, but it has two significant disadvantages that prevent it from being useful for our purposes: it requires the entire technical apparatus of sheaf theory and sheaf cohomology, which would take us too far afield; and although it produces an isomorphism between de Rham and singular cohomology, it is not easy to see that the isomorphism is given specifically by integration. Nonetheless, because the technique leads to other important applications in such fields as differential geometry, algebraic geometry, algebraic topology, and complex analysis, it is worth taking some time and effort to study it if you get the opportunity.

Problems

18-1. Suppose M is an oriented smooth manifold and ω is a closed p-form on M.
  (a) Show that ω is exact if and only if the integral of ω over every smooth p-cycle is zero.
  (b) Suppose $H_p(M)$ is generated by the homology classes of finitely many smooth p-cycles $\{c_1, \ldots, c_m\}$. Define real numbers $P_1(\omega), \ldots, P_m(\omega)$, called the *periods of* ω with respect to this set of generators, by $P_i(\omega) = \int_{c_i} \omega$. Show that ω is exact if and only if all of its periods are zero. [Remark: if you look back now at Problem 11-17, you will see that it is essentially proving the same theorem in the special case of a 1-form on $\mathbb{T}^n$.]

18-2. If G is a group, the *commutator subgroup of* G, denoted by $[G, G]$, is the smallest normal subgroup containing all elements of the form $g_1 g_2 g_1^{-1} g_2^{-1}$ for $g_1, g_2 \in G$; and the *abelianization of* G, denoted by $\mathrm{Ab}(G)$, is the quotient group $G/[G, G]$. Suppose M is a connected smooth manifold and $q \in M$. It can be shown that there is a group homomorphism from $\pi_1(M, q)$ to $H_1(M)$ that sends the homotopy class of a loop γ to the homology class of the 1-cycle determined by γ, and this map descends to an isomorphism from $\mathrm{Ab}\big(\pi_1(M, q)\big)$ to $H_1(M)$ (see [LeeTM, Thm. 13.14]). Use this result together with the de Rham theorem to prove that the map $\Phi \colon H^1_{\mathrm{dR}}(M) \to \mathrm{Hom}\big(\pi_1(M, q), \mathbb{R}\big)$ of Theorem 17.17 is an isomorphism.

18-3. Let M be a smooth n-manifold and suppose $S \subseteq M$ is an oriented compact embedded p-dimensional submanifold. A *smooth triangulation of* S is a smooth p-chain $c = \sum_i \sigma_i$ in M with the following properties:
  (i) Each $\sigma_i \colon \Delta_p \to S$ is a smooth orientation-preserving embedding.
  (ii) If $i \neq j$, then $\sigma_i(\mathrm{Int}\, \Delta_p) \cap \sigma_j(\mathrm{Int}\, \Delta_p) = \varnothing$.
  (iii) $S = \bigcup_i \sigma_i(\Delta_p)$.
  (iv) $\partial c = 0$.
  (It can be shown that every oriented compact embedded submanifold admits a smooth triangulation, but we will not use that fact; see [Mun66] for a proof.) Two oriented compact embedded p-dimensional submanifolds $S, S' \subseteq M$ are said to be *homologous* if there exist smooth triangulations c for S and c' for S' such that $c - c'$ is a boundary.
  (a) Show that for any smooth triangulation c of S and any smooth p-form ω on M, we have $\int_c \omega = \int_S \omega$.

(b) Show that if ω is closed and S, S' are homologous, then $\int_S \omega = \int_{S'} \omega$.

18-4. Suppose (M, g) is a Riemannian n-manifold. A smooth p-form ω on M is called a **calibration** if ω is closed and $\omega_x(v_1, \ldots, v_p) \leq 1$ whenever $(v_1, \ldots, v_p)$ are orthonormal vectors in some tangent space $T_x M$. An oriented embedded p-dimensional submanifold $S \subseteq M$ is said to be **calibrated** if there is a calibration ω such that the pullback $\iota_S^* \omega$ is the volume form for the induced Riemannian metric on S. Suppose $S \subseteq M$ is a smoothly triangulated calibrated compact submanifold. Prove that the volume of S (with respect to the induced Riemannian metric) is less than or equal to that of any other submanifold homologous to S (see Problem 18-3). [Remark: calibrations were invented in 1982 by Reese Harvey and Blaine Lawson [HL82]; they have become increasingly important in recent years because in many situations a calibration is the only known way of proving that a given submanifold is volume-minimizing in its homology class.]

18-5. Let $D \subseteq \mathbb{R}^3$ be the surface obtained by revolving the circle $(r - 2)^2 + z^2 = 1$ around the z-axis, with the induced Riemannian metric (see Example 13.18(a)), and let $C \subseteq D$ be the "inner circle" defined by $C = \{(x, y, z) : z = 0, x^2 + y^2 = 1\}$. Show that C is calibrated, and therefore is the shortest curve in its homology class.

18-6. For any smooth manifold M, let $H_c^p(M)$ denote the pth compactly supported de Rham cohomology group of M.

    (a) Given an open subset $U \subseteq M$, let $\iota : U \hookrightarrow M$ denote the inclusion map, and define a linear map $\iota_\# : \Omega_c^p(U) \to \Omega_c^p(M)$ by extending each compactly supported form to be zero on $M \smallsetminus U$. Show that $d \circ \iota_\# = \iota_\# \circ d$, and so $\iota_\#$ induces a linear map on compactly supported cohomology, denoted by $\iota_* : H_c^p(U) \to H_c^p(M)$.

    (b) MAYER–VIETORIS WITH COMPACT SUPPORTS: Suppose M is a smooth manifold and $U, V \subseteq M$ are open subsets whose union is M. Prove that for each nonnegative integer p, there is a linear map $\delta_* : H_c^p(M) \to H_c^{p+1}(U \cap V)$ such that the following sequence is exact:

$$\cdots \xrightarrow{\delta_*} H_c^p(U \cap V) \xrightarrow{i_* \oplus (-j_*)} H_c^p(U) \oplus H_c^p(V)$$

$$\xrightarrow{k_* + l_*} H_c^p(M) \xrightarrow{\delta_*} H_c^{p+1}(U \cap V) \xrightarrow{i_* \oplus (-j_*)} \cdots,$$

where i, j, k, l are the inclusion maps as in (17.6).

    (c) Let $H_c^p(M)^*$ denote the algebraic dual space to $H_c^p(M)$, that is, the vector space of all linear maps from $H_c^p(M)$ to $\mathbb{R}$. Show that the following sequence is also exact:

$$\cdots \xrightarrow{(\delta_*)^*} H_c^p(M)^* \xrightarrow{(k_*)^* \oplus (l_*)^*} H_c^p(U)^* \oplus H_c^p(V)^*$$

$$\xrightarrow{(i_*)^* - (j_*)^*} H_c^p(U \cap V)^* \xrightarrow{(\delta_*)^*} H_c^{p-1}(M)^* \xrightarrow{(k_*)^* \oplus (l_*)^*} \cdots.$$

$$\tag{18.16}$$

18-7. THE POINCARÉ DUALITY THEOREM: Let M be an oriented smooth n-manifold. Define a map PD: $\Omega^p(M) \to \Omega_c^{n-p}(M)^*$ by

$$PD(\omega)(\eta) = \int_M \omega \wedge \eta.$$

(a) Show that PD descends to a linear map (still denoted by the same symbol) PD: $H_{dR}^p(M) \to H_c^{n-p}(M)^*$.

(b) Show that PD is an isomorphism for each p. [Hint: imitate the proof of the de Rham theorem, with "de Rham manifold" replaced by "PD manifold." You will need Lemma 17.27 and Problem 18-6.]

18-8. Let M be a compact smooth n-manifold.

(a) Show that all de Rham groups of M are finite-dimensional. [Hint: for the orientable case, use Poincaré duality to show that $H_{dR}^p(M) \cong H_{dR}^p(M)^{**}$, and use the result of Problem 11-2. For the nonorientable case, use Lemma 17.33.]

(b) Show that if M is orientable, then $\dim H_{dR}^p(M) = \dim H_{dR}^{n-p}(M)$ for all p.

18-9. Let M be a smooth n-manifold all of whose de Rham groups are finite-dimensional. (Problem 18-8 shows that this is always the case when M is compact.) The *Euler characteristic of M* is the number

$$\chi(M) = \sum_{p=0}^{n} (-1)^p \dim H_{dR}^p(M).$$

Show that $\chi(M)$ is a homotopy invariant of M, and $\chi(M) = 0$ when M is compact, orientable, and odd-dimensional.

Chapter 19
Distributions and Foliations

Suppose V is a nonvanishing vector field on a smooth manifold M. The results of Chapter 9 imply that each integral curve of V is a smooth immersion, and that locally the images of the integral curves fit together nicely like parallel lines in Euclidean space. The fundamental theorem on flows tells us that these curves are determined by the knowledge of their velocity vectors.

In this chapter we explore an important generalization of this idea to higher-dimensional submanifolds. The general setup is this: suppose M is a smooth manifold, and we are given a k-dimensional subbundle of TM, called a *distribution* on M. Is there a k-dimensional submanifold (called an *integral manifold* of the distribution) whose tangent space at each point is the given subspace? The answer in this case is more complicated than in the case of vector fields: there is a nontrivial necessary condition, called *involutivity*, that must be satisfied by the distribution.

In the first section of the chapter, we define involutivity and give examples of both involutive and noninvolutive distributions. Next, we show how the involutivity condition can be rephrased in terms of differential forms.

The main theorem of this chapter, the *Frobenius theorem*, tells us that involutivity is also sufficient for the existence of an integral manifold through each point. We prove the Frobenius theorem in two forms. First, we prove a local form, which says that a neighborhood of every point is filled up with integral manifolds, fitting together nicely like parallel affine subspaces of $\mathbb{R}^n$. Then we prove a global form, which says that associated with each involutive distribution is a partition of the entire manifold into immersed integral manifolds fitting together nicely, called a *foliation*.

In the next section, we apply the theory of foliations to prove an important result in the theory of Lie groups. We already know that to each Lie subgroup of a Lie group G, there corresponds a Lie subalgebra of $\mathrm{Lie}(G)$. Using the theory of foliations, we prove that the correspondence also goes the other way: every Lie subalgebra of $\mathrm{Lie}(G)$ corresponds to some Lie subgroup of G.

At the end of the chapter, we give a few applications of the Frobenius theorem to partial differential equations.

J.M. Lee, *Introduction to Smooth Manifolds*, Graduate Texts in Mathematics 218,   490
DOI 10.1007/978-1-4419-9982-5_19, © Springer Science+Business Media New York 2013

Distributions and Involutivity

Let M be a smooth manifold. A *distribution on M of rank k* is a rank-k subbundle of TM. It is called a *smooth distribution* if it is a smooth subbundle. Distributions are also sometimes called *tangent distributions* (especially if there is any opportunity for confusion with the use of the term *distribution* for generalized functions in analysis), *k-plane fields*, or *tangent subbundles*.

Often a rank-k distribution is described by specifying for each $p \in M$ a linear subspace $D_p \subseteq T_p M$ of dimension k, and letting $D = \bigcup_{p \in M} D_p$. It then follows from the local frame criterion for subbundles (Lemma 10.32) that D is a smooth distribution if and only if each point of M has a neighborhood U on which there are smooth vector fields $X_1, \ldots, X_k \colon U \to TM$ such that $X_1|_q, \ldots, X_k|_q$ form a basis for D_q at each $q \in U$. In this case, we say that D is the distribution (*locally*) *spanned by the vector fields $X_1, \ldots, X_k$*.

Integral Manifolds and Involutivity

Suppose $D \subseteq TM$ is a smooth distribution. A nonempty immersed submanifold $N \subseteq M$ is called an *integral manifold of D* if $T_p N = D_p$ at each point $p \in N$. The main question we want to address in this chapter is that of the existence of integral manifolds.

Before we proceed with the general theory, let us describe some examples of distributions and integral manifolds that you should keep in mind.

Example 19.1 (Distributions and Integral Manifolds).

(a) If V is a nowhere-vanishing smooth vector field on a manifold M, then V spans a smooth rank-1 distribution on M (see Example 10.33(a)). The image of any integral curve of V is an integral manifold of D.

(b) In $\mathbb{R}^n$, the vector fields $\partial/\partial x^1, \ldots, \partial/\partial x^k$ span a smooth distribution of rank k. The k-dimensional affine subspaces parallel to $\mathbb{R}^k$ are integral manifolds.

(c) Let R be the distribution on $\mathbb{R}^n \smallsetminus \{0\}$ spanned by the unit radial vector field $x^i \partial/\partial x^i$, and let $R^\perp$ be its orthogonal complement bundle (see Lemma 10.35). Then $R^\perp$ is a smooth rank-$(n-1)$ distribution on $\mathbb{R}^n \smallsetminus \{0\}$. Through each point $x \in \mathbb{R}^n \smallsetminus \{0\}$, the sphere of radius $|x|$ around 0 is an integral manifold of $R^\perp$.

(d) Let D be the smooth distribution on $\mathbb{R}^3$ spanned by the following vector fields:

$$X = \frac{\partial}{\partial x} + y\frac{\partial}{\partial z}, \qquad Y = \frac{\partial}{\partial y}.$$

(See Fig. 19.1.) It turns out that D has no integral manifolds. To get an idea why, suppose N is an integral manifold through the origin. Because X and Y are tangent to N, any integral curve of X or Y that starts in N has to stay in N, at least for a short time (Problem 9-2). Thus, N contains an open subset of the x-axis (which is an integral curve of X). It also contains, for each sufficiently small x, an open subset of the line parallel to the y-axis and passing through

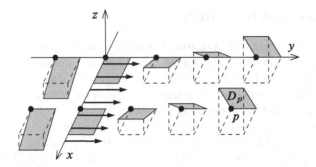

Fig. 19.1 A smooth distribution with no integral manifolds

$(x, 0, 0)$ (which is an integral curve of Y). Therefore, N contains an open subset of the (x, y)-plane. However, the tangent plane to the (x, y)-plane at any point p off of the x-axis is not equal to D_p. Therefore, no such integral manifold exists.   //

The last example shows that in general, integral manifolds may fail to exist. Suppose D is a smooth distribution on M. We say that D is ***involutive*** if given any pair of smooth local sections of D (i.e., smooth vector fields X, Y defined on an open subset of M such that $X_p, Y_p \in D_p$ for each p), their Lie bracket is also a local section of D. The next proposition shows that the involutivity condition can be expressed concisely in terms of lie algebras.

Proposition 19.2. *Let $D \subseteq TM$ be a smooth distribution, and let $\Gamma(D) \subseteq \mathfrak{X}(M)$ denote the space of smooth global sections of D. Then D is involutive if and only if $\Gamma(D)$ is a Lie subalgebra of $\mathfrak{X}(M)$.*

Proof. If D is involutive, the definition implies that $\Gamma(D)$ is closed under Lie brackets. Because it is also a linear subspace of $\mathfrak{X}(M)$, it is a Lie subalgebra.

Conversely, suppose $\Gamma(D)$ is a Lie subalgebra of $\mathfrak{X}(M)$, and let X, Y be smooth local sections of D over an open subset $U \subseteq M$. Given $p \in M$, let $\psi \in C^\infty(M)$ be a bump function that is identically 1 on a neighborhood of p and supported in U. Then ψX and ψY are smooth global sections of D, so their Lie bracket is also a section of D by hypothesis. This Lie bracket is $[\psi X, \psi Y] = \psi^2 [X, Y] + \psi(X\psi)Y - \psi(Y\psi)X$, which is equal to $[X, Y]$ in a neighborhood of p. Thus, $[X, Y]_p \in D_p$ for each $p \in U$, so D is involutive.   □

A smooth distribution D on M is said to be ***integrable*** if each point of M is contained in an integral manifold of D.

Proposition 19.3. *Every integrable distribution is involutive.*

Proof. Let $D \subseteq TM$ be an integrable distribution. Suppose X and Y are smooth local sections of D defined on some open subset $U \subseteq M$. Let p be any point in U, and let N be an integral manifold of D containing p. The fact that X and Y are sections of D means that X and Y are tangent to N. By Corollary 8.32, $[X, Y]$ is also tangent to N, and therefore $[X, Y]_p \in D_p$. Since this is true at each $p \in U$, it follows that D is involutive. $\qquad\square$

Note, for example, that the distribution D of Example 19.1(d) is not involutive, because $[X, Y] = -\partial/\partial z$, which is not a section of D.

The next lemma shows that the involutivity condition does not have to be checked for every pair of smooth vector fields, just those of a smooth local frame in a neighborhood of each point.

Lemma 19.4 (Local Frame Criterion for Involutivity). *Let $D \subseteq TM$ be a distribution. If in a neighborhood of every point of M there exists a smooth local frame $(V_1, \dots, V_k)$ for D such that $[V_i, V_j]$ is a section of D for each $i, j = 1, \dots, k$, then D is involutive.*

Proof. Suppose the hypothesis holds, and suppose X and Y are smooth local sections of D over some open subset $U \subseteq M$. Given $p \in U$, choose a smooth local frame $(V_1, \dots, V_k)$ satisfying the hypothesis in a neighborhood of p, and write $X = X^i V_i$ and $Y = Y^i V_i$ in that neighborhood. Then, using (8.11),

$$[X, Y] = \left[X^i V_i, Y^j V_j\right] = X^i Y^j [V_i, V_j] + X^i \left(V_i Y^j\right) V_j - Y^j \left(V_j X^i\right) V_i.$$

It follows from the hypothesis that this last expression is a section of D. $\qquad\square$

Involutivity and Differential Forms

Differential forms yield an alternative way to describe distributions and involutivity.

Lemma 19.5 (1-Form Criterion for Smooth Distributions). *Suppose M is a smooth n-manifold and $D \subseteq TM$ is a distribution of rank k. Then D is smooth if and only if each point $p \in M$ has a neighborhood U on which there are smooth 1-forms $\omega^1, \dots, \omega^{n-k}$ such that for each $q \in U$,*

$$D_q = \operatorname{Ker} \omega^1\big|_q \cap \cdots \cap \operatorname{Ker} \omega^{n-k}\big|_q. \tag{19.1}$$

Proof. First suppose that there exist such forms $\omega^1, \dots, \omega^{n-k}$ in a neighborhood of each point. The assumption (19.1) together with the fact that D has rank k implies that the forms $\omega^1, \dots, \omega^{n-k}$ are independent on U for dimensional reasons. By Proposition 10.15, we can complete them to a smooth coframe $(\omega^1, \dots, \omega^n)$ on a (possibly smaller) neighborhood of each point. If $(E_1, \dots, E_n)$ is the dual frame, it is easy to check that D is locally spanned by $E_{n-k+1}, \dots, E_n$, so it is smooth by the local frame criterion.

Conversely, suppose D is smooth. In a neighborhood of any $p \in M$, there are smooth vector fields $Y_1, \dots, Y_k$ spanning D. By Proposition 10.15 again, we can complete these vector fields to a smooth local frame $(Y_1, \dots, Y_n)$ for M in a neighborhood of p. With the dual coframe denoted by $(\varepsilon^1, \dots, \varepsilon^n)$, it follows easily that D is characterized locally by

$$D_q = \operatorname{Ker} \varepsilon^{k+1}\big|_q \cap \cdots \cap \operatorname{Ker} \varepsilon^n\big|_q. \qquad\square$$

If D is a rank-k distribution on a smooth n-manifold M, any $n - k$ linearly independent 1-forms $\omega^1, \ldots, \omega^{n-k}$ defined on an open subset $U \subseteq M$ and satisfying (19.1) for each $q \in U$ are called **local defining forms for D**. More generally, if $0 \le p \le n$, we say that a p-form $\omega \in \Omega^p(M)$ **annihilates D** if $\omega(X_1, \ldots, X_p) = 0$ whenever $X_1, \ldots, X_p$ are local sections of D. (In the case $p = 0$, only the zero function annihilates D.)

Lemma 19.6. *Suppose M is a smooth n-manifold and D is a smooth rank-k distribution on M. Let $\omega^1, \ldots, \omega^{n-k}$ be smooth local defining forms for D over an open subset $U \subseteq M$. A smooth p-form η defined on U annihilates D if and only if it can be expressed in the form*

$$\eta = \sum_{i=1}^{n-k} \omega^i \wedge \beta^i \tag{19.2}$$

for some smooth $(p-1)$-forms $\beta^1, \ldots, \beta^{n-k}$ on U.

Remark. In the case $p = 1$, the β^i's are smooth functions, and we interpret a wedge product with a smooth function to be ordinary multiplication. Thus, in this case, the lemma just says that η is a linear combination of the ω^i's with smooth coefficients.

Proof. It is easy to check that any form η that satisfies (19.2) in a neighborhood of each point annihilates D. Conversely, suppose η annihilates D on U. In a neighborhood of each point we can complete the $(n - k)$-tuple $(\omega^1, \ldots, \omega^{n-k})$ to a smooth local coframe $(\omega^1, \ldots, \omega^n)$ for M (Proposition 10.15). If $(E_1, \ldots, E_n)$ is the dual frame, then D is locally spanned by $E_{n-k+1}, \ldots, E_n$. In terms of this coframe, any $\eta \in \Omega^p(M)$ can be written locally in a unique way as

$$\eta = {\sum_I}' \eta_I \omega^{i_1} \wedge \cdots \wedge \omega^{i_p},$$

where the coefficients η_I are determined by $\eta_I = \eta(E_{i_1}, \ldots, E_{i_p})$. Thus, η annihilates D in U if and only if $\eta_I = 0$ whenever $n - k + 1 \le i_1 < \cdots < i_p \le n$, in which case η can be written locally as

$$\eta = {\sum_{I : i_1 \le n-k}}' \eta_I \omega^{i_1} \wedge \cdots \wedge \omega^{i_p} = \sum_{i_1=1}^{n-k} \omega^{i_1} \wedge \left({\sum_{I'}}' \eta_{i_1 I'} \omega^{i_2} \wedge \cdots \wedge \omega^{i_p} \right),$$

where we have written $I' = (i_2, \ldots, i_p)$. This holds in a neighborhood of each point of U; patching together with a partition of unity, we obtain a similar expression on all of U.   □

When expressed in terms of differential forms, the involutivity condition translates into a statement about exterior derivatives.

Theorem 19.7 (1-Form Criterion for Involutivity). *Suppose $D \subseteq TM$ is a smooth distribution. Then D is involutive if and only if the following condition is* *satisfied:*

> If η is any smooth 1-form that annihilates D on an open subset
> $U \subseteq M$, then $d\eta$ also annihilates D on U. $\qquad$ (19.3)

Proof. First, assume that D is involutive, and suppose η is a smooth 1-form that annihilates D on $U \subseteq M$. Then for any smooth local sections X, Y of D, formula (14.28) for $d\eta$ gives

$$d\eta(X, Y) = X(\eta(Y)) - Y(\eta(X)) - \eta([X, Y]).$$

The hypothesis implies that each of the terms on the right-hand side is zero on U.

Conversely, suppose D satisfies (19.3), and suppose X and Y are smooth local sections of D. If $\omega^1, \ldots, \omega^{n-k}$ are smooth local defining forms for D, then (14.28) shows that for each $i = 1, \ldots, n - k$,

$$\omega^i([X, Y]) = X(\omega^i(Y)) - Y(\omega^i(X)) - d\omega^i(X, Y) = 0,$$

which implies that $[X, Y]$ takes its values in D. Thus D is involutive. $\qquad\square$

Just like the Lie bracket condition for involutivity, the exterior derivative condition need only be checked for a particular set of smooth defining forms in a neighborhood of each point, as the next proposition shows.

Proposition 19.8 (Local Coframe Criterion for Involutivity). *Let D be a smooth distribution of rank k on a smooth n-manifold M, and let $\omega^1, \ldots, \omega^{n-k}$ be smooth defining forms for D on an open subset $U \subseteq M$. The following are equivalent:*

(a) *D is involutive on U.*
(b) *$d\omega^1, \ldots, d\omega^{n-k}$ annihilate D.*
(c) *There exist smooth 1-forms $\{\alpha_j^i : i, j = 1, \ldots, n - k\}$ such that*

$$d\omega^i = \sum_{j=1}^{n-k} \omega^j \wedge \alpha_j^i, \quad \text{for each } i = 1, \ldots, n - k.$$

▶ **Exercise 19.9.** Prove the preceding proposition.

With a bit more algebraic terminology, there is an elegant way to express the involutivity condition in terms of differential forms. Recall that we have defined the graded algebra of smooth differential forms on a smooth n-manifold M as $\Omega^*(M) = \Omega^0(M) \oplus \cdots \oplus \Omega^n(M)$ (see p. 360). An *ideal in $\Omega^*(M)$* is a linear subspace $\mathcal{I} \subseteq \Omega^*(M)$ that is closed under wedge products with arbitrary elements of $\Omega^*(M)$; that is, $\omega \in \mathcal{I}$ implies $\eta \wedge \omega \in \mathcal{I}$ for every $\eta \in \Omega^*(M)$.

Now suppose D is a smooth distribution on a smooth n-manifold M. Let $\mathcal{I}^p(D) \subseteq \Omega^p(M)$ denote the space of smooth p-forms that annihilate D, and let $\mathcal{I}(D) = \mathcal{I}^0(D) \oplus \cdots \oplus \mathcal{I}^n(D) \subseteq \Omega^*(M)$.

▶ **Exercise 19.10.** For any smooth distribution $D \subseteq TM$, show that $\mathcal{I}(D)$ is an ideal in $\Omega^*(M)$.

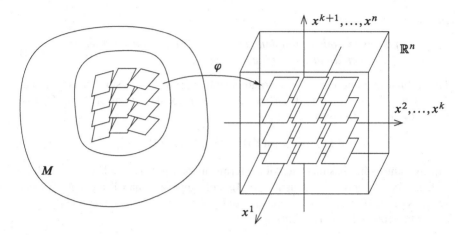

Fig. 19.2 A flat chart for a distribution

Any ideal of the form $\mathcal{I}(D)$ for some smooth distribution D is sometimes called a *Pfaffian system*. An ideal $\mathcal{I} \subseteq \Omega^*(M)$ is said to be a *differential ideal* if $d(\mathcal{I}) \subseteq \mathcal{I}$, that is, if $\omega \in \mathcal{I}$ implies $d\omega \in \mathcal{I}$.

Proposition 19.11 (Differential Ideal Criterion for Involutivity). *Let M be a smooth manifold. A smooth distribution $D \subseteq TM$ is involutive if and only if $\mathcal{I}(D)$ is a differential ideal in $\Omega^*(M)$.*

Proof. Problem 19-1.   □

The Frobenius Theorem

In Example 19.1, all of the distributions we defined except the last one had the property that there was an integral manifold through each point. Moreover, locally these submanifolds all "fit together" nicely like parallel affine subspaces of $\mathbb{R}^n$. Given a rank-k distribution $D \subseteq TM$, let us say that a smooth coordinate chart (U, φ) on M is *flat for D* if $\varphi(U)$ is a cube in $\mathbb{R}^n$, and at points of U, D is spanned by the first k coordinate vector fields $\partial/\partial x^1, \ldots, \partial/\partial x^k$ (Fig. 19.2). In any such chart, each slice of the form $x^{k+1} = c^{k+1}, \ldots, x^n = c^n$ for constants $c^{k+1}, \ldots, c^n$ is an integral manifold of D. This is the nicest possible local situation for integral manifolds. We say that a distribution $D \subseteq TM$ is *completely integrable* if there exists a flat chart for D in a neighborhood of each point of M. Obviously, every completely integrable distribution is integrable and therefore involutive. In summary,

$$\text{completely integrable} \Rightarrow \text{integrable} \Rightarrow \text{involutive}.$$

The next theorem is the main result of this chapter, and indeed one of the central theorems in smooth manifold theory. It says that the implications above are actually equivalences:

$$\text{completely integrable} \Leftrightarrow \text{integrable} \Leftrightarrow \text{involutive}.$$

Theorem 19.12 (Frobenius). *Every involutive distribution is completely integrable.*

Proof. The canonical form theorem for commuting vector fields (Theorem 9.46) implies that any distribution locally spanned by independent smooth *commuting* vector fields is completely integrable, because the coordinate chart whose existence is guaranteed by that theorem is flat (after shrinking the domain if necessary so the image is a cube). Thus, it suffices to show that every involutive distribution is locally spanned by independent smooth commuting vector fields.

Let D be an involutive distribution of rank k on an n-dimensional manifold M, and let $p \in M$. Since complete integrability is a local question, by passing to a smooth coordinate neighborhood of p, we may replace M by an open subset $U \subseteq \mathbb{R}^n$, and choose a smooth local frame $X_1, \ldots, X_k$ for D. By reordering the coordinates if necessary, we may assume that D_p is complementary to the subspace of $T_p\mathbb{R}^n$ spanned by $\left(\partial/\partial x^{k+1}|_p, \ldots, \partial/\partial x^n|_p \right)$.

Let $\pi \colon \mathbb{R}^n \to \mathbb{R}^k$ be the projection onto the first k coordinates, $\pi\left(x^1, \ldots, x^n\right) = \left(x^1, \ldots, x^k\right)$ (Fig. 19.3). This induces a smooth bundle homomorphism $d\pi \colon T\mathbb{R}^n \to T\mathbb{R}^k$, which can be written

$$d\pi\left(\sum_{i=1}^n v^i \left.\frac{\partial}{\partial x^i}\right|_q \right) = \sum_{i=1}^k v^i \left.\frac{\partial}{\partial x^i}\right|_{\pi(q)}.$$

(Notice that the summation on the right-hand side is only over $i = 1, \ldots, k$.) Because $d\pi|_D$ is the composition of the inclusion $D \hookrightarrow TU$ followed by $d\pi$, it is a smooth bundle homomorphism. Thus, the matrix entries of $d\pi|_{D_q}$ with respect to the frames $\left(X_i|_q\right)$ and $\left(\partial/\partial x^j|_{\pi(q)}\right)$ are smooth functions of q.

By our choice of coordinates, $D_p \subseteq T_p\mathbb{R}^n$ is complementary to the kernel of $d\pi_p$, so the restriction of $d\pi_p$ to D_p is bijective. By continuity, therefore, the same is true of $d\pi|_{D_q}$ for q in a neighborhood of p, and the matrix entries of $\left(d\pi|_{D_q}\right)^{-1} \colon T_{\pi(q)}\mathbb{R}^k \to D_q$ are also smooth functions of q. Define a new smooth local frame $V_1, \ldots, V_k$ for D in a neighborhood of p by

$$V_i|_q = \left(d\pi|_{D_q}\right)^{-1} \left.\frac{\partial}{\partial x^i}\right|_{\pi(q)}. \tag{19.4}$$

The theorem will be proved if we can show that $[V_i, V_j] = 0$ for all i, j.

First observe that V_i and $\partial/\partial x^i$ are π-related, because (19.4) implies that

$$\left.\frac{\partial}{\partial x^i}\right|_{\pi(q)} = \left(d\pi|_{D_q}\right)V_i|_q = d\pi_q\left(V_i|_q\right).$$

Therefore, by the naturality of Lie brackets,

$$d\pi_q\left([V_i, V_j]_q\right) = \left[\frac{\partial}{\partial x^i}, \frac{\partial}{\partial x^j} \right]_{\pi(q)} = 0.$$

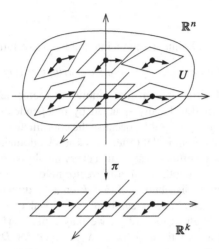

Fig. 19.3 Proof of the Frobenius theorem

Since involutivity of D implies that $[V_i, V_j]$ takes its values in D, and $d\pi$ is injective on each fiber of D, this implies that $[V_i, V_j]_q = 0$ for each q, thus completing the proof. □

For later use in our treatment of overdetermined partial differential equations, we note the following easy corollary to the proof.

Corollary 19.13. *Suppose M is a smooth manifold, D is an involutive rank-k distribution on M, and $S \subseteq M$ is a codimension-k embedded submanifold. If $p \in S$ is a point such that $T_p S$ is complementary to D_p, then there is a flat chart $\left(U, (s^i)\right)$ for D centered at p in which $S \cap U$ is the slice $s^1 = \cdots = s^k = 0$.*

Proof. The proof of the theorem showed that locally D is spanned by k commuting vector fields, and then the corollary follows from Theorem 9.46. □

As is often the case, embedded in the proof of the Frobenius theorem is a technique for finding integral manifolds. The idea is to use a coordinate projection to find commuting vector fields spanning the same distribution, and then use the technique of Example 9.47 to find a flat chart. Here is an example.

Example 19.14. Let $D \subseteq T\mathbb{R}^3$ be the distribution spanned by the vector fields

$$X = x\frac{\partial}{\partial x} + \frac{\partial}{\partial y} + x(y+1)\frac{\partial}{\partial z},$$

$$Y = \frac{\partial}{\partial x} + y\frac{\partial}{\partial z}.$$

The computation of Example 8.27 showed that

$$[X, Y] = -\frac{\partial}{\partial x} - y\frac{\partial}{\partial z} = -Y,$$

so D is involutive. Let us try to find a flat chart in a neighborhood of the origin. Since D is complementary to the span of $\partial/\partial z$, the coordinate projection $\pi \colon \mathbb{R}^3 \to \mathbb{R}^2$ given by $\pi(x, y, z) = (x, y)$ induces an isomorphism $d\pi|_{D_{(x,y,z)}} \colon D_{(x,y,z)} \to T_{(x,y)}\mathbb{R}^2$ for each $(x, y, z) \in \mathbb{R}^3$. The proof of the Frobenius theorem shows that if we can find smooth local sections V, W of D that are π-related to $\partial/\partial x$ and $\partial/\partial y$, respectively, they will be commuting vector fields spanning D. It is easy to check that V, W have this property if and only if they take their values in D and are of the form

$$V = \frac{\partial}{\partial x} + u(x, y, z)\frac{\partial}{\partial z},$$

$$W = \frac{\partial}{\partial y} + v(x, y, z)\frac{\partial}{\partial z},$$

for some smooth real-valued functions u, v. A bit of linear algebra shows that the vector fields

$$V = Y = \frac{\partial}{\partial x} + y\frac{\partial}{\partial z},$$

$$W = X - xY = \frac{\partial}{\partial y} + x\frac{\partial}{\partial z},$$

do the trick. The flows of these vector fields are easily found by solving the two systems of ODEs. Sparing the details, we find that the flow of V is

$$\alpha_t(x, y, z) = (x + t, y, z + ty), \tag{19.5}$$

and that of W is

$$\beta_t(x, y, z) = (x, y + t, z + tx). \tag{19.6}$$

Thus, by the procedure of Example 9.47, we can define the inverse Φ of our coordinate map by starting on the z-axis and flowing out along these two flows in succession:

$$\Phi(u, v, w) = \alpha_u \circ \beta_v(0, 0, w) = \alpha_u(0, v, w) = (u, v, w + uv).$$

The flat coordinates we seek are given by inverting the map $(x, y, z) = \Phi(u, v, w) = (u, v, w + uv)$, to yield

$$(u, v, w) = \Phi^{-1}(x, y, z) = (x, y, z - xy).$$

It follows that the integral manifolds of D are the level sets of $w(x, y, z) = z - xy$. (Since the flat chart we have constructed is actually a global chart in this case, this describes all of the integral manifolds, not just the ones near the origin.)   //

▶ **Exercise 19.15.** Verify that the flows of V and W are given by (19.5) and (19.6), respectively, and that the level sets of $z - xy$ are integral manifolds of D.

The next proposition is one of the most important consequences of the Frobenius theorem.

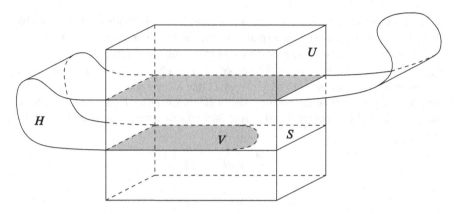

Fig. 19.4 The local structure of an integral manifold

Proposition 19.16 (Local Structure of Integral Manifolds). *Let D be an involutive distribution of rank k on a smooth manifold M, and let $(U, (x^i))$ be a flat chart for D. If H is any integral manifold of D, then $H \cap U$ is a union of countably many disjoint open subsets of parallel k-dimensional slices of U, each of which is open in H and embedded in M.*

Proof. Let H be an integral manifold of D. Because the inclusion map $\iota\colon H \hookrightarrow M$ is continuous, $H \cap U = \iota^{-1}(U)$ is open in H, and thus consists of a countable disjoint union of connected components, each of which is open in H.

Let V be any component of $H \cap U$ (Fig. 19.4). We show first that V is contained in a single slice. Since $dx^{k+1}, \dots, dx^n$ are local defining forms for D, it follows that the pullbacks of these 1-forms to V are identically zero. Because V is connected, this implies that $x^{k+1}, \dots, x^n$ are all constant on V, so V lies in a single slice S.

Because S is embedded in M, the inclusion map $V \hookrightarrow M$ is also smooth as a map into S by Corollary 5.30. The inclusion $V \hookrightarrow S$ is thus an injective smooth immersion between manifolds of the same dimension, and therefore a local diffeomorphism, an open map, and a homeomorphism onto an open subset of S. The inclusion map $V \hookrightarrow M$ is a composition of the smooth embeddings $V \hookrightarrow S \hookrightarrow M$, so it is a smooth embedding. $\qquad\square$

The preceding proposition implies the following important result about integral manifolds, which we will use in our study of Lie subgroups at the end of this chapter. Recall that a smooth submanifold $H \subseteq M$ is said to be *weakly embedded in M* if every smooth map $F\colon N \to M$ whose image lies in H is smooth as a map from N to H. (See Chapter 5.)

Theorem 19.17. *Every integral manifold of an involutive distribution is weakly embedded.*

Proof. Let M be a smooth n-manifold, let $H \subseteq M$ be an integral manifold of an involutive rank-k distribution D on M, and suppose $F\colon N \to M$ is a smooth map such that $F(N) \subseteq H$. Let $p \in N$ be arbitrary, and set $q = F(p) \in H$. Let $(y^1, \ldots, y^n)$ be flat coordinates for D on a neighborhood U of q, and let (x^i) be smooth coordinates for N on a connected neighborhood B of p such that $F(B) \subseteq U$. With the coordinate representation of F written as

$$(y^1, \ldots, y^n) = (F^1(x), \ldots, F^n(x)),$$

the fact that $F(B) \subseteq H \cap U$ means that the coordinate functions $F^{k+1}, \ldots, F^n$ take on only countably many values. Because B is connected, the intermediate value theorem implies that these coordinate functions are constant, and thus $F(B)$ lies in a single slice $S \subseteq U$. Because $S \cap H$ is an open subset of H that is embedded in M, it follows that $F|_B$ is smooth from B into $S \cap H$, and thus by composition, $F|_B : B \to (S \cap H) \hookrightarrow H$ is smooth into H. $\qquad \square$

Foliations

When we put together all of the maximal integral manifolds of an involutive rank-k distribution on a smooth manifold M, we obtain a partition of M into k-dimensional submanifolds that "fit together" locally like the slices in a flat chart.

To express more precisely what we mean by "fitting together," we need to extend our notion of a flat chart slightly. Let M be a smooth n-manifold, and let $\mathcal{F}$ be any collection of k-dimensional submanifolds of M. A smooth chart (U, φ) for M is said to be *flat for $\mathcal{F}$* if $\varphi(U)$ is a cube in $\mathbb{R}^n$, and each submanifold in $\mathcal{F}$ intersects U in either the empty set or a countable union of k-dimensional slices of the form $x^{k+1} = c^{k+1}, \ldots, x^n = c^n$. We define a *foliation of dimension k on M* to be a collection $\mathcal{F}$ of disjoint, connected, nonempty, immersed k-dimensional submanifolds of M (called the *leaves of the foliation*), whose union is M, and such that in a neighborhood of each point $p \in M$ there exists a flat chart for $\mathcal{F}$.

Example 19.18 (Foliations).

(a) The collection of all k-dimensional affine subspaces of $\mathbb{R}^n$ parallel to $\mathbb{R}^k \times \{0\}$ is a k-dimensional foliation of $\mathbb{R}^n$.

(b) The collection of open rays of the form $\{\lambda x : \lambda > 0\}$ as x ranges over $\mathbb{R}^n \smallsetminus \{0\}$ is a 1-dimensional foliation of $\mathbb{R}^n \smallsetminus \{0\}$.

(c) The collection of all spheres centered at 0 is an $(n-1)$-dimensional foliation of $\mathbb{R}^n \smallsetminus \{0\}$.

(d) If M and N are connected smooth manifolds, the collection of subsets of the form $M \times \{q\}$ as q ranges over points in N forms a foliation of $M \times N$, each of whose leaves is diffeomorphic to M. For example, the collection of all circles of the form $\mathbb{S}^1 \times \{q\} \subseteq \mathbb{T}^2$ for $q \in \mathbb{S}^1$ yields a foliation of the torus $\mathbb{T}^2$ (Fig. 19.5(a)). A different foliation of $\mathbb{T}^2$ is given by the collection of circles of the form $\{p\} \times \mathbb{S}^1$ (Fig. 19.5(b)).

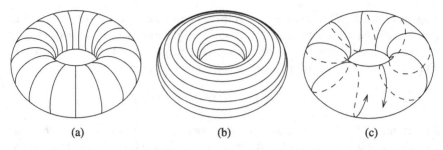

Fig. 19.5 Foliations of the torus

(e) If α is a fixed real number, the images of all curves of the form

$$\gamma_\theta(t) = \left(e^{it}, e^{i(\alpha t + \theta)}\right)$$

as θ ranges over $\mathbb{R}$ form a 1-dimensional foliation of the torus (Fig. 19.5(c)). If α is rational, each leaf is an embedded circle; whereas if α is irrational, each leaf is dense (see Example 4.20 and Problem 4-4).

(f) The collection of connected components of the curves in the (y, z)-plane defined by the following equations is a foliation of $\mathbb{R}^2$ (Fig. 19.6(a)):

$$z = \sec y + c, \quad c \in \mathbb{R};$$
$$y = \left(k + \tfrac{1}{2}\right)\pi, \quad k \in \mathbb{Z}.$$

(g) If we revolve the curves of the previous example around the z-axis, we obtain a 2-dimensional foliation of $\mathbb{R}^3$ in which some of the leaves are diffeomorphic to disks and some are diffeomorphic to cylinders (Fig. 19.6(b)). //

The main fact about foliations is that they are in one-to-one correspondence with involutive distributions. One direction, expressed in the next proposition, is an easy consequence of the definition.

Proposition 19.19. *Let $\mathcal{F}$ be a foliation on a smooth manifold M. The collection of tangent spaces to the leaves of $\mathcal{F}$ forms an involutive distribution on M.*

▶ **Exercise 19.20.** Prove Proposition 19.19.

The Frobenius theorem allows us to conclude the following converse, which is much more profound. By the way, it is worth noting that this result is one of the two primary reasons why the notion of immersed submanifold has been defined. (The other is for the study of Lie subgroups.)

Theorem 19.21 (Global Frobenius Theorem). *Let D be an involutive distribution on a smooth manifold M. The collection of all maximal connected integral manifolds of D forms a foliation of M.*

The theorem will be an easy consequence of the following lemma.

Lemma 19.22. *Suppose $D \subseteq TM$ is an involutive distribution, and let $\{N_\alpha\}_{\alpha \in A}$ be any collection of connected integral manifolds of D with a point in common. Then*

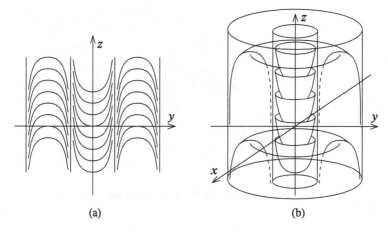

Fig. 19.6 Foliations of $\mathbb{R}^2$ and $\mathbb{R}^3$

$N = \bigcup_\alpha N_\alpha$ *has a unique smooth manifold structure making it into a connected integral manifold of* D.

Proof. If we can construct a topology and smooth manifold structure making N into an integral manifold of D, then Theorem 5.33 shows that the topology and smooth structure are uniquely determined, because integral manifolds are weakly embedded.

To construct the topology, first we need to show that $N_\alpha \cap N_\beta$ is open in N_α and in N_β for each $\alpha, \beta \in A$. Let $q \in N_\alpha \cap N_\beta$ be arbitrary, and choose a flat chart for D on a neighborhood W of q (Fig. 19.7). Let V_α, V_β denote the components of $N_\alpha \cap W$ and $N_\beta \cap W$, respectively, containing q. By Proposition 19.16, V_α and V_β are open subsets of single slices with the subspace topology, and since both contain q, they both must lie in the same slice S. Thus $V_\alpha \cap V_\beta$ is open in S and also in both N_α and N_β, so q has a neighborhood in N_α and a neighborhood in N_β contained in $N_\alpha \cap N_\beta$.

Define a topology on N by declaring a subset $U \subseteq N$ to be open if and only if $U \cap N_\alpha$ is open in N_α for each α. Using the result of the previous paragraph, it is easy to check that this is a topology and that each N_α is an open subspace of N. With this topology, N is locally Euclidean of dimension k, because each $q \in N$ has a coordinate neighborhood V in some N_α, and V is an open subset of N because N_α is open in N. Moreover, the inclusion map $N \hookrightarrow M$ is continuous: for any open subset $U \subseteq M$, $U \cap N$ is open in N because $U \cap N_\alpha$ is open in N_α for each α.

To see that N is Hausdorff, let q, q' be distinct points of N. There are disjoint open subsets $U, U' \subseteq M$ containing q and q', respectively, and because inclusion $N \hookrightarrow M$ is continuous, $N \cap U$ and $N \cap U'$ are disjoint open subsets of N containing q and q'.

Next we show that N is second-countable. We can cover M with countably many flat charts for D, say $\{W_i\}$. It suffices to show that $N \cap W_i$ is contained in a countable union of slices for each i, because any open subset of a single slice is second-

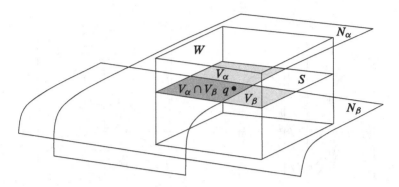

Fig. 19.7 A union of integral manifolds

countable, and thus N can be expressed as a union of countably many subsets, each of which is second-countable and open in N.

Let p_0 be a point contained in N_α for every α. Let us say that a slice S of some W_k is **accessible from p_0** if there is a finite sequence of indices $i_1, \ldots, i_m$ and for each i_j a slice $S_{i_j} \subseteq W_{i_j}$ with the properties that $p \in S_{i_1}$, $S_{i_m} = S$, and $S_{i_j} \cap S_{i_{j+1}} \neq \varnothing$ for each $j = 1, \ldots, m-1$ (Fig. 19.8).

Let W_k be one of our countable collection of flat charts, and suppose $S \subseteq W_k$ is a slice that contains a point $q \in N$. Then q is contained in one of the integral manifolds N_α. Because p_0 is also in N_α, there is a continuous path $\gamma : [0, 1] \to N_\alpha$ connecting p_0 and q. Since $\gamma([0, 1])$ is compact, there exist finitely many numbers $0 = t_0 < t_1 < \cdots < t_m = 1$ such that for each $j = 1, \ldots, m$, the set $\gamma([t_{j-1}, t_j])$ is contained in one of the flat charts W_{i_j}. Since $\gamma([t_{j-1}, t_j])$ is connected, it is contained in a single component of $W_{i_j} \cap N_\alpha$ and therefore in a single slice $S_{i_j} \subseteq W_{i_j}$. For each $j = 1, \ldots, m-1$, the slices S_{i_j} and $S_{i_{j+1}}$ have the point $\gamma(t_j)$ in common, so it follows that the slice S is accessible from p_0.

This shows that every slice of some W_k containing a point of N is accessible from p_0. To complete the proof of second-countability, we just note that each S_{i_j} is itself an integral manifold, and therefore it meets at most countably many slices of $W_{i_{j+1}}$ by Proposition 19.16; thus, there are only countably many slices accessible from p_0. Therefore, N is a topological manifold of dimension k. It is connected because it is a union of connected subspaces with a point in common.

To construct a smooth structure on N, we define an atlas consisting of all charts of the form $(S \cap N, \psi)$, where S is a single slice of some flat chart, and $\psi : S \to \mathbb{R}^k$ is the map whose coordinate representation in the flat chart is projection onto the first k coordinates: $\psi\left(x^1, \ldots, x^k, x^{k+1}, \ldots, x^n\right) = \left(x^1, \ldots, x^k\right)$. Because any slice is an embedded submanifold, its smooth structure is uniquely determined, and thus whenever two such slices S, S' overlap the transition map $\psi' \circ \psi^{-1}$ is smooth. With respect to this smooth structure, the inclusion map $N \hookrightarrow M$ is a smooth immersion (because it is a smooth embedding on each slice), and the tangent space to N at each point $q \in N$ is equal to D_q (because this is true for slices). $\square$

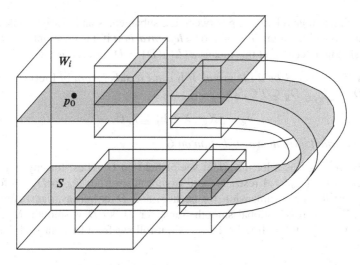

Fig. 19.8 A slice S accessible from p_0

Proof of the global Frobenius theorem. For each $p \in M$, let L_p be the union of all connected integral manifolds of D containing p. By the preceding lemma, L_p is a connected integral manifold of D containing p, and it is clearly maximal. If any two such maximal integral manifolds L_p and $L_{p'}$ intersect, their union $L_p \cup L_{p'}$ is an integral manifold containing both p and p', so by maximality $L_p = L_{p'}$. Thus, the various maximal connected integral manifolds are either disjoint or identical.

If (U, φ) is any flat chart for D, then $L_p \cap U$ is a countable union of open subsets of slices by Proposition 19.16. For any such slice S, if $L_p \cap S$ is neither empty nor all of S, then $L_p \cup S$ is a connected integral manifold properly containing L_p, which contradicts the maximality of L_p. Therefore, $L_p \cap U$ is precisely a countable union of slices, so the collection $\{L_p : p \in M\}$ is the desired foliation.   □

Suppose M is a smooth manifold and $\Phi \colon M \to M$ is a diffeomorphism. A distribution D on M is said to be **Φ-invariant** if $d\Phi(D) = D$; or more precisely if for each $x \in M$, $d\Phi_x(D_x) = D_{\Phi(x)}$. Similarly, a foliation $\mathscr{F}$ on M is said to be **Φ-invariant** if for each leaf L of $\mathscr{F}$, the submanifold $\Phi(L)$ is also a leaf of $\mathscr{F}$.

Proposition 19.23. *Let M be a smooth manifold and $\Phi \colon M \to M$ be a diffeomorphism. Suppose D is an involutive distribution on M and $\mathscr{F}$ is the foliation it determines. Then D is Φ-invariant if and only if $\mathscr{F}$ is Φ-invariant.*

Proof. Problem 19-9.   □

Lie Subalgebras and Lie Subgroups

Foliations have profound applications to the theory of Lie groups. Here we present two such applications; we will see many more in the next two chapters. Both rely on the following simple relationship between Lie subalgebras and distributions. A distribution D on a Lie group G is said to be **left-invariant** if it is invariant under every left translation. (Recall that this means $d(L_g)(D) = D$ for each $g \in G$.)

Lemma 19.24. *Let G be a Lie group. If $\mathfrak{h}$ is a Lie subalgebra of* $\mathrm{Lie}(G)$, *then the subset $D = \bigcup_{g \in G} D_g \subseteq TG$, where*

$$D_g = \{X_g : X \in \mathfrak{h}\} \subseteq T_g G, \tag{19.7}$$

is a left-invariant involutive distribution on G.

Proof. Each $X \in \mathfrak{h}$ is a left-invariant vector field on G. Thus, for any $g, g' \in G$, the differential $d(L_{g'g^{-1}})$ restricts to an isomorphism from D_g to $D_{g'}$. It follows that D_g has the same dimension for each g, and D is left-invariant. Any basis $(X_1, \ldots, X_k)$ for $\mathfrak{h}$ is a global smooth frame for D, so D is smooth. Moreover, because $[X_i, X_j] \in \mathfrak{h}$ for all $i, j \in \{1, \ldots, k\}$, it follows from Lemma 19.4 that D is involutive. $\qquad\qquad\square$

Theorem 19.25 (Lie Subgroups Are Weakly Embedded). *Every Lie subgroup is an integral manifold of an involutive distribution, and therefore is a weakly embedded submanifold.*

Proof. Suppose G is a Lie group and $H \subseteq G$ is a Lie subgroup. Theorem 8.46 shows that the Lie algebra of H is canonically isomorphic to the Lie subalgebra $\mathfrak{h} = \iota_*(\mathrm{Lie}(H)) \subseteq \mathrm{Lie}(G)$, where $\iota \colon H \hookrightarrow G$ is inclusion. Let $D \subseteq TG$ be the involutive distribution determined by $\mathfrak{h}$ as in Lemma 19.24. It follows from the definitions that at each point $h \in H$, the tangent space $T_h H$ is equal to D_h, so H is an integral manifold of D. It then follows from Theorem 19.17 that H is weakly embedded in G. $\qquad\qquad\square$

Theorem 19.26 (The Lie Subgroup Associated with a Lie Subalgebra). *Suppose G is a Lie group and $\mathfrak{g}$ is its Lie algebra. If $\mathfrak{h}$ is any Lie subalgebra of $\mathfrak{g}$, then there is a unique connected Lie subgroup of G whose Lie algebra is $\mathfrak{h}$.*

Proof. Suppose $\mathfrak{h}$ is a Lie subalgebra of $\mathfrak{g}$. Let $D \subseteq TG$ be the involutive distribution defined by (19.7). Let $\mathcal{H}$ denote the foliation determined by D, and for any $g \in G$, let $\mathcal{H}_g$ denote the leaf of $\mathcal{H}$ containing g (Fig. 19.9). Because D is left-invariant, it follows from Proposition 19.23 that each left translation takes leaves to leaves: for any $g, g' \in G$, we have $L_g(\mathcal{H}_{g'}) = \mathcal{H}_{gg'}$.

Define $H = \mathcal{H}_e$, the leaf containing the identity. We will show that H is the desired Lie subgroup.

First, to see that H is a subgroup, observe that for any $h, h' \in H$,

$$hh' = L_h(h') \in L_h(H) = L_h(\mathcal{H}_e) = \mathcal{H}_h = H.$$

Similarly,

$$h^{-1} = h^{-1}e \in L_{h^{-1}}(\mathcal{H}_e) = L_{h^{-1}}(\mathcal{H}_h) = \mathcal{H}_{h^{-1}h} = H.$$

To show that H is a Lie group, we need to show that the map $\mu \colon (h, h') \mapsto hh'^{-1}$ is smooth as a map from $H \times H$ to H. Because $H \times H$ is a submanifold of $G \times G$,

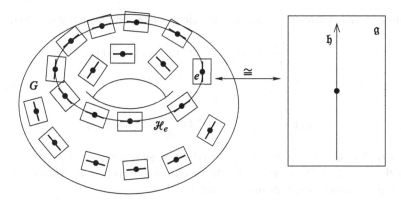

Fig. 19.9 Finding a subgroup whose Lie algebra is $\mathfrak{h}$

it is immediate that $\mu \colon H \times H \to G$ is smooth. Since H is an integral manifold of an involutive distribution, Theorem 19.17 shows that it is weakly embedded, so μ is also smooth as a map into H.

The fact that H is a leaf of $\mathcal{H}$ implies that the Lie algebra of H is $\mathfrak{h}$, because the tangent space to H at the identity is $D_e = \{X_e : X \in \mathfrak{h}\}$. To see that H is the unique connected subgroup with Lie algebra $\mathfrak{h}$, suppose $\widetilde{H}$ is any other connected subgroup with the same Lie algebra. Any such Lie subgroup is easily seen to be an integral manifold of D, so by maximality of $H = \mathcal{H}_e$, we must have $\widetilde{H} \subseteq H$. On the other hand, if U is the domain of a flat chart for D containing the identity, then by Proposition 19.16, $\widetilde{H} \cap U$ is a union of open subsets of slices. Since the slice containing e is an open subset of H, this implies that $\widetilde{H}$ contains a neighborhood of the identity in H. Since any neighborhood of the identity generates H (Proposition 7.14), this implies that $\widetilde{H} = H$. $\qquad\square$

Overdetermined Systems of Partial Differential Equations

The partial differential equations we considered in Chapter 9 were all single equations for one unknown function. In some applications, it is necessary to consider systems of PDEs that are ***overdetermined***, which means that there are more equations than unknown functions. In general, overdetermined systems have solutions only if they satisfy certain compatibility conditions. For some first-order systems, the compatibility condition can be interpreted as a statement about involutivity of a distribution, and the Frobenius theorem can be used to prove local existence and uniqueness of solutions.

First, we consider certain linear systems. Suppose W is an open subset of $\mathbb{R}^n$ and m is a positive integer less than or equal to n. Consider the following system of

m linear partial differential equations for a single unknown function $u \in C^\infty(W)$:

$$a_1^1(x)\frac{\partial u}{\partial x^1}(x) + \cdots + a_1^n(x)\frac{\partial u}{\partial x^n}(x) = f_1(x),$$

$$\vdots \tag{19.8}$$

$$a_m^1(x)\frac{\partial u}{\partial x^1}(x) + \cdots + a_m^n(x)\frac{\partial u}{\partial x^n}(x) = f_m(x),$$

where $\left(a_i^j\right)$ is an $n \times m$ matrix of smooth real-valued functions and $f_1, \ldots, f_m$ are smooth real-valued functions on W. The case $m = 1$ is covered by Theorem 9.51, so this discussion is useful primarily when $m > 1$.

If we let A_i denote the vector field $a_i^j \, \partial/\partial x^j$, the system (19.8) can be written more succinctly as $A_1 u = f_1, \ldots, A_m u = f_m$. To avoid redundant or degenerate systems of equations, we assume that the matrix $\left(a_i^j\right)$ has rank m at each point of W, or equivalently that the vector fields $A_1, \ldots, A_m$ are linearly independent. The following theorem is an analogue of Theorem 9.51 for the overdetermined case.

Theorem 19.27. *Let $W \subseteq \mathbb{R}^n$ be an open subset and let m be an integer such that $1 \le m \le n$. Suppose we are given an embedded codimension-m submanifold $S \subseteq W$, a linearly independent m-tuple of smooth vector fields $(A_1, \ldots, A_m)$ on W whose span is complementary to $T_p S$ at each $p \in S$, and functions $f_1, \ldots, f_m \in C^\infty(W)$. Suppose also that there are smooth functions $c_{ij}^k \in C^\infty(W)$ for $i, j, k = 1, \ldots, m$ such that the following compatibility conditions are satisfied:*

$$[A_i, A_j] = c_{ij}^k A_k, \tag{19.9}$$

$$A_i f_j - A_j f_i = c_{ij}^k f_k. \tag{19.10}$$

(In these expressions, k is implicitly summed from 1 to m.) Then for each $p \in S$, there is a neighborhood U of p such that for every $\varphi \in C^\infty(S \cap U)$, there exists a unique solution $u \in C^\infty(U)$ to the following overdetermined Cauchy problem:

$$A_i u = f_i \quad \text{for } i = 1, \ldots, m, \tag{19.11}$$

$$u\big|_{S \cap U} = \varphi. \tag{19.12}$$

Proof. Let D be the distribution on W spanned by $A_1, \ldots, A_m$, and let $p \in S$ be arbitrary. It follows from (19.9) that D is involutive, so by Corollary 19.13, on some neighborhood U of p there is a flat chart for D centered at p that is also a slice chart for S. Label the coordinates in this chart as $(v, w) = \left(v^1, \ldots, v^m, w^1, \ldots, w^{n-m}\right)$, so that $S \cap U$ is the slice where $v^1 = \cdots = v^m = 0$, and each $w = $ constant slice is an integral manifold of D in U, which we denote by H_w. Because (19.11)–(19.12) is a coordinate-independent statement, we can replace A_i and f_i by their coordinate representations in U, solve the equation there, and then use the inverse coordinate transformation to convert the solution back to the original coordinates.

Because $\mathrm{span}\left(A_1|_q, \ldots, A_m|_q\right) = \mathrm{span}\left(\partial/\partial v^1|_q, \ldots, \partial/\partial v^m|_q\right)$ for each $q \in U$, the n-tuple $\left(A_1, \ldots, A_m, \partial/\partial w^1, \ldots, \partial/\partial w^{n-m}\right)$ is a smooth local frame for U. Let

$(\alpha^1, \ldots, \alpha^m, \beta^1, \ldots, \beta^{n-m})$ denote the dual coframe, and define a smooth 1-form $\omega \in \Omega^1(U)$ by $\omega = f_k \alpha^k$ (with the implied summation from 1 to m). The system of equations (19.11) is satisfied if and only if $du(A_i) = \omega(A_i)$ for $i = 1, \ldots, m$, which is equivalent to saying that the pullback of $du - \omega$ to each H_w is equal to zero.

Using formula (14.28) for the exterior derivative together with (19.9), we obtain

$$d\alpha^k(A_i, A_j) = A_i(\alpha^k(A_j)) - A_j(\alpha^k(A_i)) - \alpha^k([A_i, A_j]) = -c_{ij}^k$$

for each $i, j, k = 1, \ldots, m$. It then follows from (19.10) that

$$d\omega(A_i, A_j) = (df_k \wedge \alpha^k + f_k d\alpha^k)(A_i, A_j)$$

$$= (A_i f_k)\delta_j^k - (A_j f_k)\delta_i^k - f_k c_{ij}^k = A_i f_j - A_j f_i - f_k c_{ij}^k = 0.$$

Since $(A_1, \ldots, A_m)$ restricts to a frame on each integral manifold H_w, this shows that the pullback of ω to each H_w is closed.

Given $\varphi \in C^\infty(U \cap S)$, let $u = u_0 + u_1$, where $u_0, u_1 \in C^\infty(U)$ are defined by

$$u_0(v, w) = \varphi(0, w),$$

$$u_1(v, w) = \int_0^1 \omega_k(tv, w)v^k \, dt,$$

and $\omega_k dv^k$ is the coordinate expression for ω.

Recall that a flat chart is cubical by definition, and thus star-shaped, so the integral is well defined for all $(v, w) \in U$, and differentiation under the integral sign shows that u_1 is a smooth function of (v, w). Because $u_0|_{S \cap U} = \varphi$ and $u_1|_{S \cap U} = 0$, it follows that u satisfies the initial condition (19.12).

The function u_0 satisfies $A_1 u_0 = \cdots = A_m u_0 = 0$ because it is independent of the v-coordinates. On the other hand, for each fixed w, u_1 is the potential function on H_w for $\iota_{H_w}^* \omega$ given by formula (11.24). The proof of Theorem 11.49 shows that $\iota_{H_w}^* du = \iota_{H_w}^* \omega$ for each w. It follows that $A_k u = A_k(u_1) = f_k$ for each $k = 1, \ldots, m$, so u is a solution to (19.11) as well.

To prove uniqueness, suppose $\tilde{u}$ is any other solution to (19.11)–(19.12) on U, and let $\psi = u - \tilde{u}$. Then $A_k \psi = 0$ for each k, so ψ is independent of v. It follows that $\psi(v, w) = \psi(0, w)$, which is zero because u and $\tilde{u}$ satisfy (19.12). □

Next we apply the Frobenius theorem to a class of nonlinear overdetermined PDEs. These are equations for a vector-valued function $u = (u^1, \ldots, u^m)$ that express all first partial derivatives of u in terms of the independent variables and the values of u. We explain it first in the case of a single real-valued function u of two independent variables (x, y), in which case the notation is considerably simpler.

Suppose we seek a solution u to the system

$$\frac{\partial u}{\partial x}(x, y) = \alpha(x, y, u(x, y)),$$

$$\frac{\partial u}{\partial y}(x, y) = \beta(x, y, u(x, y)),$$

(19.13)

where α and β are smooth real-valued functions defined on some open subset $W \subseteq \mathbb{R}^3$. This is an overdetermined system of (possibly nonlinear) first-order PDEs. (In fact, almost any pair of smooth first-order partial differential equations for one unknown function of two variables can be put into this form, at least locally, simply by solving the two equations for $\partial u / \partial x$ and $\partial u / \partial y$. Whether this can be done in principle is a question that is completely answered by the implicit function theorem; whether it can be done in practice depends on the specific equations and how clever you are.)

To determine the compatibility conditions that α and β must satisfy for solvability of (19.13), assume u is a smooth solution on some open subset of $\mathbb{R}^2$. Because $\partial^2 u / \partial x \partial y = \partial^2 u / \partial y \partial x$, (19.13) implies

$$\frac{\partial}{\partial y}\left(\alpha\left(x, y, u(x, y)\right)\right) = \frac{\partial}{\partial x}\left(\beta\left(x, y, u(x, y)\right)\right)$$

and therefore by the chain rule

$$\frac{\partial \alpha}{\partial y} + \beta \frac{\partial \alpha}{\partial z} = \frac{\partial \beta}{\partial x} + \alpha \frac{\partial \beta}{\partial z}. \tag{19.14}$$

This is true at a point $(x, y, z) \in W$ provided there is a smooth solution u with $u(x, y) = z$. In particular, (19.14) is a necessary condition for (19.13) to have a solution in a neighborhood of each point (x_0, y_0) with freely specified initial value $u(x_0, y_0) = z_0$. Using the Frobenius theorem, we can show that this condition is sufficient.

Proposition 19.28. *Suppose α and β are smooth real-valued functions defined on some open subset $W \subseteq \mathbb{R}^3$ and satisfying (19.14) there. For each $(x_0, y_0, z_0) \in W$, there exist a neighborhood U of (x_0, y_0) in $\mathbb{R}^2$ and a unique smooth function $u \colon U \to \mathbb{R}$ satisfying (19.13) and $u(x_0, y_0) = z_0$.*

Proof. The idea of the proof is that the system (19.13) determines the partial derivatives of u in terms of its values, and therefore determines the tangent plane to the graph of u at each point in terms of the coordinates of the point on the graph. This collection of tangent planes defines a smooth rank-2 distribution on W (Fig. 19.10), and (19.14) is equivalent to the involutivity condition for this distribution.

If there is a solution u on an open subset $U \subseteq \mathbb{R}^2$, the map $F \colon U \to W$ given by

$$F(x, y) = \left(x, y, u(x, y)\right)$$

is a smooth global parametrization of the graph $\Gamma(u) \subseteq U \times \mathbb{R}$. At any point $p = F(x, y)$, the tangent space $T_p \Gamma(u)$ is spanned by the vectors

$$dF\left(\left.\frac{\partial}{\partial x}\right|_{(x,y)}\right) = \left.\frac{\partial}{\partial x}\right|_p + \frac{\partial u}{\partial x}(x, y) \left.\frac{\partial}{\partial z}\right|_p,$$

$$dF\left(\left.\frac{\partial}{\partial y}\right|_{(x,y)}\right) = \left.\frac{\partial}{\partial y}\right|_p + \frac{\partial u}{\partial y}(x, y) \left.\frac{\partial}{\partial z}\right|_p.$$

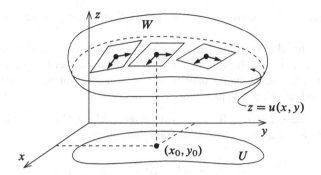

Fig. 19.10 Solving for u by finding its graph

The system (19.13) is satisfied if and only if

$$dF\left(\frac{\partial}{\partial x}\bigg|_{(x,y)}\right) = \frac{\partial}{\partial x}\bigg|_p + \alpha(x,y,u(x,y))\frac{\partial}{\partial z}\bigg|_p,$$

$$dF\left(\frac{\partial}{\partial y}\bigg|_{(x,y)}\right) = \frac{\partial}{\partial y}\bigg|_p + \beta(x,y,u(x,y))\frac{\partial}{\partial z}\bigg|_p. \tag{19.15}$$

Let X and Y be the vector fields

$$X = \frac{\partial}{\partial x} + \alpha(x,y,z)\frac{\partial}{\partial z},$$

$$Y = \frac{\partial}{\partial y} + \beta(x,y,z)\frac{\partial}{\partial z} \tag{19.16}$$

on W, and let D be the distribution on W spanned by X and Y. Because (19.15) says that $T_p\Gamma(u)$ is spanned by X_p and Y_p, a necessary condition for the system (19.13) to be satisfied is that $\Gamma(u)$ be an integral manifold of D. On the other hand, this condition is also sufficient: if $\Gamma(u)$ is an integral manifold, then $dF(\partial/\partial x)$ and $dF(\partial/\partial y)$ must both be linear combinations of X and Y, and comparing $\partial/\partial x$ and $\partial/\partial y$ components shows that this can happen only if (19.15) holds.

A straightforward computation using (19.14) shows that $[X, Y] \equiv 0$, so given any point $p = (x_0, y_0, z_0) \in W$, there is an integral manifold N of D containing p. Let $\Phi\colon V \to \mathbb{R}$ be a defining function for N on some neighborhood V of p; for example, we could take Φ to be the third coordinate function in a flat chart. The tangent space to N at each point $p \in N$ (namely D_p) is equal to the kernel of $d\Phi_p$. Since $\partial/\partial z|_p \notin D_p$ at any point p, this implies that $\partial\Phi/\partial z \neq 0$ at p, so by the implicit function theorem N is the graph of a smooth function $z = u(x,y)$ in some neighborhood of p. You can verify easily that u is a solution to the problem. Uniqueness follows immediately from Proposition 19.16. $\qquad\square$

As in several cases we have seen before, the proof of Proposition 19.28 actually contains a procedure for finding solutions to (19.13): find flat coordinates (u, v, w) for the distribution spanned by the vector fields X and Y defined by (19.16), and solve the equation $w = $ constant for $z = u(x, y)$. Some examples are given in Problem 19-13.

There is a straightforward generalization of this result to higher dimensions. The general statement of the theorem is a bit complicated, but verifying the necessary conditions in specific examples usually just amounts to computing mixed partial derivatives and applying the chain rule.

Proposition 19.29. *Suppose W is an open subset of $\mathbb{R}^n \times \mathbb{R}^m$, and $\alpha = \left(\alpha_j^i\right)\colon W \to M(m \times n, \mathbb{R})$ is a smooth matrix-valued function satisfying*

$$\frac{\partial \alpha_j^i}{\partial x^k} + \alpha_k^l \frac{\partial \alpha_j^i}{\partial z^l} = \frac{\partial \alpha_k^i}{\partial x^j} + \alpha_j^l \frac{\partial \alpha_k^i}{\partial z^l} \quad \text{for all } i, j, k,$$

where we denote a point in $\mathbb{R}^n \times \mathbb{R}^m$ by $(x, z) = \left(x^1, \dots, x^n, z^1, \dots, z^m\right)$. For any $(x_0, z_0) \in W$, there is a neighborhood U of x_0 in $\mathbb{R}^n$ and a unique smooth function $u\colon U \to \mathbb{R}^m$ such that $u(x_0) = z_0$ and the Jacobian of u satisfies

$$\frac{\partial u^i}{\partial x^j}\left(x^1, \dots, x^n\right) = \alpha_j^i\left(x^1, \dots, x^n, u^1(x), \dots, u^m(x)\right).$$

▶ **Exercise 19.30.** Prove Proposition 19.29.

Problems

19-1. Prove Proposition 19.11 (a smooth distribution is involutive if and only if it determines a differential ideal).

19-2. Let D be a smooth distribution of rank k on a smooth n-manifold M, and suppose $\omega^1, \dots, \omega^{n-k}$ are smooth local defining forms for D on an open subset $U \subseteq M$. Show that D is involutive on U if and only if the following identity holds for each $i = 1, \dots, n - k$:

$$d\omega^i \wedge \omega^1 \wedge \cdots \wedge \omega^{n-k} = 0.$$

(*Used on p. 582.*)

19-3. Let ω be a smooth 1-form on a smooth manifold M. A smooth positive function μ on some open subset $U \subseteq M$ is called an ***integrating factor*** for ω if $\mu\omega$ is exact on U. Prove the following statements:
   (a) If ω is nowhere-vanishing, then ω admits an integrating factor in a neighborhood of each point if and only if $d\omega \wedge \omega \equiv 0$.
   (b) If $\dim M = 2$, then every nonvanishing smooth 1-form admits an integrating factor in a neighborhood of each point.

19-4. Let $U \subseteq \mathbb{R}^3$ be the subset where all three coordinates are positive, and let D be the distribution on U spanned by the vector fields

$$X = y\frac{\partial}{\partial z} - z\frac{\partial}{\partial y}, \qquad Y = z\frac{\partial}{\partial x} - x\frac{\partial}{\partial z}.$$

Find an explicit global flat chart for D on U.

19-5. Let D be the distribution on $\mathbb{R}^3$ spanned by

$$X = \frac{\partial}{\partial x} + yz\frac{\partial}{\partial z}, \qquad Y = \frac{\partial}{\partial y}.$$

    (a) Find an integral submanifold of D passing through the origin.

    (b) Is D involutive? Explain your answer in light of part (a).

19-6. Let D be an involutive distribution on a smooth manifold M, and let $\gamma: J \to M$ be a smooth curve. Prove the following statements.

    (a) If H is an integral manifold of D, and the image of γ is contained in H, then $\gamma'(t)$ is in $T_{\gamma(t)}H \subseteq T_{\gamma(t)}M$ for all $t \in J$.

    (b) Conversely, if $\gamma'(t)$ lies in D for all t, then the image of γ is contained in a single leaf of the foliation determined by D.

[Remark: compare this to the result of Problem 5-19.]

19-7. Let D be an involutive distribution on a smooth manifold M, and let N be a connected integral manifold of D. Show that if N is a closed subset of M, then it is a maximal connected integral manifold and is therefore a leaf of the foliation determined by D. (*Used on p. 545.*)

19-8. Suppose M and N are smooth manifolds and $F: M \to N$ is a smooth submersion. Show that the connected components of the nonempty level sets of F form a foliation of M.

19-9. Prove Proposition 19.23 (invariant distributions vs. invariant foliations).

19-10. Let M and N be smooth manifolds. Suppose $\mathcal{F}$ is a foliation on M of codimension k, and $\varphi: N \to M$ is a smooth map. Show that if φ is transverse to each leaf of $\mathcal{F}$, then there is a unique codimension-k foliation $\varphi^*\mathcal{F}$ on N, called the ***pullback of*** $\mathcal{F}$, such that φ maps each leaf of $\varphi^*\mathcal{F}$ into a single leaf of $\mathcal{F}$.

19-11. Consider the following system of PDEs for $u \in C^\infty(\mathbb{R}^3)$:

$$x\frac{\partial u}{\partial x} + \frac{\partial u}{\partial y} + x(y+1)\frac{\partial u}{\partial z} = xy,$$

$$\frac{\partial u}{\partial x} + y\frac{\partial u}{\partial z} = y - 1.$$

Find a solution u to this system satisfying $u(0,0,z) = z$. [Hint: look at Example 19.14.]

19-12. Consider the following system of PDEs for $u \in C^\infty(\mathbb{R}^4)$:

$$\frac{\partial u}{\partial w} + x\frac{\partial u}{\partial y} + 2(w + xy)\frac{\partial u}{\partial z} = 0,$$

$$\frac{\partial u}{\partial x} - w\frac{\partial u}{\partial y} + 2(x - wy)\frac{\partial u}{\partial z} = 0.$$

    (a) Show that there do not exist two solutions u^1, u^2 with linearly independent differentials on any open subset of $\mathbb{R}^4$.

(b) Show that in a neighborhood of each point, there exists a solution with nonvanishing differential.

19-13. Of the systems of partial differential equations below, determine which ones have solutions $z(x, y)$ (or, for part (c), $z(x, y)$ and $w(x, y)$) in a neighborhood of the origin for arbitrary positive values of $z(0, 0)$ (respectively, $z(0, 0)$ and $w(0, 0)$).

(a) $\dfrac{\partial z}{\partial x} = z \cos y; \qquad \dfrac{\partial z}{\partial y} = -z \log z \tan y.$

(b) $\dfrac{\partial z}{\partial x} = e^{xz}; \qquad \dfrac{\partial z}{\partial y} = xe^{yz}.$

(c) $\dfrac{\partial z}{\partial x} = z; \qquad \dfrac{\partial z}{\partial y} = w; \qquad \dfrac{\partial w}{\partial x} = w; \qquad \dfrac{\partial w}{\partial y} = z.$

Chapter 20
The Exponential Map

In this chapter we apply the tools of flows, Lie derivatives, and foliations to delve deeper into the relationships between Lie groups and Lie algebras.

In the first section we define *one-parameter subgroups* of a Lie group G, which are just Lie group homomorphisms from $\mathbb{R}$ to G, and show that there is a one-to-one correspondence between elements of $\mathrm{Lie}(G)$ and one-parameter subgroups of G.

Next we introduce the focal point of our study, which is a canonical smooth map from the Lie algebra into the group called the *exponential map*. It maps lines through the origin in $\mathrm{Lie}(G)$ to one-parameter subgroups of G.

As our first major application of the exponential map, we prove the *closed subgroup theorem*, which says that every topologically closed subgroup of a Lie group is actually an embedded Lie subgroup.

Next we prove a higher-dimensional generalization of the fundamental theorem on flows. Instead of a single smooth vector field generating an action of $\mathbb{R}$, we consider a finite-dimensional family of vector fields and ask when they generate an action of some Lie group. The main theorem is that if G is a simply connected Lie group, then any Lie algebra homomorphism from $\mathrm{Lie}(G)$ into the set of complete vector fields on M generates a smooth action of G on M.

Finally, in the last two sections, we bring together all of these results to deepen our understanding of the correspondence between Lie groups and Lie algebras. First, we prove that there is a one-to-one correspondence between isomorphism classes of finite-dimensional Lie algebras and isomorphism classes of simply connected Lie groups; and then we show that for any Lie group G, connected normal subgroups of G correspond to *ideals* in the Lie algebra of G, which are subspaces that are stable under bracketing with arbitrary elements of the algebra. This is an excellent illustration of the fundamental philosophy of Lie theory: as much as possible, we use the Lie group/Lie algebra correspondence to translate group-theoretic questions about a Lie group into linear-algebraic questions about its Lie algebra.

One-Parameter Subgroups and the Exponential Map

Suppose G is a Lie group. Since left-invariant vector fields are naturally defined in terms of the group structure of G, one might reasonably expect to find some relationship between the group law for the flow of a left-invariant vector field and group multiplication in G. We begin by exploring this relationship.

One-Parameter Subgroups

A ***one-parameter subgroup of*** G is defined to be a Lie group homomorphism $\gamma \colon \mathbb{R} \to G$, with $\mathbb{R}$ considered as a Lie group under addition. By this definition, a one-parameter subgroup is *not* a Lie subgroup of G, but rather a homomorphism into G. (However, the *image* of a one-parameter subgroup is a Lie subgroup when endowed with a suitable smooth manifold structure; see Problem 20-1.)

Theorem 20.1 (Characterization of One-Parameter Subgroups). *Let G be a Lie group. The one-parameter subgroups of G are precisely the maximal integral curves of left-invariant vector fields starting at the identity.*

Proof. First suppose γ is the maximal integral curve of some left-invariant vector field $X \in \mathrm{Lie}(G)$ starting at the identity. Because left-invariant vector fields are complete (Theorem 9.18), γ is defined on all of $\mathbb{R}$. Left-invariance means that X is L_g-related to itself for every $g \in G$, so Proposition 9.6 implies that L_g takes integral curves of X to integral curves of X. Applying this with $g = \gamma(s)$ for some $s \in \mathbb{R}$, we conclude that the curve $t \mapsto L_{\gamma(s)}\big(\gamma(t)\big)$ is an integral curve starting at $\gamma(s)$. But the translation lemma (Lemma 9.4) implies that $t \mapsto \gamma(s+t)$ is also an integral curve with the same initial point, so they are equal:

$$\gamma(s)\gamma(t) = \gamma(s+t).$$

This says precisely that $\gamma \colon \mathbb{R} \to G$ is a one-parameter subgroup.

Conversely, suppose $\gamma \colon \mathbb{R} \to G$ is a one-parameter subgroup, and let $X = \gamma_*(d/dt) \in \mathrm{Lie}(G)$, treating d/dt as a left-invariant vector field on $\mathbb{R}$. Since $\gamma(0) = e$, we just have to show that γ is an integral curve of X. Recall that $\gamma_*(d/dt)$ is defined as the unique left-invariant vector field on G that is γ-related to d/dt (see Theorem 8.44). Therefore, for any $t_0 \in \mathbb{R}$,

$$\gamma'(t_0) = d\gamma_{t_0}\left(\frac{d}{dt}\bigg|_{t_0} \right) = X_{\gamma(t_0)},$$

so γ is an integral curve of X.   □

Given $X \in \mathrm{Lie}(G)$, the one-parameter subgroup determined by X in this way is called the ***one-parameter subgroup generated by*** X. Because left-invariant vector fields are uniquely determined by their values at the identity, it follows that each one-parameter subgroup is uniquely determined by its initial velocity in $T_e G$, and thus there are one-to-one correspondences

$$\{\text{one-parameter subgroups of } G\} \longleftrightarrow \mathrm{Lie}(G) \longleftrightarrow T_e G.$$

The one-parameter subgroups of $GL(n, \mathbb{R})$ are not hard to compute explicitly.

Proposition 20.2. *For any* $A \in \mathfrak{gl}(n, \mathbb{R})$, *let*

$$e^A = \sum_{k=0}^{\infty} \frac{1}{k!} A^k = I_n + A + \frac{1}{2} A^2 + \cdots. \tag{20.1}$$

This series converges to an invertible matrix $e^A \in GL(n, \mathbb{R})$, *and the one-parameter subgroup of* $GL(n, \mathbb{R})$ *generated by* $A \in \mathfrak{gl}(n, \mathbb{R})$ *is* $\gamma(t) = e^{tA}$.

Proof. First, we verify convergence. From Exercise B.48, matrix multiplication satisfies $|AB| \leq |A||B|$, where the norm is the Frobenius norm on $\mathfrak{gl}(n, \mathbb{R})$. It follows by induction that $|A^k| \leq |A|^k$. The Weierstrass M-test then shows that (20.1) converges uniformly on any bounded subset of $\mathfrak{gl}(n, \mathbb{R})$, by comparison with the series $\sum_k (1/k!) c^k = e^c$.

Fix $A \in \mathfrak{gl}(n, \mathbb{R})$. Under our identification of $\mathfrak{gl}(n, \mathbb{R})$ with $\mathrm{Lie}(GL(n, \mathbb{R}))$, the matrix A corresponds to the left-invariant vector field A^L given by (8.15). Thus, the one-parameter subgroup generated by A is an integral curve of A^L on $GL(n, \mathbb{R})$, and therefore satisfies the ODE initial value problem

$$\gamma'(t) = A^L\big|_{\gamma(t)}, \qquad \gamma(0) = I_n.$$

Using (8.15), the condition for γ to be an integral curve can be rewritten as

$$\dot{\gamma}_k^i(t) = \gamma_j^i(t) A_k^j,$$

or in matrix notation

$$\gamma'(t) = \gamma(t) A.$$

We will show that $\gamma(t) = e^{tA}$ satisfies this equation. Since $\gamma(0) = I_n$, this implies that γ is the unique integral curve of A^L starting at the identity and is therefore the desired one-parameter subgroup.

To see that γ is differentiable, we note that differentiating the series (20.1) formally term by term yields the result

$$\gamma'(t) = \sum_{k=1}^{\infty} \frac{k}{k!} t^{k-1} A^k = \left(\sum_{k=1}^{\infty} \frac{1}{(k-1)!} t^{k-1} A^{k-1} \right) A = \gamma(t) A.$$

Since the differentiated series converges uniformly on bounded sets (because apart from the additional factor of A, it is the same series!), the term-by-term differentiation is justified. A similar computation shows that $\gamma'(t) = A\gamma(t)$. By smoothness of solutions to ODEs, γ is a smooth curve.

It remains only to show that $\gamma(t)$ is invertible for all t, so that γ actually takes its values in $GL(n, \mathbb{R})$. If we let $\sigma(t) = \gamma(t)\gamma(-t) = e^{tA} e^{-tA}$, then σ is a smooth curve in $\mathfrak{gl}(n, \mathbb{R})$, and by the previous computation and the product rule it satisfies

$$\sigma'(t) = \big(\gamma(t) A \big) \gamma(-t) - \gamma(t) \big(A\gamma(-t) \big) = 0.$$

It follows that σ is the constant curve $\sigma(t) \equiv \sigma(0) = I_n$, which is to say that $\gamma(t)\gamma(-t) = I_n$. Substituting $-t$ for t, we obtain $\gamma(-t)\gamma(t) = I_n$, which shows that $\gamma(t)$ is invertible and $\gamma(t)^{-1} = \gamma(-t)$.   □

Next we would like to compute the one-parameter subgroups of $\mathrm{GL}(n, \mathbb{R})$, such as $\mathrm{O}(n)$. To do so, we need the following result.

Proposition 20.3. *Suppose G is a Lie group and $H \subseteq G$ is a Lie subgroup. The one-parameter subgroups of H are precisely those one-parameter subgroups of G whose initial velocities lie in $T_e H$.*

Proof. Let $\gamma \colon \mathbb{R} \to H$ be a one-parameter subgroup. Then the composite map

$$\mathbb{R} \xrightarrow{\gamma} H \hookrightarrow G$$

is a Lie group homomorphism and thus a one-parameter subgroup of G, which clearly satisfies $\gamma'(0) \in T_e H$.

Conversely, suppose $\gamma \colon \mathbb{R} \to G$ is a one-parameter subgroup whose initial velocity lies in $T_e H$. Let $\widetilde{\gamma} \colon \mathbb{R} \to H$ be the one-parameter subgroup of H with the same initial velocity $\widetilde{\gamma}'(0) = \gamma'(0) \in T_e H \subseteq T_e G$. As in the preceding paragraph, by composing with the inclusion map, we can also consider $\widetilde{\gamma}$ as a one-parameter subgroup of G. Since γ and $\widetilde{\gamma}$ are both one-parameter subgroups of G with the same initial velocity, they must be equal.   □

Example 20.4. If H is a Lie subgroup of $\mathrm{GL}(n, \mathbb{R})$, the preceding proposition shows that the one-parameter subgroups of H are precisely the maps of the form $\gamma(t) = e^{tA}$ for $A \in \mathfrak{h}$, where $\mathfrak{h} \subseteq \mathfrak{gl}(n, \mathbb{R})$ is the subalgebra corresponding to $\mathrm{Lie}(H)$ as in Theorem 8.46. For example, taking $H = \mathrm{O}(n)$, this shows that the one-parameter subgroups of $\mathrm{O}(n)$ are the maps of the form $\gamma(t) = e^{tA}$ for an arbitrary skew-symmetric matrix A. In particular, this shows that the exponential of any skew-symmetric matrix is orthogonal.   //

The Exponential Map

In the preceding section we saw that the matrix exponential maps $\mathfrak{gl}(n, \mathbb{R})$ to $\mathrm{GL}(n, \mathbb{R})$ and takes each line through the origin to a one-parameter subgroup. This has a powerful generalization to arbitrary Lie groups.

Given a Lie group G with Lie algebra $\mathfrak{g}$, we define a map $\exp \colon \mathfrak{g} \to G$, called the *exponential map of G*, as follows: for any $X \in \mathfrak{g}$, we set

$$\exp X = \gamma(1),$$

where γ is the one-parameter subgroup generated by X, or equivalently the integral curve of X starting at the identity (Fig. 20.1). The following proposition shows that, like the matrix exponential, this map sends the line through X to the one-parameter subgroup generated by X.

Proposition 20.5. *Let G be a Lie group. For any $X \in \mathrm{Lie}(G)$, $\gamma(s) = \exp sX$ is the one-parameter subgroup of G generated by X.*

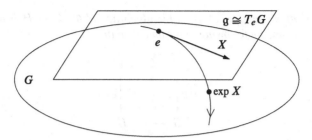

Fig. 20.1 The exponential map

Proof. Let $\gamma: \mathbb{R} \to G$ be the one-parameter subgroup generated by X, which is the integral curve of X starting at e. For any fixed $s \in \mathbb{R}$, it follows from the rescaling lemma (Lemma 9.3) that $\tilde{\gamma}(t) = \gamma(st)$ is the integral curve of sX starting at e, so

$$\exp sX = \tilde{\gamma}(1) = \gamma(s). \qquad \square$$

Here are two simple but important examples.

Example 20.6. The results of the preceding section show that the exponential map of $GL(n, \mathbb{R})$ (or any Lie subgroup of it) is given by $\exp A = e^A$. This, obviously, is the reason for the term *exponential map*.   //

Example 20.7. If V is a finite-dimensional real vector space, a choice of basis for V yields isomorphisms $GL(V) \cong GL(n, \mathbb{R})$ and $\mathfrak{gl}(V) \cong \mathfrak{gl}(n, \mathbb{R})$. The analysis of the $GL(n, \mathbb{R})$ case then shows that the exponential map of $GL(V)$ can be written in the form

$$\exp A = \sum_{k=0}^{\infty} \frac{1}{k!} A^k, \tag{20.2}$$

where we consider $A \in \mathfrak{gl}(V)$ as a linear map from V to itself, and $A^k = A \circ \cdots \circ A$ is the k-fold composition of A with itself.   //

Proposition 20.8 (Properties of the Exponential Map). *Let G be a Lie group and let $\mathfrak{g}$ be its Lie algebra.*

(a) *The exponential map is a smooth map from $\mathfrak{g}$ to G.*
(b) *For any $X \in \mathfrak{g}$ and $s, t \in \mathbb{R}$, $\exp(s + t)X = \exp sX \exp tX$.*
(c) *For any $X \in \mathfrak{g}$, $(\exp X)^{-1} = \exp(-X)$.*
(d) *For any $X \in \mathfrak{g}$ and $n \in \mathbb{Z}$, $(\exp X)^n = \exp(nX)$.*
(e) *The differential $(d \exp)_0 : T_0\mathfrak{g} \to T_eG$ is the identity map, under the canonical identifications of both $T_0\mathfrak{g}$ and T_eG with $\mathfrak{g}$ itself.*
(f) *The exponential map restricts to a diffeomorphism from some neighborhood of 0 in $\mathfrak{g}$ to a neighborhood of e in G.*

(g) *If H is another Lie group, $\mathfrak{h}$ is its Lie algebra, and $\Phi\colon G \to H$ is a Lie group homomorphism, the following diagram commutes*:

$$
\begin{array}{ccc}
\mathfrak{g} & \xrightarrow{\ \Phi_*\ } & \mathfrak{h} \\[2pt]
\exp \Big\downarrow & & \Big\downarrow \exp \\[2pt]
G & \xrightarrow[\ \Phi\ ]{} & H.
\end{array}
\qquad (20.3)
$$

(h) *The flow θ of a left-invariant vector field X is given by $\theta_t = R_{\exp tX}$ (right multiplication by $\exp tX$).*

Proof. In this proof, for any $X \in \mathfrak{g}$ we let $\theta_{(X)}$ denote the flow of X. To prove (a), we need to show that the expression $\theta_{(X)}^{(e)}(1)$ depends smoothly on X, which amounts to showing that the flow varies smoothly as the vector field varies. This is a situation not covered by the fundamental theorem on flows, but we can reduce it to that theorem by the following simple trick. Define a vector field Ξ on the product manifold $G \times \mathfrak{g}$ by

$$
\Xi_{(g,X)} = (X_g, 0) \in T_g G \oplus T_X \mathfrak{g} \cong T_{(g,X)}(G \times \mathfrak{g}).
$$

(See Fig. 20.2.) To see that Ξ is a smooth vector field, choose any basis $(X_1, \ldots, X_k)$ for $\mathfrak{g}$, and let (x^i) be the corresponding global coordinates for $\mathfrak{g}$, defined by $(x^i) \leftrightarrow x^i X_i$. Let (w^i) be any smooth local coordinates for G. If $f \in C^\infty(G \times \mathfrak{g})$ is arbitrary, then locally we can write

$$
\Xi f\left(w^i, x^i\right) = x^j X_j f\left(w^i, x^i\right),
$$

where each vector field X_j differentiates f only in the w^i-directions. Since this depends smoothly on (w^i, x^i), it follows from Proposition 8.14 that Ξ is smooth. It is easy to verify that the flow Θ of Ξ is given by

$$
\Theta_t(g, X) = \left(\theta_{(X)}(t, g), X\right).
$$

By the fundamental theorem on flows, Θ is smooth. Since $\exp X = \pi_G\left(\Theta_1(e, X)\right)$, where $\pi_G\colon G \times \mathfrak{g} \to G$ is the projection, it follows that $\exp$ is smooth.

Next, (b) and (c) follow immediately from Proposition 20.5, because $t \mapsto \exp tX$ is a group homomorphism from $\mathbb{R}$ to G. Then (d) for nonnegative n follows from (b) by induction, and for negative n it follows from (c).

To prove (e), let $X \in \mathfrak{g}$ be arbitrary, and let $\sigma\colon \mathbb{R} \to \mathfrak{g}$ be the curve $\sigma(t) = tX$. Then $\sigma'(0) = X$, and Proposition 20.5 implies

$$
(d\exp)_0(X) = (d\exp)_0\left(\sigma'(0)\right) = (\exp\circ\sigma)'(0) = \left.\frac{d}{dt}\right|_{t=0} \exp tX = X.
$$

Part (f) then follows immediately from (e) and the inverse function theorem.

Next, to prove (g) we need to show that $\exp(\Phi_* X) = \Phi(\exp X)$ for every $X \in \mathfrak{g}$. In fact, we will show that for all $t \in \mathbb{R}$,

$$
\exp(t\Phi_* X) = \Phi(\exp tX).
$$

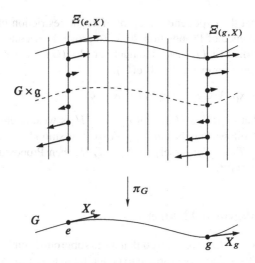

Fig. 20.2 Proof that the exponential map is smooth

The left-hand side is, by Proposition 20.5, the one-parameter subgroup generated by $\Phi_* X$. Thus, if we put $\sigma(t) = \Phi(\exp tX)$, it suffices to show that $\sigma \colon \mathbb{R} \to H$ is a Lie group homomorphism satisfying $\sigma'(0) = (\Phi_* X)_e$. It is a Lie group homomorphism because it is the composition of the homomorphisms Φ and $t \mapsto \exp tX$. The initial velocity is computed as follows:

$$\sigma'(0) = \frac{d}{dt}\Big|_{t=0} \Phi(\exp tX) = d\Phi_0 \left(\frac{d}{dt}\Big|_{t=0} \exp tX \right) = d\Phi_0(X_e) = (\Phi_* X)_e.$$

Finally, to show that $(\theta_{(X)})_t = R_{\exp tX}$, we use the fact that for any $g \in G$, the left multiplication map L_g takes integral curves of X to integral curves of X. Thus, the map $t \mapsto L_g(\exp tX)$ is the integral curve starting at g, which means it is equal to $\theta_{(X)}^{(g)}(t)$. It follows that

$$R_{\exp tX}(g) = g \exp tX = L_g(\exp tX) = \theta_{(X)}^{(g)}(t) = (\theta_{(X)})_t (g). \qquad \square$$

It is important to notice that Proposition 20.8(b) does not imply $\exp(X + Y) = (\exp X)(\exp Y)$ for arbitrary X, Y in the Lie algebra. In fact, for connected groups, this is true only when the group is abelian (see Problem 20-8).

The exponential map yields the following alternative characterization of the Lie subalgebra of a subgroup. We will use this later in the chapter when we study normal subgroups.

Proposition 20.9. *Let G be a Lie group, and let $H \subseteq G$ be a Lie subgroup. With $\mathrm{Lie}(H)$ considered as a subalgebra of $\mathrm{Lie}(G)$ in the usual way, the exponential map of H is the restriction to $\mathrm{Lie}(H)$ of the exponential map of G, and*

$$\mathrm{Lie}(H) = \left\{ X \in \mathrm{Lie}(G) : \exp tX \in H \text{ for all } t \in \mathbb{R} \right\}.$$

Proof. The fact that the exponential map of H is the restriction of that of G is an immediate consequence of Proposition 20.3. To prove the second assertion, by the way we have identified $\mathrm{Lie}(H)$ as a subalgebra of $\mathrm{Lie}(G)$, we need to establish the following equivalence for every $X \in \mathrm{Lie}(G)$:

$$\exp t X \in H \quad \text{for all } t \in \mathbb{R} \quad \Leftrightarrow \quad X_e \in T_e H.$$

Assume first that $\exp t X \in H$ for all t. Since H is weakly embedded in G by Theorem 19.25, it follows that the curve $t \mapsto \exp t X$ is smooth as a map into H, and thus $X_e = \gamma'(0) \in T_e H$. Conversely, if $X_e \in T_e H$, then Proposition 20.3 implies that $\exp t X \in H$ for all t. $\qquad\square$

The Closed Subgroup Theorem

Recall that in Theorem 7.21 we showed that a Lie subgroup is embedded if and only if it is closed. In this section, we use the exponential map to prove a much stronger form of that theorem, showing that if a subgroup of a Lie group is topologically a closed subset, then it is actually an embedded Lie subgroup.

We begin with a simple result that shows how group multiplication in G is reflected "to first order" in the vector space structure of its Lie algebra.

Proposition 20.10. *Let G be a Lie group and let $\mathfrak{g}$ be its Lie algebra. For any $X, Y \in \mathfrak{g}$, there is a smooth function $Z: (-\varepsilon, \varepsilon) \to \mathfrak{g}$ for some $\varepsilon > 0$ such that the following identity holds for all $t \in (-\varepsilon, \varepsilon)$:*

$$(\exp t X)(\exp t Y) = \exp\big(t(X + Y) + t^2 Z(t)\big). \tag{20.4}$$

Proof. Since the exponential map is a diffeomorphism on some neighborhood of the origin in $\mathfrak{g}$, there is some $\varepsilon > 0$ such that the map $\varphi: (-\varepsilon, \varepsilon) \to \mathfrak{g}$ defined by

$$\varphi(t) = \exp^{-1}(\exp t X \, \exp t Y)$$

is smooth. It obviously satisfies $\varphi(0) = 0$ and

$$\exp t X \, \exp t Y = \exp \varphi(t).$$

Observe that we can write φ as the composition

$$\mathbb{R} \xrightarrow{\ e_X \times e_Y\ } G \times G \xrightarrow{\ m\ } G \xrightarrow{\ \exp^{-1}\ } \mathfrak{g},$$

where $e_X(t) = \exp t X$ and $e_Y(t) = \exp t Y$. The result of Problem 7-2 shows that $dm_{(e,e)}(X, Y) = X + Y$ for $X, Y \in T_e G$, which implies

$$\varphi'(0) = \big((d \exp)_0\big)^{-1} \big(e_X'(0) + e_Y'(0)\big) = X + Y.$$

Therefore, Taylor's theorem yields

$$\varphi(t) = t(X + Y) + t^2 Z(t)$$

for some smooth function Z. $\qquad\square$

Corollary 20.11. *Under the hypotheses of the preceding proposition,*

$$\lim_{n\to\infty} \left(\left(\exp\frac{t}{n}X \right) \left(\exp\frac{t}{n}Y \right) \right)^n = \exp t(X+Y). \qquad (20.5)$$

Proof. Formula (20.4) implies that for any $t \in \mathbb{R}$ and any sufficiently large $n \in \mathbb{Z}$,

$$\left(\exp\frac{t}{n}X \right) \left(\exp\frac{t}{n}Y \right) = \exp\left(\frac{t}{n}(X+Y) + \frac{t^2}{n^2}Z\left(\frac{t}{n}\right) \right),$$

and then Proposition 20.8(d) yields

$$\left(\left(\exp\frac{t}{n}X \right) \left(\exp\frac{t}{n}Y \right) \right)^n = \left(\exp\left(\frac{t}{n}(X+Y) + \frac{t^2}{n^2}Z\left(\frac{t}{n}\right) \right) \right)^n$$

$$= \exp\left(t(X+Y) + \frac{t^2}{n}Z\left(\frac{t}{n}\right) \right).$$

Fixing t and taking the limit as $n \to \infty$, we obtain (20.5).   □

Theorem 20.12 (Closed Subgroup Theorem). *Suppose G is a Lie group and $H \subseteq G$ is a subgroup that is also a closed subset of G. Then H is an embedded Lie subgroup.*

Proof. By Proposition 7.11, it suffices to show that H is an embedded submanifold of G. We begin by identifying a subspace of $\mathrm{Lie}(G)$ that will turn out to be the Lie algebra of H.

Let $\mathfrak{g} = \mathrm{Lie}(G)$, and define a subset $\mathfrak{h} \subseteq \mathfrak{g}$ by

$$\mathfrak{h} = \{X \in \mathfrak{g} : \exp tX \in H \text{ for all } t \in \mathbb{R}\}.$$

We need to show that $\mathfrak{h}$ is a linear subspace of $\mathfrak{g}$. It is obvious from the definition that $\mathfrak{h}$ is closed under scalar multiplication: if $X \in \mathfrak{h}$, then $tX \in \mathfrak{h}$ for all $t \in \mathbb{R}$. Suppose $X, Y \in \mathfrak{h}$, and let $t \in \mathbb{R}$ be arbitrary. Then $\exp\big((t/n)X\big)$ and $\exp\big((t/n)Y\big)$ are in H for each positive integer n, and because H is a closed subgroup of G, (20.5) implies that $\exp t(X+Y) \in H$. Thus $X + Y \in \mathfrak{h}$, so $\mathfrak{h}$ is a subspace.

Next we show that there is a neighborhood U of the origin in $\mathfrak{g}$ on which the exponential map of G is a diffeomorphism, and which has the property that

$$\exp(U \cap \mathfrak{h}) = (\exp U) \cap H. \qquad (20.6)$$

(See Fig. 20.3.) This will enable us to construct a slice chart for H in a neighborhood of the identity, and we will then use left translation to get a slice chart in a neighborhood of any point of H.

If $U \subseteq \mathfrak{g}$ is any neighborhood of 0 on which $\exp$ is a diffeomorphism, then $\exp(U \cap \mathfrak{h}) \subseteq (\exp U) \cap H$ by definition of $\mathfrak{h}$. So to find a neighborhood satisfying (20.6), all we need to do is to show that U can be chosen small enough that $(\exp U) \cap H \subseteq \exp(U \cap \mathfrak{h})$. Assume this is not possible.

Choose a linear subspace $\mathfrak{b} \subseteq \mathfrak{g}$ that is complementary to $\mathfrak{h}$, so that $\mathfrak{g} = \mathfrak{h} \oplus \mathfrak{b}$ as vector spaces. By the result of Problem 20-3, the map $\Phi : \mathfrak{h} \oplus \mathfrak{b} \to G$ given

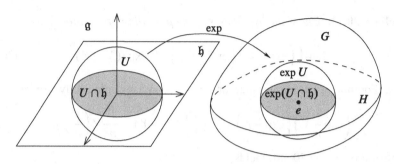

Fig. 20.3 A neighborhood used to construct a slice chart for H

by $\Phi(X, Y) = \exp X \exp Y$ is a diffeomorphism in some neighborhood of $(0, 0)$. Choose neighborhoods U_0 of 0 in $\mathfrak{g}$ and $\widetilde{U}_0$ of $(0, 0)$ in $\mathfrak{h} \oplus \mathfrak{b}$ such that both $\exp|_{U_0}$ and $\Phi|_{\widetilde{U}_0}$ are diffeomorphisms onto their images. Let $\{U_i\}$ be a countable neighborhood basis for $\mathfrak{g}$ at 0 (e.g., a countable sequence of coordinate balls whose radii approach zero). If we set $V_i = \exp(U_i)$ and $\widetilde{U}_i = \Phi^{-1}(V_i)$, then $\{V_i\}$ and $\{\widetilde{U}_i\}$ are neighborhood bases for G at e and $\mathfrak{h} \oplus \mathfrak{b}$ at $(0, 0)$, respectively. By discarding finitely many terms at the beginning of the sequence, we may assume that $U_i \subseteq U_0$ and $\widetilde{U}_i \subseteq \widetilde{U}_0$ for each i.

Our assumption implies that for each i, there exists $h_i \in (\exp U_i) \cap H$ such that $h_i \notin \exp(U_i \cap \mathfrak{h})$. This means $h_i = \exp Z_i$ for some $Z_i \in U_i$. Because $\exp(U_i) = \Phi(\widetilde{U}_i)$, we can also write

$$h_i = \exp X_i \exp Y_i$$

for some $(X_i, Y_i) \in \widetilde{U}_i$. If Y_i were zero, then we would have $\exp Z_i = \exp X_i \in \exp(\mathfrak{h})$; but because $\exp$ is injective on U_0, this implies $X_i = Z_i \in U_i \cap \mathfrak{h}$, which contradicts our assumption that $h_i \notin \exp(U_i \cap \mathfrak{h})$ (Fig. 20.4). Since $\{\widetilde{U}_i\}$ is a neighborhood basis, $Y_i \to 0$ as $i \to \infty$. Observe that $\exp X_i \in H$ by definition of $\mathfrak{h}$, so it follows that $\exp Y_i = (\exp X_i)^{-1} h_i \in H$ as well.

Choose an inner product on $\mathfrak{b}$ and let $|\cdot|$ denote the norm associated with this inner product. If we define $c_i = |Y_i|$, then we have $c_i \to 0$ as $i \to \infty$. The sequence $(c_i^{-1} Y_i)$ lies on the unit sphere in $\mathfrak{b}$, so replacing it by a subsequence we may assume that $c_i^{-1} Y_i \to Y \in \mathfrak{b}$, with $|Y| = 1$ by continuity. In particular, $Y \neq 0$. We will show that $\exp t Y \in H$ for all $t \in \mathbb{R}$, which implies that $Y \in \mathfrak{h}$. Since $\mathfrak{h} \cap \mathfrak{b} = \{0\}$, this is a contradiction.

Let $t \in \mathbb{R}$ be arbitrary, and for each i, let n_i be the greatest integer less than or equal to t/c_i. Then

$$\left| n_i - \frac{t}{c_i} \right| \leq 1,$$

which implies

$$|n_i c_i - t| \leq c_i \to 0,$$

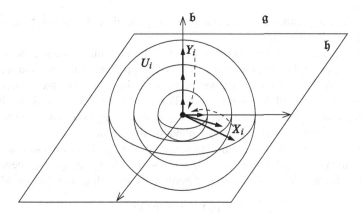

Fig. 20.4 Proof of the closed subgroup theorem

so $n_i c_i \to t$. Thus,

$$n_i Y_i = (n_i c_i)\left(c_i^{-1} Y_i\right) \to t Y,$$

which implies $\exp n_i Y_i \to \exp t Y$ by continuity. But $\exp n_i Y_i = (\exp Y_i)^{n_i} \in H$, so the fact that H is closed implies $\exp t Y \in H$. This completes the proof of the existence of U satisfying (20.6).

Choose any linear isomorphism $E \colon \mathfrak{g} \to \mathbb{R}^m$ that sends $\mathfrak{h}$ to $\mathbb{R}^k$. The composite map $\varphi = E \circ \exp^{-1} \colon \exp U \to \mathbb{R}^m$ is then a smooth chart for G, and $\varphi\big((\exp U) \cap H\big) = E(U \cap \mathfrak{h})$ is the slice obtained by setting the last $m - k$ coordinates equal to zero. Moreover, if $h \in H$ is arbitrary, the left translation map L_h is a diffeomorphism from $\exp U$ to a neighborhood of h. Since H is a subgroup, $L_h(H) = H$, and so

$$L_h\big((\exp U) \cap H\big) = L_h(\exp U) \cap H,$$

and $\varphi \circ L_h^{-1}$ is easily seen to be a slice chart for H in a neighborhood of h. Thus, H is an embedded submanifold of G, hence a Lie subgroup. $\qquad\square$

The following corollary summarizes the results of the closed subgroup theorem, Proposition 7.11, and Theorem 7.21.

Corollary 20.13. *If G is a Lie group and H is any subgroup of G, the following are equivalent:*

(a) *H is closed in G.*
(b) *H is an embedded submanifold of G.*
(c) *H is an embedded Lie subgroup of G.* $\qquad\square$

Infinitesimal Generators of Group Actions

In Chapter 9, we showed that a complete vector field on a manifold generates an action of $\mathbb{R}$ on the manifold. In this section, using the Frobenius theorem and properties of the exponential map, we show how to generalize this notion to actions of higher-dimensional groups.

To begin, we need to specify what we mean by an "infinitesimal generator" of a Lie group action. For reasons that will become apparent, in this section we work primarily with right actions. Afterwards, we will show how the theory has to be modified in the case of left actions (see Theorem 20.18). Because $\mathbb{R}$ is abelian, global flows can be considered either as left actions or as right actions, so everything in this section applies to global flows without modification.

Suppose we are given a smooth right action of a Lie group G on a smooth manifold M, which we denote either by $\theta \colon M \times G \to M$ or by $(p, g) \mapsto p \cdot g$, depending on context. Each element $X \in \mathrm{Lie}(G)$ determines a smooth global flow on M:

$$(t, p) \mapsto p \cdot \exp t X.$$

(It is a flow because $\exp(0X) = e$ and $(\exp sX)(\exp tX) = \exp(s + t)X$.) Let $\widehat{X} \in \mathfrak{X}(M)$ be the infinitesimal generator of this flow, so for each $p \in M$,

$$\widehat{X}_p = \frac{d}{dt}\bigg|_{t=0} p \cdot \exp t X. \qquad (20.7)$$

Thus we obtain a map $\widehat{\theta} \colon \mathrm{Lie}(G) \to \mathfrak{X}(M)$, defined by $\widehat{\theta}(X) = \widehat{X}$.

There is a useful alternative characterization of $\widehat{X}$ in terms of the orbit map $\theta^{(p)} \colon G \to M$ defined by $\theta^{(p)}(g) = p \cdot g$. Since $\gamma(t) = \exp t X$ is a smooth curve in G whose initial velocity is $\gamma'(0) = X_e$, it follows from Corollary 3.25 that for each $p \in M$ we have

$$d\left(\theta^{(p)}\right)_e (X_e) = \left(\theta^{(p)} \circ \gamma\right)'(0) = \frac{d}{dt}\bigg|_{t=0} p \cdot \exp t X = \widehat{X}_p. \qquad (20.8)$$

Lemma 20.14. *Suppose G is a Lie group and θ is a smooth right action of G on a smooth manifold M. For any $X \in \mathrm{Lie}(G)$ and $p \in M$, the vector fields X and $\widehat{\theta}(X)$ are $\theta^{(p)}$-related.*

Proof. Let $X \in \mathrm{Lie}(G)$ and $p \in M$ be arbitrary, and write $\widehat{X} = \widehat{\theta}(X)$. Note that the group law $p \cdot gg' = (p \cdot g) \cdot g'$ translates to

$$\theta^{(p)} \circ L_g(g') = \theta^{(p \cdot g)}(g'). \qquad (20.9)$$

Let $g \in G$ be arbitrary, and write $q = p \cdot g = \theta^{(p)}(g)$. Then (20.9) yields $\theta^{(p)} \circ L_g = \theta^{(q)}$. Using this together with (20.8) and the fact that X is left-invariant, we obtain

$$\widehat{X}_q = d\left(\theta^{(q)}\right)_e (X_e) = d\left(\theta^{(p)}\right)_g \circ d(L_g)_e (X_e) = d\left(\theta^{(p)}\right)_g (X_g),$$

which proves the claim. □

Theorem 20.15. *Suppose G is a Lie group and θ is a smooth right action of G on a smooth manifold M. Then the map $\widehat{\theta} \colon \mathrm{Lie}(G) \to \mathfrak{X}(M)$ defined above is a Lie algebra homomorphism.*

Proof. For each $p \in M$, it follows from (20.8) that $\widehat{X}_p$ depends linearly on X, so $\widehat{\theta}$ is a linear map. Given $p \in M$, Lemma 20.14 together with the naturality of Lie brackets implies that $[X, Y]$ is $\theta^{(p)}$-related to $\left[\widehat{X}, \widehat{Y}\right]$. This means, in particular, that

$$\left[\widehat{X}, \widehat{Y}\right]_p = d\left(\theta^{(p)}\right)_e([X, Y]_e) = \widehat{[X, Y]}_p.$$

Since every point of M is in the image of some orbit map, we conclude that $\left[\widehat{\theta}(X), \widehat{\theta}(Y)\right] = \widehat{\theta}([X, Y])$ as claimed. $\qquad\square$

The Lie algebra homomorphism $\widehat{\theta}\colon \mathrm{Lie}(G) \to \mathfrak{X}(M)$ defined above is known as the ***infinitesimal generator of θ***. More generally, if $\mathfrak{g}$ is an arbitrary finite-dimensional Lie algebra, any Lie algebra homomorphism $\widehat{\theta}\colon \mathfrak{g} \to \mathfrak{X}(M)$ is called a (***right***) ***$\mathfrak{g}$-action on M***. A $\mathfrak{g}$-action $\widehat{\theta}$ is said to be ***complete*** if for every $X \in \mathfrak{g}$, the vector field $\widehat{\theta}(X)$ is complete.

Just as every complete vector field generates an $\mathbb{R}$-action, the next theorem shows that, at least for simply connected groups, every complete Lie algebra action generates a Lie group action.

Theorem 20.16 (Fundamental Theorem on Lie Algebra Actions). *Let M be a smooth manifold, let G be a simply connected Lie group, and let $\mathfrak{g} = \mathrm{Lie}(G)$. Suppose $\widehat{\theta}\colon \mathfrak{g} \to \mathfrak{X}(M)$ is a complete $\mathfrak{g}$-action on M. Then there is a unique smooth right G-action on M whose infinitesimal generator is $\widehat{\theta}$.*

Proof. We begin by defining a distribution D on $G \times M$; we will show that D is involutive, and then each leaf will turn out to be the graph of an orbit map $\theta^{(p)}\colon G \to M$. For brevity, given $X \in \mathfrak{g}$, we use the notation $\widehat{X}$ for $\widehat{\theta}(X) \in \mathfrak{X}(M)$.

Define D as follows: for each $X \in \mathfrak{g}$, define a smooth vector field $\widetilde{X}$ on $G \times M$ by

$$\widetilde{X}_{(g,p)} = \left(X_g, \widehat{X}_p\right) \in T_g G \oplus T_p M \cong T_{(g,p)}(G \times M).$$

In the notation of Problem 8-17, this is $\widetilde{X} = X \oplus \widehat{X}$. Then for each $(g, p) \in G \times M$, let $D_{(g,p)}$ be the set of all vectors of the form $\widetilde{X}_{(g,p)}$ as X ranges over $\mathfrak{g}$. If $X_1, \ldots, X_k$ is a basis for $\mathfrak{g}$, then the smooth vector fields $\widetilde{X}_1, \ldots, \widetilde{X}_k$ are independent and span D, so D is a smooth distribution whose rank is equal to the dimension of G. To see that it is involutive, note that Problem 8-17 and the fact that $\widehat{\theta}$ is a Lie algebra homomorphism imply

$$\left[\widetilde{X}_i, \widetilde{X}_j\right] = [X_i, X_j] \oplus \left[\widehat{X}_i, \widehat{X}_j\right] = [X_i, X_j] \oplus \widehat{[X_i, X_j]} = \widetilde{[X_i, X_j]}.$$

Let $\mathcal{S}$ denote the foliation determined by D, and for each $(g, p) \in G \times M$, let $\mathcal{S}_{(g,p)}$ denote the leaf of $\mathcal{S}$ containing (g, p).

Next we show that D is invariant under a certain G-action on $G \times M$. Combining the natural action of G on itself by left translation with the trivial action of G on M, we get a left action of G on $G \times M$ given by

$$\psi_g(g', p) = (gg', p).$$

A straightforward computation shows

$$d(\psi_g)_{(g',p)}\big(\widetilde{X}_{(g',p)}\big) = d(\psi_g)_{(g',p)}\big(X_{g'}, \widehat{X}_p\big) = \big(d(L_g)_{g'}(X_{g'}), \widehat{X}_p\big)$$
$$= \big(X_{gg'}, \widehat{X}_p\big) = \widetilde{X}_{(gg',p)},$$

so D is invariant under ψ_g for each $g \in G$. It follows that ψ_g takes leaves of $\mathcal{S}$ to leaves of $\mathcal{S}$ (Proposition 19.23).

Let $\pi_G \colon G \times M \to G$ and $\pi_M \colon G \times M \to M$ denote the projections. Let $p \in M$ be arbitrary, let $S_p = \mathcal{S}_{(e,p)} \subseteq G \times M$ denote the leaf containing (e, p), and let $\Pi_p = \pi_G|_{S_p} \colon S_p \to G$. We will show that Π_p is a smooth covering map. To begin with, at each point $(g, q) \in S_p$, $d(\Pi_p)_{(g,p)}\big(\widetilde{X}_{(g,p)}\big) = X_g$ for all $X \in \mathfrak{g}$, so Π_p is a smooth submersion, and for dimensional reasons it is a local diffeomorphism.

To show that Π_p is a covering map, choose a connected neighborhood U of e in G small enough that the exponential map of G is a diffeomorphism from some neighborhood V of 0 in $\mathfrak{g}$ onto U, and for any $g \in G$, consider the neighborhood $gU = \{gh : h \in U\}$ of g. We will show that gU is evenly covered by constructing local sections. For each $q \in M$ such that (g, q) is in the fiber $\Pi_p^{-1}(g)$, define a map $\sigma_q \colon gU \to G \times M$ by

$$\sigma_q(g \exp X) = \big(g \exp X, \eta_{(\widehat{X})}(1, q)\big),$$

where $X \in V$ and $\eta_{(\widehat{X})}$ denotes the flow of $\widehat{X}$. It follows immediately from the definition that σ_q is smooth and satisfies $\pi_G \circ \sigma_q = \mathrm{Id}_{gU}$, so to show that σ_q is a local section of Π_p, it suffices to show that it takes its values in S_p. A straightforward computation shows that $\gamma(t) = \big(g \exp tX, \eta_{(\widehat{X})}(t, q)\big)$ is an integral curve of $\widetilde{X}$ starting at (g, q), from which it follows easily that $\sigma_q(g \exp X) = \gamma(1) \in S_p$. It is smooth because it is a local section of the local diffeomorphism Π_p.

For each $(g, q) \in \Pi_p^{-1}(g)$, the set $\sigma_q(gU)$ is a connected open subset of S_p, which is mapped diffeomorphically onto gU by Π_p. To complete the proof that Π_p is a covering map, we need only prove that every point in $\Pi_p^{-1}(gU)$ is in exactly one such set. First suppose $(g', q') \in \Pi_p^{-1}(gU)$. Then $\Pi_p(g', q') \in gU$ means that $g' = g \exp X$ for some $X \in V$. If we let $q = \eta_{(\widehat{X})}(-1, q')$, then the group law for $\eta_{(\widehat{X})}$ implies that $q' = \eta_{(\widehat{X})}(1, q)$ and therefore $(g', q') = \sigma_q(g \exp X)$. On the other hand, suppose two such sets $\sigma_q(gU)$ and $\sigma_{q'}(gU)$ intersect nontrivially. Then for some $X, X' \in V$, we have $\big(g \exp X, \eta_{(\widehat{X})}(1, q)\big) = \big(g \exp X', \eta_{(\widehat{X'})}(1, q')\big)$, which implies that $X = X'$ and therefore $\eta_{(\widehat{X})}(1, q) = \eta_{(\widehat{X})}(1, q')$; then flowing back along the integral curve of $\widehat{X}$ for time -1 shows that $q = q'$. This completes the proof that Π_p is a smooth covering map. Because we are assuming G is simply connected, Π_p is actually a diffeomorphism.

Now for each $p \in M$, define $\theta^{(p)} \colon G \to M$ by $\theta^{(p)} = \pi_M \circ \Pi_p^{-1}$ (so S_p is the graph of $\theta^{(p)}$), and define an action of G on M by $p \cdot g = \theta^{(p)}(g)$. This is equivalent to declaring that $p \cdot g = q$ if and only if $\mathcal{S}_{(e,p)} = \mathcal{S}_{(g,q)}$. To show that this is an action, assuming $p \cdot g = q$ and $q \cdot g' = r$, we need to show that $p \cdot gg' = r$.

Equivalently, assuming that $\mathcal{S}_{(e,p)} = \mathcal{S}_{(g,q)}$ and $\mathcal{S}_{(e,q)} = \mathcal{S}_{(g',r)}$, we need to show that $\mathcal{S}_{(e,p)} = \mathcal{S}_{(gg',r)}$. This follows from ψ-invariance:

$$\mathcal{S}_{(e,p)} = \mathcal{S}_{(g,q)} = \psi_g\left(\mathcal{S}_{(e,q)}\right) = \psi_g\left(\mathcal{S}_{(g',r)}\right) = \mathcal{S}_{(gg',r)}.$$

It remains to show that the action is smooth, that $\hat{\theta}$ is its infinitesimal generator, and that it is the unique such action. For $g = \exp X$ near the identity, the discussion above shows that the action can be expressed as

$$p \cdot g = \theta^{(p)}(\exp X) = \pi_M \circ \sigma_p(\exp X) = \eta_{(\hat{X})}(1, p). \tag{20.10}$$

An argument analogous to the one we used to prove smoothness of the exponential map (with $\Xi_{(p,X)} = (\hat{X}_g, 0)$ on $M \times \mathfrak{g}$) shows that this depends smoothly on X and p and thus on g and p. But since any neighborhood of the identity generates G (Proposition 7.14), every element of G can be expressed as a finite product of elements of the form $\exp X$ for $X \in V$, so it follows that $(p, g) \mapsto p \cdot g$ can be written as a finite composition of smooth maps. The fact that the infinitesimal generator of the action is $\hat{\theta}$ is an immediate consequence of (20.10). Uniqueness is left as an exercise.   □

▶ **Exercise 20.17.** Prove that the action constructed in the previous proof is the unique one that has $\hat{\theta}$ as its infinitesimal generator.

Left Actions

The situation for left actions is similar, but with a slight twist. Let G be a Lie group and M be a smooth manifold. If $\theta: G \times M \to M$ is a smooth left action of G on M, define the ***infinitesimal generator of*** θ as the map $\hat{\theta}: \mathrm{Lie}(G) \to \mathfrak{X}(M)$ given by $\hat{\theta}(X) = \hat{X}$, where

$$\hat{X}_p = \frac{d}{dt}\bigg|_{t=0} \left((\exp tX) \cdot p\right) = d\left(\theta^{(p)}\right)_e(X_e), \tag{20.11}$$

and $\theta^{(p)}: G \to M$ is the orbit map $\theta^{(p)}(g) = g \cdot p$.

We have the following analogue of Theorems 20.15 and 20.16 for left actions.

Theorem 20.18. *Suppose G is a Lie group and M is a smooth manifold.*

(a) *If θ is a smooth left action of G on M, the map $\hat{\theta}: \mathrm{Lie}(G) \to \mathfrak{X}(M)$ defined by (20.11) is an **antihomomorphism** (a linear map satisfying $\hat{\theta}([X, Y]) = -[\hat{\theta}(X), \hat{\theta}(Y)]$ for all $X, Y \in \mathrm{Lie}(G)$).*

(b) *Conversely, if G is simply connected, every antihomomorphism $\hat{\theta}: \mathrm{Lie}(G) \to \mathfrak{X}(M)$ such that $\hat{\theta}(X)$ is complete for each $X \in \mathrm{Lie}(G)$ is the infinitesimal generator of a unique left G-action.*

Proof. Problem 20-15.   □

Because of this theorem, for a finite-dimensional Lie algebra $\mathfrak{g}$ and a smooth manifold M, a *left* $\mathfrak{g}$-*action on* M is defined as an antihomomorphism from $\mathfrak{g}$ to $\mathfrak{X}(M)$.

The Lie Correspondence

Many of our results about Lie groups show how essential properties of a Lie group are reflected in its Lie algebra, and vice versa. This raises a natural question: To what extent is the correspondence between Lie groups and Lie algebras (or at least between their isomorphism classes) one-to-one?

We have already seen in Chapter 8 that the assignment that sends a Lie group to its Lie algebra and a Lie group homomorphism to its induced Lie algebra homomorphism is a functor from the category of Lie groups to the category of finite-dimensional Lie algebras. Because functors take isomorphisms to isomorphisms, it follows that isomorphic Lie groups have isomorphic Lie algebras. The converse is easily seen to be false: both $\mathbb{R}^n$ and $\mathbb{T}^n$ have n-dimensional abelian Lie algebras, which are obviously isomorphic to each other, but $\mathbb{R}^n$ and $\mathbb{T}^n$ are certainly not isomorphic Lie groups. However, as we will see in this section, if we restrict our attention to simply connected Lie groups, then we do obtain a one-to-one correspondence.

In order to prove this correspondence, we need a way to construct an isomorphism between simply connected Lie groups when we are given an isomorphism between their algebras. Theorem 8.44 showed that every Lie group homomorphism gives rise to a Lie algebra homomorphism. Using the fundamental theorem on Lie algebra actions, we can prove the following partial converse.

Theorem 20.19. *Suppose G and H are Lie groups with G simply connected, and let $\mathfrak{g}$ and $\mathfrak{h}$ be their Lie algebras. For any Lie algebra homomorphism $\varphi\colon \mathfrak{g} \to \mathfrak{h}$, there is a unique Lie group homomorphism $\Phi\colon G \to H$ such that $\Phi_* = \varphi$.*

Proof. The Lie algebra homomorphism $\varphi\colon \mathfrak{g} \to \mathfrak{h} \subseteq \mathfrak{X}(H)$ is, in particular, a complete $\mathfrak{g}$-action on H (since every left-invariant vector field is complete). Thus, by Theorem 20.16, there is a unique smooth right G-action $\theta\colon H \times G \to H$ for which φ is the infinitesimal generator. Let us use the notation $\widehat{X} = \varphi(X)$ for $X \in \mathfrak{g}$, and the notation $h \cdot g = \theta(h, g)$ for $h \in H$ and $g \in G$. Define a smooth map $\Phi\colon G \to H$ by $\Phi(g) = e \cdot g$ (where e is the identity in H). We will show that Φ is the desired homomorphism.

Lemma 20.14 shows that for each $h \in H$ and each $X \in \mathfrak{g}$, the vector fields X and $\widehat{X}$ are $\theta^{(h)}$-related. By Proposition 9.6, $\theta^{(h)}$ takes integral curves of X to integral curves of $\widehat{X}$. Therefore, $t \mapsto h \cdot \exp tX$ is the integral curve of $\widehat{X}$ starting at h.

On the other hand, because $\widehat{X}$ is a left-invariant vector field on H, left translation in H takes integral curves of $\widehat{X}$ to integral curves of $\widehat{X}$. For any $h \in H$ and $X \in \mathfrak{g}$, therefore,

$$L_h(e \cdot \exp tX) = h \cdot \exp tX. \tag{20.12}$$

Applying this with $h = e \cdot g$ for some $g \in G$, we get

$$(e \cdot g)(e \cdot \exp tX) = (e \cdot g) \cdot \exp tX = e \cdot g \exp tX.$$

(The last equality follows from the fact that θ is an action.) Rewritten in terms of Φ, this says

$$\Phi(g)\Phi(\exp tX) = \Phi(g \exp tX).$$

Since G is connected, it is generated by the image of the exponential map by Proposition 7.14, so this implies that Φ is a homomorphism.

To see that $\Phi_* = \varphi$, let $X \in \mathfrak{g}$ be arbitrary. The fact that φ is the infinitesimal generator of θ means

$$\varphi(X)\big|_e = \frac{d}{dt}\bigg|_{t=0} (e \cdot \exp tX) = \frac{d}{dt}\bigg|_{t=0} \Phi(\exp tX) = d\Phi_e(X_e).$$

Since Φ_* is determined by the action of $d\Phi_e$, this implies $\Phi_* X = \varphi(X)$.

The proof is completed by invoking Problem 20-17, which shows that Φ is the unique homomorphism with this property. $\qquad\square$

Corollary 20.20. *If G and H are simply connected Lie groups with isomorphic Lie algebras, then G and H are isomorphic.*

Proof. Let $\mathfrak{g}$, $\mathfrak{h}$ be the Lie algebras of G and H, respectively, and let $\varphi \colon \mathfrak{g} \to \mathfrak{h}$ be a Lie algebra isomorphism between them. By the preceding theorem, there are Lie group homomorphisms $\Phi \colon G \to H$ and $\Psi \colon H \to G$ satisfying $\Phi_* = \varphi$ and $\Psi_* = \varphi^{-1}$. Both the identity map of G and the composition $\Psi \circ \Phi$ are Lie group homomorphisms from G to itself whose induced Lie algebra homomorphisms are equal to the identity, so the uniqueness part of Theorem 20.19 implies that $\Psi \circ \Phi = \mathrm{Id}_G$. Similarly, $\Phi \circ \Psi = \mathrm{Id}_H$, so Φ is a Lie group isomorphism. $\qquad\square$

Now we are ready for our main theorem.

Theorem 20.21 (The Lie Correspondence). *There is a one-to-one correspondence between isomorphism classes of finite-dimensional Lie algebras and isomorphism classes of simply connected Lie groups, given by associating each simply connected Lie group with its Lie algebra.*

Proof. We need to show that the functor that sends a simply connected Lie group to its Lie algebra is both surjective and injective up to isomorphism. Injectivity is precisely the content of Corollary 20.20.

To prove surjectivity, suppose $\mathfrak{g}$ is any finite-dimensional Lie algebra. By Corollary 8.50 to Ado's theorem, we may replace $\mathfrak{g}$ by an isomorphic Lie subalgebra $\mathfrak{g}_0 \subseteq \mathfrak{gl}(n, \mathbb{R})$. By Theorem 19.26, there is a connected Lie subgroup $G_0 \subseteq \mathrm{GL}(n, \mathbb{R})$ that has $\mathfrak{g}_0$ as its Lie algebra. If G is the universal covering group of G_0, Problem 8-27 shows that $\mathrm{Lie}(G) \cong \mathrm{Lie}(G_0) \cong \mathfrak{g}_0 \cong \mathfrak{g}$. $\qquad\square$

In the next chapter, we will see what happens when we remove the restriction to simply connected groups (see Theorem 21.32).

Lie's Fundamental Theorems

As the name of the previous theorem suggests, a version of the Lie correspondence theorem was proved in the nineteenth century by Sophus Lie. However, since global topological notions such as manifolds and simple connectivity had not yet been formulated, what he was able to prove was essentially a local version of the theorem.

Instead of considering Lie groups as abstract objects, Lie worked with vector fields on open subsets of Euclidean space, and the (local) group actions they generate. Define a **local Lie group** to be an open subset U in some finite-dimensional vector space V, together with an element $e \in U$ and smooth maps $m : U \times U \to V$ (multiplication) and $i : U \to V$ (inversion), satisfying the following identities for all g, h, k sufficiently close to e that both sides are defined:

$$m\big(g, m(h,k)\big) = m\big(m(g,h), k\big) \quad \text{(associativity)};$$
$$m(e,g) = g = m(g,e) \quad \text{(identity)};$$
$$m\big(i(g), g\big) = e = m\big(g, i(g)\big) \quad \text{(inverses)}.$$

The left translation map $L_g : U \to V$ is defined just as for ordinary Lie groups, and a vector field $X \in \mathfrak{X}(U)$ is said to be **left-invariant** if $d(L_g)_{g'}(X_{g'}) = X_{m(g,g')}$ for all $g, g' \in U$ such that $m(g, g') \in U$. Two local Lie groups (U, e, m, i) and (U', e', m', i') are said to be **locally isomorphic** if there is a diffeomorphism from a neighborhood of e in U to a neighborhood of e' in U' that takes e to e', m to m', and i to i', whenever the respective operations are defined. A **local (left or right) action of a local Lie group** on an open subset $W \subseteq \mathbb{R}^n$ is defined like an ordinary action, except that $g \cdot x$ (or $x \cdot g$) is required to be defined only for (g, x) in a neighborhood of $\{e\} \times W$ in $U \times W$. A coordinate neighborhood of the identity in any Lie group is a local Lie group, and any smooth action of a Lie group on a smooth manifold restricts to a local action on any sufficiently small coordinate neighborhood.

Theorem 20.22 (The Fundamental Theorems of Sophus Lie).

  (i) FIRST FUNDAMENTAL THEOREM: *The set of left-invariant vector fields on a local Lie group is a finite-dimensional Lie algebra under Lie bracket, and two local Lie groups with isomorphic Lie algebras are locally isomorphic.*

  (ii) SECOND FUNDAMENTAL THEOREM: *Given an open subset $W \subseteq \mathbb{R}^n$, there is a one-to-one correspondence between smooth right actions of local Lie groups on W and finite-dimensional Lie subalgebras of $\mathfrak{X}(W)$.*

  (iii) THIRD FUNDAMENTAL THEOREM: *Given any finite-dimensional abstract Lie algebra $\mathfrak{g}$, there exists a local Lie group whose algebra of left-invariant vector fields is isomorphic to $\mathfrak{g}$.*

It is an interesting exercise to see if you can adapt the techniques of this chapter to prove these theorems. (See Problem 20-19.)

Normal Subgroups

Normal subgroups (those that are invariant under conjugation) play a central role in abstract group theory: they are the only subgroups whose quotients have group structures, and the only subgroups that are kernels of group homomorphisms.

For Lie groups, the following criterion for normality is useful. It says that for a connected Lie group, normality need only be checked for elements that are in the image of the exponential map, because such elements generate the group.

Lemma 20.23. *Let G be a connected Lie group, and let $H \subseteq G$ be a connected Lie subgroup. Let $\mathfrak{g}$ and $\mathfrak{h}$ denote the Lie algebras of G and H, respectively. Then H is normal in G if and only if*

$$(\exp X)(\exp Y)\big(\exp(-X)\big) \in H \quad \text{for all } X \in \mathfrak{g} \text{ and } Y \in \mathfrak{h}. \tag{20.13}$$

Proof. Note that $\exp(-X) = (\exp X)^{-1}$. Thus if H is normal, then (20.13) holds by definition. Conversely, suppose (20.13) holds, and choose open subsets $V \subseteq \mathfrak{g}$ containing 0 and $U \subseteq G$ containing the identity such that $\exp\colon V \to U$ is a diffeomorphism. Since the exponential map of H is the restriction of that of G, after shrinking V if necessary, we may assume that the restriction of $\exp$ to $V \cap \mathfrak{h}$ is a diffeomorphism from $V \cap \mathfrak{h}$ to a neighborhood U_0 of the identity in H. Shrinking V still further, we may assume also that $X \in V$ if and only if $-X \in V$. Then (20.13) implies that $ghg^{-1} \in H$ whenever $g \in U$ and $h \in U_0$.

Since every element of H can be written as a finite product $h = h_1 \cdots h_m$ with $h_1, \ldots, h_m \in U_0$ (Proposition 7.14), it follows that for any $g \in U$ and $h \in H$ we have

$$ghg^{-1} = gh_1 \cdots h_m g^{-1} = \big(gh_1 g^{-1}\big) \cdots \big(gh_m g^{-1}\big) \in H. \tag{20.14}$$

Similarly, any $g \in G$ can be written $g = g_1 \cdots g_k$ with $g_1, \ldots, g_k \in U$, so it follows by induction on k that $ghg^{-1} \in H$ for all $g \in G$ and $h \in H$. $\square$

Our next goal is a theorem that expresses a deep relationship between Lie groups and their Lie algebras. If $\mathfrak{g}$ is a Lie algebra, a linear subspace $\mathfrak{h} \subseteq \mathfrak{g}$ is called an ***ideal in*** $\mathfrak{g}$ if $[X, Y] \in \mathfrak{h}$ whenever $X \in \mathfrak{g}$ and $Y \in \mathfrak{h}$ (see Problem 8-31). Because ideals are kernels of Lie algebra homomorphisms and normal subgroups are kernels of Lie group homomorphisms, it should not be surprising that there is a connection between ideals and normal subgroups. The key to analyzing this connection is the adjoint representation, which we study next.

The Adjoint Representation

Let G be a Lie group and $\mathfrak{g}$ be its Lie algebra. For any $g \in G$, the conjugation map $C_g\colon G \to G$ given by $C_g(h) = ghg^{-1}$ is a Lie group homomorphism (see Example 7.4(f)). We let $\mathrm{Ad}(g) = (C_g)_*\colon \mathfrak{g} \to \mathfrak{g}$ denote its induced Lie algebra homomorphism.

Proposition 20.24 (The Adjoint Representation). *If G is a Lie group with Lie algebra* $\mathfrak{g}$, *the map* $\mathrm{Ad}\colon G \to \mathrm{GL}(\mathfrak{g})$ *is a Lie group representation, called the **adjoint representation of** G.*

Proof. Because $C_{g_1 g_2} = C_{g_1} \circ C_{g_2}$ for any $g_1, g_2 \in G$, it follows immediately that $\mathrm{Ad}(g_1 g_2) = \mathrm{Ad}(g_1) \circ \mathrm{Ad}(g_2)$, and $\mathrm{Ad}(g)$ is invertible with inverse $\mathrm{Ad}(g^{-1})$.

To see that Ad is smooth, let $C\colon G \times G \to G$ be the smooth map defined by $C(g,h) = ghg^{-1}$. Let $X \in \mathfrak{g}$ and $g \in G$ be arbitrary. Then $\mathrm{Ad}(g)X$ is the left-invariant vector field whose value at $e \in G$ is

$$\left((C_g)_* X\right)_e = \frac{d}{dt}\bigg|_{t=0} C_g(\exp tX) = \frac{d}{dt}\bigg|_{t=0} C(g, \exp tX) = dC_{(g,e)}(0, X_e),$$

where we are regarding $(0, X_e)$ as an element of $T_{(g,e)}(G \times G) \cong T_g G \oplus T_e G$. Because $dC\colon T(G \times G) \to TG$ is a smooth bundle homomorphism by Example 10.28(a), this expression depends smoothly on g and X. Smooth coordinates on $\mathrm{GL}(\mathfrak{g})$ are obtained by choosing a basis (E_i) for $\mathfrak{g}$ and using matrix entries with respect to this basis as coordinates. If (ε^j) is the dual basis, the matrix entries of $\mathrm{Ad}(g)\colon \mathfrak{g} \to \mathfrak{g}$ are given by $\left(\mathrm{Ad}(g)\right)_i^j = \varepsilon^j\left(\mathrm{Ad}(g)E_i\right)$. The computation above with $X = E_i$ shows that these are smooth functions of g. $\square$

There is also an adjoint representation for Lie algebras. Given a finite-dimensional Lie algebra $\mathfrak{g}$, for each $X \in \mathfrak{g}$, define a map $\mathrm{ad}(X)\colon \mathfrak{g} \to \mathfrak{g}$ by $\mathrm{ad}(X)Y = [X, Y]$.

Proposition 20.25. *For any Lie algebra* $\mathfrak{g}$, *the map* $\mathrm{ad}\colon \mathfrak{g} \to \mathfrak{gl}(\mathfrak{g})$ *is a Lie algebra representation, called the **adjoint representation of** $\mathfrak{g}$.*

▶ **Exercise 20.26.** Prove the preceding proposition.

Using the exponential map, we can show that these two representations are intimately related.

Theorem 20.27. *Let G be a Lie group, let $\mathfrak{g}$ be its Lie algebra, and let* $\mathrm{Ad}\colon G \to \mathrm{GL}(\mathfrak{g})$ *be the adjoint representation of G. The induced Lie algebra representation* $\mathrm{Ad}_*\colon \mathfrak{g} \to \mathfrak{gl}(\mathfrak{g})$ *is given by* $\mathrm{Ad}_* = \mathrm{ad}$.

Proof. Let $X \in \mathfrak{g}$ be arbitrary. Then $\mathrm{Ad}_* X$ is determined by its value at the identity, which we can interpret as an element of $\mathfrak{gl}(\mathfrak{g})$, the set of all linear maps from $\mathfrak{g}$ to itself. Because $t \mapsto \exp tX$ is a smooth curve in G whose velocity vector at $t = 0$ is X_e, we can compute the action of $\mathrm{Ad}_* X$ on an element $Y \in \mathfrak{g}$ by

$$(\mathrm{Ad}_* X)Y = \left(\frac{d}{dt}\bigg|_{t=0} \mathrm{Ad}(\exp tX)\right) Y = \frac{d}{dt}\bigg|_{t=0} \left(\mathrm{Ad}(\exp tX)Y\right).$$

As an element of $\mathfrak{g}$, $\mathrm{Ad}(\exp tX)Y$ is a left-invariant vector field on G, and thus is itself determined by its value at the identity. Using the fact that $\mathrm{Ad}(g) = (C_g)_* =$

$(R_{g^{-1}})_* \circ (L_g)_*$, its value at $e \in G$ can be computed as

$$
\begin{aligned}
\big(\mathrm{Ad}(\exp tX)Y\big)_e &= d(R_{\exp(-tX)}) \circ d(L_{\exp tX})(Y_e) \\
&= d(R_{\exp(-tX)})(Y_{\exp tX}).
\end{aligned}
\tag{20.15}
$$

Recall from Proposition 20.8(h) that the flow of X is given by $\theta_t(g) = R_{\exp tX}(g)$. Therefore, (20.15) can be rewritten as

$$
\big(\mathrm{Ad}(\exp tX)Y\big)_e = d(\theta_{-t})(Y_{\theta_t(e)}).
$$

Taking the derivative with respect to t and setting $t = 0$, we obtain

$$
\big((\mathrm{Ad}_* X)Y\big)_e = \frac{d}{dt}\bigg|_{t=0} d(\theta_{-t})(Y_{\theta_t(e)}) = (\mathcal{L}_X Y)_e = [X, Y]_e.
$$

Since $(\mathrm{Ad}_* X)Y$ is determined by its value at e, this completes the proof. $\qquad\square$

Ideals and Normal Subgroups

Now we are in a position to prove the main theorem of this section.

Theorem 20.28 (Ideals and Normal Subgroups). *Let G be a connected Lie group, and suppose $H \subseteq G$ is a connected Lie subgroup. Then H is a normal subgroup of G if and only if $\mathrm{Lie}(H)$ is an ideal in $\mathrm{Lie}(G)$.*

Proof. Write $\mathfrak{g} = \mathrm{Lie}(G)$ and $\mathfrak{h} = \mathrm{Lie}(H)$, considering $\mathfrak{h}$ as a Lie subalgebra of $\mathfrak{g}$. For any $g \in G$, the commutative diagram (20.3) applied to the Lie group homomorphism $C_g(h) = ghg^{-1}$ yields

$$
\begin{array}{ccc}
\mathfrak{g} & \xrightarrow{\ \mathrm{Ad}(g)\ } & \mathfrak{g} \\
{\scriptstyle\exp}\big\downarrow & & \big\downarrow{\scriptstyle\exp} \\
G & \xrightarrow[\ C_g\ ]{} & G.
\end{array}
\tag{20.16}
$$

Suppose that $\mathfrak{h}$ is an ideal. Applying (20.16) to $Y \in \mathfrak{h}$ with $g = \exp X$, we obtain

$$
\exp\big(\mathrm{Ad}(\exp X)Y\big) = C_{\exp X}(\exp Y) = (\exp X)(\exp Y)\big(\exp(-X)\big). \tag{20.17}
$$

On the other hand, applying (20.3) to the homomorphism $\mathrm{Ad}\colon G \to \mathrm{GL}(\mathfrak{g})$ and noting that $\mathrm{Ad}_* = \mathrm{ad}$ by Theorem 20.27, we obtain

$$
\mathrm{Ad}(\exp X) = \exp(\mathrm{ad}\, X). \tag{20.18}
$$

Formula (20.2) for the exponential map of the group $\mathrm{GL}(\mathfrak{g})$ reads

$$
\mathrm{Ad}(\exp X)Y = \big(\exp(\mathrm{ad}\, X)\big)Y = \sum_{k=0}^{\infty} \frac{1}{k!}(\mathrm{ad}\, X)^k Y. \tag{20.19}
$$

Whenever $X \in \mathfrak{g}$ and $Y \in \mathfrak{h}$, we have $(\operatorname{ad} X)Y = [X, Y] \in \mathfrak{h}$, and by induction $(\operatorname{ad} X)^k Y \in \mathfrak{h}$ for all k. Therefore, (20.19) implies that $\operatorname{Ad}(\exp X)Y \in \mathfrak{h}$, and so (20.17) implies that $(\exp X)(\exp Y)(\exp(-X)) \in \exp \mathfrak{h} \subseteq H$. By Lemma 20.23, H is normal.

Conversely, suppose H is normal. Given $X \in \mathfrak{g}$ and $Y \in \mathfrak{h}$, note that (20.16) applied to sY with $g = \exp t X$ implies

$$\exp\big(\operatorname{Ad}(\exp t X)sY\big) = (\exp t X)(\exp sY)(\exp t X)^{-1} \in H.$$

Since $\operatorname{Ad}(\exp t X)$ is linear over $\mathbb{R}$, it follows that

$$\exp\big(s \operatorname{Ad}(\exp t X)Y\big) = \exp\big(\operatorname{Ad}(\exp t X)sY\big),$$

which we have just shown to be in H for all s, so $\operatorname{Ad}(\exp t X)Y \in \mathfrak{h}$ by Proposition 20.9. From the proof of Theorem 20.27, we have

$$\frac{d}{dt}\bigg|_{t=0} \operatorname{Ad}(\exp t X)Y = [X, Y],$$

and therefore $[X, Y] \in \mathfrak{h}$, so $\mathfrak{h}$ is an ideal. $\square$

Problems

20-1. Let G be a Lie group.
  (a) Show that the images of one-parameter subgroups in G are precisely the connected Lie subgroups of dimension less than or equal to 1.
  (b) Show that the image of every one-parameter subgroup is isomorphic as a Lie group to one of the following: $\mathbb{R}$, $\mathbb{S}^1$, or the trivial group $\{e\}$.

20-2. Compute the exponential maps of the abelian Lie groups $\mathbb{R}^n$ and $\mathbb{T}^n$.

20-3. Let G be a Lie group, and suppose $A, B \subseteq \mathfrak{g}$ are complementary linear subspaces of $\operatorname{Lie}(G)$. Show that the map $A \oplus B \to G$ given by $(X, Y) \mapsto \exp X \exp Y$ is a diffeomorphism from some neighborhood of $(0, 0)$ in $A \oplus B$ to a neighborhood of e in G. (*Used on p. 523.*)

20-4. Show that the matrix exponential satisfies the identity

$$\det e^A = e^{\operatorname{tr} A}.$$

[Hint: apply Proposition 20.8(g) to det: $\operatorname{GL}(n, \mathbb{R}) \to \mathbb{R}^*$.]

20-5. Let a, b, c be real numbers, and let A, B, and C be the following elements of $\mathfrak{gl}(3, \mathbb{R})$:

$$A = \begin{pmatrix} a & 0 & 0 \\ 0 & b & 0 \\ 0 & 0 & c \end{pmatrix}; \qquad B = \begin{pmatrix} 0 & a & b \\ 0 & 0 & c \\ 0 & 0 & 0 \end{pmatrix}; \qquad C = \begin{pmatrix} 0 & 1 & 0 \\ -1 & 0 & 0 \\ 0 & 0 & 0 \end{pmatrix}.$$

Give explicit formulas (not infinite series) for the one-parameter subgroups of $\operatorname{GL}(3, \mathbb{R})$ generated by A, B, and C.

20-6. This problem shows that the exponential map of a connected Lie group need not be surjective.

    (a) Suppose $A \in SL(n, \mathbb{R})$ is of the form e^B for some $B \in \mathfrak{gl}(n, \mathbb{R})$. Show that A has a square root in $SL(n, \mathbb{R})$, i.e., a matrix $C \in SL(n, \mathbb{R})$ such that $C^2 = A$.

    (b) Let

$$A = \begin{pmatrix} -\frac{1}{2} & 0 \\ 0 & -2 \end{pmatrix}.$$

Show that $\exp \colon \mathfrak{sl}(2, \mathbb{R}) \to SL(2, \mathbb{R})$ is not surjective, by showing that A is not in its image. [Remark: in the next chapter, Problem 21-25 will show that $SL(2, \mathbb{R})$ is connected.]

20-7. Let G be a connected Lie group and let $\mathfrak{g}$ be its Lie algebra.

    (a) For any $X, Y \in \mathfrak{g}$, show that $[X, Y] = 0$ if and only if

$$\exp tX \exp sY = \exp sY \exp tX \quad \text{for all } s, t \in \mathbb{R}.$$

    (b) Show that G is abelian if and only if $\mathfrak{g}$ is abelian.

    (c) Give a counterexample to (b) when G is not connected.

20-8. Suppose G is a Lie group. Prove that $\exp(X + Y) = (\exp X)(\exp Y)$ for all $X, Y \in \mathrm{Lie}(G)$ if and only if the identity component of G is abelian. [Hint: for the "if" direction, prove that $t \mapsto (\exp tX)(\exp tY)$ is a 1-parameter subgroup. For the "only if" direction, use Problem 20-7.]

20-9. Extend the result of Proposition 20.10 by showing that under the same hypotheses there is a smooth function $\widehat{Z} \colon (-\varepsilon, \varepsilon) \to \mathfrak{g}$ such that

$$(\exp tX)(\exp tY) = \exp\left(t(X + Y) + \tfrac{1}{2}t^2[X, Y] + t^3 \widehat{Z}(t)\right). \quad (20.20)$$

[Remark: there is an explicit formula, known as the *Baker–Campbell–Hausdorff formula*, for all of the terms in the Taylor series of the map $\varphi \colon (-\varepsilon, \varepsilon) \to \mathfrak{g}$ that satisfies $\exp tX \exp tY = \exp \varphi(t)$. Formula (20.20) gives the first two terms in this series. See [Var84] for the full formula.]

20-10. Suppose G is a Lie group and S is a Lie subgroup of G. Show that the closure of S is also a Lie subgroup. Conclude that every Lie subgroup of G is either a properly embedded submanifold of G, or a dense subset of a properly embedded submanifold. [Remark: this shows that the subgroup $S \subseteq \mathbb{T}^3$ of Exercise 7.20—a dense subgroup of a properly embedded subgroup—is typical of nonembedded Lie subgroups.]

20-11. Let G and H be Lie groups.

    (a) Show that every continuous homomorphism $\gamma \colon \mathbb{R} \to H$ is smooth. [Hint: let $V \subseteq \mathrm{Lie}(H)$ be a neighborhood of 0 such that the exponential map is a diffeomorphism from $2V = \{2X : X \in V\}$ to $\exp(2V)$. Choose t_0 small enough that $\gamma(t) \in \exp(V)$ whenever $|t| \le t_0$, and let X_0 be the element of V such that $\gamma(t_0) = \exp X_0$. Show that $\gamma(qt_0) = \exp(qX_0)$ whenever q is a dyadic rational, i.e., a number of the form $m/2^n$ for $m, n \in \mathbb{Z}$.]

(b) Show that every continuous homomorphism $F\colon G \to H$ is smooth. [Hint: show that there is a map $\varphi\colon \mathrm{Lie}(G) \to \mathrm{Lie}(H)$ such that the following diagram commutes:

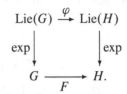

Then use Corollary 20.11 to show that φ is linear.]
(c) Show that with the given topology on G, there is only one smooth structure that makes G into a Lie group.
   (*Used on p. 556.*)

20-12. Let G be a Lie group. Show that the infinitesimal generator of the action of G on itself by right translation is the inclusion map $\mathrm{Lie}(G) \hookrightarrow \mathfrak{X}(G)$.

20-13. Let $\mathfrak{g}$ be a finite-dimensional Lie algebra and let M be a smooth manifold. A Lie algebra action $\widehat{\theta}\colon \mathfrak{g} \to \mathfrak{X}(M)$ is said to be ***transitive*** if for every $p \in M$, the vectors of the form $\widehat{X}_p$ for $\widehat{X} \in \widehat{\theta}(\mathfrak{g})$ span T_pM. Show that a smooth right action of a Lie group G on a connected smooth manifold M is transitive if and only if its infinitesimal generator is transitive. [Hint: show that if the Lie algebra action is transitive, then every orbit is open.]

20-14. Let M be a smooth manifold, and suppose $\mathfrak{g}$ is a finite-dimensional Lie subalgebra of $\mathfrak{X}(M)$ consisting only of complete vector fields. Show that there is a smooth right action of a Lie group G on M such that $\mathfrak{g}$ is the image of its infinitesimal generator. Determine such an action for the Lie subalgebra of $\mathfrak{X}\left(\mathbb{R}^3\right)$ described in Problem 8-20.

20-15. Prove Theorem 20.18 (the fundamental theorem on left Lie algebra actions). [Hint: use the one-to-one correspondence between left actions and right actions given by $g \cdot p = p \cdot g^{-1}$.]

20-16. Let G be a simply connected Lie group and let $\mathfrak{g}$ be its Lie algebra. Show that every representation of $\mathfrak{g}$ is of the form $\rho_*\colon \mathfrak{g} \to \mathfrak{gl}(V)$ for some representation $\rho\colon G \to \mathrm{GL}(V)$ of G.

20-17. Suppose G is a connected Lie group, H is any Lie group, and $\Phi, \Psi\colon G \to H$ are Lie group homomorphisms such that $\Phi_* = \Psi_*\colon \mathrm{Lie}(G) \to \mathrm{Lie}(H)$. Prove that $\Phi = \Psi$. (*Used on p. 531.*)

20-18. If C and D are categories, a covariant functor $\mathscr{F}\colon \mathsf{C} \to \mathsf{D}$ is called an ***equivalence of categories*** if every object of D is isomorphic to $\mathscr{F}(X)$ for some $X \in \mathrm{Ob}(\mathsf{C})$, and the map $\mathscr{F}\colon \mathrm{Hom}_{\mathsf{C}}(X, Y) \to \mathrm{Hom}_{\mathsf{D}}(\mathscr{F}(X), \mathscr{F}(Y))$ is bijective for each pair of objects $X, Y \in \mathrm{Ob}(\mathsf{C})$. Show that the assignment $G \mapsto \mathrm{Lie}(G)$, $\varphi \mapsto \varphi_*$ is an equivalence of categories between the category SLie of simply connected Lie groups and the category lie of finite-dimensional Lie algebras.

20-19. Prove Theorem 20.22 (Lie's fundamental theorems).

20-20. Let G be a connected Lie group and let $\mathfrak{g}$ be its Lie algebra. Prove that the kernel of Ad: $G \to GL(\mathfrak{g})$ is the *center of G*, that is, the set of elements of G that commute with every element of G.

20-21. Show that the adjoint representation of $GL(n, \mathbb{R})$ is given by $\mathrm{Ad}(A)Y = AYA^{-1}$ for $A \in GL(n, \mathbb{R})$ and $Y \in \mathfrak{gl}(n, \mathbb{R})$. Show that it is not faithful.

20-22. If $\mathfrak{g}$ is a Lie algebra, the *center of* $\mathfrak{g}$ is the set of all $X \in \mathfrak{g}$ such that $[X, Y] = 0$ for all $Y \in \mathfrak{g}$. Suppose G is a connected Lie group. Show that the center of Lie(G) is the Lie algebra of the center of G.

Chapter 21
Quotient Manifolds

In Chapter 4, we showed that surjective smooth submersions play a role in smooth manifold theory that is strongly parallel to the role of quotient maps in topology. But one question we did not address there was which quotients of smooth manifolds are themselves smooth manifolds.

In general, the class of all quotient spaces is far too broad to admit a good general theory. But there is one class of quotients about which quite a lot can be said: those resulting from smooth Lie group actions. This is one of the most useful applications of Lie groups to smooth manifold theory.

There are many examples of smooth Lie group actions whose quotients are not manifolds, so the class of all smooth Lie group actions is still too broad; we need to impose some additional conditions on an action to ensure that we get a nice quotient space. In this chapter, we explore a pair of conditions that, taken together, ensure that a group action has a well-behaved quotient space. The first condition is that the action be *free* (meaning that the group acts without fixed points); the second condition is that it be *proper* (which means roughly that each compact subset is moved away from itself by most elements of the group). The main theorem of the chapter is the *quotient manifold theorem*, which asserts that a Lie group acting smoothly, freely, and properly on a smooth manifold yields a quotient space with a natural smooth manifold structure.

In the first section of the chapter, we explore what freeness and properness mean for group actions, and study some examples to help understand why these are reasonable conditions to require. Then in the second section, we prove the quotient manifold theorem.

After the proof, we explore two significant special classes of Lie group actions. First we study actions by discrete groups, which under suitable conditions yield covering maps. Then we study *homogeneous spaces*, which are smooth manifolds endowed with smooth transitive Lie group actions; we show that they are equivalent to Lie groups modulo closed subgroups. Finally, at the end of the chapter we describe a number of applications to the theory of Lie groups.

J.M. Lee, *Introduction to Smooth Manifolds*, Graduate Texts in Mathematics 218,
DOI 10.1007/978-1-4419-9982-5_21, © Springer Science+Business Media New York 2013

Quotients of Manifolds by Group Actions

Suppose we are given an action of a group G on a topological space M, which we write either as $\theta\colon G \times M \to M$ or as $(g, p) \mapsto g \cdot p$. (For definiteness, let us assume that G acts on the left; similar considerations apply to right actions.) Recall that the *orbit* of a point $p \in M$ is the set of images of p under all elements of the group:

$$G \cdot p = \{g \cdot p : g \in G\}.$$

Define a relation on M by setting $p \sim q$ if there exists $g \in G$ such that $g \cdot p = q$. This is an equivalence relation, whose equivalence classes are exactly the orbits of G in M. The set of orbits is denoted by M/G; with the quotient topology it is called the *orbit space* of the action. It is of great importance to determine conditions under which an orbit space is a smooth manifold.

It is important to be aware that some authors use the notation $G \backslash M$ for an orbit space determined by a left action of G on M, reserving M/G for one determined by a right action. We use only the latter notation, relying on the context, not the notation, to distinguish between the two cases.

Here is a simple but important property of orbit spaces.

Lemma 21.1. *For any continuous action of a topological group G on a topological space M, the quotient map $\pi\colon M \to M/G$ is an open map.*

Proof. For any $g \in G$ and any subset $U \subseteq M$, we define a set $g \cdot U \subseteq M$ by

$$g \cdot U = \{g \cdot x : x \in U\}.$$

If $U \subseteq M$ is open, then $\pi^{-1}\big(\pi(U)\big)$ is equal to the union of all sets of the form $g \cdot U$ as g ranges over G. Since $p \mapsto g \cdot p$ is a homeomorphism, each such set is open, and therefore $\pi^{-1}\big(\pi(U)\big)$ is open in M. Because π is a quotient map, this implies that $\pi(U)$ is open in M/G, and therefore π is an open map. $\qquad\square$

It is easy to construct smooth actions by Lie groups on smooth manifolds whose orbit spaces are themselves manifolds, and others whose orbit spaces are not. Here are a few examples.

Example 21.2 (Orbit Spaces of Smooth Lie Group Actions).

(a) Let G be any group and let M be any smooth manifold. The trivial action (given by $g \cdot p = p$ for all $g \in G$ and $p \in M$) has one-point sets as orbits and $M/G = M$, so the orbit space is a smooth manifold for silly reasons.

(b) The simplest nontrivial example to keep in mind is the action of $\mathbb{R}^k$ on $\mathbb{R}^k \times \mathbb{R}^n$ by translation in the $\mathbb{R}^k$ factor: $v \cdot (x, y) = (v + x, y)$. The orbits are the affine subspaces parallel to $\mathbb{R}^k$, and the orbit space $\big(\mathbb{R}^k \times \mathbb{R}^n\big)/\mathbb{R}^k$ is homeomorphic to $\mathbb{R}^n$. The quotient map $\pi\colon \mathbb{R}^k \times \mathbb{R}^n \to \mathbb{R}^n$ is a smooth submersion.

(c) The circle group $\mathbb{S}^1$ acts on the plane $\mathbb{C}$ by complex multiplication: $z \cdot w = zw$. The orbits are circles centered at the origin and the singleton $\{0\}$. The orbit space is homeomorphic to $[0, \infty)$, as you can see by applying Theorem A.31 to the continuous map $f\colon \mathbb{C} \to [0, \infty)$ given by $f(z) = |z|$, which is a quotient map that makes the same identifications as the projection $\pi \colon \mathbb{C} \to \mathbb{C}/\mathbb{S}^1$. Thus, the orbit space is not a manifold.

(d) An even more dramatic example of how an orbit space can fail to be a manifold is given by the natural action of $\mathrm{GL}(n, \mathbb{R})$ on $\mathbb{R}^n$ by matrix multiplication. In this case, there are two orbits, $\{0\}$ and $\mathbb{R}^n \smallsetminus \{0\}$, and the only open subsets in the quotient topology are the empty set, the whole set, and the singleton $\{[\mathbb{R}^n \smallsetminus \{0\}]\}$. This orbit space is not even Hausdorff, let alone a manifold.

(e) The restriction of the natural action of $\mathrm{GL}(n, \mathbb{R})$ on $\mathbb{R}^n$ to $\mathrm{O}(n) \times \mathbb{R}^n \to \mathbb{R}^n$ defines a smooth left action of $\mathrm{O}(n)$ on $\mathbb{R}^n$. In this case, the orbits are the origin and the spheres centered at the origin. To see why, note that any orthogonal linear transformation preserves norms, so $\mathrm{O}(n)$ takes the sphere of radius R to itself; on the other hand, any nonzero vector of length R can be taken to any other by an orthogonal matrix. (If v and v' are such vectors, complete $v/|v|$ and $v'/|v'|$ to orthonormal bases and let A and A' be the orthogonal matrices whose columns are these orthonormal bases; then it is easy to check that $A'A^{-1}$ takes v to v'.) As in (c), the orbit space is homeomorphic to $[0, \infty)$.

(f) If we delete the origin from each of the three preceding examples, we obtain orbit spaces that are manifolds: the quotient of $\mathbb{C} \smallsetminus \{0\}$ by $\mathbb{S}^1$ is homeomorphic to $\mathbb{R}^+$, as is the quotient of $\mathbb{R}^n \smallsetminus \{0\}$ by $\mathrm{O}(n)$; and the quotient of $\mathbb{R}^n \smallsetminus \{0\}$ by $\mathrm{GL}(n, \mathbb{R})$ is a single point.

(g) Further restricting the natural action to $\mathrm{O}(n) \times \mathbb{S}^{n-1} \to \mathbb{S}^{n-1}$, we obtain an action of $\mathrm{O}(n)$ on $\mathbb{S}^{n-1}$. It is smooth by Corollary 5.30, because $\mathbb{S}^{n-1}$ is an embedded submanifold of $\mathbb{R}^n$. This action is transitive (recall that this means the only orbit is $\mathbb{S}^{n-1}$ itself), so the quotient space is a singleton. //

In (c), (d), and (e) above, the problematic point is the origin. In each case, the origin is the only point that is fixed by every element of the group. Recall that in Chapter 7, we defined a *free action* to be one for which every isotropy group is trivial. In Example 21.2, the action of $\mathbb{R}^k$ on $\mathbb{R}^n$ in part (b) is free, as is the action of $\mathbb{S}^1$ on $\mathbb{C} \smallsetminus \{0\}$ described in (f); the other examples are not. Of course, freeness is not necessary for an action to have a smooth manifold quotient, as Example 21.2(a) shows. Nor is it sufficient by itself, as the next example shows.

Example 21.3. Let α be an irrational number, and let $\mathbb{R}$ act on $\mathbb{T}^2 = \mathbb{S}^1 \times \mathbb{S}^1$ by

$$t \cdot (w, z) = \left(e^{2\pi i t} w, e^{2\pi i \alpha t} z\right).$$

This is a smooth action, and the arguments of Example 4.20 and Problem 4-4 can be adapted easily to show that it is free and has dense orbits. This means that the only saturated open subsets of $\mathbb{T}^2$ are $\varnothing$ and $\mathbb{T}^2$, so the orbit space $\mathbb{T}^2/\mathbb{R}$ has the trivial topology. In particular, it is not Hausdorff and therefore not a manifold. //

To avoid pathological cases such as these, we need to introduce one more restriction on our group actions. A continuous left action of a Lie group G on a manifold M is said to be a ***proper action*** if the map $G \times M \to M \times M$ given by $(g, p) \mapsto (g \cdot p, p)$ is a proper map. Note that this is generally a weaker condition than requiring that the map $G \times M \to M$ defining the action be a proper map (see Problem 21-1).

Proposition 21.4. *If a Lie group acts continuously and properly on a manifold, then the orbit space is Hausdorff.*

Proof. Suppose G is a Lie group acting continuously and properly on a manifold M. Let $\Theta\colon G \times M \to M \times M$ be the proper map $\Theta(g, p) = (g \cdot p, p)$, and let $\pi\colon M \to M/G$ be the quotient map. Define the ***orbit relation*** $\mathcal{O} \subseteq M \times M$ by

$$\mathcal{O} = \Theta(G \times M) = \{(g \cdot p, p) \in M \times M : p \in M,\ g \in G\}.$$

(It is called the orbit relation because $(q, p) \in \mathcal{O}$ if and only if p and q are in the same G-orbit.) Since proper continuous maps are closed (Theorem A.57), it follows that $\mathcal{O}$ is a closed subset of $M \times M$. Because π is an open map by Lemma 21.1, the result of Exercise A.36 shows that M/G is Hausdorff.   □

It is not always easy to tell whether a given action is proper. The next proposition gives two alternative characterizations of proper actions that are often useful.

Proposition 21.5 (Characterizations of Proper Actions). *Let M be a manifold, and let G be a Lie group acting continuously on M. The following are equivalent.*

(a) *The action is proper.*
(b) *If (p_i) is a sequence in M and (g_i) is a sequence in G such that both (p_i) and $(g_i \cdot p_i)$ converge, then a subsequence of (g_i) converges.*
(c) *For every compact subset $K \subseteq M$, the set $G_K = \{g \in G : (g \cdot K) \cap K \neq \varnothing\}$ is compact.*

Proof. Throughout this proof, let $\Theta\colon G \times M \to M \times M$ denote the map $\Theta(g, p) = (g \cdot p, p)$; thus, the action is proper if and only if Θ is a proper map. We will prove (a) $\Rightarrow$ (b) $\Rightarrow$ (c) $\Rightarrow$ (a).

Assume first that Θ is proper, and let (p_i), (g_i) be sequences satisfying the hypotheses of (b). Let U and V be precompact neighborhoods of the points $p = \lim_i p_i$ and $q = \lim_i (g_i \cdot p_i)$, respectively. The assumption means that the points $\Theta(g_i, p_i)$ all lie in the compact set $\overline{V} \times \overline{U}$ when i is large enough, so a subsequence of $((g_i, p_i))$ converges in $G \times M$. In particular, this means that a subsequence of (g_i) converges in G, and therefore (b) holds.

Assume next that (b) holds, and let K be a compact subset of M. To show that G_K is compact, suppose (g_i) is any sequence of points in G_K. This means that for each i, there exists $p_i \in (g_i \cdot K) \cap K$, which is to say that $p_i \in K$ and $g_i^{-1} \cdot p_i \in K$. After passing to a subsequence, we may assume that (p_i) converges, and then passing to a subsequence of that, we may assume also that $(g_i^{-1} \cdot p_i)$ converges. By (b), there is a subsequence (g_{i_k}) such that $(g_{i_k}^{-1})$ converges, which implies that (g_{i_k}) also converges. Since each subsequence of G_K has a convergent subsequence, G_K is compact.

Finally, assume that (c) holds. Suppose $L \subseteq M \times M$ is compact, and let $K = \pi_1(L) \cup \pi_2(L) \subseteq M$, where $\pi_1, \pi_2\colon M \times M \to M$ are the projections onto the first and second factors, respectively. Then

$$\Theta^{-1}(L) \subseteq \Theta^{-1}(K \times K) = \{(g, p) : g \cdot p \in K \text{ and } p \in K\} \subseteq G_K \times K.$$

Since $\Theta^{-1}(L)$ is closed by continuity, it is a closed subset of the compact set $G_K \times K$ and is therefore compact. This shows that the action is proper. □

Corollary 21.6. *Every continuous action by a compact Lie group on a manifold is proper.*

Proof. If (p_i) and (g_i) are sequences satisfying the hypotheses of Proposition 21.5(b), then a subsequence of (g_i) converges, for the simple reason that *every* sequence in G has a convergent subsequence. □

Proposition 21.7 (Orbits of Proper Actions). *Suppose θ is a proper smooth action of a Lie group G on a smooth manifold M. For any point $p \in M$, the orbit map $\theta^{(p)} \colon G \to M$ is a proper map, and thus the orbit $G \cdot p = \theta^{(p)}(G)$ is closed in M. If in addition $G_p = \{e\}$, then $\theta^{(p)}$ is a smooth embedding, and the orbit is a properly embedded submanifold.*

Proof. If $K \subseteq M$ is compact, then $(\theta^{(p)})^{-1}(K)$ is closed in G by continuity, and since it is contained in $G_{K \cup \{p\}}$, it is compact by Proposition 21.5. Therefore, $\theta^{(p)}$ is a proper map, which implies that $G \cdot p = \theta^{(p)}(G)$ is closed by Theorem A.57. The final statement of the theorem then follows from Propositions 7.26 and 4.22. □

The preceding results yield some simple necessary conditions for an action to be proper.

Corollary 21.8. *If a Lie group G acts properly on a manifold M, then each orbit is a closed subset of M, and each isotropy group is compact.*

Proof. The first statement follows immediately from Proposition 21.7, and the second from Proposition 21.5, using the fact that the isotropy group of a point $p \in M$ is the set G_K for $K = \{p\}$. □

Example 21.9. We can see in two ways that the action of $\mathbb{R}^+$ on $\mathbb{R}^n$ given by

$$t \cdot (x^1, \dots, x^n) = (tx^1, \dots, tx^n) \tag{21.1}$$

is not proper: the isotropy group of the origin is all of $\mathbb{R}^+$, which is not compact; and the orbits of other points are open rays, which are not closed in $\mathbb{R}^n$. //

The Quotient Manifold Theorem

In this section, we prove that smooth, free, and proper group actions always yield smooth manifolds as orbit spaces. The basic idea of the proof is that if G acts smoothly, freely, and properly on M, the set of orbits forms a foliation of M whose leaves are embedded submanifolds diffeomorphic to G. Flat charts for the foliation can then be used to construct coordinates on the orbit space.

Theorem 21.10 (Quotient Manifold Theorem). *Suppose G is a Lie group acting smoothly, freely, and properly on a smooth manifold M. Then the orbit space M/G is a topological manifold of dimension equal to $\dim M - \dim G$, and has a unique* *smooth structure with the property that the quotient map* $\pi: M \to M/G$ *is a smooth submersion.*

Proof. Before we get started, let us establish some notation. Throughout the proof, we assume without loss of generality that G acts on the left. Let $\mathfrak{g}$ denote the Lie algebra of G, and write $k = \dim G$, $m = \dim M$, and $n = m - k$. Let $\theta: G \times M \to M$ denote the action and $\Theta: G \times M \to M \times M$ the proper map $\Theta(g, p) = (g \cdot p, p)$.

First, we take care of the easy part: the uniqueness of the smooth structure. Suppose M/G has two different smooth structures such that $\pi: M \to M/G$ is a smooth submersion. Let $(M/G)_1$ and $(M/G)_2$ denote M/G with the first and second smooth structures, respectively. By Theorem 4.29, the identity map is smooth from $(M/G)_1$ to $(M/G)_2$:

$$
\begin{array}{ccc}
 & M & \\
\pi \swarrow & & \searrow \pi \\
(M/G)_1 & \underset{\mathrm{Id}}{\longrightarrow} & (M/G)_2.
\end{array}
$$

The same argument shows that it is also smooth in the opposite direction, so the two smooth structures are identical; this proves uniqueness.

Now we have to show that M/G is a topological manifold and construct a smooth structure for it. The main tools are certain special coordinate charts for M. Let us say that a smooth chart (U, φ) for M is **adapted to the G-action** if it is a cubical chart with coordinate functions $(x, y) = (x^1, \ldots, x^k, y^1, \ldots, y^n)$, such that each G-orbit intersects U either in the empty set or in a single slice of the form $(y^1, \ldots, y^n) = (c^1, \ldots, c^n)$. The heart of the proof is the following claim: *For each $p \in M$, there exists an adapted chart centered at p.*

To prove the claim, note first that the G-orbits are properly embedded submanifolds of M diffeomorphic to G by Proposition 21.7. In fact, we will show that the orbits are integral manifolds of a smooth distribution on M.

Define a subset $D \subseteq TM$ by

$$
D = \bigcup_{p \in M} D_p, \quad \text{where } D_p = T_p(G \cdot p).
$$

Because every point is contained in exactly one orbit, and the orbits are submanifolds of dimension k, each D_p has dimension k. To see that D is a smooth distribution, for each $X \in \mathfrak{g}$ let $\widehat{X}$ be the vector field on M defined by (20.11) (the infinitesimal generator of the flow $(t, p) \mapsto (\exp tX) \cdot p$). If $(X_1, \ldots, X_k)$ is a basis for $\mathfrak{g}$, then $(\widehat{X}_1, \ldots, \widehat{X}_k)$ is a global frame for D, so D is smooth. Each point is contained in a G-orbit, which is an integral manifold of D, so D is involutive. Because the G-orbits are closed, Problem 19-7 implies that each connected component of an orbit is a leaf of the foliation determined by D.

Let $p \in M$ be arbitrary, and let (U, φ) be a smooth chart for M centered at p that is flat for D with coordinate functions $(x, y) = (x^1, \ldots, x^k, y^1, \ldots, y^n)$, so each G-orbit intersects U either in the empty set or in a countable union of $y = \text{constant}$

slices. To complete the proof of the claim, we need to show that we can find a cubical subset $U_0 \subseteq U$ centered at p that intersects each G-orbit in at most a single slice.

Assume there is no such subset U_0. For each positive integer i, let U_i be the cubical subset of U consisting of points whose coordinates are all less than $1/i$ in absolute value. Let Y be the n-dimensional submanifold of M consisting of points in U whose coordinate representations are of the form $(0, y)$, and for each i let $Y_i = U_i \cap Y$. Since each k-slice of U_i intersects Y_i in exactly one point, our assumption implies that for each i there exist distinct points $p_i, p_i' \in Y_i$ that are in the same orbit, which is to say that $g_i \cdot p_i = p_i'$ for some $g_i \in G$. By our choice of $\{Y_i\}$, both sequences (p_i) and $(p_i' = g_i \cdot p_i)$ converge to p. Because G acts properly, Proposition 21.5(b) shows that we may pass to a subsequence and assume that $g_i \to g \in G$. By continuity, therefore,

$$g \cdot p = \lim_{i \to \infty} g_i \cdot p_i = \lim_{i \to \infty} p_i' = p.$$

Since G acts freely, this implies $g = e$.

Let $\theta^Y \colon G \times Y \to M$ be the restriction of the G-action to $G \times Y$. Note that $G \times Y$ and M both have dimension $k + n = m$. The restriction of θ^Y to $\{e\} \times Y$ is just the inclusion map $Y \hookrightarrow M$, and its restriction to $G \times \{p\}$ is the orbit map $\theta^{(p)}$ (if we make the obvious identifications $\{e\} \times Y \approx Y$ and $G \times \{p\} \approx G$). Since both of these are embeddings, and $T_p M = T_p(G \cdot p) \oplus T_p Y$, it follows that $d\left(\theta^Y\right)_{(e,p)}$ is an isomorphism. Thus, there is a neighborhood W of (e, p) in $G \times Y$ such that $\theta^Y|_W$ is a diffeomorphism onto its image and hence injective. However, this contradicts the fact that $\theta^Y(g_i, p_i) = p_i' = \theta^Y(e, p_i')$ as soon as i is large enough that (g_i, p_i) and (e, p_i') are in W, because we are assuming $p_i \neq p_i'$. This completes the proof of the claim that for each $p \in M$ there exists an adapted chart centered at p.

Now we prove that M/G, with the quotient topology, is a topological n-manifold. It is Hausdorff by Proposition 21.4. If $\{B_i\}$ is a countable basis for the topology of M, then $\{\pi(B_i)\}$ is a countable collection of open subsets of M/G (because π is an open map), and it is easy to check that it is a basis for the topology of M/G. Thus, M/G is second-countable.

To show that M/G is locally Euclidean, let $q = \pi(p)$ be an arbitrary point of M/G, and let (U, φ) be an adapted chart for M centered at p, with $\varphi(U)$ equal to an open cube in $\mathbb{R}^k \times \mathbb{R}^n$, which we write as $\varphi(U) = U' \times U''$, where U' and U'' are open cubes in $\mathbb{R}^k$ and $\mathbb{R}^n$, respectively. Let $V = \pi(U)$ (Fig. 21.1), which is an open subset of M/G because π is an open map. With the coordinate functions of φ denoted by $(x^1, \ldots, x^k, y^1, \ldots, y^n)$ as before, let $Y \subseteq U$ be the submanifold $\{x^1 = \cdots = x^k = 0\}$. Note that $\pi \colon Y \to V$ is bijective by the definition of an adapted chart. Moreover, if W is an open subset of Y, then

$$\pi(W) = \pi\left(\{(x, y) : (0, y) \in W\}\right),$$

which is open in M/G, and thus $\pi|_Y \colon Y \to V$ is a homeomorphism. Let $\sigma \colon V \to Y \subseteq U$ be the map $\sigma = (\pi|_Y)^{-1}$; it is a local section of π.

Define a map $\eta \colon V \to U''$ by sending the equivalence class of a point (x, y) to y; this is well defined by the definition of an adapted chart. Formally, $\eta = \pi'' \circ \varphi \circ \sigma$,

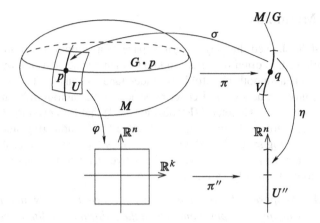

Fig. 21.1 A coordinate chart for M/G

where $\pi'' \colon U' \times U'' \to U'' \subseteq \mathbb{R}^n$ is the projection onto the second factor. Because σ is a homeomorphism from V to Y and $\pi'' \circ \varphi$ is a homeomorphism from Y to U'', it follows that η is a homeomorphism. This shows that M/G is locally Euclidean, and thus completes the proof that it is a topological n-manifold.

Finally, we need to show that M/G has a smooth structure such that π is a smooth submersion. We use the atlas consisting of all charts (V, η) as constructed in the preceding paragraph. With respect to any such chart for M/G and the corresponding adapted chart for M, π has the coordinate representation $\pi(x, y) = y$, which is certainly a smooth submersion. Thus we need only show that any two such charts for M/G are smoothly compatible.

Let (U, φ) and $(\widetilde{U}, \widetilde{\varphi})$ be two adapted charts for M, and let (V, η) and $(\widetilde{V}, \widetilde{\eta})$ be the corresponding charts for M/G. First consider the case in which the two adapted charts are both centered at the same point $p \in M$. Write the adapted coordinates as (x, y) and $(\widetilde{x}, \widetilde{y})$. The fact that the coordinates are adapted to the G-action means that two points with the same y-coordinate are in the same orbit, and therefore also have the same $\widetilde{y}$-coordinate. This means that the transition map between these coordinates can be written $(\widetilde{x}, \widetilde{y}) = (A(x, y), B(y))$, where A and B are smooth maps defined on some neighborhood of the origin. The transition map $\widetilde{\eta} \circ \eta^{-1}$ is just $\widetilde{y} = B(y)$, which is clearly smooth.

In the general case, suppose (U, φ) and $(\widetilde{U}, \widetilde{\varphi})$ are adapted charts for M, and $p \in U$, $\widetilde{p} \in \widetilde{U}$ are points such that $\pi(p) = \pi(\widetilde{p})$. After modifying both charts by adding constant vectors, we can assume that they are centered at p and $\widetilde{p}$, respectively. Since p and $\widetilde{p}$ are in the same orbit, there is an element $g \in G$ such that $g \cdot p = \widetilde{p}$. Because $\theta_g \colon M \to M$ is a diffeomorphism taking orbits to orbits, it follows that $\widetilde{\varphi}' = \widetilde{\varphi} \circ \theta_g$ is another adapted chart centered at p. Moreover, $\widetilde{\sigma}' = \theta_g^{-1} \circ \widetilde{\sigma}$ is the local section corresponding to $\widetilde{\varphi}'$, and therefore $\widetilde{\eta}' = \pi'' \circ \widetilde{\varphi}' \circ \widetilde{\sigma}' = \pi'' \circ \widetilde{\varphi} \circ \theta_g \circ \theta_g^{-1} \circ \widetilde{\sigma} = \pi'' \circ \widetilde{\varphi} \circ \widetilde{\sigma} = \widetilde{\eta}$. Thus we are back in the situation of the preceding paragraph, and the two charts are smoothly compatible. $\qquad\square$

Covering Manifolds

Proposition 4.40 showed that any covering space of a smooth manifold is itself a smooth manifold. It is often important to know when a space *covered by* a smooth manifold is itself a smooth manifold. To understand the answer to this question, we need to study the action on a covering space by the automorphism group of the covering. We saw in Chapter 7 (Proposition 7.23) that for any smooth covering $\pi\colon E \to M$, the automorphism group is a discrete Lie group acting smoothly and freely on the covering space E. Below, we will show that the action is also proper. Before proceeding, it is useful to have an alternative characterization of properness for free actions of discrete Lie groups.

Lemma 21.11. *Suppose a discrete Lie group Γ acts continuously and freely on a manifold E. The action is proper if and only if the following conditions both hold:*

(i) *Every point $p \in E$ has a neighborhood U such that for each $g \in \Gamma$, $(g \cdot U) \cap U = \varnothing$ unless $g = e$.*
(ii) *If $p, p' \in E$ are not in the same Γ-orbit, there exist neighborhoods V of p and V' of p' such that $(g \cdot V) \cap V' = \varnothing$ for all $g \in \Gamma$.*

Proof. First, suppose that the action is free and proper, and let $\pi\colon E \to E/\Gamma$ denote the quotient map. By Proposition 21.4, E/Γ is Hausdorff. If $p, p' \in E$ are not in the same orbit, we can choose disjoint neighborhoods W of $\pi(p)$ and W' of $\pi(p')$, and then $V = \pi^{-1}(W)$ and $V' = \pi^{-1}(W')$ satisfy the conclusion of condition (ii).

To prove (i), let $p \in E$, and let V be a precompact neighborhood of p. By Proposition 21.5, the set $\Gamma_{\bar V}$ is a compact subset of Γ, and hence finite because Γ is discrete. Write $\Gamma_{\bar V} = \{e, g_1, \dots, g_m\}$. Shrinking V if necessary, we may assume that $g_i^{-1} \cdot p \notin \bar V$ (which implies $p \notin g_i \cdot \bar V$) for $i = 1, \dots, m$. Then the open subset

$$U = V \smallsetminus \left(g_1 \cdot \bar V \cup \cdots \cup g_m \cdot \bar V \right)$$

satisfies the conclusion of (i).

Conversely, assume that (i) and (ii) hold. Suppose (g_i) is a sequence in Γ and (p_i) is a sequence in E such that $p_i \to p$ and $g_i \cdot p_i \to p'$. If p and p' are in different orbits, there exist neighborhoods V of p and V' of p' as in (ii); but for large enough i, we have $p_i \in V$ and $g_i \cdot p_i \in V'$, which contradicts the fact that $(g_i \cdot V) \cap V' = \varnothing$. Thus, p and p' are in the same orbit, so there exists $g \in \Gamma$ such that $g \cdot p = p'$. This implies $g^{-1} g_i \cdot p_i \to p$. Choose a neighborhood U of p as in (i), and let i be large enough that p_i and $g^{-1} g_i \cdot p_i$ are both in U. Because $\left(g^{-1} g_i \cdot U \right) \cap U \neq \varnothing$, it follows that $g^{-1} g_i = e$. So $g_i = g$ when i is large enough, which certainly converges. By Proposition 21.5(b), the action is proper. $\square$

In the literature, a continuous discrete group action satisfying condition (i) of this lemma (or conditions (i) and (ii), or something closely related to them) has traditionally been called ***properly discontinuous***. We avoid this terminology, because its meaning is not universally agreed upon, and because it leads to such oxymoronic expressions as "continuous properly discontinuous actions." Instead, following Allan Hatcher [Hat02], we call a continuous action satisfying (i) a ***covering space action***, and simply refer to actions satisfying (i) and (ii) as "free and proper actions." It can be shown that any covering space action on a topological space yields a covering map, though the quotient space need not be Hausdorff (see, e.g., [LeeTM, Chap. 12]).

Proposition 21.12. *Let M be a smooth manifold, and let $\pi: E \to M$ be a smooth covering map. With the discrete topology, the automorphism group $\mathrm{Aut}_\pi(E)$ acts smoothly, freely, and properly on E.*

Proof. We already showed in Proposition 7.23 that the action is smooth and free. To show it is proper, we will show that it satisfies conditions (i) and (ii) of Lemma 21.11. First, if $p \in E$ is arbitrary, choose $W \subseteq M$ to be an evenly covered neighborhood of $\pi(p)$. If U is the component of $\pi^{-1}(W)$ containing p, then it is easy to check that U satisfies (i). Second, if $p, p' \in E$ are in different orbits, then just as in the proof Lemma 21.11, we can choose disjoint neighborhoods W of $\pi(p)$ and W' of $\pi(p')$, and it follows that $V = \pi^{-1}(W)$ and $V' = \pi^{-1}(W')$ satisfy (ii). $\qquad\square$

The quotient manifold theorem yields an important partial converse to the preceding proposition.

Theorem 21.13. *Suppose E is a connected smooth manifold and Γ is a discrete Lie group acting smoothly, freely, and properly on E. Then the orbit space E/Γ is a topological manifold and has a unique smooth structure such that $\pi: E \to E/\Gamma$ is a smooth normal covering map.*

Proof. It follows from the quotient manifold theorem that E/Γ has a unique smooth manifold structure such that π is a smooth submersion. Because a smooth covering map is in particular a smooth submersion, any other smooth manifold structure on E making π into a smooth covering map must be equal to this one. Because $\dim E/\Gamma = \dim E - \dim \Gamma = \dim E$, π is a local diffeomorphism. Thus, to prove the theorem, it suffices to show that π is a normal covering map.

Let $p \in E$. By Lemma 21.11, p has a neighborhood U in E satisfying

$$(g \cdot U) \cap U = \varnothing \quad \text{for all } g \in \Gamma \text{ except } g = e. \tag{21.2}$$

Shrinking U if necessary, we may assume it is connected. Let $V = \pi(U)$, which is open in E/Γ by Lemma 21.1. Because $\pi^{-1}(V)$ is the union of the disjoint connected open subsets $g \cdot U$ for $g \in \Gamma$, to show that π is a covering map we need only show that π is a homeomorphism from each such set onto V. For each $g \in \Gamma$, the following diagram commutes:

$$
\begin{array}{ccc}
U & \xrightarrow{\ \ g\ \ } & g \cdot U \\
 & \searrow{\scriptstyle \pi} \quad \swarrow{\scriptstyle \pi} & \\
 & V. &
\end{array}
$$

Since $g: U \to g \cdot U$ is a homeomorphism (in fact, a diffeomorphism), it suffices to show that $\pi: U \to V$ is a homeomorphism. We already know that it is surjective, continuous, and open. To see that it is injective, suppose $\pi(q) = \pi(q')$ for

$q, q' \in U$, which means that $q' = g \cdot q$ for some $g \in \Gamma$. By (21.2), this can happen only if $g = e$, which is to say that $q = q'$. This completes the proof that π is a smooth covering map. Because elements of Γ act as automorphisms of π, and Γ acts transitively on fibers by definition, the covering is normal. $\square$

Example 21.14 (Proper Discrete Group Actions).

(a) The discrete Lie group $\mathbb{Z}^n$ acts smoothly and freely on $\mathbb{R}^n$ by translations (Example 7.22(e)). If (x_i) and (m_i) are sequences in $\mathbb{R}^n$ and $\mathbb{Z}^n$, respectively, such that $x_i \to x$ and $m_i + x_i \to y$, then $m_i \to y - x$, so the action is proper by Proposition 21.5(b). The orbit space $\mathbb{R}^n/\mathbb{Z}^n$ is homeomorphic to the n-torus $\mathbb{T}^n$, and Theorem 21.13 says that there is a unique smooth structure on $\mathbb{T}^n$ making the quotient map into a smooth covering map. To verify that this smooth structure on $\mathbb{T}^n$ is the same as the one we defined previously (thinking of $\mathbb{T}^n$ as the product manifold $\mathbb{S}^1 \times \cdots \times \mathbb{S}^1$), we just check that the covering map $\mathbb{R}^n \to \mathbb{T}^n$ given by $(x^1, \dots, x^n) \mapsto (e^{2\pi i x^1}, \dots, e^{2\pi i x^n})$ is a local diffeomorphism with respect to the product smooth structure on $\mathbb{T}^n$, and makes the same identifications as the quotient map $\mathbb{R}^n \to \mathbb{R}^n/\mathbb{Z}^n$; thus Theorem 4.31 implies that $\mathbb{R}^n/\mathbb{Z}^n$ is diffeomorphic to $\mathbb{T}^n$.

(b) The two-element group $\{\pm 1\}$ acts on $\mathbb{S}^n$ by multiplication. This action is smooth and free, and it is proper because the group is compact. This defines a smooth structure on $\mathbb{S}^n/\{\pm 1\}$. In fact, this orbit space is diffeomorphic to $\mathbb{RP}^n$ with the smooth structure we defined in Chapter 1, which can be seen as follows. Let $q \colon \mathbb{S}^n \to \mathbb{RP}^n$ be the smooth covering map defined in Example 2.13(f) (see also Problem 4-10). This map makes the same identifications as the quotient map $\pi \colon \mathbb{S}^n \to \mathbb{S}^n/\{\pm 1\}$. By Theorem 4.31, therefore, $\mathbb{S}^n/\{\pm 1\}$ is diffeomorphic to $\mathbb{RP}^n$.   //

Homogeneous Spaces

Some of the most interesting group action are transitive ones. A smooth manifold endowed with a transitive smooth action by a Lie group G is called a ***homogeneous G-space*** (or a ***homogeneous space*** or ***homogeneous manifold*** if it is not important to specify the group).

In most examples of homogeneous spaces, the group action preserves some extra structure on the manifold (such as a Riemannian metric, a distribution, a vector field, a differential form, or a foliation), and the fact that the action is transitive means that this structure "looks the same" everywhere on the manifold. Often, homogeneous spaces are models for various kinds of geometries, and as such they play a central role in many areas of differential geometry.

Here are some important examples of homogeneous spaces.

Example 21.15 (Homogeneous Spaces).

(a) The natural action of $O(n)$ on $\mathbb{S}^{n-1}$ is transitive, as we observed in Example 21.2(g). Thus, $\mathbb{S}^{n-1}$ is a homogeneous space of $O(n)$.

(b) The natural action of $O(n)$ restricts to a smooth action of $SO(n)$ on $\mathbb{S}^{n-1}$. When $n = 1$, this action is trivial because $SO(1)$ is the trivial group. But if $n > 1$, then $SO(n)$ acts transitively on $\mathbb{S}^{n-1}$. To see this, it suffices to show that for any $v \in \mathbb{S}^{n-1}$, there is a matrix $A \in SO(n)$ taking the first standard basis vector e_1 to v. Since $O(n)$ acts transitively, there is a matrix $A \in O(n)$ taking e_1 to v. Either $\det A = 1$, in which case $A \in SO(n)$, or $\det A = -1$, in which case the matrix obtained by multiplying the second column of A by -1 is in $SO(n)$ and takes e_1 to v. Thus for $n \geq 2$, $\mathbb{S}^{n-1}$ is also a homogeneous space of $SO(n)$.

(c) The Euclidean group $E(n)$ defined in Example 7.32 acts on $\mathbb{R}^n$ by rigid motions. Because any point in $\mathbb{R}^n$ can be taken to any other by a translation, $E(n)$ acts transitively on $\mathbb{R}^n$, so $\mathbb{R}^n$ is a homogeneous $E(n)$-space.

(d) The group $SL(2,\mathbb{R})$ acts smoothly and transitively on the upper half-plane $\mathbb{U} = \{z \in \mathbb{C} : \operatorname{Im} z > 0\}$ by the formula

$$\begin{pmatrix} a & b \\ c & d \end{pmatrix} \cdot z = \frac{az + b}{cz + d}.$$

The resulting complex-analytic diffeomorphisms from $\mathbb{U}$ to itself are called **Möbius transformations**.

(e) For $n \geq 1$, the natural action of $GL(n,\mathbb{C})$ on $\mathbb{C}^n$ restricts to natural smooth actions of both $U(n)$ and $SU(n)$ on $\mathbb{S}^{2n-1}$, identified with the set of unit vectors in $\mathbb{C}^n$. The next exercise shows when these actions are transitive.   //

▶ **Exercise 21.16.** Show that the natural action of $U(n)$ on $\mathbb{S}^{2n-1}$ is transitive for all $n \geq 1$, and that of $SU(n)$ is transitive for all $n \geq 2$.

Next we describe a construction that can be used to generate a great number of homogeneous spaces, as quotients of Lie groups by closed subgroups. Let G be a Lie group and $H \subseteq G$ be a Lie subgroup. A subset of G of the form

$$gH = \{gh : h \in H\}$$

for some $g \in G$ is called a **left coset of H**. The left cosets form a partition of G, and the quotient space determined by this partition (i.e., the set of left cosets with the quotient topology) is called the **left coset space of G modulo H**, and is denoted by G/H. Two elements $g_1, g_2 \in G$ are in the same left coset of H if and only if $g_1^{-1} g_2 \in H$; in this case we write $g_1 \equiv g_2 \pmod{H}$ and say g_1 and g_2 are **congruent modulo H**.

Theorem 21.17 (Homogeneous Space Construction Theorem). *Let G be a Lie group and let H be a closed subgroup of G. The left coset space G/H is a topological manifold of dimension equal to $\dim G - \dim H$, and has a unique smooth structure such that the quotient map $\pi : G \to G/H$ is a smooth submersion. The left action of G on G/H given by*

$$g_1 \cdot (g_2 H) = (g_1 g_2) H \tag{21.3}$$

turns G/H into a homogeneous G-space.

Proof. If we let H act on G by right translation, then $g_1, g_2 \in G$ are in the same H-orbit if and only if $g_1 h = g_2$ for some $h \in H$, which is the same as saying that g_1 and g_2 are in the same coset of H. In other words, the orbit space determined by the *right* action of H on G is precisely the *left* coset space G/H.

The subgroup H is a properly embedded Lie subgroup of G by the closed subgroup theorem, and the H-action on G is smooth because it is simply the restriction of the multiplication map of G. It is a free action because $gh = g$ implies $h = e$. To see that it is proper, we use Proposition 21.5(b). Suppose (g_i) is a convergent sequence in G and (h_i) is a sequence in H such that $(g_i h_i)$ converges in G. By continuity, $h_i = g_i^{-1}(g_i h_i)$ converges to a point in G, and since H is closed in G and has the subspace topology, it follows that (h_i) converges in H.

The quotient manifold theorem now implies that G/H has a unique smooth manifold structure such that the quotient map $\pi\colon G \to G/H$ is a smooth submersion. Since a product of smooth submersions is a smooth submersion, it follows that $\mathrm{Id}_G \times \pi\colon G \times G \to G \times G/H$ is also a smooth submersion. Consider the following diagram:

$$
\begin{array}{ccc}
G \times G & \xrightarrow{\ m\ } & G \\
{\scriptstyle \mathrm{Id}_G \times \pi}\downarrow & & \downarrow{\scriptstyle \pi} \\
G \times G/H & \xrightarrow[\theta]{} & G/H,
\end{array}
$$

where m is group multiplication and θ is the action of G on G/H given by (21.3). It is straightforward to check that $\pi \circ m$ is constant on the fibers of $\mathrm{Id}_G \times \pi$, and therefore θ is well defined and smooth by Theorem 4.30. It is immediate from the definition that θ satisfies the conditions for a group action. Finally, given any two points $g_1 H, g_2 H \in G/H$, the element $g_2 g_1^{-1} \in G$ satisfies $(g_2 g_1^{-1}) \cdot g_1 H = g_2 H$, so the action is transitive. $\square$

The homogeneous spaces constructed in this theorem turn out to be of central importance because, as the next theorem shows, every homogeneous space is equivalent to one of this type.

Theorem 21.18 (Homogeneous Space Characterization Theorem). *Let G be a Lie group, let M be a homogeneous G-space, and let p be any point of M. The isotropy group G_p is a closed subgroup of G, and the map $F\colon G/G_p \to M$ defined by $F(g G_p) = g \cdot p$ is an equivariant diffeomorphism.*

Proof. For simplicity, let us write $H = G_p$. Note that H is closed by continuity, because $H = (\theta^{(p)})^{-1}(p)$, where $\theta^{(p)}\colon G \to M$ is the orbit map.

To see that F is well defined, assume that $g_1 H = g_2 H$, which means that $g_2 = g_1 h$ for some $h \in H$. Then

$$F(g_2 H) = g_2 \cdot p = g_1 h \cdot p = g_1 \cdot p = F(g_1 H).$$

Also, F is equivariant, because

$$F(g' g H) = (g' g) \cdot p = g' \cdot F(g H).$$

It is smooth because it is obtained from the orbit map $\theta^{(p)} \colon G \to M$ by passing to the quotient (see Theorem 4.30).

Next, we show that F is bijective. Given any point $q \in M$, by transitivity there is a group element $g \in G$ such that $g \cdot p = q$, and thus $F(gH) = q$. On the other hand, if $F(g_1 H) = F(g_2 H)$, then $g_1 \cdot p = g_2 \cdot p$ implies $g_1^{-1} g_2 \cdot p = p$, so $g_1^{-1} g_2 \in H$, which implies $g_1 H = g_2 H$. Thus, F is an equivariant smooth bijection, so it is a diffeomorphism by the equivariant rank theorem. $\qquad\Box$

This theorem shows that the study of homogeneous spaces can be reduced to the largely algebraic problem of understanding quotients of Lie groups by closed subgroups. If we are given a smooth manifold M together with a transitive action by a Lie group G, we can always use the preceding theorem to identify M equivariantly with a coset space of the form G/H, and thus use all of the machinery that is available for analyzing quotient spaces, such as Theorems 4.29–4.31 about surjective smooth submersions.

Because of this identification, some authors *define* a homogeneous space to be a quotient manifold of the form G/H, where G is a Lie group and H is a closed subgroup of G. One advantage of that definition is that it makes explicit the intimate relationship between the homogeneous space and the Lie group that acts transitively on it. However, a disadvantage is that it suggests there is something special about the identity coset of H (the image of $e \in G$ under the quotient map), while the essence of homogeneous spaces is that every point "looks the same" as every other.

Applying the characterization theorem to the examples of transitive group actions we discussed earlier, we see that some familiar spaces are diffeomorphic to quotients of Lie groups by closed subgroups.

Example 21.19 (Homogeneous Spaces Revisited).

(a) Consider again the natural action of $O(n)$ on $\mathbb{S}^{n-1}$. If we choose our base point in $\mathbb{S}^{n-1}$ to be the north pole $N = (0, \ldots, 0, 1)$, then it is easy to check that the isotropy group is $O(n-1)$, thought of as orthogonal transformations of $\mathbb{R}^n$ that fix the last variable. Thus $\mathbb{S}^{n-1}$ is diffeomorphic to the quotient manifold $O(n)/O(n-1)$.

(b) For the action of $SO(n)$ on $\mathbb{S}^{n-1}$, the isotropy group is $SO(n-1)$, so $\mathbb{S}^{n-1}$ is also diffeomorphic to $SO(n)/SO(n-1)$ when $n \geq 2$.

(c) Because the Euclidean group $E(n)$ acts smoothly and transitively on $\mathbb{R}^n$, and the isotropy group of the origin is the subgroup $O(n) \subseteq E(n)$, $\mathbb{R}^n$ is diffeomorphic to $E(n)/O(n)$.

(d) Next, consider the transitive action of $SL(2, \mathbb{R})$ on the upper half-plane by Möbius transformations. Direct computation shows that the isotropy group of the point $i \in \mathbb{U}$ consists of matrices of the form $\left(\begin{smallmatrix} a & b \\ -b & a \end{smallmatrix} \right)$ with $a^2 + b^2 = 1$. This subgroup is exactly $SO(2) \subseteq SL(2, \mathbb{R})$, so the characterization theorem gives rise to a diffeomorphism $\mathbb{U} \approx SL(2, \mathbb{R})/SO(2)$.

(e) By virtue of the result of Exercise 21.16, we obtain diffeomorphisms $\mathbb{S}^{2n-1} \approx U(n)/U(n-1)$ for all $n \geq 1$ and $\mathbb{S}^{2n-1} \approx SU(n)/SU(n-1)$ for all $n \geq 2$. $\quad$ //

Because homogeneous spaces have such rich structures, it is natural to wonder whether *every* smooth manifold can be realized as a quotient of a Lie group modulo a closed subgroup, or equivalently, whether every smooth manifold admits a transitive Lie group action. The answer is emphatically no, because there are strong topological restrictions. In the first place, a simple necessary condition for a disconnected manifold is that all of its connected components be diffeomorphic to each other (see Problem 21-11). Much more significantly, a remarkable 2005 theorem by G.D. Mostow [Mos05] shows that a compact homogeneous space must have nonnegative Euler characteristic (see Problem 18-9 for the definition of the Euler characteristic).

Sets with Transitive Group Actions

A highly useful application of the homogeneous space characterization theorem is to put smooth manifold structures on *sets* that admit transitive Lie group actions.

Theorem 21.20. *Suppose X is a set, and we are given a transitive action of a Lie group G on X such that for some point $p \in X$, the isotropy group G_p is closed in G. Then X has a unique smooth manifold structure with respect to which the given action is smooth. With this structure, $\dim X = \dim G - \dim G_p$.*

Proof. Theorem 21.17 shows that G/G_p is a smooth manifold of dimension equal to $\dim G - \dim G_p$. The map $F \colon G/G_p \to X$ defined by $F(gG_p) = g \cdot p$ is an equivariant bijection by exactly the same argument as we used in the proof of the characterization theorem, Theorem 21.18. (That part of the proof did not use the assumption that M was a manifold or that the action was smooth.) If we define a topology and smooth structure on X by declaring F to be a diffeomorphism, then the given action of G on X is smooth because it can be written $(g, x) \mapsto F\big(g \cdot F^{-1}(x)\big)$.

If $\widetilde{X}$ denotes the set X with any smooth manifold structure such that the given action is smooth, then the homogeneous space characterization theorem shows that the map F is also an equivariant diffeomorphism from G/G_p to $\widetilde{X}$, so the topology and smooth structure of $\widetilde{X}$ are equal to those constructed above.   □

Example 21.21 (Grassmannians). Let $G_k(\mathbb{R}^n)$ denote the Grassmannian of k-dimensional subspaces of $\mathbb{R}^n$ as in Example 1.36. The general linear group $\mathrm{GL}(n, \mathbb{R})$ acts transitively on $G_k(\mathbb{R}^n)$: given two subspaces A and A', choose bases for both subspaces and extend them to bases for $\mathbb{R}^n$, and then the linear transformation taking the first basis to the second also takes A to A'. The isotropy group of the subspace $\mathbb{R}^k \subseteq \mathbb{R}^n$ is

$$H = \left\{ \begin{pmatrix} A & B \\ 0 & D \end{pmatrix} : A \in \mathrm{GL}(k, \mathbb{R}), \ D \in \mathrm{GL}(n-k, \mathbb{R}), \ B \in \mathrm{M}(k \times (n-k), \mathbb{R}) \right\},$$

which is easily seen to be closed in $\mathrm{GL}(n, \mathbb{R})$ (why?). Therefore, $G_k(\mathbb{R}^n)$ has a unique smooth manifold structure making the natural $\mathrm{GL}(n, \mathbb{R})$ action smooth. Problem 21-12 shows that this is the same structure we defined in Example 1.36. //

Example 21.22 (Flag Manifolds). Let V be a real vector space of dimension $n > 1$, and let $K = (k_1, \ldots, k_m)$ be a finite sequence of integers satisfying $0 < k_1 < \cdots < k_m < n$. A *flag in V of type K* is a nested sequence of linear subspaces $S_1 \subseteq S_2 \subseteq \cdots \subseteq S_m \subseteq V$, with $\dim S_i = k_i$ for each i. The set of all flags of type K in V is denoted by $F_K(V)$. (For example, if $K = (k)$, then $F_K(V)$ is the Grassmannian $G_k(V)$.) It is not hard to show that $GL(V)$ acts transitively on $F_K(V)$ with a closed subgroup as isotropy group (see Problem 21-16), so $F_K(V)$ has a unique smooth manifold structure making it into a homogeneous $GL(V)$-space. With this structure, $F_K(V)$ is called a *flag manifold*. //

Applications to Lie Theory

The quotient manifold theorem has a wealth of applications to the theory of Lie groups. In this section we describe a few of them.

Quotient Groups

There are two reasons why normal subgroups are important in abstract group theory: on the one hand, they are the only subgroups whose quotients have group structures; and on the other hand, they are the only subgroups that can occur as kernels of group homomorphisms. These facts are summarized in the following two fundamental results from group theory.

Theorem 21.23 (Quotient Theorem for Abstract Groups). *Suppose G is any group and $K \subseteq G$ is a normal subgroup. Then the set G/K of left cosets is a group with multiplication given by $(g_1 K)(g_2 K) = (g_1 g_2)K$, and the projection $\pi \colon G \to G/K$ sending each element of G to its coset is a surjective homomorphism whose kernel is K.*

Theorem 21.24 (First Isomorphism Theorem for Abstract Groups). *If G and H are groups and $F \colon G \to H$ is a group homomorphism, then the kernel of F is a normal subgroup of G, the image of F is a subgroup of H, and F descends to a group isomorphism $\widetilde{F} \colon G/\operatorname{Ker} F \to \operatorname{Im} F$. If F is surjective, then $G/\operatorname{Ker} F$ is isomorphic to H.*

▶ **Exercise 21.25.** Prove these two theorems (or look them up in any abstract algebra text such as [Hun97] or [Her75]).

For Lie groups, we have the following smooth analogues of these results.

Theorem 21.26 (Quotient Theorem for Lie Groups). *Suppose G is a Lie group and $K \subseteq G$ is a closed normal subgroup. Then G/K is a Lie group, and the quotient map $\pi \colon G \to G/K$ is a surjective Lie group homomorphism whose kernel is K.*

Proof. By the homogeneous space construction theorem, G/K is a smooth manifold and π is a smooth submersion; and by Theorem 21.23, G/K is a group and

$\pi: G \to G/K$ is a surjective homomorphism with kernel K. Thus, the only thing that needs to be verified is that multiplication and inversion in G/K are smooth, both of which follow easily from Theorem 4.30. $\qquad\square$

Theorem 21.27 (First Isomorphism Theorem for Lie Groups). *If $F: G \to H$ is a Lie group homomorphism, then the kernel of F is a closed normal Lie subgroup of G, the image of F has a unique smooth manifold structure making it into a Lie subgroup of H, and F descends to a Lie group isomorphism $\widetilde{F}: G/\operatorname{Ker} F \to \operatorname{Im} F$. If F is surjective, then $G/\operatorname{Ker} F$ is smoothly isomorphic to H.*

Proof. By Theorem 21.24, $\operatorname{Ker} F$ is a normal subgroup, $\operatorname{Im} F$ is a subgroup, and F descends to a group isomorphism $\widetilde{F}: G/\operatorname{Ker} F \to \operatorname{Im} F$. By continuity, $\operatorname{Ker} F$ is closed in G, so it follows from Theorem 21.26 that $G/\operatorname{Ker} F$ is a Lie group and the projection $\pi: G \to G/\operatorname{Ker} F$ is a surjective Lie group homomorphism. Because π is surjective and has constant rank, it is a smooth submersion by the global rank theorem, so the characteristic property of surjective smooth submersions (Theorem 4.29) guarantees that $\widetilde{F}$ is smooth. Since $\operatorname{Im} F$ is the image of the injective Lie group homomorphism $\widetilde{F}$, Proposition 7.17 shows that it has a smooth manifold structure with respect to which it is a Lie subgroup of H and $\widetilde{F}: G/\operatorname{Ker} F \to \operatorname{Im} F$ is a Lie group isomorphism. The uniqueness of the smooth structure on $\operatorname{Im} F$ follows from Problem 20-11. The last statement follows immediately just by substituting $H = \operatorname{Im} F$. $\qquad\square$

These results are particularly significant when we apply them to a ***discrete subgroup***, that is, a subgroup that is a discrete space in the subspace topology.

Proposition 21.28. *Every discrete subgroup of a Lie group is a closed Lie subgroup of dimension zero.*

Proof. Let G be a Lie group and $\Gamma \subseteq G$ be a discrete subgroup. With the subspace topology, Γ is a countable discrete space and thus a zero-dimensional Lie group. By the closed subgroup theorem, Γ is a closed Lie subgroup of G if and only if it is a closed subset. A discrete subset is closed if and only if it has no limit points in G, so assume for the sake of contradiction that g is a limit point of Γ in G. By discreteness, there is a neighborhood U of e in G such that $U \cap \Gamma = \{e\}$, and then Problem 7-6 shows that there is a smaller neighborhood V of e such that $g_1 g_2^{-1} \in U$ whenever $g_1, g_2 \in V$. Then $Vg = \{hg : h \in V\}$ is a neighborhood of g, and because G is Hausdorff, Vg contains infinitely many points of Γ. Let γ_1, γ_2 be two distinct points in $\Gamma \cap Vg$. Then $\gamma_1 g^{-1}, \gamma_2 g^{-1} \in V$, which implies

$$\gamma_1 \gamma_2^{-1} = \left(\gamma_1 g^{-1}\right)\left(\gamma_2 g^{-1}\right)^{-1} \in U.$$

Since $U \cap \Gamma = \{e\}$, this implies $\gamma_1 \gamma_2^{-1} = e$, so $\gamma_1 = \gamma_2$, contradicting our assumption that γ_1 and γ_2 are distinct. $\qquad\square$

Theorem 21.29 (Quotients of Lie Groups by Discrete Subgroups). *If G is a connected Lie group and $\Gamma \subseteq G$ is a discrete subgroup, then G/Γ is a smooth manifold and the quotient map $\pi: G \to G/\Gamma$ is a smooth normal covering map.*

Proof. The proof of Theorem 21.17 showed that Γ acts smoothly, freely, and properly on G on the right, and its quotient—which is the coset space G/Γ—is a smooth manifold. The theorem is then an immediate consequence of Theorem 21.13. □

Example 21.30. Let C be a cube centered at the origin in $\mathbb{R}^3$. The set of orientation-preserving orthogonal transformations of $\mathbb{R}^3$ that take C to itself is a finite subgroup $\Gamma \subseteq SO(3)$, and $SO(3)/\Gamma$ is a connected smooth 3-manifold with finite fundamental group and with $\mathbb{S}^3$ as its universal covering space (see Problem 21-21). Similar examples are obtained from the symmetry groups of other regular polyhedra, such as a regular tetrahedron, octahedron, dodecahedron, or icosahedron. //

Combining the results of Theorems 21.27 and 21.29, we obtain the following important characterization of homomorphisms with discrete kernels.

Theorem 21.31 (Homomorphisms with Discrete Kernels). *Let G and H be connected Lie groups. For any Lie group homomorphism $F: G \to H$, the following are equivalent:*

(a) *F is surjective and has discrete kernel.*
(b) *F is a smooth covering map.*
(c) *F is a local diffeomorphism.*
(d) *The induced homomorphism $F_*: \mathrm{Lie}(G) \to \mathrm{Lie}(H)$ is an isomorphism.*

Proof. We will show that (a) $\Rightarrow$ (b) $\Rightarrow$ (c) $\Rightarrow$ (a) and (c) $\Leftrightarrow$ (d). First, assume that F is surjective with discrete kernel $\Gamma \subseteq G$. Then Theorem 21.29 implies that the quotient map $\pi: G \to G/\Gamma$ is a smooth covering map, and Theorem 21.27 shows that F descends to a Lie group isomorphism $\widetilde{F}: G/\Gamma \to H$. This means that $F = \widetilde{F} \circ \pi$, which is a composition of a smooth covering map followed by a diffeomorphism and therefore is itself a smooth covering map. This proves that (a) $\Rightarrow$ (b). The next implication, (b) $\Rightarrow$ (c), follows from Proposition 4.33.

Under the assumption that F is a local diffeomorphism, each level set is an embedded 0-dimensional submanifold by the submersion level set theorem, so $\mathrm{Ker}\, F$ is discrete. Since a local diffeomorphism is an open map, $F(G)$ is an open subgroup of H, and thus by Proposition 7.15, it is all of H. This shows that (c) $\Rightarrow$ (a).

The implication (c) $\Rightarrow$ (d) is the content of Problem 8-27. Conversely, if F_* is an isomorphism, the inverse function theorem implies F is a local diffeomorphism in a neighborhood of $e \in G$. Because Lie group homomorphisms have constant rank, this means that rank $F = \dim G = \dim H$, which implies that F is a local diffeomorphism everywhere, and thus (d) $\Rightarrow$ (c). □

This theory allows us to give a precise description of all the connected Lie groups with a given Lie algebra. (For the disconnected case, see Problem 21-22.)

Theorem 21.32. *Let $\mathfrak{g}$ be a finite-dimensional Lie algebra. The connected Lie groups whose Lie algebras are isomorphic to $\mathfrak{g}$ are (up to isomorphism) precisely those of the form G/Γ, where G is the simply connected Lie group with Lie algebra $\mathfrak{g}$, and Γ is a discrete normal subgroup of G.*

Proof. Given $\mathfrak{g}$, by Theorem 20.21 there exists a simply connected Lie group G with Lie algebra isomorphic to $\mathfrak{g}$. Suppose H is any other connected Lie group whose Lie algebra is isomorphic to $\mathfrak{g}$, and let $\varphi\colon \mathrm{Lie}(G) \to \mathrm{Lie}(H)$ be a Lie algebra isomorphism. Theorem 20.19 guarantees that there is a Lie group homomorphism $\Phi\colon G \to H$ such that $\Phi_* = \varphi$. Because φ is an isomorphism, Theorem 21.31 implies that Φ is surjective and its kernel is a discrete normal subgroup of G. It follows that H is isomorphic to $G/\operatorname{Ker}\Phi$ by Theorem 21.27. $\qquad\square$

Connectivity of Lie Groups

Another application of homogeneous space theory is to identify the connected components of many familiar Lie groups. The key result is the following proposition.

Proposition 21.33. *Suppose a topological group G acts continuously, freely, and properly on a topological space M. If G and M/G are connected, then M is connected.*

Proof. Assume for the sake of contradiction that G and M/G are connected but M is not. Then there are disjoint nonempty open subsets $U, V \subseteq M$ whose union is M. Each G-orbit in M is the image of G under an orbit map $\theta^{(p)}\colon G \to M$; since G is connected, each orbit must lie entirely in one set U or V.

By Lemma 21.1, $\pi(U)$ and $\pi(V)$ are both open in M/G. If $\pi(U) \cap \pi(V)$ were not empty, some G-orbit in M would contain points of both U and V, which we have just shown is impossible. Thus, $\pi(U)$ and $\pi(V)$ are disjoint nonempty open subsets of M/G whose union is M/G, contradicting the assumption that M/G is connected. $\qquad\square$

Proposition 21.34. *For each $n \geq 1$, the Lie groups $\mathrm{SO}(n)$, $\mathrm{U}(n)$, and $\mathrm{SU}(n)$ are connected. The group $\mathrm{O}(n)$ has exactly two components, one of which is $\mathrm{SO}(n)$.*

Proof. First, we prove by induction on n that $\mathrm{SO}(n)$ is connected. For $n = 1$ this is obvious, because $\mathrm{SO}(1)$ is the trivial group. Now suppose we have shown that $\mathrm{SO}(n-1)$ is connected for some $n \geq 2$. Because the homogeneous space $\mathrm{SO}(n)/\mathrm{SO}(n-1)$ is diffeomorphic to $\mathbb{S}^{n-1}$ and therefore is connected, Proposition 21.33 and the induction hypothesis imply that $\mathrm{SO}(n)$ is connected. A similar argument applies to $\mathrm{U}(n)$ and $\mathrm{SU}(n)$, using the facts that $\mathrm{U}(n)/\mathrm{U}(n-1) \approx \mathrm{SU}(n)/\mathrm{SU}(n-1) \approx \mathbb{S}^{2n-1}$.

As we noted in Example 7.28, $\mathrm{O}(n)$ is equal to the union of the two open subsets $\mathrm{O}^+(n)$ and $\mathrm{O}^-(n)$ consisting of orthogonal matrices with determinant $+1$ and -1, respectively. By the argument in the preceding paragraph, $\mathrm{O}^+(n) = \mathrm{SO}(n)$ is connected. On the other hand, if A is any orthogonal matrix whose determinant is -1, then left translation L_A is a diffeomorphism from $\mathrm{O}^+(n)$ to $\mathrm{O}^-(n)$, so $\mathrm{O}^-(n)$ is connected as well. Therefore, the components of $\mathrm{O}(n)$ are $\mathrm{O}^+(n)$ and $\mathrm{O}^-(n)$. $\quad\square$

Determining the components of the general linear groups is a bit more involved. Let $\mathrm{GL}^+(n,\mathbb{R})$ and $\mathrm{GL}^-(n,\mathbb{R})$ denote the open subsets of $\mathrm{GL}(n,\mathbb{R})$ consisting of matrices with positive determinant and negative determinant, respectively.

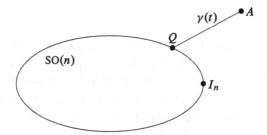

Fig. 21.2 Proof that $GL^+(n,\mathbb{R})$ is connected

Proposition 21.35. *The connected components of* $GL(n,\mathbb{R})$ *are* $GL^+(n,\mathbb{R})$ *and* $GL^-(n,\mathbb{R})$.

Proof. By continuity of the determinant function, $GL^+(n,\mathbb{R})$ and $GL^-(n,\mathbb{R})$ are nonempty, disjoint, open subsets of $GL(n,\mathbb{R})$ whose union is $GL(n,\mathbb{R})$, so all we need to prove is that both subsets are connected. We begin by showing that $GL^+(n,\mathbb{R})$ is connected. It suffices to show that it is path-connected, which will follow once we show that for any $A \in GL^+(n,\mathbb{R})$, there is a continuous path in $GL^+(n,\mathbb{R})$ from A to the identity matrix I_n.

Let $A \in GL^+(n,\mathbb{R})$ be arbitrary, and let $(A_1,\dots,A_n)$ denote the columns of A, considered as vectors in $\mathbb{R}^n$. The Gram–Schmidt algorithm (Proposition B.40) shows that there is an orthonormal basis $(Q_1,\dots,Q_n)$ for $\mathbb{R}^n$ with the property that $\operatorname{span}(Q_1,\dots,Q_k) = \operatorname{span}(A_1,\dots,A_k)$ for each $k = 1,\dots,n$. Thus, we can write

$$A_1 = R_1^1 Q_1,$$

$$A_2 = R_2^1 Q_1 + R_2^2 Q_2,$$

$$\vdots$$

$$A_n = R_n^1 Q_1 + R_n^2 Q_2 + \cdots + R_n^n Q_n,$$

for some constants R_i^j. Replacing each Q_i by $-Q_i$ if necessary, we may assume that each number R_i^i (no summation) is positive. In matrix notation, this is equivalent to $A = QR$, where Q is orthogonal and R is upper triangular with positive entries on the diagonal. Since the determinant of R is the product of its diagonal entries and $(\det Q)(\det R) = \det A > 0$, it follows that $Q \in SO(n)$. (This **QR decomposition** plays an important role in numerical linear algebra.)

Let $R_t = tI_n + (1-t)R$ for $t \in [0,1]$. It is immediate that each matrix R_t is upper triangular with positive diagonal entries, so $R_t \in GL^+(n,\mathbb{R})$. Therefore, the path $\gamma\colon [0,1] \to GL^+(n,\mathbb{R})$ given by $\gamma(t) = QR_t$ satisfies $\gamma(0) = A$ and $\gamma(1) = Q \in SO(n)$ (Fig. 21.2). Because $SO(n)$ is connected, there is a path in $SO(n)$ from Q to the identity matrix. This shows that $GL^+(n,\mathbb{R})$ is path-connected. Any matrix B with $\det B < 0$ yields a diffeomorphism $L_B\colon GL^+(n,\mathbb{R}) \to GL^-(n,\mathbb{R})$, so $GL^-(n,\mathbb{R})$ is connected as well. This completes the proof. $\square$

Problems

21-1. Suppose a Lie group G acts continuously on a manifold M. Show that if the map $G \times M \to M$ defining the action is proper, then the action is a proper action. Give a counterexample to show that the converse need not be true.

21-2. Let X be the orbit space $\mathbb{R}^n / \mathbb{R}^+$, where the action of $\mathbb{R}^+$ is given by (21.1). Show that X has an open subset homeomorphic to $\mathbb{S}^{n-1}$, and a point that belongs to every nonempty closed subset.

21-3. Show that the Hopf action of $\mathbb{S}^1$ on $\mathbb{S}^{2n+1}$ (see Problem 7-11) is free and proper, and that the orbit space $\mathbb{S}^{2n+1} / \mathbb{S}^1$ is diffeomorphic to $\mathbb{CP}^n$. [Hint: consider the restriction of the natural quotient map $\mathbb{C}^{n+1} \smallsetminus \{0\} \to \mathbb{CP}^n$ to $\mathbb{S}^{2n+1}$. The quotient map $\mathbb{S}^{2n+1} \to \mathbb{CP}^n$ is known as the *Hopf map*.]

21-4. Let M be a smooth n-manifold, and suppose V is a smooth vector field on M such that every integral curve of V is periodic with the same period (see Problem 9-1). Define an equivalence relation on M by saying $p \sim q$ if p and q are in the image of the same integral curve of V. Let $M/\sim$ be the quotient space, and let $\pi \colon M \to M/\sim$ be the quotient map. Show that $M/\sim$ is a topological $(n-1)$-manifold and has a unique smooth structure such that π is a smooth submersion.

21-5. Prove the following partial converse to the quotient manifold theorem: if a Lie group G acts smoothly and freely on a smooth manifold M, and the orbit space M/G has a smooth manifold structure such that the quotient map $\pi \colon M \to M/G$ is a smooth submersion, then G acts properly. [Hint: first show that for any smooth local section $\sigma \colon U \to M$ of π, the map $(g, x) \mapsto g \cdot \sigma(x)$ is a diffeomorphism from $G \times U$ to $\pi^{-1}(U)$.]

21-6. Suppose a Lie group G acts smoothly, freely, and properly on a smooth manifold M. Show that M is the total space of a smooth fiber bundle with base M/G, model fiber G, and projection equal to the quotient map $\pi \colon M \to M/G$. [Hint: see the hint for Problem 21-5.] Conclude from Problem 10-19 that M is compact if and only if both G and M/G are compact. [Remark: any fiber bundle obtained in this way is called a *principal G-bundle*.]

21-7. Let Γ be a discrete Lie group acting smoothly, freely, and properly on a connected smooth manifold M. Show that a Riemannian metric g on M is the pullback of a metric on M/Γ by the quotient map $\pi \colon M \to M/\Gamma$ if and only if Γ acts by isometries of g (i.e., $x \mapsto \gamma \cdot x$ is an isometry for each $\gamma \in \Gamma$).

21-8. Let M be a smooth manifold, and let $\pi \colon E \to M$ be a smooth vector bundle over M. Suppose Γ is a discrete Lie group acting smoothly, freely, and properly on both E and M. Suppose further that π is Γ-equivariant, and for each $p \in M$ and each $g \in \Gamma$, the map from E_p to $E_{g \cdot p}$ given by $v \mapsto g \cdot v$ is linear. Show that E/Γ can be given the structure of a smooth vector bundle over M/Γ in such a way that the following diagram commutes:

$$
\begin{array}{ccc}
E & \longrightarrow & E/\Gamma \\
\downarrow & & \downarrow \\
M & \longrightarrow & M/\Gamma.
\end{array}
$$

(The horizontal maps are quotient maps, and the vertical ones are bundle projections.)

21-9. Show that the action of $\mathbb{Z}$ on $\mathbb{R}^2$ given by $n \cdot (x, y) = \left(x + n, (-1)^n y\right)$ is smooth, free, and proper. Use this together with the result of Problem 21-8 to give a simple proof that the Möbius bundle of Example 10.3 is a non-trivial smooth rank-1 vector bundle over $\mathbb{S}^1$, and to prove that the quotient map $q \colon \mathbb{R}^2 \to E$ used to define E is a smooth normal covering map whose covering automorphism group is the given $\mathbb{Z}$-action on $\mathbb{R}^2$.

21-10. Let $\mathbb{CP}^n$ denote n-dimensional complex projective space (Problem 1-9). Show that the natural action of $U(n + 1)$ on $\mathbb{C}^{n+1} \smallsetminus \{0\}$ descends to a smooth, transitive action on $\mathbb{CP}^n$, so $\mathbb{CP}^n$ is a homogeneous $U(n + 1)$-space. Choose a point and identify the isotropy group.

21-11. This problem shows that a disconnected smooth manifold is homogeneous if and only if its components are all diffeomorphic to one another and one of them is homogeneous.
  (a) Suppose M is a homogeneous G-space. Show that all of its components are diffeomorphic to one another, and there is an open subgroup $G_0 \subseteq G$ such that each component is a homogeneous G_0-space.
  (b) Conversely, suppose M is a smooth manifold all of whose components are diffeomorphic to each other, and there is a Lie group G_0 that acts smoothly and transitively on one of its components M_0. Show that the direct product group $G_0 \times \mathbb{Z}$ acts smoothly and transitively on M. [Hint: first, show that M is diffeomorphic to a product of M_0 with a countable discrete space.]

21-12. Show that the smooth structure on the Grassmannian $G_k(\mathbb{R}^n)$ defined in Example 21.21 is the same as the one defined in Example 1.36.

21-13. Let V be a finite-dimensional real vector space. Prove that the Grassmannian $G_k(V)$ is compact for each k. [Hint: show that it is a quotient space of a compact Lie group.]

21-14. Let V be an n-dimensional real vector space, and let $G_k(V)$ be the Grassmannian of k-dimensional subspaces of V for some integer k with $0 < k < n$. Let $\mathbb{P}(\Lambda^k(V))$ denote the projectivization of $\Lambda^k(V)$ (see Problem 2-11). Define a map $\rho \colon G_k(V) \to \mathbb{P}(\Lambda^k(V))$ by

$$
\rho(S) = [v_1 \wedge \cdots \wedge v_k] \quad \text{if } S = \operatorname{span}(v_1, \ldots, v_k).
$$

Show that ρ is well defined, and is a smooth embedding whose image is the set of all equivalence classes of nonzero decomposable elements of $\Lambda^k(V)$. (It is called the **Plücker embedding**.)

21-15. The set of k-dimensional complex-linear subspaces of $\mathbb{C}^n$ is denoted by $G_k(\mathbb{C}^n)$. Show that $G_k(\mathbb{C}^n)$ has a unique smooth manifold structure making it into a homogeneous $GL(n, \mathbb{C})$-space, with the action of $GL(n, \mathbb{C})$ induced from its usual action on $\mathbb{C}^n$. Show that $G_k(\mathbb{C}^n)$ is compact, and determine its dimension.

21-16. Let $F_K(V)$ be the set of flags of type K in a finite-dimensional vector space V as in Example 21.22. Show that $GL(V)$ acts transitively on $F_K(V)$, and that the isotropy group of a particular flag is a closed subgroup of $GL(V)$, and conclude that $F_K(V)$ has a unique smooth manifold structure such that the action is smooth. What is $\dim F_K(V)$? For which K is $F_K(V)$ compact?

21-17. Suppose a Lie group acts smoothly (but not necessarily properly or freely) on a smooth manifold M. Show that each orbit is an immersed submanifold of M, which is embedded if the action is proper.

21-18. The *center* of a group G is the set of all elements that commute with every element of G; a subgroup of G is said to be *central* if it is contained in the center of G. Show that every discrete normal subgroup of a connected Lie group is central. [Hint: use the result of Problem 7-8.]

21-19. Use the results of Theorem 7.7 and Problem 21-18 to show that the fundamental group of every Lie group is abelian. You may use without proof the fact that if $\pi\colon \widetilde{G} \to G$ is a universal covering map, then the automorphism group $\mathrm{Aut}_\pi(\widetilde{G})$ is isomorphic to $\pi_1(G, e)$ (see [LeeTM, Chap. 12]).

21-20. (a) Let G be a connected abelian Lie group. Prove that the exponential map of G is a surjective Lie group homomorphism from $\mathrm{Lie}(G)$ (considered as a Lie group under vector addition) to G.

     (b) Show that every connected abelian Lie group is isomorphic to $\mathbb{R}^k \times \mathbb{T}^l$ for some nonnegative integers k and l.

21-21. Show that $SO(3)$ is isomorphic to $SU(2)/\{\pm I_2\}$ and diffeomorphic to $\mathbb{R}\mathbb{P}^3$, as follows:

     (a) Let $\mathcal{S} \subseteq \mathbb{H}$ denote the group of unit quaternions, and let $E \subseteq \mathbb{H}$ be the subspace of imaginary quaternions (see Problems 7-22, 7-23, and 8-6). For any $q \in \mathcal{S}$, show that the linear map $\mathbb{H} \to \mathbb{H}$ given by $v \mapsto qvq^*$ takes E to itself and preserves the inner product $\langle v, w \rangle = \frac{1}{2}(v^*w + w^*v)$ on E.

     (b) For each $q \in \mathcal{S}$, let $\rho(q)$ be the matrix representation of the map $v \mapsto qvq^*$ with respect to the basis $(\hat{\imath}, \hat{\jmath}, \Bbbk)$ for E. Show that $\rho(q) \in SO(3)$ and that the map $\rho\colon \mathcal{S} \to SO(3)$ is a surjective Lie group homomorphism whose kernel is $\{\pm 1\}$.

     (c) Prove the result.

21-22. If G and H are Lie groups, and there exists a surjective Lie group homomorphism from G to H with kernel G_0, we say that G is an *extension of G_0 by H*. Let $\mathfrak{g}$ be any finite-dimensional Lie algebra, and show that the disconnected Lie groups whose Lie algebras are isomorphic to $\mathfrak{g}$ are precisely the extensions of the connected ones by discrete Lie groups. [Hint: see Proposition 7.15.] (*Used on p. 557.*)

21-23. Show that $GL^+(n,\mathbb{R})$ is diffeomorphic to $SO(n) \times \mathbb{R}^{n(n+1)/2}$. [Hint: use the QR decomposition introduced in Proposition 21.35 to construct a diffeomorphism $SO(n) \times T^+(n,\mathbb{R}) \approx GL^+(n,\mathbb{R})$, where $T^+(n,\mathbb{R})$ is the Lie group of $n \times n$ upper triangular real matrices with positive diagonal entries.]

21-24. Show that $GL(n,\mathbb{C})$ is diffeomorphic to $U(n) \times \mathbb{R}^{n^2}$, and thus is connected. [Hint: argue as in Problem 21-23, but use the Hermitian dot product $z \cdot w = \sum_i z^i \overline{w^i}$ in place of the Euclidean dot product.]

21-25. Show that $SL(n,\mathbb{R})$ and $SL(n,\mathbb{C})$ are diffeomorphic to $SO(n) \times \mathbb{R}^{n(n+1)/2-1}$ and $SU(n) \times \mathbb{R}^{n^2-1}$ respectively. Conclude that $SL(n,\mathbb{R})$ and $SL(n,\mathbb{C})$ are connected and $SL(2,\mathbb{C})$ is simply connected. (See Proposition 21.34 and Problem 7-16.)

21-26. By the result of Problem 21-25, $SL(2,\mathbb{R})$ is diffeomorphic to $SO(2) \times \mathbb{R}^2 \approx \mathbb{S}^1 \times \mathbb{R}^2$, and therefore its fundamental group is isomorphic to $\mathbb{Z}$. Let $\widetilde{SL(2,\mathbb{R})}$ denote the universal covering group of $SL(2,\mathbb{R})$ (see Theorem 7.7). Show that $\widetilde{SL(2,\mathbb{R})}$ does not admit a faithful representation, as follows. Suppose $\rho \colon \widetilde{SL(2,\mathbb{R})} \to GL(V)$ is any representation. By choosing a basis for V over $\mathbb{R}$, we might as well replace $GL(V)$ with $GL(n,\mathbb{R})$ for some n. Then ρ induces a Lie algebra homomorphism $\rho_* \colon \mathfrak{sl}(2,\mathbb{R}) \to \mathfrak{gl}(n,\mathbb{R})$. Define a map $\varphi \colon \mathfrak{sl}(2,\mathbb{C}) \to \mathfrak{gl}(n,\mathbb{C})$ by

$$\varphi(A + iB) = \rho_* A + i\rho_* B, \quad A, B \in \mathfrak{sl}(2,\mathbb{R}).$$

(a) Show that φ is a Lie algebra homomorphism.

(b) Show there is a Lie group homomorphism $\Phi \colon SL(2,\mathbb{C}) \to GL(n,\mathbb{C})$ such that $\Phi_* = \varphi$ and the following diagram commutes:

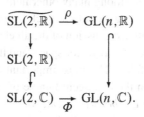

[Hint: use Problem 21-25.]

(c) Show that ρ is not faithful.

Chapter 22
Symplectic Manifolds

In this final chapter we introduce a new kind of geometric structure on manifolds, called a *symplectic structure*, which is superficially similar to a Riemannian metric but turns out to have profoundly different properties. It is simply a choice of a closed, nondegenerate 2-form. The motivation for the definition may not be evident at first, but it will emerge gradually as we see how these properties are used. For now, suffice it to say that nondegeneracy is important because it yields a tangent-cotangent isomorphism like that provided by a Riemannian metric, and closedness is important because it leads to a deep relationship between smooth functions and flows (see the discussions of Hamiltonian vector fields and Poisson brackets later in this chapter). Symplectic structures have surprisingly varied applications in mathematics and physics, including partial differential equations, differential topology, and classical mechanics, among many other fields.

In this chapter, we can give only a quick overview of the subject of symplectic geometry. We begin with a discussion of the algebra of nondegenerate alternating 2-tensors on a finite-dimensional vector space, and then turn our attention to symplectic structures on manifolds. The most important example is a canonically defined symplectic structure on the cotangent bundle of each smooth manifold. We give a proof of the important *Darboux theorem*, which shows that every symplectic form can be put into canonical form locally by a choice of smooth coordinates, so, unlike the situation for Riemannian metrics, there is no local obstruction to "flatness" of symplectic structures.

Then we explore one of the most important applications of symplectic structures. Any smooth real-valued function on a symplectic manifold gives rise to a canonical system of ordinary differential equations called a *Hamiltonian system*. These systems are central to the study of classical mechanics.

After treating Hamiltonian systems, we give a brief introduction to an odd-dimensional analogue of symplectic structures, called *contact structures*. Then at the end of the chapter, we show how symplectic and contact geometry can be used to construct solutions to first-order partial differential equations.

J.M. Lee, *Introduction to Smooth Manifolds*, Graduate Texts in Mathematics 218,
DOI 10.1007/978-1-4419-9982-5_22, © Springer Science+Business Media New York 2013

Symplectic Tensors

We begin with linear algebra. A 2-covector ω on a finite-dimensional vector space V is said to be ***nondegenerate*** if the linear map $\hat{\omega} \colon V \to V^*$ defined by $\hat{\omega}(v) = v \lrcorner \omega$ is invertible.

▶ **Exercise 22.1.** Show that the following are equivalent for 2-covector ω on a finite-dimensional vector space V:

(a) ω is nondegenerate.
(b) For each nonzero $v \in V$, there exists $w \in V$ such that $\omega(v, w) \neq 0$.
(c) In terms of some (hence every) basis, the matrix (ω_{ij}) representing ω is nonsingular.

A nondegenerate 2-covector is called a ***symplectic tensor***. A vector space V endowed with a specific symplectic tensor is called a ***symplectic vector space***. (A symplectic tensor is also often called a "symplectic form," because it is in particular a bilinear form. But to avoid confusion, we reserve that name for something slightly different, to be defined below.)

Example 22.2. Let V be a vector space of dimension $2n$, with a basis denoted by $(A_1, B_1, \ldots, A_n, B_n)$. Let $(\alpha^1, \beta^1, \ldots, \alpha^n, \beta^n)$ denote the corresponding dual basis for V^*, and let $\omega \in \Lambda^2(V^*)$ be the 2-covector defined by

$$\omega = \sum_{i=1}^{n} \alpha^i \wedge \beta^i. \tag{22.1}$$

Note that the action of ω on basis vectors is given by

$$\omega(A_i, A_j) = \omega(B_i, B_j) = 0, \qquad \omega(A_i, B_j) = -\omega(B_j, A_i) = \delta_{ij}. \tag{22.2}$$

Suppose $v = a^i A_i + b^i B_i \in V$ satisfies $\omega(v, w) = 0$ for all $w \in V$. Then $0 = \omega(v, B_i) = a^i$ and $0 = \omega(v, A_i) = -b^i$, which implies that $v = 0$. Thus ω is nondegenerate, and so is a symplectic tensor. //

It is interesting to consider the special case in which $\dim V = 2$. In this case, using the notation of the preceding example, $\alpha^1 \wedge \beta^1$ is nondegenerate, and every 2-covector is a multiple of $\alpha^1 \wedge \beta^1$. Thus every nonzero 2-covector on a 2-dimensional vector space is symplectic.

If (V, ω) is a symplectic vector space and $S \subseteq V$ is any linear subspace, we define the ***symplectic complement of*** S, denoted by $S^\perp$, to be the subspace

$$S^\perp = \{v \in V : \omega(v, w) = 0 \text{ for all } w \in S\}.$$

As the notation suggests, the symplectic complement is analogous to the orthogonal complement in an inner product space. Just as in the inner product case, the dimension of $S^\perp$ is the codimension of S, as the next lemma shows.

Lemma 22.3. *Let* (V, ω) *be a symplectic vector space. For any linear subspace* $S \subseteq V$, *we have* $\dim S + \dim S^\perp = \dim V$.

Proof. Let $S \subseteq V$ be a subspace, and define a linear map $\Phi: V \to S^*$ by $\Phi(v) = (v \lrcorner \omega)|_S$, or equivalently

$$\Phi(v)(w) = \omega(v, w) \quad \text{for } v \in V, \ w \in S.$$

Suppose φ is an arbitrary element of S^*, and let $\tilde{\varphi} \in V^*$ be any extension of φ to a linear functional on all of V. Since the map $\hat{\omega}: V \to V^*$ defined by $v \mapsto v \lrcorner \omega$ is an isomorphism, there exists $v \in V$ such that $v \lrcorner \omega = \tilde{\varphi}$. It follows that $\Phi(v) = \varphi$, and therefore Φ is surjective. By the rank–nullity law, $S^\perp = \text{Ker } \Phi$ has dimension equal to $\dim V - \dim S^* = \dim V - \dim S$. $\qquad\square$

▶ **Exercise 22.4.** Let (V, ω) be a symplectic vector space and $S \subseteq V$ be a linear subspace. Show that $(S^\perp)^\perp = S$.

Symplectic complements differ from orthogonal complements in one important respect: although it is always true that $S \cap S^\perp = \{0\}$ in an inner product space, this need not be true in a symplectic vector space. Indeed, if S is 1-dimensional, the fact that ω is alternating forces $\omega(v, v) = 0$ for every $v \in S$, so $S \subseteq S^\perp$. Carrying this idea a little further, a linear subspace $S \subseteq V$ is said to be

- **symplectic** if $S \cap S^\perp = \{0\}$;
- **isotropic** if $S \subseteq S^\perp$;
- **coisotropic** if $S \supseteq S^\perp$;
- **Lagrangian** if $S = S^\perp$.

Proposition 22.5. *Let (V, ω) be a symplectic vector space, and let $S \subseteq V$ be a linear subspace.*

(a) *S is symplectic if and only if $S^\perp$ is symplectic.*
(b) *S is symplectic if and only if $\omega|_S$ is nondegenerate.*
(c) *S is isotropic if and only if $\omega|_S = 0$.*
(d) *S is coisotropic if and only if $S^\perp$ is isotropic.*
(e) *S is Lagrangian if and only if $\omega|_S = 0$ and $\dim S = \frac{1}{2} \dim V$.*

Proof. Problem 22-1. $\qquad\square$

▶ **Exercise 22.6.** Let (V, ω) be the symplectic vector space of dimension $2n$ described in Example 22.2, and let k be an integer such that $0 \le k \le n$.

(a) Show that $\text{span}(A_1, B_1, \ldots, A_k, B_k)$ is symplectic.
(b) Show that $\text{span}(A_1, \ldots, A_k)$ is isotropic.
(c) Show that $\text{span}(A_1, \ldots, A_n, B_1, \ldots, B_k)$ is coisotropic.
(d) Show that $\text{span}(A_1, \ldots, A_n)$ is Lagrangian.
(e) If $n \ge 3$, which of the four kinds of subspace, if any, is $\text{span}(A_1, A_2, B_1)$?

The symplectic tensor ω defined in Example 22.2 turns out to be the prototype of all symplectic tensors, as the next proposition shows. This can be viewed as a symplectic version of the Gram–Schmidt algorithm.

Proposition 22.7 (Canonical Form for a Symplectic Tensor). *Let ω be a symplectic tensor on an m-dimensional vector space V. Then V has even dimension $m = 2n$, and there exists a basis for V in which ω has the form (22.1).*

Proof. The tensor ω has the form (22.1) with respect to a basis $(A_1, B_1, \ldots, A_n, B_n)$ if and only if its action on basis vectors is given by (22.2). We prove the theorem by induction on $m = \dim V$ by showing that there is a basis with this property.

For $m = 0$ there is nothing to prove. Suppose (V, ω) is a symplectic vector space of dimension $m \geq 1$, and assume that the proposition is true for all symplectic vector spaces of dimension less than m. Let A_1 be any nonzero vector in V. Since ω is nondegenerate, there exists $B_1 \in V$ such that $\omega(A_1, B_1) \neq 0$. Multiplying B_1 by a constant if necessary, we may assume that $\omega(A_1, B_1) = 1$. Because ω is alternating, B_1 cannot be a multiple of A_1, so the set $\{A_1, B_1\}$ is linearly independent, and hence $\dim V \geq 2$.

Let $S \subseteq V$ be the span of $\{A_1, B_1\}$. Then $\dim S^{\perp} = m - 2$ by Lemma 22.3. Since $\omega|_S$ is nondegenerate, by Proposition 22.5 it follows that S is symplectic, and thus $S^{\perp}$ is also symplectic. By induction, $S^{\perp}$ is even-dimensional and there is a basis $(A_2, B_2, \ldots, A_n, B_n)$ for $S^{\perp}$ such that (22.2) is satisfied for $2 \leq i, j \leq n$. It follows easily that $(A_1, B_1, A_2, B_2, \ldots, A_n, B_n)$ is the required basis for V. $\square$

Because of this proposition, if (V, ω) is a symplectic vector space, a basis $(A_1, B_1, \ldots, A_n, B_n)$ for V is called a **symplectic basis** if (22.2) holds, which is equivalent to ω being given by (22.1) in terms of the dual basis. The proposition then says that every symplectic vector space has a symplectic basis.

This leads to another useful criterion for 2-covector to be nondegenerate. For an alternating tensor ω, the notation ω^k denotes the k-fold wedge product $\omega \wedge \cdots \wedge \omega$.

Proposition 22.8. *Suppose V is a $2n$-dimensional vector space and $\omega \in \Lambda^2(V^*)$. Then ω is a symplectic tensor if and only if $\omega^n \neq 0$.*

Proof. Suppose first that ω is a symplectic tensor. Let (A_i, B_i) be a symplectic basis for V, and write $\omega = \sum_i \alpha^i \wedge \beta^i$ in terms of the dual coframe. Then $\omega^n = \sum_I \alpha^{i_1} \wedge \beta^{i_1} \wedge \cdots \wedge \alpha^{i_n} \wedge \beta^{i_n}$, where $I = (i_1, \ldots, i_n)$ ranges over all multi-indices of length n. Any term in this sum for which I has a repeated index is zero because $\alpha^i \wedge \alpha^i = 0$. The surviving terms are those for which I is a permutation of $(1, \ldots, n)$, and these terms are all equal to each other because 2-forms commute with each other under wedge product. Thus

$$\omega^n = n!\left(\alpha^1 \wedge \beta^1 \wedge \cdots \wedge \alpha^n \wedge \beta^n\right) \neq 0.$$

Conversely, suppose ω is degenerate. Then there is a nonzero vector $v \in V$ such that $v \lrcorner \omega = \hat{\omega}(v) = 0$. Since interior multiplication by v is an antiderivation, this implies $v \lrcorner (\omega^n) = n(v \lrcorner \omega) \wedge \omega^{n-1} = 0$. We can extend v to a basis $(E_1, E_2, \ldots, E_{2n})$ for V with $E_1 = v$, and then $\omega^n(E_1, \ldots, E_{2n}) = 0$, which implies $\omega^n = 0$. $\square$

Symplectic Structures on Manifolds

Now let us turn to a smooth manifold M. A **nondegenerate 2-form** on M is a 2-form ω such that ω_p is a nondegenerate 2-covector for each $p \in M$. A **symplectic** *form* on M is a closed nondegenerate 2-form. A smooth manifold endowed with a specific choice of symplectic form is called a *symplectic manifold*. A choice of symplectic form is also sometimes called a *symplectic structure*.

Proposition 22.7 implies that a symplectic manifold must be even-dimensional. However, not all even-dimensional smooth manifolds admit symplectic structures. For example, Proposition 22.8 shows that if ω is a symplectic form on a $2n$-manifold, then ω^n is a nonvanishing $2n$-form, so every symplectic manifold is orientable. In addition, a necessary homological condition is described in Problem 22-5. It implies, in particular, that $\mathbb{S}^2$ is the only sphere that admits a symplectic structure.

If (M_1, ω_1) and (M_2, ω_2) are symplectic manifolds, a diffeomorphism $F: M_1 \to M_2$ satisfying $F^*\omega_2 = \omega_1$ is called a *symplectomorphism*. The study of properties of symplectic manifolds that are invariant under symplectomorphisms is known as *symplectic geometry* or *symplectic topology*.

Example 22.9 (Symplectic Manifolds).

(a) With standard coordinates on $\mathbb{R}^{2n}$ denoted by $(x^1, \dots, x^n, y^1, \dots, x^n)$, the 2-form

$$\omega = \sum_{i=1}^{n} dx^i \wedge dy^i$$

is symplectic: it is obviously closed, and it is nondegenerate because its value at each point is the symplectic tensor of Example 22.2. This is called the *standard symplectic form on* $\mathbb{R}^{2n}$. (In formulas involving the standard symplectic form, like those involving the Euclidean inner product, it is usually necessary to insert explicit summation signs, because the summation index i appears twice in the upper position.)

(b) Suppose Σ is any orientable smooth 2-manifold and ω is a nonvanishing smooth 2-form on Σ. Then ω is closed because $d\omega$ is a 3-form on a 2-manifold. Moreover, as we observed just after Example 22.2, in two dimensions every nonvanishing 2-form is nondegenerate, so (Σ, ω) is a symplectic manifold.   //

Suppose (M, ω) is a symplectic manifold. An (immersed or embedded) submanifold $S \subseteq M$ is said to be a *symplectic, isotropic, coisotropic,* or *Lagrangian submanifold* if T_pS (thought of as a subspace of T_pM) has the corresponding property at each point $p \in S$. More generally, a smooth immersion (or embedding) $F: N \to M$ is said to have one of these properties if the subspace $dF_p(T_pN) \subseteq T_{F(p)}M$ has the corresponding property for every $p \in N$. Thus a submanifold is symplectic (isotropic, etc.) if and only if its inclusion map has the same property.

▶ **Exercise 22.10.** Suppose (M, ω) is a symplectic manifold and $F: N \to M$ is a smooth immersion. Show that F is isotropic if and only if $F^*\omega = 0$, and F is symplectic if and only if $F^*\omega$ is a symplectic form.

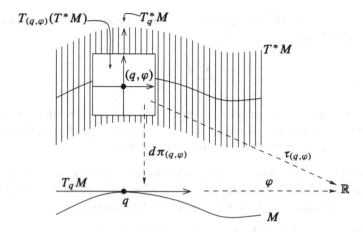

Fig. 22.1 The tautological 1-form on T^*M

The Canonical Symplectic Form on the Cotangent Bundle

The most important symplectic manifolds are total spaces of cotangent bundles, which carry canonical symplectic structures that we now define. First, there is a natural 1-form τ on the total space of T^*M, called the ***tautological 1-form***, defined as follows. A point in T^*M is a covector $\varphi \in T_q^*M$ for some $q \in M$; we denote such a point by the notation (q, φ). The natural projection $\pi: T^*M \to M$ is then just $\pi(q, \varphi) = q$, and its pointwise pullback at q is a linear map $d\pi_{(q,\varphi)}^*: T_q^*M \to T_{(q,\varphi)}^*(T^*M)$. We define $\tau \in \Omega^1(T^*M)$ (a 1-form on the total space of T^*M) by

$$\tau_{(q,\varphi)} = d\pi_{(q,\varphi)}^*\varphi. \tag{22.3}$$

(See Fig. 22.1.) In other words, the value of τ at $(q, \varphi) \in T^*M$ is the pullback with respect to π of the covector φ itself. If v is a tangent vector in $T_{(q,\varphi)}(T^*M)$, then

$$\tau_{(q,\varphi)}(v) = \varphi\big(d\pi_{(q,\varphi)}(v)\big).$$

Proposition 22.11. *Let M be a smooth manifold. The tautological 1-form τ is smooth, and $\omega = -d\tau$ is a symplectic form on the total space of T^*M.*

Proof. Let (x^i) be smooth coordinates on M, and let (x^i, ξ_i) denote the corresponding natural coordinates on T^*M as defined on p. 277. Recall that the coordinates of $(q, \varphi) \in T^*M$ are defined to be (x^i, ξ_i), where (x^i) is the coordinate representation of q, and $\xi_i \, dx^i$ is the coordinate representation of φ. In terms of these coordinates, the projection $\pi: T^*M \to M$ has the coordinate expression $\pi(x, \xi) = x$. This implies that $d\pi^*(dx^i) = dx^i$, so the coordinate expression for τ is

$$\tau_{(x,\xi)} = d\pi_{(x,\xi)}^*\left(\xi_i \, dx^i\right) = \xi_i \, dx^i. \tag{22.4}$$

It follows immediately that τ is smooth, because its component functions in these coordinates are linear.

Let $\omega = -d\tau \in \Omega^2(T^*M)$. Clearly, ω is closed, because it is exact. Moreover, in natural coordinates, (22.4) yields

$$\omega = \sum_i dx^i \wedge d\xi_i.$$

Under the identification of an open subset of T^*M with an open subset of $\mathbb{R}^{2n}$ by means of these coordinates, ω corresponds to the standard symplectic form on $\mathbb{R}^{2n}$ (with ξ_i substituted for y^i). It follows that ω is symplectic. $\qquad\square$

The symplectic form defined in this proposition is called the **canonical symplectic form on T^*M**. One of its many uses is in giving the following somewhat more "geometric" interpretation of what it means for a 1-form to be closed.

Proposition 22.12. *Let M be a smooth manifold, and let σ be a smooth 1-form on M. Thought of as a smooth map from M to T^*M, σ is a smooth embedding, and σ is closed if and only if its image $\sigma(M)$ is a Lagrangian submanifold of T^*M.*

Proof. Throughout this proof we need to remember that $\sigma\colon M \to T^*M$ is playing two roles: on the one hand, it is a 1-form on M, and on the other hand, it is a smooth map between manifolds. Since they are literally the same map, we do not use different notations to distinguish between them; but you should be careful to think about which role σ is playing at each step of the argument.

In terms of smooth local coordinates (x^i) for M and corresponding natural coordinates (x^i, ξ_i) for T^*M, the map $\sigma\colon M \to T^*M$ has the coordinate representation

$$\sigma\left(x^1, \ldots, x^n\right) = \left(x^1, \ldots, x^n, \sigma_1(x), \ldots, \sigma_n(x)\right),$$

where $\sigma_i\, dx^i$ is the coordinate representation of σ as a 1-form. It follows immediately that σ is a smooth immersion, and it is injective because $\pi \circ \sigma = \mathrm{Id}_M$. To show that it is an embedding, it suffices by Proposition 4.22 to show that it is a proper map. This in turn follows from the fact that π is a left inverse for σ, by Proposition A.53.

Because $\sigma(M)$ is n-dimensional, it is Lagrangian if and only if it is isotropic, which is the case if and only if $\sigma^*\omega = 0$. The pullback of the tautological form τ under σ is

$$\sigma^*\tau = \sigma^*\left(\xi_i\, dx^i\right) = \sigma_i\, dx^i = \sigma.$$

This can also be seen somewhat more invariantly from the computation

$$(\sigma^*\tau)_p(v) = \tau_{\sigma(p)}\left(d\sigma_p(v)\right) = \sigma_p\left(d\pi_{\sigma(p)} \circ d\sigma_p(v)\right) = \sigma_p(v),$$

which follows from the definition of τ and the fact that $\pi \circ \sigma = \mathrm{Id}_M$. Therefore,

$$\sigma^*\omega = -\sigma^*d\tau = -d(\sigma^*\tau) = -d\sigma.$$

It follows that σ is a Lagrangian embedding if and only if $d\sigma = 0$. $\qquad\square$

The Darboux Theorem

Our next theorem is one of the most fundamental results in symplectic geometry. It is a nonlinear analogue of the canonical form for a symplectic tensor given in Proposition 22.7. It illustrates the most dramatic difference between symplectic structures and Riemannian metrics: unlike the Riemannian case, there is no local obstruction to a symplectic structure being locally equivalent to the standard flat model.

Theorem 22.13 (Darboux). *Let (M, ω) be a $2n$-dimensional symplectic manifold. For any $p \in M$, there are smooth coordinates $(x^1, \ldots, x^n, y^1, \ldots, y^n)$ centered at p in which ω has the coordinate representation*

$$\omega = \sum_{i=1}^{n} dx^i \wedge dy^i. \tag{22.5}$$

We will prove the theorem below. Any coordinates satisfying (22.5) theorem are called **Darboux coordinates**, **symplectic coordinates**, or **canonical coordinates**. Obviously, the standard coordinates $(x^1, \ldots, x^n, y^1, \ldots, y^n)$ on $\mathbb{R}^{2n}$ are Darboux coordinates. The proof of Proposition 22.11 showed that the natural coordinates (x^i, ξ_i) are Darboux coordinates for T^*M with its canonical symplectic structure.

The Darboux theorem was first proved (in a slightly different form) by Gaston Darboux in 1882, in connection with his work on ordinary differential equations arising in classical mechanics. The proof we give was discovered in 1971 by Alan Weinstein [Wei71], based on a technique due to Jürgen Moser [Mos65]. A more elementary—but less elegant—proof is outlined in Problem 22-19.

Weinstein's proof of the Darboux theorem is based on the theory of time-dependent flows (see Theorem 9.48). Before we carry out the proof, we need some preliminary results regarding such flows.

First, recall that Proposition 12.36 shows how to use Lie derivatives to compute the derivative of a tensor field under a flow. We need the following generalization of that result to the case of time-dependent flows.

Proposition 22.14. *Let M be a smooth manifold. Suppose $V : J \times M \to TM$ is a smooth time-dependent vector field and $\psi : \mathcal{E} \to M$ is its time-dependent flow. For any smooth covariant tensor field $A \in \mathcal{T}^k(M)$ and any $(t_1, t_0, p) \in \mathcal{E}$,*

$$\left. \frac{d}{dt} \right|_{t=t_1} \left(\psi_{t,t_0}^* A \right)_p = \left(\psi_{t_1,t_0}^* \left(\mathcal{L}_{V_{t_1}} A \right) \right)_p. \tag{22.6}$$

Proof. First, assume $t_1 = t_0$. In this case, ψ_{t_0,t_0} is the identity map of M, so we need to prove

$$\left. \frac{d}{dt} \right|_{t=t_0} \left(\psi_{t,t_0}^* A \right)_p = \left(\mathcal{L}_{V_{t_0}} A \right)_p. \tag{22.7}$$

We begin with the special case in which $A = f$ is a smooth 0-tensor field:

$$\left. \frac{d}{dt} \right|_{t=t_0} \left(\psi_{t,t_0}^* f \right)(p) = \left. \frac{\partial}{\partial t} \right|_{t=t_0} f\left(\psi(t, t_0, p) \right) = V\left(t_0, \psi(t_0, t_0, p) \right) f$$

$$= (\mathscr{L}_{V_{t_0}} f)(p).$$

Next consider an exact 1-form $A = df$. In any smooth local coordinates (x^i), the function $\psi_{t,t_0}^* f(x) = f(\psi(t, t_0, x))$ depends smoothly on all $n + 1$ variables $(t, x^1, \ldots, x^n)$. Thus, the operator d/dt (which is more properly written as $\partial/\partial t$ in this situation) commutes with each of the partial derivatives $\partial/\partial x^i$ when applied to $\psi_{t,t_0}^* f$. In particular, this means that the exterior derivative operator d commutes with $\partial/\partial t$, and so

$$\frac{d}{dt}\bigg|_{t=t_0} (\psi_{t,t_0}^* df)_p = \frac{\partial}{\partial t}\bigg|_{t=t_0} d\,(\psi_{t,t_0}^* f)_p = d\left(\frac{\partial}{\partial t}\bigg|_{t=t_0} (\psi_{t,t_0}^* f)\right)_p$$

$$= d(\mathscr{L}_{V_{t_0}} f)_p = (\mathscr{L}_{V_{t_0}} df)_p.$$

Thus, the result is proved for 0-tensors and for exact 1-forms.

Now suppose that $A = B \otimes C$ for some smooth covariant tensor fields B and C, and assume that the proposition is true for B and C. (We include the possibility that B or C has rank 0, in which case the tensor product is just ordinary multiplication.) By the product rule for Lie derivatives (Proposition 12.32(c)), the right-hand side of (22.7) satisfies

$$(\mathscr{L}_{V_{t_0}}(B \otimes C))_p = (\mathscr{L}_{V_{t_0}} B)_p \otimes C_p + B_p \otimes (\mathscr{L}_{V_{t_0}} C)_p.$$

On the other hand, by an argument entirely analogous to that in the proof of Proposition 12.32, the left-hand side satisfies a similar product rule:

$$\frac{d}{dt}\bigg|_{t=t_0} (\psi_{t,t_0}^* (B \otimes C))_p = \left(\frac{d}{dt}\bigg|_{t=t_0} (\psi_{t,t_0}^* B)_p\right) \otimes C_p$$

$$+ B_p \otimes \left(\frac{d}{dt}\bigg|_{t=t_0} (\psi_{t,t_0}^* C)_p\right).$$

This shows that (22.7) holds for $A = B \otimes C$, provided it holds for B and C. The case of arbitrary tensor fields now follows by induction, using the fact that any smooth covariant tensor field can be written locally as a sum of tensor fields of the form $A = f\, dx^{i_1} \otimes \cdots \otimes dx^{i_k}$.

To handle arbitrary t_1, we use Theorem 9.48(d), which shows that $\psi_{t,t_0} = \psi_{t,t_1} \circ \psi_{t_1,t_0}$ wherever the right-hand side is defined. Therefore, because the linear map $d(\psi_{t_1,t_0})_p^* : T^k(T_{\psi_{t_1,t_0}(p)}^* M) \to T^k(T_p^* M)$ does not depend on t,

$$\frac{d}{dt}\bigg|_{t=t_1} (\psi_{t,t_0}^* A)_p = \frac{d}{dt}\bigg|_{t=t_1} d(\psi_{t_1,t_0})_p^* \circ d(\psi_{t,t_1})_{\psi_{t_1,t_0}(p)}^* (A_{\psi_{t,t_0}(p)})$$

$$= d(\psi_{t_1,t_0})_p^* \frac{d}{dt}\bigg|_{t=t_1} d(\psi_{t,t_1})_{\psi_{t_1,t_0}(p)}^* (A_{\psi_{t,t_1} \circ \psi_{t_1,t_0}(p)})$$

$$= (\psi_{t_1,t_0}^* (\mathscr{L}_{V_{t_1}} A))_p.$$

$\square$

A *smooth time-dependent tensor field* on a smooth manifold M is a smooth map $A\colon J \times M \to T^k T^* M$, where $J \subseteq \mathbb{R}$ is an interval, satisfying $A(t, p) \in T^k(T_p^* M)$ for each $(t, p) \in J \times M$.

Proposition 22.15. *Let M be a smooth manifold and $J \subseteq \mathbb{R}$ be an open interval. Suppose $V\colon J \times M$ is a smooth time-dependent vector field on M, $\psi\colon \mathcal{E} \to M$ is its time-dependent flow, and $A\colon J \times M \to T^k T^* M$ is a smooth time-dependent tensor field on M. Then for any $(t_1, t_0, p) \in \mathcal{E}$,*

$$\frac{d}{dt}\bigg|_{t=t_1} \left(\theta_{t,t_0}^* A_t\right)_p = \left(\theta_{t_1,t_0}^* \left(\mathcal{L}_{V_{t_1}} A_{t_1} + \frac{d}{dt}\bigg|_{t=t_1} A_t\right)\right)_p. \tag{22.8}$$

Proof. For sufficiently small $\varepsilon > 0$, consider the smooth map $F\colon (t_1 - \varepsilon, t_1 + \varepsilon) \times (t_1 - \varepsilon, t_1 + \varepsilon) \to T^k(T_p^* M)$ defined by

$$F(u, v) = \left(\theta_{u,t_0}^* A_v\right)_p = d\left(\theta_{u,t_0}\right)_p^* \left(A_v\big|_{\theta_{u,t_0}(p)}\right).$$

Since F takes its values in the finite-dimensional vector space $T^k(T_p^* M)$, we can apply the chain rule together with Proposition 22.14 to conclude that

$$\frac{d}{dt}\bigg|_{t=t_1} F(t, t) = \frac{\partial F}{\partial u}(t_1, t_1) + \frac{\partial F}{\partial v}(t_1, t_1)$$

$$= \left(\theta_{t_1,t_0}^* \left(\mathcal{L}_{V_{t_1}} A_{t_1}\right)\right)_p + \frac{\partial}{\partial v}\bigg|_{v=t_1} d\left(\theta_{t_1,t_0}\right)_p^* \left(A_v\big|_{\theta_{t_1,t_0}(p)}\right).$$

Just as in the proof of Proposition 22.14, the linear map $d\left(\theta_{t_1,t_0}\right)_p^*$ commutes past $\partial/\partial v$, yielding (22.8). $\square$

Proof of the Darboux theorem. Let ω_0 denote the given symplectic form on M, and let $p_0 \in M$ be arbitrary. The theorem will be proved if we can find a smooth coordinate chart (U_0, φ) centered at p_0 such that $\varphi^* \omega_1 = \omega_0$, where $\omega_1 = \sum_{i=1}^n dx^i \wedge dy^i$ is the standard symplectic form on $\mathbb{R}^{2n}$. Since this is a local question, by choosing smooth coordinates $\left(x^1, \ldots, x^n, y^1, \ldots, y^n\right)$ in a neighborhood of p_0, we may replace M with an open ball $U \subseteq \mathbb{R}^{2n}$. Proposition 22.7 shows that we can arrange by a linear change of coordinates that $\omega_0|_{p_0} = \omega_1|_{p_0}$.

Let $\eta = \omega_1 - \omega_0$. Because η is closed, the Poincaré lemma (Theorem 17.14) shows that we can find a smooth 1-form α on U such that $d\alpha = -\eta$. By subtracting a constant-coefficient (and thus closed) 1-form from α, we may assume without loss of generality that $\alpha_{p_0} = 0$.

For each $t \in \mathbb{R}$, define a closed 2-form ω_t on U by

$$\omega_t = \omega_0 + t\eta = (1 - t)\omega_0 + t\omega_1.$$

Let J be a bounded open interval containing $[0, 1]$. Because $\omega_t|_{p_0} = \omega_0|_{p_0}$ is non-degenerate for all t, a simple compactness argument shows that there is some neighborhood $U_1 \subseteq U$ of p_0 such that ω_t is nondegenerate on U_1 for all $t \in \bar{J}$. Because of this nondegeneracy, the smooth bundle homomorphism $\hat{\omega}_t : TU_1 \to T^*U_1$ defined by $\hat{\omega}_t(X) = X \lrcorner \omega_t$ is an isomorphism for each $t \in \bar{J}$.

Define a smooth time-dependent vector field $V : J \times U_1 \to TU_1$ by $V_t = \hat{\omega}_t^{-1}\alpha$, or equivalently

$$V_t \lrcorner \omega_t = \alpha.$$

Our assumption that $\alpha_{p_0} = 0$ implies that $V_t|_{p_0} = 0$ for each t. If $\theta : \mathcal{E} \to U_1$ denotes the time-dependent flow of V, it follows that $\theta(t, 0, p_0) = p_0$ for all $t \in J$, so $J \times \{0\} \times \{p_0\} \subseteq \mathcal{E}$. Because $\mathcal{E}$ is open in $J \times J \times M$ and $[0, 1]$ is compact, there is some neighborhood U_0 of p_0 such that $[0, 1] \times \{0\} \times U_0 \subseteq \mathcal{E}$.

For each $t_1 \in [0, 1]$, it follows from Proposition 22.15 that

$$\left. \frac{d}{dt} \right|_{t=t_1} (\theta_{t,0}^*\omega_t) = \theta_{t_1,0}^* \left(\mathcal{L}_{V_{t_1}}\omega_{t_1} + \left. \frac{d}{dt} \right|_{t=t_1} \omega_t \right)$$

$$= \theta_{t_1,0}^* \left(V_{t_1} \lrcorner d\omega_{t_1} + d(V_{t_1} \lrcorner \omega_{t_1}) + \eta \right)$$

$$= \theta_{t_1,0}^* (0 + d\alpha + \eta) = 0.$$

Therefore, $\theta_{t,0}^*\omega_t = \theta_{0,0}^*\omega_0 = \omega_0$ for all t. In particular, $\theta_{1,0}^*\omega_1 = \omega_0$. It follows from Theorem 9.48(c) that $\theta_{1,0}$ is a diffeomorphism onto its image, so it is a coordinate map. Because $\theta_{1,0}(p_0) = p_0 = 0$, these coordinate are centered at p_0. $\square$

Hamiltonian Vector Fields

One of the most useful constructions on symplectic manifolds is a symplectic analogue of the gradient, defined as follows. Suppose (M, ω) is a symplectic manifold. For any smooth function $f \in C^\infty(M)$, we define the **Hamiltonian vector field of f** to be the smooth vector field X_f defined by

$$X_f = \hat{\omega}^{-1}(df),$$

where $\hat{\omega} : TM \to T^*M$ is the bundle isomorphism determined by ω. Equivalently,

$$X_f \lrcorner \omega = df,$$

or for any vector field Y,

$$\omega(X_f, Y) = df(Y) = Yf.$$

In any Darboux coordinates, X_f can be computed explicitly as follows. Writing

$$X_f = \sum_{i=1}^n \left(a^i \frac{\partial}{\partial x^i} + b^i \frac{\partial}{\partial y^i} \right)$$

for some coefficient functions (a^i, b^i) to be determined, we compute

$$X_f \lrcorner \omega = \sum_{j=1}^n \left(a^j \frac{\partial}{\partial x^j} + b^j \frac{\partial}{\partial y^j} \right) \lrcorner \sum_{i=1}^n dx^i \wedge dy^i = \sum_{i=1}^n \left(a^i \, dy^i - b^i \, dx^i \right).$$

On the other hand,

$$df = \sum_{i=1}^{n} \left(\frac{\partial f}{\partial x^i} dx^i + \frac{\partial f}{\partial y^i} dy^i \right).$$

Setting these two expressions equal to each other, we find that $a^i = \partial f / \partial y^i$ and $b^i = -\partial f / \partial x^i$, which yields the following formula for the Hamiltonian vector field of f in Darboux coordinates:

$$X_f = \sum_{i=1}^{n} \left(\frac{\partial f}{\partial y^i} \frac{\partial}{\partial x^i} - \frac{\partial f}{\partial x^i} \frac{\partial}{\partial y^i} \right). \tag{22.9}$$

This formula holds, in particular, on $\mathbb{R}^{2n}$ with its standard symplectic form.

Although the definition of the Hamiltonian vector field is formally analogous to that of the gradient on a Riemannian manifold, Hamiltonian vector fields differ from gradients in some very significant ways, as the next lemma shows.

Proposition 22.16 (Properties of Hamiltonian Vector Fields). *Let (M, ω) be a symplectic manifold and let $f \in C^\infty(M)$.*

(a) *f is constant along each integral curve of X_f.*
(b) *At each regular point of f, the Hamiltonian vector field X_f is tangent to the level set of f.*

Proof. Both assertions follow from the fact that

$$X_f f = df(X_f) = \omega(X_f, X_f) = 0$$

because ω is alternating. $\qquad\square$

A smooth vector field X on M is said to be ***symplectic*** if ω is invariant under the flow of X. It is said to be ***Hamiltonian*** (or ***globally Hamiltonian***) if there exists a function $f \in C^\infty(M)$ such that $X = X_f$, and ***locally Hamiltonian*** if each point p has a neighborhood on which X is Hamiltonian. Clearly, every globally Hamiltonian vector field is locally Hamiltonian.

Proposition 22.17 (Hamiltonian and Symplectic Vector Fields). *Let (M, ω) be a symplectic manifold. A smooth vector field on M is symplectic if and only if it is locally Hamiltonian. Every locally Hamiltonian vector field on M is globally Hamiltonian if and only if $H^1_{dR}(M) = 0$.*

Proof. By Theorem 12.37, a smooth vector field X is symplectic if and only if $\mathcal{L}_X \omega = 0$. Using Cartan's magic formula, we compute

$$\mathcal{L}_X \omega = d(X \lrcorner \omega) + X \lrcorner (d\omega) = d(X \lrcorner \omega). \tag{22.10}$$

Therefore, X is symplectic if and only if the 1-form $X \lrcorner \omega$ is closed. On the one hand, if X is locally Hamiltonian, then in a neighborhood of each point there is a real-valued function f such that $X = X_f$, so $X \lrcorner \omega = X_f \lrcorner \omega = df$, which is certainly closed. Conversely, if X is symplectic, then by the Poincaré lemma each point $p \in M$ has a neighborhood U on which the closed 1-form $X \lrcorner \omega$ is exact. This means there is a smooth real-valued function f defined on U such that $X \lrcorner \omega = df$; because ω is nondegenerate, this implies that $X = X_f$ on U.

Now suppose M is a smooth manifold with $H^1_{\mathrm{dR}}(M) = 0$. If X is a locally Hamiltonian vector field, then it is symplectic, so (22.10) shows that $X \lrcorner \omega$ is closed. The hypothesis then implies that there is a function $f \in C^\infty(M)$ such that $X \lrcorner \omega = df$. This means that $X = X_f$, so X is globally Hamiltonian. Conversely, suppose every locally Hamiltonian vector field is globally Hamiltonian. Let η be a closed 1-form, and let X be the vector field $X = \hat{\omega}^{-1}\eta$. Then (22.10) shows that $\mathcal{L}_X \omega = d\eta = 0$, so X is symplectic and therefore locally Hamiltonian. By hypothesis, there is a global smooth real-valued function f such that $X = X_f$, and then unwinding the definitions, we find that $\eta = df$. $\square$

A symplectic manifold (M, ω) together with a smooth function $H \in C^\infty(M)$ is called a **Hamiltonian system**. The function H is called the **Hamiltonian** of the system; the flow of the Hamiltonian vector field X_H is called its **Hamiltonian flow**, and the integral curves of X_H are called the **trajectories** or **orbits** of the system. In Darboux coordinates, formula (22.9) implies that the orbits are those curves $\gamma(t) = (x^i(t), y^i(t))$ that satisfy

$$\dot{x}^i(t) = \frac{\partial H}{\partial y^i}(x(t), y(t)),$$

$$\dot{y}^i(t) = -\frac{\partial H}{\partial x^i}(x(t), y(t))$$

$$\text{(22.11)}$$

(with dots denoting ordinary derivatives of component functions with respect to t). These are called **Hamilton's equations**.

Hamiltonian systems play a central role in classical mechanics. We illustrate how they arise with a simple example.

Example 22.18 (The n-Body Problem). Consider n physical particles moving in space, and suppose their masses are $m_1, \ldots, m_n$. For many purposes, an effective model of such a system is obtained by idealizing the particles as points in $\mathbb{R}^3$, which we denote by $\mathbf{q}_1, \ldots, \mathbf{q}_n$. Writing the coordinates of $\mathbf{q}_k$ at time t as $(q_k^1(t), q_k^2(t), q_k^3(t))$, we can represent the evolution of the system over time by a curve in $\mathbb{R}^{3n}$:

$$q(t) = (q_1^1(t), q_1^2(t), q_1^3(t), \ldots, q_n^1(t), q_n^2(t), q_n^3(t)).$$

The **collision set** is the subset $\mathcal{C} \subseteq \mathbb{R}^{3n}$ where two or more particles occupy the same position in space:

$$\mathcal{C} = \{q \in \mathbb{R}^{3n} : \mathbf{q}_k = \mathbf{q}_l \text{ for some } k \neq l\}.$$

We consider only motions with no collisions, so we are interested in curves in the open subset $Q = \mathbb{R}^{3n} \smallsetminus \mathcal{C}$.

Suppose the particles are acted upon by forces that depend only on the positions of all the particles in the system. (A typical example is gravitational forces.)

If we denote the components of the net force on the kth particle by $\mathbf{F}_k(q) = \left(F_k^1(q), F_k^2(q), F_k^3(q)\right)$, then Newton's second law of motion asserts that the particles' motion satisfies $m_k \ddot{\mathbf{q}}_k(t) = \mathbf{F}_k(\mathbf{q}(t))$ for each k, which translates into the $3n \times 3n$ system of second-order ODEs

$$m_k \ddot{q}_k^1(t) = F_k^1(q(t)),$$
$$m_k \ddot{q}_k^2(t) = F_k^2(q(t)),$$
$$m_k \ddot{q}_k^3(t) = F_k^3(q(t)), \quad k = 1, \dots, n.$$

(There is no implied summation in these equations.)

This can be written in a more compact form if we relabel the $3n$ position coordinates as $q(t) = \left(q^1(t), \dots, q^{3n}(t)\right)$ and the $3n$ components of the forces as $F(q) = \left(F_1(q), \dots, F_{3n}(q)\right)$, and let $M = (M_{ij})$ denote the $3n \times 3n$ diagonal matrix whose diagonal entries are $(m_1, m_1, m_1, m_2, m_2, m_2, \dots, m_n, m_n, m_n)$. Then Newton's second law can be written

$$M_{ij} \ddot{q}^j(t) = F_i(q(t)). \tag{22.12}$$

(Here the summation convention is in force.) We assume that the forces depend smoothly on q, so we can interpret $F(q) = (F_1(q), \dots, F_{3n}(q))$ as the components of a smooth covector field on Q. We assume further that the forces are conservative, which by the results of Chapter 11 is equivalent to the existence of a smooth function $V \in C^\infty(Q)$ (called the *potential energy* of the system) such that $F = -dV$.

Under the physically reasonable assumption that all of the masses are positive, the matrix M is positive definite, and thus can be interpreted as a (constant-coefficient) Riemannian metric on Q. It therefore defines a smooth bundle isomorphism $\widehat{M} : TQ \to T^*Q$. If we denote the natural coordinates on TQ by $\left(q^i, v^i\right)$ and those on T^*Q by $\left(q^i, p_i\right)$, then $M(v, w) = M_{ij} v^i w^j$, and $\widehat{M}$ has the coordinate representation

$$\left(q^i, p_i\right) = \widehat{M}\left(q^i, v^i\right) = \left(q^i, M_{ij} v^j\right).$$

If $q'(t) = \left(\dot{q}^1(t), \dots, \dot{q}^{3n}(t)\right)$ is the velocity vector of the system of particles at time t, then the covector $p(t) = \widehat{M}\left(q'(t)\right)$ is given by the formula

$$p_i(t) = M_{ij} \dot{q}^j(t). \tag{22.13}$$

To give this equation a physical interpretation, we can revert to our original labeling of the coordinates and write

$$p(t) = \left(p_1^1, p_1^2, p_1^3, \dots, p_n^1, p_n^2, p_n^3\right),$$

and then $\mathbf{p}_k(t) = \left(p_k^1(t), p_k^2(t), p_k^3(t)\right) = m_k \dot{\mathbf{q}}_k(t)$ is interpreted as the *momentum* of the kth particle at time t.

Using (22.13), we see that a curve $q(t)$ in Q satisfies Newton's second law (22.12) if and only if the curve $\gamma(t) = \left(q(t), p(t)\right)$ in T^*Q satisfies the first-order system of ODEs

$$\dot{q}^i(t) = M^{ij} p_j(t),$$

$$\dot{p}_i(t) = -\frac{\partial V}{\partial q^i}(q(t)), \tag{22.14}$$

where (M^{ij}) is the inverse of the matrix of (M_{ij}). Define a function $E \in C^\infty(T^*Q)$, called the **total energy** of the system, by

$$E(q, p) = V(q) + K(p),$$

where V is the potential energy introduced above, and K is the **kinetic energy**, defined by

$$K(p) = \tfrac{1}{2} M^{ij} p_i p_j.$$

Since (q^i, p_i) are Darboux coordinates on T^*Q, a comparison of (22.14) with (22.11) shows that (22.14) is precisely Hamilton's equations for the Hamiltonian flow of E. The fact that E is constant along the trajectories of its own Hamiltonian flow is known as the **law of conservation of energy**.   //

An elaboration of the same technique can be applied to virtually any classical dynamical system in which the forces are conservative. For example, if the positions of a system of particles are subject to constraints, as are the constituent particles of a rigid body, for example, then the configuration space is typically a submanifold of $\mathbb{R}^{3n}$ rather than an open subset. Under very general hypotheses, the equations of motion of such a system can be formulated as a Hamiltonian system on the cotangent bundle of the constraint manifold. For much more on Hamiltonian systems, see [AM78].

Poisson Brackets

Hamiltonian vector fields allow us to define an operation on real-valued functions on a symplectic manifold M similar to the Lie bracket of vector fields. Given $f, g \in C^\infty(M)$, we define their **Poisson bracket** $\{f, g\} \in C^\infty(M)$ by any of the following equivalent formulas:

$$\{f, g\} = \omega(X_f, X_g) = df(X_g) = X_g f. \tag{22.15}$$

Two functions are said to **Poisson commute** if their Poisson bracket is zero.

The geometric interpretation of the Poisson bracket is evident from the characterization $\{f, g\} = X_g f$: it is a measure of the rate of change of f along the Hamiltonian flow of g. In particular, f and g Poisson commute if and only if f is constant along the Hamiltonian flow of g.

Using (22.9), we can readily compute the Poisson bracket of two functions f, g in Darboux coordinates:

$$\{f, g\} = \sum_{i=1}^n \frac{\partial f}{\partial x^i} \frac{\partial g}{\partial y^i} - \frac{\partial f}{\partial y^i} \frac{\partial g}{\partial x^i}. \tag{22.16}$$

Proposition 22.19 (Properties of the Poisson Bracket). *Suppose (M, ω) is a symplectic manifold, and $f, g, h \in C^\infty(M)$.*

(a) BILINEARITY: $\{f, g\}$ *is linear over $\mathbb{R}$ in f and in g.*
(b) ANTISYMMETRY: $\{f, g\} = -\{g, f\}$.
(c) JACOBI IDENTITY: $\{\{f, g\}, h\} + \{\{g, h\}, f\} + \{\{h, f\}, g\} = 0$.
(d) $X_{\{f,g\}} = -[X_f, X_g]$.

Proof. Parts (a) and (b) are obvious from the characterization $\{f, g\} = \omega\left(X_f, X_g\right)$ together with the fact that $X_f = \hat{\omega}^{-1}(df)$ depends linearly on f. Because of the nondegeneracy of ω, to prove (d), it suffices to show that the following holds for every vector field Y:

$$\omega\left(X_{\{f,g\}}, Y\right) + \omega\left([X_f, X_g], Y\right) = 0. \tag{22.17}$$

On the one hand, note that $\omega\left(X_{\{f,g\}}, Y\right) = d(\{f, g\})(Y) = Y\{f, g\} = YX_g\, f$. On the other hand, because Hamiltonian vector fields are symplectic, the Lie derivative formula of Corollary 12.33 yields

$$\begin{aligned}
0 &= \left(\mathcal{L}_{X_g}\omega\right)(X_f, Y) \\
&= X_g\left(\omega(X_f, Y)\right) - \omega\left([X_g, X_f], Y\right) - \omega\left(X_f, [X_g, Y]\right). \tag{22.18}
\end{aligned}$$

The first and third terms on the right-hand side can be simplified as follows:

$$\begin{aligned}
X_g\left(\omega(X_f, Y)\right) &= X_g\left(df(Y)\right) = X_g Y f. \\
\omega\left(X_f, [X_g, Y]\right) &= df\left([X_g, Y]\right) = [X_g, Y]f = X_g Y f - Y X_g f \\
&= X_g Y f - \omega\left(X_{\{f,g\}}, Y\right).
\end{aligned}$$

Inserting these into (22.18), we obtain (22.17).

Finally, (c) follows from (d), (b), and (22.15):

$$\begin{aligned}
\{f, \{g, h\}\} &= X_{\{g,h\}}\, f = -[X_g, X_h]f = -X_g X_h f + X_h X_g f \\
&= -X_g\{f, h\} + X_h\{f, g\} = -\{\{f, h\}, g\} + \{\{f, g\}, h\} \\
&= -\{g, \{h, f\}\} - \{h, \{f, g\}\}. \qquad \square
\end{aligned}$$

The following corollary is immediate.

Corollary 22.20. *If (M, ω) is a symplectic manifold, the vector space $C^\infty(M)$ is a Lie algebra under the Poisson bracket.* $\square$

If (M, ω, H) is a Hamiltonian system, any function $f \in C^\infty(M)$ that is constant on every integral curve of X_H is called a **conserved quantity** of the system. Conserved quantities turn out to be deeply related to symmetries, as we now show.

A smooth vector field V on M is called an **infinitesimal symmetry** of (M, ω, H) if both ω and H are invariant under the flow of V.

Proposition 22.21. *Let* (M, ω, H) *be a Hamiltonian system.*

(a) *A function* $f \in C^\infty(M)$ *is a conserved quantity if and only if* $\{f, H\} = 0$.
(b) *The infinitesimal symmetries of* (M, ω, H) *are precisely the symplectic vector fields* V *that satisfy* $VH = 0$.
(c) *If* θ *is the flow of an infinitesimal symmetry and* γ *is a trajectory of the system, then for each* $s \in \mathbb{R}$, $\theta_s \circ \gamma$ *is also a trajectory on its domain of definition.*

Proof. Problem 22-18.   □

The following theorem, first proved (in a somewhat different form) by Emmy Noether in 1918 [Noe71], has had a profound influence on both physics and mathematics. It shows that for many Hamiltonian systems, there is a one-to-one correspondence between conserved quantities (modulo constants) and infinitesimal symmetries.

Theorem 22.22 (Noether's Theorem). *Let* (M, ω, H) *be a Hamiltonian system. If* f *is any conserved quantity, then its Hamiltonian vector field is an infinitesimal symmetry. Conversely, if* $H^1_{\mathrm{dR}}(M) = 0$, *then each infinitesimal symmetry is the Hamiltonian vector field of a conserved quantity, which is unique up to addition of a function that is constant on each component of* M.

Proof. Suppose f is a conserved quantity. Proposition 22.21 shows that $\{f, H\} = 0$. This in turn implies that $X_f H = \{H, f\} = 0$, so H is constant along the flow of X_f. Since ω is invariant along the flow of any Hamiltonian vector field by Proposition 22.17, this shows that X_f is an infinitesimal symmetry.

Now suppose that M is a smooth manifold with $H^1_{\mathrm{dR}}(M) = 0$. Let V be an infinitesimal symmetry of (M, ω, H). Then V is symplectic by definition, and globally Hamiltonian by Proposition 22.17. Writing $V = X_f$, the fact that H is constant along the flow of V implies that $\{H, f\} = X_f H = VH = 0$, so f is a conserved quantity. If $\widetilde{f}$ is any other function that satisfies $X_{\widetilde{f}} = V = X_f$, then

$$d(\widetilde{f} - f) = (X_{\widetilde{f}} - X_f) \lrcorner\, \omega = 0,$$ so $\widetilde{f} - f$ must be constant on each component of M.   □

There is one conserved quantity that every Hamiltonian system possesses: the Hamiltonian H itself. The infinitesimal symmetry corresponding to it, of course, generates the Hamiltonian flow of the system, which describes how the system evolves over time. Since H is typically interpreted as the total energy of the system (as in Example 22.18), one usually says that the symmetry corresponding to conservation of energy is "translation in the time variable."

Hamiltonian Flowouts

Hamiltonian vector fields are powerful tools for constructing isotropic and Lagrangian submanifolds. Because Lagrangian submanifolds of T^*M correspond to closed 1-forms (Proposition 22.12), which in turn correspond locally to differentials of functions, such constructions have numerous applications in PDE theory. We will see one such application later in this chapter.

Theorem 22.23 (Hamiltonian Flowout Theorem). *Suppose (M, ω) is a symplectic manifold, $H \in C^\infty(M)$, Γ is an embedded isotropic submanifold of M that is contained in a single level set of H, and the Hamiltonian vector field X_H is nowhere tangent to Γ. If S is a flowout from Γ along X_H, then S is also isotropic and contained in the same level set of H.*

Proof. Let θ be the flow of X_H. Recall from Theorem 9.20 that the flowout is parametrized by the restriction of θ to a neighborhood $\mathcal{O}_\delta$ of $\{0\} \times \Gamma$ in $\mathbb{R} \times \Gamma$. First consider a point $p \in \Gamma \subseteq S$. If we choose a basis $E_1, \ldots, E_k$ for $T_p\Gamma$, then T_pS is spanned by $(X_H|_p, E_1, \ldots, E_k)$. The assumption that Γ is isotropic implies that $\omega_p(E_i, E_j) = 0$ for all i and j. On the other hand, by definition of the Hamiltonian vector field,

$$\omega_p(X_H|_p, E_j) = dH_p(E_j) = 0,$$

because E_j is tangent to Γ, which is contained in a level set of H. This shows that the restriction of ω to T_pS is zero when $p \in \Gamma$.

Any other point $p' \in S$ is of the form $p' = \theta_t(p)$ for some $(t, p) \in \mathcal{O}_\delta \subseteq \mathbb{R} \times \Gamma$. Because θ_t is a local diffeomorphism that maps a neighborhood of p in S to a neighborhood of p' in S, its differential takes T_pS isomorphically onto $T_{p'}S$. Thus, for any vectors $v, w \in T_{p'}S$, there are vectors $\hat{v}, \hat{w} \in T_pS$ such that $d(\theta_t)_p(\hat{v}) = v$ and $d(\theta_t)_p(\hat{w}) = w$. Moreover, because X_H is a symplectic vector field, its flow preserves ω. Therefore,

$$\omega_{p'}(v, w) = \omega_{p'}\big(d(\theta_t)_p(\hat{v}), d(\theta_t)_p(\hat{w})\big) = (\theta_t^* \omega)_p(\hat{v}, \hat{w}) = \omega_p(\hat{v}, \hat{w}) = 0.$$

It follows that S is isotropic. By Proposition 22.16, H is constant along each integral curve of X_H, so S is contained in the same level set of H as Γ. $\square$

Contact Structures

As we have seen, symplectic manifolds must be even-dimensional; but there is a closely related structure called a *contact structure* that one can define on odd-dimensional manifolds. It also has important applications in geometry and analysis. In this section, we introduce the main elements of contact geometry.

Suppose M is a smooth manifold of odd dimension $2n + 1$. A **contact form** on M is a nonvanishing smooth 1-form θ with the property that for each $p \in M$, the restriction of $d\theta_p$ to the subspace $\text{Ker } \theta_p \subseteq T_pM$ is nondegenerate, which is to say it is a symplectic tensor. A **contact structure on M** is a smooth distribution $H \subseteq TM$ of rank $2n$ with the property that any smooth local defining form θ for H is a contact form. A **contact manifold** is a smooth manifold M together with a contact structure on M. If (M, H) is a contact manifold, any (local or global) defining form for H is called a **contact form for H**. It was proved in 1971 by Jean Martinet [Mar71] that every oriented compact smooth 3-manifold admits a contact structure; but the question of which higher-dimensional manifolds admit contact structures is still unresolved.

Proposition 22.24. *A smooth 1-form θ on a $(2n+1)$-manifold is a contact form if and only if $\theta \wedge d\theta^n$ is nonzero everywhere on M, where $d\theta^n$ represents the n-fold wedge product $d\theta \wedge \cdots \wedge d\theta$.*

▶ **Exercise 22.25.** Prove the preceding proposition.

▶ **Exercise 22.26.** Suppose H is a contact structure on a smooth manifold M. Show that if θ_1 and θ_2 are any two local contact forms for H, then on their common domain there is a smooth nonvanishing function f such that $\theta_2 = f\theta_1$.

It follows from the result of Problem 19-2 that a codimension-1 distribution $H \subseteq TM$ is integrable if and only if any local defining form θ satisfies $\theta \wedge d\theta \equiv 0$. If H is a contact structure, by contrast, not only is $\theta \wedge d\theta$ nonzero everywhere on M, but it remains nonzero after taking $n-1$ more wedge products with $d\theta$. Thus, a contact structure is, in a sense, a "maximally nonintegrable distribution."

Example 22.27 (Contact Forms).

(a) On $\mathbb{R}^{2n+1}$ with coordinates $\left(x^1,\ldots,x^n,y^1,\ldots,y^n,z\right)$, define a 1-form θ by

$$\theta = dz - \sum_{i=1}^{n} y^i \, dx^i, \tag{22.19}$$

and let $H \subseteq T\mathbb{R}^{2n+1}$ be the rank-$2n$ distribution annihilated by θ. Then $d\theta = \sum_{i=1}^{n} dx^i \wedge dy^i$. If we define vector fields $\{X_i, Y_i : i = 1,\ldots,n\}$ by

$$X_i = \frac{\partial}{\partial x^i} + y^i \frac{\partial}{\partial z}, \qquad Y_i = \frac{\partial}{\partial y^i},$$

then $\left(X_i, Y_i\right)$ is a smooth frame for H, and it is straightforward to check that it satisfies $d\theta\left(X_i, X_j\right) = d\theta\left(Y_i, Y_j\right) = 0$ and $d\theta\left(X_i, Y_j\right) = \delta_{ij}$. It follows just as in Example 22.2 that $d\theta|_H$ is nondegenerate, so θ is a contact form. Theorem 22.31 below shows that every contact form can be put into this form locally by a change of coordinates.

(b) Let M be a smooth n-manifold, and define a smooth 1-form θ on the $(2n+1)$-manifold $\mathbb{R} \times T^*M$ by $\theta = dz - \tau$, where z is the standard coordinate on $\mathbb{R}$ and τ is the tautological 1-form on T^*M. In terms of natural coordinates $\left(x^i, \xi_i\right)$ for T^*M, the form θ has the coordinate representation

$$\theta = dz - \sum_{i=1}^{n} \xi_i \, dx^i,$$

so it is a contact form by the same argument as in part (a).

(c) On $\mathbb{R}^{2n+2}$ with coordinates $\left(x^1,\ldots,x^{n+1},y^1,\ldots,y^{n+1}\right)$, consider the 1-form

$$\Theta = \sum_{i=1}^{n+1} \left(x^i \, dy^i - y^i \, dx^i\right).$$

The *standard contact form on* $\mathbb{S}^{2n+1}$ is the smooth 1-form $\theta = \iota^*\Theta$, where $\iota\colon \mathbb{S}^{2n+1} \hookrightarrow \mathbb{R}^{2n+2}$ is inclusion. To see that θ is indeed a contact form, note first that $d\Theta = 2\sum_{i=1}^{n+1} dx^i \wedge dy^i$ is a symplectic form on $\mathbb{R}^{2n+2}$. Consider the following vector fields on $\mathbb{R}^{2n+1} \smallsetminus \{0\}$:

$$N = x^j \frac{\partial}{\partial x^j} + y^j \frac{\partial}{\partial y^j}, \qquad T = x^j \frac{\partial}{\partial y^j} - y^j \frac{\partial}{\partial x^j}.$$

A computation shows that N is normal to $\mathbb{S}^{2n+1}$ (with respect to the Euclidean metric) and T is tangent to it. Let $S \subseteq T\left(\mathbb{R}^{2n+2} \smallsetminus \{0\}\right)$ denote the subbundle spanned by N and T, and let $S^\perp$ denote its symplectic complement with respect to $d\Theta$. For each $p \in \mathbb{S}^{2n+1}$, $S_p^\perp$ is the set of vectors $X \in T_p\mathbb{R}^{2n+2}$ such that $d\Theta_p\left(N_p, X_p\right) = d\Theta_p\left(T_p, X_p\right) = 0$. We compute

$$N \lrcorner\, d\Theta = 2\sum_{i=1}^{n+1}\left(x^i\, dy^i - y^i\, dx^i\right) = 2\Theta,$$

$$T \lrcorner\, d\Theta = 2\sum_{i=1}^{n+1}\left(x^i\, dx^i + y^i\, dy^i\right) = d\left(|x|^2 + |y|^2\right).$$

It follows that $S_p^\perp$ is the common kernel of Θ_p and $d\left(|x|^2 + |y|^2\right)_p$, which is $\operatorname{Ker}\Theta_p \cap T_p\mathbb{S}^{2n+1} = \operatorname{Ker}\theta_p$. Because $d\Theta(N, T) = |x|^2 + |y|^2 \neq 0$ on $\mathbb{R}^{2n+1} \smallsetminus \{0\}$, S_p is a symplectic subspace of $T_p\mathbb{R}^{2n+2}$, and thus $\operatorname{Ker}\theta_p = S_p^\perp$ is also a symplectic subspace by Proposition 22.5(a). Because the restriction of $d\theta_p$ to $\operatorname{Ker}\theta_p$ is the same as the restriction of $d\Theta_p$, it is nondegenerate, so θ is a contact form. //

Theorem 22.28 (The Reeb Field). *Let (M, H) be a contact manifold, and suppose θ is a contact form for H. There is a unique vector field $T \in \mathfrak{X}(M)$, called the **Reeb field of** θ, that satisfies the following two conditions:*

$$T \lrcorner\, d\theta = 0, \qquad \theta(T) = 1. \tag{22.20}$$

Proof. Define a smooth bundle homomorphism $\Phi\colon TM \to T^*M$ by $\Phi(X) = X \lrcorner\, d\theta$, and for each $p \in M$, let Φ_p denote the linear map $\Phi|_{T_pM}\colon T_pM \to T_p^*M$. The fact that $d\theta_p$ restricts to a nondegenerate 2-tensor on H_p implies that $\Phi_p|_{H_p}$ is injective, so Φ_p has rank at least $2n$ (where $2n + 1$ is the dimension of M). On the other hand, Φ_p cannot have rank $2n + 1$, because then $d\theta_p$ would be nondegenerate, which is impossible because T_pM is odd-dimensional. Thus Φ_p has rank exactly $2n$, so $\dim \operatorname{Ker}\Phi_p = 1$. Since $\operatorname{Ker}\Phi_p$ is not contained in $H_p = \operatorname{Ker}\theta_p$, there is a unique vector $T_p \in \operatorname{Ker}\Phi_p$ satisfying $\theta_p(T_p) = 1$. This shows that there is a unique rough vector field T satisfying (22.20).

To see that T is smooth, note that $\operatorname{Ker}\Phi$ is a smooth rank-1 subbundle of TM by Theorem 10.34. Given $p \in M$, let X be any smooth nonvanishing section of $\operatorname{Ker}\Phi$ on a neighborhood of p. Because $\theta(X) \neq 0$, we can write the Reeb field locally as $T = \theta(X)^{-1}X$, which is also smooth. $\square$

▶ **Exercise 22.29.** Show that the Reeb fields of the three contact forms described in Example 22.27 are as follows:

(a) $T = \dfrac{\partial}{\partial z}$;   (b) $T = \dfrac{\partial}{\partial t}$;   (c) $T = \left(x^j \dfrac{\partial}{\partial y^j} - y^j \dfrac{\partial}{\partial x^j} \right) \Big|_{\mathbb{S}^{2n+1}}$.

▶ **Exercise 22.30.** Let θ be a contact form and T be its Reeb field. Show that $\mathscr{L}_T \theta = 0$.

Many of the constructs that we described for symplectic manifolds have analogues in contact geometry. We begin with an analogue of the Darboux theorem.

Theorem 22.31 (Contact Darboux Theorem). *Suppose θ is a contact form on a $(2n + 1)$-dimensional manifold M. For each $p \in M$, there are smooth coordinates $\left(x^1, \dots, x^n, y^1, \dots, y^n, z \right)$ centered at p in which θ has the form (22.19).*

Proof. Let $p \in M$ be arbitrary. Let $\left(U, \left(u^i \right) \right)$ be a smooth coordinate cube centered at p in which the Reeb field of θ has the form $T = \partial / \partial u^1$, and let $Y \subseteq U$ be the slice defined by $u^1 = 0$. Because T is nowhere tangent to Y, it follows that the pullback of $d\theta$ to Y is a symplectic form. After shrinking U and Y if necessary, we can find Darboux coordinates $\left(x^1, \dots, x^n, y^1, \dots, y^n \right)$ for Y centered at p, and extend them to U by requiring them to be independent of u^1 (or equivalently, to be constant on the integral curves of T). Let α be the 1-form $\sum_i y^i \, dx^i$ on U, so the pullbacks of $d\theta$ and $-d\alpha$ to Y agree. Because $T \lrcorner d\theta = T \lrcorner d\alpha = 0$, it follows that $d\theta + d\alpha = 0$ at points of Y. Then $\mathscr{L}_T \theta = \mathscr{L}_T \alpha = 0$ implies that $d(\theta + \alpha)$ is invariant under the flow of T, so in fact $d(\theta + \alpha) = 0$ on all of U. By the Poincaré lemma, there is a smooth function z on U such that $dz = \theta + \alpha$; by subtracting a constant, we may arrange that $z(p) = 0$. Because $dz_p(T_p) = \theta_p(T_p) = 1$, it follows that $\{ dx^i|_p, dy^i|_p, dz|_p \}$ are linearly independent, so Problem 11-6 shows that there is a neighborhood of p on which $\left(x^1, \dots, x^n, y^1, \dots, y^n, z \right)$ are the coordinates we seek. $\qquad\square$

The next proposition describes a contact analogue of Hamiltonian vector fields.

Proposition 22.32. *Suppose (M, H) is a contact manifold and θ is a contact form for H. For any function $f \in C^\infty(M)$, there is a unique vector field X_f, called the **contact Hamiltonian vector field of** f, that satisfies $\theta(X_f) = f$ and $(X_f \lrcorner d\theta)|_H = -df|_H$.*

Proof. Suppose $f \in C^\infty(M)$. Because the restriction of $d\theta$ to H is nondegenerate, there is a unique smooth vector field $B \in \Gamma(H)$ such that $B \lrcorner d\theta|_H = df|_H$. If we set $X_f = fT - B$, where T is the Reeb field for θ, then it is easy to check that the required conditions are satisfied. $\qquad\square$

Suppose (M, H) is a contact manifold. A smooth vector field $X \in \mathfrak{X}(M)$ is called a **contact vector field** if its flow θ preserves the contact structure, in the sense that $d(\theta_t)_p(H_p) = H_{\theta_t(p)}$ for all (t, p) in the domain of θ.

Theorem 22.33 (Characterization of Contact Vector Fields). *If (M, H) is a contact manifold and θ is a contact form for H, then a smooth vector field on M is a contact vector field if and only if it is a contact Hamiltonian vector field.*

Proof. Problem 22-21. □

If (M, H) is a contact manifold, a smooth submanifold $S \subseteq M$ is said to be *isotropic* if $TS \subseteq H$, or equivalently if $\iota^*\theta = 0$ for any contact form θ, where $\iota: S \hookrightarrow M$ is inclusion. If $S \subseteq M$ is isotropic, then $\iota^* d\theta = d(\iota^*\theta) = 0$. This implies that for each $p \in S$, the tangent space $T_p S$ is an isotropic subspace of the symplectic vector space H_p, and thus its dimension cannot be any larger than n, where $2n + 1$ is the dimension of M. An isotropic submanifold of the maximum possible dimension n is called a *Legendrian submanifold*.

The next theorem is a contact analogue of the Hamiltonian flowout theorem, and is proved in much the same way. It is the main tool for constructing solutions of fully nonlinear PDEs (see Theorem 22.39 below).

Theorem 22.34 (Contact Flowout Theorem). *Suppose (M, H) is a contact manifold, $F \in C^\infty(M)$, Γ is an embedded isotropic submanifold of M that is contained in the zero set of F, and the contact Hamiltonian vector field X_F is nowhere tangent to Γ. If S is a flowout from Γ along X_F, then S is also isotropic and contained in the zero set of H.*

Proof. Problem 22-23. □

Nonlinear First-Order PDEs

In Chapter 9, we discussed first-order partial differential equations, and showed how to use the theory of flows to solve them in the linear and quasilinear cases. In this section, we show how to use symplectic and contact geometry to solve fully nonlinear first-order equations (i.e., equations that are not quasilinear).

We begin with a somewhat special case. A first-order partial differential equation that involves only the first derivatives of the unknown function but not the values of the function itself is called a *Hamilton–Jacobi equation*. Such an equation for an unknown function $u(x^1, \ldots, x^n)$ can be written in the form

$$F\left(x^1, \ldots, x^n, \frac{\partial u}{\partial x^1}(x), \ldots, \frac{\partial u}{\partial x^n}(x)\right) = 0, \qquad (22.21)$$

where F is a smooth function defined on an open subset of $\mathbb{R}^{2n}$. (The terminology regarding Hamilton–Jacobi equations is not universally agreed upon. Some authors reserve the term *Hamilton–Jacobi equation* for the special case of an equation of the form

$$\frac{\partial u}{\partial x^1} + H\left(x^1, \ldots, x^n, \frac{\partial u}{\partial x^2}(x), \ldots, \frac{\partial u}{\partial x^n}(x)\right) = 0. \qquad (22.22)$$

The implicit function theorem shows that an equation of the general form (22.21) can be locally rewritten in this special form if and only if the partial derivative of F with respect to its $(n + 1)$st variable is nonzero. On the other hand, other authors use the term *eikonal equation* to refer to any equation of the form (22.21).

We reserve that term for another special case to which it was originally applied; see Problem 22-24.)

More generally, if M is a smooth manifold, a Hamilton–Jacobi equation on M is given by a smooth real-valued function F defined on an open subset $W \subseteq T^*M$, and a solution to the equation is a smooth real-valued function u defined on an open subset $U \subseteq M$ such that the image of du lies in the zero set of F:

$$F\big(x, du(x)\big) = 0 \quad \text{for all } x \in U. \tag{22.23}$$

(We write the covector $du_x \in T_x^*M$ as $\big(x, du(x)\big)$, in order to be more consistent with the coordinate representation (22.21) of the equation.) We are interested in solving a Cauchy problem for (22.23): given an embedded hypersurface $S \subseteq M$ and a smooth function $\varphi \colon S \to \mathbb{R}$, we wish to find a smooth function u defined on a neighborhood of S in M and satisfying (22.23) together with the initial condition

$$u\big|_S = \varphi. \tag{22.24}$$

Just as in Chapter 9, in order to obtain solutions we need to assume that the problem is of a type called *noncharacteristic*; we will describe what this means below.

Because Equation (22.23) involves only du, not u itself, we look first for a closed 1-form α satisfying $F\big(x, \alpha(x)\big) \equiv 0$; then the Poincaré lemma guarantees that locally $\alpha = du$ for some function u, which then satisfies (22.23). By Proposition 22.12, it suffices to construct a Lagrangian submanifold of T^*M that is the image of a 1-form and is contained in $F^{-1}(0)$. The key to finding such a submanifold is the Hamiltonian flowout theorem (Theorem 22.23): after identifying an appropriate isotropic embedded initial submanifold $\Gamma \subseteq T^*M$, we will construct the image of α as the flowout from Γ along the Hamiltonian field of F.

The first challenge is to construct an appropriate initial submanifold $\Gamma \subseteq T^*M$. The image of $d\varphi$ will not do, because it lies in T^*S, not T^*M (and there is no canonical way to identify T^*S as a subset of T^*M). Thus, we must first look for an appropriate section of the restricted bundle $T^*M|_S$, that is, a smooth map $\sigma \colon S \to T^*M$ such that $\sigma(x) \in T_x^*M$ for each $x \in S$. This will be the value of du along S for our eventual solution u. Thus, we should expect that it matches $d\varphi$ when restricted to vectors tangent to S, and that it satisfies the PDE at points of S. In summary, we require σ to satisfy the following conditions:

$$\sigma(x)\big|_{T_xS} = d\varphi(x) \quad \text{for all } x \in S, \tag{22.25}$$

$$F\big(x, \sigma(x)\big) = 0 \qquad \text{for all } x \in S. \tag{22.26}$$

To find such a σ, at least locally, begin by extending φ to a smooth function $\widetilde{\varphi}$ in a neighborhood of S and choosing a smooth local defining function ψ for S. Since σ must agree with $d\varphi$ when restricted to TS, and the annihilator of TS at each point is spanned by $d\psi$, the only possibility for σ is a section of the form $\sigma = d\widetilde{\varphi} + f\, d\psi$ for some unknown real-valued function f defined in a neighborhood of S. You can then insert this into the equation $F\big(x, \sigma(x)\big) = 0$, and attempt to solve for the values of f along S.

The Cauchy problem (22.23)–(22.24) is said to be **noncharacteristic** if there exists a smooth section $\sigma \in \Gamma\left(T^*M|_S\right)$ satisfying (22.25)–(22.26), with the additional property that if $\left(x^i\right)$ are any local coordinates on M and $\left(x^1,\ldots,x^n,\xi_1,\ldots,\xi_n\right)$ are the corresponding natural coordinates on T^*M, the following vector field along S is nowhere tangent to S:

$$A^\sigma\big|_x = \frac{\partial F}{\partial \xi_1}\left(x,\sigma(s)\right)\frac{\partial}{\partial x^1} + \cdots + \frac{\partial F}{\partial \xi_n}\left(x,\sigma(s)\right)\frac{\partial}{\partial x^n}. \qquad (22.27)$$

(As we will see in the proof of the next theorem, A^σ is actually globally defined as a vector field along S, and does not depend on the choice of coordinates.) When this condition is satisfied, we can solve the Cauchy problem.

Theorem 22.35 (The Cauchy Problem for a Hamilton–Jacobi Equation). *Suppose M is a smooth manifold, $W \subseteq T^*M$ is an open subset, $F\colon W \to \mathbb{R}$ is a smooth function, $S \subseteq M$ is an embedded hypersurface, and $\varphi\colon S \to \mathbb{R}$ is a smooth function. If the Cauchy problem (22.23)–(22.24) is noncharacteristic, then for each $p \in S$ there is a smooth solution defined on some neighborhood of p in M.*

Proof. Given $\sigma\colon S \to T^*M|_S$ satisfying (22.25)–(22.26), let $\Gamma \subseteq W$ be the image of σ. Then Γ is an embedded submanifold of dimension $n-1$, where $n = \dim M$. In order to apply the Hamiltonian flowout theorem, we need to check first that Γ is isotropic with respect to the canonical symplectic structure ω on T^*M. Since $\sigma\colon S \to T^*M$ is a smooth embedding whose image is Γ, this is equivalent to showing that $\sigma^*\omega = 0$. Let $\pi\colon T^*M \to M$ be the projection; then $\pi \circ \sigma$ is equal to the inclusion $\iota\colon S \hookrightarrow M$. If τ denotes the tautological 1-form on T^*M, the defining equation (22.3) for τ implies

$$(\sigma^*\tau)(p) = d\sigma_p^*\left(d\pi_{\sigma(p)}^*\sigma(p)\right) = d(\pi \circ \sigma)_p^*\sigma(p) = d\iota_p^*\sigma(p) = d\varphi(p).$$

Thus $\sigma^*\tau = d\varphi$, and it follows that $\sigma^*\omega = \sigma^*(-d\tau) = -d(\sigma^*\tau) = -d(d\varphi) = 0$. Thus Γ is isotropic.

Next we need to check that the Hamiltonian vector field X_F is nowhere tangent to Γ. This follows from the noncharacteristic condition just as in the proof of Theorem 9.53: because $\pi\colon T^*M \to M$ restricts to a diffeomorphism from Γ to S, if X_F were tangent to Γ at some point $\left(p,\sigma(p)\right) \in \Gamma$, then $d\pi\left(X_F|_{(p,\sigma(p))}\right)$ would be tangent to S at p. Using (22.9) in natural coordinates $\left(x^i,\xi_i\right)$ on T^*M (which are Darboux coordinates for the canonical symplectic form), we find that

$$X_F = \frac{\partial F}{\partial \xi_1}\frac{\partial}{\partial x^1} + \cdots + \frac{\partial F}{\partial \xi_n}\frac{\partial}{\partial x^n} - \frac{\partial F}{\partial x^1}\frac{\partial}{\partial \xi_1} - \cdots - \frac{\partial F}{\partial x^n}\frac{\partial}{\partial \xi_n}.$$

Thus $d\pi\left(X_F|_{(p,\sigma(p))}\right) = A^\sigma|_p$, so the assumption that the Cauchy problem is noncharacteristic guarantees that X_F is nowhere tangent to Γ. (This calculation also shows that A^σ is well defined independently of coordinates, because it is the pushforward of X_F from points of Γ.)

Let $\mathcal{S}$ be a flowout from Γ along X_F. The Hamiltonian flowout theorem guarantees that $\mathcal{S}$ is an n-dimensional isotropic—and therefore Lagrangian—submanifold of T^*M contained in $F^{-1}(0)$. Using the result of Problem 22-11, we conclude that it will be the image of a closed 1-form on a neighborhood of p provided that it is transverse to the fiber of π at $(p, \sigma(p))$. Once again, we use the fact that $T_{(p,\sigma(p))}S$ is spanned by $T_{(p,\sigma(p))}\Gamma$ and $X_F|_{(p,\sigma(p))}$. Because $d\pi$ maps $X_F|_{(p,\sigma(p))}$ to $A^\sigma|_p$ and maps $T_{(p,\sigma(p))}\Gamma$ isomorphically onto T_pS, the noncharacteristic assumption guarantees that $T_{(p,\sigma(p))}T^*M = T_{(p,\sigma(p))}S \oplus \mathrm{Ker}\, d\pi_{(p,\sigma(p))}$, and thus S intersects the fiber transversely at $(p, \sigma(p))$. By Problem 22-11, there is a closed 1-form α defined on a neighborhood U of p whose graph is an open subset of S. Because the image of σ is contained in S, it follows that

$$\alpha(x) = \sigma(x) \quad \text{for } x \in S \cap U. \tag{22.28}$$

By the Poincaré lemma, after shrinking U further if necessary, we can find a smooth function $u \colon U \to \mathbb{R}$ such that $du = \alpha$. Because $S \subseteq F^{-1}(0)$, we conclude that u satisfies (22.23). To ensure that the initial condition is also satisfied, shrink U further so that $S \cap U$ is connected. By adding a constant to u, we may arrange that $u(p) = \varphi(p)$. Then for any $x \in S$, it follows from (22.25) and (22.28) that

$$du(x)\big|_{T_xS} = \alpha(x)\big|_{T_xS} = \sigma(x)\big|_{T_xS} = d\varphi(x).$$

Because $S \cap U$ is connected, this means that $u - \varphi$ is constant on $S \cap U$. Since this difference vanishes at p, it vanishes identically, so (22.24) is satisfied on $S \cap U$. $\square$

Note that we did not claim any uniqueness in this theorem. In Cauchy problems for fully nonlinear equations, even local uniqueness can fail. For example, consider the following Cauchy problem in the plane:

$$\left(\frac{\partial u}{\partial x} \right)^2 = 1, \qquad u(0, y) = 0.$$

This is noncharacteristic, as you can check. Both $u(x, y) = x$ and $u(x, y) = -x$ are solutions to this problem, but they are not equal in any open subset. The problem here is that there are two possible choices for the initial 1-form σ (namely, $\sigma = dx$ and $\sigma = -dx$), and they yield different initial manifolds Γ and therefore different solutions to the Cauchy problem. As Problem 22-25 shows, once σ is chosen, local uniqueness holds just as in the quasilinear case.

Example 22.36 (A Hamilton–Jacobi Equation). Consider the following Cauchy problem in the plane:

$$\frac{\partial u}{\partial x} - \left(\frac{\partial u}{\partial y} \right)^2 = 0, \qquad u(0, y) = y^2.$$

The corresponding function on $T^*\mathbb{R}^2$ is $F(x, y, \xi, \eta) = \xi - \eta^2$, where we use (x, y, ξ, η) to denote natural coordinates on $T^*\mathbb{R}^2$ associated with (x, y).

To check that the problem is noncharacteristic, we need to find a suitable 1-form σ along the initial manifold $S = \{(x, y) : x = 0\}$. Since x is a defining function for S, we can write $\sigma = d(y^2) + f(y)\, dx = 2y\, dy + f(y)\, dx$ and solve the equation

$F(0, y, f(y), 2y) = f(y) - (2y)^2 = 0$ to obtain $f(y) = 4y^2$, and thus we can set $\sigma(y) = 2y\, dy + 4y^2\, dx$. The vector field A^σ is given by

$$A^\sigma\big|_{(x,y)} = \frac{\partial}{\partial x} - 4y\frac{\partial}{\partial y},$$

which is nowhere tangent to S.

The initial curve S can be parametrized by $X(s) = (0, s)$, and therefore the initial curve $\Gamma \subseteq T^*\mathbb{R}^2$ (the image of σ) can be parametrized by $\widetilde{X}(s) = (0, s, 4s^2, 2s)$. The Hamiltonian field of F is

$$X_F\big|_{(x,y,\xi,\eta)} = \frac{\partial}{\partial x} - 2\eta\frac{\partial}{\partial y},$$

and it is an easy matter to solve the corresponding system of ODEs with initial conditions $(x, y, \xi, \eta) = (0, s, 4s^2, 2s)$ to obtain the following parametrization of S:

$$\Psi(t, s) = (t, s - 4st, 4s^2, 2s).$$

Solving $(x, y) = (t, s - 4st)$ for (t, s) and inserting into the formulas for (ξ, η), we find that S is the image of the following 1-form:

$$\alpha = \frac{4y^2}{(1 - 4x)^2}\, dx + \frac{2y}{1 - 4x}\, dy.$$

This is indeed a closed 1-form, and using the procedure sketched at the end of Chapter 11 we find that $\alpha = du$ on the set $\{(x, y) : x < 1/4\}$, where

$$u(x, y) = \frac{y^2}{1 - 4x}.$$

In principle, we might have to add a constant to u to satisfy the initial condition, but in this case $u(0, y) = y^2$ already, so this is the solution to our Cauchy problem.   //

General Nonlinear Equations

Finally, we show how the preceding method can be adapted to solve arbitrary first-order PDEs by using contact geometry in place of symplectic geometry. For this purpose, we introduce one last geometric construction. If M is a smooth manifold, the *1-jet bundle of M* is the smooth vector bundle $J^1M = \mathbb{R} \times T^*M \to M$, whose fiber at $x \in M$ is $\mathbb{R} \times T_x^*M$. (It is the Whitney sum of a trivial $\mathbb{R}$-bundle with T^*M.) If $u: M \to \mathbb{R}$ is a smooth function, the *1-jet of u* is the section $j^1u: M \to J^1M$ defined by $j^1u(x) = (u(x), du(x))$. A point in the fiber of J^1M over $x \in M$ can be viewed as a first-order Taylor polynomial at x of a smooth function on M, represented invariantly as the values of the function and its differential at x. (One can also define higher-order jet bundles that give invariant representations of higher-order Taylor polynomials. They are useful for studying higher-order PDEs, but we do not pursue them here.)

The *canonical contact form* on $J^1 M$ is the 1-form $\theta = dz - \tau$ defined in Example 22.27(b). A smooth (local or global) section $\eta \colon M \to J^1 M$ is said to be *Legendrian* if its image is a Legendrian submanifold of $J^1 M$, or equivalently if $\eta^* \theta = 0$. The next proposition is a contact analogue of Proposition 22.12.

Proposition 22.37. *Let M be a smooth manifold. A smooth local section of $J^1 M$ is the 1-jet of a smooth function if and only if it is Legendrian.*

▶ **Exercise 22.38.** Prove the preceding proposition.

The 1-jet bundle provides the most general setting in which to consider first-order partial differential equations. If M is a smooth manifold, a first-order PDE for a function $u \colon M \to \mathbb{R}$ can be viewed as a real-valued function F on the 1-jet bundle of M, and a solution is a function whose 1-jet takes its values in the zero set of F.

Let M be a smooth manifold, and suppose we are given a function $F \in C^\infty(W)$ on some open subset $W \subseteq J^1 M$, a smooth hypersurface $S \subseteq M$, and a smooth function $\varphi \colon S \to \mathbb{R}$. We wish to solve the following Cauchy problem for u:

$$F\big(x, u(x), du(x)\big) \equiv 0, \tag{22.29}$$

$$u\big|_S = \varphi. \tag{22.30}$$

This problem is said to be *noncharacteristic* if there exists a smooth section $\sigma \in \Gamma\big(T^* M|_S\big)$ taking its values in W and satisfying

$$\sigma(x)\big|_{T_x S} = d\varphi(x) \quad \text{for all } x \in S, \tag{22.31}$$

$$F\big(x, \varphi(x), \sigma(x)\big) = 0 \qquad \text{for all } x \in S, \tag{22.32}$$

and such that the following vector field along S is nowhere tangent to S:

$$A^{\varphi,\sigma}\big|_x = \frac{\partial F}{\partial \xi_1}\big(x, \varphi(x), \sigma(x)\big) \frac{\partial}{\partial x^1} + \cdots + \frac{\partial F}{\partial \xi_n}\big(x, \varphi(x), \sigma(x)\big) \frac{\partial}{\partial x^n}. \tag{22.33}$$

The proof of the next theorem is very similar to that of Theorem 22.35, but uses the contact flowout theorem instead of the Hamiltonian one.

Theorem 22.39 (The General First-Order Cauchy Problem). *Suppose M is a smooth manifold, $W \subseteq J^1 M$ is an open subset, $F \colon W \to \mathbb{R}$ is a smooth function, $S \subseteq M$ is an embedded hypersurface, and $\varphi \colon S \to \mathbb{R}$ is a smooth function. If the Cauchy problem (22.29)–(22.30) is noncharacteristic, then for each $p \in S$ there is a smooth solution on some neighborhood of p in M.*

Proof. Problem 22-26.   □

Problems

22-1. Prove Proposition 22.5 (properties of symplectic, isotropic, coisotropic, and Lagrangian subspaces).

22-2. Let (V, ω) be a symplectic vector space of dimension $2n$. Show that for each symplectic, isotropic, coisotropic, or Lagrangian subspace $S \subseteq V$, there exists a symplectic basis (A_i, B_i) for V with the following property:
(a) If S is symplectic, $S = \text{span}(A_1, B_1, \ldots, A_k, B_k)$ for some k.
(b) If S is isotropic, $S = \text{span}(A_1, \ldots, A_k)$ for some k.
(c) If S is coisotropic, $S = \text{span}(A_1, \ldots, A_n, B_1, \ldots, B_k)$ for some k.
(d) If S is Lagrangian, $S = \text{span}(A_1, \ldots, A_n)$.

22-3. The **real symplectic group** is the subgroup $\text{Sp}(2n, \mathbb{R}) \subseteq \text{GL}(2n, \mathbb{R})$ consisting of all $2n \times 2n$ matrices that leave the standard symplectic tensor $\omega = \sum_{i=1}^{n} dx^i \wedge dy^i$ invariant, that is, the set of invertible linear maps $Z: \mathbb{R}^{2n} \to \mathbb{R}^{2n}$ such that $\omega(Zx, Zy) = \omega(x, y)$ for all $x, y \in \mathbb{R}^{2n}$.
(a) Show that a matrix Z is in $\text{Sp}(2n, \mathbb{R})$ if and only if it takes the standard basis to a symplectic basis.
(b) Show that $Z \in \text{Sp}(2n, \mathbb{R})$ if and only if $Z^T J Z = J$, where J is the $2n \times 2n$ block diagonal matrix

$$J = \begin{pmatrix} j & \cdots & 0 \\ \vdots & \ddots & \vdots \\ 0 & \cdots & j \end{pmatrix},$$

with copies of the 2×2 block $j = \begin{pmatrix} 0 & 1 \\ -1 & 0 \end{pmatrix}$ along the main diagonal, and zeros elsewhere.
(c) Show that $\text{Sp}(2n, \mathbb{R})$ is an embedded Lie subgroup of $\text{GL}(2n, \mathbb{R})$, and determine its dimension.
(d) Determine the Lie algebra of $\text{Sp}(2n, \mathbb{R})$ as a subalgebra of $\mathfrak{gl}(2n, \mathbb{R})$.
(e) Is $\text{Sp}(2n, \mathbb{R})$ compact?

22-4. Let (M, ω) be a symplectic manifold, and suppose $F: N \to M$ is a smooth map such that $F^*\omega$ is symplectic. Show that F is a smooth immersion.

22-5. Suppose (M, ω) is a $2n$-dimensional compact symplectic manifold.
(a) Show that ω^n (the n-fold wedge product of ω with itself) is not exact.
(b) Show that $H_{dR}^{2p}(M) \neq 0$ for $p = 1, \ldots, n$.
(c) Show that $\mathbb{S}^2$ is the only sphere that admits a symplectic structure.

22-6. Prove that $\mathbb{R}^{2n}$ (with its standard symplectic structure) does not have any compact symplectic submanifolds.

22-7. Let (M, ω) and $(\widetilde{M}, \widetilde{\omega})$ be symplectic manifolds. Define a 2-form Ω on $M \times \widetilde{M}$ by $\Omega = \pi^*\omega - \widetilde{\pi}^*\widetilde{\omega}$, where $\pi: M \times \widetilde{M} \to M$ and $\widetilde{\pi}: M \times \widetilde{M} \to \widetilde{M}$ are the projections.
(a) Show that Ω is symplectic.
(b) Show that a diffeomorphism $F: M \to \widetilde{M}$ is a symplectomorphism if and only if its graph $\Gamma(F) = \{(x, y) \in M \times \widetilde{M} : y = F(x)\}$ is a Lagrangian submanifold of $(M \times \widetilde{M}, \Omega)$.

22-8. Suppose (M, ω) is a symplectic manifold and $S \subseteq M$ is a coisotropic submanifold. An immersed submanifold $N \subseteq S$ is said to be **characteristic** if

$T_p N = (T_p S)^\perp$ for each $p \in N$. Show that there is a foliation of S by connected characteristic submanifolds of S whose dimension is equal to the codimension of S in M.

22-9. Considering $\mathbb{R}^{2n}$ as a symplectic manifold with its standard symplectic structure $\omega = \sum_i dx^i \wedge dy^i$, let $\Lambda_n \subseteq G_n\left(\mathbb{R}^{2n}\right)$ denote the set of Lagrangian subspaces of $\mathbb{R}^{2n}$.
   (a) Show that the real symplectic group $\mathrm{Sp}(2n, \mathbb{R})$ acts transitively on Λ_n (see Problem 22-3).
   (b) Show that Λ_n has a unique smooth manifold structure such that the action of $\mathrm{Sp}(2n, \mathbb{R})$ is smooth, and determine its dimension.
   (c) Is Λ_n compact?

22-10. Show that the canonical symplectic form on the cotangent bundle is invariant under diffeomorphisms, in the following sense: suppose Q and $\tilde{Q}$ are smooth manifolds and $F: Q \to \tilde{Q}$ is a diffeomorphism. Let $dF^*: T^*\tilde{Q} \to T^*Q$ be the smooth bundle homomorphism described in Problem 11-8. Show that dF^* is a symplectomorphism when both T^*Q and $T^*\tilde{Q}$ are endowed with their canonical symplectic forms.

22-11. Let Q be a smooth manifold, and let S be an embedded Lagrangian submanifold of the total space of T^*Q. Prove the following statements.
   (a) If S is transverse to the fiber of T^*Q at a point $q \in T^*Q$, then there exist a neighborhood V of q in S and a neighborhood U of $\pi(q)$ in Q such that V is the image of a smooth closed 1-form defined on U.
   (b) S is the image of a globally defined smooth closed 1-form on Q if and only if S intersects each fiber transversely in exactly one point.
   (Cf. Theorem 6.32 and Corollary 6.33.) (*Used on p. 588.*)

22-12. Let M be a smooth manifold of dimension at least 1. Show that there is no 1-form σ on M such that the tautological form $\tau \in \Omega^1(T^*M)$ is equal to the pullback $\pi^*\sigma$.

22-13. Let M be a smooth manifold and let $S \subseteq M$ be an embedded submanifold. Define the ***conormal bundle of*** S to be the subset $N^*S \subseteq T^*M$ defined by

$$N^*S = \left\{(q, \eta) \in T^*M : q \in S, \ \eta\big|_{T_q S} \equiv 0\right\}.$$

Show that N^*S is a smooth subbundle of $T^*M|_S$, and an embedded Lagrangian submanifold of T^*M (with respect to the canonical symplectic structure on T^*M).

22-14. Prove the following global version of the Darboux theorem, due to Moser [Mos65]: Let M be a compact smooth manifold, and let ω_0 be a symplectic form on M. Suppose there is a smooth time-dependent 1-form $\alpha: [0, 1] \times M \to T^*M$ such that $\omega_t = \omega_0 + d\alpha_t$ is symplectic for each $t \in [0, 1]$. Show that there is a diffeomorphism $F: M \to M$ such that $F^*\omega_1 = \omega_0$.

22-15. Using the same technique as in the proof of Theorem 22.13, prove the following theorem of Moser [Mos65]: If M is an oriented compact smooth

n-manifold, $n \geq 1$, and ω_0, ω_1 are smooth orientation forms on M such that $\int_M \omega_0 = \int_M \omega_1$, then there exists a diffeomorphism $F \colon M \to M$ such that $F^*\omega_1 = \omega_0$. [Hint: for any orientation form ω, show that the map $\beta \colon \mathfrak{X}(M) \to \Omega^{n-1}(M)$ defined by $\beta(X) = X \lrcorner\, \omega$ as in (16.11) is a bundle isomorphism.]

22-16. Let (M, ω) be a symplectic manifold. Let $\mathcal{S}(M) \subseteq \mathfrak{X}(M)$ denote the set of symplectic vector fields on M, and $\mathcal{H}(M) \subseteq \mathfrak{X}(M)$ the set of Hamiltonian vector fields.

   (a) Show that $\mathcal{S}(M)$ is a Lie subalgebra of $\mathfrak{X}(M)$, and $\mathcal{H}(M)$ is a Lie subalgebra of $\mathcal{S}(M)$.

   (b) Show that the map from $\mathcal{S}(M)$ to $\Omega^1(M)$ given by $X \mapsto X \lrcorner\, \omega$ descends to a vector space isomorphism between $\mathcal{S}(M)/\mathcal{H}(M)$ and $H^1_{\mathrm{dR}}(M)$.

22-17. Consider the 2-*body problem*, that is, the Hamiltonian system (T^*Q, ω, E) described in Example 22.18 in the special case $n = 2$. Suppose that the potential energy V depends only on the distance between the particles. More precisely, suppose that $V(q) = v\big(r(q)\big)$ for some smooth function $v \colon (0, \infty) \to \mathbb{R}$, where

$$r(q) = |\mathbf{q}_1 - \mathbf{q}_2| = \sqrt{\left(q_1^1 - q_2^1\right)^2 + \left(q_1^2 - q_2^2\right)^2 + \left(q_1^3 - q_2^3\right)^2}.$$

   (a) Let $\mathbf{u} = \left(u^1, u^2, u^3\right)$ be a unit vector in $\mathbb{R}^3$, and show that the function $P \colon T^*Q \to \mathbb{R}$ defined by

$$P(q, p) = \mathbf{u} \cdot \mathbf{p}_1 + \mathbf{u} \cdot \mathbf{p}_2$$
$$= u^1 p_1^1 + u^2 p_1^2 + u^3 p_1^3 + u^1 p_2^1 + u^2 p_2^2 + u^3 p_2^3$$

   is a conserved quantity (called the ***total linear momentum in the u-direction***), and that the corresponding infinitesimal symmetry generates translations in the **u**-direction:

$$\theta_t(q, p) = (\mathbf{q}_1 + t\mathbf{u}, \mathbf{q}_2 + t\mathbf{u}, \mathbf{p}_1, \mathbf{p}_2)$$
$$= \big(q_1^1 + tu^1, q_1^2 + tu^2, q_1^3 + tu^3, q_2^1 + tu^1, q_2^2 + tu^2,$$
$$q_2^3 + tu^3, p_1^1, p_1^2, p_1^3, p_2^1, p_2^2, p_2^3\big).$$

   (b) Show that the function $L \colon T^*Q \to \mathbb{R}$ defined by

$$L(q, p) = q_1^1 p_1^2 - q_1^2 p_1^1 + q_2^1 p_2^2 - q_2^2 p_2^1$$

   is a conserved quantity (called the ***total angular momentum about the z-axis***), and find the flow of the corresponding infinitesimal symmetry. Explain what this has to do with rotational symmetry.

22-18. Prove Proposition 22.21 (properties of conserved quantities and infinitesimal symmetries).

22-19. This problem outlines a different proof of the Darboux theorem. Let (M, ω) be a $2n$-dimensional symplectic manifold and $p \in M$.

(a) Show that smooth coordinates $(x^1, \ldots, x^n, y^1, \ldots, y^n)$ on an open subset $U \subseteq M$ are Darboux coordinates if and only if their Poisson brackets satisfy

$$\{x^i, y^j\} = \delta^{ij}; \qquad \{x^i, x^j\} = \{y^i, y^j\} = 0. \qquad (22.34)$$

(b) Prove the following statement by induction on k: for each $k = 0, \ldots, n$, there are smooth functions $(x^1, \ldots, x^k, y^1, \ldots, y^k)$ vanishing at p and satisfying (22.34) in a neighborhood of p such that the $2k$-tuple of 1-forms $(dx^1, \ldots, dx^k, dy^1, \ldots, dy^k)$ is linearly independent at p. When $k = n$, this proves the theorem. [Hint: for the inductive step, assuming that $(x^1, \ldots, x^k, y^1, \ldots, y^k)$ have been found, find smooth coordinates $(u^1, \ldots, u^{2n})$ such that

$$\frac{\partial}{\partial u^i} = X_{x^i}, \qquad \frac{\partial}{\partial u^{i+k}} = X_{y^i}, \quad i = 1, \ldots, k,$$

and let $y^{k+1} = u^{2k+1}$. Then find new coordinates $(v^1, \ldots, v^{2n})$ with

$$\frac{\partial}{\partial v^i} = X_{x^i}, \quad i = 1, \ldots, k,$$

$$\frac{\partial}{\partial v^{i+k}} = X_{y^i}, \quad i = 1, \ldots, k+1,$$

and let $x^{k+1} = v^{2k+1}$.]

22-20. Suppose (M, H) is a contact manifold of dimension $2n + 1$. Show that if n is odd, then M is orientable, while if n is even, then M is orientable if and only if there exists a global contact form for H.

22-21. Prove Theorem 22.33 (characterization of contact vector fields).

22-22. Suppose (M, H) is a contact manifold and X is a smooth vector field on M. Prove that X is the Reeb field of some contact form for H if and only if it is a contact vector field that takes no values in H.

22-23. Prove Theorem 22.34 (the contact flowout theorem).

22-24. The classical **eikonal equation** for a real-valued function u on an open subset $U \subseteq \mathbb{R}^n$ is

$$\sum_{i=1}^{n} \left(\frac{\partial u}{\partial x^i} \right)^2 = f(x), \qquad (22.35)$$

where f is a given smooth real-valued function on u. It plays an important role in the theory of optics. (The word "eikonal" stems from the Greek word for "image," the same root from which our word "icon" is derived.) In the special case $f(x) \equiv 1$, find an explicit solution u to (22.35) on an open subset of $\mathbb{R}^n$ with $u = 0$ on the unit sphere.

22-25. Suppose (22.24) is a noncharacteristic initial condition for a Hamilton–Jacobi equation (22.23). For any choice of $\sigma\colon S \to T^*\mathbb{R}^n$ satisfying (22.25), (22.26), and (22.27), and any $p \in S$, show that there is a neighborhood U of p on which there is a *unique* solution to the Cauchy problem (22.23)–(22.24) satisfying $du(x) = \sigma(x)$ for $x \in S$.

22-26. Prove Theorem 22.39 (solution to the general first-order Cauchy problem).

Appendix A
Review of Topology

This book is written for readers who have already completed a rigorous course in basic topology, including an introduction to the fundamental group and covering maps. A convenient source for this material is [LeeTM], which covers all the topological ideas we need, and uses notations and conventions that are compatible with those in the present book. But almost any other good topology text would do as well, such as [Mun00, Sie92, Mas89]. In this appendix we state the most important definitions and results, with most of the proofs left as exercises. If you have had sufficient exposure to topology, these exercises should be straightforward, although you might want to look a few of them up in the topology texts listed above.

Topological Spaces

We begin with the definitions. Let X be a set. A *topology on X* is a collection $\mathcal{T}$ of subsets of X, called *open subsets*, satisfying

(i) X and $\varnothing$ are open.
(ii) The union of any family of open subsets is open.
(iii) The intersection of any finite family of open subsets is open.

A pair $(X, \mathcal{T})$ consisting of a set X together with a topology $\mathcal{T}$ on X is called a *topological space*. Ordinarily, when the topology is understood, one omits mention of it and simply says "X is a topological space."

There are a host of constructions and definitions associated with topological spaces. Here we summarize the ones that are most important for this book.

Suppose X is a topological space, $p \in X$, and $S \subseteq X$.

- A *neighborhood of p* is an open subset containing p. Similarly, a *neighborhood of the set S* is an open subset containing S. (Be warned that some authors use the word "neighborhood" in the more general sense of a subset containing an open subset containing p or S.)
- S is said to be *closed* if $X \smallsetminus S$ is open (where $X \smallsetminus S$ denotes the *set difference* $\{x \in X : x \notin S\}$).

- The *interior of* S, denoted by Int S, is the union of all open subsets of X contained in S.
- The *exterior of* S, denoted by Ext S, is the union of all open subsets of X contained in $X \smallsetminus S$.
- The *closure of* S, denoted by $\bar{S}$, is the intersection of all closed subsets of X containing S.
- The *boundary of* S, denoted by ∂S, is the set of all points of X that are in neither Int S nor Ext S.
- A point $p \in S$ is said to be an *isolated point of* S if p has a neighborhood $U \subseteq X$ such that $U \cap S = \{p\}$.
- A point $p \in X$ (not necessarily in S) is said to be a *limit point of* S if every neighborhood of p contains at least one point of S other than p.
- S is said to be *dense in* X if $\bar{S} = X$, or equivalently if every nonempty open subset of X contains at least one point of S.
- S is said to be *nowhere dense in* X if $\bar{S}$ contains no nonempty open subset.

The most important concepts of topology are continuous maps and convergent sequences, which we define next. Let X and Y be topological spaces.

- A map $F: X \to Y$ is said to be *continuous* if for every open subset $U \subseteq Y$, the preimage $F^{-1}(U)$ is open in X.
- A continuous bijective map $F: X \to Y$ with continuous inverse is called a *homeomorphism*. If there exists a homeomorphism from X to Y, we say that X and Y are *homeomorphic*.
- A continuous map $F: X \to Y$ is said to be a *local homeomorphism* if every point $p \in X$ has a neighborhood $U \subseteq X$ such that $F(U)$ is open in Y and F restricts to a homeomorphism from U to $F(U)$.
- Given a sequence $(p_i)_{i=1}^{\infty}$ of points in X and a point $p \in X$, the sequence is said to *converge to* p if for every neighborhood U of p, there exists a positive integer N such that $p_i \in U$ for all $i \geq N$. In this case, we write $p_i \to p$ or $\lim_{i \to \infty} p_i = p$.

▶ **Exercise A.1.** Let $F: X \to Y$ be a map between topological spaces. Prove that each of the following properties is equivalent to continuity of F:

(a) For every subset $A \subseteq X$, $F(\bar{A}) \subseteq \overline{F(A)}$.
(b) For every subset $B \subseteq Y$, $F^{-1}(\text{Int } B) \subseteq \text{Int } F^{-1}(B)$.

▶ **Exercise A.2.** Let X, Y, and Z be topological spaces. Show that the following maps are continuous:

(a) The *identity map* $\text{Id}_X: X \to X$, defined by $\text{Id}_X(x) = x$ for all $x \in X$.
(b) Any *constant map* $F: X \to Y$ (i.e., a map such that $F(x) = F(y)$ for all $x, y \in X$).
(c) Any composition $G \circ F$ of continuous maps $F: X \to Y$ and $G: Y \to Z$.

▶ **Exercise A.3.** Let X and Y be topological spaces. Suppose $F: X \to Y$ is continuous and $p_i \to p$ in X. Show that $F(p_i) \to F(p)$ in Y.

The most important examples of topological spaces, from which most of our examples of manifolds are built in one way or another, are described below.

Example A.4 (Discrete Spaces). If X is an arbitrary set, the *discrete topology* on X is the topology defined by declaring every subset of X to be open. Any space that has the discrete topology is called a *discrete space*.   //

Example A.5 (Metric Spaces). A *metric space* is a set M endowed with a *distance function* (also called a *metric*) $d: M \times M \to \mathbb{R}$ (where $\mathbb{R}$ denotes the set of real numbers) satisfying the following properties for all $x, y, z \in M$:

(i) POSITIVITY: $d(x, y) \geq 0$, with equality if and only if $x = y$.
(ii) SYMMETRY: $d(x, y) = d(y, x)$.
(iii) TRIANGLE INEQUALITY: $d(x, z) \leq d(x, y) + d(y, z)$.

If M is a metric space, $x \in M$, and $r > 0$, the *open ball of radius r around x* is the set

$$B_r(x) = \{ y \in M : d(x, y) < r \},$$

and the *closed ball of radius r* is

$$\bar{B}_r(x) = \{ y \in M : d(x, y) \leq r \}.$$

The *metric topology on M* is defined by declaring a subset $S \subseteq M$ to be open if for every point $x \in S$, there is some $r > 0$ such that $B_r(x) \subseteq S$.   //

If M is a metric space and S is any subset of M, the restriction of the distance function to pairs of points in S turns S into a metric space and thus also a topological space. We use the following standard terminology for metric spaces:

- A subset $S \subseteq M$ is *bounded* if there exists a positive number R such that $d(x, y) \leq R$ for all $x, y \in S$.
- If S is a nonempty bounded subset of M, the *diameter of S* is the number $\operatorname{diam} S = \sup\{ d(x, y) : x, y \in S \}$.
- A sequence of points $(x_i)_{i=1}^{\infty}$ in M is a *Cauchy sequence* if for every $\varepsilon > 0$, there exists an integer N such that $i, j \geq N$ implies $d(x_i, x_j) < \varepsilon$.
- A metric space M is said to be *complete* if every Cauchy sequence in M converges to a point of M.

Example A.6 (Euclidean Spaces). For each integer $n \geq 1$, the set $\mathbb{R}^n$ of ordered n-tuples of real numbers is called *n-dimensional Euclidean space*. We denote a point in $\mathbb{R}^n$ by $(x^1, \ldots, x^n)$, (x^i), or x; the numbers x^i are called the *components* or *coordinates of x*. (When n is small, we often use more traditional names such as (x, y, z) for the coordinates.) Notice that we write the coordinates of a point $(x^1, \ldots, x^n) \in \mathbb{R}^n$ with superscripts, not subscripts as is usually done in linear algebra and calculus books, so as to be consistent with the Einstein summation convention, explained in Chapter 1. By convention, $\mathbb{R}^0$ is the one-element set $\{0\}$.

For each $x \in \mathbb{R}^n$, the *Euclidean norm of x* is the nonnegative real number

$$|x| = \sqrt{\left(x^1\right)^2 + \cdots + \left(x^n\right)^2},$$

and for $x, y \in \mathbb{R}^n$, the ***Euclidean distance function*** is defined by

$$d(x, y) = |x - y|.$$

This distance function turns $\mathbb{R}^n$ into a complete metric space. The resulting metric topology on $\mathbb{R}^n$ is called the ***Euclidean topology***. //

Example A.7 (Complex Euclidean Spaces). We also sometimes have occasion to work with complex Euclidean spaces. We consider the set $\mathbb{C}$ of complex numbers, as a set, to be simply $\mathbb{R}^2$, with the complex number $x + iy$ corresponding to $(x, y) \in \mathbb{R}^2$. For any positive integer n, the ***n-dimensional complex Euclidean space*** is the set $\mathbb{C}^n$ of ordered n-tuples of complex numbers. It becomes a topological space when identified with $\mathbb{R}^{2n}$ via the correspondence

$$\left(x^1 + iy^1, \ldots, x^n + iy^n\right) \leftrightarrow \left(x^1, y^1, \ldots, x^n, y^n\right).$$ //

Example A.8 (Subsets of Euclidean Spaces). Every subset of $\mathbb{R}^n$ or $\mathbb{C}^n$ becomes a metric space, and thus a topological space, when endowed with the Euclidean metric. Whenever we mention such a subset, it is always assumed to have this metric topology unless otherwise specified. It is a complete metric space if and only if it is a closed subset of $\mathbb{R}^n$. Here are some standard subsets of Euclidean spaces that we work with frequently:

- The ***unit interval*** is the subset $I \subseteq \mathbb{R}$ defined by
$$I = [0, 1] = \{x \in \mathbb{R} : 0 \le x \le 1\}.$$

- The (***open***) ***unit ball of dimension n*** is the subset $\mathbb{B}^n \subseteq \mathbb{R}^n$ defined by
$$\mathbb{B}^n = \{x \in \mathbb{R}^n : |x| < 1\}.$$

- The ***closed unit ball of dimension n*** is the subset $\bar{\mathbb{B}}^n \subseteq \mathbb{R}^n$ defined by
$$\bar{\mathbb{B}}^n = \{x \in \mathbb{R}^n : |x| \le 1\}.$$

  The terms (***open***) ***unit disk*** and ***closed unit disk*** are commonly used for $\mathbb{B}^2$ and $\bar{\mathbb{B}}^2$, respectively.

- For $n \ge 0$, the (***unit***) ***n-sphere*** is the subset $\mathbb{S}^n \subseteq \mathbb{R}^{n+1}$ defined by
$$\mathbb{S}^n = \{x \in \mathbb{R}^{n+1} : |x| = 1\}.$$

  Sometimes it is useful to think of an odd-dimensional sphere $\mathbb{S}^{2n+1}$ as a subset of $\mathbb{C}^{n+1}$, by means of the usual identification of $\mathbb{C}^{n+1}$ with $\mathbb{R}^{2n+2}$.

- The (***unit***) ***circle*** is the 1-sphere $\mathbb{S}^1$, considered either as a subset of $\mathbb{R}^2$ or as a subset of $\mathbb{C}$. //

Hausdorff Spaces

Topological spaces allow us to describe a wide variety of concepts of "spaces." But for the purposes of manifold theory, arbitrary topological spaces are far too general, because they can have some unpleasant properties, as the next exercise illustrates.

▶ **Exercise A.9.** Let X be any set. Show that $\{X, \varnothing\}$ is a topology on X, called the *trivial topology*. Show that when X is endowed with this topology, every sequence in X converges to every point of X, and every map from a topological space into X is continuous.

To avoid pathological cases like this, which result when X does not have sufficiently many open subsets, we often restrict our attention to topological spaces satisfying the following special condition. A topological space X is said to be a *Hausdorff space* if for every pair of distinct points $p, q \in X$, there exist disjoint open subsets $U, V \subseteq X$ such that $p \in U$ and $q \in V$.

▶ **Exercise A.10.** Show that every metric space is Hausdorff in the metric topology.

▶ **Exercise A.11.** Let X be a Hausdorff space. Show that each finite subset of X is closed, and that each convergent sequence in X has a unique limit.

Bases and Countability

Suppose X is a topological space. A collection $\mathcal{B}$ of open subsets of X is said to be a *basis for the topology of X* (plural: *bases*) if every open subset of X is the union of some collection of elements of $\mathcal{B}$.

More generally, suppose X is merely a set, and $\mathcal{B}$ is a collection of subsets of X satisfying the following conditions:

(i) $X = \bigcup_{B \in \mathcal{B}} B$.
(ii) If $B_1, B_2 \in \mathcal{B}$ and $x \in B_1 \cap B_2$, then there exists $B_3 \in \mathcal{B}$ such that $x \in B_3 \subseteq B_1 \cap B_2$.

Then the collection of all unions of elements of $\mathcal{B}$ is a topology on X, called the *topology generated by $\mathcal{B}$*, and $\mathcal{B}$ is a basis for this topology.

If X is a topological space and $p \in X$, a *neighborhood basis at p* is a collection $\mathcal{B}_p$ of neighborhoods of p such that every neighborhood of p contains at least one $B \in \mathcal{B}_p$.

A set is said to be *countably infinite* if it admits a bijection with the set of positive integers, and *countable* if it is finite or countably infinite. A topological space X is said to be *first-countable* if there is a countable neighborhood basis at each point, and *second-countable* if there is a countable basis for its topology. Since a countable basis for X contains a countable neighborhood basis at each point, second-countability implies first-countability.

The next lemma expresses the most important properties of first-countable spaces. To say that a sequence is *eventually in a subset* means that all but finitely many terms of the sequence are in the subset.

Lemma A.12 (Sequence Lemma). *Let X be a first-countable space, let $A \subseteq X$ be any subset, and let $x \in X$.*

(a) $x \in \bar{A}$ *if and only if x is a limit of a sequence of points in A.*
(b) $x \in \operatorname{Int} A$ *if and only if every sequence in X converging to x is eventually in A.*

(c) *A is closed in X if and only if A contains every limit of every convergent sequence of points in A.*

(d) *A is open in X if and only if every sequence in X converging to a point of A is eventually in A.*

▶ **Exercise A.13.** Prove the sequence lemma.

▶ **Exercise A.14.** Show that every metric space is first-countable.

▶ **Exercise A.15.** Show that the set of all open balls in $\mathbb{R}^n$ whose radii are rational and whose centers have rational coordinates is a countable basis for the Euclidean topology, and thus $\mathbb{R}^n$ is second-countable.

One of the most important properties of second-countable spaces is expressed in the following proposition. Let X be a topological space. A **cover of X** is a collection $\mathcal{U}$ of subsets of X whose union is X; it is called an **open cover** if each of the sets in $\mathcal{U}$ is open. A **subcover of $\mathcal{U}$** is a subcollection of $\mathcal{U}$ that is still a cover.

Proposition A.16. *Let X be a second-countable topological space. Every open cover of X has a countable subcover.*

Proof. Let $\mathcal{B}$ be a countable basis for X, and let $\mathcal{U}$ be an arbitrary open cover of X. Let $\mathcal{B}' \subseteq \mathcal{B}$ be the collection of basis open subsets $B \in \mathcal{B}$ such that $B \subseteq U$ for some $U \in \mathcal{U}$. For each $B \in \mathcal{B}'$, choose a particular set $U_B \in \mathcal{U}$ containing B. The collection $\{U_B : B \in \mathcal{B}'\}$ is countable, so it suffices to show that it covers X. Given a point $x \in X$, there is some $V \in \mathcal{U}$ containing x, and because $\mathcal{B}$ is a basis there exists $B \in \mathcal{B}$ such that $x \in B \subseteq V$. This implies, in particular, that $B \in \mathcal{B}'$, and therefore $x \in B \subseteq U_B$. $\square$

Subspaces, Products, Disjoint Unions, and Quotients

Subspaces

Probably the simplest way to obtain new topological spaces from old ones is by taking subsets of other spaces. If X is a topological space and $S \subseteq X$ is an arbitrary subset, we define the **subspace topology on S** (sometimes called the **relative topology**) by declaring a subset $U \subseteq S$ to be open in S if and only if there exists an open subset $V \subseteq X$ such that $U = V \cap S$. A subset of S that is open or closed in the subspace topology is sometimes said to be **relatively open** or **relatively closed in S**, to make it clear that we do not mean open or closed as a subset of X. Any subset of X endowed with the subspace topology is said to be a **subspace of X**. Whenever we treat a subset of a topological space as a space in its own right, we always assume that it has the subspace topology unless otherwise specified.

If X and Y are topological spaces, a continuous injective map $F: X \to Y$ is called a **topological embedding** if it is a homeomorphism onto its image $F(X) \subseteq Y$ in the subspace topology.

The most important properties of the subspace topology are summarized in the next proposition.

Proposition A.17 (Properties of the Subspace Topology). *Let X be a topological space and let S be a subspace of X.*

(a) CHARACTERISTIC PROPERTY: *If Y is a topological space, a map $F: Y \to S$ is continuous if and only if the composition $\iota_S \circ F: Y \to X$ is continuous, where $\iota_S: S \hookrightarrow X$ is the inclusion map (the restriction of the identity map of X to S).*

(b) *The subspace topology is the unique topology on S for which the characteristic property holds.*

(c) *A subset $K \subseteq S$ is closed in S if and only if there exists a closed subset $L \subseteq X$ such that $K = L \cap S$.*

(d) *The inclusion map $\iota_S: S \hookrightarrow X$ is a topological embedding.*

(e) *If Y is a topological space and $F: X \to Y$ is continuous, then $F|_S: S \to Y$ (the restriction of F to S) is continuous.*

(f) *If $\mathcal{B}$ is a basis for the topology of X, then $\mathcal{B}_S = \{B \cap S : B \in \mathcal{B}\}$ is a basis for the subspace topology on S.*

(g) *If X is Hausdorff, then so is S.*

(h) *If X is first-countable, then so is S.*

(i) *If X is second-countable, then so is S.*

▶ **Exercise A.18.** Prove the preceding proposition.

If X and Y are topological spaces and $F: X \to Y$ is a continuous map, part (e) of the preceding proposition guarantees that the restriction of F to every subspace of X is continuous (in the subspace topology). We can also ask the converse question: If we know that the restriction of F to certain subspaces of X is continuous, is F itself continuous? The next two propositions express two somewhat different answers to this question.

Lemma A.19 (Continuity Is Local). *Continuity is a local property, in the following sense: if $F: X \to Y$ is a map between topological spaces such that every point $p \in X$ has a neighborhood U on which the restriction $F|_U$ is continuous, then F is continuous.*

Lemma A.20 (Gluing Lemma for Continuous Maps). *Let X and Y be topological spaces, and suppose one of the following conditions holds:*

(a) *$B_1, \dots, B_n$ are finitely many closed subsets of X whose union is X.*

(b) *$\{B_i\}_{i \in A}$ is a collection of open subsets of X whose union is X.*

Suppose that for all i we are given continuous maps $F_i: B_i \to Y$ that agree on overlaps: $F_i|_{B_i \cap B_j} = F_j|_{B_i \cap B_j}$. Then there exists a unique continuous map $F: X \to Y$ whose restriction to each B_i is equal to F_i.

▶ **Exercise A.21.** Prove the two preceding lemmas.

▶ **Exercise A.22.** Let X be a topological space, and suppose X admits a countable open cover $\{U_i\}$ such that each set U_i is second-countable in the subspace topology. Show that X is second-countable.

Product Spaces

Next we consider finite products of topological spaces. If $X_1, \ldots, X_k$ are (finitely many) sets, their ***Cartesian product*** is the set $X_1 \times \cdots \times X_k$ consisting of all ordered k-tuples of the form $(x_1, \ldots, x_k)$ with $x_i \in X_i$ for each i. The ***i th projection map*** is the map $\pi_i \colon X_1 \times \cdots \times X_k \to X_i$ defined by $\pi_i(x_1, \ldots, x_k) = x_i$.

Suppose $X_1, \ldots, X_k$ are topological spaces. The collection of all subsets of $X_1 \times \cdots \times X_k$ of the form $U_1 \times \cdots \times U_k$, where each U_i is open in X_i, forms a basis for a topology on $X_1 \times \cdots \times X_k$, called the ***product topology***. Endowed with this topology, a finite product of topological spaces is called a ***product space***. Any open subset of the form $U_1 \times \cdots \times U_k \subseteq X_1 \times \cdots \times X_k$, where each U_i is open in X_i, is called a ***product open subset***. (A slightly different definition is required for products of infinitely many spaces, but we need only the finite case. See [LeeTM] for more about infinite product spaces.)

Proposition A.23 (Properties of the Product Topology). *Suppose $X_1, \ldots, X_k$ are topological spaces, and let $X_1 \times \cdots \times X_k$ be their product space.*

(a) CHARACTERISTIC PROPERTY: *If B is a topological space, a map $F \colon B \to X_1 \times \cdots \times X_k$ is continuous if and only if each of its component functions $F_i = \pi_i \circ F \colon B \to X_i$ is continuous.*
(b) *The product topology is the unique topology on $X_1 \times \cdots \times X_k$ for which the characteristic property holds.*
(c) *Each projection map $\pi_i \colon X_1 \times \cdots \times X_k \to X_i$ is continuous.*
(d) *Given any continuous maps $F_i \colon X_i \to Y_i$ for $i = 1, \ldots, k$, the **product map** $F_1 \times \cdots \times F_k \colon X_1 \times \cdots \times X_k \to Y_1 \times \cdots \times Y_k$ is continuous, where*

$$F_1 \times \cdots \times F_k(x_1, \ldots, x_k) = \big(F_1(x_1), \ldots, F_k(x_k)\big).$$

(e) *If S_i is a subspace of X_i for $i = 1, \ldots, n$, the product topology and the subspace topology on $S_1 \times \cdots \times S_n \subseteq X_1 \times \cdots \times X_n$ coincide.*
(f) *For any $i \in \{1, \ldots, k\}$ and any choices of points $a_j \in X_j$ for $j \neq i$, the map $x \mapsto (a_1, \ldots, a_{i-1}, x, a_{i+1}, \ldots, a_k)$ is a topological embedding of X_i into the product space $X_1 \times \cdots \times X_k$.*
(g) *If $\mathcal{B}_i$ is a basis for the topology of X_i for $i = 1, \ldots, k$, then the collection*

$$\mathcal{B} = \{B_1 \times \cdots \times B_k : B_i \in \mathcal{B}_i\}$$

is a basis for the topology of $X_1 \times \cdots \times X_k$.
(h) *Every finite product of Hausdorff spaces is Hausdorff.*
(i) *Every finite product of first-countable spaces is first-countable.*
(j) *Every finite product of second-countable spaces is second-countable.*

▶ **Exercise A.24.** Prove the preceding proposition.

Disjoint Union Spaces

Another simple way of building new topological spaces is by taking disjoint unions of other spaces. From a set-theoretic point of view, the disjoint union is defined as follows. If $(X_\alpha)_{\alpha \in A}$ is an indexed family of sets, their *disjoint union* is the set

$$\coprod_{\alpha \in A} X_\alpha = \{(x,\alpha) : \alpha \in A, \ x \in X_\alpha\}.$$

For each α, there is a canonical injective map $\iota_\alpha \colon X_\alpha \to \coprod_{\alpha \in A} X_\alpha$ given by $\iota_\alpha(x) = (x,\alpha)$, and the images of these maps for different values of α are disjoint. Typically, we implicitly identify X_α with its image in the disjoint union, thereby viewing X_α as a subset of $\coprod_{\alpha \in A} X_\alpha$. The α in the notation (x,α) should be thought of as a "tag" to indicate which set x comes from, so that the subsets corresponding to different values of α are disjoint, even if some or all of the original sets X_α were identical.

Given an indexed family of topological spaces $(X_\alpha)_{\alpha \in A}$, we define the *disjoint union topology* on $\coprod_{\alpha \in A} X_\alpha$ by declaring a subset of $\coprod_{\alpha \in A} X_\alpha$ to be open if and only if its intersection with each X_α is open in X_α.

Proposition A.25 (Properties of the Disjoint Union Topology). *Suppose* $(X_\alpha)_{\alpha \in A}$ *is an indexed family of topological spaces, and* $\coprod_{\alpha \in A} X_\alpha$ *is endowed with the disjoint union topology.*

(a) CHARACTERISTIC PROPERTY: *If Y is a topological space, a map*

$$F \colon \coprod_{\alpha \in A} X_\alpha \to Y$$

is continuous if and only if $F \circ \iota_\alpha \colon X_\alpha \to Y$ is continuous for each $\alpha \in A$.
(b) *The disjoint union topology is the unique topology on* $\coprod_{\alpha \in A} X_\alpha$ *for which the characteristic property holds.*
(c) *A subset of* $\coprod_{\alpha \in A} X_\alpha$ *is closed if and only if its intersection with each X_α is closed.*
(d) *Each injection* $\iota_\alpha \colon X_\alpha \to \coprod_{\alpha \in A} X_\alpha$ *is a topological embedding.*
(e) *Every disjoint union of Hausdorff spaces is Hausdorff.*
(f) *Every disjoint union of first-countable spaces is first-countable.*
(g) *Every disjoint union of countably many second-countable spaces is second-countable.*

▶ **Exercise A.26.** Prove the preceding proposition.

Quotient Spaces and Quotient Maps

If X is a topological space, Y is a set, and $\pi \colon X \to Y$ is a surjective map, the *quotient topology on Y determined by π* is defined by declaring a subset $U \subseteq Y$ to be open if and only if $\pi^{-1}(U)$ is open in X. If X and Y are topological spaces, a map $\pi \colon X \to Y$ is called a *quotient map* if it is surjective and continuous and Y has the quotient topology determined by π.

The following construction is the most common way of producing quotient maps. A relation $\sim$ on a set X is called an *equivalence relation* if it is *reflexive* ($x \sim x$ for all $x \in X$), *symmetric* ($x \sim y$ implies $y \sim x$), and *transitive* ($x \sim y$ and $y \sim z$ imply

$x \sim z$). If $R \subseteq X \times X$ is any relation on X, then the intersection of all equivalence relations on X containing R is an equivalence relation, called the **equivalence relation generated by R**. If $\sim$ is an equivalence relation on X, then for each $x \in X$, the **equivalence class of x**, denoted by $[x]$, is the set of all $y \in X$ such that $y \sim x$. The set of all equivalence classes is a **partition of X**: a collection of disjoint nonempty subsets whose union is X.

Suppose X is a topological space and $\sim$ is an equivalence relation on X. Let $X/\sim$ denote the set of equivalence classes in X, and let $\pi \colon X \to X/\sim$ be the natural projection sending each point to its equivalence class. Endowed with the quotient topology determined by π, the space $X/\sim$ is called the **quotient space** (or **identification space**) **of X determined by $\sim$**. For example, suppose X and Y are topological spaces, $A \subseteq Y$ is a closed subset, and $f \colon A \to X$ is a continuous map. The relation $a \sim f(a)$ for all $a \in A$ generates an equivalence relation on $X \amalg Y$, whose quotient space is denoted by $X \cup_f Y$ and called an **adjunction space**. It is said to be formed by **attaching Y to X along f**.

If $\pi \colon X \to Y$ is a map, a subset $U \subseteq X$ is said to be **saturated with respect to π** if U is the entire preimage of its image: $U = \pi^{-1}\big(\pi(U)\big)$. Given $y \in Y$, the **fiber of π over y** is the set $\pi^{-1}(y)$. Thus, a subset of X is saturated if and only if it is a union of fibers.

Theorem A.27 (Properties of Quotient Maps). *Let $\pi \colon X \to Y$ be a quotient map.*

(a) CHARACTERISTIC PROPERTY: *If B is a topological space, a map $F \colon Y \to B$ is continuous if and only if $F \circ \pi \colon X \to B$ is continuous.*

(b) *The quotient topology is the unique topology on Y for which the characteristic property holds.*

(c) *A subset $K \subseteq Y$ is closed if and only if $\pi^{-1}(K)$ is closed in X.*

(d) *If π is injective, then it is a homeomorphism.*

(e) *If $U \subseteq X$ is a saturated open or closed subset, then the restriction $\pi|_U \colon U \to \pi(U)$ is a quotient map.*

(f) *Any composition of π with another quotient map is again a quotient map.*

▶ **Exercise A.28.** Prove the preceding theorem.

▶ **Exercise A.29.** Let X and Y be topological spaces, and suppose that $F \colon X \to Y$ is a surjective continuous map. Show that the following are equivalent:

(a) F is a quotient map.

(b) F takes saturated open subsets to open subsets.

(c) F takes saturated closed subsets to closed subsets.

The next two properties of quotient maps play important roles in topology, and have equally important generalizations in smooth manifold theory (see Chapter 4).

Theorem A.30 (Passing to the Quotient). *Suppose $\pi \colon X \to Y$ is a quotient map, B is a topological space, and $F \colon X \to B$ is a continuous map that is constant on the fibers of π (i.e., $\pi(p) = \pi(q)$ implies $F(p) = F(q)$). Then there exists a unique continuous map $\widetilde{F} \colon Y \to B$ such that $F = \widetilde{F} \circ \pi$.*

Proof. The existence and uniqueness of $\widetilde{F}$ follow from set-theoretic considerations, and its continuity is an immediate consequence of the characteristic property of the quotient topology. $\square$

Theorem A.31 (Uniqueness of Quotient Spaces). *If $\pi_1 \colon X \to Y_1$ and $\pi_2 \colon X \to Y_2$ are quotient maps that are constant on each other's fibers (i.e., $\pi_1(p) = \pi_1(q)$ if and only if $\pi_2(p) = \pi_2(q)$), then there exists a unique homeomorphism $\varphi \colon Y_1 \to Y_2$ such that $\varphi \circ \pi_1 = \pi_2$.*

Proof. Applying the preceding theorem to the quotient map $\pi_1 \colon X \to Y_1$, we see that π_2 passes to the quotient, yielding a continuous map $\widetilde{\pi}_2 \colon Y_1 \to Y_2$ satisfying $\widetilde{\pi}_2 \circ \pi_1 = \pi_2$. Applying the same argument with the roles of π_1 and π_2 reversed, there is a continuous map $\widetilde{\pi}_1 \colon Y_2 \to Y_1$ satisfying $\widetilde{\pi}_1 \circ \pi_2 = \pi_1$. Together, these identities imply that $\widetilde{\pi}_2 \circ \widetilde{\pi}_1 \circ \pi_2 = \pi_2$. Applying Theorem A.30 again with π_2 playing the roles of both π and F, we see that both $\widetilde{\pi}_2 \circ \widetilde{\pi}_1$ and Id_{Y_2} are obtained from π_2 by passing to the quotient, so the uniqueness assertion of Theorem A.30 implies that $\widetilde{\pi}_2 \circ \widetilde{\pi}_1 = \mathrm{Id}_{Y_2}$. A similar argument shows that $\widetilde{\pi}_1 \circ \widetilde{\pi}_2 = \mathrm{Id}_{Y_1}$, so that $\widetilde{\pi}_2$ is the desired homeomorphism. $\square$

Open and Closed Maps

A map $F \colon X \to Y$ (continuous or not) is said to be an ***open map*** if for every open subset $U \subseteq X$, the image set $F(U)$ is open in Y, and a ***closed map*** if for every closed subset $K \subseteq X$, the image $F(K)$ is closed in Y. Continuous maps may be open, closed, both, or neither, as can be seen by examining simple examples involving subsets of the plane.

▶ **Exercise A.32.** Suppose $X_1, \ldots, X_k$ are topological spaces. Show that each projection $\pi_i \colon X_1 \times \cdots \times X_k \to X_i$ is an open map.

▶ **Exercise A.33.** Let $(X_\alpha)_{\alpha \in A}$ be an indexed family of topological spaces. Show that each injection $\iota_\alpha \colon X_\alpha \to \coprod_{\alpha \in A} X_\alpha$ is both open and closed.

▶ **Exercise A.34.** Show that every local homeomorphism is an open map.

▶ **Exercise A.35.** Show that every bijective local homeomorphism is a homeomorphism.

▶ **Exercise A.36.** Suppose $q \colon X \to Y$ is an open quotient map. Prove that Y is Hausdorff if and only if the set $\mathcal{R} = \{(x_1, x_2) : q(x_1) = q(x_2)\}$ is closed in $X \times X$.

▶ **Exercise A.37.** Let X and Y be topological spaces, and let $F \colon X \to Y$ be a map. Prove the following:

(a) F is closed if and only if for every $A \subseteq X$, $F\left(\overline{A}\right) \supseteq \overline{F(A)}$.
(b) F is open if and only if for every $B \subseteq Y$, $F^{-1}(\mathrm{Int}\, B) \supseteq \mathrm{Int}\, F^{-1}(B)$.

The most important classes of continuous maps in topology are the homeomorphisms, quotient maps, and topological embeddings. Obviously, it is necessary for a map to be bijective in order for it to be a homeomorphism, surjective for it to be a quotient map, and injective for it to be a topological embedding. However, even when a continuous map is known to satisfy one of these necessary set-theoretic conditions, it is not always easy to tell whether it has the desired topological property. One simple sufficient condition is that it be either an open or a closed map, as the next theorem shows.

Theorem A.38. *Suppose X and Y are topological spaces, and $F: X \to Y$ is a continuous map that is either open or closed.*

(a) *If F is surjective, then it is a quotient map.*
(b) *If F is injective, then it is a topological embedding.*
(c) *If F is bijective, then it is a homeomorphism.*

Proof. Suppose first that F is surjective. If it is open, it certainly takes saturated open subsets to open subsets. Similarly, if it is closed, it takes saturated closed subsets to closed subsets. Thus it is a quotient map by Exercise A.29.

Now suppose F is open and injective. Then $F: X \to F(X)$ is bijective, so $F^{-1}: F(X) \to X$ exists by elementary set-theoretic considerations. If $U \subseteq X$ is open, then $(F^{-1})^{-1}(U) = F(U)$ is open in Y by hypothesis, and therefore is also open in $F(X)$ by definition of the subspace topology on $F(X)$. This proves that F^{-1} is continuous, so that F is a homeomorphism onto its image. If F is closed, the same argument goes through with "open" replaced by "closed" (using the characterization of closed subsets of $F(X)$ given in Proposition A.17(c)). This proves part (b), and part (c) is just the special case of (b) in which $F(X) = Y$. $\qquad\square$

Connectedness and Compactness

A topological space X is said to be **disconnected** if it has two disjoint nonempty open subsets whose union is X, and it is **connected** otherwise. Equivalently, X is connected if and only if the only subsets of X that are both open and closed are $\varnothing$ and X itself. If X is any topological space, a **connected subset of** X is a subset that is a connected space when endowed with the subspace topology. For example, the nonempty connected subsets of $\mathbb{R}$ are the **singletons** (one-element sets) and the **intervals**, which are the subsets $J \subseteq \mathbb{R}$ containing more than one point and having the property that whenever $a, b \in J$ and $a < c < b$, it follows that $c \in J$ as well.

A maximal connected subset of X (i.e., a connected subset that is not properly contained in any larger connected subset) is called a **component** (or **connected component) of** X.

Proposition A.39 (Properties of Connected Spaces). *Let X and Y be topological spaces.*

(a) *If $F: X \to Y$ is continuous and X is connected, then $F(X)$ is connected.*
(b) *Every connected subset of X is contained in a single component of X.*

(c) *A union of connected subspaces of X with a point in common is connected.*
(d) *The components of X are disjoint nonempty closed subsets whose union is X, and thus they form a partition of X.*
(e) *If S is a subset of X that is both open and closed, then S is a union of components of X.*
(f) *Every finite product of connected spaces is connected.*
(g) *Every quotient space of a connected space is connected.*

▶ **Exercise A.40.** Prove the preceding proposition.

Closely related to connectedness is *path connectedness*. If X is a topological space and $p, q \in X$, a ***path in X from p to q*** is a continuous map $f : I \to X$ (where $I = [0, 1]$) such that $f(0) = p$ and $f(1) = q$. If for every pair of points $p, q \in X$ there exists a path in X from p to q, then X is said to be ***path-connected***. The ***path components of X*** are its maximal path-connected subsets.

Proposition A.41 (Properties of Path-Connected Spaces).

(a) *Proposition A.39 holds with "connected" replaced by "path-connected" and "component" by "path component" throughout.*
(b) *Every path-connected space is connected.*

▶ **Exercise A.42.** Prove the preceding proposition.

For most topological spaces we treat in this book, including all manifolds, connectedness and path connectedness turn out to be equivalent. The link between the two concepts is provided by the following notion. A topological space is said to be ***locally path-connected*** if it admits a basis of path-connected open subsets.

Proposition A.43 (Properties of Locally Path-Connected Spaces). *Let X be a locally path-connected topological space.*

(a) *The components of X are open in X.*
(b) *The path components of X are equal to its components.*
(c) *X is connected if and only if it is path-connected.*
(d) *Every open subset of X is locally path-connected.*

▶ **Exercise A.44.** Prove the preceding proposition.

A topological space X is said to be ***compact*** if every open cover of X has a finite subcover. A ***compact subset*** of a topological space is one that is a compact space in the subspace topology. For example, it is a consequence of the Heine–Borel theorem that a subset of $\mathbb{R}^n$ is compact if and only if it is closed and bounded.

Proposition A.45 (Properties of Compact Spaces). *Let X and Y be topological spaces.*

(a) *If $F : X \to Y$ is continuous and X is compact, then $F(X)$ is compact.*
(b) *If X is compact and $f : X \to \mathbb{R}$ is continuous, then f is bounded and attains its maximum and minimum values on X.*

(c) *Any union of finitely many compact subspaces of X is compact.*
(d) *If X is Hausdorff and K and L are disjoint compact subsets of X, then there exist disjoint open subsets $U, V \subseteq X$ such that $K \subseteq U$ and $L \subseteq V$.*
(e) *Every closed subset of a compact space is compact.*
(f) *Every compact subset of a Hausdorff space is closed.*
(g) *Every compact subset of a metric space is bounded.*
(h) *Every finite product of compact spaces is compact.*
(i) *Every quotient of a compact space is compact.*

▶ **Exercise A.46.** Prove the preceding proposition.

For maps between metric spaces, there are several variants of continuity that are useful, especially in the context of compact spaces. Suppose (M_1, d_1) and (M_2, d_2) are metric spaces, and $F\colon M_1 \to M_2$ is a map. Then F is said to be ***uniformly continuous*** if for every $\varepsilon > 0$, there exists $\delta > 0$ such that for all $x, y \in M_1$, $d_1(x, y) < \delta$ implies $d_2\big(F(x), F(y)\big) < \varepsilon$. It is said to be ***Lipschitz continuous*** if there is a constant C such that $d_2\big(F(x), F(y)\big) \leq C d_1(x, y)$ for all $x, y \in M_1$. Any such C is called a ***Lipschitz constant for F***. We say that F is ***locally Lipschitz continuous*** if every point $x \in M_1$ has a neighborhood on which F is Lipschitz continuous. (To emphasize the distinction, Lipschitz continuous functions are sometimes called ***uniformly*** or ***globally Lipschitz continuous***.)

▶ **Exercise A.47.** For maps between metric spaces, show that Lipschitz continuous ⇒ uniformly continuous ⇒ continuous, and Lipschitz continuous ⇒ locally Lipschitz continuous ⇒ continuous. (Exercise A.49 below shows that these implications are not reversible.)

Proposition A.48. *Suppose (M_1, d_1) and (M_2, d_2) are metric spaces and $F\colon M_1 \to M_2$ is a map. Let K be any compact subset of M_1.*

(a) *If F is continuous, then $F|_K$ is uniformly continuous.*
(b) *If F is locally Lipschitz continuous, then $F|_K$ is Lipschitz continuous.*

Proof. First we prove (a). Assume F is continuous, and let $\varepsilon > 0$ be given. For each $x \in K$, by continuity there is a positive number $\delta(x)$ such that $d_1(x, y) < 2\delta \Rightarrow d_2\big(F(x), F(y)\big) < \varepsilon/2$. Because the open balls $\{B_{\delta(x)}(x) : x \in K\}$ cover K, by compactness there are finitely many points $x_1, \dots, x_n \in K$ such that $K \subseteq B_{\delta(x_1)}(x_1) \cup \cdots \cup B_{\delta(x_n)}(x_n)$. Let $\delta = \min\{\delta(x_1), \dots, \delta(x_n)\}$. Suppose $x, y \in K$ satisfy $d_1(x, y) < \delta$. There is some i such that $x \in B_{\delta(x_i)}(x_i)$, and then the triangle inequality implies that x and y both lie in $B_{2\delta(x_i)}(x_i)$. It follows that

$$d_2\big(F(x), F(y)\big) \leq d_2\big(F(x), F(x_i)\big) + d_2\big(F(x_i), F(y)\big) < \varepsilon/2 + \varepsilon/2 = \varepsilon.$$

Next we prove (b). Assume F is locally Lipschitz continuous. Because F is continuous, Proposition A.45 shows that $F(K)$ is compact and therefore bounded. Let $D = \operatorname{diam} F(K)$. For each $x \in K$, there is a positive number $\delta(x)$ such that F is Lipschitz continuous on $B_{2\delta(x)}(x)$, with Lipschitz constant $C(x)$. By compactness, there are points $x_1, \dots, x_n \in K$ such that $K \subseteq B_{\delta(x_1)}(x_1) \cup \cdots \cup B_{\delta(x_n)}(x_n)$.

Let $C = \max\{C(x_1), \ldots, C(x_n)\}$ and $\delta = \min\{\delta(x_1), \ldots, \delta(x_n)\}$, and let $x, y \in K$ be arbitrary. On the one hand, if $d_1(x, y) < \delta$, then by the same argument as in the preceding paragraph, x and y lie in one of the balls on which F is Lipschitz continuous, so $d_2(F(x), F(y)) \le C d_1(x, y)$. On the other hand, if $d_1(x, y) \ge \delta$, then $d_2(F(x), F(y)) \le D \le (D/\delta) d_1(x, y)$. Therefore, $\max\{C, D/\delta\}$ is a Lipschitz constant for F on K. $\qquad\square$

▶ **Exercise A.49.** Let $f, g \colon [0, \infty) \to \mathbb{R}$ be defined by $f(x) = \sqrt{x}$ and $g(x) = x^2$. Show that f is uniformly continuous but not locally or globally Lipschitz continuous, and g is locally Lipschitz continuous but not uniformly continuous or globally Lipschitz continuous.

For manifolds, subsets of manifolds, and most other spaces we work with, there are two other equivalent formulations of compactness that are frequently useful. Proofs of the next proposition can be found in [LeeTM, Chap. 4], [Mun00, Chap. 3], and [Sie92, Chap. 7].

Proposition A.50 (Equivalent Formulations of Compactness). *Suppose M is a second-countable Hausdorff space or a metric space. The following are equivalent.*

(a) *M is compact.*
(b) *Every infinite subset of M has a limit point in M.*
(c) *Every sequence in M has a convergent subsequence in M.*

▶ **Exercise A.51.** Show that every compact metric space is complete.

The next lemma expresses one of the most useful properties of compact spaces.

Lemma A.52 (Closed Map Lemma). *Suppose X is a compact space, Y is a Hausdorff space, and $F \colon X \to Y$ is a continuous map.*

(a) *F is a closed map.*
(b) *If F is surjective, it is a quotient map.*
(c) *If F is injective, it is a topological embedding.*
(d) *If F is bijective, it is a homeomorphism.*

Proof. By virtue of Theorem A.38, the last three assertions follow from the first, so we need only prove that F is closed. Suppose $K \subseteq X$ is a closed subset. Then part (e) of Proposition A.45 implies that K is compact; part (a) of that proposition implies that $F(K)$ is compact; and part (f) implies that $F(K)$ is closed in Y. $\qquad\square$

If X and Y are topological spaces, a map $F \colon X \to Y$ (continuous or not) is said to be ***proper*** if for every compact set $K \subseteq Y$, the preimage $F^{-1}(K)$ is compact. Here are some useful sufficient conditions for a map to be proper.

Proposition A.53 (Sufficient Conditions for Properness). *Suppose X and Y are topological spaces, and $F \colon X \to Y$ is a continuous map.*

(a) *If X is compact and Y is Hausdorff, then F is proper.*
(b) *If F is a closed map with compact fibers, then F is proper.*
(c) *If F is a topological embedding with closed image, then F is proper.*

(d) *If Y is Hausdorff and F has a continuous left inverse (i.e., a continuous map $G: Y \to X$ such that $G \circ F = \mathrm{Id}_X$), then F is proper.*

(e) *If F is proper and $A \subseteq X$ is a subset that is saturated with respect to F, then $F|_A: A \to F(A)$ is proper.*

▶ **Exercise A.54.** Prove the preceding proposition.

Locally Compact Hausdorff Spaces

In general, the topological spaces whose properties are most familiar are those whose topologies are induced by metrics; such a topological space is said to be *metrizable*. However, when studying manifolds, it is often quite inconvenient to exhibit a metric that generates a manifold's topology. Fortunately, as shown in Chapter 1, manifolds belong to another class of spaces with similarly nice properties, the *locally compact Hausdorff spaces*. In this section, we review some of the properties of these spaces.

A topological space X is said to be ***locally compact*** if every point has a neighborhood contained in a compact subset of X. If X is Hausdorff, this property has two equivalent formulations that are often more useful, as the next exercise shows. A subset of X is said to be ***precompact in X*** if its closure in X is compact.

▶ **Exercise A.55.** For a Hausdorff space X, show that the following are equivalent:

(a) X is locally compact.
(b) Each point of X has a precompact neighborhood.
(c) X has a basis of precompact open subsets.

▶ **Exercise A.56.** Prove that every open or closed subspace of a locally compact Hausdorff space is itself a locally compact Hausdorff space.

The next result can be viewed as a generalization of the closed map lemma (Lemma A.52).

Theorem A.57 (Proper Continuous Maps Are Closed). *Suppose X is a topological space and Y is a locally compact Hausdorff space. Then every proper continuous map $F: X \to Y$ is closed.*

Proof. Let $K \subseteq X$ be a closed subset. To show that $F(K)$ is closed in Y, we show that it contains all of its limit points. Let y be a limit point of $F(K)$, and let U be a precompact neighborhood of y. Then y is also a limit point of $F(K) \cap \bar{U}$. Because F is proper, $F^{-1}(\bar{U})$ is compact, which implies that $K \cap F^{-1}(\bar{U})$ is compact. Because F is continuous, $F(K \cap F^{-1}(\bar{U})) = F(K) \cap \bar{U}$ is compact and therefore closed in Y. In particular, $y \in F(K) \cap \bar{U} \subseteq F(K)$, so $F(K)$ is closed. $\qquad\square$

Here is an important property of locally compact Hausdorff spaces, which is also shared by complete metric spaces. For a proof, see [LeeTM, Chap. 4].

Theorem A.58 (Baire Category Theorem). *In a locally compact Hausdorff space or a complete metric space, every countable union of nowhere dense sets has empty interior.*

Corollary A.59. *In a locally compact Hausdorff space or a complete metric space, every nonempty countable closed subset contains at least one isolated point.*

Proof. Assume X is such a space. Let $A \subseteq X$ be a nonempty countable closed subset, and assume that A has no isolated points. The fact that A is closed in X means that A itself is either a locally compact Hausdorff space or a complete metric space. For each $a \in A$, the singleton $\{a\}$ is nowhere dense in A: it is closed in A because A is Hausdorff, and it contains no nonempty open subset because A has no isolated points. Since A is a countable union of singletons, the Baire category theorem implies that A has empty interior in A, which is a contradiction. $\qquad\square$

If we add the hypothesis of second-countability to a locally compact Hausdorff space, we can prove even more. A sequence $(K_i)_{i=1}^{\infty}$ of compact subsets of a topological space X is called an ***exhaustion of X by compact sets*** if $X = \bigcup_i K_i$ and $K_i \subseteq \operatorname{Int} K_{i+1}$ for each i.

Proposition A.60. *A second-countable, locally compact Hausdorff space admits an exhaustion by compact sets.*

Proof. Let X be such a space. Because X is a locally compact Hausdorff space, it has a basis of precompact open subsets; since it is second-countable, it is covered by countably many such sets. Let $(U_i)_{i=1}^{\infty}$ be such a countable cover. Beginning with $K_1 = \bar{U}_1$, assume by induction that we have constructed compact sets $K_1, \ldots, K_k$ satisfying $U_j \subseteq K_j$ for each j and $K_{j-1} \subseteq \operatorname{Int} K_j$ for $j \geq 2$. Because K_k is compact, there is some m_k such that $K_k \subseteq U_1 \cup \cdots \cup U_{m_k}$. If we let $K_{k+1} = \bar{U}_1 \cup \cdots \cup \bar{U}_{m_k}$, then K_{k+1} is a compact set whose interior contains K_k. Moreover, by increasing m_k if necessary, we may assume that $m_k \geq k + 1$, so that $U_{k+1} \subseteq K_{k+1}$. By induction, we obtain the required exhaustion. $\qquad\square$

Homotopy and the Fundamental Group

If X and Y are topological spaces and $F_0, F_1 \colon X \to Y$ are continuous maps, a ***homotopy from F_0 to F_1*** is a continuous map $H \colon X \times I \to Y$ satisfying

$$H(x, 0) = F_0(x),$$

$$H(x, 1) = F_1(x),$$

for all $x \in X$. If there exists a homotopy from F_0 to F_1, we say that ***F_0 and F_1 are homotopic***, and write $F_0 \simeq F_1$. If the homotopy satisfies $H(x, t) = F_0(x) = F_1(x)$ for all $t \in I$ and all x in some subset $A \subseteq X$, the maps F_0 and F_1 are said to be ***homotopic relative to A***. Both "homotopic" and "homotopic relative to A" are equivalence relations on the set of all continuous maps from X to Y.

The most important application of homotopies is to paths. Suppose X is a topological space. Two paths $f_0, f_1 \colon I \to X$ are said to be ***path-homotopic***, denoted symbolically by $f_0 \sim f_1$, if they are homotopic relative to $\{0, 1\}$. Explicitly, this means that there is a continuous map $H: I \times I \to X$ satisfying

$$
\begin{aligned}
H(s,0) &= f_0(s), & s &\in I; \\
H(s,1) &= f_1(s), & s &\in I; \\
H(0,t) &= f_0(0) = f_1(0), & t &\in I; \\
H(1,t) &= f_0(1) = f_1(1), & t &\in I.
\end{aligned}
$$

For any given points $p, q \in X$, path homotopy is an equivalence relation on the set of all paths from p to q. The equivalence class of a path f is called its **path class**, and is denoted by $[f]$.

Given two paths $f, g: I \to X$ such that $f(1) = g(0)$, their **product** is the path $f \cdot g: I \to X$ defined by

$$
f \cdot g(s) = \begin{cases} f(2s), & 0 \le s \le \frac{1}{2}; \\ g(2s-1), & \frac{1}{2} \le s \le 1. \end{cases}
$$

If $f \sim f'$ and $g \sim g'$, it is not hard to show that $f \cdot g \sim f' \cdot g'$. Therefore, it makes sense to define the product of the path classes $[f]$ and $[g]$ by $[f] \cdot [g] = [f \cdot g]$. Although multiplication of paths is not associative, it is associative up to path homotopy: $([f] \cdot [g]) \cdot [h] = [f] \cdot ([g] \cdot [h])$. When we need to consider products of three or more actual paths (as opposed to path classes), we adopt the convention that such products are to be evaluated from left to right: $f \cdot g \cdot h = (f \cdot g) \cdot h$.

If X is a topological space and q is a point in X, a **loop in X based at q** is a path in X from q to q, that is, a continuous map $f: I \to X$ such that $f(0) = f(1) = q$. The set of path classes of loops based at q is denoted by $\pi_1(X, q)$. Equipped with the product described above, it is a group, called the **fundamental group of X based at q**. The identity element of this group is the path class of the **constant path** $c_q(s) \equiv q$, and the inverse of $[f]$ is the path class of the **reverse path** $\bar{f}(s) = f(1-s)$.

It can be shown that for path-connected spaces, the fundamental groups based at different points are isomorphic. If X is path-connected and for some (hence every) $q \in X$, the fundamental group $\pi_1(X, q)$ is the trivial group consisting of $[c_q]$ alone, we say that X is **simply connected**. This means that every loop is path-homotopic to a constant path.

▶ **Exercise A.61.** Let X be a path-connected topological space. Show that X is simply connected if and only if every pair of paths in X with the same starting and ending points are path-homotopic.

A key feature of the homotopy relation is that it is preserved by composition, as the next proposition shows.

Proposition A.62. *If $F_0, F_1: X \to Y$ and $G_0, G_1: Y \to Z$ are continuous maps with $F_0 \simeq F_1$ and $G_0 \simeq G_1$, then $G_0 \circ F_0 \simeq G_1 \circ F_1$. Similarly, if $f_0, f_1: I \to X$ are path-homotopic and $F: X \to Y$ is a continuous map, then $F \circ f_0 \sim F \circ f_1$.*

▶ **Exercise A.63.** Prove the preceding proposition.

Thus if $F: X \to Y$ is a continuous map, for each $q \in X$ we obtain a well-defined map $F_*: \pi_1(X, q) \to \pi_1(Y, F(q))$ by setting

$$F_*[f] = [F \circ f].$$

Proposition A.64. *If X and Y are topological spaces and $F: X \to Y$ is a continuous map, then $F_*: \pi_1(X, q) \to \pi_1(Y, F(q))$ is a group homomorphism, known as the **homomorphism induced by F**.*

Proposition A.65 (Properties of the Induced Homomorphism).

(a) *Let $F: X \to Y$ and $G: Y \to Z$ be continuous maps. Then for each $q \in X$, $(G \circ F)_* = G_* \circ F_*: \pi_1(X, q) \to \pi_1(Z, G(F(q)))$.*

(b) *For each space X and each $q \in X$, the homomorphism induced by the identity map $\mathrm{Id}_X: X \to X$ is the identity map of $\pi_1(X, q)$.*

(c) *If $F: X \to Y$ is a homeomorphism, then $F_*: \pi_1(X, q) \to \pi_1(Y, F(q))$ is an isomorphism. Thus, homeomorphic spaces have isomorphic fundamental groups.*

▶ **Exercise A.66.** Prove the two preceding propositions.

▶ **Exercise A.67.** A subset $U \subseteq \mathbb{R}^n$ is said to be **star-shaped** if there is a point $c \in U$ such that for each $x \in U$, the line segment from c to x is contained in U. Show that every star-shaped set is simply connected.

Proposition A.68 (Fundamental Groups of Spheres).

(a) *$\pi_1(\mathbb{S}^1, (1, 0))$ is the infinite cyclic group generated by the path class of the loop $\omega: I \to \mathbb{S}^1$ given by $\omega(s) = (\cos 2\pi s, \sin 2\pi s)$.*

(b) *If $n > 1$, $\mathbb{S}^n$ is simply connected.*

Proposition A.69 (Fundamental Groups of Product Spaces). *Suppose $X_1, \ldots, X_k$ are topological spaces, and let $p_i: X_1 \times \cdots \times X_k \to X_i$ denote the ith projection map. For any points $q_i \in X_i$, $i = 1, \ldots, k$, define a map*

$$P: \pi_1(X_1 \times \cdots \times X_k, (q_1, \ldots, q_k)) \to \pi_1(X_1, q_1) \times \cdots \times \pi_1(X_k, q_k)$$

by

$$P[f] = (p_{1*}[f], \ldots, p_{k*}[f]).$$

Then P is an isomorphism.

▶ **Exercise A.70.** Prove the two preceding propositions.

A continuous map $F: X \to Y$ between topological spaces is said to be a **homotopy equivalence** if there is a continuous map $G: Y \to X$ such that $F \circ G \simeq \mathrm{Id}_Y$ and $G \circ F \simeq \mathrm{Id}_X$. Such a map G is called a **homotopy inverse for F**. If there exists a homotopy equivalence between X and Y, the two spaces are said to be **homotopy equivalent**. For example, the inclusion map $\iota: \mathbb{S}^{n-1} \hookrightarrow \mathbb{R}^n \smallsetminus \{0\}$ is a homotopy equivalence with homotopy inverse $r(x) = x/|x|$, because $r \circ \iota = \mathrm{Id}_{\mathbb{S}^{n-1}}$ and $\iota \circ r$ is homotopic to the identity map of $\mathbb{R}^n \smallsetminus \{0\}$ via the straight-line homotopy $H(x, t) = tx + (1 - t)x/|x|$.

Theorem A.71 (Homotopy Invariance). *If $F \colon X \to Y$ is a homotopy equivalence, then for each $p \in X$, $F_* \colon \pi_1(X, p) \to \pi_1(Y, F(p))$ is an isomorphism.*

For a proof, see any of the topology texts mentioned at the beginning of this appendix.

Covering Maps

Suppose E and X are topological spaces. A map $\pi \colon E \to X$ is called a *covering map* if E and X are connected and locally path-connected, π is surjective and continuous, and each point $p \in X$ has a neighborhood U that is *evenly covered by π*, meaning that each component of $\pi^{-1}(U)$ is mapped homeomorphically onto U by π. In this case, X is called the *base of the covering*, and E is called a *covering space of X*. If U is an evenly covered subset of X, the components of $\pi^{-1}(U)$ are called the *sheets of the covering over U*.

Some immediate consequences of the definition should be noted. First, it follows from Proposition A.43 that E and X are actually path-connected. Second, suppose $U \subseteq X$ is any evenly covered open subset. Because $\pi^{-1}(U)$ is open in E, it is locally path-connected, and therefore its components are open subsets of $\pi^{-1}(U)$ and thus also of E. Because U is the homeomorphic image of any one of the components of $\pi^{-1}(U)$, each of which is path-connected, it follows that evenly covered open subsets are path-connected.

▶ **Exercise A.72.** Show that every covering map is a local homeomorphism, an open map, and a quotient map.

▶ **Exercise A.73.** Show that an injective covering map is a homeomorphism.

▶ **Exercise A.74.** Show that all fibers of a covering map have the same cardinality, called the *number of sheets of the covering*.

▶ **Exercise A.75.** Show that a covering map is a proper map if and only if it is finite-sheeted.

▶ **Exercise A.76.** Show that every finite product of covering maps is a covering map.

The main properties of covering maps that we need are summarized in the next four propositions. For proofs, you can consult [LeeTM, Chaps. 11 and 12], [Mun00, Chaps. 9 and 13], or [Sie92, Chap. 14].

If $\pi \colon E \to X$ is a covering map and $F \colon B \to X$ is a continuous map, a *lift of F* is a continuous map $\widetilde{F} \colon B \to E$ such that $\pi \circ \widetilde{F} = F$:

Proposition A.77 (Lifting Properties of Covering Maps). *Suppose* $\pi: E \to X$ *is a covering map.*

(a) UNIQUE LIFTING PROPERTY: *If B is a connected space and $F: B \to X$ is a continuous map, then any two lifts of F that agree at one point are identical.*
(b) PATH LIFTING PROPERTY: *If $f: I \to X$ is a path, then for any point $e \in E$ such that $\pi(e) = f(0)$, there exists a unique lift $\tilde{f}_e: I \to E$ of f such that $\tilde{f}(0) = e$.*
(c) MONODROMY THEOREM: *If $f, g: I \to X$ are path-homotopic paths and $\tilde{f}_e, \tilde{g}_e: I \to E$ are their lifts starting at the same point $e \in E$, then $\tilde{f}_e$ and $\tilde{g}_e$ are path-homotopic and $\tilde{f}_e(1) = \tilde{g}_e(1)$.*

Proposition A.78 (Lifting Criterion). *Suppose $\pi: E \to X$ is a covering map, Y is a connected and locally path-connected space, and $F: Y \to X$ is a continuous map. Let $y \in Y$ and $e \in E$ be such that $\pi(e) = F(y)$. Then there exists a lift $\tilde{F}: Y \to E$ of F satisfying $\tilde{F}(y) = e$ if and only if $F_*\big(\pi_1(Y, y)\big) \subseteq \pi_*\big(\pi_1(E, e)\big)$.*

Proposition A.79 (Coverings of Simply Connected Spaces). *If X is a simply connected space, then every covering map $\pi: E \to X$ is a homeomorphism.*

A topological space is said to be *locally simply connected* if it admits a basis of simply connected open subsets.

Proposition A.80 (Existence of a Universal Covering Space). *If X is a connected and locally simply connected topological space, there exists a simply connected topological space $\tilde{X}$ and a covering map $\pi: \tilde{X} \to X$. If $\hat{\pi}: \hat{X} \to X$ is any other simply connected covering of X, there is a homeomorphism $\varphi: \tilde{X} \to \hat{X}$ such that $\hat{\pi} \circ \varphi = \pi$.*

The simply connected covering space $\tilde{X}$ whose existence and uniqueness (up to homeomorphism) are guaranteed by this proposition is called the *universal covering space of X*.

Appendix B
Review of Linear Algebra

For the basic properties of vector spaces and linear maps, you can consult almost any linear algebra book that treats vector spaces abstractly, such as [FIS03]. Here we just summarize the main points, with emphasis on those aspects that are most important for the study of smooth manifolds.

Vector Spaces

Let $\mathbb{R}$ denote the field of real numbers. A *vector space over $\mathbb{R}$* (or *real vector space*) is a set V endowed with two operations: *vector addition* $V \times V \to V$, denoted by $(v, w) \mapsto v + w$, and *scalar multiplication* $\mathbb{R} \times V \to V$, denoted by $(a, v) \mapsto av$, satisfying the following properties:

(i) V is an abelian group under vector addition.
(ii) Scalar multiplication satisfies the following identities:

$$a(bv) = (ab)v \quad \text{for all } v \in V \text{ and } a, b \in \mathbb{R};$$
$$1v = v \qquad \text{for all } v \in V.$$

(iii) Scalar multiplication and vector addition are related by the following distributive laws:

$$(a + b)v = av + bv \quad \text{for all } v \in V \text{ and } a, b \in \mathbb{R};$$
$$a(v + w) = av + aw \quad \text{for all } v, w \in V \text{ and } a \in \mathbb{R}.$$

This definition can be generalized in two directions. First, replacing $\mathbb{R}$ by an arbitrary field $\mathbb{F}$ everywhere, we obtain the definition of a *vector space over $\mathbb{F}$*. In particular, we sometimes have occasion to consider vector spaces over $\mathbb{C}$, called *complex vector spaces*. Unless we specify otherwise, all vector spaces are assumed to be real.

Second, if $\mathbb{R}$ is replaced by a commutative ring $\mathcal{R}$, this becomes the definition of a *module over $\mathcal{R}$* (or *$\mathcal{R}$-module*). For example, if $\mathbb{Z}$ denotes the ring of integers, it is straightforward to check that modules over $\mathbb{Z}$ are just abelian groups under addition.

The elements of a vector space are usually called **vectors**. When it is necessary to distinguish them from vectors, elements of the underlying field (which is $\mathbb{R}$ unless otherwise specified) are called **scalars**.

Let V be a vector space. A subset $W \subseteq V$ that is closed under vector addition and scalar multiplication is itself a vector space with the same operations, and is called a **subspace of** V. To avoid confusion with the use of the word "subspace" in topology, we sometimes use the term **linear subspace** for a subspace of a vector space in this sense, and **topological subspace** for a subset of a topological space endowed with the subspace topology.

A finite sum of the form $\sum_{i=1}^{k} a^i v_i$, where a^i are scalars and $v_i \in V$, is called a **linear combination of the vectors** $v_1, \ldots, v_k$. (The reason we write the coefficients a^i with superscripts instead of subscripts is to be consistent with the Einstein summation convention, explained in Chapter 1.) If S is an arbitrary subset of V, the set of all linear combinations of elements of S is called the **span of** S and is denoted by $\mathrm{span}(S)$; it is easily seen to be the smallest subspace of V containing S. If $V = \mathrm{span}(S)$, we say that S **spans** V. By convention, a linear combination of no elements is considered to sum to zero, and the span of the empty set is $\{0\}$.

If p and q are points of V, the **line segment from p to q** is the set $\{(1-t)p + tq : 0 \le t \le 1\}$. A subset $B \subseteq V$ is said to be **convex** if for every two points $p, q \in B$, the line segment from p to q is contained in B.

Bases and Dimension

Suppose V is a vector space. A subset $S \subseteq V$ is said to be **linearly dependent** if there exists a linear relation of the form $\sum_{i=1}^{k} a^i v_i = 0$, where $v_1, \ldots, v_k$ are distinct elements of S and at least one of the coefficients a^i is nonzero; S is said to be **linearly independent** otherwise. In other words, S is linearly independent if and only if the only linear combination of distinct elements of S that sums to zero is the one in which all the scalar coefficients are zero. Note that every set containing the zero vector is linearly dependent. By convention, the empty set is considered to be linearly independent.

It is frequently important to work with ordered k-tuples of vectors in V; such a k-tuple is denoted by $(v_1, \ldots, v_k)$ or (v_i), with parentheses instead of braces to distinguish it from the (unordered) set of elements $\{v_1, \ldots, v_k\}$. When we consider ordered k-tuples, linear dependence takes on a slightly different meaning. We say that $(v_1, \ldots, v_k)$ is a **linearly dependent k-tuple** if there are scalars $(a^1, \ldots, a^k)$, not all zero, such that $\sum_{i=1}^{k} a^i v_i = 0$; it is a **linearly independent k-tuple** otherwise. The only difference between a *linearly independent set* and a *linearly independent k-tuple* is that the latter cannot have repeated vectors. For example if $v \in V$ is a nonzero vector, the ordered pair (v, v) is linearly dependent, while the set $\{v, v\} = \{v\}$ is linearly independent. On the other hand, if $(v_1, \ldots, v_k)$ is any linearly independent k-tuple, then the set $\{v_1, \ldots, v_k\}$ is also linearly independent.

▶ **Exercise B.1.** Let V be a vector space. Prove the following statements.

(a) If $S \subseteq V$ is linearly independent, then every subset of S is linearly independent.

(b) If $S \subseteq V$ is linearly dependent or spans V, then every subset of V that properly contains S is linearly dependent.

(c) A subset $S \subseteq V$ containing more than one element is linearly dependent if and only if some element $v \in S$ can be expressed as a linear combination of elements of $S \smallsetminus \{v\}$.

(d) If $(v_1, \ldots, v_k)$ is a linearly dependent k-tuple in V with $v_1 \neq 0$, then some v_i can be expressed as a linear combination of the *preceding* vectors $(v_1, \ldots, v_{i-1})$.

A *basis for V* (plural: *bases*) is a subset $S \subseteq V$ that is linearly independent and spans V. If S is a basis for V, every element of V has a *unique* expression as a linear combination of elements of S. If V has a finite basis, then V is said to be *finite-dimensional*, and otherwise it is *infinite-dimensional*. The trivial vector space $\{0\}$ is finite-dimensional, because it has the empty set as a basis.

If V is finite-dimensional, an *ordered basis for V* is a basis endowed with a specific ordering of the basis vectors, or equivalently a linearly independent n-tuple (E_i) that spans V. For most purposes, ordered bases are more useful than unordered bases, so we always assume, often without comment, that each basis comes with a given ordering.

If $(E_1, \ldots, E_n)$ is an (ordered) basis for V, each vector $v \in V$ has a unique expression as a linear combination of basis vectors:

$$v = \sum_{i=1}^{n} v^i E_i.$$

The numbers v^i are called the *components of v* with respect to this basis, and the ordered n-tuple $(v^1, \ldots, v^n)$ is called its *basis representation*. (Here is an example of a definition that requires an ordered basis.)

Lemma B.2. *Let V be a vector space. If V is spanned by a set of n vectors, then every subset of V containing more than n vectors is linearly dependent.*

Proof. Suppose $\{v_1, \ldots, v_n\}$ is an n-element set that spans V. To prove the lemma, it suffices to show that every set containing exactly $n + 1$ vectors is linearly dependent. Let $\{w_1, \ldots, w_{n+1}\}$ be such a set. If any of the w_i's is zero, then clearly the set is dependent, so we might as well assume they are all nonzero. By Exercise B.1(b), the set $\{w_1, v_1, \ldots, v_n\}$ is linearly dependent, and thus so is the ordered $(n+1)$-tuple $(w_1, v_1, \ldots, v_n)$. By Exercise B.1(d), one of the vectors v_j can be written as a linear combination of $\{w_1, v_1, \ldots, v_{j-1}\}$, and thus the set $\{w_1, v_1, \ldots, v_{j-1}, v_{j+1}, \ldots, v_n\}$ still spans V. Renumbering the v_i's if necessary, we may assume that the set $\{w_1, v_2, \ldots, v_n\}$ spans V.

Now suppose by induction that $\{w_1, w_2, \ldots, w_{k-1}, v_k, \ldots, v_n\}$ spans V. As before, the $(n + 1)$-tuple $(w_1, w_2, \ldots, w_{k-1}, w_k, v_k, \ldots, v_n)$ is linearly dependent, so one of the vectors in this list can be written as a linear combination of the preceding ones. If one of the w_i's can be so written, then the set $\{w_1, \ldots, w_{n+1}\}$ is dependent and we are done. Otherwise, one of the v_j's can be so written, and after reordering we may assume that $\{w_1, w_2, \ldots, w_k, v_{k+1}, \ldots, v_n\}$ still spans V. Continuing by induction, by the time we get to $k = n$, if we have not already shown that the w_i's are dependent, we conclude that the set $\{w_1, \ldots, w_n\}$ spans V. But this means that the set $\{w_1, \ldots, w_{n+1}\}$ is linearly dependent by Exercise B.1(b).   □

Proposition B.3. *If V is a finite-dimensional vector space, all bases for V contain the same number of elements.*

Proof. If $\{E_1, \ldots, E_n\}$ is a basis for V with n elements, then Lemma B.2 implies that every set containing more than n elements is linearly dependent, so no basis can have more than n elements. On the other hand, if there were a basis containing fewer than n elements, then Lemma B.2 would imply that $\{E_1, \ldots, E_n\}$ is linearly dependent, which is a contradiction.   □

Because of the preceding proposition, if V is a finite-dimensional vector space, it makes sense to define the ***dimension of V***, denoted by $\dim V$, to be the number of elements in any basis.

▶ **Exercise B.4.** Suppose V is a finite-dimensional vector space.

(a) Show that every set that spans V contains a basis, and every linearly independent subset of V is contained in a basis.
(b) Show that every subspace $S \subseteq V$ is finite-dimensional and satisfies $\dim S \leq \dim V$, with equality if and only if $S = V$.
(c) Show that $\dim V = 0$ if and only if $V = \{0\}$.

▶ **Exercise B.5.** Suppose V is an infinite-dimensional vector space.

(a) Use Zorn's lemma to show that every linearly independent subset of V is contained in a basis.
(b) Show that any two bases for V have the same cardinality. [Hint: assume that S and T are bases such that S has larger cardinality than T. Each element of T can be expressed as a linear combination of elements of S, and the hypothesis guarantees that some element of S does not appear in any of the expressions for elements of T. Show that this element can be expressed as a linear combination of other elements of S, contradicting the hypothesis that S is linearly independent.]

If S is a subspace of a finite-dimensional vector space V, we define the ***codimension of S in V*** to be $\dim V - \dim S$. By virtue of Exercise B.4(b), the codimension of S is always nonnegative, and is zero if and only if $S = V$. A (***linear***) ***hyperplane*** is a linear subspace of codimension 1.

Example B.6 (Euclidean Spaces). For each integer $n \geq 0$, $\mathbb{R}^n$ is a real vector space under the usual operations of vector addition and scalar multiplication:

$$\left(x^1, \ldots, x^n\right) + \left(y^1, \ldots, y^n\right) = \left(x^1 + y^1, \ldots, x^n + y^n\right),$$
$$a\left(x^1, \ldots, x^n\right) = \left(ax^1, \ldots, ax^n\right).$$

There is a natural basis $(e_1, \ldots, e_n)$ for $\mathbb{R}^n$, called the ***standard basis***, where $e_i = (0, \ldots, 1, \ldots, 0)$ is the vector with a 1 in the ith place and zeros elsewhere; thus

$\mathbb{R}^n$ has dimension n, as one would expect. Any element $x \in \mathbb{R}^n$ can be written $(x^1, \ldots, x^n) = \sum_{i=1}^n x^i e_i$, so its components with respect to the standard basis are just its coordinates $(x^1, \ldots, x^n)$.   //

Example B.7 (Complex Euclidean Spaces). With scalar multiplication and vector addition defined just as in the real case, the n-dimensional complex Euclidean space $\mathbb{C}^n$ becomes a complex vector space. Because the vectors $(e_1, \ldots, e_n)$, defined as above, form a basis for $\mathbb{C}^n$ over $\mathbb{C}$, it follows that $\mathbb{C}^n$ has dimension n as a complex vector space.

By restricting scalar multiplication to real scalars, we can also consider $\mathbb{C}^n$ as a *real vector space*. In this case, it is straightforward to check that the vectors $(e_1, ie_1, \ldots, e_n, ie_n)$ form a basis for $\mathbb{C}^n$ over $\mathbb{R}$, so $\mathbb{C}^n$ has dimension $2n$ when considered as a real vector space.   //

If S and T are subspaces of a vector space V, the notation $S + T$ denotes the set of all vectors of the form $v + w$, where $v \in S$ and $w \in T$. It is easily seen to be a subspace of V, and in fact is the subspace spanned by $S \cup T$. If $S + T = V$ and $S \cap T = \{0\}$, then V is said to be the (**internal**) **direct sum of S and T**, and we write $V = S \oplus T$. Two linear subspaces $S, T \subseteq V$ are said to be **complementary subspaces** if $V = S \oplus T$. In this case, every vector in V has a *unique* expression as a sum of an element of S plus an element of T.

▶ **Exercise B.8.** Suppose S and T are subspaces of a finite-dimensional vector space V.

(a) Show that $S \cap T$ is a subspace of V.
(b) Show that $\dim(S + T) = \dim S + \dim T - \dim(S \cap T)$.
(c) Suppose $V = S + T$. Show that $V = S \oplus T$ if and only if $\dim V = \dim S + \dim T$.

▶ **Exercise B.9.** Let V be a finite-dimensional vector space. Show that every subspace $S \subseteq V$ has a complementary subspace in V. In fact, given an arbitrary basis $(E_1, \ldots, E_n)$ for V, show that there is some subset $\{i_1, \ldots, i_k\}$ of the integers $\{1, \ldots, n\}$ such that $\text{span}(E_{i_1}, \ldots, E_{i_k})$ is a complement to S. [Hint: choose a basis $(F_1, \ldots, F_m)$ for S, and apply Exercise B.1(d) to the ordered $(m + n)$-tuple $(F_1, \ldots, F_m, E_1, \ldots, E_n)$.]

Suppose $S \subseteq V$ is a linear subspace. Any subset of V of the form

$$v + S = \{v + w : w \in S\}$$

for some fixed $v \in V$ is called an **affine subspace of V parallel to S**. If S is a linear hyperplane, then any affine subspace parallel to S is called an **affine hyperplane**.

▶ **Exercise B.10.** Let V be a vector space, and let $v + S$ be an affine subspace of V parallel to S.

(a) Show that $v + S$ is a linear subspace if an only if it contains 0, which is true if and only if $v \in S$.
(b) Show that $v + S = \tilde{v} + \tilde{S}$ if and only if $S = \tilde{S}$ and $v - \tilde{v} \in S$.

Because of part (b) of the preceding exercise, we can unambiguously define the *dimension of v + S* to be the dimension of S.

For each vector $v \in V$, the affine subspace $v + S$ is also called the *coset of S determined by v*. The set V/S of cosets of S is called the *quotient of V by S*.

▶ **Exercise B.11.** Suppose V is a vector space and S is a linear subspace of V. Define vector addition and scalar multiplication of cosets by

$$(v + S) + (w + S) = (v + w) + S,$$

$$c(v + S) = (cv) + S.$$

(a) Show that the quotient V/S is a vector space under these operations.
(b) Show that if V is finite-dimensional, then $\dim V/S = \dim V - \dim S$.

Linear Maps

Let V and W be real vector spaces. A map $T \colon V \to W$ is *linear* if $T(av + bw) = aTv + bTw$ for all vectors $v, w \in V$ and all scalars a, b. (Because of the close connection between linear maps and matrix multiplication described below, we generally write the action of a linear map T on a vector v as Tv without parentheses, unless parentheses are needed for grouping.) In the special case $W = \mathbb{R}$, a linear map from V to $\mathbb{R}$ is usually called a *linear functional on V*.

If $T \colon V \to W$ is a linear map, the *kernel* or *null space of T*, denoted by $\operatorname{Ker} T$ or $T^{-1}(0)$, is the set $\{v \in V : Tv = 0\}$, and the *image of T*, denoted by $\operatorname{Im} T$ or $T(V)$, is the set $\{w \in W : w = Tv \text{ for some } v \in V\}$.

One simple but important example of a linear map arises in the following way. Given a subspace $S \subseteq V$ and a complementary subspace T, there is a unique linear map $\pi \colon V \to S$ defined by

$$\pi(v + w) = v \quad \text{for } v \in S, \ w \in T.$$

This map is called the *projection onto S with kernel T*.

If V and W are vector spaces, a bijective linear map $T \colon V \to W$ is called an *isomorphism*. In this case, there is a unique inverse map $T^{-1} \colon W \to V$, and the following computation shows that T^{-1} is also linear:

$$
\begin{aligned}
aT^{-1}v + bT^{-1}w &= T^{-1}T\left(aT^{-1}v + bT^{-1}w\right) \\
&= T^{-1}\left(aTT^{-1}v + bTT^{-1}w\right) \quad \text{(by linearity of } T\text{)} \\
&= T^{-1}(av + bw).
\end{aligned}
$$

For this reason, a bijective linear map is also said to be *invertible*. If there exists an isomorphism $T \colon V \to W$, then V and W are said to be *isomorphic*.

Example B.12. Let V be an n-dimensional real vector space, and $(E_1, \dots, E_n)$ be an ordered basis for V. Define a map $E \colon \mathbb{R}^n \to V$ by

$$E\left(x^1, \dots, x^n\right) = x^1 E_1 + \cdots + x^n E_n.$$

Then E is linear and bijective, so it is an isomorphism, called the **basis isomorphism** determined by this basis. Thus, every n-dimensional real vector space is isomorphic to $\mathbb{R}^n$.   //

▶ **Exercise B.13.** Let V and W be vector spaces, and let $(E_1, \dots, E_n)$ be a basis for V. For any n elements $w_1, \dots, w_n \in W$, show that there is a unique linear map $T: V \to W$ satisfying $T(E_i) = w_i$ for $i = 1, \dots, n$.

▶ **Exercise B.14.** Let $S: V \to W$ and $T: W \to X$ be linear maps.

(a) Show that $\operatorname{Ker} S$ and $\operatorname{Im} S$ are subspaces of V and W, respectively.
(b) Show that S is injective if and only if $\operatorname{Ker} S = \{0\}$.
(c) Show that if S is an isomorphism, then $\dim V = \dim W$ (in the sense that these dimensions are either both infinite or both finite and equal).
(d) Show that if S and T are both injective or both surjective, then $T \circ S$ has the same property.
(e) Show that if $T \circ S$ is surjective, then T is surjective; give an example to show that S might not be.
(f) Show that if $T \circ S$ is injective, then S is injective; give an example to show that T might not be.

▶ **Exercise B.15.** Suppose V is a vector space and S is a subspace of V, and let $\pi: V \to V/S$ denote the projection defined by $\pi(v) = v + S$.

(a) Show that π is a surjective linear map with kernel equal to S.
(b) Given a linear map $T: V \to W$, show that there exists a linear map $\tilde{T}: V/S \to W$ such that $\tilde{T} \circ \pi = T$ if and only if $S \subseteq \operatorname{Ker} T$.

If V and W are vector spaces, a map $F: V \to W$ is called an **affine map** if it can be written in the form $F(v) = w + Tv$ for some linear map $T: V \to W$ and some fixed $w \in W$.

▶ **Exercise B.16.** Suppose $F: V \to W$ is an affine map. Show that $F(V)$ is an affine subspace of W, and the sets $F^{-1}(z)$ for $z \in W$ are parallel affine subspaces of V.

▶ **Exercise B.17.** Suppose V is a finite-dimensional vector space. Show that every affine subspace of V is of the form $F^{-1}(z)$ for some affine map $F: V \to W$ and some $z \in W$.

Now suppose V and W are finite-dimensional vector spaces with ordered bases $(E_1, \dots, E_n)$ and $(F_1, \dots, F_m)$, respectively. If $T: V \to W$ is a linear map, the **matrix of T** with respect to these bases is the $m \times n$ matrix

$$A = \left(A^i_j \right) = \begin{pmatrix} A^1_1 & \cdots & A^1_n \\ \vdots & \ddots & \vdots \\ A^m_1 & \cdots & A^m_n \end{pmatrix}$$

whose jth column consists of the components of TE_j with respect to the basis (F_i):

$$TE_j = \sum_{i=1}^{m} A^i_j F_i.$$

By linearity, the action of T on an arbitrary vector $v = \sum_j v^j E_j$ is then given by

$$T\left(\sum_{j=1}^{n} v^j E_j\right) = \sum_{i=1}^{m}\sum_{j=1}^{n} A^i_j v^j F_i.$$

If we write the components of a vector with respect to a basis as a column matrix, then the matrix representation of $w = Tv$ is given by matrix multiplication:

$$\begin{pmatrix} w^1 \\ \vdots \\ w^m \end{pmatrix} = \begin{pmatrix} A^1_1 & \cdots & A^1_n \\ \vdots & \ddots & \vdots \\ A^m_1 & \cdots & A^m_n \end{pmatrix} \begin{pmatrix} v^1 \\ \vdots \\ v^n \end{pmatrix},$$

or, more succinctly,

$$w^i = \sum_{j=1}^{n} A^i_j v^j.$$

Insofar as possible, we denote the row index of a matrix by a superscript and the column index by a subscript, so that A^i_j represents the element in the ith row and jth column. Thus the entry in the ith row and jth column of a matrix product AB is given by

$$(AB)^i_j = \sum_{k=1}^{n} A^i_k B^k_j.$$

The composition of two linear maps is represented by the product of their matrices. Provided we use the same basis for both the domain and the codomain, the identity map on an n-dimensional vector space is represented by the **$n \times n$ identity matrix**, which we denote by I_n; it is the matrix with ones on the main diagonal (where the row number equals the column number) and zeros elsewhere.

The set $M(m \times n, \mathbb{R})$ of all $m \times n$ real matrices is easily seen to be a real vector space of dimension mn. (In fact, by stringing out the matrix entries in a single row, we can identify it in a natural way with $\mathbb{R}^{mn}$.) Similarly, because $\mathbb{C}$ is a real vector space of dimension 2, the set $M(m \times n, \mathbb{C})$ of $m \times n$ complex matrices is a real vector space of dimension $2mn$. When $m = n$, we abbreviate the spaces of $n \times n$ square real and complex matrices by $M(n, \mathbb{R})$ and $M(n, \mathbb{C})$, respectively. In this case, matrix multiplication gives these spaces additional algebraic structure. If V, W, and Z are vector spaces, a map $B \colon V \times W \to Z$ is said to be **bilinear** if it is linear in each variable separately when the other is held fixed:

$$B(a_1 v_1 + a_2 v_2, w) = a_1 B(v_1, w) + a_2 B(v_2, w),$$

$$B(v, a_1 w_1 + a_2 w_2) = a_1 B(v, w_1) + a_2 B(v, w_2).$$

An **algebra** (over $\mathbb{R}$) is a real vector space V endowed with a bilinear product map $V \times V \to V$. The algebra is said to be **commutative** or **associative** if the bilinear product has that property.

▶ **Exercise B.18.** Show that matrix multiplication turns both $M(n, \mathbb{R})$ and $M(n, \mathbb{C})$ into associative algebras over $\mathbb{R}$. Show that they are noncommutative unless $n = 1$.

Suppose A is an $n \times n$ matrix. If there is a matrix B such that $AB = BA = I_n$, then A is said to be **invertible** or **nonsingular**; it is **singular** otherwise.

▶ **Exercise B.19.** Suppose A is an $n \times n$ matrix. Prove the following statements.

(a) If A is nonsingular, then there is a *unique* $n \times n$ matrix B such that $AB = BA = I_n$. This matrix is denoted by A^{-1} and is called the **inverse of** A.
(b) If A is the matrix of a linear map $T: V \to W$ with respect to some bases for V and W, then T is invertible if and only if A is invertible, in which case A^{-1} is the matrix of T^{-1} with respect to the same bases.
(c) If B is an $n \times n$ matrix such that either $AB = I_n$ or $BA = I_n$, then A is nonsingular and $B = A^{-1}$.

Because $\mathbb{R}^n$ comes equipped with the standard basis (e_i), we can unambiguously identify linear maps from $\mathbb{R}^n$ to $\mathbb{R}^m$ with $m \times n$ real matrices, and we often do so without further comment.

Change of Basis

In this book we often need to be concerned with how various objects transform when we change bases. Suppose (E_i) and $(\widetilde{E}_j)$ are two bases for a finite-dimensional real vector space V. Then each basis can be written uniquely in terms of the other, so there is an invertible matrix B, called the **transition matrix** between the two bases, such that

$$E_i = \sum_{j=1}^{n} B_i^j \widetilde{E}_j, \qquad \widetilde{E}_j = \sum_{i=1}^{n} \left(B^{-1}\right)_j^i E_i. \tag{B.1}$$

Now suppose V and W are finite-dimensional vector spaces and $T: V \to W$ is a linear map. With respect to bases (E_i) for the domain V and (F_j) for the codomain W, the map T is represented by some matrix $A = \left(A_j^i\right)$. If $(\widetilde{E}_i)$ and $(\widetilde{F}_j)$ are any other choices of bases for V and W, respectively, let B and C denote the transition matrices satisfying (B.1) and

$$F_i = \sum_{j=1}^{m} C_i^j \widetilde{F}_j, \qquad \widetilde{F}_j = \sum_{i=1}^{m} \left(C^{-1}\right)_j^i F_i.$$

Then a straightforward computation shows that the matrix $\widetilde{A}$ representing T with respect to the new bases is related to A by

$$\widetilde{A}_j^i = \sum_{k,l} C_l^i A_k^l \left(B^{-1}\right)_j^k,$$

or, in matrix notation,

$$\widetilde{A} = CAB^{-1}.$$

In particular, if T is a map from V to itself, we usually use the same basis for the domain and the codomain. In this case, if A denotes the matrix of T with respect to (E_i), and $\widetilde{A}$ is its matrix with respect to $(\widetilde{E}_i)$, we have

$$\widetilde{A} = BAB^{-1}. \tag{B.2}$$

If V and W are real vector spaces, the set $\mathrm{L}(V;W)$ of linear maps from V to W is a real vector space under the operations

$$(S+T)v = Sv + Tv; \qquad (cT)v = c(Tv).$$

If $\dim V = n$ and $\dim W = m$, then each choice of bases for V and W gives us a map $\mathrm{L}(V;W) \to \mathrm{M}(m \times n, \mathbb{R})$, by sending every linear map to its matrix with respect to the chosen bases. This map is easily seen to be linear and bijective, so $\dim \mathrm{L}(V;W) = \dim \mathrm{M}(m \times n, \mathbb{R}) = mn$.

If $T: V \to W$ is a linear map between finite-dimensional spaces, the dimension of $\operatorname{Im} T$ is called the **rank of T**, and the dimension of $\operatorname{Ker} T$ is called its **nullity**. The following theorem shows that, up to choices of bases, a linear map is completely determined by its rank together with the dimensions of its domain and codomain.

Theorem B.20 (Canonical Form for a Linear Map). *Suppose V and W are finite-dimensional vector spaces, and $T: V \to W$ is a linear map of rank r. Then there are bases for V and W with respect to which T has the following matrix representation (in block form):*

$$\begin{pmatrix} I_r & 0 \\ 0 & 0 \end{pmatrix}.$$

Proof. Choose bases $(F_1, \ldots, F_r)$ for $\operatorname{Im} T$ and $(K_1, \ldots, K_k)$ for $\operatorname{Ker} T$. Extend (F_j) arbitrarily to a basis $(F_1, \ldots, F_m)$ for W. By definition of the image, there are vectors $E_1, \ldots, E_r \in V$ such that $TE_i = F_i$ for $i = 1, \ldots, r$. We will show that $(E_1, \ldots, E_r, K_1, \ldots, K_k)$ is a basis for V; once we know this, it follows easily that T has the desired matrix representation.

Suppose first that $\sum_i a^i E_i + \sum_j b^j K_j = 0$. Applying T to this equation yields $\sum_{i=1}^r a^i F_i = 0$, which implies that all the coefficients a^i are zero. Then it follows also that all the b^j's are zero because the K_j's are linearly independent. Therefore, the $(r+k)$-tuple $(E_1, \ldots, E_r, K_1, \ldots, K_k)$ is linearly independent.

To show that these vectors span V, let $v \in V$ be arbitrary. We can express $Tv \in \operatorname{Im} T$ as a linear combination of $(F_1, \ldots, F_r)$:

$$Tv = \sum_{i=1}^r c^i F_i.$$

If we put $w = \sum_i c^i E_i \in V$, it follows that $Tw = Tv$, so $z = v - w \in \operatorname{Ker} T$. Writing $z = \sum_j d^j K_j$, we obtain

$$v = w + z = \sum_{i=1}^r c^i E_i + \sum_{j=1}^k d^j K_j,$$

so the $(r+k)$-tuple $(E_1, \ldots, E_r, K_1, \ldots, K_k)$ does indeed span V. $\qquad\square$

This theorem says that every linear map can be put into a particularly nice diagonal form by appropriate choices of bases in the domain and codomain. However, it is important to be aware of what the theorem does *not* say: if $T: V \to V$ is a linear map from a finite-dimensional vector space to itself, it might not be possible to represent T by a diagonal matrix with respect to the same basis for the domain and codomain.

The next result is central in applications of linear algebra to smooth manifold theory; it is a corollary to the proof of the preceding theorem.

Corollary B.21 (Rank-Nullity Law). *Suppose* $T: V \to W$ *is a linear map between finite-dimensional vector spaces. Then*

$$\dim V = \operatorname{rank} T + \operatorname{nullity} T = \dim(\operatorname{Im} T) + \dim(\operatorname{Ker} T).$$

Proof. The preceding proof showed that V has a basis consisting of $k + r$ elements, where $k = \dim(\operatorname{Ker} T)$ and $r = \dim(\operatorname{Im} T)$. $\qquad\square$

▶ **Exercise B.22.** Suppose V, W, X are finite-dimensional vector spaces, and $S: V \to W$ and $T: W \to X$ are linear maps. Prove the following statements.

(a) rank $S \leq \dim V$, with equality if and only if S is injective.
(b) rank $S \leq \dim W$, with equality if and only if S is surjective.
(c) If $\dim V = \dim W$ and S is either injective or surjective, then it is an isomorphism.
(d) $\operatorname{rank}(T \circ S) \leq \operatorname{rank} S$, with equality if and only if $\operatorname{Im} S \cap \operatorname{Ker} T = \{0\}$.
(e) $\operatorname{rank}(T \circ S) \leq \operatorname{rank} T$, with equality if and only if $\operatorname{Im} S + \operatorname{Ker} T = W$.
(f) If S is an isomorphism, then $\operatorname{rank}(T \circ S) = \operatorname{rank} T$.
(g) If T is an isomorphism, then $\operatorname{rank}(T \circ S) = \operatorname{rank} S$.

Let A be an $m \times n$ matrix. The ***transpose of*** A is the $n \times m$ matrix A^T obtained by interchanging the rows and columns of A: $(A^T)^j_i = A^i_j$. A square matrix A is said to be ***symmetric*** if $A = A^T$ and ***skew-symmetric*** if $A = -A^T$.

▶ **Exercise B.23.** Show that if A and B are matrices of dimensions $m \times n$ and $n \times k$, respectively, then $(AB)^T = B^T A^T$.

The ***rank*** of an $m \times n$ matrix A is defined to be the rank of the corresponding linear map from $\mathbb{R}^n$ to $\mathbb{R}^m$. Because the columns of A, thought of as vectors in $\mathbb{R}^m$, are the images of the standard basis vectors under this linear map, the rank of A can also be thought of as the dimension of the span of its columns, and is sometimes called its ***column rank***. Analogously, we define the ***row rank of*** A to be the dimension of the span of its rows, thought of similarly as vectors in $\mathbb{R}^n$.

Proposition B.24. *The row rank of a matrix is equal to its column rank.*

Proof. Let A be an $m \times n$ matrix. Because the row rank of A is equal to the column rank of A^T, we must show that $\operatorname{rank} A = \operatorname{rank} A^T$.

Suppose the (column) rank of A is k. Thought of as a linear map from $\mathbb{R}^n$ to $\mathbb{R}^m$, A factors through $\operatorname{Im} A$ as follows:

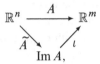

where $\tilde{A}$ is just the map A with its codomain restricted to $\operatorname{Im} A$, and ι is the inclusion of $\operatorname{Im} A$ into $\mathbb{R}^m$. Choosing a basis for the k-dimensional subspace $\operatorname{Im} A$, we can write this as a matrix equation $A = BC$, where B and C are the matrices of ι and $\tilde{A}$ with respect to the standard bases in $\mathbb{R}^n$ and $\mathbb{R}^m$ and the chosen basis in $\operatorname{Im} A$. Taking transposes, we obtain $A^T = C^T B^T$, from which it follows that $\operatorname{rank} A^T \le \operatorname{rank} B^T$. Since B^T is a $k \times m$ matrix, its column rank is at most k, which shows that $\operatorname{rank} A^T \le \operatorname{rank} A$. Reversing the roles of A and A^T and using the fact that $(A^T)^T = A$, we conclude that $\operatorname{rank} A = \operatorname{rank} A^T$. $\square$

Suppose $A = \left(A_j^i \right)$ is an $m \times n$ matrix. If we choose integers $1 \le i_1 < \cdots < i_k \le m$ and $1 \le j_1 < \cdots < j_l \le n$, we obtain a $k \times l$ matrix whose entry in the pth row and qth column is $A_{j_q}^{i_p}$:

$$
\begin{pmatrix}
A_{j_1}^{i_1} & \cdots & A_{j_l}^{i_1} \\
\vdots & \ddots & \vdots \\
A_{j_1}^{i_k} & \cdots & A_{j_l}^{i_k}
\end{pmatrix}.
$$

Such a matrix is called a **submatrix of** A. Looking at submatrices gives a convenient criterion for checking the rank of a matrix.

Proposition B.25. *Suppose A is an $m \times n$ matrix. Then* $\operatorname{rank} A \ge k$ *if and only if some $k \times k$ submatrix of A is nonsingular.*

Proof. By definition, $\operatorname{rank} A \ge k$ if and only if A has at least k linearly independent columns, which is equivalent to A having some $m \times k$ submatrix with rank k. But by Proposition B.24, an $m \times k$ submatrix has rank k if and only if it has k linearly independent rows. Thus A has rank at least k if and only if it has an $m \times k$ submatrix with k linearly independent rows, if and only if it has a $k \times k$ submatrix that is nonsingular. $\square$

The Determinant

There are a number of ways of defining the determinant of a square matrix, each of which has advantages in different contexts. The definition we give here, while not particularly intuitive, is the simplest to state and fits nicely with our treatment of alternating tensors in Chapter 14.

If X is a set, a **permutation of** X is a bijective map from X to itself. The set of all permutations of X is a group under composition. A **transposition** is a permutation that interchanges two elements and leaves all the others fixed.

We let S_n denote the group of permutations of the set $\{1,\dots,n\}$, called the **symmetric group on n elements**. The properties of S_n that we need are summarized in the following proposition; proofs can be found in any good undergraduate algebra text such as [Hun97] or [Her75].

Proposition B.26 (Properties of the Symmetric Group).

(a) *Every element of S_n can be expressed as a composition of finitely many transpositions.*

(b) *For each $\sigma \in S_n$, the parity (evenness or oddness) of the number of factors in any decomposition of σ as a product of transpositions is independent of the choice of decomposition. We say that σ is an **even permutation** if every such decomposition has an even number of factors, and an **odd permutation** otherwise.*

(c) *For each $\sigma \in S_n$, define the **sign of** σ to be the number*

$$\operatorname{sgn}\sigma = \begin{cases} +1 & \text{if } \sigma \text{ is even,} \\ -1 & \text{if } \sigma \text{ is odd.} \end{cases}$$

If $n \geq 2$, $\operatorname{sgn}\colon S_n \to \{\pm 1\}$ is a surjective group homomorphism, where we consider $\{\pm 1\}$ as a group under multiplication.

▶ **Exercise B.27.** Prove (or look up) Proposition B.26.

If $A = \left(A^i_j\right)$ is an $n \times n$ (real or complex) matrix, the **determinant of A** is defined by the expression

$$\det A = \sum_{\sigma \in S_n} (\operatorname{sgn}\sigma) A^{\sigma(1)}_1 \cdots A^{\sigma(n)}_n. \tag{B.3}$$

For simplicity, we assume throughout this section that our matrices are real. The statements and proofs, however, hold equally well in the complex case. In our study of Lie groups we also have occasion to consider determinants of complex matrices.

Although the determinant is defined as a function of matrices, it is also useful to think of it as a function of n vectors in $\mathbb{R}^n$: if $A_1,\dots,A_n \in \mathbb{R}^n$, we interpret $\det(A_1,\dots,A_n)$ to mean the determinant of the matrix whose columns are $(A_1,\dots,A_n)$:

$$\det(A_1,\dots,A_n) = \det \begin{pmatrix} A^1_1 & \cdots & A^1_n \\ \vdots & \ddots & \vdots \\ A^n_1 & \cdots & A^n_n \end{pmatrix}.$$

It is obvious from the defining formula (B.3) that the function $\det\colon \mathbb{R}^n \times \cdots \times \mathbb{R}^n \to \mathbb{R}$ so defined is **multilinear**, which means that it is linear as a function of each vector when all the other vectors are held fixed.

Proposition B.28 (Properties of the Determinant). *Let A be an $n \times n$ matrix.*

(a) *If one column of A is multiplied by a scalar c, the determinant is multiplied by the same scalar:*

$$\det(A_1,\dots,cA_i,\dots,A_n) = c\det(A_1,\dots,A_i,\dots,A_n).$$

(b) *The determinant changes sign when two columns are interchanged*:

$$\det(A_1,\ldots,A_q,\ldots,A_p,\ldots,A_n) = -\det(A_1,\ldots,A_p,\ldots,A_q,\ldots,A_n). \quad \text{(B.4)}$$

(c) *The determinant is unchanged by adding a scalar multiple of one column to any other column*:

$$\det(A_1,\ldots,A_i,\ldots,A_j+cA_i,\ldots,A_n) = \det(A_1,\ldots,A_i,\ldots,A_j\ldots,A_n).$$

(d) *For every scalar c, $\det(cA) = c^n \det A$.*
(e) *If any two columns of A are identical, then $\det A = 0$.*
(f) *If A has a column of zeros, then $\det A = 0$.*
(g) $\det A^T = \det A$.
(h) $\det I_n = 1$.
(i) *If A is singular, then $\det A = 0$.*

Proof. Part (a) is part of the definition of multilinearity, and (d) follows immediately from (a). Part (f) also follows from (a), because a matrix with a column of zeros is unchanged when that column is multiplied by zero, so $\det A = 0(\det A) = 0$. To prove (b), suppose $p < q$ and let $\tau \in S_n$ be the transposition that interchanges p and q, leaving all other indices fixed. Then the left-hand side of (B.4) is equal to

$$\det(A_1,\ldots,A_q,\ldots,A_p,\ldots,A_n)$$
$$= \sum_{\sigma \in S_n} (\mathrm{sgn}\,\sigma) A_1^{\sigma(1)} \cdots A_q^{\sigma(p)} \cdots A_p^{\sigma(q)} \cdots A_n^{\sigma(n)}$$
$$= \sum_{\sigma \in S_n} (\mathrm{sgn}\,\sigma) A_1^{\sigma(1)} \cdots A_p^{\sigma(q)} \cdots A_q^{\sigma(p)} \cdots A_n^{\sigma(n)}$$
$$= \sum_{\sigma \in S_n} (\mathrm{sgn}\,\sigma) A_1^{\sigma(\tau(1))} \cdots A_n^{\sigma(\tau(n))}$$
$$= -\sum_{\sigma \in S_n} \left(\mathrm{sgn}(\sigma\tau)\right) A_1^{\sigma(\tau(1))} \cdots A_n^{\sigma(\tau(n))}$$
$$= -\sum_{\eta \in S_n} (\mathrm{sgn}\,\eta) A_1^{\eta(1)} \cdots A_n^{\eta(n)}$$
$$= -\det(A_1,\ldots,A_p,\ldots,A_q,\ldots,A_n),$$

where the next-to-last line follows by substituting $\eta = \sigma\tau$ and noting that η runs over all elements of S_n as σ does. Part (e) is then an immediate consequence of (b), and (c) follows by multilinearity:

$$\det(A_1,\ldots,A_i,\ldots,A_j+cA_i,\ldots,A_n)$$
$$= \det(A_1,\ldots,A_i,\ldots,A_j\ldots,A_n) + c\det(A_1,\ldots,A_i,\ldots,A_i\ldots,A_n)$$
$$= \det(A_1,\ldots,A_i,\ldots,A_j\ldots,A_n) + 0.$$

Part (g) follows directly from the definition of the determinant:

$$\det A^T = \sum_{\sigma \in S_n} (\operatorname{sgn} \sigma) A^1_{\sigma(1)} \cdots A^n_{\sigma(n)}$$

$$= \sum_{\sigma \in S_n} (\operatorname{sgn} \sigma) A^{\sigma^{-1}(\sigma(1))}_{\sigma(1)} \cdots A^{\sigma^{-1}(\sigma(n))}_{\sigma(n)}$$

$$= \sum_{\sigma \in S_n} (\operatorname{sgn} \sigma) A^{\sigma^{-1}(1)}_{1} \cdots A^{\sigma^{-1}(n)}_{n}$$

$$= \sum_{\eta \in S_n} (\operatorname{sgn} \eta) A^{\eta(1)}_{1} \cdots A^{\eta(n)}_{n} = \det A.$$

In the third line we have used the fact that multiplication is commutative, and the numbers $\{A^{\sigma^{-1}(\sigma(1))}_{\sigma(1)}, \ldots, A^{\sigma^{-1}(\sigma(n))}_{\sigma(n)}\}$ are just $\{A^{\sigma^{-1}(1)}_{1}, \ldots, A^{\sigma^{-1}(n)}_{n}\}$ in a different order; and the fourth line follows by substituting $\eta = \sigma^{-1}$ and noting that $\operatorname{sgn} \sigma^{-1} = \operatorname{sgn} \sigma$. Similarly, (h) follows from the definition, because when A is the identity matrix, for each σ except the identity permutation there is some j such that $A^{\sigma(j)}_{j} = 0$.

Finally, to prove (i), suppose A is singular. Then, as a linear map from $\mathbb{R}^n$ to $\mathbb{R}^n$, A has rank less than n by parts (a) and (b) of Exercise B.22. Thus the columns of A are linearly dependent, so at least one column can be written as a linear combination of the others: $A_j = \sum_{i \neq j} c^i A_i$. The result then follows from (e) and the multilinearity of det. ◻

The operations on matrices described in parts (a), (b), and (c) of the preceding proposition (multiplying one column by a scalar, interchanging two columns, and adding a multiple of one column to another) are called *elementary column operations*. Part of the proposition, therefore, describes precisely how a determinant is affected by elementary column operations. If we define *elementary row operations* analogously, the fact that the determinant of A^T is equal to that of A implies that the determinant behaves similarly under elementary row operations.

Since the columns of an $n \times n$ matrix A are the images of the standard basis vectors under the linear map from $\mathbb{R}^n$ to itself that A defines, elementary column operations correspond to changes of basis in the domain. Thus each elementary column operation on a matrix A can be realized by multiplying A on the right by a suitable matrix. For example, multiplying the ith column by c is achieved by multiplying A by the matrix E_c that is equal to the identity matrix except for a c in the (i, i) position:

$$\begin{pmatrix} A^1_1 & \ldots & A^1_i & \ldots & A^1_n \\ \vdots & & \vdots & & \vdots \\ A^j_1 & \ldots & A^j_i & \ldots & A^j_n \\ \vdots & & \vdots & & \vdots \\ A^n_1 & \ldots & A^n_i & \ldots & A^n_n \end{pmatrix} \begin{pmatrix} 1 & \ldots & 0 & \ldots & 0 \\ & \ddots & & & \\ & & c & & \\ & & & \ddots & \\ 0 & \ldots & 0 & \ldots & 1 \end{pmatrix} = \begin{pmatrix} A^1_1 & \ldots & cA^1_i & \ldots & A^1_n \\ \vdots & & \vdots & & \vdots \\ A^j_1 & \ldots & cA^j_i & \ldots & A^j_n \\ \vdots & & \vdots & & \vdots \\ A^n_1 & \ldots & cA^n_i & \ldots & A^n_n \end{pmatrix}.$$

▶ **Exercise B.29.** Show that replacing one column of a matrix by c times that same column is equivalent to multiplying on the right by a matrix whose determinant is c; interchanging two columns is equivalent to multiplying on the right by a matrix whose determinant is -1; and adding a multiple of one column to another is equivalent to multiplying on the right by a matrix of determinant 1. Matrices of these three types are called *elementary matrices*.

▶ **Exercise B.30.** Suppose A is a nonsingular $n \times n$ matrix.

(a) Show that A can be reduced to the identity I_n by a sequence of elementary column operations.
(b) Show that A is equal to a product of elementary matrices.

Elementary matrices form a key ingredient in the proof of the following theorem, which is arguably the deepest and most important property of the determinant.

Theorem B.31. *If A and B are $n \times n$ matrices, then*

$$\det(AB) = (\det A)(\det B).$$

Proof. If B is singular, then rank $B < n$, which implies that rank $AB < n$. Therefore both $\det B$ and $\det AB$ are zero by Proposition B.28(i). On the other hand, parts (a), (b), and (c) of Proposition B.28 combined with Exercise B.29 show that the theorem is true when B is an elementary matrix. Finally, if B is an arbitrary nonsingular matrix, then B can be written as a product of elementary matrices by Exercise B.30, and the result follows by induction on the number of elementary matrices in such a product. □

Corollary B.32. *If A is a nonsingular $n \times n$ matrix, then $\det A \neq 0$ and $\det(A^{-1}) = (\det A)^{-1}$.*

Proof. Just note that $1 = \det I_n = \det(AA^{-1}) = (\det A)(\det A^{-1})$. □

Corollary B.33. *A square matrix is singular if and only if its determinant is zero.*

Proof. One direction follows from Proposition B.28(i), and the other from Corollary B.32. □

Corollary B.34. *Suppose A and B are $n \times n$ matrices and B is nonsingular. Then $\det(BAB^{-1}) = \det A$.*

Proof. This is just a computation using Theorem B.31 and Corollary B.32:

$$\det(BAB^{-1}) = (\det B)(\det A)(\det B^{-1})$$
$$= (\det B)(\det A)(\det B)^{-1}$$
$$= \det A.$$
□

The last corollary allows us to extend the definition of the determinant to linear maps on arbitrary finite-dimensional vector spaces. Suppose V is an n-dimensional vector space and $T: V \to V$ is a linear map. With respect to a choice of basis for V, T is represented by an $n \times n$ matrix. As we observed above, the matrices

A and $\widetilde{A}$ representing T with respect to two different bases are related by $\widetilde{A} = BAB^{-1}$ for some nonsingular matrix B (see (B.2)). It follows from Corollary B.34, therefore, that $\det \widetilde{A} = \det A$. Thus, we can make the following definition: for each linear map $T: V \to V$ from a finite-dimensional vector space to itself, we define the *determinant of T* to be the determinant of any matrix representation of T (using the same basis for the domain and codomain).

For actual computations of determinants, the formula in the following proposition is usually more useful than the definition.

Proposition B.35 (Expansion by Minors). *Let A be an $n \times n$ matrix, and for each i, j let M_i^j denote the $(n-1) \times (n-1)$ submatrix obtained by deleting the ith column and jth row of A. For any fixed i between 1 and n inclusive,*

$$\det A = \sum_{j=1}^{n} (-1)^{i+j} A_i^j \det M_i^j. \tag{B.5}$$

Proof. It is useful to consider first a special case: suppose A is an $n \times n$ matrix that has the block form

$$A = \begin{pmatrix} B & 0 \\ C & 1 \end{pmatrix}, \tag{B.6}$$

where B is an $(n-1) \times (n-1)$ matrix and C is a $1 \times (n-1)$ row matrix. Then in the defining formula (B.3) for $\det A$, the factor $A_n^{\sigma(n)}$ is equal to 1 when $\sigma(n) = n$ and zero otherwise, so in fact the only terms that are nonzero are those in which $\sigma \in S_{n-1}$, thought of as the subgroup of S_n consisting of elements that permute $\{1, \dots, n-1\}$ and leave n fixed. Thus the determinant of A simplifies to

$$\det A = \sum_{\sigma \in S_{n-1}} (\operatorname{sgn}\sigma) A_1^{\sigma(1)} \cdots A_{n-1}^{\sigma(n-1)} = \det B.$$

Now let A be arbitrary, and fix $i \in \{1, \dots, n\}$. For each $j = 1, \dots, n$, let X_i^j denote the matrix obtained by replacing the ith column of A by the basis vector e_j. Since the determinant is a multilinear function of its columns,

$$\det A = \det\left(A_1, \dots, A_{i-1}, \sum_{j=1}^{n} A_i^j e_j, A_{i+1}, \dots, A_n\right)$$

$$= \sum_{j=1}^{n} A_i^j \det(A_1, \dots, A_{i-1}, e_j, A_{i+1}, \dots, A_n)$$

$$= \sum_{j=1}^{n} A_i^j \det X_i^j. \tag{B.7}$$

On the other hand, by interchanging columns $n - i$ times and then interchanging rows $n - j$ times, we can transform X_i^j to a matrix of the form (B.6) with $B = M_i^j$.

Therefore, by the observation in the preceding paragraph,

$$\det X_i^j = (-1)^{n-i+n-j} \det M_i^j = (-1)^{i+j} \det M_i^j.$$

Inserting this into (B.7) completes the proof.   □

Each determinant $\det M_i^j$ is called a **minor of** A, and (B.5) is called the **expansion of** $\det A$ **by minors along the** ith **column**. Since $\det A = \det A^T$, there is an analogous expansion along any row. The factor $(-1)^{i+j} \det M_i^j$ multiplying A_i^j in (B.5) is called the **cofactor of** A_i^j, and is denoted by cof_i^j.

Proposition B.36 (Cramer's Rule). *If A is a nonsingular $n \times n$ matrix, then A^{-1} is equal to $1/(\det A)$ times the transposed cofactor matrix of A. Thus, the entry in the ith row and jth column of A^{-1} is*

$$\left(A^{-1}\right)_j^i = \frac{1}{\det A}\mathrm{cof}_i^j = \frac{1}{\det A}(-1)^{i+j} \det M_i^j. \tag{B.8}$$

Proof. Let B_j^i denote the expression on the right-hand side of (B.8). Then

$$\sum_{j=1}^n B_j^i A_k^j = \frac{1}{\det A} \sum_{j=1}^n (-1)^{i+j} A_k^j \det M_i^j. \tag{B.9}$$

When $k = i$, the summation on the right-hand side is precisely the expansion of $\det A$ by minors along the ith column, so the right-hand side of (B.9) is equal to 1. On the other hand, if $k \neq i$, the summation is equal to the determinant of the matrix obtained by replacing the ith column of A by the kth column. Since this matrix has two identical columns, its determinant is zero. Thus (B.9) is equivalent to the matrix equation $BA = I_n$, where B is the matrix (B_j^i). By Exercise B.19(c), therefore, $B = A^{-1}$.   □

A square matrix $A = \left(A_j^i\right)$ is said to be **upper triangular** if $A_j^i = 0$ for $i > j$ (i.e., the only nonzero entries are on and above the main diagonal). Determinants of upper triangular matrices are particularly easy to compute.

Proposition B.37. *If A is an upper triangular $n \times n$ matrix, then the determinant of A is the product of its diagonal entries:*

$$\det A = A_1^1 \cdots A_n^n.$$

Proof. When $n = 1$, this is trivial. So assume the result is true for $(n-1) \times (n-1)$ matrices, and let A be an upper triangular $n \times n$ matrix. In the expansion of $\det A$ by minors along the first column, there is only one nonzero entry, namely $A_1^1 \det M_1^1$. By induction, $\det M_1^1 = A_2^2 \cdots A_n^n$, which proves the proposition.   □

Suppose X is an $(m+k) \times (m+k)$ matrix. We say that X is **block upper triangular** if X has the form

$$X = \begin{pmatrix} A & B \\ 0 & C \end{pmatrix} \tag{B.10}$$

for some matrices A, B, C of sizes $m \times m$, $m \times k$, and $k \times k$, respectively.

Proposition B.38. *If X is the block upper triangular matrix given by (B.10), then* $\det X = (\det A)(\det C)$.

Proof. If A is singular, then the columns of both A and X are linearly dependent, which implies that $\det X = 0 = (\det A)(\det C)$. So let us assume that A is nonsingular.

Consider first the following special case:

$$X = \begin{pmatrix} I_m & 0 \\ 0 & C \end{pmatrix}.$$

Expanding by minors along the first column and using induction on m, we conclude easily that $\det X = \det C$ in this case. A similar argument shows that

$$\det \begin{pmatrix} A & 0 \\ 0 & I_k \end{pmatrix} = \det A.$$

In the general case, a straightforward computation yields the factorization

$$\begin{pmatrix} A & B \\ 0 & C \end{pmatrix} = \begin{pmatrix} A & 0 \\ 0 & I_k \end{pmatrix} \begin{pmatrix} I_m & 0 \\ 0 & C \end{pmatrix} \begin{pmatrix} I_m & A^{-1}B \\ 0 & I_k \end{pmatrix}. \tag{B.11}$$

By the preceding observations, the determinants of the first two factors are equal to $\det A$ and $\det C$, respectively; and the third factor is upper triangular, so its determinant is 1 by Proposition B.37. The result then follows from Theorem B.31. □

Inner Products and Norms

If V is a real vector space, an ***inner product on V*** is a map $V \times V \to \mathbb{R}$, usually written $(v, w) \mapsto \langle v, w \rangle$, that satisfies the following conditions:

(i) SYMMETRY:

$$\langle v, w \rangle = \langle w, v \rangle;$$

(ii) BILINEARITY:

$$\langle av + a'v', w \rangle = a\langle v, w \rangle + a'\langle v', w \rangle,$$
$$\langle v, bw + b'w' \rangle = b\langle v, w \rangle + b'\langle v, w' \rangle;$$

(iii) POSITIVE DEFINITENESS:

$$\langle v, v \rangle \geq 0, \quad \text{with equality if and only if } v = 0.$$

A vector space endowed with a specific inner product is called an ***inner product space***. The standard example is, of course, $\mathbb{R}^n$ with its ***Euclidean dot product***:

$$\langle x, y \rangle = x \cdot y = \sum_{i=1}^{n} x^i y^i.$$

Suppose V is an inner product space. For each $v \in V$, the **length of** v is the nonnegative real number $|v| = \sqrt{\langle v, v \rangle}$. A **unit vector** is a vector of length 1. If $v, w \in V$ are nonzero vectors, the **angle between** v **and** w is defined to be the unique $\theta \in [0, \pi]$ satisfying

$$\cos \theta = \frac{\langle v, w \rangle}{|v| \, |w|}.$$

Two vectors $v, w \in V$ are said to be **orthogonal** if $\langle v, w \rangle = 0$; this means that either one of the vectors is zero, or the angle between them is $\pi/2$.

▶ **Exercise B.39.** Let V be an inner product space. Show that the length function associated with the inner product satisfies

$$|v| > 0, \qquad v \in V, \ v \neq 0,$$
$$|cv| = |c| |v|, \qquad c \in \mathbb{R}, \ v \in V,$$
$$|v + w| \leq |v| + |w|, \qquad v, w \in V,$$

and the **Cauchy–Schwarz inequality**:

$$|\langle v, w \rangle| \leq |v| |w|, \qquad v, w \in V.$$

Suppose V is a finite-dimensional inner product space. A basis $(E_1, \dots, E_n)$ for V is said to be **orthonormal** if each E_i is a unit vector and E_i is orthogonal to E_j when $i \neq j$.

Proposition B.40 (The Gram–Schmidt Algorithm). *Let V be an inner product space of dimension $n \geq 1$. Then V has an orthonormal basis. In fact, if $(E_1, \dots, E_n)$ is an arbitrary basis for V, there is an orthonormal basis $(\tilde{E}_1, \dots, \tilde{E}_n)$ with the property that*

$$\operatorname{span}(\tilde{E}_1, \dots, \tilde{E}_k) = \operatorname{span}(E_1, \dots, E_k) \quad \text{for } k = 1, \dots, n. \tag{B.12}$$

Proof. The proof is by induction on $n = \dim V$. If $n = 1$, there is only one basis element E_1, and then $\tilde{E}_1 = E_1/|E_1|$ is an orthonormal basis.

Suppose the result is true for inner product spaces of dimension $n - 1$, and let V have dimension n. Then $W = \operatorname{span}(E_1, \dots, E_{n-1})$ is an $(n-1)$-dimensional inner product space with the inner product restricted from V, so there is an orthonormal basis $(\tilde{E}_1, \dots, \tilde{E}_{n-1})$ satisfying (B.12) for $k = 1, \dots, n-1$. Define $\tilde{E}_n$ by

$$\tilde{E}_n = \frac{E_n - \sum_{i=1}^{n-1} \langle E_n, \tilde{E}_i \rangle \tilde{E}_i}{\left| E_n - \sum_{i=1}^{n-1} \langle E_n, \tilde{E}_i \rangle \tilde{E}_i \right|}. \tag{B.13}$$

A computation shows that $(\tilde{E}_1, \dots, \tilde{E}_n)$ is the desired orthonormal basis for V. $\square$

▶ **Exercise B.41.** For $w = (w^1, \dots, w^n)$ and $z = (z^1, \dots, z^n) \in \mathbb{C}^n$, define the **Hermitian dot product** by $w \cdot z = \sum_{j=1}^n w^j \overline{z^j}$, where, for any complex number $z = x + iy$, the notation $\bar{z}$ denotes the **complex conjugate**: $\bar{z} = x - iy$. A basis $(E_1, \dots, E_n)$ for $\mathbb{C}^n$ (over $\mathbb{C}$) is said to be *orthonormal* if $E_i \cdot E_i = 1$ and $E_i \cdot E_j = 0$ for $i \neq j$. Show that the statement and proof of Proposition B.40 hold for the Hermitian dot product.

An isomorphism $T\colon V \to W$ between inner product spaces is called a ***linear isometry*** if it takes the inner product of V to that of W:

$$\langle Tv, Tw \rangle = \langle v, w \rangle \quad \text{for all } v, w \in V.$$

▶ **Exercise B.42.** Show that every linear isometry between inner product spaces is a homeomorphism that preserves lengths, angles, and orthogonality, and takes orthonormal bases to orthonormal bases.

▶ **Exercise B.43.** Given any basis (E_i) for a finite-dimensional vector space V, show that there is a unique inner product on V for which (E_i) is orthonormal.

▶ **Exercise B.44.** Suppose V is a finite-dimensional inner product space and $E\colon \mathbb{R}^n \to V$ is the basis map determined by some orthonormal basis. Show that E is a linear isometry when $\mathbb{R}^n$ is endowed with the Euclidean inner product.

The preceding exercise shows that finite-dimensional inner product spaces are geometrically indistinguishable from the Euclidean space of the same dimension.

If V is a finite-dimensional inner product space and $S \subseteq V$ is a subspace, the ***orthogonal complement of S in V*** is the set

$$S^\perp = \{ v \in V : \langle v, w \rangle = 0 \text{ for all } w \in S \}.$$

▶ **Exercise B.45.** Let V be a finite-dimensional inner product space and let $S \subseteq V$ be a subspace. Show that $S^\perp$ is a subspace and $V = S \oplus S^\perp$.

Thanks to the result of the preceding exercise, for any subspace S of an inner product space V, there is a natural projection $\pi\colon V \to S$ with kernel $S^\perp$. This is called the ***orthogonal projection of V onto S***.

Norms

If V is a real vector space, a ***norm on V*** is a function from V to $\mathbb{R}$, written $v \mapsto |v|$, satisfying the following properties.

  (i) POSITIVITY: $|v| \geq 0$ for all $v \in V$, with equality if and only if $v = 0$.
 (ii) HOMOGENEITY: $|cv| = |c|\,|v|$ for all $c \in \mathbb{R}$ and $v \in V$.
(iii) TRIANGLE INEQUALITY: $|v + w| \leq |v| + |w|$ for all $v, w \in V$.

A vector space together with a specific choice of norm is called a ***normed linear space***. Exercise B.39 shows that the length function associated with any inner product is a norm; thus, in particular, every finite-dimensional vector space possesses many norms. Given a norm on V, the distance function $d(v, w) = |v - w|$ turns V into a metric space, yielding a topology on V called the ***norm topology***.

Example B.46 (Euclidean Spaces). Endowed with the ***Euclidean norm*** defined by

$$|x| = \sqrt{x \cdot x}, \tag{B.14}$$

$\mathbb{R}^n$ is a normed linear space, whose norm topology is exactly the Euclidean topology described in Appendix A. //

Example B.47 (The Frobenius Norm on Matrices). The vector space $M(m \times n, \mathbb{R})$ of $m \times n$ real matrices has a natural Euclidean inner product, obtained by identifying a matrix with a point in $\mathbb{R}^{mn}$:

$$A \cdot B = \sum_{i,j} A_j^i B_j^i.$$

This yields a norm on matrices, called the **Frobenius norm**:

$$|A| = \sqrt{\sum_{i,j} \left(A_j^i\right)^2}. \tag{B.15}$$

Whenever we use a norm on matrices, it is always this one.   //

▶ **Exercise B.48.** For any matrices $A \in M(m \times n, \mathbb{R})$ and $B \in M(n \times k, \mathbb{R})$, show that

$$|AB| \le |A|\,|B|.$$

Two norms $|\cdot|_1$ and $|\cdot|_2$ on a vector space V are said to be **equivalent** if there are positive constants c, C such that

$$c|v|_1 \le |v|_2 \le C|v|_1 \quad \text{for all } v \in V.$$

▶ **Exercise B.49.** Show that equivalent norms determine the same topology.

▶ **Exercise B.50.** Show that any two norms on a finite-dimensional vector space are equivalent. [Hint: first do the case in which $V = \mathbb{R}^n$ and one of the norms is the Euclidean norm, and consider the restriction of the other norm to the unit sphere.]

The preceding exercise shows that finite-dimensional normed linear spaces of the same dimension are topologically indistinguishable from one another. Thus, any such space automatically inherits all the usual topological properties of Euclidean space, such as compactness of closed and bounded subsets.

If V and W are normed linear spaces, a linear map $T: V \to W$ is said to be **bounded** if there exists a positive constant C such that

$$|Tv| \le C|v| \quad \text{for all } v \in V.$$

▶ **Exercise B.51.** Show that a linear map between normed linear spaces is continuous if and only if it is bounded. [Hint: to show that continuity of T implies boundedness, first show that there exists $\delta > 0$ such that $|x| < \delta \Rightarrow |T(x)| < 1$.]

▶ **Exercise B.52.** Show that every linear map between finite-dimensional normed linear spaces is bounded and therefore continuous.

Direct Products and Direct Sums

If $V_1, \dots, V_k$ are real vector spaces, their **direct product** is the vector space whose underlying set is the Cartesian product $V_1 \times \cdots \times V_k$, with addition and scalar multiplication defined componentwise:

$$(v_1, \dots, v_k) + (v_1', \dots, v_k') = (v_1 + v_1', \dots, v_k + v_k'),$$

$$c(v_1, \ldots, v_k) = (cv_1, \ldots, cv_k).$$

The basic example is the Euclidean space $\mathbb{R}^n = \mathbb{R} \times \cdots \times \mathbb{R}$.

For some applications (chiefly in our treatment of de Rham cohomology in Chapters 17 and 18), it is important to generalize this to an infinite number of vector spaces. For this discussion, we turn to the general setting of modules over a commutative ring $\mathcal{R}$. Linear maps between $\mathcal{R}$-modules are defined exactly as for vector spaces: if V and W are $\mathcal{R}$-modules, a map $F\colon V \to W$ is said to be **$\mathcal{R}$-linear** if $F(av + bw) = aF(v) + bF(w)$ for all $a, b \in \mathcal{R}$ and $v, w \in V$. If V is an $\mathcal{R}$-module, a subset $S \subseteq V$ is called a **submodule of V** if it is closed under addition and scalar multiplication, so it is itself an $\mathcal{R}$-module. Throughout the rest of this section we assume that $\mathcal{R}$ is a fixed commutative ring. In all of our applications, $\mathcal{R}$ will be either the field $\mathbb{R}$ of real numbers, in which case the modules are real vector spaces and the linear maps are the usual ones, or the ring of integers $\mathbb{Z}$, in which case the modules are abelian groups and the linear maps are group homomorphisms.

If $(V_\alpha)_{\alpha \in A}$ is an arbitrary indexed family of sets, their **Cartesian product**, denoted by $\prod_{\alpha \in A} V_\alpha$, is defined as the set of functions $v\colon A \to \bigcup_{\alpha \in A} V_\alpha$ with the property that $v(\alpha) \in V_\alpha$ for each α. Thanks to the axiom of choice, the Cartesian product of a nonempty indexed family of nonempty sets is nonempty. If v is an element of the Cartesian product, we usually denote the value of v at $\alpha \in A$ by v_α instead of $v(\alpha)$; the element v itself is usually denoted by $(v_\alpha)_{\alpha \in A}$, or just (v_α) if the index set is understood. This can be thought of as an indexed family of elements of the sets V_α, or an "A-tuple." For each $\beta \in A$, we have a canonical **projection map** $\pi_\beta\colon \prod_{\alpha \in A} V_\alpha \to V_\beta$, defined by

$$\pi_\beta\big((v_\alpha)_{\alpha \in A}\big) = v_\beta.$$

Now suppose that $(V_\alpha)_{\alpha \in A}$ is an indexed family of $\mathcal{R}$-modules. The **direct product** of the family is the set $\prod_{\alpha \in A} V_\alpha$, made into an $\mathcal{R}$-module by defining addition and scalar multiplication as follows:

$$(v_\alpha) + (v'_\alpha) = (v_\alpha + v'_\alpha),$$
$$c(v_\alpha) = (cv_\alpha).$$

The zero element of this module is the A-tuple with $v_\alpha = 0$ for every α. It is easy to check that each projection map π_β is $\mathcal{R}$-linear.

Proposition B.53 (Characteristic Property of the Direct Product). *Let $(V_\alpha)_{\alpha \in A}$ be an indexed family of $\mathcal{R}$-modules. Given an $\mathcal{R}$-module W and a family of $\mathcal{R}$-linear maps $G_\alpha\colon W \to V_\alpha$, there exists a unique $\mathcal{R}$-linear map $G\colon W \to \prod_{\alpha \in A} V_\alpha$ such that $\pi_\alpha \circ G = G_\alpha$ for each $\alpha \in A$.*

▶ **Exercise B.54.** Prove the preceding proposition.

Complementary to direct products is the notion of direct sums. Given an indexed family $(V_\alpha)_{\alpha \in A}$ as above, we define the **direct sum** of the family to be the submodule of their direct product consisting of A-tuples $(v_\alpha)_{\alpha \in A}$ with the property that $v_\alpha = 0$

for all but finitely many α. The direct sum is denoted by $\bigoplus_{\alpha \in A} V_\alpha$, or in the case of a finite family by $V_1 \oplus \cdots \oplus V_k$. For finite families of modules, the direct product and the direct sum are identical.

For each $\beta \in A$, there is a canonical $\mathcal{R}$-linear injection $\iota_\beta \colon V_\beta \to \bigoplus_{\alpha \in A} V_\alpha$, defined by letting $\iota_\beta(v)$ be the A-tuple $(v_\alpha)_{\alpha \in A}$ with $v_\beta = v$ and $v_\alpha = 0$ for $\alpha \neq \beta$. In the case of a finite direct sum, this just means $\iota_\beta(v) = (0, \ldots, 0, v, 0, \ldots, 0)$, with v in position β.

Proposition B.55 (Characteristic Property of the Direct Sum). *Let $(V_\alpha)_{\alpha \in A}$ be an indexed family of $\mathcal{R}$-modules. Given an $\mathcal{R}$-module W and a family of $\mathcal{R}$-linear maps $G_\alpha \colon V_\alpha \to W$, there exists a unique $\mathcal{R}$-linear map $G \colon \bigoplus_{\alpha \in A} V_\alpha \to W$ such that $G \circ \iota_\alpha = G_\alpha$ for each $\alpha \in A$.*

▶ **Exercise B.56.** Prove the preceding proposition.

If W is an $\mathcal{R}$-module and $(V_\alpha)_{\alpha \in A}$ is a family of subspaces of W, then the characteristic property applied to the inclusions $\iota_\alpha \colon V_\alpha \hookrightarrow W$ guarantees the existence of a canonical $\mathcal{R}$-linear map $\bigoplus_\alpha V_\alpha \to W$ that restricts to inclusion on each V_α. This map is an isomorphism precisely when the V_α's are chosen so that every element of W has a unique expression as a finite linear combination $\sum_\alpha c_\alpha v_\alpha$ with $v_\alpha \in V_\alpha$ for each α. In this case, we can naturally *identify* W with $\bigoplus_\alpha V_\alpha$, and we say that W is the ***internal direct sum*** of the submodules $\{V_\alpha\}$, extending the terminology we introduced earlier for two complementary subspaces of a vector space. A direct sum of an abstract family of modules is sometimes called their ***external direct sum*** to distinguish it from an internal direct sum.

If V and W are $\mathcal{R}$-modules, the set $\operatorname{Hom}_{\mathcal{R}}(V, W)$ of all $\mathcal{R}$-linear maps from V to W is an $\mathcal{R}$-module under pointwise addition and scalar multiplication:

$$(F + G)(v) = F(v) + G(v),$$

$$(a F)(v) = a F(v).$$

(If V and W are real vector spaces, then $\operatorname{Hom}_{\mathcal{R}}(V, W)$ is just the space $\mathrm{L}(V; W)$ of $\mathbb{R}$-linear maps; if they are abelian groups, then $\mathbb{Z}$-linear maps are group homomorphisms, and we usually write $\operatorname{Hom}(V, W)$ instead of $\operatorname{Hom}_{\mathbb{Z}}(V, W)$.) Our last proposition is used in the proof of the de Rham theorem in Chapter 18.

Proposition B.57. *Let $(V_\alpha)_{\alpha \in A}$ be an indexed family of $\mathcal{R}$-modules. For each $\mathcal{R}$-module W, there is a canonical isomorphism*

$$\operatorname{Hom}_{\mathcal{R}}\left(\bigoplus_{\alpha \in A} V_\alpha, W \right) \cong \prod_{\alpha \in A} \operatorname{Hom}_{\mathcal{R}}(V_\alpha, W).$$

Proof. Define a map $\Phi \colon \operatorname{Hom}_{\mathcal{R}}\left(\bigoplus_{\alpha \in A} V_\alpha, W \right) \to \prod_{\alpha \in A} \operatorname{Hom}_{\mathcal{R}}(V_\alpha, W)$ by setting $\Phi(F) = (F_\alpha)_{\alpha \in A}$, where $F_\alpha = F \circ \iota_\alpha$.

To prove that Φ is surjective, suppose $(F_\alpha)_{\alpha \in A}$ is an arbitrary element of $\prod_{\alpha \in A} \operatorname{Hom}_{\mathcal{R}}(V_\alpha, W)$. This just means that for each α, F_α is an $\mathcal{R}$-linear map from V_α to W. The characteristic property of the direct sum then guarantees the existence of an $\mathcal{R}$-linear map $F: \bigoplus_{\alpha \in A} V_\alpha \to W$ satisfying $F \circ \iota_\alpha = F_\alpha$ for each α, which is equivalent to $\Phi(F) = (F_\alpha)_{\alpha \in A}$.

To prove that Φ is injective, suppose that $\Phi(F) = (F_\alpha)_{\alpha \in A} = 0$. By definition of the zero element of the direct product, this means that $F_\alpha = F \circ \iota_\alpha$ is the zero homomorphism for each α. By the uniqueness assertion in Proposition B.55, this implies that F itself is the zero homomorphism. $\square$

Appendix C
Review of Calculus

In this appendix we summarize the main results from multivariable calculus and real analysis that are needed in this book. For details on most of the ideas touched on here, you can consult [Apo74], [Rud76], or [Str00].

Total and Partial Derivatives

For maps between (open subsets of) finite-dimensional vector spaces, the most general notion of derivative is the total derivative.

Let V, W be finite-dimensional vector spaces, which we may assume to be endowed with norms. If $U \subseteq V$ is an open subset and $a \in U$, a map $F \colon U \to W$ is said to be *differentiable at a* if there exists a linear map $L \colon V \to W$ such that

$$\lim_{v \to 0} \frac{|F(a + v) - F(a) - Lv|}{|v|} = 0. \tag{C.1}$$

The norm in the numerator of this expression is that of W, while the norm in the denominator is that of V. Because all norms on a finite-dimensional vector space are equivalent (Exercise B.49), the definition is independent of both choices of norms.

▶ **Exercise C.1.** Suppose $F \colon U \to W$ is differentiable at $a \in U$. Show that the linear map L satisfying (C.1) is unique.

If F is differentiable at a, the linear map L satisfying (C.1) is denoted by $DF(a)$ and is called the *total derivative of F at a*. Condition (C.1) can also be written

$$F(a + v) = F(a) + DF(a)v + R(v), \tag{C.2}$$

where the remainder term $R(v) = F(a + v) - F(a) - DF(a)v$ satisfies $|R(v)|/|v| \to 0$ as $v \to 0$. Thus the total derivative represents the "best linear approximation" to $F(a + v) - F(a)$ near a.

▶ **Exercise C.2.** Suppose V, W, X are finite-dimensional vector spaces, $U \subseteq V$ is an open subset, a is a point in U, and $F, G \colon U \to W$ and $f, g \colon U \to \mathbb{R}$ are maps. Prove the following statements.

J.M. Lee, *Introduction to Smooth Manifolds*, Graduate Texts in Mathematics 218, DOI 10.1007/978-1-4419-9982-5, © Springer Science+Business Media New York 2013

(a) If F is differentiable at a, then it is continuous at a.

(b) If F is a constant map, then F is differentiable at a and $DF(a) = 0$.

(c) If F and G are differentiable at a, then $F + G$ is also, and

$$D(F + G)(a) = DF(a) + DG(a).$$

(d) If f and g are differentiable at a, then fg is also, and

$$D(fg)(a) = f(a)Dg(a) + g(a)Df(a).$$

(e) If f and g are differentiable at a and $g(a) \neq 0$, then f/g is differentiable at a, and

$$D(f/g)(a) = \frac{g(a)Df(a) - f(a)Dg(a)}{g(a)^2}.$$

(f) If $T : V \to W$ is a linear map, then T is differentiable at every point $v \in V$, with total derivative equal to T itself: $DT(v) = T$.

(g) If $B : V \times W \to X$ is a bilinear map, then B is differentiable at every point $(v, w) \in V \times W$, and

$$DB(v, w)(x, y) = B(v, y) + B(x, w).$$

Proposition C.3 (The Chain Rule for Total Derivatives). *Suppose* V, W, X *are finite-dimensional vector spaces,* $U \subseteq V$ *and* $\tilde{U} \subseteq W$ *are open subsets, and* $F : U \to \tilde{U}$ *and* $G : \tilde{U} \to X$ *are maps. If* F *is differentiable at* $a \in U$ *and* G *is differentiable at* $F(a) \in \tilde{U}$, *then* $G \circ F$ *is differentiable at* a, *and*

$$D(G \circ F)(a) = DG\big(F(a)\big) \circ DF(a).$$

Proof. Let $A = DF(a)$ and $B = DG\big(F(a)\big)$. We need to show that

$$\lim_{v \to 0} \frac{|G(F(a + v)) - G(F(a)) - BAv|}{|v|} = 0. \tag{C.3}$$

Let us write $b = F(a)$ and $w = F(a + v) - F(a)$. With these substitutions, we can rewrite the quotient in (C.3) as

$$\frac{|G(b + w) - G(b) - BAv|}{|v|} = \frac{|G(b + w) - G(b) - Bw + Bw - BAv|}{|v|}$$

$$\leq \frac{|G(b + w) - G(b) - Bw|}{|v|} + \frac{|B(w - Av)|}{|v|}. \tag{C.4}$$

Since A and B are linear, Exercise B.52 shows that there are constants C, C' such that $|Ax| \leq C|x|$ for all $x \in V$, and $|By| \leq C'|y|$ for all $y \in W$. The differentiability of F at a means that for any $\varepsilon > 0$, we can ensure that

$$|w - Av| = |F(a + v) - F(a) - Av| \leq \varepsilon|v|$$

as long as v lies in a small enough neighborhood of 0. Moreover, as $v \to 0$, $|w| = |F(a + v) - F(a)| \to 0$ by continuity of F. Therefore, the differentiability of G at b means that by making $|v|$ even smaller if necessary, we can also achieve

$$|G(b + w) - G(b) - Bw| \leq \varepsilon|w|.$$

Putting all of these estimates together, we see that for $|v|$ sufficiently small, (C.4) is bounded by

$$\varepsilon \frac{|w|}{|v|} + C' \frac{|w - Av|}{|v|} = \varepsilon \frac{|w - Av + Av|}{|v|} + C' \frac{|w - Av|}{|v|}$$

$$\leq \varepsilon \frac{|w - Av|}{|v|} + \varepsilon \frac{|Av|}{|v|} + C' \frac{|w - Av|}{|v|}$$

$$\leq \varepsilon^2 + \varepsilon C + C'\varepsilon,$$

which can be made as small as desired.   □

Partial Derivatives

Now we specialize to maps between Euclidean spaces. Suppose $U \subseteq \mathbb{R}^n$ is open and $f: U \to \mathbb{R}$ is a real-valued function. For any $a = (a^1, \dots, a^n) \in U$ and any $j \in \{1, \dots, n\}$, the *jth partial derivative of f at a* is defined to be the ordinary derivative of f with respect to x^j while holding the other variables fixed:

$$\frac{\partial f}{\partial x^j}(a) = \lim_{h \to 0} \frac{f(a^1, \dots, a^j + h, \dots, a^n) - f(a^1, \dots, a^j, \dots, a^n)}{h}$$

$$= \lim_{h \to 0} \frac{f(a + he_j) - f(a)}{h},$$

if the limit exists.

More generally, for a vector-valued function $F: U \to \mathbb{R}^m$, we can write the coordinates of $F(x)$ as $F(x) = (F^1(x), \dots, F^m(x))$. This defines m functions $F^1, \dots, F^m: U \to \mathbb{R}$ called the *component functions of F*. The partial derivatives of F are defined simply to be the partial derivatives $\partial F^i / \partial x^j$ of its component functions. The matrix $(\partial F^i / \partial x^j)$ of partial derivatives is called the *Jacobian matrix of F*, and its determinant is called the *Jacobian determinant of F*.

If $F: U \to \mathbb{R}^m$ is a function for which each partial derivative exists at each point in U and the functions $\partial F^i / \partial x^j: U \to \mathbb{R}$ so defined are all continuous, then F is said to be of *class C^1* or *continuously differentiable*. If this is the case, we can differentiate the functions $\partial F^i / \partial x^j$ to obtain *second-order partial derivatives*

$$\frac{\partial^2 F^i}{\partial x^k \partial x^j} = \frac{\partial}{\partial x^k} \left(\frac{\partial F^i}{\partial x^j} \right),$$

if they exist. Continuing this way leads to higher-order partial derivatives: the *partial derivatives of F of order k* are the (first) partial derivatives of those of order $k - 1$, when they exist.

In general, if $U \subseteq \mathbb{R}^n$ is an open subset and $k \geq 0$, a function $F: U \to \mathbb{R}^m$ is said to be of *class C^k* or *k times continuously differentiable* if all the partial derivatives of F of order less than or equal to k exist and are continuous functions on U. (Thus a function of class C^0 is just a continuous function.) Because existence and continuity of derivatives are local properties, clearly F is C^k if and only if it has that property in a neighborhood of each point in U.

A function that is of class C^k for every $k \geq 0$ is said to be of *class C^∞*, *smooth*, or *infinitely differentiable*. If U and V are open subsets of Euclidean spaces, a function $F: U \to V$ is called a *diffeomorphism* if it is smooth and bijective and its inverse function is also smooth.

One consequence of the chain rule is worth noting.

Proposition C.4. *Suppose $U \subseteq \mathbb{R}^n$ and $V \subseteq \mathbb{R}^m$ are open subsets and $F: U \to V$ is a diffeomorphism. Then $m = n$, and for each $a \in U$, the total derivative $DF(a)$ is invertible, with $DF(a)^{-1} = D(F^{-1})(F(a))$.*

Proof. Because $F^{-1} \circ F = \mathrm{Id}_U$, the chain rule implies that for each $a \in U$,

$$\mathrm{Id}_{\mathbb{R}^n} = D(\mathrm{Id}_U)(a) = D(F^{-1} \circ F)(a) = D(F^{-1})(F(a)) \circ DF(a). \qquad \text{(C.5)}$$

Similarly, $F \circ F^{-1} = \mathrm{Id}_V$ implies that $DF(a) \circ D(F^{-1})(F(a))$ is the identity on $\mathbb{R}^m$. This implies that $DF(a)$ is invertible with inverse $D(F^{-1})(F(a))$, and therefore $m = n$. $\qquad \square$

We sometimes need to consider smoothness of functions whose domains are subsets of $\mathbb{R}^n$ that are not open. If $A \subseteq \mathbb{R}^n$ is an *arbitrary* subset, a function $F: A \to \mathbb{R}^m$ is said to be *smooth on A* if it admits a smooth extension to an open neighborhood of each point, or more precisely, if for every $x \in A$, there exist an open subset $U_x \subseteq \mathbb{R}^n$ containing x and a smooth function $\tilde{F}: U_x \to \mathbb{R}^m$ that agrees with F on $U_x \cap A$. The notion of diffeomorphism extends to arbitrary subsets in the obvious way: given arbitrary subsets $A, B \subseteq \mathbb{R}^n$, a *diffeomorphism from A to B* is a smooth bijective map $f: A \to B$ with smooth inverse.

We are especially concerned with real-valued functions, that is, functions whose codomain is $\mathbb{R}$. If $U \subseteq \mathbb{R}^n$ is open, the set of all real-valued functions of class C^k on U is denoted by $C^k(U)$, and the set of all smooth real-valued functions by $C^\infty(U)$. Sums, constant multiples, and products of functions are defined pointwise: for $f, g: U \to \mathbb{R}$ and $c \in \mathbb{R}$,

$$(f + g)(x) = f(x) + g(x),$$
$$(cf)(x) = c(f(x)),$$
$$(fg)(x) = f(x)g(x).$$

▶ **Exercise C.5.** Let $U \subseteq \mathbb{R}^n$ be an open subset, and suppose $f, g \in C^\infty(U)$ and $c \in \mathbb{R}$.

(a) Show that $f + g$, cf, and fg are smooth.
(b) Show that these operations turn $C^\infty(U)$ into a commutative ring and a commutative and associative algebra over $\mathbb{R}$ (see p. 624).
(c) Show that if g never vanishes on U, then f/g is smooth.

The following important result shows that for most interesting functions, the order in which we take partial derivatives is irrelevant. For a proof, see [Apo74, Rud76, Str00].

Proposition C.6 (Equality of Mixed Partial Derivatives). *If U is an open subset of $\mathbb{R}^n$ and $F\colon U \to \mathbb{R}^m$ is a function of class C^2, then the mixed second-order partial derivatives of F do not depend on the order of differentiation:*

$$\frac{\partial^2 F^i}{\partial x^j \partial x^k} = \frac{\partial^2 F^i}{\partial x^k \partial x^j}.$$

Corollary C.7. *If $F\colon U \to \mathbb{R}^m$ is smooth, then the mixed partial derivatives of F of any order are independent of the order of differentiation.* $\square$

Next we study the relationship between total and partial derivatives. Suppose $U \subseteq \mathbb{R}^n$ is open and $F\colon U \to \mathbb{R}^m$ is differentiable at $a \in U$. As a linear map between Euclidean spaces $\mathbb{R}^n$ and $\mathbb{R}^m$, $DF(a)$ can be identified with an $m \times n$ matrix. The next proposition identifies that matrix as the Jacobian of F.

Proposition C.8. *Let $U \subseteq \mathbb{R}^n$ be open, and suppose $F\colon U \to \mathbb{R}^m$ is differentiable at $a \in U$. Then all of the partial derivatives of F at a exist, and $DF(a)$ is the linear map whose matrix is the Jacobian of F at a:*

$$DF(a) = \left(\frac{\partial F^j}{\partial x^i}(a) \right).$$

Proof. Let $B = DF(a)$, and for $v \in \mathbb{R}^n$ small enough that $a + v \in U$, let $R(v) = F(a + v) - F(a) - Bv$. The fact that F is differentiable at a implies that each component of the vector-valued function $R(v)/|v|$ goes to zero as $v \to 0$. The ith partial derivative of F^j at a, if it exists, is

$$\frac{\partial F^j}{\partial x^i}(a) = \lim_{t \to 0} \frac{F^j(a + te_i) - F^j(a)}{t} = \lim_{t \to 0} \frac{B^j_i t + R^j(te_i)}{t}$$

$$= B^j_i + \lim_{t \to 0} \frac{R^j(te_i)}{t}.$$

The norm of the quotient on the right above is $\left| R^j(te_i) \right| / |te_i|$, which approaches zero as $t \to 0$. It follows that $\partial F^j / \partial x^i(a)$ exists and is equal to B^j_i as claimed. $\square$

▶ **Exercise C.9.** Suppose $U \subseteq \mathbb{R}^n$ is open. Show that a function $F\colon U \to \mathbb{R}^m$ is differentiable at $a \in U$ if and only if each of its component functions $F^1, \ldots, F^m$ is differentiable at a. Show that if this is the case, then

$$DF(a) = \begin{pmatrix} DF^1(a) \\ \vdots \\ DF^m(a) \end{pmatrix}.$$

The preceding exercise implies that for an open interval $J \subseteq \mathbb{R}$, a map $\gamma\colon J \to \mathbb{R}^m$ is differentiable if and only if its component functions are differentiable in the sense of one-variable calculus.

The next proposition gives the most important sufficient condition for differentiability; in particular, it shows that all of the usual functions of elementary calculus are differentiable. For a proof, see [Apo74, Rud76, Str00].

Proposition C.10. *Let $U \subseteq \mathbb{R}^n$ be open. If $F : U \to \mathbb{R}^m$ is of class C^1, then it is differentiable at each point of U.*

For functions between Euclidean spaces, the chain rule can be rephrased in terms of partial derivatives.

Corollary C.11 (The Chain Rule for Partial Derivatives). *Let $U \subseteq \mathbb{R}^n$ and $\tilde{U} \subseteq \mathbb{R}^m$ be open subsets, and let $x = (x^1, \dots, x^n)$ denote the standard coordinates on U and $y = (y^1, \dots, y^m)$ those on $\tilde{U}$.*

(a) *A composition of C^1 functions $F : U \to \tilde{U}$ and $G : \tilde{U} \to \mathbb{R}^p$ is again of class C^1, with partial derivatives given by*

$$\frac{\partial (G^i \circ F)}{\partial x^j}(x) = \sum_{k=1}^{m} \frac{\partial G^i}{\partial y^k}(F(x)) \frac{\partial F^k}{\partial x^j}(x).$$

(b) *If F and G are smooth, then $G \circ F$ is smooth.*

▶ **Exercise C.12.** Prove Corollary C.11.

From the chain rule and induction one can derive formulas for the higher partial derivatives of a composite function as needed, provided the functions in question are sufficiently differentiable.

▶ **Exercise C.13.** Suppose $A \subseteq \mathbb{R}^n$ and $B \subseteq \mathbb{R}^m$ are arbitrary subsets, and $F : A \to \mathbb{R}^m$ and $G : B \to \mathbb{R}^p$ are smooth maps (in the sense that they have smooth extensions in a neighborhood of each point) such that $F(A) \subseteq B$. Show that $G \circ F : A \to \mathbb{R}^p$ is smooth.

Now suppose $f : U \to \mathbb{R}$ is a smooth real-valued function on an open subset $U \subseteq \mathbb{R}^n$, and $a \in U$. For each vector $v \in \mathbb{R}^n$, we define the ***directional derivative of f in the direction v at a*** to be the number

$$D_v f(a) = \frac{d}{dt}\bigg|_{t=0} f(a + tv). \tag{C.6}$$

(This definition makes sense for any vector v; we do not require v to be a unit vector as one sometimes does in elementary calculus.)

Since $D_v f(a)$ is the ordinary derivative of the composite function $t \mapsto a + tv \mapsto f(a + tv)$, by the chain rule it can be written more concretely as

$$D_v f(a) = \sum_{i=1}^{n} v^i \frac{\partial f}{\partial x^i}(a) = Df(a)v.$$

The fundamental theorem of calculus expresses one well-known relationship between integrals and derivatives. Another is that integrals of smooth functions can be differentiated under the integral sign. A precise statement is given in the next theorem; this is not the best that can be proved, but it is more than sufficient for our purposes. For a proof, see [Apo74, Rud76, Str00].

Theorem C.14 (Differentiation Under an Integral Sign). *Let $U \subseteq \mathbb{R}^n$ be an open subset, let $a, b \in \mathbb{R}$, and let $f : U \times [a,b] \to \mathbb{R}$ be a continuous function such that the partial derivatives $\partial f / \partial x^i : U \times [a,b] \to \mathbb{R}$ exist and are continuous on $U \times [a,b]$ for $i = 1, \ldots, n$. Define $F : U \to \mathbb{R}$ by*

$$F(x) = \int_a^b f(x,t) \, dt.$$

Then F is of class C^1, and its partial derivatives can be computed by differentiating under the integral sign:

$$\frac{\partial F}{\partial x^i}(x) = \int_a^b \frac{\partial f}{\partial x^i}(x,t) \, dt.$$

You are probably familiar with Taylor's theorem, which shows how a sufficiently smooth function can be approximated near a point by a polynomial. We need a version of Taylor's theorem in several variables that gives an explicit integral form for the remainder term. In order to express it concisely, it helps to introduce some shorthand notation. For any m-tuple $I = (i_1, \ldots, i_m)$ of indices with $1 \le i_j \le n$, we let $|I| = m$ denote the number of indices in I, and

$$\partial_I = \frac{\partial^m}{\partial x^{i_1} \cdots \partial x^{i_m}},$$
$$(x - a)^I = \left(x^{i_1} - a^{i_1}\right) \cdots \left(x^{i_m} - a^{i_m}\right).$$

Theorem C.15 (Taylor's Theorem). *Let $U \subseteq \mathbb{R}^n$ be an open subset, and let $a \in U$ be fixed. Suppose $f \in C^{k+1}(U)$ for some $k \ge 0$. If W is any convex subset of U containing a, then for all $x \in W$,*

$$f(x) = P_k(x) + R_k(x), \tag{C.7}$$

*where P_k is the **kth-order Taylor polynomial of f at a**, defined by*

$$P_k(x) = f(a) + \sum_{m=1}^{k} \frac{1}{m!} \sum_{I:|I|=m} \partial_I f(a)(x-a)^I, \tag{C.8}$$

*and R_k is the **kth remainder term**, given by*

$$R_k(x) = \frac{1}{k!} \sum_{I:|I|=k+1} (x-a)^I \int_0^1 (1-t)^k \partial_I f\left(a + t(x-a)\right) dt. \tag{C.9}$$

Proof. For $k = 0$ (where we interpret P_0 to mean $f(a)$), this is just the fundamental theorem of calculus applied to the function $u(t) = f\left(a + t(x-a)\right)$, together with the chain rule. Assuming the result holds for some k, integration by parts applied to the integral in the remainder term yields

$$\int_0^1 (1-t)^k \partial_I f\left(a + t(x-a)\right) dt$$

$$= \left[-\frac{(1-t)^{k+1}}{k+1} \partial_I f \big(a + t(x-a)\big) \right]_{t=0}^{t=1}$$

$$+ \int_0^1 \frac{(1-t)^{k+1}}{k+1} \frac{\partial}{\partial t} \big(\partial_I f \big(a + t(x-a)\big) \big) \, dt$$

$$= \frac{1}{k+1} \partial_I f(a)$$

$$+ \frac{1}{k+1} \sum_{j=1}^n (x^j - a^j) \int_0^1 (1-t)^{k+1} \frac{\partial}{\partial x^j} \partial_I f \big(a + t(x-a)\big) \, dt.$$

When we insert this into (C.7), we obtain the analogous formula with k replaced by $k+1$. $\qquad\square$

Corollary C.16. *Suppose $U \subseteq \mathbb{R}^n$ is an open subset, $a \in U$, and $f \in C^{k+1}(U)$ for some $k \geq 0$. If W is a convex subset of U containing a on which all of the $(k+1)$st partial derivatives of f are bounded in absolute value by a constant M, then for all $x \in W$,*

$$\big| f(x) - P_k(x) \big| \leq \frac{n^{k+1} M}{(k+1)!} |x - a|^{k+1},$$

where P_k is the kth Taylor polynomial of f at a, defined by (C.8).

Proof. There are n^{k+1} terms on the right-hand side of (C.9), and each term is bounded in absolute value by $(1/(k+1)!)|x-a|^{k+1} M$. $\qquad\square$

Multiple Integrals

In this section we give a brief review of some basic facts regarding multiple integrals in $\mathbb{R}^n$. For our purposes, the Riemann integral is more than sufficient. Readers who are familiar with the theory of Lebesgue integration are free to interpret all of our integrals in the Lebesgue sense, because the two integrals are equal for the types of functions we consider. For more details on the aspects of integration theory described here, you can consult [Apo74, Rud76, Str00].

A *closed rectangle* in $\mathbb{R}^n$ is a product set of the form $[a^1, b^1] \times \cdots \times [a^n, b^n]$, for real numbers $a^i < b^i$. Analogously, an *open rectangle* is a set of the form $(a^1, b^1) \times \cdots \times (a^n, b^n)$. If A is a rectangle of either type, the *volume of A*, denoted by $\mathrm{Vol}(A)$, is defined to be the product of the lengths of its component intervals:

$$\mathrm{Vol}(A) = (b^1 - a^1) \cdots (b^n - a^n). \tag{C.10}$$

A rectangle is called a *cube* if all of its side lengths $(b^i - a^i)$ are equal.

Given a closed interval $[a, b] \subseteq \mathbb{R}$, a *partition of $[a, b]$* is a finite sequence $P = (a_0, \ldots, a_k)$ of real numbers such that $a = a_0 < a_1 < \cdots < a_k = b$. Each of the intervals $[a_{i-1}, a_i]$ for $i = 1, \ldots, k$ is called a *subinterval of P*. Similarly, if $A = [a^1, b^1] \times \cdots \times [a^n, b^n]$ is a closed rectangle, a *partition of A* is an n-tuple $P =$

$(P_1, \ldots, P_n)$, where each P_i is a partition of $[a^i, b^i]$. Each rectangle of the form $I_1 \times \cdots \times I_n$, where I_j is a subinterval of P_j, is called a **subrectangle of P**. Clearly, A is the union of all the subrectangles in any partition, and distinct subrectangles intersect only on their boundaries.

Suppose $A \subseteq \mathbb{R}^n$ is a closed rectangle and $f : A \to \mathbb{R}$ is a bounded function. For each partition P of A, we define the **lower sum of f with respect to P** by

$$L(f, P) = \sum_j \left(\inf_{R_j} f \right) \mathrm{Vol}(R_j),$$

where the sum is over all the subrectangles R_j of P. Similarly, the **upper sum** is

$$U(f, P) = \sum_j \left(\sup_{R_j} f \right) \mathrm{Vol}(R_j).$$

The lower sum with respect to P is obviously less than or equal to the upper sum with respect to the same partition. In fact, more is true.

Lemma C.17. *Let $A \subseteq \mathbb{R}^n$ be a closed rectangle, and let $f : A \to \mathbb{R}$ be a bounded function. For any pair of partitions P and P' of A,*

$$L(f, P) \leq U(f, P').$$

Proof. Write $P = (P_1, \ldots, P_n)$ and $P' = (P'_1, \ldots, P'_n)$, and let Q be the partition $Q = (P_1 \cup P'_1, \ldots, P_n \cup P'_n)$. Each subrectangle of P or P' is a union of finitely many subrectangles of Q. An easy computation shows that

$$L(f, P) \leq L(f, Q) \leq U(f, Q) \leq U(f, P'),$$

from which the result follows. $\square$

The **lower integral of f over A** is

$$\underline{\int}_A f \, dV = \sup\{L(f, P) : P \text{ is a partition of } A\},$$

and the **upper integral** is

$$\overline{\int}_A f \, dV = \inf\{U(f, P) : P \text{ is a partition of } A\}.$$

Clearly, both numbers exist, because f is bounded, and Lemma C.17 implies that the lower integral is less than or equal to the upper integral.

If $f : A \to \mathbb{R}$ is a bounded function whose upper and lower integrals are equal, we say that f is **(Riemann) integrable over A**, and their common value, denoted by

$$\int_A f \, dV,$$

is called the *integral of f over A*. The "dV" in this notation, like the "dx" in the notation for single integrals, has no meaning on its own; it is just a "closing bracket" for the integral sign. Other common notations are

$$\int_A f \quad \text{or} \quad \int_A f \, dx^1 \cdots dx^n \quad \text{or} \quad \int_A f\left(x^1, \ldots, x^n\right) dx^1 \cdots dx^n.$$

In $\mathbb{R}^2$, the symbol dV is often replaced by dA.

There is a simple criterion for a bounded function to be Riemann integrable. It is based on the following notion. A subset $X \subseteq \mathbb{R}^n$ is said to have *measure zero* if for every $\delta > 0$, there exists a countable cover of X by open rectangles $\{C_i\}$ such that $\sum_i \text{Vol}(C_i) < \delta$. (Those who are familiar with the theory of Lebesgue measure will notice that this is equivalent to the condition that the Lebesgue measure of X be equal to zero.)

Proposition C.18 (Properties of Sets of Measure Zero).

(a) *If $X \subseteq \mathbb{R}^n$ has measure zero and $x_0 \in \mathbb{R}^n$, then the translated subset $x_0 + X = \{x_0 + a : a \in X\}$ also has measure zero.*

(b) *Every subset of a set of measure zero in $\mathbb{R}^n$ has measure zero.*

(c) *A countable union of sets of measure zero in $\mathbb{R}^n$ has measure zero.*

(d) *If $k < n$, then every subset of $\mathbb{R}^k$ (viewed as the set of points $x \in \mathbb{R}^n$ with $x^{k+1} = \cdots = x^n = 0$) has measure zero in $\mathbb{R}^n$.*

▶ **Exercise C.19.** Prove Proposition C.18.

Part (d) of this proposition illustrates that having measure zero is a property of a set in relation to a particular Euclidean space containing it, not of a set in and of itself. For example, an open interval in the x-axis has measure zero as a subset of $\mathbb{R}^2$, but not when considered as a subset of $\mathbb{R}^1$. For this reason, we sometimes say that a subset of $\mathbb{R}^n$ has *n-dimensional measure zero* if we wish to emphasize that it has measure zero as a subset of $\mathbb{R}^n$.

The following proposition gives a sufficient condition for a function to be integrable. It shows, in particular, that every bounded continuous function is integrable.

Proposition C.20 (Lebesgue's Integrability Criterion). *Let $A \subseteq \mathbb{R}^n$ be a closed rectangle, and let $f : A \to \mathbb{R}$ be a bounded function. If the set*

$$S = \{x \in A : f \text{ is not continuous at } x\}$$

has measure zero, then f is integrable.

Proof. Suppose the set S has measure zero, and let $\varepsilon > 0$ be given. By definition of measure zero sets, S can be covered by a countable collection of open rectangles $\{C_i\}$, the sum of whose volumes is less than ε. For each $q \in A \smallsetminus S$, since f is continuous at q, there is an open rectangle D_q centered at q such that $|f(x) - f(q)| < \varepsilon$ for all $x \in D_q \cap A$. By shrinking D_q a little, we can arrange that the same inequality holds for all $x \in \bar{D}_q \cap A$. This implies $\sup_{\bar{D}_q} f - \inf_{\bar{D}_q} f \le 2\varepsilon$.

The collection of all rectangles of the form C_i or D_q is an open cover of A. By compactness, finitely many of them cover A. Let us relabel these rectangles as $\{C_1, \ldots, C_k, D_1, \ldots, D_l\}$. Replacing each C_i or D_j by its intersection with Int A, we may assume that each $\bar{C}_i$ and each $\bar{D}_j$ is contained in A.

Since there are only finitely many rectangles $\{C_i, D_j\}$, there is a partition P of A with the property that each $\bar{C}_i$ or $\bar{D}_j$ is equal to a union of subrectangles of P. (Just use the union of all the endpoints of the component intervals of the rectangles C_i and D_j to define the partition.) We can divide the subrectangles of P into two disjoint sets $\mathcal{C}$ and $\mathcal{D}$ such that every subrectangle in $\mathcal{C}$ is contained in $\bar{C}_i$ for some i, and every subrectangle in $\mathcal{D}$ is contained in $\bar{D}_j$ for some j. Then

$$
U(f, P) - L(f, P)
$$

$$
= \sum_i \left(\sup_{R_i} f \right) \mathrm{Vol}(R_i) - \sum_i \left(\inf_{R_i} f \right) \mathrm{Vol}(R_i)
$$

$$
= \sum_{R_i \in \mathcal{C}} \left(\sup_{R_i} f - \inf_{R_i} f \right) \mathrm{Vol}(R_i) + \sum_{R_i \in \mathcal{D}} \left(\sup_{R_i} f - \inf_{R_i} f \right) \mathrm{Vol}(R_i)
$$

$$
\leq \left(\sup_A f - \inf_A f \right) \sum_{R_i \in \mathcal{C}} \mathrm{Vol}(R_i) + 2\varepsilon \sum_{R_i \in \mathcal{D}} \mathrm{Vol}(R_i)
$$

$$
\leq \left(\sup_A f - \inf_A f \right) \varepsilon + 2\varepsilon \, \mathrm{Vol}(A).
$$

It follows that

$$
\overline{\int_A} f \, dV - \underline{\int_A} f \, dV \leq \left(\sup_A f - \inf_A f \right) \varepsilon + 2\varepsilon \, \mathrm{Vol}(A),
$$

which can be made as small as desired by taking ε sufficiently small. This implies that the upper and lower integrals of f are equal, so f is integrable. □

In fact, Lebesgue's criterion is both necessary and sufficient for Riemann integrability, but we do not need that.

Now suppose $D \subseteq \mathbb{R}^n$ is an arbitrary bounded set, and $f : D \to \mathbb{R}$ is a bounded function. Define $f_D : \mathbb{R}^n \to \mathbb{R}$ by

$$
f_D(x) = \begin{cases} f(x), & x \in D, \\ 0, & x \in \mathbb{R}^n \smallsetminus D. \end{cases} \tag{C.11}
$$

If the integral

$$
\int_A f_D \, dV \tag{C.12}
$$

exists for some closed rectangle A containing D, then f is said to be **integrable over D**. The integral (C.12) is denoted by $\int_D f \, dV$ and called the **integral of f**

over **D**. It is easy to check that both the integrability of f and the value of the integral are independent of the rectangle chosen.

In practice, we are interested only in integrals of bounded continuous functions. However, since we sometimes need to integrate them over domains other than rectangles, it is necessary to consider also integrals of discontinuous functions such as the function f_D defined by (C.11). The main reason for proving Proposition C.20 is that it allows us to give a simple description of domains on which all bounded continuous functions are integrable.

A subset $D \subseteq \mathbb{R}^n$ is called a ***domain of integration*** if D is bounded and ∂D has n-dimensional measure zero. It follows from Proposition C.18 that every open or closed rectangle is a domain of integration, and a finite union of domains of integration is again a domain of integration.

Proposition C.21. *If $D \subseteq \mathbb{R}^n$ is a domain of integration, then every bounded continuous real-valued function on D is integrable over D.*

Proof. Let $f : D \to \mathbb{R}$ be bounded and continuous, let $f_D : \mathbb{R}^n \to \mathbb{R}$ be the function defined by (C.11), and let A be a closed rectangle containing D. To prove the theorem, we need only show that the set of points in A where f_D is discontinuous has measure zero.

If $x \in \text{Int } D$, then $f_D = f$ on a neighborhood of x, so f_D is continuous at x. Similarly, if $x \in \mathbb{R}^n \smallsetminus \bar{D}$, then $f_D \equiv 0$ on a neighborhood of x, so again f is continuous at x. Thus the set of points where f_D is discontinuous is contained in ∂D, and therefore has measure zero. $\square$

Of course, if D is compact, then the assumption that f is bounded in the preceding proposition is superfluous.

If D is a domain of integration, the ***volume of*** **D** is defined to be

$$\text{Vol}(D) = \int_D 1 \, dV. \tag{C.13}$$

The integral on the right-hand side is often abbreviated $\int_D dV$.

The next two propositions collect some basic facts about volume and integrals of continuous functions.

Proposition C.22 (Properties of Volume). *Let $D \subseteq \mathbb{R}^n$ be a domain of integration.*

(a) *If D is an open or closed rectangle, then the two definitions (C.10) and (C.13) of $\text{Vol}(D)$ agree.*

(b) $\text{Vol}(D) \geq 0$, *with equality if and only if D has measure zero.*

(c) *If $D_1, \ldots, D_k$ are domains of integration whose union is D, then*

$$\text{Vol}(D) \leq \text{Vol}(D_1) + \cdots + \text{Vol}(D_k),$$

with equality if and only if $D_i \cap D_j$ has measure zero for each $i \neq j$.

(d) *If D_1 is a domain of integration contained in D, then $\text{Vol}(D_1) \leq \text{Vol}(D)$, with equality if and only if $D \smallsetminus D_1$ has measure zero.*

Proposition C.23 (Properties of Integrals). *Let $D \subseteq \mathbb{R}^n$ be a domain of integration, and let $f, g: D \to \mathbb{R}$ be continuous and bounded.*

(a) *For any $a, b \in \mathbb{R}$,*

$$\int_D (af + bg) \, dV = a \int_D f \, dV + b \int_D g \, dV.$$

(b) *If D has measure zero, then $\int_D f \, dV = 0$.*
(c) *If $D_1, \dots, D_k$ are domains of integration whose union is D and whose pairwise intersections have measure zero, then*

$$\int_D f \, dV = \int_{D_1} f \, dV + \cdots + \int_{D_k} f \, dV.$$

(d) *If $f \geq 0$ on D, then $\int_D f \, dV \geq 0$, with equality if and only if $f \equiv 0$ on $\operatorname{Int} D$.*
(e) *$(\inf_D f) \operatorname{Vol}(D) \leq \int_D f \, dV \leq (\sup_D f) \operatorname{Vol}(D)$.*
(f) *$\left| \int_D f \, dV \right| \leq \int_D |f| \, dV$.*

▶ **Exercise C.24.** Prove Propositions C.22 and C.23.

Corollary C.25. *A set of measure zero in $\mathbb{R}^n$ contains no nonempty open subset.*

Proof. Assume for the sake of contradiction that $D \subseteq \mathbb{R}^n$ has measure zero and contains a nonempty open subset U. Then U contains a nonempty open rectangle, which has positive volume and therefore does not have measure zero by Proposition C.22. But this contradicts the fact that every subset of D has measure zero by Proposition C.18. $\qquad \square$

There are two more fundamental properties of multiple integrals that we need. The proofs are too involved to be included in this summary, but you can look them up in [Apo74, Rud76, Str00] if you are interested. Each of these theorems can be stated in various ways, some stronger than others. The versions we give here are quite sufficient for our applications.

Theorem C.26 (Change of Variables). *Suppose D and E are open domains of integration in $\mathbb{R}^n$, and $G: \bar{D} \to \bar{E}$ is smooth map that restricts to a diffeomorphism from D to E. For every continuous function $f: \bar{E} \to \mathbb{R}$,*

$$\int_E f \, dV = \int_D (f \circ G) \, |\det DG| \, dV.$$

Theorem C.27 (Fubini's Theorem). *Let $A = [a^1, b^1] \times \cdots \times [a^n, b^n]$ be a closed rectangle in $\mathbb{R}^n$, and let $f: A \to \mathbb{R}$ be continuous. Then*

$$\int_A f \, dV = \int_{a^n}^{b^n} \left(\cdots \left(\int_{a^1}^{b^1} f(x^1, \dots, x^n) \, dx^1 \right) \cdots \right) dx^n,$$

and the same is true if the variables in the iterated integral on the right-hand side are reordered in any way.

Integrals of Vector-Valued Functions

If $D \subseteq \mathbb{R}^n$ is a domain of integration and $F\colon D \to \mathbb{R}^k$ is a bounded continuous vector-valued function, we define the integral of F over D to be the vector in $\mathbb{R}^k$ obtained by integrating F component by component:

$$\int_D F \, dV = \left(\int_D F^1 \, dV, \dots, \int_D F^k \, dV \right).$$

The analogues of parts (a)–(c) of Proposition C.23 obviously hold for vector-valued integrals, just by applying them to each component. Part (f) holds as well, but requires a bit more work to prove.

Proposition C.28. *Suppose $D \subseteq \mathbb{R}^n$ is a domain of integration and $F\colon D \to \mathbb{R}^k$ is a bounded continuous vector-valued function. Then*

$$\left| \int_D F \, dV \right| \le \int_D |F| \, dV. \tag{C.14}$$

Proof. Let G denote the vector $\int_D F \, dV \in \mathbb{R}^k$. If $G = 0$, then (C.14) obviously holds, so we may as well assume that $G \ne 0$. We compute

$$|G|^2 = \sum_{i=1}^k (G^i)^2 = \sum_{i=1}^k G^i \int_D F^i \, dV = \sum_{i=1}^k \int_D G^i F^i \, dV = \int_D (G \cdot F) \, dV.$$

Applying Proposition C.23(f) to the scalar integral $\int_D (G \cdot F) \, dV$, we obtain

$$|G|^2 \le \int_D |G \cdot F| \, dV \le \int_D |G||F| \, dV = |G| \int_D |F| \, dV.$$

Dividing both sides of the inequality above by $|G|$ yields (C.14). $\square$

As an application of (C.14), we prove an important estimate for the local behavior of a C^1 function in terms of its total derivative.

Proposition C.29 (Lipschitz Estimate for C^1 Functions). *Let $U \subseteq \mathbb{R}^n$ be an open subset, and suppose $F\colon U \to \mathbb{R}^m$ is of class C^1. Then F is Lipschitz continuous on every compact convex subset $K \subseteq U$. The Lipschitz constant can be taken to be $\sup_{x \in K} |DF(x)|$.*

Proof. Since $|DF(x)|$ is a continuous function of x, it is bounded on the compact set K. (The norm here is the Frobenius norm on matrices defined in (B.15).) Let $M = \sup_{x \in K} |DF(x)|$. For arbitrary $a, b \in K$, we have $a + t(b - a) \in K$ for all $t \in I$ because K is convex. By the fundamental theorem of calculus applied to each component of F, together with the chain rule,

$$F(b) - F(a) = \int_0^1 \frac{d}{dt} F\big(a + t(b - a)\big) \, dt$$

$$= \int_0^1 DF\big(a + t(b - a)\big)(b - a)\, dt.$$

Therefore, by (C.14) and Exercise B.48,

$$\big|F(b) - F(a)\big| \le \int_0^1 \big|DF\big(a + t(b - a)\big)\big|\, |b - a|\, dt$$

$$\le \int_0^1 M\, |b - a|\, dt = M\, |b - a|. \qquad \Box$$

Corollary C.30. *If* $U \subseteq \mathbb{R}^n$ *is an open subset and* $F: U \to \mathbb{R}^m$ *is of class* C^1, *then* f *is locally Lipschitz continuous.*

Proof. Each point of U is contained in a ball whose closure is contained in U, and Proposition C.29 shows that the restriction of F to such a ball is Lipschitz continuous. $\qquad \Box$

Sequences and Series of Functions

Let $S \subseteq \mathbb{R}^n$, and suppose we are given functions $f: S \to \mathbb{R}^m$ and $f_i: S \to \mathbb{R}^m$ for each integer $i \ge 1$. The sequence $(f_i)_{i=1}^{\infty}$ is said to **converge pointwise to** f if for each $a \in S$ and each $\varepsilon > 0$, there exists an integer N such that $i \ge N$ implies $|f_i(a) - f(a)| < \varepsilon$. The sequence is said to **converge uniformly to** f if N can be chosen independently of the point a: for each $\varepsilon > 0$ there exists N such that $i \ge N$ implies $|f_i(a) - f(a)| < \varepsilon$ for all $a \in S$. The sequence is **uniformly Cauchy** if for every $\varepsilon > 0$ there exists N such that $i, j \ge N$ implies $|f_i(a) - f_j(a)| < \varepsilon$ for all $a \in S$.

Theorem C.31 (Properties of Uniform Convergence). *Let* $S \subseteq \mathbb{R}^n$, *and suppose* $f_i: S \to \mathbb{R}^m$ *is continuous for each integer* $i \ge 1$.

(a) *If* $f_i \to f$ *uniformly, then* f *is continuous.*
(b) *If the sequence* $(f_i)_{i=1}^{\infty}$ *is uniformly Cauchy, then it converges uniformly to a continuous function.*
(c) *If* $f_i \to f$ *uniformly and* S *is a compact domain of integration, then*

$$\lim_{i \to \infty} \int_S f_i\, dV = \int_S f\, dV.$$

(d) *If* S *is open, each* f_i *is of class* C^1, $f_i \to f$ *pointwise, and* $(\partial f_i / \partial x^j)$ *converges uniformly on* S *as* $i \to \infty$, *then* $\partial f / \partial x^j$ *exists on* S *and*

$$\frac{\partial f}{\partial x^j} = \lim_{i \to \infty} \frac{\partial f_i}{\partial x^j}.$$

For a proof, see [Apo74, Rud76, Str00].

Given an infinite series of (real-valued or vector-valued) functions $\sum_{i=0}^{\infty} f_i$ on $S \subseteq \mathbb{R}^n$, one says *the series converges pointwise* if the corresponding sequence of partial sums converges pointwise to some function f:

$$f(x) = \lim_{N \to \infty} \sum_{i=0}^{N} f_i(x) \quad \text{for all } x \in S.$$

We say *the series converges uniformly* if its partial sums do so.

Proposition C.32 (Weierstrass M-test). *Suppose $S \subseteq \mathbb{R}^n$, and $f_i \colon S \to \mathbb{R}^k$ are functions. If there exist positive real numbers M_i such that $\sup_S |f_i| \leq M_i$ and $\sum_i M_i$ converges, then $\sum_i f_i$ converges uniformly on S.*

▶ **Exercise C.33.** Prove Proposition C.32.

The Inverse and Implicit Function Theorems

The last two theorems in this appendix are central results about smooth functions. They say that under certain hypotheses, the local behavior of a smooth function is modeled by the behavior of its total derivative.

Theorem C.34 (Inverse Function Theorem). *Suppose U and V are open subsets of $\mathbb{R}^n$, and $F \colon U \to V$ is a smooth function. If $DF(a)$ is invertible at some point $a \in U$, then there exist connected neighborhoods $U_0 \subseteq U$ of a and $V_0 \subseteq V$ of $F(a)$ such that $F|_{U_0} \colon U_0 \to V_0$ is a diffeomorphism.*

The proof of this theorem is based on an elementary result about metric spaces, which we describe first.

Let X be a metric space. A map $G \colon X \to X$ is said to be a *contraction* if there is a constant $\lambda \in (0, 1)$ such that $d(G(x), G(y)) \leq \lambda d(x, y)$ for all $x, y \in X$. Clearly, every contraction is continuous. A *fixed point* of a map $G \colon X \to X$ is a point $x \in X$ such that $G(x) = x$.

Lemma C.35 (Contraction Lemma). *Let X be a nonempty complete metric space. Every contraction $G \colon X \to X$ has a unique fixed point.*

Proof. Uniqueness is immediate, for if x and x' are both fixed points of G, the contraction property implies $d(x, x') = d(G(x), G(x')) \leq \lambda d(x, x')$, which is possible only if $x = x'$.

To prove the existence of a fixed point, let x_0 be an arbitrary point in X, and define a sequence $(x_n)_{n=0}^{\infty}$ inductively by $x_{n+1} = G(x_n)$. For any $i \geq 1$ we have $d(x_i, x_{i+1}) = d(G(x_{i-1}), G(x_i)) \leq \lambda d(x_{i-1}, x_i)$, and therefore by induction

$$d(x_i, x_{i+1}) \leq \lambda^i d(x_0, x_1).$$

If N is a positive integer and $j \geq i \geq N$,

$$d(x_i, x_j) \leq d(x_i, x_{i+1}) + d(x_{i+1}, x_{i+2}) + \cdots + d(x_{j-1}, x_j)$$

$$\leq \left(\lambda^i + \cdots + \lambda^{j-1} \right) d(x_0, x_1)$$

$$\leq \lambda^i \left(\sum_{n=0}^{\infty} \lambda^n \right) d(x_0, x_1)$$

$$\leq \lambda^N \frac{1}{1-\lambda} d(x_0, x_1).$$

Since this last expression can be made as small as desired by choosing N large, the sequence (x_n) is Cauchy and therefore converges to a limit $x \in X$. Because G is continuous,

$$G(x) = G\left(\lim_{n \to \infty} x_n \right) = \lim_{n \to \infty} G(x_n) = \lim_{n \to \infty} x_{n+1} = x,$$

so x is the desired fixed point. $\square$

Proof of the inverse function theorem. We begin by making some simple modifications to the function F to streamline the proof. First, the function F_1 defined by

$$F_1(x) = F(x + a) - F(a)$$

is smooth on a neighborhood of 0 and satisfies $F_1(0) = 0$ and $DF_1(0) = DF(a)$; clearly, F is a diffeomorphism on a connected neighborhood of a if and only if F_1 is a diffeomorphism on a connected neighborhood of 0. Second, the function $F_2 = DF_1(0)^{-1} \circ F_1$ is smooth on the same neighborhood of 0 and satisfies $F_2(0) = 0$ and $DF_2(0) = I_n$; and if F_2 is a diffeomorphism in a neighborhood of 0, then so is F_1 and therefore also F. Henceforth, replacing F by F_2, we assume that F is defined in a neighborhood U of 0, $F(0) = 0$, and $DF(0) = I_n$. Because the determinant of $DF(x)$ is a continuous function of x, by shrinking U if necessary, we may assume that $DF(x)$ is invertible for each $x \in U$.

Let $H(x) = x - F(x)$ for $x \in U$. Then $DH(0) = I_n - I_n = 0$. Because the matrix entries of $DH(x)$ are continuous functions of x, there is a number $\delta > 0$ such that $\bar{B}_\delta(0) \subseteq U$ and for all $x \in \bar{B}_\delta(0)$, we have $|DH(x)| \leq \frac{1}{2}$. If $x, x' \in \bar{B}_\delta(0)$, the Lipschitz estimate for smooth functions (Proposition C.29) implies

$$|H(x') - H(x)| \leq \tfrac{1}{2}|x' - x|. \tag{C.15}$$

In particular, taking $x' = 0$, this implies

$$|H(x)| \leq \tfrac{1}{2}|x|. \tag{C.16}$$

Since $x' - x = F(x') - F(x) + H(x') - H(x)$, it follows that

$$|x' - x| \leq |F(x') - F(x)| + |H(x') - H(x)| \leq |F(x') - F(x)| + \tfrac{1}{2}|x' - x|,$$

and rearranging gives

$$|x' - x| \leq 2|F(x') - F(x)| \tag{C.17}$$

for all $x, x' \in \bar{B}_\delta(0)$. In particular, this shows that F is injective on $\bar{B}_\delta(0)$.

Now let $y \in B_{\delta/2}(0)$ be arbitrary. We will show that there exists a unique point $x \in B_\delta(0)$ such that $F(x) = y$. Let $G(x) = y + H(x) = y + x - F(x)$, so that $G(x) = x$ if and only if $F(x) = y$. If $|x| \leq \delta$, (C.16) implies

$$|G(x)| \leq |y| + |H(x)| < \frac{\delta}{2} + \frac{1}{2}|x| \leq \delta, \tag{C.18}$$

so G maps $\bar{B}_\delta(0)$ to itself. It then follows from (C.15) that $|G(x) - G(x')| = |H(x) - H(x')| \leq \frac{1}{2}|x - x'|$, so G is a contraction. Since $\bar{B}_\delta(0)$ is a complete metric space (see Example A.6), the contraction lemma implies that G has a unique fixed point $x \in \bar{B}_\delta(0)$. From (C.18), $|x| = |G(x)| < \delta$, so in fact $x \in B_\delta(0)$, thus proving the claim.

Let $V_0 = B_{\delta/2}(0)$ and $U_0 = B_\delta(0) \cap F^{-1}(V_0)$. Then U_0 is open in $\mathbb{R}^n$, and the argument above shows that $F : U_0 \to V_0$ is bijective, so $F^{-1} : V_0 \to U_0$ exists. Substituting $x = F^{-1}(y)$ and $x' = F^{-1}(y')$ into (C.17) shows that F^{-1} is continuous. Thus $F : U_0 \to V_0$ is a homeomorphism, and it follows that U_0 is connected because V_0 is.

The only thing that remains to be proved is that F^{-1} is smooth. If we knew it were smooth, Proposition C.4 would imply that $D(F^{-1})(y) = DF(x)^{-1}$, where $x = F^{-1}(y)$. We begin by showing that F^{-1} is differentiable at each point of V_0, with total derivative given by this formula.

Let $y \in V_0$ be arbitrary, and set $x = F^{-1}(y)$ and $L = DF(x)$. We need to show that

$$\lim_{y' \to y} \frac{F^{-1}(y') - F^{-1}(y) - L^{-1}(y' - y)}{|y' - y|} = 0.$$

Given $y' \in V_0 \smallsetminus \{y\}$, write $x' = F^{-1}(y') \in U_0 \smallsetminus \{x\}$. Then

$$\frac{F^{-1}(y') - F^{-1}(y) - L^{-1}(y' - y)}{|y' - y|}$$

$$= L^{-1}\left(\frac{L(x' - x) - (y' - y)}{|y' - y|}\right)$$

$$= \frac{|x' - x|}{|y' - y|}L^{-1}\left(-\frac{F(x') - F(x) - L(x' - x)}{|x' - x|}\right).$$

The factor $|x' - x|/|y' - y|$ above is bounded thanks to (C.17), and because L^{-1} is linear and therefore bounded (Exercise B.52), the norm of the second factor is bounded by a constant multiple of

$$\frac{|F(x') - F(x) - L(x' - x)|}{|x' - x|}. \tag{C.19}$$

As $y' \to y$, it follows that $x' \to x$ by continuity of F^{-1}, and then (C.19) goes to zero because $L = DF(x)$ and F is differentiable. This completes the proof that F^{-1} is differentiable.

By Proposition C.8, the partial derivatives of F^{-1} are defined at each point $y \in V_0$. Observe that the formula $D(F^{-1})(y) = DF(F^{-1}(y))^{-1}$ implies that the matrix-valued function $y \mapsto D(F^{-1})(y)$ can be written as the composition

$$y \xrightarrow{F^{-1}} F^{-1}(x) \xrightarrow{DF} DF(F^{-1}(y)) \xrightarrow{i} DF(F^{-1}(y))^{-1}, \qquad \text{(C.20)}$$

where i is matrix inversion. In this composition, F^{-1} is continuous; DF is smooth because its component functions are the partial derivatives of F; and i is smooth because Cramer's rule expresses the entries of an inverse matrix as rational functions of the entries of the matrix. Because $D(F^{-1})$ is a composition of continuous functions, it is continuous. Thus the partial derivatives of F^{-1} are continuous, so F^{-1} is of class C^1.

Now assume by induction that we have shown that F^{-1} is of class C^k. This means that each of the functions in (C.20) is of class C^k. Because $D(F^{-1})$ is a composition of C^k functions, it is itself C^k; this implies that the partial derivatives of F^{-1} are of class C^k, so F^{-1} itself is of class C^{k+1}. Continuing by induction, we conclude that F^{-1} is smooth. $\qquad \square$

Corollary C.36. *Suppose $U \subseteq \mathbb{R}^n$ is an open subset, and $F : U \to \mathbb{R}^n$ is a smooth function whose Jacobian determinant is nonzero at every point in U.*

(a) *F is an open map.*
(b) *If F is injective, then $F : U \to F(U)$ is a diffeomorphism.*

Proof. For each $a \in U$, the fact that the Jacobian determinant of F is nonzero implies that $DF(a)$ is invertible, so the inverse function theorem implies that there exist open subsets $U_a \subseteq U$ containing a and $V_a \subseteq F(U)$ containing $F(a)$ such that F restricts to a diffeomorphism $F|_{U_a} : U_a \to V_a$. In particular, this means that each point of $F(U)$ has a neighborhood contained in $F(U)$, so $F(U)$ is open. If $U_0 \subseteq U$ is an arbitrary open subset, the same argument with U replaced by U_0 shows that $F(U_0)$ is also open; this proves (a). If in addition F is injective, then the inverse map $F^{-1} : F(U) \to U$ exists for set-theoretic reasons; on a neighborhood of each point $F(a) \in F(U)$ it is equal to the inverse of $F|_{U_a}$, so it is smooth. $\qquad \square$

The next two examples illustrate the use of the preceding corollary.

Example C.37 (Polar Coordinates). As you know from calculus, *polar coordinates* (r, θ) in the plane are defined implicitly by the relations $x = r \cos \theta$, $y = r \sin \theta$. The map $F : (0, \infty) \times \mathbb{R} \to \mathbb{R}^2$ defined by $F(r, \theta) = (r \cos \theta, r \sin \theta)$ is smooth and has Jacobian determinant equal to r, which is nonzero everywhere on the domain. Thus, Corollary C.36 shows that the restriction of F to any open subset on which it is injective is a diffeomorphism onto its image. One such subset is $\{(r, \theta) : r > 0, -\pi < \theta < \pi\}$, which is mapped bijectively by F onto the complement of the nonpositive part of the x-axis. $\qquad /\!/$

Example C.38 (Spherical Coordinates). Similarly, *spherical coordinates* on $\mathbb{R}^3$ are the functions (ρ, φ, θ) defined by the relations

$$x = \rho \sin \varphi \cos \theta,$$

$$y = \rho \sin \varphi \sin \theta,$$

$$z = \rho \cos \varphi.$$

Geometrically, ρ is the distance from the origin, φ is the angle from the positive z-axis, and θ is the angle from the $x > 0$ half of the (x, z)-plane. If we define $G: (0, \infty) \times (0, \pi) \times \mathbb{R} \to \mathbb{R}^3$ by $G(\rho, \varphi, \theta) = (\rho \sin \varphi \cos \theta, \rho \sin \varphi \sin \theta, \rho \cos \varphi)$, a computation shows that the Jacobian determinant of G is $\rho^2 \sin \varphi \neq 0$. Thus, the restriction of G to any open subset on which it is injective is a diffeomorphism onto its image. One such subset is

$$\{(\rho, \varphi, \theta) : \rho > 0,\ 0 < \varphi < \pi, -\pi < \theta < \pi\}.$$

Notice how much easier it is to argue this way than to try to construct an inverse map explicitly out of inverse trigonometric functions. //

▶ **Exercise C.39.** Verify the claims in the preceding two examples.

The next result is one of the most important consequences of the inverse function theorem. It gives conditions under which a level set of a smooth function is locally the graph of a smooth function.

Theorem C.40 (Implicit Function Theorem). *Let $U \subseteq \mathbb{R}^n \times \mathbb{R}^k$ be an open subset, and let $(x, y) = (x^1, \ldots, x^n, y^1, \ldots, y^k)$ denote the standard coordinates on U. Suppose $\Phi: U \to \mathbb{R}^k$ is a smooth function, $(a, b) \in U$, and $c = \Phi(a, b)$. If the $k \times k$ matrix*

$$\left(\frac{\partial \Phi^i}{\partial y^j}(a, b) \right)$$

is nonsingular, then there exist neighborhoods $V_0 \subseteq \mathbb{R}^n$ of a and $W_0 \subseteq \mathbb{R}^k$ of b and a smooth function $F: V_0 \to W_0$ such that $\Phi^{-1}(c) \cap (V_0 \times W_0)$ is the graph of F, that is, $\Phi(x, y) = c$ for $(x, y) \in V_0 \times W_0$ if and only if $y = F(x)$.

Proof. Consider the smooth function $\Psi: U \to \mathbb{R}^n \times \mathbb{R}^k$ defined by $\Psi(x, y) = (x, \Phi(x, y))$. Its total derivative at (a, b) is

$$D\Psi(a, b) = \begin{pmatrix} I_n & 0 \\ \dfrac{\partial \Phi^i}{\partial x^j}(a, b) & \dfrac{\partial \Phi^i}{\partial y^j}(a, b) \end{pmatrix},$$

which is nonsingular because it is block lower triangular and the two blocks on the main diagonal are nonsingular. Thus by the inverse function theorem there exist connected neighborhoods U_0 of (a, b) and Y_0 of (a, c) such that $\Psi: U_0 \to Y_0$ is a diffeomorphism. Shrinking U_0 and Y_0 if necessary, we may assume that $U_0 = V \times W$ is a product neighborhood.

Writing $\Psi^{-1}(x, y) = (A(x, y), B(x, y))$ for some smooth functions A and B, we compute

$$(x, y) = \Psi(\Psi^{-1}(x, y)) = \Psi(A(x, y), B(x, y))$$

$$= (A(x, y), \Phi(A(x, y), B(x, y))). \tag{C.21}$$

Comparing the first components in this equation, we find that $A(x, y) = x$, so Ψ^{-1} has the form $\Psi^{-1}(x, y) = \big(x, B(x, y)\big)$.

Now let $V_0 = \{x \in V : (x, c) \in Y_0\}$ and $W_0 = W$, and define $F \colon V_0 \to W_0$ by $F(x) = B(x, c)$. Comparing the second components in (C.21) yields

$$c = \Phi\big(x, B(x, c)\big) = \Phi\big(x, F(x)\big)$$

whenever $x \in V_0$, so the graph of F is contained in $\Phi^{-1}(c)$. Conversely, suppose $(x, y) \in V_0 \times W_0$ and $\Phi(x, y) = c$. Then $\Psi(x, y) = \big(x, \Phi(x, y)\big) = (x, c)$, so

$$(x, y) = \Psi^{-1}(x, c) = \big(x, B(x, c)\big) = \big(x, F(x)\big),$$

which implies that $y = F(x)$. This completes the proof. $\square$

Appendix D
Review of Differential Equations

The theory of ordinary differential equations (ODEs) underlies much of the study of smooth manifolds. In this appendix, we review both the theoretical and the practical aspects of the subject. Since we need to work only with first-order equations and systems, we concentrate our attention on those. For more detail, consult any good ODE textbook, such as [BR89] or [BD09].

Existence, Uniqueness, and Smoothness

Here is the general setting in which ODEs appear in this book: we are given n real-valued continuous functions $V^1, \dots, V^n$ defined on some open subset $W \subseteq \mathbb{R}^{n+1}$, and the goal is to find differentiable real-valued functions $y^1, \dots, y^n$ solving the following *initial value problem*:

$$\dot{y}^i(t) = V^i\big(t, y^1(t), \dots, y^n(t)\big), \quad i = 1, \dots, n, \tag{D.1}$$

$$y^i(t_0) = c^i, \qquad\qquad\qquad i = 1, \dots, n, \tag{D.2}$$

where $(t_0, c^1, \dots, c^n)$ is an arbitrary point in W. (Here and elsewhere in the book, we use a dot to denote an ordinary derivative with respect to t whenever convenient, primarily when there are superscripts that would make the prime notation cumbersome.)

The fundamental fact about ordinary differential equations is that for smooth equations, there always exists a unique solution to the initial value problem, at least for a short time, and the solution is a smooth function of the initial conditions as well as time. The existence and uniqueness parts of this theorem are proved in most ODE textbooks, but the smoothness part is often omitted. Because this result is so fundamental to smooth manifold theory, we give a complete proof here.

Most of our applications of the theory are confined to the following special case: if the functions V^i on the right-hand side of (D.1) do not depend explicitly on t, the system is said to be *autonomous*; otherwise, it is *nonautonomous*. We begin by stating and proving our main theorem in the autonomous case. Afterwards, we show how the general case follows from this one.

J.M. Lee, *Introduction to Smooth Manifolds*, Graduate Texts in Mathematics 218, DOI 10.1007/978-1-4419-9982-5, © Springer Science+Business Media New York 2013

Theorem D.1 (Fundamental Theorem for Autonomous ODEs). *Suppose $U \subseteq \mathbb{R}^n$ is open, and $V: U \to \mathbb{R}^n$ is a smooth vector-valued function. Consider the initial value problem*

$$\dot{y}^i(t) = V^i\big(y^1(t),\ldots,y^n(t)\big), \quad i = 1,\ldots,n, \tag{D.3}$$

$$y^i(t_0) = c^i, \qquad\qquad\qquad i = 1,\ldots,n, \tag{D.4}$$

for arbitrary $t_0 \in \mathbb{R}$ and $c = (c^1,\ldots,c^n) \in U$.

(a) EXISTENCE: *For any $t_0 \in \mathbb{R}$ and $x_0 \in U$, there exist an open interval J_0 containing t_0 and an open subset $U_0 \subseteq U$ containing x_0 such that for each $c \in U_0$, there is a C^1 map $y: J_0 \to U$ that solves (D.3)–(D.4).*

(b) UNIQUENESS: *Any two differentiable solutions to (D.3)–(D.4) agree on their common domain.*

(c) SMOOTHNESS: *Let J_0 and U_0 be as in (a), and let $\theta: J_0 \times U_0 \to U$ be the map defined by $\theta(t,x) = y(t)$, where $y: J_0 \to U$ is the unique solution to (D.3) with initial condition $y(t_0) = x$. Then θ is smooth.*

The existence, uniqueness, and smoothness parts of this theorem will be proved separately below. The following comparison theorem is useful in the proofs to follow.

Theorem D.2 (ODE Comparison Theorem). *Let $J \subseteq \mathbb{R}$ be an open interval, and suppose the differentiable function $u: J \to \mathbb{R}^n$ satisfies the following differential inequality for all $t \in J$:*

$$|u'(t)| \le f\big(|u(t)|\big),$$

where $f: [0,\infty) \to [0,\infty)$ is Lipschitz continuous. If for some $t_0 \in J$, $v: [0,\infty) \to [0,\infty)$ is a differentiable real-valued function satisfying the initial-value problem

$$v'(t) = f\big(v(t)\big),$$

$$v(0) = |u(t_0)|,$$

then the following inequality holds for all $t \in J$:

$$|u(t)| \le v\big(|t - t_0|\big). \tag{D.5}$$

Proof. Assume first that $t_0 = 0$, and let $J^+ = \{t \in J : t \ge 0\}$. We begin by proving that (D.5) holds for all $t \in J^+$. On the open subset of J^+ where $|u(t)| > 0$, $|u(t)|$ is a differentiable function of t, and the Cauchy–Schwarz inequality shows that

$$\frac{d}{dt}|u(t)| = \frac{d}{dt}\big(u(t) \cdot u(t)\big)^{1/2} = \frac{1}{2}\big(u(t) \cdot u(t)\big)^{-1/2}\big(2u(t) \cdot u'(t)\big)$$

$$\le \frac{1}{2}|u(t)|^{-1}\big(2\,|u(t)|\,|u'(t)|\big) = |u'(t)| \le f\big(|u(t)|\big).$$

Let A be a Lipschitz constant for f, and consider the continuous function $w: J^+ \to \mathbb{R}$ defined by

$$w(t) = e^{-At}\big(|u(t)| - v(t)\big).$$

Then $w(0) = 0$, and (D.5) for $t \in J^+$ is equivalent to $w(t) \leq 0$.

At any $t \in J^+$ such that $w(t) > 0$ (and therefore $|u(t)| > v(t) \geq 0$), w is differentiable and satisfies

$$w'(t) = -Ae^{-At}\left(|u(t)| - v(t)\right) + e^{-At}\frac{d}{dt}\left(|u(t)| - v(t)\right)$$
$$\leq -Ae^{-At}\left(|u(t)| - v(t)\right) + e^{-At}\left(f\left(|u(t)|\right) - f\left(v(t)\right)\right)$$
$$\leq 0,$$

where the last inequality follows from the Lipschitz estimate for f.

Now suppose there is some $t_1 \in J^+$ such that $w(t_1) > 0$. Let

$$\tau = \sup\{t \in [0, t_1] : w(t) \leq 0\}.$$

Then $w(\tau) = 0$ by continuity, and $w(t) > 0$ for $t \in (\tau, t_1]$. Since w is continuous on $[\tau, t_1]$ and differentiable on (τ, t_1), the mean value theorem implies that there must exist $t \in (\tau, t_1)$ such that $w(t) > 0$ and $w'(t) > 0$. But this contradicts the calculation above, which showed that $w'(t) \leq 0$ whenever $w(t) > 0$, thus proving that $w(t) \leq 0$ for all $t \in J^+$.

Now, the result for $t \leq 0$ follows easily by substituting $-t$ for t in the argument above. Finally, for the general case in which $t_0 \neq 0$, we simply apply the above argument to the function $\tilde{u}(t) = u(t + t_0)$ on the interval $\tilde{J} = \{t : t + t_0 \in J\}$. $\quad\square$

Remark. In the statement of the comparison theorem, we have assumed for simplicity that both f and v are defined for all nonnegative t, but these hypotheses can be weakened: the proof goes through essentially without modification as long as v is defined on an interval $[0, b)$ large enough that $J \subseteq (t_0 - b, t_0 + b)$, and f is defined on some interval that contains $|u(t)|$ and $v(t)$ for all $t \in J$.

Theorem D.3 (Existence of ODE Solutions). *Let $U \subseteq \mathbb{R}^n$ be an open subset, and suppose $V : U \to \mathbb{R}^n$ is locally Lipschitz continuous. Let $(t_0, x_0) \in \mathbb{R} \times U$ be given. There exist an open interval $J_0 \subseteq \mathbb{R}$ containing t_0, an open subset $U_0 \subseteq U$ containing x_0, and for each $c \in U_0$, a C^1 map $y : J_0 \to U$ satisfying the initial value problem (D.3)–(D.4).*

Proof. By shrinking U if necessary, we may assume that V is Lipschitz continuous on U. We begin by showing that the system (D.3)–(D.4) is equivalent to a certain integral equation. Suppose y is any solution to (D.3)–(D.4) on some interval J_0 containing t_0. Because y is differentiable, it is continuous, and then the fact that the right-hand side of (D.3) is a continuous function of t implies that y is of class C^1. Integrating (D.3) with respect to t and applying the fundamental theorem of calculus shows that y satisfies the following integral equation:

$$y^i(t) = c^i + \int_{t_0}^t V^i\left(y(s)\right) ds. \tag{D.6}$$

Conversely, if $y : J_0 \to U$ is a continuous map satisfying (D.6), then the fundamental theorem of calculus implies that y satisfies (D.3)–(D.4) and therefore is actually of class C^1.

This motivates the following definition. Suppose J_0 is an open interval containing t_0. For any continuous map $y\colon J_0 \to U$, we define a new map $Iy\colon J_0 \to \mathbb{R}^n$ by

$$Iy(t) = c + \int_{t_0}^t V(y(s))\, ds. \tag{D.7}$$

Then we are led to seek a fixed point for I in a suitable metric space of maps.

Let C be a Lipschitz constant for V, so that

$$\left|V(y) - V(\tilde{y})\right| \le C\left|y - \tilde{y}\right|, \quad y, \tilde{y} \in U. \tag{D.8}$$

Given $t_0 \in \mathbb{R}$ and $x_0 \in U$, choose $r > 0$ small enough that $\bar{B}_r(x_0) \subseteq U$. Let M be the supremum of $|V|$ on the compact set $\bar{B}_r(x_0)$. Choose $\delta > 0$ and $\varepsilon > 0$ small enough that

$$\delta < \frac{r}{2}, \qquad \varepsilon < \min\left(\frac{r}{2M}, \frac{1}{C}\right),$$

and set $J_0 = (t_0 - \varepsilon, t_0 + \varepsilon) \subseteq \mathbb{R}$ and $U_0 = B_\delta(x_0) \subseteq U$. For any $c \in U_0$, let $\mathcal{M}_c$ denote the set of all continuous maps $y\colon J_0 \to \bar{B}_r(x_0)$ satisfying $y(t_0) = c$. We define a metric on $\mathcal{M}_c$ by

$$d(y, \tilde{y}) = \sup_{t \in J_0} \left|y(t) - \tilde{y}(t)\right|.$$

Any sequence of maps in $\mathcal{M}_c$ that is Cauchy in this metric is uniformly Cauchy, and thus converges to a continuous limit y. Clearly, the conditions that y take its values in $\bar{B}_r(x_0)$ and $y(t_0) = c$ are preserved in the limit. Therefore, $\mathcal{M}_c$ is a complete metric space.

We wish to define a map $I\colon \mathcal{M}_c \to \mathcal{M}_c$ by formula (D.7). The first thing we need to verify is that I really does map $\mathcal{M}_c$ into itself. It is clear from the definition that $Iy(t_0) = c$ and Iy is continuous (in fact, it is differentiable by the fundamental theorem of calculus). Thus, we need only check that Iy takes its values in $\bar{B}_r(x_0)$. If $y \in \mathcal{M}_c$, then for any $t \in J_0$,

$$|Iy(t) - x_0| = \left|c + \int_{t_0}^t V(y(s))\, ds - x_0\right|$$

$$\le |c - x_0| + \int_{t_0}^t \left|V(y(s))\right| ds$$

$$< \delta + M\varepsilon < r$$

by our choice of δ and ε.

Next we check that I is a contraction (see p. 657). If $y, \tilde{y} \in \mathcal{M}_c$, then

$$d(Iy, I\tilde{y}) = \sup_{t \in J_0} \left|\int_{t_0}^t V(y(s))\, ds - \int_{t_0}^t V(\tilde{y}(s))\, ds\right|$$

$$\le \sup_{t \in J_0} \int_{t_0}^t \left|V(y(s)) - V(\tilde{y}(s))\right| ds$$

$$\leq \sup_{t \in J_0} \int_{t_0}^{t} C |y(s) - \tilde{y}(s)| \, ds \leq C \varepsilon d(y, \tilde{y}).$$

Because we have chosen ε so that $C\varepsilon < 1$, this shows that I is a contraction. By the contraction lemma (Lemma C.35), I has a fixed point $y \in \mathcal{M}_c$, which is a solution to (D.6) and thus also to (D.3)–(D.4). $\square$

Theorem D.4 (Uniqueness of ODE Solutions). *Let $U \subseteq \mathbb{R}^n$ be an open subset, and suppose $V : U \to \mathbb{R}^n$ is locally Lipschitz continuous. For any $t_0 \in \mathbb{R}$ and $c \in U$, any two differentiable solutions to (D.3)–(D.4) are equal on their common domain.*

Proof. Suppose first that $y, \tilde{y} : J_0 \to U$ are two differentiable functions that both satisfy (D.3) on the same open interval $J_0 \subseteq \mathbb{R}$, but not necessarily with the same initial conditions. Let J_1 be a bounded open interval containing t_0 such that $\bar{J}_1 \subseteq J_0$. The union of $y(\bar{J}_1)$ and $\tilde{y}(\bar{J}_1)$ is a compact subset of U, and Proposition A.48(b) shows that there is a Lipschitz constant C for V on that set. Thus

$$\left| \frac{d}{dt}\left(\tilde{y}(t) - y(t) \right) \right| = \left| V\left(\tilde{y}(t) \right) - V\left(y(t) \right) \right| \leq C \left| \tilde{y}(t) - y(t) \right|.$$

Applying the ODE comparison theorem (Theorem D.2) with $u(t) = \tilde{y}(t) - y(t)$, $f(v) = Cv$, and $v(t) = e^{Ct} |\tilde{y}(t_0) - y(t_0)|$, we conclude that

$$|\tilde{y}(t) - y(t)| \leq e^{C|t - t_0|} |\tilde{y}(t_0) - y(t_0)|, \quad t \in \bar{J}_1. \tag{D.9}$$

Thus, $y(t_0) = \tilde{y}(t_0)$ implies $y \equiv \tilde{y}$ on all of $\bar{J}_1$. Since every point of J_0 is contained in some such subinterval J_1, it follows that $y \equiv \tilde{y}$ on all of J_0. $\square$

Theorem D.5 (Smoothness of ODE Solutions). *Suppose $U \subseteq \mathbb{R}^n$ is an open subset and $V : U \to \mathbb{R}^n$ is locally Lipschitz continuous. Suppose also that $U_0 \subseteq U$ is an open subset, $J_0 \subseteq \mathbb{R}$ is an open interval containing t_0, and $\theta : J_0 \times U_0 \to U$ is a map such that for each $x \in U_0$, $y(t) = \theta(t, x)$ solves the initial value problem (D.3)–(D.4) with initial condition $c = x$. If V is of class C^k for some $k \geq 0$, then so is θ.*

Proof. Let $(t_1, x_1) \in J_0 \times U_0$ be arbitrary. It suffices to prove that θ is C^k on some neighborhood of (t_1, x_1). We prove this claim by induction on k.

Let J_1 be a bounded open interval containing t_0 and t_1 such that $\bar{J}_1 \subseteq J_0$. Because the restriction of θ to $J_0 \times \{x_1\}$ is an integral curve of V, it is continuous and therefore the set $K = \theta(\bar{J}_1 \times \{x_1\})$ is compact. Thus, there exists $c > 0$ such that $\bar{B}_{2c}(y) \subseteq U$ for every $y \in K$. Let $W = \bigcup_{y \in K} B_c(y)$, so that W is a precompact neighborhood of K in U. The restriction of V to $\overline{W}$ is bounded by compactness, and is Lipschitz continuous by Proposition A.48(b). Let C be a Lipschitz constant for V on $\overline{W}$, and define constants M and T by

$$M = \sup_{\overline{W}} |V|, \qquad T = \sup_{t \in \bar{J}_1} |t - t_0|.$$

For any $x, \tilde{x} \in W$, both $t \mapsto \theta(t, x)$ and $t \mapsto \theta(t, \tilde{x})$ are integral curves of V for $t \in J_1$. As long as both curves stay in W, (D.9) implies

$$|\theta(t, x) - \theta(t, \tilde{x})| \le e^{CT} |\tilde{x} - x|. \tag{D.10}$$

Choose $r > 0$ such that $2re^{CT} < c$, and let $U_1 = B_r(x_1)$ and $U_2 = B_{2r}(x_1)$. We will prove that θ maps $\bar{J}_1 \times \bar{U}_2$ into W. Assume not, which means there is some $(t_2, x_2) \in \bar{J}_1 \times \bar{U}_2$ such that $\theta(t_2, x_2) \notin W$; for simplicity, assume $t_2 > t_0$. Let τ be the infimum of times $t > t_0$ in J_1 such that $\theta(t, x_2) \notin \overline{W}$. By continuity, this means $\theta(\tau, x_2) \in \partial W$. But because both $\theta(t, x_1)$ and $\theta(t, x_2)$ are in W for $t \in [t_0, \tau]$, (D.10) yields $|\theta(\tau, x_2) - \theta(\tau, x_1)| \le 2re^{CT} < c$, which means that $\theta(\tau, x_2) \in W$, a contradiction. This proves the claim.

For the $k = 0$ step, we need to show that θ is continuous on $\bar{J}_1 \times \bar{U}_1$. It follows from (D.10) that it is Lipschitz continuous there as a function of x. We need to show that it is jointly continuous in (t, x).

Let $(t, x) \in \bar{J}_1 \times \bar{U}_1$ be arbitrary. Since every solution to the initial value problem satisfies the integral equation (D.6), we find that

$$\theta^i(t, x) = x^i + \int_{t_0}^{t} V^i(\theta(s, x)) \, ds, \tag{D.11}$$

and therefore (assuming for simplicity that $t_1 \ge t$),

$$|\theta(t_1, x_1) - \theta(t, x)| \le |x_1 - x| + \left| \int_{t_0}^{t_1} V(\theta(s, x_1)) \, ds - \int_{t_0}^{t} V(\theta(s, x)) \, ds \right|$$

$$\le |x_1 - x| + \int_{t_0}^{t} |V(\theta(s, x_1)) - V(\theta(s, x))| \, ds$$

$$+ \int_{t}^{t_1} |V(\theta(s, x_1))| \, ds$$

$$\le |x_1 - x| + C \int_{t_0}^{t} |\theta(s, x_1) - \theta(s, x)| \, ds + \int_{t}^{t_1} M \, ds$$

$$\le |x_1 - x| + CTe^{CT} |x_1 - x| + M |t_1 - t|.$$

It follows that θ is continuous at (t_1, x_1).

Next we tackle the $k = 1$ step, which is the hardest part of the proof. Suppose that V is of class C^1, and let $\bar{J}_1, \bar{U}_1$ be defined as above. Expressed in terms of θ, the initial value problem (D.3)–(D.4) with $c = x$ reads

$$\frac{\partial \theta^i}{\partial t}(t, x) = V^i(\theta(t, x)),$$

$$\theta^i(t_0, x) = x^i. \tag{D.12}$$

Because we know that θ is continuous by the argument above, this shows that $\partial \theta^i / \partial t$ is continuous. We will prove that for each j, $\partial \theta^i / \partial x^j$ exists and is continuous on $\bar{J}_1 \times \bar{U}_1$.

For any real number h such that $0 < |h| < r$ and any indices $i, j \in \{1, \ldots, n\}$, we let $(\Delta_h)^i_j \colon \bar{J}_1 \times \bar{U}_1 \to \mathbb{R}$ be the difference quotient

$$(\Delta_h)^i_j(t, x) = \frac{\theta^i(t, x + he_j) - \theta^i(t, x)}{h}.$$

Then $\partial \theta^i / \partial x^j(t, x) = \lim_{h \to 0}(\Delta_h)^i_j(t, x)$ if the limit exists. In fact, we will show that $(\Delta_h)^i_j$ converges uniformly on $\bar{J}_1 \times \bar{U}_1$ as $h \to 0$, from which it follows that $\partial \theta^i / \partial x^j$ exists and is continuous there, because it is a uniform limit of continuous functions. Let $\Delta_h \colon \bar{J}_1 \times \bar{U}_1 \to M(n, \mathbb{R})$ be the matrix-valued function whose matrix entries are $(\Delta_h)^i_j(t, x)$. Note that (D.10) implies $|(\Delta_h)^i_j(t, x)| \le e^{CT}$ for each i and j, and thus

$$|\Delta_h(t, x)| \le n e^{CT}, \tag{D.13}$$

where the norm on the left-hand side is the Frobenius norm on matrices.

Let us compute the derivative of $(\Delta_h)^i_j$ with respect to t:

$$\begin{aligned}
\frac{\partial}{\partial t}(\Delta_h)^i_j(t, x) &= \frac{1}{h}\left(\frac{\partial \theta^i}{\partial t}(t, x + he_j) - \frac{\partial \theta^i}{\partial t}(t, x)\right) \\
&= \frac{1}{h}\left(V^i(\theta(t, x + he_j)) - V^i(\theta(t, x))\right).
\end{aligned} \tag{D.14}$$

The mean value theorem applied to the C^1 function

$$u(s) = V^i\left((1 - s)\theta(t, x) + s\theta(t, x + he_j)\right)$$

implies that there is some $c \in (0, 1)$ such that $u(1) - u(0) = u'(c)$. If we substitute $y_0 = (1 - c)\theta(t, x) + c\theta(t, x + he_j)$ (a point on the line segment between $\theta(t, x)$ and $\theta(t, x + he_j)$), this becomes

$$V^i\left(\theta(t, x + he_j)\right) - V^i\left(\theta(t, x)\right) = \sum_{k=1}^n \frac{\partial V^i}{\partial y^k}(y_0)\left(\theta^k(t, x + he_j) - \theta^k(t, x)\right)$$

$$= h\sum_{k=1}^n \frac{\partial V^i}{\partial y^k}(y_0)(\Delta_h)^k_j(t, x).$$

Inserting this into (D.14) yields

$$\frac{\partial}{\partial t}(\Delta_h)^i_j(t, x) = \sum_{k=1}^n \frac{\partial V^i}{\partial y^k}(y_0)(\Delta_h)^k_j(t, x).$$

Thus for any sufficiently small nonzero real numbers $h, \tilde{h}$,

$$\frac{\partial}{\partial t}\left((\Delta_h)^i_j(t,x) - (\Delta_{\tilde{h}})^i_j(t,x)\right)$$

$$= \sum_{k=1}^n \frac{\partial V^i}{\partial y^k}(y_0)(\Delta_h)^k_j(t,x) - \sum_{k=1}^n \frac{\partial V^i}{\partial y^k}(\tilde{y}_0)(\Delta_{\tilde{h}})^k_j(t,x)$$

$$= \sum_{k=1}^n \frac{\partial V^i}{\partial y^k}(y_0)\left((\Delta_h)^k_j(t,x) - (\Delta_{\tilde{h}})^k_j(t,x)\right)$$

$$+ \sum_{k=1}^n \left(\frac{\partial V^i}{\partial y^k}(y_0) - \frac{\partial V^i}{\partial y^k}(\tilde{y}_0)\right)(\Delta_{\tilde{h}})^k_j(t,x), \tag{D.15}$$

where $\tilde{y}_0$ is defined similarly to y_0, but with $\tilde{h}$ in place of h.

Now let $\varepsilon > 0$ be given. Because the continuous functions $\partial V^i/\partial y^k$ are uniformly continuous on $\bar{U}_1$ (Proposition A.48(a)), there exists $\delta > 0$ such that the following inequality holds whenever $|y_1 - y_2| < \delta$:

$$\left|\frac{\partial V^i}{\partial y^k}(y_1) - \frac{\partial V^i}{\partial y^k}(y_2)\right| < \varepsilon.$$

Suppose $|h|$ and $|\tilde{h}|$ are both less than $\delta e^{-CT}/n$. Then we have

$$|y_0 - \theta(t,x)| = c|\theta(t,x+he_j) - \theta(t,x)| \le c|h||\Delta_h(t,x)| < \delta, \tag{D.16}$$

and similarly $|\tilde{y}_0 - \theta(t,x)| < \delta$, so

$$\left|\frac{\partial V^i}{\partial y^k}(y_0) - \frac{\partial V^i}{\partial y^k}(\tilde{y}_0)\right|$$

$$\le \left|\frac{\partial V^i}{\partial y^k}(y_0) - \frac{\partial V^i}{\partial y^k}(\theta(t,x))\right| + \left|\frac{\partial V^i}{\partial y^k}(\theta(t,x)) - \frac{\partial V^i}{\partial y^k}(\tilde{y}_0)\right| < 2\varepsilon. \tag{D.17}$$

Inserting (D.17) and (D.13) into (D.15), we find that the matrix-valued function $\Delta_h - \Delta_{\tilde{h}}$ satisfies the following differential inequality:

$$\left|\frac{\partial}{\partial t}\left(\Delta_h(t,x) - \Delta_{\tilde{h}}(t,x)\right)\right| \le E|\Delta_h(t,x) - \Delta_{\tilde{h}}(t,x)| + 2\varepsilon n e^{CT},$$

where E is the supremum of $|DV|$ on $\bar{U}_1$. Note that $\theta^i(t_0,x) = x^i$ implies that $(\Delta_h)^i_j$ satisfies the following initial condition:

$$(\Delta_h)^i_j(t_0,x) = \frac{\theta^i(t_0,x+he_j) - \theta^i(t_0,x)}{h} = \frac{(x^i + h\delta^i_j) - x^i}{h} = \delta^i_j. \tag{D.18}$$

Thus, $\Delta_h(t_0,x) - \Delta_{\tilde{h}}(t_0,x) = 0$, and we can apply the ODE comparison theorem with $f(v) = Ev + B$ and $v(t) = (B/E)(e^{Et} - 1)$ (where $B = 2\varepsilon n e^{CT}$) to conclude that

$$\left| \Delta_h(t,x) - \Delta_{\tilde{h}}(t,x) \right| \leq \frac{2\varepsilon n e^{CT}}{E}\left(e^{E|t-t_0|} - 1\right) \leq \frac{2\varepsilon n e^{CT}}{E}\left(e^{ET} - 1\right). \quad \text{(D.19)}$$

Since the expression on the right can be made as small as desired by choosing h and $\tilde{h}$ sufficiently small, this shows that for each i and j, and any sequence $h_k \to 0$, the sequence of functions $\left((\Delta_{h_k})^i_j\right)_{k=1}^\infty$ is uniformly Cauchy and therefore uniformly convergent to a continuous limit function. It follows easily from (D.19) that the limit is independent of the choice of (h_k), so the limit is in fact equal to $\lim_{h\to 0}(\Delta_h)^i_j(t,x)$, which is $\partial\theta^i/\partial x^j(t,x)$ by definition. This shows that θ^i has continuous first partial derivatives, and completes the proof of the $k = 1$ case.

Now assume that the theorem is true for some $k \geq 1$, and suppose V is of class C^{k+1}. By the inductive hypothesis, θ is of class C^k, and therefore by (D.12), $\partial\theta^i/\partial t$ is also C^k. We can differentiate under the integral sign in (D.11) to obtain

$$\frac{\partial\theta^i}{\partial x^j}(t,x) = \delta^i_j + \sum_{k=1}^n \int_{t_0}^t \frac{\partial V^i}{\partial y^k}(\theta(s,x))\frac{\partial\theta^k}{\partial x^j}(s,x)\,ds.$$

By the fundamental theorem of calculus, this implies that $\partial\theta^i/\partial x^j$ satisfies the differential equation

$$\frac{\partial}{\partial t}\frac{\partial\theta^i}{\partial x^j}(t,x) = \sum_{k=1}^n \frac{\partial V^i}{\partial y^k}(\theta(t,x))\frac{\partial\theta^k}{\partial x^j}(t,x).$$

Consider the following initial value problem for the $n + n^2$ unknown functions (α^i, β^i_j):

$$\dot{\alpha}^i(t) = V^i(\alpha(t)), \qquad\qquad \alpha^i(t_0) = a^i,$$

$$\dot{\beta}^i_j(t) = \sum_{k=1}^n \frac{\partial V^i}{\partial y^k}(\alpha(t))\beta^k_j(t), \quad \beta^i_j(t_0) = b^i_j.$$

The functions on the right-hand side of this system are C^k functions of (α^i, β^i_j), so the inductive hypothesis implies that its solutions are C^k functions of (t, a^i, b^i_j). The discussion in the preceding paragraph shows that $\alpha^i(t) = \theta^i(t,x)$ and $\beta^i_j(t) = \partial\theta^i/\partial x^j(t,x)$ solve this system with initial conditions $a^i = x^i, b^i_j = \delta^i_j$. This shows that $\partial\theta^i/\partial x^j$ is a C^k function of (t,x), so θ itself is of class C^{k+1}, thus completing the induction. □

Proof of the fundamental theorem. Suppose $U \subseteq \mathbb{R}^n$ is open and $V\colon U \to \mathbb{R}^n$ is smooth. Let $t_0 \in \mathbb{R}$ and $x_0 \in U$ be arbitrary. Because V is smooth, it is locally Lipschitz continuous by Corollary C.30, so the theorems of this appendix apply. Theorem D.3 shows that there exist neighborhoods J_0 of t_0 and U_0 of x_0 such that for each $c \in U_0$, there is a C^1 solution $y\colon J_0 \to U$ to (D.3)–(D.4). Uniqueness of solutions is an immediate consequence of Theorem D.4. Finally, Theorem D.5 shows that the solution is C^k for every k as a function of (t,c), so it is smooth. □

Nonautonomous Systems

Many applications of ODEs require the consideration of nonautonomous systems. In this section we show how the main theorem can be extended to cover the nonautonomous case.

Theorem D.6 (Fundamental Theorem for Nonautonomous ODEs). *Let $J \subseteq \mathbb{R}$ be an open interval and $U \subseteq \mathbb{R}^n$ be an open subset, and let $V : J \times U \to \mathbb{R}^n$ be a smooth vector-valued function.*

(a) EXISTENCE: *For any $s_0 \in J$ and $x_0 \in U$, there exist an open interval $J_0 \subseteq J$ containing s_0 and an open subset $U_0 \subseteq U$ containing x_0, such that for each $t_0 \in J_0$ and $c = (c^1, \ldots, c^n) \in U_0$, there is a C^1 map $y \colon J_0 \to U$ that solves (D.1)–(D.2).*

(b) UNIQUENESS: *Any two differentiable solutions to (D.1)–(D.2) agree on their common domain.*

(c) SMOOTHNESS: *Let J_0 and U_0 be as in (a), and define a map $\theta \colon J_0 \times J_0 \times U_0 \to U$ by letting $\theta(t, t_0, c) = y(t)$, where $y \colon J_0 \to U$ is the unique solution to (D.1)–(D.2). Then θ is smooth.*

Proof. Consider the following autonomous initial value problem for the $n + 1$ functions $y^0, \ldots, y^n$:

$$
\begin{aligned}
\dot{y}^0(t) &= 1; \\
\dot{y}^i(t) &= V^i\big(y^0(t), y^1(t), \ldots, y^n(t)\big), \quad i = 1, \ldots, n; \\
y^0(t_0) &= t_0; \\
y^i(t_0) &= c^i, \qquad\qquad\qquad\qquad\qquad\qquad i = 1, \ldots, n.
\end{aligned}
\tag{D.20}
$$

Any solution to (D.20) satisfies $y^0(t) = t$ for all t, and therefore $(y^1, \ldots, y^n)$ solves the nonautonomous system (D.1)–(D.2); and conversely, any solution to (D.1)–(D.2) yields a solution to (D.20) by setting $y^0(t) = t$. Theorem D.1 guarantees that there is an open interval $J_0 \subseteq \mathbb{R}$ containing s_0 and an open subset $W_0 \subseteq J \times U$ containing (s_0, x_0), such that for any $(t_0, c) \in W_0$ there exists a unique solution to (D.20) defined for $t \in J_0$, and the solution depends smoothly on (t, t_0, c). Shrinking J_0 and W_0 if necessary, we may assume that $J_0 \subseteq J$ and $W_0 = J_0 \times U_0$ for some open subset $U_0 \subseteq U$. The result follows.   □

Simple Solution Techniques

To get the most out of this book, you need to be able to find explicit solutions to a few differential equations and systems of differential equations. You have probably learned a variety of solution techniques for such equations. The following simple types of equations are more than adequate for the needs of this book.

Separable Equations

A first-order differential equation for a single function $y(t)$ that can be written in the form

$$y'(t) = f(y(t))g(t),$$

where f and g are continuous functions with f nonvanishing, is said to be **separable**. Any separable equation can be solved (at least in principle) by dividing through by $f(y(t))$, integrating both sides, and using substitution to transform the left-hand integral:

$$\frac{y'(t)}{f(y(t))} = g(t),$$

$$\int \frac{y'(t)\,dt}{f(y(t))} = \int g(t)\,dt,$$

$$\int \frac{dy}{f(y)} = \int g(t)\,dt.$$

If the resulting indefinite integrals can be computed explicitly, the result is a relation involving y and t that can (again, in principle) be solved for y. The constant of integration can then be adjusted to achieve the desired initial condition for y. Separable equations include those of the form $y'(t) = g(t)$ as a trivial special case, which can be solved by direct integration.

2×2 Constant-Coefficient Linear Systems

A system of the form

$$x'(t) = ax(t) + by(t),$$
$$y'(t) = cx(t) + dy(t), \tag{D.21}$$

where a, b, c, d are constants, can be written in matrix notation as $Z'(t) = AZ(t)$, where

$$Z(t) = \begin{pmatrix} x(t) \\ y(t) \end{pmatrix}, \qquad A = \begin{pmatrix} a & b \\ c & d \end{pmatrix}.$$

The set of solutions to this type of system always forms a 2-dimensional vector space over $\mathbb{R}$. Once two linearly independent solutions have been found, every other solution is a linear combination of these, and the constants can be adjusted to match any initial conditions.

It is always possible to find at least one (perhaps complex-valued) solution of the form $Z(t) = e^{\lambda t} Z_0$, where λ is an eigenvalue of A and Z_0 is a corresponding eigenvector. If A has two distinct eigenvalues, then there are two such solutions, and they span the solution space. (If the initial conditions are real, then the corresponding solution is real.) On the other hand, if A has only one eigenvalue, there are two cases. If $A - \lambda I_2 \neq 0$, then there is a vector Z_1 (called a **generalized eigenvector**) such that $(A - \lambda I_2)Z_1 = Z_0$, and a second linearly independent solution is given by $Z(t) = e^{\lambda t}(t Z_0 + Z_1)$. Otherwise, $A = \lambda I_2$, and the two equations in (D.21) are uncoupled and can be solved independently.

Partially Uncoupled Systems

If one of the differential equations in (D.1), say the equation for $\dot{y}^i(t)$, involves none of the dependent variables other than $y^i(t)$, then one can attempt to solve that equation first and substitute the solution into the other equations, thus obtaining a system with fewer unknown functions, which might be solvable by one of the methods above.

▶ **Exercise D.7.** Solve the following initial value problems.

(a) $x'(t) = x(t)^2$; $\quad x(0) = x_0$.

(b) $x'(t) = \dfrac{1}{x(t)}$; $\quad x(0) = x_0 > 0$.

(c) $x'(t) = y$; $\quad x(0) = x_0$;

$\quad\ y'(t) = 1$; $\quad y(0) = y_0$.

(d) $x'(t) = x$; $\quad x(0) = x_0$;

$\quad\ y'(t) = 2y$; $\quad y(0) = y_0$.

(e) $x'(t) = -y$; $\quad x(0) = x_0$;

$\quad\ y'(t) = x$; $\quad\ y(0) = y_0$.

(f) $x'(t) = -x(t) + y(t)$; $\quad x(0) = x_0$;

$\quad\ y'(t) = -x(t) - y(t)$; $\quad y(0) = y_0$.

(g) $x'(t) = 1$; $\quad\quad\quad\quad\ x(0) = x_0$;

$\quad\ y'(t) = \dfrac{1}{1 + x(t)^2}$; $\quad y(0) = y_0$.

References

[AM78]   Abraham, Ralph, Marsden, Jerrold E.: Foundations of Mechanics, 2nd edn. Benjamin/ Cummings, Reading (1978)

[Apo74] Apostol, Tom M.: Mathematical Analysis, 2nd edn. Addison–Wesley, Reading (1974)

[Bae02] Baez, John C.: The octonions. Bull. Am. Math. Soc. (N.S.) **39**(2), 145–205 (2002)

[BR89]   Birkhoff, Garrett, Rota, Gian-Carlo: Ordinary Differential Equations, 4th edn. Wiley, New York (1989)

[Boo86] Boothby, William M.: An Introduction to Differentiable Manifolds and Riemannian Geometry, 2nd edn. Academic Press, Orlando (1986)

[MB58]   Bott, Raoul, Milnor, John: On the parallelizability of the spheres. Bull. Am. Math. Soc. (N.S.) **64**, 87–89 (1958)

[BD09]   Boyce, William E., DiPrima, Richard C.: Elementary Differential Equations and Boundary Value Problems, 9th edn. Wiley, New York (2009)

[Bre93] Bredon, Glen E.: Topology and Geometry. Springer, New York (1993)

[Cha06] Chavel, Isaac: Riemannian Geometry: A Modern Introduction. Cambridge Studies in Advanced Mathematics, vol. 98, 2nd edn. Cambridge University Press, Cambridge (2006)

[dC92]   do Carmo, Manfredo Perdigão: Riemannian Geometry. Birkhäuser, Boston (1992). Translated from the second Portuguese edition by Francis Flaherty

[DK90]   Donaldson, S.K., Kronheimer, P.B.: The Geometry of Four-Manifolds. Clarendon Press, New York (1990)

[Eva98] Evans, Lawrence C.: Partial Differential Equations. Am. Math. Soc., Providence (1998)

[Fol95] Folland, Gerald B.: Introduction to Partial Differential Equations, 2nd edn. Princeton University Press, Princeton (1995)

[FQ90]   Freedman, Michael, Quinn, Frank: Topology of 4-Manifolds. Princeton University Press, Princeton (1990)

[FIS03] Friedberg, Stephen H., Insel, Arnold J., Spence, Lawrence E.: Linear Algebra, 4th edn. Prentice–Hall, Upper Saddle River (2003)

[Gil95] Gilkey, Peter B.: Invariance Theory, the Heat Equation, and the Atiyah–Singer Index Theorem, 2nd edn. CRC Press, Boca Raton (1995)

[HL82]   Harvey, Reese, Lawson, H. Blaine Jr.: Calibrated geometries. Acta Math.-Djursholm **148**, 47–157 (1982)

[Hat02] Hatcher, Allen: Algebraic Topology. Cambridge University Press, Cambridge (2002)

[HE73]   Hawking, S.W., Ellis, G.F.R.: The Large Scale Structure of Space–Time. Cambridge University Press, London (1973)

[Her75] Herstein, Israel N.: Topics in Algebra, 2nd edn. Wiley, New York (1975)

[Hör90]   Hörmander, Lars: The Analysis of Linear Partial Differential Operators I: Distribution Theory and Fourier Analysis, 2nd edn. Springer, Berlin (1990)

[Hun97]   Hungerford, Thomas W.: Abstract Algebra: An Introduction, 2nd edn. Saunders College Publishing, Fort Worth (1997)

[Joh91]   John, Fritz: Partial Differential Equations. Applied Mathematical Sciences, vol. 1, 4th edn. Springer, New York (1991)

[Ker58]   Kervaire, Michel A.: Non-parallelizability of the n sphere for $n > 7$. Proc. Natl. Acad. Sci. USA **44**, 280–283 (1958)

[Ker60]   Kervaire, Michel A.: A manifold which does not admit any differentiable structure. Comment. Math. Helv. **34**, 257–270 (1960)

[KM63]   Kervaire, Michel A., Milnor, John W.: Groups of homotopy spheres: I. Ann. Math. **77**, 504–537 (1963)

[KN69]   Kobayashi, Shoshichi, Nomizu, Katsumi: Foundations of Differential Geometry, vols. I & II. Wiley, New York (1969)

[LeeJeff09] Lee, Jeffrey M.: Manifolds and Differential Geometry. Am. Math. Soc., Providence (2009)

[LeeRM]   Lee, John M.: Riemannian Manifolds: An Introduction to Curvature. Springer, New York (1997)

[LeeTM]   Lee, John M.: Introduction to Topological Manifolds, 2nd edn. Springer, New York (2011)

[Mar71]   Martinet, J.: Formes de contact sur les variétés de dimension 3. In: Proc. Liverpool Singularities Symp. II, pp. 142–163 (1971) (French)

[Mas89]   Massey, William S.: Algebraic Topology: An Introduction. Springer, New York (1989)

[Mil56]   Milnor, John W.: On manifolds homeomorphic to the 7-sphere. Ann. Math. **64**, 399–405 (1956)

[Mil63]   Milnor, John W.: Morse Theory. Princeton University Press, Princeton (1963)

[Moi77]   Moise, Edwin E.: Geometric Topology in Dimensions 2 and 3. Springer, New York (1977)

[Mos65]   Moser, Jürgen: On the volume elements on a manifold. Trans. Am. Math. Soc. **120**, 286–294 (1965)

[Mos05]   Mostow, G.D.: A structure theorem for homogeneous spaces. Geom. Dedic. **114**, 87–102 (2005)

[Mun60]   Munkres, James R.: Obstructions to the smoothing of piecewise differentiable homeomorphisms. Ann. Math. **72**, 521–554 (1960)

[Mun66]   Munkres, James R.: Elementary Differential Topology. Princeton University Press, Princeton (1966)

[Mun84]   Munkres, James R.: Elements of Algebraic Topology. Addison–Wesley, Menlo Park (1984)

[Mun00]   Munkres, James R.: Topology, 2nd edn. Prentice–Hall, Upper Saddle River (2000)

[Nes03]   Nestruev, Jet: Smooth Manifolds and Observables. Springer, New York (2003)

[Noe71]   Noether, Emmy: Invariant variation problems. Transp. Theory Stat. **1**(3), 186–207 (1971). Translated from the German (Nachr. Akad. Wiss. Gött. Math.-Phys. Kl. II 1918, 235–257)

[Osb82]   Osborn, Howard: Vector Bundles, vol. 1. Academic Press Inc. [Harcourt Brace Jovanovich Publishers], New York (1982)

[Pet06]   Petersen, Peter: Riemannian Geometry, 2nd edn. Springer, New York (2006)

[Rud76]   Rudin, Walter: Principles of Mathematical Analysis, 3rd edn. McGraw–Hill, New York (1976)

[Sha97]   Sharpe, R.W.: Differential Geometry: Cartan's Generalization of Klein's Erlangen Program. Springer, New York (1997)

[Sie92]   Sieradski, Allan J.: An Introduction to Topology and Homotopy. PWS-Kent, Boston (1992)

[Sma62]   Smale, S.: On the structure of manifolds. Am. J. Math. **84**, 387–399 (1962)

[Spi99]   Spivak, Michael: A Comprehensive Introduction to Differential Geometry, vols. 1–5, 3rd edn. Publish or Perish, Berkeley (1999)

[Str00]   Strichartz, Robert S.: The Way of Analysis, revised edn. Jones & Bartlett, Boston (2000)

[Var84]   Varadarajan, V.S.: Lie Groups, Lie Algebras, and Their Representations. Springer, New York (1984)

[Wal65]   Wall, C.T.C.: All 3-manifolds imbed in 5-space. Bull. Am. Math. Soc. (N.S.) **71**, 564–567 (1965)

[War83]   Warner, Frank W.: Foundations of Differentiable Manifolds and Lie Groups. Springer, New York (1983)

[Wei71]   Weinstein, Alan: Symplectic manifolds and their Lagrangian submanifolds. Adv. Math. **6**, 329–346 (1971)

[Whi36]   Whitney, Hassler: Differentiable manifolds. Ann. Math. **37**, 645–680 (1936)

[Whi44a]   Whitney, Hassler: The self-intersections of a smooth n-manifold in $2n$-space. Ann. Math. **45**, 220–246 (1944)

[Whi44b]   Whitney, Hassler: The singularities of a smooth n-manifold in $(2n-1)$-space. Ann. Math. **45**, 247–293 (1944)

Notation Index

Subject Index